21st Century Nanoscience – A Handbook

21ˢᵗ Century Nanoscience – A Handbook

Design Strategies for Synthesis and Fabrication (Volume Two)

Edited by

Klaus D. Sattler

CRC Press
Taylor & Francis Group
Boca Raton London New York

CRC Press is an imprint of the
Taylor & Francis Group, an **informa** business

CRC Press
Taylor & Francis Group
6000 Broken Sound Parkway NW, Suite 300
Boca Raton, FL 33487-2742

First issued in paperback 2022

© 2020 by Taylor & Francis Group, LLC

CRC Press is an imprint of Taylor & Francis Group, an Informa business

No claim to original U.S. Government works

ISBN-13: 978-0-815-39233-0 (hbk)
ISBN-13: 978-1-03-233732-6 (pbk)
DOI: 10.1201/9780367341558

Library of Congress Cataloging-in-Publication Data

Names: Sattler, Klaus D., editor.
Title: 21st century nanoscience : a handbook / edited by Klaus D. Sattler.
Description: Boca Raton, Florida : CRC Press, [2020] | Includes bibliographical references and index. | Contents: volume 1. Nanophysics sourcebook—volume 2. Design strategies for synthesis and fabrication—volume 3. Advanced analytic methods and instrumentation—volume 5. Exotic nanostructures and quantum systems—volume 6. Nanophotonics, nanoelectronics, and nanoplasmonics—volume 7. Bioinspired systems and methods. | Summary: "This 21st Century Nanoscience Handbook will be the most comprehensive, up-to-date large reference work for the field of nanoscience. Handbook of Nanophysics, by the same editor, published in the fall of 2010, was embraced as the first comprehensive reference to consider both fundamental and applied aspects of nanophysics. This follow-up project has been conceived as a necessary expansion and full update that considers the significant advances made in the field since 2010. It goes well beyond the physics as warranted by recent developments in the field"—Provided by publisher.
Identifiers: LCCN 2019024160 (print) | LCCN 2019024161 (ebook) | ISBN 9780815384434 (v. 1 ; hardback) | ISBN 9780815392330 (v. 2 ; hardback) | ISBN 9780815384731 (v. 3 ; hardback) | ISBN 9780815355281 (v. 4 ; hardback) | ISBN 9780815356264 (v. 5 ; hardback) | ISBN 9780815356417 (v. 6 ; hardback) | ISBN 9780815357032 (v. 7 ; hardback) | ISBN 9780815357070 (v. 8 ; hardback) | ISBN 9780815357087 (v. 9 ; hardback) | ISBN 9780815357094 (v. 10 ; hardback) | ISBN 9780367333003 (v. 1 ; ebook) | ISBN 9780367341558 (v. 2 ; ebook) | ISBN 9780429340420 (v. 3 ; ebook) | ISBN 9780429347290 (v. 4 ; ebook) | ISBN 9780429347313 (v. 5 ; ebook) | ISBN 9780429351617 (v. 6 ; ebook) | ISBN 9780429351525 (v. 7 ; ebook) | ISBN 9780429351587 (v. 8 ; ebook) | ISBN 9780429351594 (v. 9 ; ebook) | ISBN 9780429351631 (v. 10 ; ebook)
Subjects: LCSH: Nanoscience—Handbooks, manuals, etc.
Classification: LCC QC176.8.N35 A22 2020 (print) | LCC QC176.8.N35 (ebook) | DDC 500—dc23
LC record available at https://lccn.loc.gov/2019024160
LC ebook record available at https://lccn.loc.gov/2019024161

Visit the Taylor & Francis Web site at
http://www.taylorandfrancis.com

and the CRC Press Web site at
http://www.crcpress.com

Contents

Editor

Klaus D. Sattler pursued his undergraduate and master's courses at the University of Karlsruhe in Germany. He received his PhD under the guidance of Professors G. Busch and H.C. Siegmann at the Swiss Federal Institute of Technology (ETH) in Zurich. For three years was a Heisenberg fellow at the University of California, Berkeley, where he initiated the first studies with a scanning tunneling microscope of atomic clusters on surfaces. Dr. Sattler accepted a position as professor of physics at the University of Hawaii, Honolulu, in 1988. In 1994, his group produced the first carbon nanocones. His current work focuses on novel nanomaterials and solar photocatalysis with nanoparticles for the purification of water. He is the editor of the sister references, *Carbon Nanomaterials Sourcebook* (2016) and *Silicon Nanomaterials Sourcebook* (2017), as well as *Fundamentals of Picoscience* (2014). Among his many other accomplishments, Dr. Sattler was awarded the prestigious Walter Schottky Prize from the German Physical Society in 1983. At the University of Hawaii, he teaches courses in general physics, solid state physics, and quantum mechanics.

Contributors

Mohd Athar
School of Chemical Sciences
Central University of Gujarat
Gandhinagar, India

Aydin Berenjian
Faculty of Science and Engineering
School of Engineering
The University of Waikato
Hamilton, New Zealand

Maria Teresa Caccamo
CNR–IPCF, Consiglio Nazionale delle
    Ricerche
Istituto per i Processi Chimico Fisici
Messina, Italy

Pietro Calandra
CNR-ISMN, Consiglio Nazionale delle
    Ricerche
Istituto Studio Materiali
    Nanostrutturati
Roma, Italy

Costas N. Costa
Laboratory of Environmental
    Catalysis
Department of Environmental Science
    and Technology
Cyprus University of Technology
Lemesos, Cyprus

Zheyuan Ding
Beijing Key Laboratory of
    Lignocellulosic Chemistry, Collage
    of Materials Science and
    Technology
Beijing Forestry University
Beijing, China

Hai M. Duong
Department of Mechanical
    Engineering
National University of Singapore
Singapore, Singapore

Sriparna Dutta
Green Chemistry Network Centre
Department of Chemistry
University of Delhi
Delhi, India

Alireza Ebrahiminezhad
Department of Medical
    Nanotechnology
School of Advanced Medical Sciences
    and Technologies
Shiraz University of Medical Sciences
Shiraz, Iran
and
Department of Pharmaceutical
    Biotechnology
School of Pharmacy and
    Pharmaceutical Sciences Research
    Center
Shiraz University of Medical Sciences
Shiraz, Iran

Xolile Fuku
Energy Materials
Council for Scientific and Industrial
    Research
Pretoria, South Africa

Enlai Gao
Department of Engineering Mechanics
School of Civil Engineering
Wuhan University
Wuhan, P. R. China

Younes Ghasemi
Department of Pharmaceutical
    Biotechnology
School of Pharmacy and
    Pharmaceutical Sciences Research
    Center
Shiraz University of Medical Sciences
Shiraz, Iran

Hiroyuki Hasegawa
Course of School Teacher Training
Faculty of Education
Department of Natural Science
    Education
Shimane University
Matsue, Japan

Matias Holz
Ilmenau University of Technology
Institute for Micro & Nanoelectronics
Ilmenau, Germany

Ahmad Fauzi Ismail
Advanced Membrane Technology
    Research Centre (AMTEC)
Universiti Teknologi Malaysia
Skudai, Malaysia

Vinod K. Jain
Department of Chemistry
School of Sciences
Gujarat University
Ahmedabad, India

Xiangzheng Jia
Department of Engineering Mechanics
School of Civil Engineering
Wuhan University
Wuhan, P. R. China

Anita R. Kongor
Department of Chemistry
School of Sciences
Gujarat University
Ahmedabad, India

Ulrich J. Krull
Department of Chemical and Physical
    Sciences
University of Toronto Mississauga
Mississauga, Canada

Vladimir Lesnyak
Physical Chemistry
TU Dresden
Dresden, Germany

Ruishan Li
Department of Engineering Mechanics
School of Civil Engineering
Wuhan University
Wuhan, P. R. China

Domenico Lombardo
CNR–IPCF, Consiglio Nazionale delle
    Ricerche
Istituto per i Processi Chimico Fisici
Messina, Italy

Yao Lu
National Laboratory of Solid State
   Microstructures
School of Physics
Nanjing University
Nanjing, P. R. China

Karan Malhotra
Department of Chemical and Physical
   Sciences
University of Toronto Mississauga
Mississauga, Ontario, Canada

Badal Kumar Mandal
Trace Elements Speciation Research
   Laboratory
Department of Chemistry, School of
   Advanced Sciences
Vellore Institute of Technology (VIT)
Vellore, India

Carlos I. Mendoza
Instituto de Investigaciones en
   Materiales
Universidad Nacional Autónoma de
   México
Ciudad de México, Mexico

Mmalewane Modibedi
Energy Materials
Council for Scientific and Industrial
   Research
Pretoria, South Africa

Antaryami Mohanta
Laboratory for Advanced Materials
   Processing
Empa–Swiss Federal Laboratories for
   Materials Science and Technology
Thun, Switzerland

Subrata Mondal
Department of Mechanical
   Engineering
National Institute of Technical
   Teachers' Training and Research
   (NITTTR) Kolkata
Kolkata, India

Hifza Najib
Department of Chemical and Physical
   Sciences
University of Toronto Mississauga
Mississauga, Ontario, Canada

Manthan K. Panchal
Department of Chemistry
School of Sciences
Gujarat University
Ahmedabad, India

Ivo W. Rangelow
Ilmenau University of Technology
Institute for Micro & Nanoelectronics
Ilmenau, Germany

Daniel Salgado-Blanco
Centro Nacional de Supercómputo
Instituto Potosino de Investigación
Científica y Tecnológica
San Luis Potosí, Mexico

Alireza Samavati
Advanced Membrane Technology
   Research Centre (AMTEC)
Universiti Teknologi Malaysia
Skudai, Malaysia

Petros G. Savva
Laboratory of Environmental
   Catalysis
Department of Environmental Science
   and Technology
Cyprus University of Technology
Lemesos, Cyprus

Mostafa Seifan
Faculty of Science and Engineering
School of Engineering
The University of Waikato
Hamilton, New Zealand

Muhammad Shahrukh
Department of Chemical and Physical
   Sciences
University of Toronto Mississauga
Mississauga, Ontario, Canada

Rakesh Kumar Sharma
Green Chemistry Network Centre
Department of Chemistry
University of Delhi
Delhi, India

Han Shui
Department of Engineering Mechanics
School of Civil Engineering
Wuhan University
Wuhan, P. R. China

Xuefeng Song
State Key Lab of Metal Matrix
   Composites
School of Materials Science and
   Engineering
Shanghai Jiao Tong University
Shanghai, P. R. China

Saeed Taghizadeh
Department of Medical Biotechnology
School of Advanced Medical Sciences
   and Technologies
Shiraz University of Medical Sciences
Shiraz, Iran

Seyedeh-Masoumeh Taghizadeh
Department of Pharmaceutical
   Biotechnology
School of Pharmacy and
   Pharmaceutical Sciences Research
   Center
Shiraz University of Medical Sciences
Shiraz, Iran

Kai Tao
State Key Lab of Metal Matrix
   Composites
School of Materials Science and
   Engineering
Shanghai Jiao Tong University
Shanghai, P. R. China

Raj K. Thareja
Indian Institute of Technology Bhilai
Raipur, India
and
Department of Physics
Indian Institute of Technology Kanpur
Kanpur, India

Thang Q. Tran
Department of Mechanical
   Engineering
National University of Singapore
Singapore, Singapore

Shuai Wang
State Key Lab of Metal Matrix
   Composites
School of Materials Science and
   Engineering
Shanghai Jiao Tong University
Shanghai, P. R. China

Xiluan Wang
Beijing Key Laboratory of
   Lignocellulosic Chemistry, College
   of Materials Science and
   Technology
Beijing Forestry University
Beijing, China

Wei Wu
National & Local Joint Engineering
   Research Center of Advanced
   Packaging Materials Developing
   Technology
Hunan University of Technology
Zhuzhou, P. R. China
and
Laboratory of Printable Functional
   Nanomaterials and Printed
   Electronics
School of Printing and Packaging
Wuhan University
Wuhan, P. R. China

Zhaohui Wu
Hunan Key Laboratory of Applied
   Environmental Photocatalysis
Changsha University
Changsha, P. R. China

Weiren Xia
National Laboratory of Solid State
   Microstructures
School of Physics
Nanjing University
Nanjing, P. R. China

Jijin Xu
State Key Lab of Metal Matrix
   Composites
School of Materials Science and
   Engineering
Shanghai Jiao Tong University
Shanghai, P. R. China

Sneha Yadav
Green Chemistry Network Centre
Department of Chemistry
University of Delhi
Delhi, India

Xiaofeng Zhao
State Key Lab of Metal Matrix
   Composites
School of Materials Science and
   Engineering
Shanghai Jiao Tong University
Shanghai, P. R. China

Xinhua Zhu
National Laboratory of Solid State
   Microstructures
School of Physics
Nanjing University
Nanjing, P. R. China

1

Large-Scale Colloidal Synthesis of Nanoparticles

Vladimir Lesnyak
TU Dresden

1.1 Introduction

Over the last few decades, research on nanoparticles (NPs) has been an intensively developing area of science inspired by a fascinating endeavor of understanding the behavior of a solid when its physical dimensions decrease to only a few hundreds/thousands of atoms. In this size regime, the quantum mechanical coupling of the constituent atoms gives rise to new interesting physical phenomena. Over the years, intensive efforts by many research groups around the world have made it possible to reach an outstanding level of control of these quantum mechanical properties by tuning the size, shape, structure, and composition of many different types of materials at the nanoscale. Furthermore, the small size of NPs, combined with their large surface-to-volume ratio, make them valuable contenders for several applications in many different fields, for instance, in (opto)electronics, sensor technologies, photovoltaics, nanophotonics, catalysis, and medicine (Choi et al., 2015; Kagan et al., 2016; Kovalenko et al., 2015; Medintz et al., 2005; Michalet et al., 2005; Panthani & Korgel, 2012; Pelaz et al., 2012; Sapsford et al., 2013; Talapin et al., 2010; Yang et al., 2016). Ligand-capped NPs, i.e., particles made of an inorganic core and an organic shell of stabilizing ligands (surfactants) (Figure 1.1b), are particularly suitable candidates for low-cost processing, as they can be employed in the form of liquid solutions, e.g., as inks in printed electronics or in the form of composites with biomolecules in biomedical applications.

Compared to a large body of fundamental research that has been devoted to nanomaterials over the years, in part fueled by the hype on nanoscience and nanotechnology, the number of applications that have made their way into consumer products has been relatively small so far. To a large extent, this is due to one simple reason – making NPs in large amounts is still difficult. For example, the reaction conditions of the colloidal synthesis have to be carefully

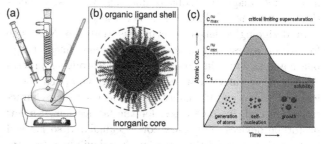

FIGURE 1.1 (a) A common setup for hot-injection synthesis of semiconductor NPs. (b) Calculated atomic structure of a 5 nm sized PbS NP capped with oleic acid (Zherebetskyy et al., 2014). (c) LaMer diagram depicting the concentration variation during the nucleation and growth of NPs (Xia et al., 2009). (Reproduced with permission from AAAS and Wiley VCH Verlag GmbH & Co.)

controlled to produce monodisperse NPs. There is a very narrow window of variability of parameters for the NP properties to remain practically the same from batch to batch. This has led to the major road block in making nanomaterials available to solve real issues and to be used to their full potential – the lack of large-scale production techniques.

Among various nanomaterials, colloidal semiconductor NPs (otherwise known as quantum dots, QDs), primarily made of metal chalcogenide compounds, have been the most studied objects. QDs exhibit extremely attractive optical features, such as widely tunable absorption spectra extending over a broad wavelength range, high photoluminescence (PL) quantum yields (QYs), narrow PL spectral bandwidth, and high thermal and photochemical stability. Above all, the possibility to tune their optical features by varying their size, shape, and composition by simple chemical methods induced the development of more refined approaches, resulting in the NPs with widely varied sizes and shapes, easily accessible from well-consolidated solution-phase synthetic schemes. Furthermore, many types of QDs have already been commercialized, and several applications have been developed. One of the latest and most fascinating examples is a line-up of Samsung QLED TVs, in which QDs are employed as color converters. Recently added members to the list of fluorescent QDs are represented by methylammonium (or cesium) lead halide perovskite NPs. Perovskites have attracted intensive interest over the last several years due to their unique properties that make them appealing for applications in photovoltaics, optoelectronics, and lighting (Akkerman et al., 2018; Kovalenko et al., 2017). Initially evolved as thin films and bulk single crystals, they have soon been developed in the nanocrystal form too, featured by their strong fluorescence (especially the $APbX_3$ compounds, with A = methylammonium, formamidinium, or Cs, X = Cl, Br, or I) (Bai et al., 2016; González-Carrero et al., 2016; González-Carrero et al., 2015; Protesescu et al., 2015). In addition to QDs, many companies all over the world have commercialized the production of noble metal and magnetic NPs, having in view their great perspective in biomedical applications, in particular, in biosensing (noble metal NPs), magnetic resonance imaging (MRI), and hyperthermia (magnetic NPs), as well as catalysis.

In designing a lab-scale synthesis route to NPs, one focuses on a desired composition with well-defined crystal phase and stability in common solvents, through a process which can be carried out in a moderately equipped chemical laboratory, and in quantities sufficient for their characterization and, in some cases, for the fabrication of a few prototype devices. The know-how accumulated so far will provide directions for the future work and will help to identify materials that could be explored more extensively, with a synthesis process that can be industrially scalable. There have been many reports on the colloidal synthesis of various NPs based on relatively simple, generalized, inexpensive methods and employing nontoxic reagents. Many of these methods have been claimed to be easily scalable to prepare at least grams of products. When seen from an industrial perspective, the fabrication requirements are unfortunately not much promising, as they involve high temperatures, or the use of vacuum and inert atmosphere as typical reaction conditions. Therefore, scaling-up of successful lab protocols to an industrial level is often not straightforward, and there is still a strong demand for simpler and inexpensive methods for large-scale production of high-quality NPs.

It is quite difficult to define exact minimal space-time production yields in the synthesis of colloidal NPs, after which a system can be considered as a large-scale production unit. It is mainly due to unspecified times and capabilities of the post-synthetic treatment of NPs, especially their purification, which is an unavoidable step of the whole process and which is highly time-consuming. Obviously, different materials require different processing and thus different times. Therefore, focusing solely on the reaction, a system with a throughput of at least 1 g/h would be evaluated as capable of large-scale production. This, however, is a very rough estimation which should not be considered literally. Also, important is the potential for upscaling which is quite obvious in the case of flow syntheses, which are discussed in

Section 1.3, although in order to make them truly large scale, a continuous, uninterrupted supply of precursors should be further developed.

Taking into account the abovementioned need for upscaling the colloidal synthesis of various NPs, in this chapter, the main methods of the large-scale production of NPs developed up to date are summarized, beginning with simplest one-batch reactions, which are widely used in a common laboratory practice, followed by more sophisticated approaches based on the continuous production of nanomaterials in flow reactors, such as microfluidic and spinning disc reactors. Prior to overviewing these synthetic approaches, the main features of the colloidal synthesis of NPs will be introduced.

1.1.1 Basics of the Colloidal Synthesis of Nanoparticles

Solution-based colloidal synthesis has emerged as a separate branch of synthetic chemistry due to the flexibility in preparing NPs of a broad range of materials and due to excellent control over their sizes and shapes, in addition to employing a relatively simple experimental equipment. Typical setup for the so-called hot-injection synthesis (discussed in more detail in Section 1.2.2) is presented in Figure 1.1a. High-quality materials can be produced by this method at surprisingly low cost. In the colloidal synthesis, suitable molecular precursors (mainly inorganic salts or organometallic compounds) are reacted in a liquid medium in the presence of stabilizing agents (also known as capping ligands). Chemical reactions between the precursors release reactive species, called monomers which are responsible for the further nucleation and the growth of NPs (Kim et al., 2013; Klimov, 2004; Kwon et al., 2007; Lee et al., 2016; Rogach, 2008; Schmid, 2010; Yang et al., 2015). A general synthetic strategy involves several consecutive steps: (i) nucleation, when a critical concentration of the active monomer species in the reaction mixture is reached; (ii) growth, when the formed nuclei do not dissolve back into monomeric units but instead keep attracting monomers from the solution phase and further enlarge in size; (iii) arresting the growth – quenching the reaction and separation of the particles from the reaction mixture, when the desired size/composition has been achieved; and (iv) post-synthetic treatments, if necessary. It is important to temporally separate nucleation and growth in order to achieve a narrow size distribution of the NPs (Murray et al., 1993; Park et al., 2007; Thanh et al., 2014). This is generally achieved by tuning the rate at which the monomer species are released from corresponding precursors. The chemical reactions which produce these monomer species can be, e.g., the decomposition of molecular precursors (in the case of QDs), the hydrolysis of salts with subsequent oxidation (for metal oxide NPs), or the reduction of metal ions (in the synthesis of noble metal NPs) (Park et al., 2007).

The pioneering work by LaMer and Dinegar defines the synthesis parameters for obtaining monodisperse colloidal particles (LaMer & Dinegar, 1950). The LaMer diagram in Figure 1.1c shows the evolution of the concentration in the formation of monodisperse NPs demarcating the three stages of (i) generation of atoms (or other active species): the monomer formation, (ii) the nucleation when the concentration of monomers reaches a certain minimal value (so-called nucleation threshold), and (iii) the growth of the formed nuclei by consumption of the monomer species in the solution. In particular, the formation of monomers needs to be controlled so that during the growth stage, their concentration should not cross the nucleation threshold, since this would create new nuclei which would have less time to grow compared to the ones formed before, a scenario that would ultimately lead to polydisperse in size NPs. This can be achieved by several handles, such as temperature, material supply, pH, etc.

A range of reactions can be selected for producing NPs with a particular composition. Most of the methods employed in a lab use batch-type reactors, which involve quite large volumes of reagents or solvents. This complicates the precise control of the synthetic parameters uniformly throughout the reaction, since in large volume systems, thermal and concentration gradients are unavoidable. Moreover, achieving rapid changes in order to follow the LaMer curve is a very challenging task. To overcome these limitations, various synthetic strategies have been developed during last years with a special focus on the continuous flow production of NPs (Chen et al., 2014; Jensen et al., 2014; Lignier et al., 2012; Nightingale & de Mello, 2010; Nightingale & deMello, 2013; Niu et al., 2015; Phillips et al., 2014). Among approaches to scalable synthesis of NPs, the microfluidic route has demonstrated a great potential, owing to a precise control over the reaction environment, resulting in various industrial chemical syntheses (Elvira et al., 2013; Jahn et al., 2008; Marre & Jensen, 2010; Myers et al., 2014; Reizman & Jensen, 2016; Song et al., 2008; Whitesides, 2006; Zhao et al., 2011). Each following section presents syntheses of nanomaterials starting from simple mono- or binary compounds followed by examples of more complex alloyed or heterostructures built of at least two different materials, such as core/shell or even more complex core/shell/shell QDs composed of three different semiconductors.

1.2   One-Batch Synthesis

The main approaches used to synthesize large amounts of NPs are schematically summarized in Figure 1.2. In the one-batch synthesis, one can define two main routes: heat-up and hot-injection. Heat-up, the simplest method, basically consists of heating a precursor (or a mixture of precursors) in high boiling solvents in the presence of appropriate ligand molecules, which control the nucleation and the growth of NPs. The hot-injection method involves instead a fast injection of one of the precursors into a hot solution of another precursor in a mixture of high boiling point solvents (or the injection of both precursors at once in a heated mixture of

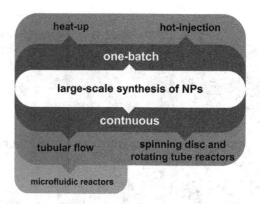

FIGURE 1.2 Main approaches used in large-scale syntheses of NPs.

solvents) upon vigorous stirring. These conditions favor a burst of homogeneous nucleation, followed by the homogeneous growth of NPs.

1.2.1   Heat-Up Method

This technique is suitable for the synthesis of different materials, such as metal (e.g., Ag) (Park et al., 2012), metal oxide, both magnetic (Fe_3O_4) and wide band gap semiconductor (ZnO, TiO_2) (Lee et al., 2013), binary II–VI, III_2–VI_3 metal chalcogenide semiconductor (CdSe, Bi_2Te_3) (Son et al., 2012; Yang et al., 2005), and more complex ternary I–III–VI_2 and even quaternary I_2–II–IV–VI_4 alloyed semiconductor, such as $CuInS_2$ (CIS) (Lu et al., 2011) and Cu_2ZnSnS_4 (CZTS) (Chesman et al., 2013; van Embden et al., 2014) NPs. Transmission electron microscopy (TEM) images of main examples of NPs prepared by this approach are presented in Figure 1.3, demonstrating its ability to produce fairly monodisperse particles not only of spherical, but also of platelet-, bullet-, and tetrahedral shapes.

Room Temperature Syntheses

Before beginning with heat-up synthesis, it is worthwhile to mention several protocols which formally do not belong to this category, as they involve reactions run at room temperature (i.e., no extra heat, except for ambient, is required) and, in some cases, even without stirring the reaction mixture. This route is especially useful for the synthesis of noble metal NPs via reduction of metal salts in solution. For example, Feng et al. reported a method to prepare thin Au nanowires with a diameter of 1.8 nm via reduction of $HAuCl_4$ by triisopropylsilane in the presence of oleylamine (OlAm) at room temperature, in which the latter acted not only as a stabilizer but also as a one-dimensional growth template, while triisopropylsilane accelerated the formation of the wires (Feng et al., 2009). This synthesis can easily be scaled up, as it requires neither heating, nor stirring, nor inert atmosphere, and involves only three chemicals. Monodisperse Au NPs were synthesized at room temperature by mixing $HAuCl_4$ with OlAm and storing the resulting solution for three days, with the formation of a pink colored precipitate (Lakshminarayana & Qing-Hua, 2009). Gold and

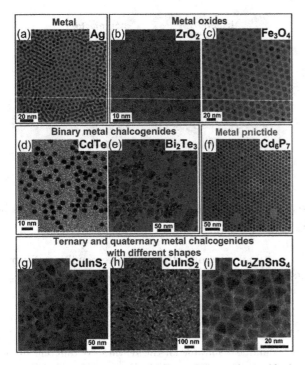

FIGURE 1.3 TEM images of different NPs synthesized by heat-up approach: (a) Ag (Park et al., 2012), (b) ZrO$_2$ (Joo et al., 2003), (c) Fe$_3$O$_4$ (Park et al., 2004), (d) CdTe (Yang et al., 2005), (e) Bi$_2$Te$_3$ (Son et al., 2012), (f) Cd$_6$P$_7$ (Miao et al., 2012), (g, h) CuInS$_2$ (Lu et al., 2011), and (i) Cu$_2$ZnSnS$_4$ (Chesman et al., 2013). (Reproduced with permission from Wiley VCH Verlag GmbH & Co, the Royal Society of Chemistry, American Chemical Society, and Nature Publishing Group.)

silver nanorods, spheroids, nanowires, platelets, and cubes of 4–50 nm were obtained employing a mixture of strong and weak reducing agents (NaBH$_4$ and ascorbic acid, respectively) in an aqueous solution of the corresponding metal salts (Jana, 2005). A fine control of the nucleation-growth kinetics and the presence of rod-like micelles templating the growing species enabled the formation of high-quality anisotropic NPs.

Syntheses at Elevated Temperatures

The heat-up synthesis of small-sized (1.7 and 3.5 nm, shown in Figure 1.3a) Ag NPs was developed by the Hyeon group, employing a highly concentrated Ag-precursor, which makes scaling-up very straightforward (Park et al., 2012). As a size controller, the authors used heating rate of the reaction mixture containing silver nitrate, OlAm (acting as a reducing agent and a surfactant), while oleic acid (OlAc) was employed as a co-surfactant and a co-solvent. In this method, ~2 g of Ag NPs was obtained from a single reaction at 180°C. Following a similar approach, but working at lower temperature, Sato et al. prepared 0.5 g of small monodisperse Pd NPs with sizes ranging from 2 to 5 nm and with a high reaction yield (93%) (Sato et al., 2011).

Monodisperse 4 nm zirconia (ZrO$_2$) NPs (Figure 1.3b) were synthesized in gram scale (5 g in a single reaction) using a nonhydrolytic sol-gel reaction between zirconium(IV)

isopropoxide and zirconium(IV) chloride or bromide under argon atmosphere at 340°C (Joo et al., 2003). Another example of uniform-sized oxide NPs is ceria (CeO$_2$) crystallized in sphere, wire, and tadpole shapes by the sol-gel reaction of cerium(III) nitrate and diphenyl ether in the presence of appropriate surfactants at 320°C, with single reaction yield of more than 10 g (Yu et al., 2005). A straightforward heat-up method was undertaken by Park et al. to prepare various monodisperse nanocrystals (iron, iron oxide, MnO, and CoO). This work demonstrated a great potential as a large-scale process, with an example of 40 g of iron oxide NPs (reaction yield >95%) obtained from a single reaction (Figure 1.3c) (Park et al., 2004). The synthesis was carried out at 250°C for 30 min in an inert atmosphere using iron-oleate complex, OlAc, oleyl alcohol, and diphenyl ether.

Among several magnetic oxides, iron oxide is biocompatible and in the NP-form is approved by the US Food and Drug Administration for use as an MRI contrast agent. As an example, uniform small-sized (<4 nm) iron oxide NPs synthesized according to the abovementioned procedure (Park et al., 2004) were applied for *in vivo* MRI and were found to be efficient T$_1$ contrast agent, showing a long circulation time, eventually resulting in a high-resolution imaging (Kim et al., 2011). A valuable alternative to reactions in organic high boiling solvents is a colloidal synthesis in aqueous media, which is more economical (as the main costs of the synthesis are represented by solvents), more environment-friendly (so-called "green" approach), and thus overall industrially attractive. By using water as a solvent, a straightforward synthesis of Fe$_3$O$_4$ nanocrystals via oxidative precipitation yielding up to 20 g from a single reaction was developed by Marciello et al. (2013). The reaction between FeSO$_4$ and KNO$_3$ acting as an oxidant in a basic media under nitrogen protection resulted in relatively uniform magnetite particles with average diameter from 20 to 30 nm, which are the optimum sizes for magnetic hyperthermia, another developing field of application of magnetic NPs.

The aqueous approach can be easily extended to the syntheses of various semiconductor metal chalcogenide NPs and apparently has a great potential for upscaling (Lesnyak et al., 2013). Mild synthetic conditions are both an advantage and a disadvantage of the method, as often they do not allow for a perfect crystallization of the NPs and usually they do not permit a direct and efficient shape control of the particles during their growth, due to temperature limitations. Therefore, the quality of the materials obtained by the synthesis in high boiling solvents is in fact unmatched by any low-temperature route at present. Here, a typical non-injection synthesis method of CdSe and CdTe (Figure 1.3d) QDs consisted of reacting selenium powder (or tributylphosphine-Te complex as a Te-precursor for CdTe NP synthesis) and cadmium myristate or cadmium octadecylphosphonate in octadecene (ODE, a high boiling liquid which is very often used as a solvent in the colloidal synthesis of different nanomaterials) (Yang et al., 2005). As another example of large-scale synthesis of metal chalcogenide NPs,

1.12 g of ultrathin Bi_2Te_3 nanoplates with thickness of 1–3 nm was obtained upon reacting bismuth dodecanethiolate and tri-*n*-octylphosphine-tellurium complex (TOP-Te) in the presence of OlAm (Figure 1.3e) (Son et al., 2012). After being sintered into nanostructured bulk pellets, this material exhibited excellent thermoelectric properties.

Based on the setup used in the aqueous synthesis of CdTe QDs introduced by Gaponik et al. (2002) and in the organometallic synthesis of InP QDs developed in the Reiss group (Li et al., 2008), Miao et al. established a hot-bubbling synthesis of monodisperse cadmium phosphide NPs (Figure 1.3f) using ex situ produced phosphine, which yielded gram-scale amounts (Miao et al., 2012). In this method, PH_3 gas, generated by the reaction between Ca_3P_2 powder and H_2SO_4, was bubbled through a hot solution of Cd-oleate in ODE, which afterwards was kept at a desired temperature (from 80°C to 250°C). Size-controlling parameters in this synthesis were the reaction time and the temperature: an increase of either parameter led to the growth of Cd_6P_7 NPs.

By employing the heat-up method, it is possible to synthesize more complex metal chalcogenide NPs, composed of three (ternary compounds) or even four (quaternary) different elements, in gram-scale amounts. Thus, monodisperse 3–8 nm sized CIS pyramidal nanocrystals, which are a promising material for solar harvesting and lighting applications, were synthesized with 80%–90% reaction yield (2.3 g in one batch) by Zhong et al. by heating a mixture of copper iodide, indium acetate, dodecanethiol (DDT), OlAc, and ODE at 200°C for 2 h (Zhong et al., 2010). In another report, wurtzite CIS NPs were obtained by a one-pot reaction of copper nitrate, indium nitrate, and DDT, with >80% yield (Lu et al., 2011). The size and the shape of these NPs were controlled by varying the Cu/In ratio and by introducing additional ligands, such as OlAm and OlAc (Figure 1.3g,h). For instance, uniform nanorods were formed starting from a Cu/In feed ratio of 2:1 with the assistance of both DDT and OlAc. They were demonstrated to assemble into large-scale superlattices on substrates by a solvent evaporation method. One more promising material for photovoltaic applications is represented by more complex quaternary CZTS NPs. Chesman et al. reported a non-injection synthesis of this material that delivered multi-gram yields (Chesman et al., 2013; van Embden et al., 2014). Their approach provides a precise control of the metal composition of the particles (Cu/Zn/Sn ratio), through the selective use of a binary sulfur precursor-ligand in the reaction mixture, which decouples the nucleation and the growth stages. Heating zinc ethyl xanthate, CuI, and $SnCl_4 \times 5H_2O$ in a mixture of DDT and OlAm (1/3) to 250°C, and then maintaining this temperature for additional 30 min under a nitrogen flow resulted in tetrahedral CZTS nanocrystals (Figure 1.3i) with approximately 80% yield.

Solvothermal and Microwave Irradiation Syntheses

The solvothermal (with hydrothermal as a particular case of it) approach utilizes solvents under elevated pressures and temperatures, either above or below their critical point, to increase the solubility of solid precursors and to speed up reactions among them in a sealed vessel. Using this method, highly crystalline hydrophilic and hydrophobic magnetic 20 nm Fe_3O_4 NPs exhibiting a room temperature magnetization as high as 84 emu/g were synthesized (Kolen'ko et al., 2014). A single reaction gave 86% (1.4 g) or 68% (0.9 g) yield of hydrophilic and hydrophobic magnetite nanocrystals, respectively. These NPs not only demonstrated super-paramagnetic behavior with notably high saturation magnetization, but also exhibited excellent performance in magnetic hyperthermia. Another example of potentially up-scalable process is the synthesis of Bi_2S_3 and Sb_2S_3 nanorods from single-source precursors under air at relatively low reaction temperatures (Lou et al., 2007).

Microwave irradiation methodology has been explored to synthesize various NPs, such as lanthanide orthophosphate nanorods (Patra et al., 2005); InGaP, InP, and CdSe NPs (Gerbec et al., 2005); rods and wires of ZnS, ZnSe, CdS, and CdSe (Panda et al., 2006); as well as rare earth oxide (M_2O_3, M = Pr, Nd, Sm, Eu, Gd, Tb, Dy), nanorods and nanoplates (Panda et al., 2007). Most of them used conventional/cooking microwave ovens, making the process rapid and simple, without the need of inert atmosphere or high pressure. The more uniform heating provided by microwave irradiation eliminates issues linked to thermal gradients, making this method promising for preparing large quantities of various nanomaterials. Also, much higher heating rates than with conventional heating can be achieved, which therefore provide an additional means of tuning reactivity. Apparently, further upscaling of this synthesis to industrial amounts requires its implementation in a continuous reaction mode by coupling, e.g., with microfluidic systems which are described in Section 1.3.1.

1.2.2 Hot-Injection Method

The intensive development of the hot-injection method for the colloidal synthesis of NPs was ignited by a pioneering work of Murray, Norris, and Bawendi on the synthesis of monodisperse CdS, CdSe, and CdTe QDs (Murray et al., 1993). On one hand, this approach provides an excellent control over the size, shape, and composition of NPs, including complex alloyed and heterostructures. On the other, its successful scale-up is challenged by significantly retarded heat and mass transport when large volumes of reaction mixtures are involved. Nevertheless, there have been reports on a large-scale hot-injection production of different materials, such as metallic Fe (Yang et al., 2007) and Cu (Lee et al., 2008), Cu_2O (Park et al., 2009), as well as various metal sulfide NPs (Zhang et al., 2012). For example, a fast injection of $Fe(CO)_5$ into a mixture of kerosene and OlAm at 180°C yielded ~2 g of 3–12 nm Fe NPs (Yang et al., 2007). The particle size was controlled by varying the reaction time, the molar ratio of OlAm to $Fe(CO)_5$, and the $Fe(CO)_5$ concentration in kerosene. Another example of metal particles prepared in large quantities by the hot-injection is represented by copper nanocrystals (Lee et al., 2008). Approximately 50 g of Cu NPs was obtained in one batch by the rapid injection of 1 L of a 1 M solution of copper sulfate in ethylene glycol into a mixture of sodium hypophosphite (as a reducing agent) and polyvinylpyrrolidone (PVP) (used as a stabilizer) in 4 L of ethylene glycol at 90°C under ambient atmosphere. The size of these copper NPs was controlled between 30 and 65 nm by varying the reaction time, temperature, and $CuSO_4$/PVP ratio. Thus produced NP colloids were further processed via ink-jet printing into metallic copper patterns on polyimide substrates. After a relatively low-temperature sintering process, the final patterns had low electrical resistivity (≥ 3.6 μΩ cm, or ≥ 2.2 times the resistivity of bulk copper).

More efforts have been devoted to the large-scale hot-injection synthesis of various metal chalcogenide NPs. A general approach to the gram-scale production (more than 30 g in a single reaction) of monodisperse metal sulfide (Cu_2S, CdS, SnS, ZnS, MnS, Ag_2S, and Bi_2S_3) NPs was developed by the group of Robinson, who employed for the first time $(NH_4)_2S$ as a sulfur precursor (Zhang et al., 2012). This method enabled high reaction yields ($\geq 87\%$) at moderate reaction temperatures, due to the high reactivity of $(NH_4)_2S$. The size tuning of the NPs was achieved by varying reaction temperature, time, and precursor ratios. Du et al. (2012) synthesized ultrathin nanosheets of different metal sulfides in gram scales with $\sim 100\%$ yield, starting from their respective metal chlorides: 1.4 g of 1.3 nm thick CuS nanosheets (reaction scheme is presented in Figure 1.4a), 1.3 g of ZnS, 1.6 g of Bi_2S_3, and 1.7 g of Sb_2S_3 nanowires (sub-2 nm diameter) were obtained from a single batch. According to this scheme, first, a lamellar complex CuCl(octylamine(OctAm), $OlAm)_x$ is formed by the reaction of corresponding compounds at 130°C, which then acts as a soft template for the nucleation and growth of ultrathin CuS nanosheets after the addition of S powder. A similar lamellar soft template, $CdCl_2(OctAm, OlAm)_2$, was employed by the Hyeon group for the large-scale synthesis of CdSe and CdTe nanosheets as thin as 1.4 nm (Son et al., 2009): about 4 g of single-layered CdSe nanosheets were prepared by reacting Se-powder dispersed in OctAm/OlAm mixture with $CdCl_2(OctAm, OlAm)_2$ complex, followed by mild sonication in chloroform.

Another important family of NPs is the cadmium chalcogenides (CdS, CdSe, and CdTe), whose large-scale hot-injection synthesis became feasible after the introduction by Peng and Peng of CdO as a Cd-precursor replacing dimethyl cadmium $Cd(CH_3)_2$ (extremely toxic, pyrophoric, expensive, unstable at room temperature, and explosive at elevated temperatures compound) (Peng & Peng, 2000). They demonstrated the synthesis of about 0.7 g of CdTe NPs in a single lab-scale reaction. In another report, a gram-scale synthesis of red-emitting CdSe QDs (using standard air-free techniques) was developed (Zhong et al., 2006). In the logic of further simplifying the synthesis conditions, Liu et al. used diesel fuel as a solvent and OlAc

as a ligand in an open-air synthesis to produce 9.6 g of CdSe QDs, thus effectively avoiding the commonly used in this process expensive solvents and toxic, air-sensitive organic phosphines (Liu et al., 2008). A fast method with minimum cost and effort to prepare core/shell QDs was developed by successive injections of precursors in one pot, according to the synthetic scheme displayed in Figure 1.4b, yielding 0.2 kg of CdSe/ZnSe QDs, which were easily purified and redispersed in various organic solvents (Kim & Lee, 2006).

The hot-injection technique has been used by several groups to obtain high-quality NPs of various materials and compositions, but translating this to a large-scale synthesis is not straightforward as the step of injecting a large quantity of a precursor/reactant in a short time, while regulating the precise reaction temperature of a large volume, is indeed a challenge. The group of Reiss solved this issue by using a peristaltic pump and a mechanical stirrer in a 2 L batch reactor (see Figure 1.4c) (Protière et al., 2011). They demonstrated an efficient approach to optimize the synthesis and prepare differently sized CdSe QDs by design-of-experiment method. The lab-scale synthesis yielded 100–200 mg, and when the reaction was scaled up by a factor of 20 using a 2 L batch reactor with mechanical stirring, 2–3 g of CdSe QDs was obtained, without sacrificing the narrow size distribution and the reaction yield. Furthermore, using the same setup, the authors synthesized ~ 1.3 g of CdSe/CdS/ZnS core/shell/shell QDs, which exhibited an excellent photoluminescence quantum yield (PLQY) of up to 81%. Yarema et al. proposed an ingenious solution to speed up the injection of a large volume of a precursor from an addition funnel by applying mild underpressure (1–10 mbars) to the flask reactor (Yarema et al., 2017). Using this method, they successfully upscaled synthetic protocols for Sn, PbS, $CsPbBr_3$, and $Cu_3In_5Se_9$ NPs by 1–2 orders of magnitude to obtain tens of grams of the products per synthesis maintaining their high quality. This work shows that fast addition of large injection volumes does not intrinsically limit upscaling of hot-injection-based colloidal syntheses.

As Cd-containing compounds are considered to be toxic, alternatives have been sought. An example is a large-scale synthesis of InPZnS alloy QDs, which yielded ~ 3 g of the product (after the work-up process) at lab scale (see Figure 1.4d) (Kim et al., 2012). In this method, a phosphorous precursor (a solution of tris(trimethylsilyl) phosphine (TMS_3P) in ODE) was slowly added to the hot mixture (210°C) of metal precursors. Such slow addition appears to be more favorable for a potential industrial production of the material than a rapid injection. Thereafter, ZnS-rich shell was grown by adding zinc acetate and DDT followed by heating at 230°C for 5 h to improve the PLQY of the QDs and their stability. The emission of these InPZnS alloy/ZnS core/shell QDs could be precisely and reproducibly tuned by controlling the ratio of DDT/palmitic acid employed as a sulfur-source and ligands.

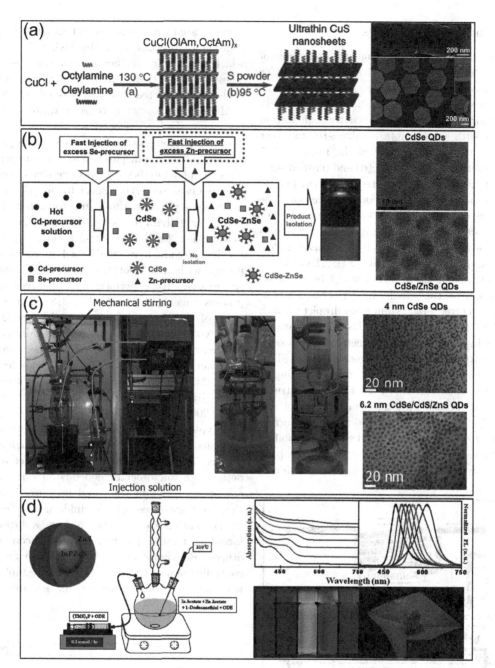

FIGURE 1.4 Examples of large-scale hot-injection syntheses of different materials. (a) Scheme of the synthesis of CuS nanosheets with corresponding electron microscopy images (Du et al., 2012). (b) Scheme of successive injection of precursors in one pot for preparing CdSe/ZnSe core/shell QDs with a photo of corresponding NP colloid and TEM images of core CdSe and core/shell CdSe/CdS QDs (Kim & Lee, 2006). (c, from left to right) Experimental setup used for the gram-scale synthesis of CdSe QDs; CdSe/CdS/ZnS core/shell/shell QDs in the 2 L reactor at the final stage of the synthesis; purification using a glass filter column retaining the precipitated QDs, while the solvent and by-products are collected in the round-bottom flask below; corresponding TEM images (Protière et al., 2011). (d) Experimental setup for the synthesis of InPZnS alloy QDs, absorption and emission spectra of alloyed InPZnS/ZnS core/shell QDs, photo of different QD colloids under UV light, and photo demonstrating the amount of the final product in a single reaction (Kim et al., 2012). (Reproduced with permission from Wiley VCH Verlag GmbH & Co, Nature Publishing Group, Springer, and American Chemical Society.)

1.3 Continuous and Multiphase Flow Synthesis

Continuous flow processes are performed in two main types of reactors: tubular and spinning disc (or tube) reactors. Although the main body of work on the continuous flow synthesis of NPs has been done on microscopic flow capillary-based reactors, there are few reports on large-scale synthesis using simpler macroscopic tubular reactors.

1.3.1   Tubular Flow Reactors

Shavel et al. demonstrated the continuous production of grams of CZTS NPs with controlled composition under open-air conditions in a macroscopic tubular reactor (Figure 1.5a) (Shavel et al., 2012). In this method, a pre-prepared precursor solution was pumped through a 1 m long bronze tube with a 3 mm internal diameter and kept at 300°C–320°C, with the flow rate in the range of 1–5 mL/min. The product NPs were collected and purified similarly to batch synthesis, i.e., washed several times by precipitation with isopropanol and redispersion in chloroform. The Suslick group proposed an original flow synthesis of Cd-chalcogenide nanocrystals, employing a tubular furnace as a reactor, the so-called continuous chemical aerosol flow synthesis (Figure 1.5b) (Bang et al., 2008; Didenko & Suslick, 2005). Although the reaction vessel in this case was a macroscopic tube, the actual reaction between two or three precursors occurred inside submicron droplets, which were generated by the ultrasonic nebulization of solutions of high boiling point liquids containing metal and chalcogen precursors in an ultrasonic transducer working at 1.7 MHz, taken from a domestic humidifier. The mist consisting of these droplets was then carried by an argon gas stream into a heated tube furnace with temperature controllable from 180°C to 400°C. In order to facilitate droplet formation, the precursor solutions were diluted with a low boiling point solvent (toluene). As the temperature of the droplets entering the furnace exceeded the boiling point of toluene,

it evaporated, leaving submicron droplets of a concentrated solution of precursors in the high boiling point solvent, which reacted by further temperature rise forming NPs. These particles then exited the furnace, and were rapidly cooled and collected in a solvent-filled bubbler.

Microfluidic reactor systems for various applications have been extensively developed within last years, with many design improvements made to avoid fouling and to increase product output (multiple channels, pile-up prevention, etc.). This method has been very successful in diverse areas of the chemical research, especially for the miniaturization of (bio)analytical methods and for synthetic (bio)organic chemistry (Elvira et al., 2013; Rodrigues et al., 2014; Whitesides, 2006). The strides made in this field over the last decade have established it into a low-cost technology, with small-footprint machinery, allowing for sensitive analyses to be performed at high resolution and short process times (Elvira et al., 2013). The various advantages of this method stem from the fact that it involves small volumes of liquids in geometrically controlled environments and is subdivided into further components, such as mixer chips, reactor coils, and analytical units (Jahn et al., 2008; Marre & Jensen, 2010; Nightingale & de Mello, 2010; Nightingale & deMello, 2013; Niu et al., 2015; Song et al., 2008; Zhao et al., 2011).

A typical flow chemistry platform comprising of the key elements is depicted in Figure 1.6, although all of them are not required at the same time (Myers et al., 2014). The design of a microreactor depends on various factors, such as operating temperature and pressure, intended application, and harshness of the fluids used. The material for its fabrication can be chosen from a broad assortment, including metals, polymers, glass, silicon, ceramics, and others (Rodrigues et al., 2014). Manually or computer-controlled pumps (A) are employed for establishing a stream with accurate flow rates for the suitable solvents. The

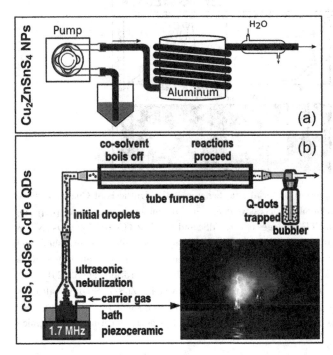

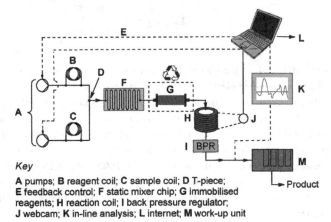

FIGURE 1.5   Examples of flow reactors for the synthesis of NPs. (a) Scheme of the setup for the synthesis of CZTS NPs (Shavel et al., 2012). (b) Setup for chemical aerosol flow synthesis of CdS, CdSe, and CdTe QDs. Inset: a photo of nebulized mist produced by the ultrasonic fountain (Didenko & Suslick, 2005). (Reproduced with permission from the American Chemical Society.)

FIGURE 1.6   Scheme of a typical micro-flow chemical synthesis platform depicting reagent pumps, mixing chips, immobilized reagent containing columns, flow reactor coils, and analytical (e.g. spectrometers) and work-up units (in-line evaporators) (Myers et al., 2014). (Reproduced with permission from Wiley VCH Verlag GmbH & Co.)

sample coils (B and C) are used to feed the precursors at proper temperatures, which can be achieved also by microwave heating. The precursor streams can be mixed at the T-piece (D) or in an in-line glass static mixer chip (F). Since most of the time the reaction has to run by constant supply of reagents, such provision can be made through cartridges packed with immobilized materials (G), which may be recycled periodically during breaks in the continuous flow process. The choice of the tubing is based on the temperature, pressure, and the nature of the liquid flowing through. The materials for this range from polyfluoroacetate to polytetrafluoroethylene (Teflon), steel, Hastelloy, and others. The reactor coil (H) needs to be carefully designed, since in this part of the flow system, almost all the "action" takes place. Consequently, the coil should be manufactured of unreactive material to withstand a wide range of temperatures, pressures, and different flow rates. The pressure can be adjusted through in-line back-pressure regulators (I). The entire system is monitored through a network of webcams (J) and in-line monitoring equipment (K; such as various spectrometers), which are computer-controlled and might be regulated remotely through the internet (L). The work-up unit (M) at the end of the system leads to the final product collection. In the reactor, converging streams of fluids flow in parallel but without turbulence, resulting in diffusion controlled mixing of compounds at the congregating fluid-interface, enabling a rapid transport of heat and mass. Additionally, the high surface-to-volume ratio of miniaturized channels and reactor coils often results in a dramatic enhancement in throughput and yields, as compared to a corresponding batch synthesis (Reyes et al., 2002). The flow systems are valuable for conducting reactions involving harsh conditions, such as superheated solvents, owing to a better control and more safety in comparison with the conventional methods. Similarly, toxic precursors and side products are confined in the system and can be removed and disposed continuously, thus avoiding a pile-up of large amounts of toxic materials.

Different groups have designed their own reactors with in-line heating, based on the requirements for the synthesis. Microfluidic production of Au, Ag, Co, Cu, SiO_2, TiO_2, CdS, CdSe, CdSe/ZnS, and CdSe/CdS core/shell NPs has been reported. Some of the setups developed are illustrated in Figure 1.7. An early report on the flow synthesis of colloidal CdS NPs in an aqueous media was published in 1992 by Fischer and Giersig (1992). In their method, two water solutions ($Cd(ClO_4)_2$ with sodium polyphosphate and NaSH with polyphosphate) were pumped into a mixing chamber in conjunction with a chromatographic column to obtain NPs, which further passed through the in-line detector for a rapid recording of their absorption spectra. One of the initial reasons why flow reactors gained interest as a NP synthesis technique was that the existing at that time methods were time-consuming and unsuitable for large-scale production, as the NP samples were polydisperse and required repetitive size-selection to obtain a desired particle size.

In a next step, in-line heating (such as oil bath shown in Figure 1.7b,c) was introduced. This helped to solve the long-standing problem of inhomogeneity in mixing and temperature control in large-scale reactors. For example, CdSe QDs were prepared at 230°C–300°C from a mixture of TOP-Se and $Cd(CH_3COO)_2$ in stearic acid, and trioctylphosphine oxide (TOPO) loaded into a glass syringe and connected to a glass capillary, one section of which was heated in an oil bath (Nakamura et al., 2002). Continuous production was demonstrated, and narrow size distribution of the NPs was achieved using a 200 µm thin capillary. In 2002, a significant step forward was made by the de Mello group in the synthesis of CdS NPs employing a continuous flow micromixer, where the inlets containing the reactants were split into separate multichannel streams before they were brought in contact for mixing (Edel et al., 2002). Later, in 2004, the same group reported the synthesis of CdSe NPs through the direct reaction of Se and $Cd(CH_3COO)_2$ dissolved in a TOPO-TOP mixture, coupled with an in-line analysis system that enabled real-time monitoring of the quality of the obtained material (Krishnadasan et al., 2004). This was a major step forward to designing intelligent feedback-driven NPs production. Likewise, OlAc/OlAm stabilized InP nanocrystals were obtained in a microfluidic reactor by in-line recording of their PL spectra (Nightingale & de Mello, 2009). As in the case of single batch reactions, the use of single-source precursors simplifies the hardware and facilitates the control of the NP synthesis in a microfluidic setup. An example of this strategy is the synthesis of CdS, CdSe, CdSe/CdS core/shell, and CdSeS alloy QDs from $Cd(S_2CNMe^nHex)_2$ and $Cd(Se_2P^iPr_2)_2$ dissolved in OlAm (Abdelhady et al., 2011).

Another breakthrough was made by the Weller group when they developed a continuous flow system that mimics the hot-injection method to produce fluorescent CdSe QDs with optional shell-coating (CdS, ZnS, or CdS/ZnS double shell) (Ness et al., 2012). The interchangeable components of the reactor made it possible to obtain the QDs of any desired configuration. In this system, nucleation and growth steps were separated in time and in space in two different parts of the system. Moreover, the reactor was able to withstand temperatures as high as 320°C, typical to hot-injection synthesis. Recently, high-temperature continuous flow synthesis of core/shell CdS/ZnS and CdSeS/ZnS as well as core/shell/shell CdSe/CdS/ZnS QDs was successfully realized in rather millifluidic than microfluidic regime (in a stainless steel coil with an inner diameter of 2.2 mm) (Naughton et al., 2015). A simple design of the reactor allowed the authors to exclude preheating or in-line mixing and to achieve high flow rates of ~10 mL/min. Moreover, high PLQYs of 50%–60% for the obtained QDs were demonstrated.

In recent years, the use of capillary-based droplet reactors has become more popular due to a series of issues with continuous flow reactors. One of them is the spread of residence times due to the viscous drag at the channel walls, leading to different fluid velocities in the flow. Another one

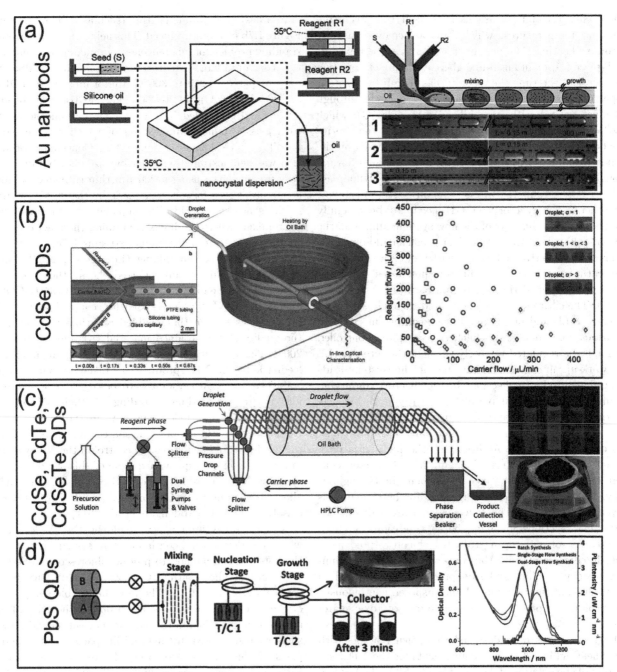

FIGURE 1.7 Examples of the microfluidic reactor systems. (a, left) Setup for the synthesis of anisotropic Au NPs: aqueous solutions of Au^{3+}, Ag^+, and CTAB (R1), ascorbic acid (R2), and Au seed NPs (S) are separately fed through individual inlets into one arm of the T-junction. Silicone oil, as the continuous liquid, is fed to the perpendicular arm. (a, right, top) Reagents get in contact at the T-junction, being dispensed into small picoliter droplets. The oil forms a thin lubricating layer around the droplets preventing contact between growing NPs and the microchannel walls. (a, right, bottom) Stereomicroscopy images of flow patterns observed in the microchannel under different flow conditions (Duraiswamy & Khan, 2009). (b, left) Setup for the synthesis of CdSe NPs comprising droplet generator, heated oil-bath, and in-line optical detector with a close-up of a droplet generation stage. (b, right) Dynamic phase diagram for ODE droplets in perfluoropolyether (PFPE), indicating the flow rate of the carrier and reagent phases with the marker shape denoting the aspect ratio of the resultant droplets (Nightingale et al., 2011). (c, left) A multichannel reactor for the synthesis of CdSe, CdTe, and alloyed CdSeTe NPs. (c, right) Photographs of the corresponding NP colloids under UV illumination and of 54.4 g of purified CdTe NP powder from 9 h production run (Nightingale et al., 2013). (d, left) Schematic of a dual-stage continuous flow reactor for the synthesis of PbS QDs from Pb-oleate in ODE (precursor A) and bis(trimethylsilyl) sulfide (TMS) in ODE (precursor B). (d, right) Absorbance and PL spectra of PbS QDs synthesized with a single-stage flow approach compared to batch synthesis and dual-stage flow setup (Pan et al., 2013).

(Continued)

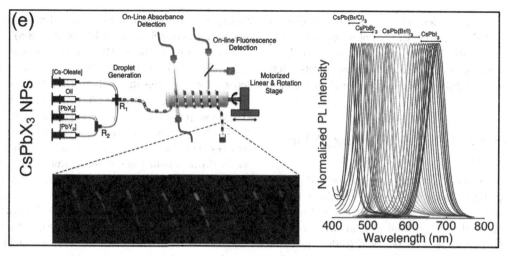

FIGURE 1.7 (CONTINUED) (e, top left) Setup integrated with online absorbance and PL detection for the synthesis and real-time characterization of CsPbX$_3$ perovskite NPs, which allows for precise tuning of the chemical payload of the formed droplets by continuously varying the ratio between the lead and cesium sources (R1) and the ratio between halides (R2). (e, bottom left) Image of the generated droplets after exiting the heating zone, showing bright PL of CsPbX$_3$ NPs under UV excitation. (e, right) Online PL spectra of CsPbX$_3$ NPs (X: Cl, Br, I, and Cl/Br and Br/I mixed halide systems) (Lignos et al., 2016). (Reproduced with permission from Wiley VCH Verlag GmbH & Co, the Royal Society of Chemistry, and American Chemical Society.)

is the deposition of NPs on the reactor's walls, eventually leading to reactor fouling. Instead, in capillary-based droplet reactors, droplets are generated commonly by mixing a precursor solution with an immiscible liquid acting as a carrier fluid. For example, an aqueous solution containing precursors and metal NP seeds was mixed with silicone oil (Figure 1.7a) (Duraiswamy & Khan, 2009). In the other case, the solvent of the droplet phase was ODE, and the carrier fluid was a perfluorinated polyether used to synthesize Ag, TiO$_2$, and CdSe NPs (Figure 1.7b) (Nightingale et al., 2011). This capillary-based process was also extended to prepare dextran-coated superparamagnetic iron oxide NPs in water, which were small, stable, and highly crystalline, with narrow size distribution (Kumar et al., 2012). The same group also developed a multichannel (five-channel prototype) droplet reactor for the production of CdSe, CdTe and alloyed CdSeTe QDs (Figure 1.7c) (Nightingale et al., 2013). In particular, by using this system the authors demonstrated large-scale production of CdTe QDs (54.4 g from a 9 h production run) at a rate of 145 g/day.

Mixing fluid and gas streams was demonstrated to generate segmented flow where droplets of liquid were chopped by gas bubbles inside the capillary. This method was successfully employed for the continuous synthesis of Pt and Pd nanostructures of varied morphologies (Sebastian et al., 2016). In this approach, O$_2$, CO, or N$_2$ were used to chop both aqueous and non-aqueous fluids containing reactants and ligands. This enabled tuning the reduction kinetics and controlling the size and the shape of the NPs. In addition to the droplet generation, separation of the nucleation and the growth steps by introduction of a dual-temperature stage was used in the synthesis of PbS QDs developed by the Sargent group on a commercially available continuous flow reactor, FlowSyn Multi-X, (Figure 1.7d)

(Pan et al., 2013). The flow rate in this setup was typically in the range of 1–3 mL/min, the nucleation temperature was varied in the range of 80°C–150°C, while the growth temperature was set between 50°C and 100°C. The authors demonstrated that the solar cell fabricated from these QDs had a performance equivalent to that made from batch synthesized particles, and was superior to the conventional single-stage flow-synthesis product. In-line real-time monitoring of the near-infrared PL spectra enabled a rapid optimization of reaction conditions resulted in high-quality PbS QDs, emitting in the range of 765–1600 nm with QYs of up to 28%, as well as PbSe QDs with PL ranging from 860 to 1600 nm (Lignos et al., 2014). The thus-produced QDs were successfully used in the fabrication of Schottky solar cells, exhibiting 3.4% power conversion efficiency.

Recently developed and already booming cesium lead halide perovskite CsPbX$_3$ nanocrystals were synthesized on a microfluidic platform by the groups of deMello and Kovalenko (Figure 1.7e) (Lignos et al., 2016). Apart from its great potential for a large-scale and reproducible production of these nanomaterials having bright and tunable PL, microfluidic system in combination with in-line PL and absorption measurements allows the rigorous and rapid mapping of the reaction parameters, including molar ratios of precursors, reaction temperatures, and reaction times. The rapid screening is especially important in this system, since it has an extremely fast reaction kinetics, which is very difficult to monitor using a corresponding batch approach. This in turn translates into enormous savings in reagent usage and screening times. Another recent example of a complex synthesis realized on a microfluidic platform with the aid of in-line monitoring of the optical properties is a scalable production of core/shell CuInS$_2$/ZnS NPs (Yashina et al., 2016). This in-line detection enabled a rapid assessment of optimal reaction parameters which resulted in the formation of NPs with high PLQYs of up to 55%, emitting in the range of 580–760 nm, within a few seconds. In this method ZnS-single-source precursor was injected into droplets containing CuInS$_2$ NP cores without their purification.

1.3.2 Spinning Disc and Rotating Tube Reactors

Spinning disc and rotating tube reactors (also called processors, SDP and RTP, respectively) are considered as a scalable, continuous, and high-throughput method for NPs synthesis (Chen et al., 2014). Basically, an SDP consists of a rotating disc with speed control and feeding jets that supply reactants to the disc, as shown in Figure 1.8a. The disc can be rotated at high speeds to obtain thin fluid films (1–200 μm) resulting in a high area-to-volume ratio. These thin films are divided into two zones: (i) the injection point where the reagents hit the disc to form a pool in its center, and (ii) the acceleration and synchronized flow area, where the fluid film initially experiences an increase in radial flow velocity and then the liquid is rotating close to the disc velocity. The shear forces and viscous drag between the moving fluid layer and the disc surface create turbulence and ripples, thus giving rise to highly efficient mixing within the thin fluid film that ensures extremely short reaction residence times and enables impulse heating and immediate subsequent cooling. Controlling the levels of mixing associated with the waves offers the possibility of governing the

transport rates within thin films and hence tuning the NP size and size distribution (Chin et al., 2008).

Two designs of the SDPs have been reported so far. The first one is relatively straightforward, with the reagents fed to the spinning disc, where they form a thin film and interact, yielding the product which is taken away to the collection point at the bottom (Figure 1.8a) (Iyer et al., 2007). In the second type, also known as a high-gravity process, the disc is placed vertically so that the product easily drops out of the reactor chamber (Figure 1.8b). Two operation modes can be employed in these reactors: a continuous mode (Tai et al., 2009a) or a recycle mode (Tai et al., 2009b), when the collected product is fed back to the disc. Using the first type of design, Ag NPs of different sizes were synthesized from silver nitrate, ascorbic acid, and soluble starch as a stabilizing agent (as its biocompatibility is favorable for biomedical application) in water at room temperature (Iyer et al., 2007). The tunability in particle size from 5 to 200 nm was achieved by varying the disc speed and the concentration of the reactants. ZnO NPs were synthesized using two feed solutions: zinc nitrate hexahydrate and PVP in ethanol, and a base in ethanol (Hartlieb et al., 2007). In this method, size control was realized by raising the disc temperature up to 80°C, although this also increased the polydispersity of the NPs. For superparamagnetic Fe$_3$O$_4$ NPs to be used in various applications such as MRI, targeted drug delivery, magnetic separation, and others, the size has to be less than 20 nm. A continuous synthesis by SDP delivered Fe$_3$O$_4$ NPs in the range of 5–10 nm, with narrow size distribution, characterized by remarkably high

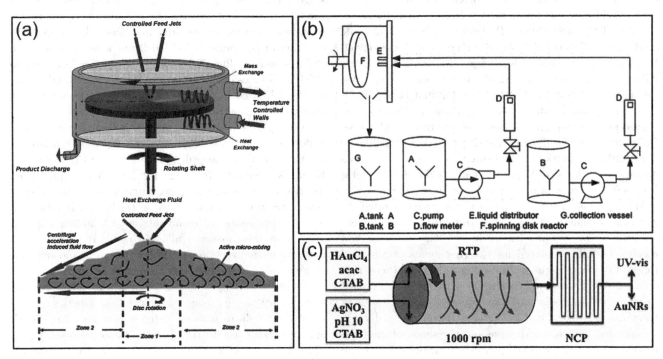

FIGURE 1.8 (a, top) Schematic representation of an SDP (Iyer et al., 2007). (a, bottom) Hydrodynamics of the fluid flow over a spinning disc surface (Chin et al., 2008). (b) High-gravity SDP system (Tai et al., 2009a). (c) Continuous flow synthesis of Au nanorods using an RTP serially with a narrow channel processor (NCP) (Bullen et al., 2011). (Reproduced with permission from the Royal Society of Chemistry, Wiley VCH Verlag GmbH & Co, and American Chemical Society.)

saturation magnetizations (Chin et al., 2008). For the first time, ammonia gas was used as a base blown on a thin film of aqueous iron salt on a disc.

By using the second type of the design in the recycle mode, Ag NPs of ≥10 nm were synthesized at room temperature by a green chemistry approach (Tai et al., 2008), in which two aqueous solutions, $AgNO_3$ with starch and NaOH with glucose (as a reducer), were fed into the center of the rotating disc (1000–4000 rpm) to form thin film and subsequently reacted, yielding the NPs. Recycle mode ensured NP yield of >56%, as the residence time on the disc was short and sometimes, one run was insufficient to reach an efficient conversion of reagents. This approach was further extended to synthesize Ag NPs of <10 nm in size by replacing starch with PVP, where the reaction yield was improved up to 85% by increasing the recycling time (Tai et al., 2009b). They also exploited this synthesis in continuous mode in order to increase the production rate and found that the yield was higher in the recycle mode (Tai et al., 2009a). Furthermore, using the same setup, spherical- and disc-shaped magnesium hydroxide NPs were synthesized (Tai et al., 2007).

RTPs are related to SDPs in generating thin films with turbulent flow, with an advantage of an extended residence time of the liquid in the reactor. While in SDPs, this time is less than a second for a 10 cm disc, in RDPs, the residence time can be controlled up to a minute depending on the size of the tube, flow rates, and viscosity of the liquid (Bullen et al., 2011). In an RTP, the centrifugal force generates dynamic thin films as the liquid moves along the rotating tube. Figure 1.8c presents an example of such reactor coupled with a narrow channel processor (NCP) which were used for the continuous seedless synthesis of Au nanorods (Bullen et al., 2011). The RTP employed in this work was 6 cm in diameter and 30 cm in length, and its rotation rate was set at 1000 rpm resulting in an average film thickness of ca. 300 μm at a residence time of 30 s. The authors used two feed solutions: $HAuCl_4$/cetyltrimethylammonium bromide (CTAB)/acetylacetonate and $AgNO_3$/CTAB/carbonate buffer. During the processing, Au NPs formed in the RTP were fed into the NCP for the further growth of the Au nanorods.

1.4 Conclusions and Outlook

In this chapter, the main methods for the colloidal synthesis of various NPs that have demonstrated large-scale production capabilities are summarized. Among the one-batch modes, the heat-up approach seems to be the most suitable for scale-up, as it does not require rapid heat and mass transport within the reaction mixture, opposite to the hot-injection method. In general, although one-batch synthesis appears to be more adaptable to existing industrial infrastructures, it still suffers from inefficient heat and mass management, especially in the case of high-temperature processes in large volumes, which eventually can lead to a poor reaction control. These issues can be surmounted

by turning the synthesis into a continuous mode employing tubular or spinning disc reactors. Especially attractive is the application of microfluidic systems, which benefit from a better control owing to a small amount of reaction mixture processed at a time, and reduced hazard potential. At the same time, this small reaction volume can be disadvantageous, as it significantly decreases throughput of the system. Here, a further development is needed to adopt microfluidic reactors to scale up amounts of NPs produced to meet industrial demands.

As a general trend in the large-scale production of NPs, one can define the use of low-cost precursors, low-energy consumption, as well as simple and environment-friendly processes. As discussed in this chapter, modern synthetic protocols, even by upscaling, ensure high quality and narrow size distribution of NPs, which has a special importance for the QDs with bright and narrow fluorescence engaged in various light-emitting applications. This allows one to eliminate additional post-synthetic treatment processes, such as size selection of the particles. However, such common steps, as the separation and the purification of the as-prepared NPs, are typically performed using batch principles rather than continuous ones. Such steps, involving the coagulation of NPs in reaction mixture, with their precipitation by centrifugation followed by repetitive washing, are presently not optimized or automated. This means that there is still plenty of room for improving the whole efficiency of the process, as, e.g., these steps too could be executed in a continuous mode. Here, regeneration and recycling of organic solvents/ligands used in the synthesis should lead to a significant reduction of the processes and materials costs, as solvents/ligands constitute a main part of the reaction mixture in the colloidal synthesis. This is the area where a valuable contribution from the chemical and process engineers is highly demanded.

For further advancing the large-scale production of NPs, one would need efficient coupling of the synthesis and the post-synthetic treatment units with a high-throughput analytical systems for in-line characterization and control of the processes and products. Directly related to this is the development of a user-friendly advanced software which would allow controlling all the production stages from one operator center. Therefore, to turn NPs to industrial and commercial lines, coordinated efforts from a wide range of specialists including synthetic and analytical chemists, chemical, process and automatization engineers, physicists, managers, etc. are highly required.

References

Abdelhady, A.L., Afzaal, M., Malik, M.A. & O'Brien, P. 2011. Flow reactor synthesis of CdSe, CdS, CdSe/CdS and CdSeS nanoparticles from single molecular precursor(s). *J. Mater. Chem.*, 21(46):18768–18775.

Akkerman, Q.A., Rainò, G., Kovalenko, M.V. & Manna, L. 2018. Genesis, challenges and opportunities for colloidal lead halide perovskite nanocrystals. *Nat. Mater.*, 17(5):394–405.

Bai, S., Yuan, Z. & Gao, F. 2016. Colloidal metal halide perovskite nanocrystals: synthesis, characterization, and applications. *J. Mater. Chem. C*, 4(18): 3898–3904.

Bang, J.H., Suh, W.H. & Suslick, K.S. 2008. Quantum dots from chemical aerosol flow synthesis: preparation, characterization, and cellular imaging. *Chem. Mater.*, 20(12):4033–4038.

Bullen, C., Latter, M.J., D'Alonzo, N.J., Willis, G.J. & Raston, C.L. 2011. A seedless approach to continuous flow synthesis of gold nanorods. *Chem. Commun.*, 47(14):4123–4125.

Chen, X., Smith, N.M., Iyer, K.S. & Raston, C.L. 2014. Controlling nanomaterial synthesis, chemical reactions and self assembly in dynamic thin films. *Chem. Soc. Rev.*, 43(5):1387–1399.

Chesman, A.S.R., Duffy, N.W., Peacock, S., et al. 2013. Non-injection synthesis of Cu_2ZnSnS_4 nanocrystals using a binary precursor and ligand approach. *RSC Adv.*, 3(4):1017–1020.

Chin, S.F., Iyer, K.S., Raston, C.L. & Saunders, M. 2008. Size selective synthesis of superparamagnetic nanoparticles in thin fluids under continuous flow conditions. *Adv. Funct. Mater.*, 18(6):922–927.

Choi, M.K., Yang, J., Kang, K., et al. 2015. Wearable red–green–blue quantum dot light-emitting diode array using high-resolution intaglio transfer printing. *Nat. Commun.*, 6:7149.

Didenko, Y.T. & Suslick, K.S. 2005. Chemical aerosol flow synthesis of semiconductor nanoparticles. *J. Am. Chem. Soc.*, 127(35):12196–12197.

Du, Y., Yin, Z., Zhu, J., et al. 2012. A general method for the large-scale synthesis of uniform ultrathin metal sulphide nanocrystals. *Nat. Commun.*, 3:1177.

Duraiswamy, S. & Khan, S.A. 2009. Droplet-based microfluidic synthesis of anisotropic metal nanocrystals. *Small*, 5(24):2828–2834.

Edel, J.B., Fortt, R., deMello, J.C. & deMello, A.J. 2002. Microfluidic routes to the controlled production of nanoparticles. *Chem. Commun.*, (10):1136–1137.

Elvira, K.S., i Solvas, X.C., Wootton, R.C.R. & deMello, A.J. 2013. The past, present and potential for microfluidic reactor technology in chemical synthesis. *Nat. Chem.*, 5(11):905–915.

Feng, H., Yang, Y., You, Y., et al. 2009. Simple and rapid synthesis of ultrathin gold nanowires, their self-assembly and application in surface-enhanced Raman scattering. *Chem. Commun.*, (15):1984–1986.

Fischer, C.H. & Giersig, M. 1992. Colloidal cadmium sulfide preparation via flow techniques: ultrasmall particles and the effect of a chromatographic column. *Langmuir*, 8(5):1475–1478.

Gaponik, N., Talapin, D.V., Rogach, A.L., et al. 2002. Thiol-capping of CdTe nanocrystals: an alternative to organometallic synthetic routes. *J. Phys. Chem. B*, 106(29):7177–7185.

Gerbec, J.A., Magana, D., Washington, A. & Strouse, G.F. 2005. Microwave-enhanced reaction rates for nanoparticle synthesis. *J. Am. Chem. Soc.*, 127(45):15791–15800.

González-Carrero, S., Galian, R.E. & Pérez-Prieto, J. 2015. Organometal halide perovskites: bulk low-dimension materials and nanoparticles. *Part. Part. Syst. Charact.*, 32(7):709–720.

González-Carrero, S., Galian, R.E. & Pérez-Prieto, J. 2016. Organic-inorganic and all-inorganic lead halide nanoparticles. *Opt. Express*, 24(2):A285–A301.

Hartlieb, K.J., Raston, C.L. & Saunders, M. 2007. Controlled scalable synthesis of ZnO nanoparticles. *Chem. Mater.*, 19(23):5453–5459.

Iyer, K.S., Raston, C.L. & Saunders, M. 2007. Continuous flow nano-technology: manipulating the size, shape, agglomeration, defects and phases of silver nanoparticles. *Lab. Chip*, 7(12):1800–1805.

Jahn, A., Reiner, J., Vreeland, W., et al. 2008. Preparation of nanoparticles by continuous-flow microfluidics. *J. Nanopart. Res.*, 10(6):925–934.

Jana, N.R. 2005. Gram-Scale synthesis of soluble, near-monodisperse gold nanorods and other anisotropic nanoparticles. *Small*, 1(8–9):875–882.

Jensen, K.F., Reizman, B.J. & Newman, S.G. 2014. Tools for chemical synthesis in microsystems. *Lab. Chip*, 14(17):3206–3212.

Joo, J., Yu, T., Kim, Y.W., et al. 2003. Multigram scale synthesis and characterization of monodisperse tetragonal zirconia nanocrystals. *J. Am. Chem. Soc.*, 125(21):6553–6557.

Kagan, C.R., Lifshitz, E., Sargent, E.H. & Talapin, D.V. 2016. Building devices from colloidal quantum dots. *Science*, 353(6302):885.

Kim, B.H., Lee, N., Kim, H., et al. 2011. Large-scale synthesis of uniform and extremely small-sized iron oxide nanoparticles for high-resolution T1 magnetic resonance imaging contrast agents. *J. Am. Chem. Soc.*, 133(32):12624–12631.

Kim, B.H., Shin, K., Kwon, S.G., et al. 2013. Sizing by weighing: characterizing sizes of ultrasmall-sized iron oxide nanocrystals using MALDI-TOF mass spectrometry. *J. Am. Chem. Soc.*, 135(7):2407–2410.

Kim, J.I. & Lee, J.K. 2006. Sub-kilogram-scale one-pot synthesis of highly luminescent and monodisperse core/shell quantum dots by the successive injection of precursors. *Adv. Funct. Mater.*, 16(16):2077–2082.

Kim, T., Kim, S.W., Kang, M. & Kim, S.-W. 2012. Large-scale synthesis of InPZnS alloy quantum dots with dodecanethiol as a composition controller. *J. Phys. Chem. Lett.*, 3(2):214–218.

Klimov, V.I. 2004. *Semiconductor and Metal Nanocrystals: Synthesis and Electronic and Optical Properties*. 500. [Online] Available from www.crcpress.com/Semiconductor-and-Metal-Nanocrystals-Synthesis-and-Electronic-and-Optical/Klimov/p/book/9780824747169.

Kolen'ko, Y.V., Bañobre-López, M., Rodríguez-Abreu, C., et al. 2014. Large-scale synthesis of colloidal

Fe$_3$O$_4$ nanoparticles exhibiting high heating efficiency in magnetic hyperthermia. *J. Phys. Chem. C,* 118(16):8691–8701.

Kovalenko, M.V., Manna, L., Cabot, A., et al. 2015. Prospects of nanoscience with nanocrystals. *ACS Nano,* 9(2):1012–1057.

Kovalenko, M.V., Protesescu, L. & Bodnarchuk, M.I. 2017. Properties and potential optoelectronic applications of lead halide perovskite nanocrystals. *Science,* 358(6364):745–750.

Krishnadasan, S., Tovilla, J., Vilar, R., deMello, A.J. & deMello, J.C. 2004. On-line analysis of CdSe nanoparticle formation in a continuous flow chip-based microreactor. *J. Mater. Chem.,* 14(17):2655–2660.

Kumar, K., Nightingale, A.M., Krishnadasan, S.H., et al. 2012. Direct synthesis of dextran-coated superparamagnetic iron oxide nanoparticles in a capillary-based droplet reactor. *J. Mater. Chem.,* 22(11):4704–4708.

Kwon, S.G., Piao, Y., Park, J., et al. 2007. Kinetics of monodisperse iron oxide nanocrystal formation by "Heating-Up" process. *J. Am. Chem. Soc.,* 129(41): 12571–12584.

Lakshminarayana, P. & Qing-Hua, X. 2009. A simple method for large scale synthesis of highly monodisperse gold nanoparticles at room temperature and their electron relaxation properties. *Nanotechnology,* 20(18):185606.

LaMer, V.K. & Dinegar, R.H. 1950. Theory, production and mechanism of formation of monodispersed hydrosols. *J. Am. Chem. Soc.,* 72(11):4847–4854.

Lee, J., Yang, J., Kwon, S.G. & Hyeon, T. 2016. Nonclassical nucleation and growth of inorganic nanoparticles. *Nat. Rev. Mater.,* 1:16034.

Lee, J., Zhang, S. & Sun, S. 2013. High-temperature solution-phase syntheses of metal-oxide nanocrystals. *Chem. Mater.,* 25(8):1293–1304.

Lee, Y., Choi, J.-r., Lee, K.J., Stott, N.E. & Kim, D. 2008. Large-scale synthesis of copper nanoparticles by chemically controlled reduction for applications of inkjet-printed electronics. *Nanotechnology,* 19(41):415604.

Lesnyak, V., Gaponik, N. & Eychmüller, A. 2013. Colloidal semiconductor nanocrystals: the aqueous approach. *Chem. Soc. Rev.,* 42(7):2905–2929.

Li, L., Protière, M. & Reiss, P. 2008. Economic synthesis of high quality InP nanocrystals using calcium phosphide as the phosphorus precursor. *Chem. Mater.,* 20(8): 2621–2623.

Lignier, P., Bellabarba, R. & Tooze, R.P. 2012. Scalable strategies for the synthesis of well-defined copper metal and oxide nanocrystals. *Chem. Soc. Rev.,* 41(5): 1708–1720.

Lignos, I., Protesescu, L., Stavrakis, S., et al. 2014. Facile droplet-based microfluidic synthesis of monodisperse IV–VI semiconductor nanocrystals with coupled in-line NIR fluorescence detection. *Chem. Mater.,* 26(9): 2975–2982.

Lignos, I., Stavrakis, S., Nedelcu, G., et al. 2016. Synthesis of cesium lead halide perovskite nanocrystals in a droplet-based microfluidic platform: fast parametric space mapping. *Nano Lett.,* 16(3):1869–1877.

Liu, J.-H., Fan, J.-B., Gu, Z., et al. 2008. Green chemistry for large-scale synthesis of semiconductor quantum dots. *Langmuir,* 24(10):5241–5244.

Lou, W., Chen, M., Wang, X. & Liu, W. 2007. Novel single-source precursors approach to prepare highly uniform Bi$_2$S$_3$ and Sb$_2$S$_3$ nanorods via a solvothermal treatment. *Chem. Mater.,* 19(4):872–878.

Lu, X., Zhuang, Z., Peng, Q. & Li, Y. 2011. Controlled synthesis of wurtzite CuInS$_2$ nanocrystals and their side-by-side nanorod assemblies. *CrystEngComm,* 13(12): 4039–4045.

Marciello, M., Connord, V., Veintemillas-Verdaguer, S., et al. 2013. Large scale production of biocompatible magnetite nanocrystals with high saturation magnetization values through green aqueous synthesis. *J. Mater. Chem. B,* 1(43):5995–6004.

Marre, S. & Jensen, K.F. 2010. Synthesis of micro and nanostructures in microfluidic systems. *Chem. Soc. Rev.,* 39(3):1183–1202.

Medintz, I.L., Uyeda, H.T., Goldman, E.R. & Mattoussi, H. 2005. Quantum dot bioconjugates for imaging, labelling and sensing. *Nat. Mater.,* 4(6):435–446.

Miao, S., Hickey, S.G., Waurisch, C., et al. 2012. Synthesis of monodisperse cadmium phosphide nanoparticles using ex-situ produced phosphine. *ACS Nano,* 6(8):7059–7065.

Michalet, X., Pinaud, F.F., Bentolila, L.A., et al. 2005. Quantum dots for live cells, in vivo imaging, and diagnostics. *Science,* 307(5709):538–544.

Murray, C.B., Norris, D.J. & Bawendi, M.G. 1993. Synthesis and characterization of nearly monodisperse CdE (E = sulfur, selenium, tellurium) semiconductor nanocrystallites. *J. Am. Chem. Soc.,* 115(19):8706–8715.

Myers, R.M., Fitzpatrick, D.E., Turner, R.M. & Ley, S.V. 2014. Flow chemistry meets advanced functional materials. *Chem. Eur. J.,* 20(39):12348–12366.

Nakamura, H., Yamaguchi, Y., Miyazaki, M., et al. 2002. Preparation of CdSe nanocrystals in a micro-flow-reactor. *Chem. Commun.,* (23):2844–2845.

Naughton, M.S., Kumar, V., Bonita, Y., Deshpande, K. & Kenis, P.J.A. 2015. High temperature continuous flow synthesis of CdSe/CdS/ZnS, CdS/ZnS, and CdSeS/ZnS nanocrystals. *Nanoscale,* 7(38):15895–15903.

Ness, D., Niehaus, J., Tran, V.-H. & Weller, H. 2012. Sustainable synthesis of semiconductor nanoparticles in a continuous flow reactor. *Mater. Res. Soc. Symp. Proc.,* 1386: mrsf11-1386-d07-02.

Nightingale, A.M., Bannock, J.H., Krishnadasan, S.H., et al. 2013. Large-scale synthesis of nanocrystals in a multichannel droplet reactor. *J. Mater. Chem. A,* 1(12):4067–4076.

Nightingale, A.M. & de Mello, J.C. 2009. Controlled synthesis of III–V quantum dots in microfluidic reactors. *ChemPhysChem,* 10(15):2612–2614.

Nightingale, A.M. & de Mello, J.C. 2010. Microscale synthesis of quantum dots. *J. Mater. Chem.*, 20(39):8454–8463.

Nightingale, A.M. & deMello, J.C. 2013. Segmented flow reactors for nanocrystal synthesis. *Adv. Mater.*, 25(13): 1813–1821.

Nightingale, A.M., Krishnadasan, S.H., Berhanu, D., et al. 2011. A stable droplet reactor for high temperature nanocrystal synthesis. *Lab Chip*, 11(7):1221–1227.

Niu, G., Ruditskiy, A., Vara, M. & Xia, Y. 2015. Toward continuous and scalable production of colloidal nanocrystals by switching from batch to droplet reactors. *Chem. Soc. Rev.*, 44(16):5806–5820.

Pan, J., El-Ballouli, A.a.O., Rollny, L., et al. 2013. Automated synthesis of photovoltaic-quality colloidal quantum dots using separate nucleation and growth stages. *ACS Nano*, 7(11):10158–10166.

Panda, A.B., Glaspell, G. & El-Shall, M.S. 2006. Microwave synthesis of highly aligned ultra narrow semiconductor rods and wires. *J. Am. Chem. Soc.*, 128(9):2790–2791.

Panda, A.B., Glaspell, G. & El-Shall, M.S. 2007. Microwave synthesis and optical properties of uniform nanorods and nanoplates of rare earth oxides. *J. Phys. Chem. C*, 111(5):1861–1864.

Panthani, M.G. & Korgel, B.A. 2012. Nanocrystals for electronics. *Annu. Rev. Chem. Biomol. Eng.*, 3(1):287–311.

Park, J., An, K., Hwang, Y., et al. 2004. Ultra-large-scale syntheses of monodisperse nanocrystals. *Nat. Mater.*, 3(12):891–895.

Park, J., Joo, J., Soon, G.K., Jang, Y. & Hyeon, T. 2007. Synthesis of monodisperse spherical nanocrystals. *Angew. Chem. Int. Ed.*, 46(25):4630–4660.

Park, J., Kwon, S.G., Jun, S.W., Kim, B.H. & Hyeon, T. 2012. Large-scale synthesis of ultra-small-sized silver nanoparticles. *ChemPhysChem*, 13(10):2540–2543.

Park, J.C., Kim, J., Kwon, H. & Song, H. 2009. Gram-scale synthesis of Cu$_2$O nanocubes and subsequent oxidation to CuO hollow nanostructures for lithium-ion battery anode materials. *Adv. Mater.*, 21(7):803–807.

Patra, C.R., Alexandra, G., Patra, S., et al. 2005. Microwave approach for the synthesis of rhabdophane-type lanthanide orthophosphate (Ln = La, Ce, Nd, Sm, Eu, Gd and Tb) nanorods under solvothermal conditions. *New J. Chem.*, 29(5):733–739.

Pelaz, B., Jaber, S., de Aberasturi, D.J., et al. 2012. The state of nanoparticle-based nanoscience and biotechnology: progress, promises, and challenges. *ACS Nano*, 6(10):8468–8483.

Peng, Z.A. & Peng, X. 2000. Formation of high-quality CdTe, CdSe, and CdS nanocrystals using CdO as precursor. *J. Am. Chem. Soc.*, 123(1):183–184.

Phillips, T.W., Lignos, I.G., Maceiczyk, R.M., deMello, A.J. & deMello, J.C. 2014. Nanocrystal synthesis in microfluidic reactors: where next? *Lab. Chip*, 14(17): 3172–3180.

Protesescu, L., Yakunin, S., Bodnarchuk, M.I., et al. 2015. Nanocrystals of cesium lead halide perovskites (CsPbX$_3$,

X = Cl, Br, and I): novel optoelectronic materials showing bright emission with wide color gamut. *Nano Lett.*, 15(6):3692–3696.

Protière, M., Nerambourg, N., Renard, O. & Reiss, P. 2011. Rational design of the gram-scale synthesis of nearly monodisperse semiconductor nanocrystals. *Nanoscale Res. Lett.*, 6:472.

Reizman, B.J. & Jensen, K.F. 2016. Feedback in flow for accelerated reaction development. *Acc. Chem. Res.*, 49(9):1786–1796.

Reyes, D.R., Iossifidis, D., Auroux, P.-A. & Manz, A. 2002. Micro total analysis systems. 1. Introduction, theory, and technology. *Anal. Chem.*, 74(12):2623–2636.

Rodrigues, T., Schneider, P. & Schneider, G. 2014. Accessing new chemical entities through microfluidic systems. *Angew. Chem. Int. Ed.*, 53(23):5750–5758.

Rogach, A.L. 2008. *Semiconductor Nanocrystal Quantum Dots: Synthesis, Assembly, Spectroscopy and Applications*.372. [Online] Available from www.springer.com/ gp/book/9783211752357.

Sapsford, K.E., Algar, W.R., Berti, L., et al. 2013. Functionalizing nanoparticles with biological molecules: developing chemistries that facilitate nanotechnology. *Chem. Rev.*, 113(3):1904–2074.

Sato, R., Kanehara, M. & Teranishi, T. 2011. Homoepitaxial size control and large-scale synthesis of highly monodisperse amine-protected palladium nanoparticles. *Small*, 7(4):469–473.

Schmid, G. 2010. *Nanoparticles: From Theory to Application*. 533. [Online] Available from www.wiley.com/en-us/Nanoparticles%3A+From+Theory+to+Application% 2C+2nd%2C+Completely+Revised+and+Updated+ Edition-p-9783527325894.

Sebastian, V., Smith, C.D. & Jensen, K.F. 2016. Shape-controlled continuous synthesis of metal nanostructures. *Nanoscale*, 8(14):7534–7543.

Shavel, A., Cadavid, D., Ibáñez, M., Carrete, A. & Cabot, A. 2012. Continuous production of Cu$_2$ZnSnS$_4$ nanocrystals in a flow reactor. *J. Am. Chem. Soc.*, 134(3):1438–1441.

Son, J.S., Choi, M.K., Han, M.-K., et al. 2012. n-Type nanostructured thermoelectric materials prepared from chemically synthesized ultrathin Bi$_2$Te$_3$ nanoplates. *Nano Lett.*, 12(2):640–647.

Son, J.S., Wen, X.-D., Joo, J., et al. 2009. Large-scale soft colloidal template synthesis of 1.4 nm thick CdSe nanosheets. *Angew. Chem. Int. Ed.*, 48(37):6861–6864.

Song, Y., Hormes, J. & Kumar, C.S.S.R. 2008. Microfluidic synthesis of nanomaterials. *Small*, 4(6):698–711.

Tai, C.Y., Tai, C.-T., Chang, M.-H. & Liu, H.-S. 2007. Synthesis of magnesium hydroxide and oxide nanoparticles using a spinning disk reactor. *Ind. Eng. Chem. Res.*, 46(17):5536–5541.

Tai, C.Y., Wang, Y.-H. & Liu, H.-S. 2008. A green process for preparing silver nanoparticles using spinning disk reactor. *AIChE Journal*, 54(2):445–452.

Tai, C.Y., Wang, Y.-H., Tai, C.-T. & Liu, H.-S. 2009a. Preparation of silver nanoparticles using a spinning disk reactor in a continuous mode. *Ind. Eng. Chem. Res.,* 48(22):10104–10109.

Tai, C.Y., Wang, Y.H., Kuo, Y.W., Chang, M.H. & Liu, H.S. 2009b. Synthesis of silver particles below 10 nm using spinning disk reactor. *Chem. Eng. Sci.,* 64(13):3112–3119.

Talapin, D.V., Lee, J.-S., Kovalenko, M.V. & Shevchenko, E.V. 2010. Prospects of colloidal nanocrystals for electronic and optoelectronic applications. *Chem. Rev.,* 110(1):389–458.

Thanh, N.T.K., Maclean, N. & Mahiddine, S. 2014. Mechanisms of nucleation and growth of nanoparticles in solution. *Chem. Rev.,* 114(15):7610–7630.

van Embden, J., Chesman, A.S.R., Della Gaspera, E., et al. 2014. $Cu_2ZnSnS_{4x}Se_{4(1-x)}$ solar cells from polar nanocrystal inks. *J. Am. Chem. Soc.,* 136(14):5237–5240.

Whitesides, G.M. 2006. The origins and the future of microfluidics. *Nature,* 442(7101):368–373.

Xia, Y., Xiong, Y., Lim, B. & Skrabalak, S.E. 2009. Shape-controlled synthesis of metal nanocrystals: simple chemistry meets complex physics? *Angew. Chem. Int. Ed.,* 48(1):60–103.

Yang, H., Ito, F., Hasegawa, D., Ogawa, T. & Takahashi, M. 2007. Facile large-scale synthesis of monodisperse Fe nanoparticles by modest-temperature decomposition of iron carbonyl. *J. Appl. Phys.,* 101(9):09J112.

Yang, J., Choi, M.K., Kim, D.-H. & Hyeon, T. 2016. Designed assembly and integration of colloidal nanocrystals for device applications. *Adv. Mater.,* 28(6):1176–1207.

Yang, J., Fainblat, R., Kwon, S.G., et al. 2015. Route to the smallest doped semiconductor: Mn^{2+}-Doped $(CdSe)_{13}$ clusters. *J. Am. Chem. Soc.,* 137(40):12776–12779.

Yang, Y.A., Wu, H., Williams, K.R. & Cao, Y.C. 2005. Synthesis of CdSe and CdTe nanocrystals without precursor injection. *Angew. Chem. Int. Ed.,* 44(41):6712–6715.

Yarema, M., Yarema, O., Lin, W.M.M., et al. 2017. Upscaling colloidal nanocrystal hot-injection syntheses via reactor underpressure. *Chem. Mater.,* 29(2):796–803.

Yashina, A., Lignos, I., Stavrakis, S., Choo, J. & deMello, A.J. 2016. Scalable production of $CuInS_2/ZnS$ quantum dots in a two-step droplet-based microfluidic platform. *J. Mater. Chem. C,* 4(26):6401–6408.

Yu, T., Joo, J., Park, Y.I. & Hyeon, T. 2005. Large-scale nonhydrolytic Sol–Gel synthesis of uniform-sized ceria nanocrystals with spherical, wire, and tadpole shapes. *Angew. Chem. Int. Ed.,* 44(45):7411–7414.

Zhang, H., Hyun, B.-R., Wise, F.W. & Robinson, R.D. 2012. A generic method for rational scalable synthesis of monodisperse metal sulfide nanocrystals. *Nano Lett.,* 12(11):5856–5860.

Zhao, C.-X., He, L., Qiao, S.Z. & Middelberg, A.P.J. 2011. Nanoparticle synthesis in microreactors. *Chem. Eng. Sci.,* 66(7):1463–1479.

Zherebetskyy, D., Scheele, M., Zhang, Y., et al. 2014. Hydroxylation of the surface of PbS nanocrystals passivated with oleic acid. *Science,* 344(6190):1380–1384.

Zhong, H., Lo, S.S., Mirkovic, T., et al. 2010. Noninjection gram-scale synthesis of monodisperse pyramidal $CuInS_2$ nanocrystals and their size-dependent properties. *ACS Nano,* 4(9):5253–5262.

Zhong, X., Feng, Y. & Zhang, Y. 2006. Facile and reproducible synthesis of red-emitting CdSe nanocrystals in amine with long-term fixation of particle size and size distribution. *J. Phys. Chem. C,* 111(2):526–531.

2

Plasma Synthesis of Nanomaterials

Antaryami Mohanta
*Empa–Swiss Federal Laboratories for
Materials Science and Technology*

Raj K. Thareja
Indian Institute of Technology Bhilai
Indian Institute of Technology Kanpur

2.1 Introduction

Nanomaterials have taken a significant place in both academia and industries due to their remarkable physical and chemical properties differing from the bulk counterpart having the same constituents (Byrappa et al. 2008, Rossetti et al. 1983). Various potential applications in several different fields such as coatings, catalysts, sensors, magnetic data storage, solar energy devices, ferro-fluids, cell labeling, special drug delivery systems, and filler in natural fiber-reinforced composites are manifestations of novel properties of the nanoparticles (Byrappa et al. 2008, Hosseini 2017). Several techniques have been developed to synthesize varieties of nanomaterials depending on various different applications. The synthesis primarily involves conversion of solid to solid, liquid to solid, or gas to solid. The nanomaterials production techniques are broadly classified into two approaches: (i) top-down approach and (ii) bottom-up approach. Typical example of top-down approach is a physical synthesis method in which bulk materials are graved to achieve nanoscale by mechanical grinding; ball milling and ion-beam milling are two most popular techniques. Although these are convenient and simple methods, the shape and size of the synthesized nanomaterials are not uniform and not reproducible. On the other hand, chemical synthesis method is an example of the bottom-up approach; it involves the controlled arrangement of small building blocks such as atomic and molecular structures to form larger structures in the nanoscale dimensions (Su and Chang 2018). The structures so synthesized have authentic size distribution and are normally reproducible. Nevertheless, chemical synthesis approach has an inherent drawback of contributing unintentional impurities to the synthesized products. The liquid containing nanomaterials produced by using chemical routes always contain other ions and reaction products that are difficult to separate from the liquid (Patel et al. 2013), and therefore, the products are not contamination free. Plasma synthesis methods have several distinct advantages including versatility and operation with reactive gases in addition to inert, reducing, oxidizing, and other atmospheres. Pulsed laser ablation (PLA) and radio frequency inductively coupled plasma (RF ICP) have proved their abilities to produce great varieties of nanomaterials with high purity (Kim et al. 2017, Shin et al. 2006). The techniques are being pursued actively in both academia and industries. PLA is a synthesis technique in which a high-power laser pulse focused onto a target material removes the surface layer of the material in the form of plasma consisting primarily of electrons, ions, and neutrals. The created plasma expands perpendicular to the target surface and collides with the species of the ambient and cools rapidly, subsequently forming nanoparticles because of supersaturation. PLA synthesis produces contaminant-free high-purity nanoparticles because of the use of high-purity target material and absence of chemical reagents as in solutions (Patel et al. 2013, Mohanta et al. 2015). RF ICP is another technique where plasma is created by inductive coupling; RF electric current and high voltage are supplied to a helical or spiral-like torch coil that ionizes the injected gas and heats to form a plasma due to the influence of electromagnetic field. Being electrodeless nature of the ICP torch in the inert gas atmosphere assures high-purity nanoparticles

(Boulos 1997). In principle, many different kinds of gas such as H_2, O_2, N_2, Ar, NH_3, and CH_4 can be used as plasma working gas. These gases assure ICP as heat source and an environment for chemical reactions (Leblanc et al. 2017). As an example, following reactions take place (Mohanta et al. 2017b) when H_2 is injected into the inductively coupled Ar plasma in a reactor:

$$Ar^+ + H_2 \rightarrow ArH^+ + H \qquad (2.1)$$

$$ArH^+ + e^- \rightarrow Ar + H^* \qquad (2.2)$$

where H^* denotes the excited state hydrogen atoms.

Various types of nanomaterials having different dimensions have been produced using the plasma-based synthesis techniques such as nanoparticles (Amendola et al. 2007), nanowires (Morales and Lieber 1998), nanorods (Okada et al. 2005), fullerenes and carbon nanotubes (Cota-Sanchez et al. 2005), nano-flakes (Pristavita et al. 2011), thin films (Gupta and Thareja 2013), and nanocomposites (Pandey et al. 2018). Therefore, it is imperative to have systematic discussion on the type of nanomaterials in order to understand the influence of various structures of different dimensions on the physical and chemical properties of the materials. The chapter is organized as follows: Section 2.2 discusses the type of nanomaterials. Three- to zero-dimensional semiconductor structures are described in Sections 2.2.1–2.2.4. In Section 2.3, the physical mechanism of PLA and plasma formation process are discussed. Section 2.4 contains the nanomaterials synthesis process by PLA. Section 2.5 describes the experimental confirmation of vapor-phase ZnO nanoparticles formation by Rayleigh scattering and photoluminescence (PL) spectroscopy, which also includes the concept of free excitons, bound exciton complexes (BECs), and exciton–exciton scattering process. Section 2.6 presents the experimental results on the synthesis of nanomaterials in liquids. Section 2.7 discusses heterogeneous colliding plasma for the synthesis of nanocomposites, and Section 2.8 is about synthesis of nanoflakes by using RF ICP reactor. To conclude the chapter, a brief summary is presented in Section 2.9.

2.2   Type of Nanomaterials

Nanomaterials exhibiting different morphologies, shapes, and sizes have been nomenclatured as nanoparticles, nano-flakes, nanowires, nanorods, nanofibres, nanoribbons, nanoflowers, nanoplates, core–shell nanostructures (NS), etc. (ISO/TS 80004-2 Part 2 2015). The term "nanoscale" corresponds to the size ranging from approximately 1 to 100 nm (ISO/TS 80004-2 Part 1 2015). Nanoparticles are nanoscale objects where all three external dimensions are in the nanoscale. Nanoparticles can be of metals, semiconductors, or insulators having distinct properties such as mechanical, electrical, magnetic, optical, and chemical different from that of molecules and corresponding bulk materials. The size of nanoparticles has strong influence on different properties due to change in surface-to-volume ratio and the

quantum confinement effect. For a spherical nanoparticle, the surface-to-volume ratio is inversely proportional to its radius (r) since surface area and volume of a sphere are given by $4\pi r^2$ and $\frac{4}{3}\pi r^3$; surface-to-volume ratio will be $\frac{3}{r}$. Thus, the dimension of surface-to-volume ratio is inverse of length (L^{-1}) which indicates that surface-to-volume ratio increases as the size of a material decreases. For simplicity, consider a single crystal material with simple cubic structure of lattice constant $\sim$4 Å that has one atom per unit cell. The volume and surface densities of atoms are therefore $\sim$1.6 $\times$ 10^{22} atoms/cm^3 and $\sim$1.0 $\times$ 10^{14} atoms/cm^2, respectively. Thus, the total number of atoms in volume and surface of a spherical particle of radius 1 cm is $\sim$7 $\times$ 10^{22} and $\sim$1.3 $\times$ 10^{15} atoms, respectively. The ratio of number of atoms at the surface to the number of atoms in the volume is about 2 $\times$ 10^{-8}. However, the total number of atoms in volume and surface of a spherical particle of radius 1 nm is $\sim$70 and $\sim$13 atoms, respectively. The ratio of number of atoms at the surface to the number of atoms in the volume is therefore $\sim$0.2. In other words, the percentage of the atoms at the surface of the nanoparticles increases with decrease in radius of the nanoparticles. The increasing number of atoms at the surface of the smaller size nanoparticles leads to increasing chemical reactivity that influences significantly the properties of a material. As a result, the inert bulk material becomes reactive when in nano form. Size-dependent surface luminescence in nanomaterials due to the influence of difference in number of atoms closer to the surface has been reported (Shalish et al. 2004). The properties of semiconductor nanoparticles strongly depend on size. Usually, semiconductors possess small nonzero band gap. PL peak energy is correlated with band gap energy in semiconductor materials. Mohanta et al. (2008) have reported blue-shift in PL peak energy with respect to that of bulk materials in vapor-phase ZnO nanoparticles and attributed this to the quantum confinement effect. They have also observed particle size-dependent emission due to inelastic exciton-exciton scattering (Mohanta et al. 2015). Based on the effective mass approximation, the relation between band gap of bulk semiconductor and that of nanoparticles is written as (Wong et al. 1998, Thareja and Mohanta 2010)

$$E_{np} = E_{bulk} + \frac{\pi^2 \hbar^2}{2r_q^2}\left(\frac{1}{m_e^*} + \frac{1}{m_h^*}\right) \qquad (2.3)$$

where

E_{np} is the band gap of nanoparticles showing the quantum confinement effect.

E_{bulk} is the band gap of the bulk material.

r_q is the radius of nanoparticles showing quantum confinement effect.

m_e^* and m_h^* are the effective masses of the electrons and the holes, respectively.

Usually, semiconductors may have exciton states below the band gap. The exciton binding energy in ZnO, a wide band gap semiconductor material belonging to II–VI group, is 60 meV indicating stable exciton states at room temperature. An exciton is an electron–hole pair bounded by

Columbic interaction. The relation between the exciton energy (E_{ex}), band gap (E_{bulk}), and nanoparticle radius (r_q) can be expressed as (Brus 1984, Kayanuma 1988)

$$E_{ex} = E_{bulk} + \frac{\pi^2 \hbar^2}{2r_q^2} \left(\frac{1}{m_e^*} + \frac{1}{m_h^*} \right) - \frac{1.786e^2}{\varepsilon r_q} - \frac{0.248 \mu e^4}{2\hbar^2 \varepsilon^2}$$
(2.4)

Equation (2.4) can be rewritten by considering Eq. (2.3) as

$$E_{ex} = E_{np} - \frac{1.786e^2}{\varepsilon r_q} - \frac{0.248 \mu e^4}{2\hbar^2 \varepsilon^2}$$
(2.5)

where

ε is the semiconductor dielectric constant.

μ is the reduced effective mass.

$$\mu = \frac{1}{\frac{1}{m_e^*} + \frac{1}{m_h^*}}$$
(2.6)

Nanofibers are the nanoscale objects having two external dimensions in the nanoscale between 1 and 100 nm where the third dimension is significantly larger (ISO/TS 80004-2 Part 2 2015). Nanofibril and nanofilament are synonyms for the term "nanofiber". Nanorods are categorized as solid nanofiber and nanotubes as hollow nanofiber. Nanowires are classified as electrical conducting or semiconducting nanofiber. The properties of the semiconducting nanowire materials strongly depend on radius. The luminescence band in ZnO nanowires usually shows ultra-violet (UV) emission from excitonic recombination near the band edge and a broad visible (green) below band gap emission which is mediated by energy transfer from band edge states to the responsible defect states (Foreman et al. 2006). The intensity ratio of green to UV emission decreases as the wire radius increases (Shalish et al. 2004). A nanoplate is a nano-object whose one external dimension is in the nanoscale and the other two dimensions are substantially larger, i.e., more than three times on the nanoscale dimension. Nanoribbons are those nanoplates whose two larger dimensions are significantly different from each other. The core–shell nanomaterials, in particular core–shell silicon–carbon nanoparticles, have attracted significant attention due to their potential applications, for their usage in high-performance anodes in lithium ion batteries. The mean silicon nanoparticle size is about 22 nm with carbon shell of thickness 1 nm in the core–shell silicon–carbon nanoparticles (Sourice et al. 2016).

Types of nanomaterials are categorized in five different heads: (i) nanostructured powder, (ii) nanocomposite, (iii) solid nanofoam, (iv) nanoporous material, and (v) fluid nanodispersion. Although powders are the solid objects consisting of very tiny discrete particles of size <1 mm (ISO3252:1999), powders and particles are not same. In addition to particles, powders contain gases and liquids usually in the form of air and water, respectively, on the particles' surface. Powders can be referred to as nanostructured powders if it comprises nanostructured agglomerates, aggregates, or nanoparticles (ISO/TS 80004-4:2011). In comparison with the bulk materials, nanopowders have

different optical, thermal, chemical, electrical, and physical properties that allow for enhanced chemical activity, faster sintering kinetics, higher electrical resistivity, superparamagnetic, microwave absorption, and localized surface plasmon resonances. Nanopowders are thus potential candidates for catalysts, sintering aids, microwave absorption, magnetic recording media, magnetic fluids, magnetic ink, rocket propellants, conducting ink/paste, permeable reactive barriers for soil decontamination, and biomarkers and biosensors applications (Chang 2013). Nanocomposites are solid materials having multiple phases in which one or more phases are in the nanoscale embedded in a metal, ceramic, or polymer matrix. Depending on the nature of reinforcing agents such as nanoparticles, nanoplatelets etc., nanocomposites are categorized as nanoparticle-reinforced composites, nanoplatelets-reinforced composites, nanofiber-reinforced composites, carbon nanotube-reinforced composites, nanocomposite fibrils, nanocomposite films, etc. (Hussain et al. 2006). A solid nanofoam is a solid matrix filled with a gaseous phase in which either phase is in the nanoscale dimension. Metallic foams such as aluminum foams have novel physical, mechanical, and electrical properties along with low density since the large portion of the volume is filled with gaseous pores that makes the materials ultra-light (Mirzaee et al. 2012). Nanoporous materials are solid with nanopores. These materials can be crystalline, amorphous, or mixture of both the structures. The nanoporous materials are divided into three broad categories according to their pore size: (i) microporous materials with 0–2 nm pores, (ii) mesoporous materials with 2–50 nm pores, and (iii) macroporous materials with pores >50 nm (Polarza and Smarsly 2002).

Several nanoporous materials are employed as heterogeneous catalysts, particularly zeolites mostly considered as solid acid catalysts for hydrocarbon cracking and isomerization (Sneddon et al. 2014). Solid nanofoam and nanoporous materials have similar definition since both contain pores. However, solid nanofoam is a solid material in which most of the volume is filled by pores, whereas nanoporous materials are materials with small fraction of pores.

Fluid nanodispersion is a heterogeneous material system in which the dispersed objects are in the nanoscale, and has several forms such as nano-suspension, nano-emulsion, liquid nanofoam, and nano-aerosol (ISO/TS 80004-4:2011). Mohanta and Thareja (2009), Mohanta et al. (2015) have reported the formation of suspended ZnO nanoparticles in vapor phase and occurrence of exciton–exciton scattering process. In order to better understand the effect of size on different physical and chemical properties, systematic discussion on materials structures in three-dimension (3D), two-dimension (2D), one-dimension (1D), and zero-dimension (0D) is imperative. Since both electrons and holes have significant influence on carrier dynamics in semiconductors, we consider semiconductor materials structures for further discussion.

2.2.1   Three-Dimensional (3D) Structure

Semiconductors are materials with nonzero small band gap having properties intermediate between insulators and conductors. At absolute zero temperature, they behave similar to insulators since the conduction band is empty and the valence band is fully occupied. However, at sufficiently higher temperature, the semiconductors become conducting due to thermal excitation of electrons from valence to conduction band leaving behind holes in the valence band. The energy band structure at the E-k extrema; that is, the conduction band minimum and the valence band maximum are approximated to parabolic (Saleh and Teich 1991, Thareja and Mohanta 2010). Under the parabolic band approximation, the carriers, i.e., electrons or holes of effective mass m_e^*or m_h^*, are free to move in the conduction or valence band, respectively, and can be treated as free carriers confined in a three-dimensional potential box. The time-independent Schrödinger wave equation for a free particle of energy E and effective mass m^* can be written as

$$-\frac{h^2}{8\pi^2 m^*}\nabla^2\psi(\vec{r}) = E\psi(\vec{r}) \qquad (2.7)$$

ψ is the wave function of the quantum system that represents the state and can be obtained as

$$\psi(\vec{r}) = Ae^{i\left(\vec{K}\cdot\vec{r}\right)} = Ae^{i(k_x x + k_y y + k_z z)} \qquad (2.8)$$

with

$$\left|\vec{K}\right| = k = \sqrt{\left(k_x^2 + k_y^2 + k_z^2\right)} = \sqrt{\frac{8\pi^2 m^*}{h^2}}\,(E)^{\frac{1}{2}} \qquad (2.9)$$

and A is an arbitrary constant.

The allowed values of $\vec{K}$ are

$$(k_x, k_y, k_z) = 0, \pm\frac{2\pi}{L}, \pm\frac{4\pi}{L}, \pm\frac{6\pi}{L}, \ldots, \pm\frac{2n\pi}{L} \qquad (2.10)$$

which are obtained using the following periodic boundary conditions in x, y and z with period L,

$$\psi\left(x + L, y, z\right) = \psi\left(x, y, z\right)$$
$$\psi\left(x, y + L, z\right) = \psi\left(x, y, z\right)$$
$$\psi\left(x, y, z + L\right) = \psi\left(x, y, z\right) \qquad (2.11)$$

Equation (2.10) indicates that in each volume element $\left(\frac{2\pi}{L}\right)^3$ of a 3D k-space, there is one allowed wave vector $\vec{K}$, and the number of states between E and $E + \mathrm{d}E$ is

$$N_s(E)\mathrm{d}E = \pi\left(\frac{2L}{h}\right)^3 (m^*)^{\frac{3}{2}}(2E)^{\frac{1}{2}}\mathrm{d}E \quad \text{for } E \geq 0 \quad (2.12)$$

If E_c is the minimum energy of the electron at the bottom of the conduction band, then the density of the states of electrons in the conduction band per unit volume (L^3) can be expressed as

$$d_c(E) = \pi\left(\frac{2}{h}\right)^3 (m_e^*)^{\frac{3}{2}}\left(2(E - E_c)\right)^{\frac{1}{2}} \quad \text{for } E \geq E_c \quad (2.13)$$

where m_e^* is the effective mass of the electron in the conduction band.

Following Eq. (2.9), energy and momentum relation for electrons in the conduction band can be written as

$$E = E_c + \frac{h^2}{8\pi^2 m_e^*}\left(k_x^2 + k_y^2 + k_z^2\right) \qquad (2.14)$$

Similarly, for valence band

$$d_v(E) = \pi\left(\frac{2}{h}\right)^3 (m_h^*)^{\frac{3}{2}}\left(2\left(E_v - E\right)\right)^{\frac{1}{2}} \quad \text{for } E \leq E_v \qquad (2.15)$$

$$E = E_v - \left(\frac{h^2}{8\pi^2 m_h^*}\left(k_x^2 + k_y^2 + k_z^2\right)\right) \qquad (2.16)$$

where $d_v(E)$ represents the density of states for holes in the valence band per unit volume, E_v represents the maximum energy of a hole at the top of the valence band, and m_h^* is the effective mass of hole in the valence band.

2.2.2   Two-Dimensional (2D) Structure

A thin layer of semiconductor medium whose thickness is comparable to or smaller than the de Broglie wavelength of carrier shows quantum confinement effect. The de Broglie wavelength (λ_{dB}) is associated with the carrier's momentum (p) as $\lambda_{dB} = h/p$, where h is the Planck's constant. A double heterojunction structure in which such thin semiconductor layer is sandwiched between two other semiconductor materials having band gap larger than that of the thin layer is known as quantum well. In quantum wells, discontinuities resulting from the heterojunction lead to the carrier confinement. For high-energy band gap materials, the relation between refractive index (n) and energy band gap (E_g) following classical oscillator theory can be expressed as (Hervé and Vandamme 1994)

$$n^2 = 1 + \left[\frac{13.6}{(E_g + 3.4)}\right]^2 \qquad (2.17)$$

The value of E_g is in eV, and the constant 13.6 is the value of the hydrogen ionization energy in eV. Figure 2.1 shows a variation of refractive index (n) with respect to the band gap energy (E_g) in eV of different III–V and II–VI group semiconductors. It shows variation for InAs (~0.36 eV) (Adachi 1999), InN (~0.65 eV) (Mohanta et al. 2011), InP (~1.347 eV) (Pavesi et al. 1991), GaAs (~1.424 eV) (Pozhar 2015), AlAs (~2.14 eV) (Yim 1971), GaP (~2.261 eV) (Lorenz et al. 1968), ZnO (~3.3 eV), GaN (~3.4 eV) (Özgür et al. 2005), AlN (~6.2 eV) (Edgar et al. 1990), MgO (~7.8 eV) (Soma and Uchino 2017), etc. Eq. (2.17) describes the refractive index and energy band gap relation in compound semiconductors used in optoelectronic devices and for wide band gap semiconductors such as GaN and ZnO. However, Eq. (2.17) does not well represent the behavior in IV–VI group semiconductor (Hervé and Vandamme 1994). It is obvious from Figure 2.1 that the refractive index decreases

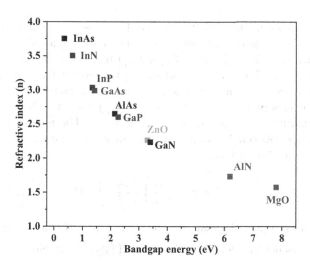

FIGURE 2.1 Variation of refractive index (n) with respect to the band gap energy (E_g).

as the band gap energy increases. An optical waveguide is a spatially nonuniform structure, which has a medium of higher value of refractive index in comparison with the surrounding medium known as cladding. As discussed above, quantum well structure has thin sandwiched layer of smaller band gap than the adjacent wider band gap materials. Thus, the refractive index of the thin layer in quantum well structure is larger than that of adjacent semiconductor materials. As a result, the quantum well structure behaves as an optical waveguide, responsible for the photon confinement in the thin sandwiched layer. Figure 2.2 shows the schematic representation of a typical GaN and AlGaN quantum well structure which has a thin layer of thickness l_x and the adjacent barrier layers. The x-direction is confined and restricted for carrier movement, whereas carriers of effective mass m^* can move freely over a large distance in other two directions (y- and z-axes) similar to the bulk semiconductor. Along x-axis, the rectangular potential well formed due to quantum well structure is approximated to a one-dimensional infinitely deep potential well with width l_x in which electrons of mass m_e^* in the conduction band and holes of mass m_h^* in the valence band are free to move. Thus, the free particle Schrödinger wave equation in a one-dimensional infinitely deep potential well can be written as

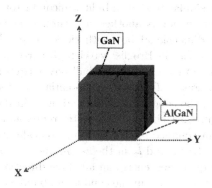

FIGURE 2.2 Schematic representation of a typical quantum well structure.

$$-\frac{h^2}{8\pi^2 m^*}\left(\frac{d^2\psi(x)}{dx^2}\right) = E\psi(x) \qquad (2.18)$$

In one-dimensional infinitely deep potential well, the wave function must vanish at the boundaries such as (i) $\psi(x = 0) = 0$ and (ii) $\psi(x = l_x) = 0$. Using these boundary conditions, we can obtain from Eq. (2.18) the following allowed energy values:

$$E_{n_x} = \frac{n_x^2 h^2}{8m^* l_x^2}, \quad n_x = 1, 2, 3, \ldots \qquad (2.19)$$

The effective mass of electron in the conduction band is $m_e^* = 0.22 m_0$ (Witowski et al. 1999) where m_0 is the rest mass of the electron. Thus, the allowed energy levels for electrons in the conduction band according to Eq. (2.19) for a GaN-based quantum well structure of well width l_x of 5 nm are 70, 275, ... meV. Since, the carriers are free to move along y- and z-directions in a quantum well similar to that in bulk semiconductor, the energy–momentum relation along y-, and z-directions can take the same form as that of Eqs. (2.14) and (2.16) for electrons and holes, respectively. Therefore, the energy–momentum relation for electrons in conduction band of a quantum well can be expressed as

$$E = (E_c + E_{n_x}) + \frac{h^2}{8\pi^2 m_e^*}\left(k_y^2 + k_z^2\right); n_x = 1, 2, 3 \ldots \quad (2.20)$$

or

$$E = E_{cn_x} + \frac{h^2}{8\pi^2 m_e^*}\left(k_y^2 + k_z^2\right); n_x = 1, 2, 3 \ldots \qquad (2.21)$$

where $E_{cn_x} = (E_c + E_{n_x})$

Similarly, for holes in valence band of a quantum well, the energy–momentum relation is written as

$$E = (E_v - E_{n_x}) - \left[\frac{h^2}{8\pi^2 m_h^*}\left(k_y^2 + k_z^2\right)\right]; n_x = 1, 2, 3 \ldots$$
$$(2.22)$$

or

$$E = E_{vn_x} - \left[\frac{h^2}{8\pi^2 m_h^*}\left(k_y^2 + k_z^2\right)\right]; n_x = 1, 2, 3 \ldots \quad (2.23)$$

where $E_{vn_x} = (E_v - E_{n_x})$

According to Eqs. (2.21) and (2.23), quantum well structure can be considered as a two-dimensional bulk semiconductor in which $E_{cn_x} [= (E_c + E_{n_x})]$ represents the energy corresponding to the bottom of the conduction band and $E_{vn_x} [= (E_v - E_{n_x})]$ corresponds to the top of the valence band for each $n_x = 1, 2, 3, \ldots$.

The number of states between E and $E + dE$ can be obtained as

$$N_{ws}(E)dE = \left(\frac{4\pi m^*}{h^2}\right) l_y l_z dE \qquad (2.24)$$

where l_y and l_z are the lengths of the quantum well along y- and z-directions, respectively.

For each value of the quantum number n_x, the density of states per unit area is $\left(\frac{4\pi m_e^*}{h^2}\right)$. Therefore,

$$d_{wc}(E) = \begin{cases} \left(\dfrac{4\pi m_e^*}{h^2}\right); & E > E_{cn_x} \\ 0 & ; \quad E < E_{cn_x} \end{cases} \qquad (2.25)$$

$$d_{wv}(E) = \begin{cases} \left(\dfrac{4\pi m_h^*}{h^2}\right); & E < E_{vn_x} \\ 0 & ; \quad E > E_{vn_x} \end{cases} \qquad (2.26)$$

where $d_{wc}(E)$ and $d_{wv}(E)$ represent the density of states per unit area for electrons in the conduction band and that for holes in the valence band, respectively.

2.2.3 One-Dimensional (1D) Structure

In aforementioned two-dimensional structures, carriers and photons are confined in one direction, i.e., along x-axis. Figure 2.3 shows a typical one-dimensional structure of a quantum wire in which carriers are confined along two directions, i.e., x- and y-axes within a distance of l_x and l_y. However, carriers are free to move along z-axis over a large distance of l_z ($\gg l_x$, l_y) in the plane of confining layer similar to that in the bulk semiconductor. More specifically, a quantum wire can be defined to be a thin wire-like structure of a semiconductor material of diameter comparable to or smaller than the de Broglie wavelength surrounded by a wider band gap semiconductor material. The thin wire can then be treated as infinitely deep two-dimensional potential well of widths l_x and l_y for carriers (electrons in the conduction band and holes in the valence band). Following Eqs. (2.20) and (2.22), the energy–momentum relation for electrons in the conduction band in a quantum wire can be expressed as

$$E = E_{cn_x n_y} + \frac{h^2 k_z^2}{8\pi^2 m_e^*}; \; n_x, n_y = 1, 2, 3 \ldots \qquad (2.27)$$

where $E_{cn_x n_y} = \left(E_c + E_{n_x} + E_{n_y}\right)$

For valence band,

$$E = E_{vn_x n_y} - \frac{h^2 k_z^2}{8\pi^2 m_h^*}; \; n_x, n_y = 1, 2, 3 \ldots \qquad (2.28)$$

where $E_{vn_x n_y} = \left(E_v - E_{n_x} - E_{n_y}\right)$

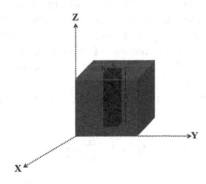

FIGURE 2.3 Schematic representation of a typical quantum wire structure.

According to Eqs. (2.27) and (2.28), a quantum wire can be regarded as one-dimensional bulk semiconductor in which $E_{cn_x n_y} \left(= E_c + E_{n_x} + E_{n_y}\right)$ represents the energy corresponding to the bottom of the conduction band and the top of the valence band is characterized by the energy $E_{vn_x n_y} \left(= E_v - E_{n_x} - E_{n_y}\right)$ for each pair of quantum numbers $(n_x, n_y) = 1, 2, 3 \ldots$. The number of states between E and $E + \mathrm{d}E$ in a quantum wire structure can be obtained as

$$N_{rs}(E)\mathrm{d}E = \left(\frac{l_z}{h}\right)\left(\frac{2m^*}{E}\right)^{\frac{1}{2}}\mathrm{d}E \qquad (2.29)$$

For each pair of quantum numbers (n_x, n_y), the density of states per unit length of the wire is $\left(\frac{2m^*}{h^2 E}\right)^{\frac{1}{2}}$. Therefore,

$$d_{rc}(E) = \begin{cases} \left(\dfrac{2m_e^*}{h^2\left(E - E_{cn_x n_y}\right)}\right)^{\frac{1}{2}}; & E > E_{cn_x n_y} \\ 0 & ; \quad E < E_{cn_x n_y} \end{cases} \qquad (2.30)$$

and

$$d_{rv}(E) = \begin{cases} \left(\dfrac{2m_h^*}{h^2\left(E_{vn_x n_y} - E\right)}\right)^{\frac{1}{2}}; & E < E_{vn_x n_y} \\ 0 & ; \quad E > E_{vn_x n_y} \end{cases} \qquad (2.31)$$

where $d_{rc}(E)$ and $d_{rv}(E)$ represent the density of states per unit length of the wire for electrons in the conduction band and that for holes in the valence band, respectively.

2.2.4 Zero-Dimensional (0D) Structure

In Sections 2.2.1–2.2.3, concept of 3D, 2D, and 1D semiconductor structures is discussed. In a 3D semiconductor structure, carriers are free to move in all three directions and are regarded as three-dimensional system. In a 2D quantum well structure, movement of carriers is restricted to a very small distance comparable to or less than the de Broglie wavelength in one direction, whereas in other two directions, carriers are allowed to move in similar manner of 3D semiconductor. In a 1D-quantum wire structure, carriers are confined in two directions within a short distance where carrier confinement is along two directions and the third direction in the plane of the confining layer allows carriers to move freely similar to that in bulk semiconductor. Furthermore, quantum dot is another structure; it consists of a small three-dimensional box with sides comparable to or smaller than the de Broglie wavelength surrounded by a wider band gap material. Such small box is equivalent to a three-dimensional infinitely deep potential well for carriers (electrons in the conduction band and holes in the valence band). Figure 2.4 shows a schematic representation of a quantum dot confined in all three directions along each side of the box—l_x, l_y, and l_z in the x-, y-, and z-axes, respectively. The quantized energy in all three directions can be expressed for electrons in the conduction band as

$$E = E_c + E_{n_x} + E_{n_y} + E_{n_z} \qquad (2.32)$$

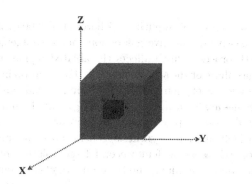

FIGURE 2.4 Schematic representation of a typical quantum dot structure.

and for holes in the valence band as

$$E = E_v - \left[E_{n_x} + E_{n_y} + E_{n_z} \right] \quad (2.33)$$

with n_x, n_y, $n_z = 1, 2, 3, \ldots$.

The density of states in quantum dot is represented by the delta functions where the carrier motion is restricted in all three directions. Since carriers are confined in all three directions to a very small distance comparable to or less than the de Broglie wavelength, quantum dots are regarded as zero-dimensional semiconductors.

2.3 Pulsed Laser Ablation

PLA has widely been used for many different applications of materials processing such as thin films deposition, cluster formation, chemical reactions, and surface modifications (synthesis of nanoclusters) (Miller and Haglan 1998). Pulsed laser deposition (PLD) is one of the versatile techniques for thin film deposition and has been used for deposition of varieties of materials such as metals, semiconductors, and ceramics depending upon the applications (Miller and Haglan 1998). For the synthesis of nanomaterials, nanosecond neodymium-doped yttrium aluminum garnet (Nd: YAG), and excimer lasers have mostly been used (Kim et al. 2017). Femtosecond lasers are also used for nanoparticle generation (Amoruso et al. 2004, Amoruso et al. 2005). The specific mechanism of absorption of laser irradiation by the materials depends on its type, laser pulse width, intensity, and wavelength (Kim et al. 2017, Brown and Arnold 2010). According to the Beer–Lambert law, the intensity of the laser radiation depends on the absorption coefficient α of the material as (Brown and Arnold 2010)

$$I(Z) = I_0 e^{-\alpha z} \quad (2.34)$$

where α is considered as constant which generally depends on wavelength and temperature, I_0 is the intensity of the incident laser radiation excluding the reflection loss from the surface, and the optical penetration depth is the inverse of absorption coefficient $\frac{1}{\alpha}$. In case of ZnO, the absorption coefficient at 355 nm is about 2×10^5 cm^{-1} which indicates an absorption depth of about 50 nm (Mohanta and Thareja 2014). In general, the laser radiation incident on semiconductors and insulators is absorbed due to transitions of valence band electrons to the conduction band or within the bands. However, inverse Bremsstrahlung mechanism is primarily responsible for the absorption of laser radiation in metals dominantly by free electrons in the conduction band (Brown and Arnold 2010), subsequently leading to heating by the transfer of energy to lattice by collision. If the thermal diffusion length, given by $l_T = 2(D\tau_p)^{1/2}$, where D is the thermal diffusion constant and τ_p the pulse width, is smaller than $1/\alpha$, the bulk will be heated down to $1/\alpha$, independent of the pulse width. This condition should be satisfied in order to have congruent evaporation of multi-elemental targets (Willmott and Huber 2000). However, in metals $1/\alpha << l_T$, absorption depths are typically 10 nm. The thermalization time, the time taken for the electron gas and the solid lattice to reach thermal balance, is 900 ps, 915 ps, 752.5 ps, and 465.7 ps, respectively, for Ag, Au, Cu, and Pb (Xu et al. 2004). This indicates that in case of femtosecond laser radiation, the laser pulse ends before the thermalization, whereas in case of nanosecond laser systems, the details of thermalization process are insignificant and the absorbed energy is directly converted to heat due to photo thermal process (Brown and Arnold 2010). This leads to heating of the ablation layer within a few picoseconds. Thus, for nanosecond laser pulses, nascent vapor cloud is ionized before the laser pulse is over. Therefore, the bulk of material under the plasma is largely screened from the remainder of the laser pulse, which is partly absorbed by the plasma as it becomes increasingly ionized (Willmott and Huber 2000). This process known as laser-supported absorption produces plasma species with high kinetic energy ranging between 1 eV and more than 100 eV. Thus, materials with small value of absorption depth (a high value of α) generate very hot species, and lower ablation yields are expected. These highly energetic plasma species provide the possibility of surface processes far from thermal equilibrium. The effective thermal diffusion length is reduced to $l_T^{eff} = 2\left(D\tau_{eff}\right)^{1/2}$ where τ_{eff} is the time needed to create an erosion plasma after the start of the laser pulse. The absorption of radiation takes place by the process of inverse bremsstrahlung with an absorption coefficient (Hughes 1975) in cm^{-1} of

$$K_v = 3.69 \times 10^8 \left\{ \frac{Z^3 n_i^2}{T^{1/2} v^3} \right\} \left[1 - \exp\left(-\frac{hv}{k_B T} \right) \right] \quad (2.35)$$

where n_i is the density of ions in cm^{-3}, T is the gas temperature in K, v is the frequency (in Hz) of incident light, k_B is Boltzmann constant, and the factor $\left[1 - \exp\left(-\frac{hv}{k_B T} \right) \right]$ gives the loss due to stimulated emission. For the Nd: YAG laser of wavelength 1.06 μm, $\left[1 - \exp\left(-\frac{hv}{k_B T} \right) \right] = 1$ for $T << 13{,}000$ K and close to $\frac{hv}{k_B T}$ for $T >> 13{,}000$ K. The vapors cloud absorbs incident radiation, becomes hotter and more ionized, and starts absorbing more efficiently. A general feature of ablation plasmas (also known as "plumes") is their high ion and electron temperatures of the order of several 1,000 K, and their high degree of ionization (Thareja et al. 2002). The high temperature affects the degree of ionization of the plume in accordance with Saha equation

(Bekefi 1976). The whole process of laser ablation is subdivided into three regimes (Singh et al. 1990):

(a) Evaporation of the surface layers of the target material due to heating by the laser pulse.

(b) The evaporated material interacts with the laser pulse leading to the formation of a high-temperature isothermal expanding plasma that primarily contains ions, electrons, and neutrals.

(c) The expansion of plasma after the termination of the laser pulse. The above two regimes (a) and (b) initiate with the interaction of the laser pulse and last up to the disappearance of the laser pulse.

The regime (a) refers to as evaporation regime, and (b) to as isothermal or laser–plasma interaction regime. After the termination of the laser pulse, plasma/plume expands adiabatically and is called as the adiabatic or plume expansion regime (c). All these processes are influenced by the target's physical and chemical properties of the material, and laser parameters (wavelength, pulse width, fluence at the target, etc.) (Singh et al. 1990, Singh and Narayan 1990).

Many researchers (Singh et al. 1990, Singh and Narayan 1990, Wood and Giles 1981, Singh and Narayan 1989, Carslaw and Jaegaer 1959) have analyzed the mechanism using nanosecond laser heating of the target material. When a laser pulse of sufficient energy is incident on the solid target, a part of the energy evaporates the surface layers of the target resulting in expanding plasma and the rest is lost due to absorption by plasma and conduction. The thickness of the target material evaporated per pulse (Δx_t) can be obtained using heat balance equation (Singh and Narayan 1990):

$$\Delta x_t = (1 - R) \left[\frac{E_L - E_{th}}{\Delta H + C_v \Delta T} \right] \quad (2.36)$$

where R, E_L, E_{th}, ΔH, C_v, and ΔT represent the reflectivity, incident laser energy, threshold laser energy, latent heat of the target material, heat capacity, and temperature rise, respectively. This equation is valid for conditions where the thermal diffusion distance is larger than the absorption length of the laser beam in the target material.

In case of femtosecond laser pulse, the ablation process may be interpreted as a direct solid–vapor or solid–plasma transition since the thermalization time is in picosecond time scale at which lattice heating results in creation of vapor and plasma phases followed by a rapid expansion. Thermal conduction during these processes is neglected in a first approximation (Chichkov et al. 1996).

2.4   Nanomaterials Synthesis by Pulsed Laser Ablation

PLA has received significant attention for thin films deposition and production of nanomaterials. It has several advantages over other synthesis techniques including the ability to produce materials with a complex stoichiometry and a narrower particle size distribution, reduced porosity, and

controlled level of impurities (Eliezer et al. 2004). Addition of a passive or reactive gas or ion source supplements the growth process, which affects the ablated plume species in the gas phase or the surface reaction. For ablation in reactive gases such as oxygen and nitrogen, simple oxide and nitride molecules are formed in the expanding ablated plume. Laser-ablated plasmas produced in vacuum or in a gas ambient are widely used for synthesis/generation of nanoparticles. Nanoparticles using femtosecond laser ablation of silicon, germanium, aluminum, and titanium targets in vacuum have been reported (Amoruso et al. 2005). The continuum emission observed from nanoclusters at time delays (1–50 μs) after the laser pulse is employed to estimate the temperature (T_p). Following the Planck's law, temperature can be calculated using the following relation (Amoruso et al. 2004):

$$I \propto \lambda^{-5} \exp \left(-\frac{hc}{\lambda k_B T_p} \right) \quad (2.37)$$

where I is the emission intensity, λ is the emission wavelength, h is the Planck's constant, k_B is the Boltzmann constant, and c is the velocity of light. The size of nanomaterials in PLA technique can be controlled by optimizing the laser wavelength, intensity, pulse duration, environment temperature, and pressure (Lee et al. 2009). The target materials used in PLA process can be metals, semiconductors, ceramics, oxides, or mixed system. Since the stoichiometry of multielement complex target in the vapor phase is maintained, PLA is used to grow superconducting super lattices and superconducting device structures.

As described in Section 2.3, in the adiabatic regime, the temperature of the plasma decreases and the occurrence of supersaturation sets in that results in the formation of the nanoparticles in vapor phase. In an ambient atmosphere, the expanding laser-ablated plume interacts hydrodynamically with the background gas. The expanding plasma plume species collide with the ambient gas species and hence lose their energy that results in eventually slowing down of the expansion. The cooling of plasma due to collision of the plasma species with the species of the ambient initiates the chemical reactions and subsequently the formation of nanoparticles.

The laser ablation process primarily involves the absorption mechanism above threshold laser fluence that depends on material properties and on laser parameters such as pulse width and wavelength. In metals, threshold fluence for ablation is between 1 and 10 J/cm^2. In inorganic insulators, it is between 0.5 and 2 J/cm^2, and in organic materials, it is between 0.1 and 1 J/cm^2 (Bäuerle 2000). The increase in laser fluence causes a change in ablation mechanism from surface vaporization to explosive boiling also known as phase explosion over certain laser fluence. Hoffman et al. (2014) have reported that the transition from the thermal ablation to the phase explosion depends on the laser wavelengths. The transition to phase explosion is indicated by the jump of the ablation rate, which takes place at lower laser fluences for shorter wavelengths. They report laser fluence of approximately 10 Jcm^{-2} at 355 nm, 25 Jcm^{-2} at 532 nm, and

55 Jcm^{-2} at 1,064 nm for carbon. The absorption coefficient (α) in direct band gap of semiconductor materials depends on the band gap energy as (Pankove 1975)

$$\alpha(hv) \propto \sqrt{\left(\frac{hc}{\lambda} - E_g\right)} \qquad (2.38)$$

where h is the Planck's constant, c is the velocity of light, λ is the laser wavelength, and E_g is the band gap energy. Equation (2.38) indicates that the absorption coefficient is larger for shorter wavelength of the laser.

In ZnO, the band gap energy is 3.3 eV which corresponds to about 375 nm. Mohanta et al. (2008), Mohanta and Thareja (2009) have synthesized ZnO nanoparticles in vapor phase by PLA using the third harmonic of an Nd: YAG laser [$\lambda = 355$ nm, pulse width of 5 ns, repetition rate = 10 Hz]. A sintered ZnO pellet of 99.99% purity was kept inside a chamber at atmospheric pressure (air). The laser beam focused onto the ZnO target ablates the material. In order to have a fresh target surface and to avoid drilling, ZnO pellet was rotated continuously by a microprocessor-controlled stepper motor (Mohanta et al. 2008). Optical emission spectroscopy (OES) is used to characterize the laser-ablated ZnO plasma. An intensified charge coupled device (ICCD) attached to a spectrometer is used to record the optical emission spectra of ZnO plasma at different time delays with respect to the ablating laser pulse and at different distances from the target surface. Figure 2.5 shows the temporal evolution of the emission spectrum in the spectral region between 460 and 490 nm dominantly containing the emission of neutral zinc (Zn I). It is obvious from Figure 2.5 that Zn I line intensity decreases with increase in time after the initiation of plasma. Similarly, the

emission intensity of Zn I lines decreases as the distance from the target surface increases (Mohanta and Thareja 2009). Moreover, electron temperature and density also decrease with increase in time with respect to the laser pulse and with the increase in distance from the target surface. Plasma expansion and three-body recombination are responsible for the decrease in electron density (Narayanan and Thareja 2004).

The density of the plasma decreases at any time t from its inner edge at a point z as (Singh and Narayan 1990, Saji et al. 2006):

$$n_e(z,t) = n_0(t)\left(1 - \frac{z}{z(t)}\right) \qquad (2.39)$$

where $n_0(t)$ is the density at the center of the laser irradiated spot ($z = 0$) at time t, the z coordinate is directed perpendicular to the target, and $Z(t)$ is the spatial coordinate of the leading edge of the plasma. The spatial dependence of electron density should be linear in accordance with Eq. (2.39). However, it follows exponential decreasing behavior (Mohanta et al. 2008). The small upward trend observed at a higher distance at 500 ns in the spatial-dependent profile of electron density is attributed to the opposition of the free expansion of plasma by the adjacent mass of the ambient (air) at the plume edge (Mohanta et al. 2008).

2.5 Experimental Confirmation of Vapor-Phase ZnO Nanoparticles Formation

The formation of nanoparticles in vapor phase due to the condensation of laser-ablated ZnO plasma is confirmed by Rayleigh scattering of visible radiation and by simultaneously observing PL from these nanoparticles.

2.5.1 Rayleigh Scattering

When an electromagnetic wave is incident on a matter, electrons of the matter are set into periodic vibration in accordance with the vibrating electric field and magnetic field of the incident radiation. The relative displacement of electrons and nuclei of neutral atoms results in electric dipoles. Rayleigh scattering is the radiation from the oscillating electrons bound in atoms, molecules, or small particles under the influence of external electromagnetic field, similar to the case of dipole radiation. The condensation of species formed in the laser-ablated plume results in the formation of nanoparticles. The scattered Rayleigh signal is not from one particle but is from an ensemble of nanoparticles. However, since the size distribution of the generated particles is narrow, the multiple scattering is neglected and a single particle scattering theory is applied to ensemble of particles (Kempkens and Uhlenbusch 2000). Assuming the incident laser radiation is polarized along the x-axis and propagating along the z-axis with propagation vector $\vec{k}$, with the electric field vector $\vec{E} = \hat{i}E_o e^{i(kz-\omega t)}$, the

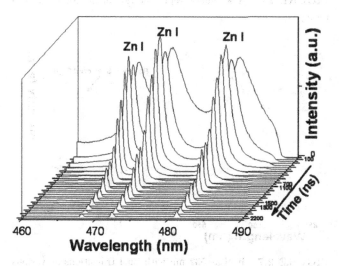

FIGURE 2.5 Temporal evolution of the spectrum containing Zn I transitions at 468 nm ($4s4d^3D_3 \rightarrow 4s4p\ ^3P_2$), 472 nm ($4s5s^3S_1 \rightarrow 4s4p^3P_1$), and 481 nm ($4s5s^3S_1 \rightarrow 4s4p^3P_2$) at a distance of 2 mm from the target surface. (Reproduced from A. Mohanta and R. K. Thareja. Rayleigh scattering from gaseous phase nanoparticles synthesized by PLA of ZnO, *Journal of Applied Physics* 106, 124909 (2009), with the permission of AIP Publishing.)

intensity I_{RS} of the Rayleigh-scattered signal (RSS) from the ensemble of particles can be expressed by the relation (Mohanta and Thareja 2009, Kempkens and Uhlenbusch 2000):

$$I_{RS} = \left[\left(\frac{c\varepsilon_0 k^4 E_0^2}{2r^2} \right) \left(\frac{\varepsilon - 1}{\varepsilon + 2} \right)^2 a^6 \sin^2 \theta \right] n \, d\tau_s$$

$I_{RS} \propto (density \, of \, particles)$

$\times \, (sixth \, power \, of \, radius \, of \, particles)$ (2.40)

In Eq. (2.40), c is the velocity of light in free space, ε_0 is the free space permittivity, a is radius of particles, ε is dielectric constant, and k is the magnitude of the propagation vector $\vec{k}$. E_0 is the amplitude of the electric field of the incident radiation, θ is the angle between the incident field (x-axis) and the direction of scattered radiation, $d\tau_s$ is the scattering volume, and n is the particle density. RSS intensity is inversely proportional to the fourth power of wavelength of incident laser radiation since $k \propto \frac{1}{\lambda}$ and directly proportional to the square of the volume of the scattering particle. It is shown that the theory holds for $a < \frac{\lambda}{20}$ (Kempkens and Uhlenbusch 2000). Following Eq. (2.40), the intensity of the scattered radiation is directly proportional to the density of the scattering particles (scatterers), and the scattering is termed as incoherent scattering. However, in case of forward scattering, the scattered signal intensity is coherent and proportional to the square of the density of the particles, n^2 (Kempkens and Uhlenbusch 2000). It follows from Eq. (2.40), the intensity of scattered radiation is zero along the direction of polarization, i.e., along x-axis, and is maximum for $\theta = 90°$. If the scatterers are spherically symmetric, i.e., for ground state atoms, the scattered signal at $\theta = 90°$ is fully polarized. Due to the presence of excited state atoms, depolarization of the RSS may occur in plasma (Meulenbroeks et al. 1992). In pulsed laser-ablated plasma, depolarization effect is nullified with increase in time and can be neglected at sufficiently large time after the initiation of plasma.

Mohanta and Thareja (2009) have considered visible radiation of wavelength 532 nm for Rayleigh scattering experiment to verify the formation of nanoparticles in vapor phase. In the experiment, third harmonic [355 nm] of an Nd: YAG laser is used to generate ZnO plasma. Second harmonic [532 nm] of another Nd: YAG laser, referred to as probe pulse, is passed through the ZnO plasma parallel to the target surface at different axial distances from the target surface at various time delays with respect to the ablating pulse (355 nm). The probe (532 nm) radiation is horizontally polarized. Thus, the scattered radiation will be maximum in the direction perpendicular to the state of polarization of the incident radiation in accordance with Eq. (2.40). The scattered radiation is imaged onto one end of a fiber placed perpendicular to the axis of both the laser beam and to the state of polarization of the incident radiation. A polarizer kept in front of the fiber after the collecting lens ensures the polarization of the radiation. The other end of the fiber is coupled with the entrance slit of a spectrograph attached with an ICCD. A schematic representation of the Rayleigh scattering experiment is shown in Figure 2.6. Since the scattered radiation is also polarized, the variation in intensity of the scattered radiation can be noticed at different states of the polarizer. The intensity of the probe pulse is kept very low in order to avoid stray scattered light that comes due to scattering from different sources other than plasma species in the interaction chamber. Figure 2.7 shows emission spectra of ZnO plasma between 450 and 550 nm obtained at different time delays of passage of the probe pulse with respect to the ablating pulse at a distance of 6 mm away and parallel to the target surface. The emission spectra contain Zn I emission lines along with the RSS at 532 nm. In the initial stage, the plasma emission predominates the Rayleigh-scattered radiation at 532 nm. However, with the increase in time delay, the plasma emission intensity decreases and the RSS intensity increases. At sufficiently large time delay, the plasma emission is completely

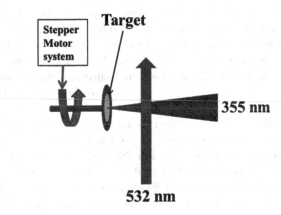

FIGURE 2.6 A schematic representation of the Rayleigh scattering experiment.

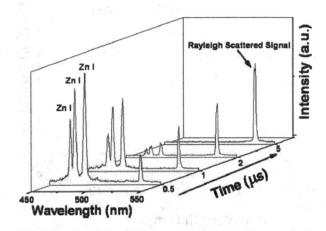

FIGURE 2.7 RSS at 532 nm with Zn I transitions at various delays of passage of the probe pulse with respect to the ablating pulse at a distance of 6 mm from the target surface. (Reproduced from A. Mohanta and R. K. Thareja. Rayleigh scattering from gaseous phase nanoparticles synthesized by PLA of ZnO, *Journal of Applied Physics* 106, 124909 (2009), with the permission of AIP Publishing.)

suppressed and the RSS only appears in the observation spectral window between 450 and 550 nm. This indicates that nanoparticles formed at much later time of the initiation of plasma due to condensation of species scatter the probe pulse at 532 nm that contributes to the increase in intensity of the scattered signal. The increased signal is manifestation of the increase in number of nanoparticles. Thus, the dynamic Rayleigh scattering could be used to confirm the formation of nanoparticles in vapor phase, synthesized by the nanosecond PLA of ZnO.

2.5.2 Photoluminescence

PL is a process in which a system is excited to a higher energy level by absorbing a photon and then spontaneously emits a photon by decaying to a lower energy level. The energy and momentum remain conserved in this process. PL spectroscopy is usually used to characterize surface, interface, and impurity levels, and to identify alloy disorder and surface roughness (Gfroerer 2000). The quality of the material is represented in terms of the intensity of PL spectra, the major advantage of this technique being for high-resistivity materials where electroluminescence would be inefficient or impractical (Gfroerer 2000). This is also useful for materials where contact or junction technology is not adequately developed. The technique is extensively used to characterize the semiconductor materials prior to the fabrication of any optoelectronic devices. The technique is simple, and very little control of environment is required. The structure of electronic energy levels of photo-excited materials is obtained by analyzing the transition energies from the PL spectrum. The PL process strongly depends on the nature of optical excitation. The PL spectroscopy technique is limited to the radiative transitions. However, it is difficult to characterize the poor luminescent indirect band gap semiconductor materials. We have looked at the PL spectrum of ZnO, an II–VI direct band gap semiconductor with 355 nm excitation.

Free Excitons

On optical excitation, an electron–hole pair is created in semiconductor material by absorption of a photon, which recombines emitting another photon. The optical transitions in semiconductors are due to both extrinsic and intrinsic effects. Intrinsic effects involve the optical transition between electrons in the conduction band and holes in the valence band including recombination of electron–hole pairs bounded by Coulomb interaction. The interaction of electron and hole via attractive Coulomb potential forms a series of hydrogen or positronium-like states below the band gap. These are called free excitons (Wannier excitons) and are characterized by the fact that the average distance between electron and hole, that is, the exciton Bohr radius is larger than the lattice constant. On the other hand, extrinsic effects in optical transitions are related to dopants or defects that create discrete electronic states in the band gap that

has strong influence in absorption and luminescence spectra. Excitons can be bound to these dopants or defects to form BECs.

A free exciton is an electron–hole pair, that is, a pair of opposite charges bounded by Coulomb potential (Pankove 1975). This indicates that electron revolves round the hole forming hydrogen-like atoms. The free-exciton energy in a semiconductor is expressed as (Mang et al. 1995)

$$E(n) = E_g - \frac{E_{e-b}}{n^2} \qquad (2.41)$$

where $E(n)$ is the exciton energy, $n = 1, 2, 3, \ldots$ is the exciton principal quantum number, E_g is the band gap energy, and E_{e-b} is the exciton binding energy. In hydrogen-like atoms, the reduced mass of nucleus and electron is equal to the mass of electron as mass of nucleus is larger in comparison with that of electron. However, the effective reduced mass of hole and electron in case of free exciton is not equal to the effective mass of the electron and is less than the effective masses of hole and electron. This is because effective masses of electron and hole are comparable, e.g., in ZnO, $m_e^* = 0.24m_0$ and $m_h^* = 0.45m_0$, where m_e^* and m_h^* are the effective masses of the electron and hole, and m_0 is the mass of the electron. The free exciton is a mobile pair and can ramble throughout the crystal. Moreover, excitonic complexes similar to positronium-like molecule can be formed by combining two free holes and two free electrons.

Such a complex has lower energy than two free excitons.

Bound Exciton Complexes

A brief discussion on free excitons is made in the previous section. However, there is a finite possibility where a free hole can combine with an electron of a neutral donor to form a positively charged excitonic ion. The electron remains bound to the donor and travels around the donor. The hole, which combined with the electron, also travels about the donor. These complexes are called BECs. On the other hand, an electron bound with a neutral acceptor is called neutral acceptor BEC. Furthermore, an exciton can be bound to an ionized donor to form ionized donor bound exciton. The abbreviations often used for ionized bound excitons, neutral donor bound excitons, and neutral acceptor bound excitons are D^+X, D^0X, and A^0X, respectively. The bound exciton does not have freedom to translate throughout the crystal. The electron and hole remain in the same unit cell. The bound excitonic transitions are observed in absorption and luminescence bands at low temperatures. In bulk ZnO, the bound excitonic transitions cover a wide range from 3.348 to 3.374 eV (Özgür et al. 2005). In case of good quality samples, the line width of bound excitons is <1 meV. At low temperatures, the luminescence spectra are dominated by the transitions of BECs. However, with increasing temperature, the bound excitons get thermalized and disappear at high temperatures.

Exciton–Exciton Scattering Process

In ZnO, exciton binding energy is 60 meV implying exciton states are stable at room temperature. At low excitation level, only excitonic emission is reflected in the PL spectra of ZnO. At intermediate excitation intensity, an emission band appears only in high-quality samples due to inelastic collision between excitons, which results in an exciton to be excited into a higher state ($n = 2, 3, 4, \ldots, \infty$), and a photon emitted with energy (Mohanta et al. 2015, Bagnall et al. 1998, Klingshirn et al. 2005):

$$P_n = E_{ex} - E_b^{ex}\left(1 - \frac{1}{n^2}\right) - \frac{3}{2}k_BT(n = 2, 3, 4\ldots, \infty)$$

$$(2.42)$$

where P_n is the photon energy, E_{ex} is the free-exciton emission energy, n is the quantum number of the envelope function, and k_BT is the thermal energy. A schematic representation of the exciton–exciton scattering process for $n = 2$ is presented in Figure 2.8. Inelastic collision between two excitons in ground state ($n = 1$) results in excitation of one exciton into the first excited state ($n = 2$) and the recombination of the second exciton with emission of a photon of energy P_2 given by the relation:

$$P_2 = E_{ex} - \left(\frac{3}{4}\right)E_b^{ex} - \frac{3}{2}k_BT \qquad (2.43)$$

The experimental geometry of the PL experiment is similar to the Rayleigh scattering experiment with the difference here the fourth harmonic (266 nm) of a Nd: YAG laser is the probe pulse. Figure 2.9 depicts the experimental scheme. At axial position very close to the target surface and at shorter time delay with respect to the initiation of plasma, the formation of nanoparticles is hindered due to high temperature (Shaikh et al. 2006). The luminescence spectra collected at 1.5 µs and at distance of 8 mm with and without passage

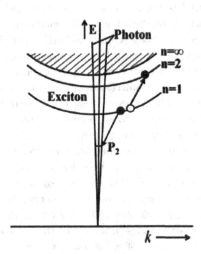

FIGURE 2.8 A schematic representation of the exciton–exciton scattering process. (Reproduced from A. Mohanta, P. Kung and R. K. Thareja. Exciton-exciton scattering in vapor-phase ZnO nanoparticles, *Applied Physics Letters* 106, 013108 (2015), with the permission of AIP Publishing.)

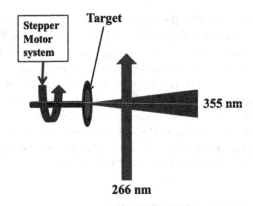

FIGURE 2.9 A schematic representation of the experimental configuration to observe PL from vapor-phase ZnO nanoparticles.

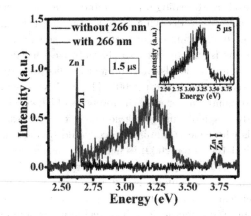

FIGURE 2.10 Luminescence spectra collected at $t = 1.5$ µs and $d = 8$ mm with and without passage of the excitation pulse (266 nm). The inset shows the PL spectrum at $t = 5$ µs and $d = 8$ mm. (Reproduced from A. Mohanta, P. Kung and R. K. Thareja. Exciton-exciton scattering in vapor-phase ZnO nanoparticles, *Applied Physics Letters* 106, 013108 (2015), with the permission of AIP Publishing.)

of the probe pulse are shown in Figure 2.10. The emission spectra obtained without passage of the probe/excitation pulse contain only the plasma emission lines which include the neutral Zn I transitions at 3.75, 3.71, 2.65 and 2.62 eV in the spectral region between 2.4 and 3.9 eV. An additional broad emission band appears peaked at about 3.22 eV along with the plasma emission in the presence of the probe pulse. This corresponds to PL emission from ZnO nanoparticles formed in the vapor phase.

Figure 2.11 shows the PL spectra of vapor-phase ZnO nanoparticles at 2 ms and at a distance of 6 mm at different intensities of the probe pulse (266 nm) at fixed intensity (2.1 GW/cm²) of the ablating pulse (355 nm). At low probe pulse intensity (1 MW/cm²), the PL spectrum is characterized by the free excitonic recombination at 3.20 eV, designated as E_1. Above 4 MW/cm², another emission band centered at 3.12 eV (designated as E_2) dominates the PL spectrum which narrows and does not shift with increasing probe pulse intensity. The E_1 peak is blue-shifted by 12 meV from the peak energy of PL of bulk ZnO obtained under similar condition which is used to determine the particle radius using

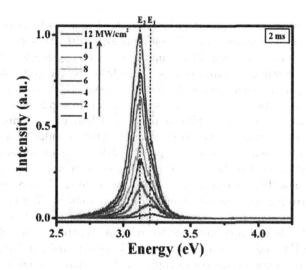

FIGURE 2.11 The PL spectra at $t = 2$ ms and $d = 6$ mm at different probe pulse intensity with fixed ablating intensity of 2.1 GW/cm^2. (Reproduced from A. Mohanta, P. Kung and R. K. Thareja. Exciton-exciton scattering in vapor-phase ZnO nanoparticles, *Applied Physics Letters* 106, 013108 (2015), with the permission of AIP Publishing.)

Eq. (2.3) that is about 13 nm. For $n = 2$ in Eq. (2.42), the energy difference between the photon energy (P_2) and E_{ex} is 84 meV, which is in good agreement with the observed spectroscopic separation between E_1 and E_2. Thus, E_2 band observed in the PL spectra at 3.12 eV in Figure 2.11 can be ascribed to the P_2 band due to exciton–exciton scattering (Mohanta et al. 2015). Exciton–exciton scattering process has also been observed in high-quality ZnO epitaxial layers (Bagnall et al. 1998). Generally, excitonic emission in PL spectra would be observed at low excitation intensity. As the excitation intensity increases, emission from biexciton transition at a different spectral energy can appear due to its different binding energy. Subsequently, emission due to exciton–exciton scattering appears. At sufficiently higher photo-generated exciton density above Mott density, excitons lose their bosonic character and electron-hole plasma forms due to phase space filling and Coulomb interactions, which shows redshift with increase in excitation intensity due to the band gap renormalization effect (Bagnall et al. 1998). It has also been reported that emission due to exciton–exciton scattering process is absent in thin ($\approx$50 nm) nanorods and is present in thick ($\approx$200 nm) nanorods (Klingshirn et al. 2005).

2.6 Nanomaterials Synthesis by Pulsed Laser Ablation in Liquids

Pulsed laser ablation in liquids (PLAL) has been employed for the synthesis of different functional materials, which involves reaction dynamics of plasma and ambient species. PLAL is a single-step process to synthesize diverse metals and alloys, ceramics, and semiconductors, and is known

to produce pure colloidal solution nanoparticles, which are free of external contamination in contrast to conventional solution-based chemical synthesis methods (Wagener et al. 2010, Mafune et al. 2003). Many different kinds of nanoparticles have been synthesized by laser ablation in different liquids. Mafune et al. (2003) have produced platinum nanoparticles with size distribution in the range of 1–7 nm by laser ablation of a platinum metal plate in an aqueous solution of sodium dodecyl sulfate (SDS). They have found that platinum nanoparticles produced in a 0.01-M SDS aqueous solution are stable. Streubel et al. (2016) have synthesized catalysis relevant metal nanoparticles such as platinum and gold with a production rate of 4 g/h by PLAL. They have overcome the major drawback of low productivity in PLAL by using a novel ultrafast high-repetition rate laser system combined with a polygon scanner having a maximum scanning speed of 500 m/s. Ganjali et al. (2015) have synthesized Ni nanoparticles with a narrow size distribution having diameter ~10 nm using PLAL in 30 mL of acetone. Stratakis et al. (2009) have reported the formation of self-organized NS on bulk Al under its ablation in air and liquids with femtosecond laser pulses. Preparation of hybrid nanomaterials has also been reported. Sajti et al. (2007) have prepared colloidal ZnO–ethanol solution by femtosecond laser ablation of a ZnO target in pure ethanol. Dye molecules were then grafted onto the ZnO nanoparticles by mixing the colloidal ZnO-ethanol solution to Tetramethylrhodamine B isothiocyante or to Rhodamine B solutions. Yan et al. (2013) have demonstrated the generation of Ag-Ag$_2$O complex NS by excimer laser ablation of Ag in water. They show that the pulsed excimer laser ablation of bulk Ag in water can generate Ag nanosheets with large {111} facets which reveals that the laser reprocessing could mediate the coalescence and growth of laser-produced clusters. They further reported that drop evaporation of the laser-produced colloidal Ag solutions can result in rice-shaped Ag-Ag$_2$O NS and their assemblies. Kumar et al. (2011) have done laser ablation of titanium metal target in deionized water ambient using nanosecond Nd: YAG pulsed laser operating in fundamental mode for the synthesis of titanium nanoparticles. They observed that synthesized nanoparticles are spherical in shape with narrow size distribution with an average size of ~15 nm. They have also investigated the growth dynamics of the nanoparticles by laser light scattering. In addition, attempts to understand nanoparticles formation mechanisms and growth dynamics have been made by considering different studies such as laser-produced plasma in confined geometry (Fabbro et al. 1990), mechanism of PLA, and cavitation bubble formation in liquid (Vogel and Venugopalan 2003). Itina (2011) have proposed major processes involved in laser ablation in liquids which affects nanoparticles formation. They have divided the laser ablation process in liquid into three stages: (i) early stage—laser energy absorption, material ejection, and plume formation; (ii) intermediate stage—plume expansion in the presence of confining liquid environment; and (iii) later stage—plume mixing with the liquid and NP coalescence/aggregation processes. Nucleation and formation of small nanoparticles start in the intermediate stage where the plume expands adiabatically. The equations governing the adiabatic expansion that involves time-dependent plasma length and pressure are given by (Itina 2011)

$$P(t) = P_0 \left[1 + \frac{\gamma + 1}{\tau}(t - \tau) \right]^{-\frac{\gamma}{\gamma+1}} \quad (2.44)$$

$$L(t) = L_0 \left[1 + \frac{\gamma + 1}{\tau}(t - \tau) \right]^{\frac{1}{\gamma+1}} \quad (2.45)$$

where P is the pressure experienced by the plasma plume, L is the length of the plasma plume, and γ is the adiabatic parameter.

2.7 Heterogeneous Colliding Plasma for Synthesis of Nanocomposites

PLD technique has been widely used to deposit thin films and NS due to ease of control of ambient pressure in a wide range irrespective of other growth parameters (Mohanta and Thareja 2008, Hayamizu et al. 1996, Okada et al. 2005). It has also been used for nanostructuring of surfaces (Pandey et al. 2018). PLD process has been classified into two categories according to the range of pressures used: (i) conventional PLD and (ii) nanoparticle-assisted PLD. The conventional PLD is mostly used for the deposition of thin films in which the background ambient pressure is relatively low in the range around 10 Pa or less, whereas higher background pressure ranging from 133 Pa to 1.3 kPa has been used in nanoparticle-assisted PLD for the synthesis of NS (Okada et al. 2005). Hartanto et al. (2004) have reported synthesis of ZnO nanorods by using nanoparticle-assisted PLD at comparatively high pressures without using a catalyst. In nanoparticle-assisted PLD, the nanoparticles are formed in vapor phase in the pathway between target and substrate due to the condensation of ablated species. The aforementioned vapor-phase nanoparticles are the major species propagating towards the substrate surface and play important role in the growth of the NS. The major inherent drawback of PLD process is the deposition of micron-sized particulates on the substrate which are ejected directly from the target surface as chunks. PLD with colliding plasma plumes can help to mitigate the debris and larger material chunks present in the plume (Pandey et al. 2016). Gupta and Thareja (2013) have comparatively investigated the ZnO thin films deposited using colliding plasma plumes and single plasma plume. They have observed that the thin films deposited using colliding plumes are of better quality than those deposited using single plume which has chunk deposition. Such observed differences in the quality have been attributed to the flux of mono-energetic plasma species with almost uniform kinetic energy and higher thermal velocity reaching the substrate from interaction/stagnation zone of colliding plasma plumes. Moreover, the dynamics of plasma plume expansion and collision of plasma with the species of ambient has significant influence on the properties of the synthesized nanocrystals. In addition to the collision of plasma with the species of the ambient which leads to condensation and nanoparticles formation in vapor phase, collision with other plasma plume can lead to new kind of alloy nanomaterials (Umezu et al. 2010). Pandey et al. (2018) have used the concept of colliding plasma plumes and considered the target comprising of concentric rings of Cu and Ti onto which two laser beams are focused to create corresponding plasma plumes in oxygen ambient. The two laser beams are produced by splitting a nanosecond Nd: YAG laser beam [pulse width of 8 ns, repetition rate of 10 Hz, and wavelength of 1064 nm] using a wedge prism. The expanding images of copper and titanium plasmas at different time delays with respect to the ablating pulse recorded by ICCD camera are shown in Figure 2.12. At lower pressures of 10^{-4} and 10^{-2} mbar, the copper and titanium plasma plumes expand linearly before colliding each other. However, the expansion of each plasma plume follows the drag force model at 10^{0} mbar (Pandey et al. 2016). Analysis of field emission scanning microscope (FESEM) images reveals that the synthesized Cu-TiO$_2$ nanocomposites have diameter in the range of 20–30 nm at 10^{-4} mbar. Particle sizes and corresponding distribution are similar to that of at 10^{-4} mbar, whereas at 10^{0} mbar, the plasma plume dynamics changes, and flattening of the stagnation layer front is initiated which can be envisaged from Figure 2.12. Consequently, the nanoparticle flux movement towards the substrate surface is reduced. Furthermore, at higher pressure, enhanced collision among plasma species decreases the agglomeration rate resulting in reduction in the size of nanoparticles (Pandey et al. 2018). The stoichiometric concentration of copper, titanium, and oxygen is found to be, respectively, Cu 70.6 wt%, Ti 19.9 wt%, and O 9.5 wt% at 10^{-4} mbar; Cu 35 wt%, Ti 33 wt%, and O 32 wt% at 10^{-4} mbar; and Cu 11.2 wt%, Ti 6.2 wt%, and O 11.2 wt% at 10^{0} mbar, and the remainder represents the contribution of the silicon substrate.

2.8 Nano-Flakes Synthesis by Using Plasma Reactor

Synthesis of nanoparticles using thermal plasma technologies is well established. RF ICP has several advantages such as large volume, high-temperature plasma zone, operating under different atmospheres, permitting synthesis of a great variety of nanoparticles, and higher production rate with lower processing cost (Jiayin et al. 2010). RF ICP torch system in inert gas atmosphere is known to produce clean plasma since it does not involve electrodes for its operation (Boulos 1991). In contrast to aforementioned PLA technique, RF thermal plasma can be employed for industrial-scale mass production of high-purity materials, as it is a continuous process to accomplish high production

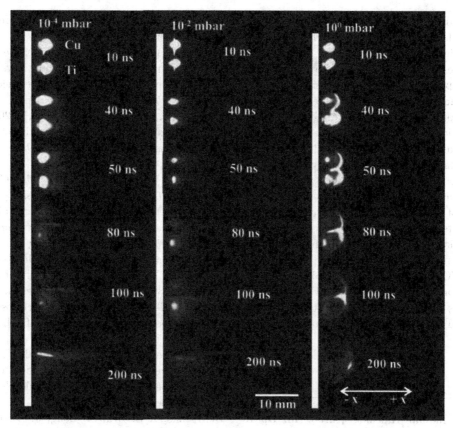

FIGURE 2.12 ICCD images of expanding copper and titanium plasmas showing stagnation at oxygen ambient pressures of 10^{-4}, 10^{-2}, and 10^0 mbar. (Reprinted by permission from RightsLink Permissions Springer Nature Customer Service Centre GmbH: [Springer Nature] *Applied Physics B: Lasers and Optics* Deposition of nanocomposite Cu-TiO$_2$ using heterogeneous colliding plasmas, P. K. Pandey, R. K. Thareja, R. P. Singh and J. T. Costello. © Springer–Verlag GmbH Germany, part of Springer Nature (2018).

volume with good synthesis control from the gaseous phase (Mohanta et al. 2017a). Great varieties of nanoparticles have been successfully synthesized using appropriate precursors in RF ICP torch system (Shin et al. 2006, Leparoux et al. 2008, Pristavita et al. 2011). Girshick et al. (1988) have conducted modeling investigation on particle nucleation and growth process in a thermal plasma reactor and proposed three routes for conversion from the gas phase to a condensed phase such as (i) homogenous nucleation, (ii) condensation to an existing particle, and (iii) heterogeneous chemistry at a particle surface. Fast high-temperature chemistry and rapid cooling that primarily occur in a plasma reactor facilitate the homogeneous nucleation. Growth of particles is then occurred by either process (ii) or (iii), and coagulation. Three main streams of gas flow into the ICP reactor chamber are used such as central, sheath, and reactive gas. Mohanta et al. (2017b) have used argon as central gas, argon and hydrogen mixture as sheath gas, and argon and methane gas mixture as reactive gas. Pristavita et al. (2011) have synthesized carbon particles with flake-like morphology called as "carbon nano-flakes" having typical dimensions of approximately 100 nm long, 50 nm wide, and 5 nm thick. Mohanta et al. (2017a, b) have performed in situ investigation of Ar/H$_2$/CH$_4$ plasma by OES to understand the growth process of graphene nano-flakes. Optical emission spectrum of Ar/H$_2$/CH$_4$ plasma was recorded during the synthesis of graphene nano-flakes which contains dominant emission from C$_2$ Swan system along with CH and C$_3$ (Mohanta et al. 2017a). The formation of C$_2$, C$_3$, and CH radicals indicates that electron impact and the dehydrogenation processes are responsible for the dissociation of CH$_4$ due to the presence of both electrons and atomic hydrogens in the plasma (Mohanta et al. 2017b, Denysenko et al. 2004). The C$_2$ transitions observed in the Ar/H$_2$/CH$_4$ plasma arise due to the energy transfer from Ar* to C$_2$H$_2$ and/or C$_2$H formed during the dissociation of CH$_4$ (Mohanta et al. 2017b, Rabeau et al. 2004).

2.9 Summary

The physical processes of plasma synthesis routes such as PLA and RF ICP have been investigated which have several distinct advantages including versatility and operation with reactive gases in addition to inert, reducing, oxidizing, and other atmospheres. PLA technique has been used to synthesize contaminant-free high-purity nanoparticles in different environments since condensation of laser-ablated plasma species leads to the formation of nanoparticles which doesn't involve chemical reagents in solutions. Rayleigh scattering and PL experiments have been used to investigate the vapor-phase nanoparticles. Exciton–exciton scattering process in vapor-phase ZnO nanoparticles has also been observed which is dependent on particle size. However, the laser ablation process is incapable of industrial-scale mass production of nanoparticles. RF ICP synthesis technique is an alternative plasma route which is capable of producing high-quantity nanoparticles for industrial applications and is better than conventional plasma due to a large volume, high-temperature plasma zone, operating under different atmospheres, permitting synthesis of a great variety of nanoparticles, and higher production rate with lower processing cost. ICP torch doesn't involve electrode for its operation leading to the generation of clean plasma that ensures synthesis of high-purity products. It is a continuous process to accomplish high production volume with good synthesis control from the gaseous phase. A wide range of nanoparticles such as Si, SiC, WC, and graphene nano-flakes have been produced using appropriate precursors in RF ICP reactor. OES technique has been used to monitor the plasma in order to understand the induction plasma synthesis process of nanoparticles.

References

Adachi, S. 1999. Indium Arsenide (InAs). In *Optical Constants of Crystalline and Amorphous Semiconductors*, 257–267. Boston, MA: Springer.

Amendola, V., Polizzi, S. and Meneghetti, M. 2007. Free silver nanoparticles synthesized by laser ablation in organic solvents and their easy functionalization. *Langmuir* 23: 6766–6770.

Amoruso, S., Ausanio, G., Bruzzese, R., Vitiello, M. and Wang, X. 2005. Femtosecond laser pulse irradiation of solid targets as a general route to nanoparticle formation in a vacuum. *Phys. Rev. B* 71: 033406–1–4.

Amoruso, S., Bruzzese, R., Spinelli, N., Velotta, R., Vitiello, M., Wang, X., Ausanio, G., Iannotti, V. and Lanotte, L. 2004. Generation of silicon nanoparticles via femtosecond laser ablation in vacuum. *Appl. Phys. Lett.* 84: 4502–4504.

Bagnall, D. M., Chen, Y. F., Zhu, Z., Yao, T., Shen, M. Y. and Goto, T. 1998. High temperature excitonic stimulated emission from ZnO epitaxial layers. *Appl. Phys. Lett.* 73: 1038–1040.

Bäuerle, D. 2000. *Laser Processing and Chemistry*. Berlin: Springer.

Bekefi, G. 1976. *Principles of Laser Plasmas*. New York: Wiley.

Boulos, M. I. 1991. Thermal plasma processing. *IEEE T. Plasma Sci.* 19: 1078–1089.

Boulos, M. I. 1997. The inductively coupled radio frequency plasma. *High Temp. Mat. Proc.* 1: 17–39.

Brown, M. S. and Arnold, C. B. 2010. Fundamentals of laser-material interaction and application to multi-scale surface modification. In *Laser Precision Microfabrication. Springer Series in Materials Science*, eds.

K. Sugioka, M. Meunier, and A. Piqu, vol 135. Berlin, Heidelberg: Springer.

Brus, L. E. 1984. Electron–electron and electron–hole interactions in small semiconductor crystallites: The size dependence of the lowest excited electronic state. *J. Chem. Phys.* 80: 4403–4409.

Byrappa, K., Ohara, S. and Adschiri, T. 2008. Nanoparticles synthesis using supercritical fluid technology – towards biomedical applications. *Adv. Drug Delivery Rev.* 60: 299–327.

Carslaw, H. S. and Jaegaer, J. C. 1959. *Conduction of Heat in Solids*. Oxford: Oxford university press.

Chang, I. 2013. Plasma synthesis of metal nanopowders. In *Advances in Powder Metallurgy: Properties, Processing and Applications*, eds. I. Chang and Y. Zhao, 69–85. Woodhead Publishing Series in Metals and Surface Engineering: Woodhead Publishing Limited, Sawston, UK.

Chichkov, B. N., Momma, C., Nolte, S., Von Alvensleben, F. and Tünnermann, A. 1996. Femtosecond, picosecond and nanosecond laser ablation of solids. *Appl. Phys. A* 63: 109–115.

Cota-Sanchez, G., Soucy, G., Huczko, A. and Lange, H. 2005. Induction plasma synthesis of fullerenes and nanotubes using carbon black-nickel particles. *Carbon* 43: 3153–3166.

Denysenko, I. B., Xu, S., Long, J. D., Rutkevych, P. P., Azarenkov, N. A. and Ostrikov, K. 2004. Inductively coupled $Ar/CH_4/H_2$ plasmas for low-temperature deposition of ordered carbon nanostructures. *J. Appl. Phys.* 95: 2713–2724.

Edgar, J. H., Yu, Z. J., Ahmed, A. U. and Rys, A. 1990. Low temperature metal-organic chemical vapor deposition of aluminum nitride with nitrogen trifluoride as the nitrogen source. *Thin Solid Films* 189: L11–L14.

Eliezer, S., Eliaz, N., Grossman, E., Fisher, D., Gouzman, I., Henis, Z., Pecker, S., Horovitz, Y., Fraenkel, M., Maman, S. and Lereah, Y. 2004. Synthesis of nanoparticles with femtosecond laser pulses. *Phys. Rev. B* 69: 144119.

Fabbro, R., Fournier, J., Ballard, P., Devaux, D. and Virmont, J. 1990. Physical study of laser-produced plasma in confined geometry. *J. Appl. Phys.* 68: 775–784.

Foreman, J. V., Li, J., Peng, H., Choi, S., Everitt, H. O. and Liu, J. 2006. Time-resolved investigation of bright visible wavelength luminescence from sulfur-doped ZnO nanowires and micropowders. *Nano Lett.* 6: 1126–1130.

Ganjali, M., Ganjali, M., Vahdatkhan, P. and Marashi, S. M. B. 2015. Synthesis of Ni nanoparticles by pulsed laser ablation method in liquid phase. *Procedia Materials Science* 11: 359–363.

Gfroerer, T. H. 2000. Photoluminescence in Analysis of Surfaces and Interfaces. In *Encyclopedia of Analytical Chemistry*, ed. R. A. Meyers, 9209–9231. Chichester: John Wiley & Sons Ltd.

Girshick, S. L., Chiu, C.-P. and McMurry, P. H. 1988. Modelling particle formation and growth in a plasma synthesis reactor. *Plasma Chem. Plasma Process.* 8: 145–157.

Gupta, S. L. and Thareja, R. K. 2013. ZnO thin film deposition using colliding plasma plumes and single plasma plume: Structural and optical properties. *J. Appl. Phys.* 114: 224903.

Hartanto, A. B., Ning, X., Nakata, Y. and Okada, T. 2004. Growth mechanism of ZnO nanorods from nanoparticles formed in a laser ablation plume. *Appl. Phys. A* 78: 299–301.

Hayamizu, S., Tabata, H., Tanaka, H. and Kawai, T. 1996. Preparation of crystallized zinc oxide films on amorphous glass substrates by pulsed laser deposition. *J. Appl. Phys.* 80: 787–791.

Hervé, P. and Vandamme, L. K. J. 1994. General relation between refractive index and energy gap in semiconductors. *Infrared Phys. Technol.* 35: 609–615.

Hoffman, J., Chrzanowska, J., Kucharski, S., Moscicki, T., Mihailescu, I. N., Ristoscu, C. and Szymanski, Z. 2014. The effect of laser wavelength on the ablation rate of carbon. *Appl. Phys. A* 117: 395–400.

Hosseini, S. B. 2017. A review: Nanomaterials as a filler in natural fiber reinforced composites. *J. Nat. Fibers* 14: 311–325. www.iso.org/obp/ui/fr/#iso:std:iso:3252:ed-4:v1:en : ISO3252:1999(en), Powder metallurgy—Vocabulary.

Hughes, T. P. 1975. *Plasma and Laser Light.* New York: Wiley.

Hussain, F., Hojjati, M., Okamoto, M. and Gorga, R. E. 2006. Review article: Polymer-matrix nanocomposites, processing, manufacturing, and application: An overview. *J. Compos. Mater.* 40: 1511–1575.

ISO/TS 80004-2:2015—Nanotechnologies—Vocabulary—Part 1: Core terms. *International Organization for Standardization.* 2015. Retrieved 2018-01-08.

ISO/TS 80004-2:2015—Nanotechnologies—Vocabulary—Part 2: Nano-objects. *International Organization for Standardization.* 2015. Retrieved 2018-01-08.

ISO/TS 80004-4:2011(en)—Nanotechnologies—Vocabulary—Part 4: Nanostructured materials, International Organization for Standardization. 2011.

Itina, T. E. 2011. On Nanoparticle formation by laser ablation in liquids. *J. Phys. Chem. C* 115: 5044–5048.

Jiayin, G., Xiaobao, F., Dolbec, R., Siwen, X., Jurewicz, J. and Boulos, M. 2010. Development of nanopowder synthesis using induction plasma. *Plasma Sci. Tech.* 12: 188–199.

Kayanuma, Y. 1988. Quantum-size effects of interacting electrons and holes in semiconductor microcrystals with spherical shape. *Phys. Rev. B* 38: 9797–9805.

Kempkens, H. and Uhlenbusch, J. 2000. Scattering diagnostics of low-temperature plasmas (Rayleigh scattering, Thomson scattering, CARS). *Plasma Sources Sci. Technol.* 9: 492.

Kim, M., Osone, S., Kim, T., Higashi, H. and Seto, T. 2017. Synthesis of nanoparticles by laser ablation: A Review. *KONA Powder Part. J.* 34: 80–90.

Klingshirn, C., Priller, H., Decker, M., BrcknerH, J., Kalt, H., Hauschild, R., Zeller, J., Waag, A., Bakin, A.,

Wehmann, H., Thonke, K., Sauer, R., Kling, R., Reuss, F. and Kirchner, Ch.. 2005. Excitonic properties of ZnO. In *Advanced in Solid State Physics*, ed. B. Kramer, vol. 45: 275–287. Berlin: Springer.

Kumar, B., Yadav, D. and Thareja, R. K. 2011. Growth dynamics of nanoparticles in laser produced plasma in liquid ambient. *J. Appl. Phys.* 110: 074903.

Leblanc, D., Dolbec, R., Guerfi, A., Jiayin, G., Hovington, P., Boulos, M. and Zaghib, K. 2017. Silicon nanopowder synthesis by inductively coupled plasma as anode for high-energy Li-ion batteries. In *Silicon Nanomaterials Sourcebook, Vol. II, Hybrid Materials, Arrays, Networks, and Devices*, ed. K. D. Sattler, 464–483. CRC Press—Taylor & Francis Group.

Lee, M.-T., Hwang, D. J., Greif, R. and Grigoropoulos, C. P. 2009. Nanocatalyst fabrication and the production of hydrogen by using photon energy. *Int. J. Hydrogen Energy* 34: 1835–1843.

Leparoux, M., Schreuder, C. and Fauchais, P. 2008. Improved plasma synthesis of Si nanopowders by quenching. *Adv. Eng. Mater.* 10: 1147–1150.

Lorenz, M. R., Pettit, G. D. and Taylor, R. C. 1968. Band gap of gallium phosphide from 0 to 900°K and light emission from diodes at high temperatures. *Phys. Rev.* 171: 876–881.

Mafune, F., Kohno, J.-y., Takeda, Y. and Kondow, T. 2003. Formation of stable platinum nanoparticles by laser ablation in water. *J. Phys. Chem. B* 107: 4218–4223.

Mang, A., Reimann, K. and Rubenacke, St. 1995. Band gaps, crystal-field splitting, spin-orbit coupling, and exciton binding energies in ZnO under hydrostatic pressure. *Solid State Commun.* 94: 251–254.

Meulenbroeks, R. F. G., Schram, D. C., Jaegers, L. J. M. and van de Sanden, M. C. M. 1992. Depolarization Rayleigh scattering as a means of molecular concentration determination in plasmas. *Phys. Rev. Lett.* 69: 1379–1382.

Miller, J. C. and Haglan, Jr., R. F. 1998. *Laser Ablation and Desorption.* New York: Academic.

Mirzaee, M., Changizi, R. and Alinejad, B. 2012. Comparison of mechanical and electrical properties of foams fabricated by the methods of sinter followed cold press and hot press. *ARPN J. Eng. Appl. Sci.* 7: 1533–1538.

Mohanta, A. and Thareja, R. K. 2008. Photoluminescence study of ZnO nanowires grown by thermal evaporation on pulsed laser deposited ZnO buffer layer. *J. Appl. Phys.* 104: 044906.

Mohanta, A. and Thareja, R. K. 2009. Rayleigh scattering from gaseous phase nanoparticles synthesized by pulsed laser ablation of ZnO. *J. Appl. Phys.* 106: 124909.

Mohanta, A. and Thareja, R. K. 2014. Photoluminescence characteristics of catalyst free ZnO nanowires. *Mater. Res. Express* 1: 015023.

Mohanta, A., Jang, D.-J., Lin, G.-T., Lin, Y.-T. and Tu, L. W. 2011. Carrier recombination dynamics in Si doped InN thin films, *J. Appl. Phys.* 110: 023703.

Mohanta, A., Kung, P. and Thareja, R. K. 2015. Exciton-exciton scattering in vapor phase ZnO nanoparticles. *Appl. Phys. Lett.* 106: 013108.

Mohanta, A., Lanfant, B., Asfaha M. and Leparoux, M. 2017a. Optical emission spectroscopic study of $Ar/H_2/CH_4$ plasma during the production of graphene nano-flakes by induction plasma synthesis. *J. Phys.: Conf. Ser.* 825: 012010.

Mohanta, A., Lanfant, B., Asfaha, M. and Leparoux, M. 2017b. Methane dissociation process in inductively coupled $Ar/H_2/CH_4$ plasma for graphene nano-flakes production. *Appl. Phys. Lett.* 110: 093109.

Mohanta, A., Singh, V. and Thareja, R. K. 2008. Photoluminescence from ZnO nanoparticles in vapor phase. *J. Appl. Phys.* 104: 064903.

Morales, A. M. and Lieber, C. M. 1998. A laser ablation method for the synthesis of crystalline semiconductor nanowires. *Science* 279: 208–211.

Narayanan, V. and Thareja, R. K. 2004. Emission spectroscopy of laser-ablated Si plasma related to nanoparticle formation. *Appl. Surf. Sci.* 222: 382–393.

Okada, T., Kawashima, K. and Ueda, M. 2005. Ultraviolet lasing and field emission characteristics of ZnO nano-rods synthesized by nano-particle-assisted pulsed-laser ablation deposition. *Appl. Phys. A: Mater. Sci. Process.* 81: 907–910.

Özgür, U., Alivov, Ya. I., Liu, C., Teke, A., Reshchikov, M. A., Doanc, S., Avrutin, V., Cho, S.-J. and Morkoçd, H. 2005. A comprehensive review of ZnO materials and devices. *J. Appl. Phys.* 98: 041301.

Pandey, P. K., Thareja, R. K. and Costello, J. T. 2016. Heterogeneous (Cu-Ti) colliding plasma dynamics. *Phys. Plasmas.* 23: 103516.

Pandey, P. K., Thareja, R. K., Singh, R. P. and Costello, J. T. 2018. Deposition of nanocomposite $Cu–TiO_2$ using heterogeneous colliding plasmas. *Appl. Phys. B: Lasers Opt.* 124: 50.

Pankove, J. I. 1975. *Optical Processes in Semiconductors.* Englewood Cliffs, NJ: Prentice-Hall, Inc.

Patel, D. N., Pandey, P. K. and Thareja, R. K. 2013. Stoichiometry of laser ablated brass nanoparticles in water and air. *Appl. Opt.* 52: 7592–7601.

Pavesi, L., Piazza, F., Rudra, A., Carlin, J. F. and Ilegems, M. 1991. Temperature dependence of the InP band gap from a photoluminescence study. *Phys. Rev. B* 44: 9052–9055.

Polarza, S. and Smarsly, B. 2002. Nanoporous materials. *J. Nanosci. Nanotechnol.* 2: 581–612.

Pozhar, L. A. 2015. Quantum dots of traditional III–V semiconductor compounds. In *Virtual Synthesis of Nanosystems by Design: From First Principles to Applications*, 113–145. Amsterdam, Netherlands: Elsevier.

Pristavita, R., Meunier, J.-L. and Berk, D. 2011. Carbon nano-flakes prod by an inductively coupled thermal plasma system for catalyst applications. *Plasma Chem. Plasma Process* 31: 393–403.

Rabeau, J. R., John, P., Wilson, J. I. B. and Fan, Y. 2004. The role of C_2 in nanocrystalline diamond growth. *J. Appl. Phys.* 96: 6724–6732.

Rossetti, R., Nakahara, S. and Brus, L. E. 1983. Quantum size effects in the redox potentials, resonance Raman spectra, and electronic spectra of CdS crystallites in aqueous solution. *J. Chem. Phys.* 79: 1086–1088.

Saji, K. J., Joshy, N. V. and Jayaraj, M. K. 2006. Optical emission spectroscopic studies on laser ablated zinc oxide plasma. *J. Appl. Phys.* 100: 043302.

Sajti, Cs. L., Giorgio, S., Khodorkovsky, V. and Marine, W. 2007. Femtosecond laser synthesized nanohybrid materials for bioapplications. *Appl. Surf. Sci.* 253: 8111–8114.

Saleh, B. E. A. and Teich, M. C. 1991. *Fundamental of Photonics.* New York: John Wiley and Sons, Inc.

Shaikh, N. M., Rashid, B., Hafeez, S., Jamil, Y. and Baig, M. A. 2006. Measurement of electron density and temperature of a laser-induced zinc plasma. *J. Phys. D: Appl. Phys.* 39: 1384–1391.

Shalish, I., Temkin, H. and Narayanamurti, V. 2004. Size-dependent surface luminescence in ZnO nanowires. *Phys. Rev. B* 69: 245401.

Shin, J. W., Miyazoe, H., Leparoux, M., Siegmann, St., Dorier, J. L. and Hollenstein, Ch. 2006. The influence of process parameters on precursor evaporation for alumina nanopowder synthesis in an inductively coupled rf thermal plasma. *Plasma Sources Sci. Technol.* 15: 441–449.

Singh, R. K. and Narayan, J. 1989. A novel method for simulating laser-solid interactions in semiconductors and layered structures. *Mater. Sci. Eng. B* 3: 217–230.

Singh, R. K. and Narayan J. 1990. Pulsed-laser evaporation technique for deposition of thin films: Physics and theoretical model. *Phys. Rev. B* 41: 8843–8859.

Singh, R. K., Holland, O. W. and Narayan, J. 1990. Theoretical model for deposition of superconducting thin films using pulsed laser evaporation technique. *J. Appl. Phys.* 68: 233–247.

Sneddon, G., Greenaway, A. and Yiu, H. H. P. 2014. The potential applications of nanoporous materials for the adsorption, separation, and catalytic conversion of carbon dioxide. *Adv. Energy Mater.* 4: 1301873 (1–19).

Soma, H. and Uchino, T. 2017. Blue and orange photoluminescence and surface band-gap narrowing in lithium-doped MgO microcrystals. *J. Phys. Chem. C* 121: 1884–1892.

Sourice, J., Bordes, A., Boulineau, A., Alpera, J. P., Franger, S., Quinsaca, A., Habert, A., Leconte, Y., De Vito, E., Porcher, W., Reynaud, C., Herlin-Boime, N. and Haon, C. 2016. Core-shell amorphous silicon-carbon nanoparticles for high performance anodes in lithium ion batteries. *J. Power Sources* 328: 527–535.

Stratakis, E., Zobra, V., Barberoglou, M., Fotakis, C. and Shafeev, G. A. 2009. Femtosecond laser writing of nanostructures on bulk Al via its ablation in air and liquids. *Appl. Surf. Sci.* 255: 5346–5350.

Streubel, R., Bendt, G. and Gökce, B. 2016. Pilot-scale synthesis of metal nanoparticles by high-speed pulsed laser ablation in liquids. *Nanotechnology* 27: 205602.

Su, S. S. and Chang I. 2018. Review of Production Routes of Nanomaterials. In *Commercialization of Nanotechnologies-A Case Study Approach*, eds. D. Brabazon, E. Pellicer, F. Zivic, J. Sort, M. Baró, N. Grujovic, N. and K.-L. Choy, 15–29. Cham, Switzerland: Springer.

Thareja, R. K. and Mohanta, A. 2010. ZnO Nanoparticles. In *Hand Book of Nanophysics, Vol. 3, Nanoparticles and Quantum dots*, ed. K. D. Sattler, 6-1–20. CRC Press—Taylor & Francis Group, Boca Raton, FL.

Thareja, R. K., Dwivedi, R. K. and Ebihara K. 2002. Interaction of ambient nitrogen gas and laser ablated carbon plume: Formation of CN. *Nucl. Instr. Meth. Phys. Res. B* 192: 301–310.

Umezu, I., Yamamoto, S. and Sugimura, A. 2010. Emission induced by collision of two plumes during pulsed laser ablation. *Appl. Phys. A* 101: 133–136.

Vogel, A. and Venugopalan, V. 2003. Mechanisms of pulsed laser ablation of biological tissues, *Chem. Rev.* 103: 577–644.

Wagener, P., Schwenke, A., Chichkov, B. N. and Barcikowski S. 2010. Pulsed laser ablation of zinc in tetrahydrofuran: Bypassing the cavitation bubble. *J. Phys. Chem. C* 114: 7618–7625.

Willmott, P. R. and Huber, J. R. 2000. Pulsed laser vaporization and deposition. *Rev. Mod. Phys.* 72: 315–328.

Witowski, A. M., Pakuła, K., Baranowski, J. M., Sadowski, M. L. and Wyder, P. 1999. Electron effective mass in hexagonal GaN. *Appl. Phys. Lett.* 75: 4154–4155.

Wong, E. M., Bonevich, J. E. and Searson, P. C. 1998. Growth kinetics of nanocrystalline ZnO particles from colloidal suspensions. *J. Phys. Chem. B* 102: 7770–7775.

Wood, R. F. and Giles, G. E. 1981. Macroscopic theory of pulsed-laser annealing. I. Thermal transport and melting. *Phys. Rev. B* 23: 2923–2942.

Xu, H.-Y., Zhang, Y.-C., Song, Y.-Q. and Chen D.-Y. 2004. Thermalization time of thin metal film heated by short pulse laser. *Chin. Phys.* 13: 1758 – 1765.

Yan, Z., Bao, R. and Chrisey, D. B. 2013. Generation of Ag-Ag$_2$O complex nanostructures by excimer laser ablation of Ag in water. *Phys. Chem. Chem. Phys.* 15: 3052–3056.

Yim, W. M. 1971. Direct and indirect optical energy gaps of AlAs. *J. Appl. Phys.* 42: 2854–2856.

3

Plant-Mediated Synthesis of Nanoparticles

Alireza Ebrahiminezhad,
Seyedeh-Masoumeh Taghizadeh,
Saeed Taghizadeh, and
Younes Ghasemi
Shiraz University of Medical Sciences

Aydin Berenjian and
Mostafa Seifan
The University of Waikato

3.1 Introduction

Nanomaterials are going to be ubiquitous in almost all sciences, technologies, and commercial products. Metallic nanoparticles are one of the most applied nanomaterials with diverse properties and functions. Unique physicochemical properties of these materials make them the arrowhead of applied nanotechnology. These particles are now being used in various fields such as biotechnology, chemical engineering (Ebrahiminezhad et al. 2017a, 2016i, Raee et al. 2018, Ranmadugala et al. 2017a, b, c, d, e, 2018, Viola et al. 2018), civil engineering (Seifan et al. 2017, 2018a, b, c) and medical sciences (Ebrahiminezhad et al. 2016a, e, b, 2012a, 2014b, 2015a). Also these compounds make a promising future in environmental sciences (Ebrahiminezhad et al. 2017e). Expansion in the applications of metallic nanoparticles resulted in the increasing demand for the production of these materials. Since now, physical and chemical processes, both in top-down and bottom-up approaches (Figure 3.1), have been well developed for the preparation of nanomaterials (Ebrahiminezhad et al. 2012b, 2013, 2014a). Bottom-up is the more accepted approach due to better control on the synthesis process and properties of the prepared nanomaterials. This approach is performed in a chemical process by

using ionic precursors of intended metal (Ebrahiminezhad et al. 2017a, Taghizadeh et al. 2017). However, the process is usually multistep and requires harsh conditions with high energy consumption. Chemical approaches also employ organic solvents and toxic chemicals which raise environmental concerns (Ebrahimi et al. 2016a, b, Iida et al. 2007, Li et al. 2009, Mandal et al. 2005, Park et al. 2004, Si et al. 2004, Sun et al. 2007, Wang et al. 2001, Yu et al. 2006). In addition, remnants of chemicals and organic solvents as contaminants can limit biological and biomedical applications of the chemically synthesized nanostructures. Production of stable and biocompatible nanoparticles is another concern in chemical synthesis. From chemical and colloidal point of view, naked nanoparticles are not stable and need to be coated or capped with biocompatible and hydrophilic compounds.

Green chemistry has emerged as a new concept in chemistry and synthesis technologies to eliminate hazardous chemicals and organic solvents especially from industrial process. This technique is based on swapping chemical compounds and organic solvents with biochemical species and aqueous matrix, respectively (Ebrahiminezhad et al. 2016a, b, c). Plants and microorganisms are the most studied sources for biologic compounds in regard to

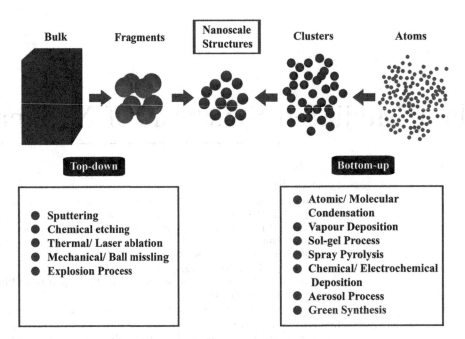

FIGURE 3.1   Top-down and bottom-up approaches in the fabrication of nanoparticles.

green synthesis. These organisms can produce biochemical compounds which are capable to synthesize and stabilize metallic nanoparticles. Since now, various microorganisms including bacteria, fungi, cyanobacteria, microalgae, and yeasts have been used for the synthesis of metallic nanoparticles (Ali et al. 2011, Chokshi et al. 2016, Ebrahiminezhad et al. 2016a, 2017h, Jena et al. 2015, Kannan et al. 2013, Kathiraven et al. 2015, Kianpour et al. 2016, Mohseniazar et al. 2011, Prasad et al. 2013, Sinha et al. 2015, Xie et al. 2007a, b). Both intracellular and secretory compounds were shown to be effective in this regard (Ebrahiminezhad et al. 2016a, Xie et al. 2007a, b). Using microorganisms for the production of nanoparticles has some major drawbacks. Most of the employed organisms are opportunistic or pathogenic, and nanoparticles that derived from such sources cannot be approved for pharmaceutical and biomedical applications. On the other hand, microbial cells usually grown in the complex and expensive media through labour-intensive and time-consuming process (Ebrahiminezhad et al. 2017h, Kianpour et al. 2016, Raïkher et al. 2010). Disposal of culture wastes such as produced biomass and waste media is the other problem.

To eliminate technical problems in microbial synthesis, plant-mediated synthesis is a promising substitution toward green and economic synthesis of nanoparticles. Now various plants, such as trees, flowers, and herbs, are being used for this purpose (Cruz et al. 2010, Ebrahiminezhad et al. 2016b, c, Mo et al. 2015). But herbs are among the widely used ones due to being cheap, abundant, ubiquitous, safe, and rich in plenty of organic and reducing compounds which are effective for nanoparticles synthesis. Most importantly, using herbs in the production process resulted in the fabrication of particles with unique properties (Ebrahiminezhad et al. 2016c, 2017e). All parts of plants can be applied in the green synthesis of nanoparticles,

considering leaf extract being the mostly used one. For instance, leaf extract of eucalyptus, black and green tea, *Lippia citriodora* (Lemon Verbena), maple (Acer sp.), *Lantana camara*, *Artemisia annua*, and *Ephedra intermedia*, and Mediterranean cypress (*Cupressus sempervirens*) were used for the synthesis of different metal nanoparticles (Ajitha et al. 2015, Basavegowda et al. 2014, Cruz et al. 2010, Ebrahiminezhad et al. 2016f, 2017e, Mo et al. 2015, Wang et al. 2014b, Xiao et al. 2015). Other parts of plants such as flower extract of *Matricaria chamomilla*, stem extract of *Desmodium gangeticum*, coffee powder extract, *Piper longum* and *Crataegus douglasii*, fruit extract, orange peel extract, *Cinnamon zeylanicum* bark extract, seed exudate from *Sterculia foetida* and *Medicago sativa*, and oil of *Plukenetia volubilis* L. were also used (Azad and Banerjee 2014, Ebrahiminezhad et al. 2016c, Ghaffari-Moghaddam and Hadi-Dabanlou 2014, Kahrilas et al. 2014, Kumar et al. 2014c, d, Lukman et al. 2011, Nadagouda and Varma 2008, Rajasekharreddy and Rani 2014, Reddy et al. 2014, Sathishkumar et al. 2009).

In this chapter, we discuss the plant-mediated green synthesis of nanoparticles in detail with focus on the molecular mechanism, effective parameters, and how to boost the synthesis reaction. Finally, there is a review on the plant-mediated synthesized nanoparticles and their characteristic features.

3.2   Mechanism of the Green Synthesis

Plant extracts are commonly rich in phytochemicals which are able to act as reducing agent and convert metal ions to metal nanoparticles. There is no exact knowledge about the effective molecules in the reduction of metal ions. But, some investigations have shown that oxygen-bearing functional groups such as hydroxyl, phenol, carboxyl, and carbonyl are the sites for entrapment and reduction of metal ions (Kozma et al. 2015). It is evident that plants with high contents in phenolic compounds are the best choice for green synthesis purposes. Phenolic compounds from *Salvia officinalis, Zataria multiflora, Melaleuca nesophila, Eucalyptus tereticornis, Rosmarinus officinalis, Mansoa alliacea*, and green tea were found to be the effective molecules in the formation of nanoparticles (Markova et al. 2014, Prasad 2016, Soliemanzadeh et al. 2016, Wang et al. 2015, 2014c). But some of these studies, especially those that examined the green tea extract for the synthesis of iron-based nanoparticles, have claimed that polyphenol compounds chelate iron ions to form nanostructures of iron complexes (Markova et al. 2014, Wang et al. 2015, 2014c).

Naked nanoparticles are chemically and colloidally unstable and should be coated or capped with a protective and hydrophilic compound (Ebrahiminezhad et al. 2012b). Also, the coating should be biocompatible if it is planned to apply nanoparticles in biological and biomedical fields. In the common chemical synthesis procedure, naked nanoparticles are coated by using biocompatible coatings such as polymers, carbohydrates, amino acids, lipoamino acids, and silica (Ebrahiminezhad et al. 2012b, 2013, 2014b, 2015a, Gholami et al. 2016, 2015, Gupta and Wells 2004, Muddineti et al. 2016, Ragaseema et al. 2012). In addition to fabricate nanoparticles, plant extracts provide stabilizing or capping agents which can engulf synthesized nanoparticles and make them colloidally and chemically stable (Ebrahiminezhad et al. 2016f, 2017c, f). Also, microparticles in the plant extract can act as a platform for the formation and stabilization of nanoparticles (Ebrahiminezhad et al. 2016g). Some evidences have shown that carbohydrates are the molecules that act in the way to engulf and stabilize nanoparticles (Ebrahiminezhad et al. 2016c, f, 2017g).

3.3 Effective Parameters in the Green Synthesis

3.3.1 Plant and Plant Extract

Synthesis of a particular nanoparticle by using various plant extract will not result in the similar particles. Various plants contain different patterns of phytochemicals which results in the production of different nanoparticles with different physicochemical characteristics. For instance, chemical structure, size, and shape of iron nanoparticles which were synthesized with various plant extracts are listed in Table 3.1. Diverse iron-based nanoparticles from zero valent to iron oxides (Fe_3O_4 and Fe_2O_3), iron oxide hydroxides (α-FeOOH and β-FeOOH), and iron complexes were synthesized by different plants. These particles have come in different shapes such as spherical, rod, triangular, cylindrical, cubic, and irregular clusters (Ebrahiminezhad et al.

2017g). Also, it is obvious that in the same plant, the pattern of phytochemicals has seasonal variations, and consequently, a particular plant cannot produce exactly same particles throughout a year (Ncube et al. 2011). This can be considered as a drawback for plant-mediated synthesis of nanoparticles that can be addressed by deposition of dried leaves.

The major advantage of plant-mediated synthesis of nanostructures is the elimination of organic solvents use. However, non-aqueous extracts can also be used for this purpose. Alcohols are the mostly used organic solvents along with the other derivatives of alcohols and organic solvents such as dichloromethane, n-hexane, n-butanol, chloroform, and ethyl acetate (Chitsazi et al. 2016, Lee et al. 2016, Nagababu and Rao 2017, Ojha et al. 2017, Sithara et al. 2017).

Employed solvent has an immense effect on the characteristic features of the prepared nanoparticles. For instance, methanolic extract of *Pulicaria gnaphalodes* provides separate spherical silver nanoparticles. But, by employing dichloromethane porous polyhedral and aggregated particles are produced (Chitsazi et al. 2016). Vélez et al. (2018) have shown that methanolic extract of *Aloe vera* produces silver nanoparticles with a size ranging from 2 to 7 nm. Authors also reported that employing aqueous extract of *Aloe vera* resulted in the larger nanoparticles with more broad size distribution (Vélez et al. 2018).

Difference in the characteristic features of the prepared particles using various solvents is due to difference in the effective compounds responsible for nanoparticles reduction and stabilization. Luis M. Carrillo-López and colleagues in 2016 find out that phenolic compounds (flavonoids and tannins) are the compounds responsible for reducing and stabilizing silver nanoparticles using methanolic extract of *Chenopodium ambrosioides*. But, in the dichloromethane and hexane extracts, the responsible compounds were mainly terpenoids including transdiol, α-terpineol, monoterpene hydroperoxides, and apiole (Carrillo-López et al. 2016). Similar differences were also reported for *Ocimum sanctum* extracts using solvents with different polarity. Methyl eugenol and β-caryophyllene were found in the hexane extract, while glycerol, phosphoric acid, succinic acid, tartaric acid, D-gluconic acid, myo-inositol, 2-methoxy-4-vinylphenol, and ferulic acid were identified in the aqueous extract (Lee et al. 2016).

As mentioned previously, aqueous extract is the most employed extract for the green synthesis of nanoparticles (Basavegowda et al. 2014, Chitsazi et al. 2016, Mo et al. 2015, Rajasekharreddy and Rani 2014, Reddy et al. 2014, Sun et al. 2014, Vivekanandhan et al. 2014). Employing aqueous extract makes the process more economical, environment-friendly, and sustainable. In this regard, the desired part of the plant should be washed to remove any possible mods and dusts, and then dried and ground to fine powder. After sieving, the powder is mixed with deionized water (1–10 w/v %) and is boiled under reflux to avoid evaporation (Cruz et al. 2010, Dubey et al. 2010a,

TABLE 3.1 Physicochemical Characteristics of Iron Nanoparticles Synthesized by Using Different Plants

Plant	Chemical Structure	Size (nm)	Shape	Reference
Hordeum vulgare	Magnetite	Up to 30	Spherical	(Makarov et al. 2014b)
Rumex acetosa		10–40		
Green tea	Magnetite	40–60 nm	Irregular cluster	(Shahwan et al. 2011)
	FeOOH			
Eucalyptus	IONPs ZVI NPs	20–80 nm	Spherical	(Wang et al. 2014a)
Sorghum **spp.**	FeOOH	50 nm	Irregular cluster	(Njagi et al. 2011)
Tangerine	IONPs	50–200 nm	Spherical	(Ehrampoush et al. 2015)
Soya bean	Magnetite	8 nm	Spherical	(Cai et al. 2010)
Watermelon	Magnetite	20 nm	Spherical	(Prasad et al. 2016)
Mimosa pudica	Magnetite	60–80 nm	Spherical	(Niraimathee et al. 2016)
Caricaya papaya	Magnetite	irregular	Irregular	(Latha and Gowri 2014)
Green tea	ZVI, IONPs, FeOOH, &	40–50 nm	Spherical	(Huang et al. 2014a)
Oolong tea	Fe$_2$O$_3$			
Black tea				
Eucalyptus	ZVI, FeOOH	20–80 nm	Quasi-spherical	(Wang et al. 2014b)
Green tea				
Omani mango	Maghemite Hematite	15 ± 2	Nano-rods	(Al-Ruqeishi et al. 2016)
Salvia officinalis	Fe$_2$O$_3$	5–25 nm	Spherical	(Wang et al. 2015)
Eucalyptus		60–20 nm		(Zhuang et al. 2015)
Pomegranate	ZVI	10–30 nm	Spherical	(Machado et al. 2013a)
Mulberry				
Cherry				
Mulberry	ZVI	5–10 nm	Spherical	(Machado et al. 2015)
Pomegranate		100 nm	Irregular	
Peach				
Pear				
Vine				
Castanea sativa	Maghemite			(Martínez-Cabanas et al. 2016)
Eucalyptus globulus				
Ulex europaeus				
Pinus pinaster				
Sapindus mukorossi	α-FeOOH	50 nm	Rod-like	(Jassal et al. 2016)
	α-Fe$_2$O$_3$			
	β-FeOH			
Eucalyptus	IONPs	80 nm	Spherical	(Cao et al. 2016)
Black tea	FeOOH	40–50 nm	Round	(Ali et al. 2016a)
	Fe$_2$O$_3$			
Mansoa alliacea	β-Fe$_2$O$_3$	18 nm		(Prasad 2016)
Andean blackberry	Magnetite	40–70 nm	Spherical	(Kumar et al. 2016a)
Amaranthus spinosus	IONPs	58–530 nm	Spherical	(Muthukumar and Matheswaran 2015)
Eucalyptus	Fe – p NPs	40–60 nm	Cubic	(Wang et al. 2013)
Green tea	Fe – p NPs	70 nm	Spherical	(Markova et al. 2014)
Eucalyptus tereticornis	Fe – p NPs	50–80 nm	Spherical	(Wang et al. 2014c)
Melaleuca nesophila				
Rosmarinus officinalis				
Shirazi thyme	ZVI & IONPs FeOOH	40–70 nm	Spherical	(Soliemanzadeh et al. 2016)
Pistachio green				
Coffee	ZVI	19.6 ± 25.8 nm	Triangular	(Kozma et al. 2015)
Green tea				
Parthenocissus tricuspidata				
Carob tree	Magnetite	5–8 nm		(Awwad and Salem 2012b)
Passiflora tripartite	Magnetite	22.3 ± 3 nm	Spherical	(Kumar et al. 2014a)
Orange	ZVI	3–300 nm	Spherical	(Machado et al. 2014)
Lime			Cylindrical	
Lemon			Irregular	
Mandarin				
Oolong tea	ZVI, Fe$_3$O$_4$ FeOOH Fe$_2$O$_3$	40–50 nm	Spherical	(Huang et al. 2014b)
Camellia sinensis	FeNPs	60 nm	Spherical	(Mystrioti et al. 2016)
Syzygium aromaticum				
Mentha spicata				
Punica granatum juice				
Red Wine				
Lawsonia inermis	FeNPs	21–32 nm		(Naseem and Farrukh 2015)
Gardenia jasminoides				
Urtica dioica	ZVI	21–71 nm	Spherical	(Ebrahiminezhad et al. 2017)
Rosa damascene	ZVI	100 nm		(Fazlzadeh et al. 2017)
Thymus vulgaris				
Urtica dioica				

Ghaffari-Moghaddam and Hadi-Dabanlou 2014, Goodarzi et al. 2014, He et al. 2013, Lukman et al. 2011, Nadagouda and Varma 2008, Njagi et al. 2011). Then, the mixture is filtered to remove plant slag and centrifuged to harvest plant microparticles. Resulted clear solution can be used as a natural source of reducing and stabilizing agents for the green synthesis of nanoparticles. This process is schematically illustrated in Figure 3.2.

3.3.2 pH of the Plant Extract

The pH of reaction solution is always critical factor in the chemical and biochemical reactions. For instance, in the chemical synthesis of magnetite nanoparticles, it is critical to raise the reaction pH near 9–11 to start the formation of magnetite nucleus and growth of magnetite nanoparticles. This extreme pH is usually obtained by using ammonium hydroxide and some other basic compounds such as

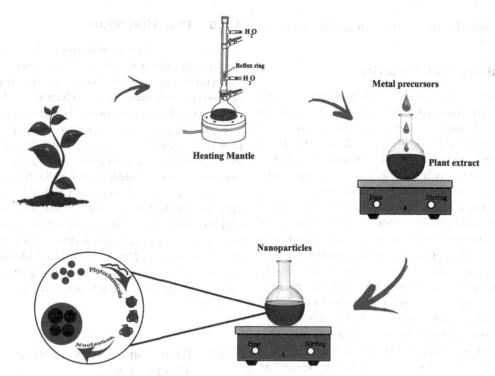

FIGURE 3.2 Schematic illustrations for the green synthesis of metallic nanoparticles.

KOH. Green synthesis is not exceptional, and the pH of the synthesis reaction has an immense effect on the prepared nanoparticles. It has been shown that plant extracts with different pH values provide particles with different characteristics. Acidic extracts provide smaller and more stable nanoparticles in contrast to less acidic or neutral extracts (Makarov et al. 2014b, Njagi et al. 2011). It is believed that capping of the prepared nanoparticles with organic acids such as oxalic acid or citric acid from plant extract resulted in the increase in the zeta potential of prepared particles. It is worth to mention that the zeta potentials more than 25 mV (negative or positive) enhance the colloidal stability of nanostructures (Ebrahiminezhad et al. 2016a).

3.3.3 Antioxidant Capacity of the Plant Extract

Antioxidant capacity of the plant and plant extract has a significant effect on the nanoparticles synthesis reaction. In the most green synthesis reactions for the fabrication of metal nanoparticles, ionic precursors should be reduced to atoms. Antioxidants are the main reducing agents in the plant extracts. Antioxidant capacity is not the same in various plants, so some plants cannot completely reduce ions to atoms and some others can do (Ebrahiminezhad et al. 2017f, Noruzi and Mousivand 2015, Wang et al. 2014a, b). In general, it is believed that plants with high antioxidant capacity are more efficient for the green synthesis of nanoparticles (Harshiny et al. 2015, Muthukumar and Matheswaran 2015). Antioxidant capacity of the plant extract is also dependent on the plant extract preparation

procedure. It has been shown that antioxidant capacity of the plant extract has a direct relation with the ratio of plant powder and solvent which is used for extraction. It means using plant powder in more weight/volume per cent ratio results in the extract with more antioxidant capacity (Harshiny et al. 2015, Machado et al. 2013a, Muthukumar and Matheswaran 2015). Harshiny et al. (2015) have shown that increase in the ration of plant powder and solvent from 5 to 20 w/v % resulted in near threefold increase in the antioxidant capacity of the extract (Harshiny et al. 2015). Showing that by increase in the plant extract concentration, the potential of the extract for the formation of nanoparticles will increase. It also needs to be considered that the overconcentrated plant extract is not suitable for the green synthesis of nanoparticles and there is an optimal value for the plant extract concentration. Ehrampoush et al. have shown that increase in the plant extract concentration from 2 to 6 w/v % resulted in a significant reduction in the particles size from 200 to 50 nm, which is an indicative for increase in the potential of plant extract for the fabrication of nanoparticles. But, further increase in the extract concentration up to 10% resulted in the agglomerated structures (Ehrampoush et al. 2015).

Antioxidant capacity of the plant extract can be evaluated by Folin–Ciocalteu method, ferric reducing antioxidant power (FRAP), and 2, 2-diphenyl-1-picryl-hydrazyl (DPPH) radical scavenging assay (Benzie and Szeto 1999, Conde et al. 2009, Pulido et al. 2000). Among phytochemicals polyphenols, amaranthine, flavonoids, and amino acids were identified as the main antioxidants with key role in the green synthesis of nanoparticles (Harshiny et al. 2015,

Machado et al. 2015, 2013a, Muthukumar and Matheswaran 2015).

3.3.4   Plant Extract Quantity

Amount of plant extract in the green synthesis reaction is another effective parameter. The ratio of plant powder and solvent that is used for green synthesis proposes can vary from 0.2 to 25 w/v % (Cruz et al. 2010, Lukman et al. 2011, Vivekanandhan et al. 2014, Yilmaz et al. 2011). But the most applied concentrations are less than 10% (Ebrahiminezhad et al. 2016b, c, f, g, Goodarzi et al. 2014, He et al. 2013, Sathishkumar et al. 2009, 2010, Song et al. 2009). At these concentrations, increase in the amount of leaf extract in the synthesis reaction usually resulted in the increase in reaction output. Ebrahiminezhad et al. have shown that increase in the amount of *Ephedra intermedia* stem extract up to 90% volume/volume ratio of the reaction mixture resulted in the increase in the formation of nanoparticles (Ebrahiminezhad et al. 2016f). Similar findings were also reported for the green synthesis reactions using leaf extract of *Lippia citriodora* (Lemon Verbena), *Sorbus aucuparia*, and *Rosa rugosa*; and fruit extract of *Tanacetum vulgare* and *Crataegus douglasii* (Cruz et al. 2010, Dubey et al. 2010a, b, c, Ghaffari-Moghaddam and Hadi-Dabanlou 2014). In addition to increase in the formation of nanoparticles, increase in the amount of plant extract has some effects on the size of the prepared particles. Increase in the amount of plant extract resulted in the increase in the total concentration of reducing and capping agents in the reaction mixture, consequently reducing the size of nanoparticles. This effect is reported for *Sorbus aucuparia*, where increase in leaf extract quantity resulted in the decrease in particle size of both synthesized silver and gold nanoparticles (Dubey et al. 2010a). In another study, steady reduction (from 16 to 13 nm) in the mean particle size is reported by increase in the leaves extract quantity from 5 to 10 mL (Cruz et al. 2010). Increase in the amount of plant extract can also reduce the particle size distribution. Significant effect is reported by Dubey and colleagues in 2010; they found that 4 mL increase in the amount of leaf extract can significantly reduce the particle size distribution from about 20 to 4 nm (Dubey et al. 2010b). As mentioned for the concentration of plant extract, there are also optimal values for the plant extract quantity in the green synthesis reaction. Ebrahiminezhad et al. (2016b) have shown that by increase in the *Alcea rosea* flower extract quantity up to 40% (volume/volume ratio of total reaction), the concentration of the fabricated particles increased. But, increase in the plant extract quantity more than 40% resulted in a significant hypochromic shift in the surface plasmon resonance (SPR) band of silver nanoparticles which is due to reduction in the concentration of fabricated particles. Also, increase in the amount of extract resulted in the appearance of second absorption peak which was due to the formation of a second population of large particles (Ebrahiminezhad et al. 2016b).

3.3.5   Reaction Time

A lot has been done to investigate the effects of reaction time on the green synthesis of nanoparticles. In general, the synthesis of nanoparticles in the bottom-up approach is divided into two main steps: (a) formation of the nuclei and (b) growth of nanostructures (Ebrahiminezhad et al. 2013). The first step is usually occurred suddenly at the beginning of synthesis reaction and then followed by the second stage in which nanostructures grow gradually. Increase in the reaction time usually resulted in the more growth of nanostructures and subsequently bigger particles formation (Ebrahiminezhad et al. 2013, 2015a). This basic premise is also approved for the green synthesis reactions. It is shown that in the green synthesis of silver nanoparticles by using *Lippia citriodora* (Lemon Verbena) leaf aqueous extract, increase in the reaction time from 3 to 24 h resulted in the 4 nm increase in mean particle size (from 16 to 20 nm) (Cruz et al. 2010).

3.3.6   Reaction and Extraction Temperature

Green synthesis reactions are usually done at room temperature, and room temperature is ideal for scale-up proposes due to energy concerns and costs (Ebrahiminezhad et al. 2016b, c, f, 2017c, e, f, g). Like all chemical reactions, reaction temperature has a significant effect on the green synthesis of nanoparticles. Increase in the reaction temperature usually resulted in the increase in nanoparticles formation. However, increase in the reaction temperature has a side effect on the characteristics features of the prepared nanoparticles. It has been shown that increase in the reaction temperature, even up to 40°C, can disturb uniformity of the particles and so the produced particles are not in tolerable quality (Ebrahiminezhad et al. 2016b, f). This side effect cannot be seen in all plants; formation of nanoparticles by using sorghum bran extract is increased by increase in the reaction temperature up to even 80°C, without resulting any negative effect on the nanoparticle properties (Njagi et al. 2011).

It also has been shown that temperature of the plant extraction process has significant effect on the synthesized nanoparticles. It has been shown that up to 50°C increase in the temperature in which plant extraction is done resulted in the increase in plant extract antioxidant capacity. But, more increase in the extraction temperature has disruptive effect on the antioxidant capacity. It is believed that high temperatures can degrade antioxidant molecules and reduce the productivity of the resulted extract (Harshiny et al. 2015, Muthukumar and Matheswaran 2015). In addition to extraction temperature, extraction time has also significant effect on the antioxidant properties of the plant extract. Heating the plant materials over a long period can disturb the antioxidant compounds. For instance, in the case of *Amaranthus dubius* leaf extract preparation, the optimal result was obtained at 45 min extraction time, and more heating resulted in the reduction of antioxidant capacity of the extract.

3.4 Assisted Green Synthesis of Nanoparticles

Green synthesis of nanoparticles can be assisted chemically or physically. In some cases, natural compounds from plant extract have no sufficient power to fabricate nanoparticles, and hence, you need an external power to complete the job. Also, it is possible that you want to fabricate nanoparticles with special characteristics that cannot be achieved simply by using a plant extract. In some cases, plant extract is used just as a source of natural coating materials and capping agents, and hence, the formation of nanoparticles should be driven by another source. This external source can be a chemical agent such as a powerful reducing agent or agents that are used for the chemical synthesis of nanoparticles. Also, fabrication of nanoparticles can be enhanced by using physical treatments, mostly irradiation. In the following, we explain these two techniques for the assisted green synthesis of nanoparticles.

3.4.1 Chemically Assisted (Semi-green) Synthesis of Nanoparticles

Some scientists have employed the benefits of plant extract in the chemical reactions. It means that they do their typical synthesis reaction in the presence of plant extract. It is believed that biochemical molecules in the plant extract can act as stabilizer for even chemically synthesized nanoparticles. This idea is not strange, because using biologic molecules as stabilizer for chemically synthesized nanoparticles is a routine approach for the fabrication of biocompatible and stable nanoparticles. In this technique, chemical reactions are conducted in the presence of a define biomolecule such as amino acids (Ebrahiminezhad et al. 2012b, 2013). The biomolecule acts as a capping agent and engulfs the prepared particles, making the particles biocompatible and stable (Ebrahiminezhad et al. 2012b, 2013, Gholami et al. 2015, Raee et al. 2018). Semi-green synthesis is similar to this approach except that in the semi-green synthesis, the chemical reaction is done in the presence of a mixture of biomolecules from plant extract. Because of the use of a chemical agent in the reaction process, the chemical-assisted green synthesis is also called semi-green synthesis (Ebrahiminezhad et al. 2017g). In practice, a typical semi-green synthesis reaction is done by the addition of a chemical agent to the mixture of plant extract and metal ions which used as precursor for nanoparticles.

Chemical agents which can be used for the semi-green synthesis of different nanoparticles vary depending on the proposed nanoparticles. For instance, in the case of iron oxide and iron oxide-hydroxide nanoparticles, the chemical synthesis is done at alkaline pH (9–11). This pH is usually achieved by using ammonium hydroxide (NH_4OH), potassium hydroxide (KOH), and sodium hydroxide (NaOH). Also, other alkaline agents such as sodium oleate and 1,6-hexanediamine ($H_2N(CH_2)_6NH_2$) were used. Hence, these compounds are applicable for the semi-green synthesis of iron oxide and iron oxide-hydroxide nanoparticles in the presence of plant extract (Iida et al. 2007, Park et al. 2004, Raee et al. 2018, Singh et al. 2014b).

Since now, semi-green synthesis of different iron-based nanoparticles has been done using a various series of plants. For instance, magnetite (Fe_3O_4) nanoparticles with superparamagnetic properties can be synthesized by employing aqueous or organic extract of *Passiflora tripartite*, Andean blackberry (*Rubus glaucus*), *Mimosa pudica*, tangerine peel extract, carob leaf, and watermelon rind powder (Awwad and Salem 2012a, Ehrampoush et al. 2015, Kumar et al. 2014a, 2016a, Prasad et al. 2016). In this regard, iron salts such as ferrous sulphate ($FeSO_4$), ferrous chloride ($FeCl_3$), and ferric chloride ($FeCl_2$) are used as iron precursor. The reaction is started by increasing the pH up to extreme alkaline using sodium hydroxide (NaOH) or ammonium hydroxide (NH_4OH) (Ehrampoush et al. 2015, Kumar et al. 2014a, 2016a, Prasad et al. 2016). Iron (III) oxide (Fe_2O_3) and iron (III) oxide-hydroxide (FeOOH) nanoparticles were also synthesized by using extracts of *Sapindus mukorossi* (raw reetha), Omani mango tree, Garlic Vine (*Mansoa alliacea*), and *Amaranthus dubius*. Ferric salts such as ferric chloride and ferric nitrate $Fe(NO_3)_3$ were used as the source of iron, and usually, sodium hydroxide was used to boost the reaction (Al-Ruqeishi et al. 2016, Harshiny et al. 2015, Jassal et al. 2016, Prasad 2016).

Semi-green synthesis was also used for the synthesis of other nanoparticles such as silver nanoparticles. In the common chemical synthesis approaches, conversion of silver ions to silver nanoparticles is done by using chemical agents such as sodium borohydride, sodium citrate, ammonium hydroxide, and ethanol (Ebrahimi et al. 2016a, b, Mehr et al. 2015, Pal et al. 2009). Prepared silver nanoparticles via these classic reactions can be stabilized by performing the reaction in the presence of plant extract. Since now, semi-green synthesized silver nanoparticles have been fabricated by using pomegranate and *Camellia sinensis* (tea plant) peel extract (Logeswari et al. 2015, Nasiriboroumand et al. 2018). Also leave extracts of *Ocimum tenuiflorum*, *Solanum trilobatum* (thoothuvalai), *Syzygium cumini* (Java plum or black plum or jamun), and *Centella asiatica* (gotu kola) were used (Logeswari et al. 2015, Nasiriboroumand et al. 2018).

3.4.2 Physically Assisted Green Synthesis

This technique employs physical treatments in the presence of plant extract to boost the green synthesis reactions. In contrast to semi-green synthesis which uses chemical agents in the presence of plant extract, physically assisted green synthesis is more green and sustainable as no chemical agent is used and no harmful waste is produced. This technique is mostly done by irradiation and so-called radiation-assisted green synthesis. Radiations by sunlight, microwave, and ultrasound waves are the most common physical treatments which have been used for the green synthesis of nanoparticles and are explained in detail in the following.

Light-Assisted Green Synthesis

Sunlight is the most used irradiation for the light-assisted green synthesis of nanoparticles. Sunlight irradiation is varying throughout a day, and hence, the characteristic feature of nanoparticles which are synthesized at different time periods of the day can be different (Prathna et al. 2014). Difference in the sunlight intensity through a day also has impacts on the speed of the reaction. Increase in the irradiation intensity can speed up the synthesis reaction. So, midday irradiation is providing the fastest reaction. It is obvious that geographical location is another significant parameter. For instance, an investigation which is performed in India demonstrated that sunlight irradiation at 12 PM increases the reaction rate more than irradiation at 10 AM and 2 PM (Prathna et al. 2014). Depending on the employed plant extract and sunlight intensity, there is an optimal exposure time for sunlight-assisted green synthesis, and long-time irradiation can disturb the quality of prepared particles. The optimal irradiation time is depending on the employed plant extract. It has been shown that, by applying lemon (*Citrus limon*) or *Polyalthia longifolia* extract for the synthesis of silver nanoparticles, it takes about 20–35 min to reach the absorption maxima. However, while using Aloe vera leaves extract, the maximum of SPR band is obtained just in 10 min (Kumar et al. 2016b, Moosa et al. 2015, Prathna et al. 2014).

Microwave-Assisted Green Synthesis

Another approach for the fast synthesis of nanoparticles using plant extract is microwave-assisted synthesis. This technique is performed simply by employing domestic microwave ovens. Microwave irradiation can increase the rate of nanoparticles formation so that a typical green synthesis reaction can perform in just few minutes or even less than 1 min (Ali et al. 2015, Joseph and Mathew 2015). But, this technique is not functional using all plant extracts. Kahrilas and his colleagues (2014) have evaluated the potential of citrus fruits (orange, grapefruit, tangelo, lemon, and lime) for the green synthesis of silver nanoparticles using microwave technology. The only successful experiment was done by using orange peel extract (Kahrilas et al. 2014). Since now, this technique was employed for the green synthesis of silver nanoparticles using various plants such as *Phyllanthus niruri* L, *Biophytum sensitivum, Elephantopus scaber*, and *Eucalyptus globulus* (Ali et al. 2015, Francis et al. 2018, Haris et al. 2017, Joseph and Mathew 2015). Copper nanoparticles were also synthesized via this technique by using potato starch, piper nigrum seeds extract, *Solanum Lycopersicum* (tomato) fruit extract, coffee powder extract, and leaves extract of black tea and *Polygonum minus* (Sirisha and Asthana 2018,

Sun et al. 2014, Suresh et al. 2014, Sutradhar et al. 2014, Tanghatari et al. 2017, Ullah et al. 2018).

The microwave-assisted synthesis of Al_2O_3 NPs by using *Cymbopogon citratus* leaf extract was proposed by Ansari et al. [64]. In this experiment, aluminium nitrate (Al $(NO_3)_3$) solution was used as the metal precursor. The aluminium salt solution was added to the aqueous extract of *C. citratus* with a ratio of 1:4 at room temperature. The mixture was then subjected to microwave irradiation at 540 W until a colour change happened. It was found that the fabricated nanoparticles ranged from 9 to 180 nm in size (Ansari et al. 2015).

Ultrasonic-Assisted Green Synthesis

Ultrasound is another physical treatment which can be used to intensify a green synthesis reaction. It has been shown that a green synthesis reaction can be performed faster when irradiated with ultrasonic waves. Kothai and Jayanthi have shown that *Camellia sinensis* extract that is fortified with lemon juice and honey can convert silver ions to silver nanoparticles in 15 min. But, if the reaction is performed under ultrasonic irradiation, it takes only 5 min for the change in colour and appearance of reddish brown colour that is an indicative for the formation of nanoparticles (Kothai and Jayanthi 2014). Similar results were reported for the fabrication of silver nanoparticles using weed plant *Lantana camara* L. leaf extract. In a typical green synthesis procedure using *Lantana camara* L. leaf extract, it takes 1 h to fabricate silver nanoparticles. Interestingly, by applying ultrasonication, the colour development and appearance of reddish brown colour can be observed within just 10 min (Manjamadha and Muthukumar 2016). This technique was also used for the synthesis of other nanoparticles such as gold nanoparticles (Babu et al. 2012).

3.5 Additives in Green Synthesis

The major disadvantage in the plant-mediated synthesis is anisotropic and non-uniform growth of the particles. Hence, one of the recent innovations to improve quality of the green synthesized nanoparticles is to employ some additional molecules (additives) such as surfactants (Khan et al. 2012a, b). The presence of various additive compounds in the reaction mixture has different effects on the synthesized nanoparticles such as creating desirable shape, capping the particles, increasing stability, and size control (Cao et al. 2016, Pandian and Palanivel 2016).

β-Cyclodextrin (βCD) is one of the well-employed molecules in this order. Zhuang et al. (2015) investigated the effect of βCD on morphology and reactivity of synthesized nanoparticles. They indicated that the addition of βCD to the reaction mixture resulted in a decrease in the particle size from 60 to 20 nm (Zhuang et al. 2015). In another study, it has been shown that Ag nanoparticles with desirable morphology can be obtained by adding suitable amount of cetyltrimethylammonium bromide (CTAB) to the synthesis reaction. In addition to shape directing role, CTAB is also able to increase the reactivity of phytochemicals to reduce Ag^+ ions (Khan et al. 2012b). All surfactants cannot play controlling role in the green synthesis reaction. For instance, sodium dodecyl sulphate (SDS) has no significant effect on the green synthesis of silver nanoparticles by using Paan (*Piper betle*) leaf petiole extract (Khan et al. 2012a).

3.6 Iron Nanoparticles

Iron-based nanoparticles are one of the most studied and employed nanostructures in various fields of science and technology. Naturally, iron has different states of oxidation with different physicochemical properties. So, nanostructures of iron in various oxidation states pose unique properties which make them suitable for specific applications. Since now, a lot has been done to fabricate iron-based nanoparticles using green synthesis approach as tabulated in Table 3.1. Attempts for the green synthesis of various iron-based nanoparticles are explained in the following sections.

3.6.1 Zero-Valent Iron Nanoparticles

Zero-valent iron (ZVI) possess incompletely filled d-orbitals whit the electron donor activity that makes them highly reactive. ZVI nanoparticles have a great potential for remediation proposes and wide applications in various environmental and industrial processes (Wang and Zhang 1997). In order to fabricate ZVI nanoparticles, we need to reduce iron ions into the zero-valent state. Since now, phytochemicals in the leaves extract from various plants such as *Rosa damascene*, *Thymus vulgaris*, *Urtica dioica*, *Azadirachta indica* (Neem), grape marc, black tea, vine leaves, *Psidium guajava* (Guava), *Cupressus sempervirens* (Mediterranean cypress), *Urtica dioica*, and 26 other different tree species have been shown to be able to do this (Ebrahiminezhad et al. 2017d, f, Fazlzadeh et al. 2017, Machado et al. 2013a, b, Pattanayak and Nayak 2013, Somchaidee and Tedsree 2018).

There are various phytochemicals in the plant extract that were reported to be capable to reduce metal ions and stabilize the nanoparticles (Mittal et al. 2013). But, reactivity of the prepared nanoparticles depends on the biochemical molecules in the plant extract. For instance, Machado et al. (2013b) have shown that extracts of black tea, grape marc, and vine leaves are able to produce ZVI nanoparticles with different reactivity capacity. Different phytochemical components in the plant extracts lead to significant impact on the morphology, reactivity, and agglomeration tendency of the prepared nanoparticles (Machado et al. 2015, 2013a). Green tea extract is also reported to be capable to synthesize spherical ZVI nanoparticles (5–15 nm) from a ferric nitrate solution. The produced nanoparticles showed a good efficiency in degradation and removal of organic contaminations such as bromothymol blue (Hoag et al. 2009).

3.6.2 Magnetite (Fe_3O_4) Nanoparticles

Magnetite nanoparticles have important multifunctional properties such as small size, high magnetism, low toxicity, and microwave absorption properties (Belachew et al. 2017). During the last years, these nanoparticles have gained immense interest by researchers from various fields such as physics, chemistry, medicine, biology, and material sciences. These particles are promising candidate for diverse medical and biological applications such as targeted drug delivery, hyperthermia, magnetic resonance imaging (MRI), cell labelling, and magnetic immobilization (Ebrahiminezhad et al. 2016i, 2015b, Yang et al. 2008).

Nowadays, the literature is replete with investigations that utilize the green synthesis as a novel method for the production of magnetite nanoparticles. In this regard, various plant extracts such as *Syzygium cumini* seed extract, plantain peel extract, Tannin extract of *Acacia mearnsii*, Rambutan peel waste extract, and fruit extract of *Cynometra ramiflora* and *Couroupita guianensis* have been used (Bishnoi et al. 2018, Khan et al. 2015, Sathishkumar et al. 2018, Venkateswarlu et al. 2013, Yuvakkumar and Hong 2014). In addition, leaves extract of *Annona squamosal*, *Albizia adianthifolia*, *Hordeum vulgare*, and *Rumex acetosa* was also employed (Makarov et al. 2014b, Sulaiman et al. 2018b, Vanitha et al. 2017).

Different plant extracts provide particles with different colloidal stability. For example, nanoparticles which synthesized by using *H. vulgare* leaf extract were colloidally unstable and were susceptible to aggregation. But, particles which produced by *Rumex acetosa* extract were highly stable. The higher stability of nanoparticles synthesized by *R. acetosa* extracts can be due to the presence of organic acids (such oxalic or citric acids) in the plant extract which can act as stabilizer on the surface of the particles (Makarov et al. 2014b).

3.6.3 Iron (III) Oxide Nanoparticles

Iron (III) oxide (Fe_2O_3) nanoparticles are one of the most interesting and potentially useful iron nanostructures. Based on the crystal structure, iron (III) oxides are divided into four different types including α-Fe_2O_3 (hematite), γ-Fe_2O_3 (maghemite), β-Fe_2O_3, and ε-Fe_2O_3. Each of these types possesses unique physicochemical features which make them suitable for specific technical and biomedical applications (Machala et al. 2011).

Nagajyothi et al. (2017) were synthesized hematite nanoparticles by using *Psoralea corylifolia* seeds extract. Obtained nanoparticles were crystalline with about 40 nm in diameter and showed efficient catalytic activity for degradation of methylene blue dye. The authors employed prepared particles for industrial wastewater treatment and as anticancer agent against renal tumour cells (Nagajyothi et al. 2017).

One comparative study reported the successful synthesis of maghemite nanoparticles by using chestnut tree (*Castanea sativa*), eucalyptus (*Eucalyptus globulus*), gorse

(*Ulex europaeus*), and pine (*Pinus pinaster*) extracts. The authors selected *E. globulus* as the best choice for green synthesis of iron (III) oxide nanoparticles (Martínez-Cabanas et al. 2016). Iron (III) oxide nanoparticles were also reported to be synthesized by using leaves extract of *Ailanthus excels* and *Anacardium occidentales* (Asoufi et al. 2018, Rufus et al. 2017). Fruit extract of *Cynometra ramiflora* and seeds extract of *Psoralea corylifolia* are the other examples (Bishnoi et al. 2018, Nagajyothi et al. 2017). Aloe vera, green tea, oolong tea, and black tea were also reported to be applicable for the synthesis of iron (III) oxide nanoparticles (Huang et al. 2014a, Mukherjee et al. 2016).

3.6.4 Iron (III) Oxide-Hydroxide (FeOOH) Nanoparticles

Iron oxyhydroxides possess four different polymorphs, namely goethite (α-FeOOH), akaganéite (β-FeOOH), lepidocrocite (γ-FeOOH), and feroxyhyte (δ-FeOOH). Nanoparticles of FeOOH are one of the most interesting candidates for air and water pollution removal (Ebrahiminezhad et al. 2017g). These nanoparticles are attracting increasing interest due to their significant properties such as biocompatibility, excellent visible light absorption, and high photostability (Jelle et al. 2016).

Iron oxyhydroxide nanoparticles can be synthesized by various physicochemical methods, but in recent years, green synthesis approaches have been developed. Since now, the utilization of *Sorghum bran*, green tea, black tea, and oolong tea extract has been reported for the synthesis of FeOOH nanoparticles (Kuang et al. 2013, Njagi et al. 2011). Also, the use of blueberry leaf extract was efficient for the synthesis of lepidocrocite (γ-FeOOH) nanoparticles (Manquián-Cerda et al. 2017).

3.6.5 Nanoparticles of Iron Complexes

In some cases, reaction between iron ions and polyphenol compounds from plant extract resulted in the formation of nanostructures of iron and organic compounds, which are called iron complex nanoparticles. In fact, polyphenols in plant extracts have redox potential which can easily reduce gold and silver ions to gold and silver nanoparticles but are not capable to reduce iron ions to ZVI as well as chelate ferric ions to produce the iron-polyphenols complex nanoparticles (Wang et al. 2015). It has been shown that polyphenols from *Eucalyptus tereticornis*, *Melaleuca nesophila*, *Rosmarinus officinalis*, *Salvia officinalis*, and green tea are able to form nanoparticles of iron complexes (Markova et al. 2014, Wang 2013, 2015, 2014c).

These nanostructures have excellent organic dye adsorption–flocculation capacity and gained applications in water purification and contaminated ground water remediation (Wang et al. 2013). In addition, iron complex nanoparticles can act as a heterogeneous Fenton-like catalyst for decolonization proposes that makes them

useful for environmental remediation (Wang et al. 2014c). Markova et al. (2014) showed that nanoparticles of iron complexes produced by green tea extract have remarkable ecotoxicological impacts on the aquatic organism (Markova et al. 2014). This toxic effect can be due to the generation of noxious radicals attributed to Fe^{2+} ions released from complexes of green tea polyphenols. Hence, the use of these materials in remediation processes where aquatic organisms are present should be considered (Markova et al. 2014).

3.7 Silver Nanoparticles

Silver (Ag) nanoparticles are one of the most studied and applied nanoparticles in science and technology. Silver has gained historical applications in the human communities since ancient time (Ebrahiminezhad et al. 2016e). As a potent antimicrobial agent, the silver cations were used in burns, wounds, and ulcer treatments. As early as 4,000 B.C.E, silver was known to the Chaldeans. Silver colloids have gained medical applications even before discovering microbial life as the causative agent for infectious disease. Romans and Egyptians were found silver compounds as functional food preservative. Also, Hippocrates and Macedonians employed silver to prevent and treat wound infections. They found silver-based compounds as potent compound not just against infections but also as a wound-healing agent (Ebrahiminezhad et al. 2016e). Discovering and application of antibiotics as a potent antimicrobial agent, in the early 1940s, with very light side effects drove ancient silver away from the spotlight. Nowadays by emerging resistance strains, silver is back in a novel structure known as "nanosilver". Vast investigations about the biological and antimicrobial properties of nanosilver, the mechanism behind it, and the fabrication of silver nanoparticles have been done (Ebrahiminezhad et al. 2016a, e). Like other nanostructures, the physical and chemical approaches were the first methods for silver nanoparticles fabrication (Ebrahimi et al. 2016a, b). But, by increase in the commercial and scientific applications of silver nanoparticles, the demand for this nanostructure was increased significantly. Hence, the production techniques were shifted to the sustainable process and so many investigations were done to fabricate silver nanoparticles in green manner using microorganisms and plants (Ebrahiminezhad et al. 2016a, b).

Since now, a lot has been done to facilitate the production of silver nanoparticles by plant-mediated synthesis (Table 3.2), and diverse parts of plants have been used for this purpose. Leaf extract is the most employed part, and the potential of various plant leaf extracts such as *Artemisia annua*, *Lippia citriodora* (Lemon Verbena), *Lantana camara*, *Zataria multiflora*, *Cupressus sempervirens* (Mediterranean Cypress), maple (Acer sp.), green and black tea, and eucalyptus has been evaluated (Ajitha et al. 2015, Basavegowda et al. 2014, Cataldo 2014, Cruz et al. 2010, Ebrahiminezhad et al. 2016g, Manjamadha and Muthukumar 2016, Mo et al. 2015, Nakhjavani et al.

TABLE 3.2 Characteristic Features of Green Synthesized Silver Nanoparticles Using Different Plant Extracts

Plant Name	Plant's Part	Precursor	Reported NPs Size	Shape	Responsible Biomolecules	References
Catharanthus roseus Linn	Leaf	AgNO$_3$	35–55 nm	Spherical	–	(Ponarulselvam et al. 2012)
Desmodium triflorum	Leaf	AgNO$_3$	5–20 nm	Spherical, oval, and elliptical	Proteins and heterocyclic compounds	(Ahmad et al. 2011)
Acalypha indica	Leaf	AgNO$_3$	20–30 nm	Spherical	Quercetin	(Krishnaraj et al. 2010)
Cinnamomum camphora	Leaf	AgNO$_3$	5–40 nm	Quasi-spherical	Polyols and heterocyclic compounds	(Huang et al. 2008)
Coleus aromaticus	Leaf	AgNO$_3$	40–50 nm	Spherical	Alcohols and polyphenols	(Vanaja and Annadurai 2013)
Coriandrum sativum	Leaf	AgNO$_3$	8–75 nm	Spherical		(Sathyavathi et al. 2010)
Gliricidia sepium	Leaf	AgNO$_3$	10–50 nm	Spherical		(Raut Rajesh et al. 2009)
Citrus limon	Leaf	AgNO$_3$	15–30 nm	Multi shape	Citric acid	(Vankar and Shukla 2012)
Lippia citriodora	Leaf	AgNO$_3$	15–30 nm	Spherical	Isoverbascosie	(Cruz et al. 2010)
Parthenium	Leaf	AgNO$_3$	30–80 nm	Irregular		(Parashar et al. 2009)
Rosa rugosa	Leaf	AgNO$_3$	Average size of 12 nm	Spherical, triangular, and hexagonal	Amines and alcohols	(Dubey et al. 2010b)
Elaeagnus indica	Leaf	AgNO$_3$	22–40 nm	Spherical	Polyphenols	(Lavanya et al. 2013)
Chenopodium album	Leaf	AgNO$_3$	Average size of ~30 nm	Spherical		(Natarajan et al. 2013)
	Leaf	AgNO$_3$	10–30 nm	Spherical		(Dwivedi and Gopal 2010)
	Leaf	AgNO$_3$	9–15 nm	Spherical		(Naik et al. 2013)
Bixa orellana	Leaf	AgNO$_3$	35–65 nm	Spherical and cubic	Phenols, tannins, and terpenoids	(Thilagam et al. 2013)
Elaeagnus latifolia	Leaf	AgNO$_3$	5–30 nm	Spherical	Polyphenols	(Phanjom et al. 2012)
Nerium oleander	Leaf	AgNO$_3$	48–67 nm	Cubic		(Suganya et al. 2012)
Ocimum bacillicum	Leaf	AgNO$_3$	58–89 nm	Spherical with a few agglomeration	Terpenoids and proteins	(Sivaranjani and Meenakshisundaram 2013)
Odina wodier	Leaf	AgNO$_3$	5–30 nm	Spherical	Unsaturated carbonyl groups	(Arunkumar et al. 2013)
Ceratonia siliqua	Leaf	AgNO$_3$	5–40 nm	Spherical	Protein	(Awwad et al. 2013)
Juglans regia L.	Leaf	AgNO$_3$	10–50 nm	Quasi-spherical		(Korbekandi et al. 2013)
Datura metel	Leaf	AgNO$_3$	6–40 nm	Spherical and ellipsoidal	Alcoholic components	(Kesharwani et al. 2009)
Euphorbia hirta	Leaf	AgNO$_3$	40–50 nm	Spherical		(Elumalai et al. 2010)
	Leaf	AgNO$_3$	–	Spherical	Proteins	(Parveen et al. 2012)
Chromolaena odorata	Leaf	AgNO$_3$	40–70 nm	Hexagonal	Flavonoids, alkaloids, and polyphenols	(Geetha et al. 2012)
Coleus amboinicus	Leaf	AgNO$_3$	25.83 ± 0.78 nm	–		(Subramanian and Suja 2012)
Sonchus asper	Leaf	AgNO$_3$	19.7–82 nm			(Verma et al. 2013)
Piper nigrum	Leaf	AgNO$_3$	21.07 nm	Spherical		(Jacob et al. 2012)
Azadirachta indica	Lalitha	AgNO$_3$	20–50 nm			(Lalitha et al. 2013)
Murraya paniculata	Leaf	AgNO$_3$		Spherical	Unsaturated carbonyl groups	(Ganesan et al. 2013)
Albizia adianthifolia	Leaf	AgNO$_3$	4–35 nm	Spherical	Saponins, proteins, and sugars	(Gengan et al. 2013)
Anacardium occidentale	Leaf	AgNO$_3$	Average size of 15.5 nm	Spherical	Proteins, aromatic amines, and polyphenols	(Sheny et al. 2011)
Annona squamosa	Leaf	AgNO$_3$	20–100 nm	Spherical	Phenols, proteins, and carbohydrates	(Vivek et al. 2012)
Chrysopogon zizanioides	Leaf	AgNO$_3$	85–110 nm	Roughly cubic	Alkaloids and phytosterols	(Arunachalam and Annamalai 2013)
Coccinia grandis	Leaf	AgNO$_3$	20–30 nm	Spherical	Alkaloids and terpenoids	(Arunachalam et al. 2012)
	Leaf	AgNO$_3$	10–80 nm	Spherical, hexahedral, oval, and truncated triangle		(Ebrahiminezhad et al. 2017c)
Zataria multiflora	Leaf	AgNO$_3$	16.3–25.4 nm	Spherical	Carbohydrate	(Ebrahiminezhad et al. 2016b)
Ficus carica	Leaf and Bark	AgNO$_3$	10–20 nm	–	Organic acids	(Singh and Bhakat 2012)
Phyllanthus amarus	Whole plant	AgNO$_3$	24 ± 8 nm size	Spherical		(Singh et al. 2014a)
Nyctanthes arbortristis	Seed	AgNO$_3$	50–80 nm	Spherical		(Basu et al. 2016)
Solanum xanthocarpum	Fruit	AgNO$_3$	45–80 nm	Spherical		(Bharath et al. 2012)
Vitis vinifera	Fruit	AgNO$_3$	30–40 nm	Spherical		(Gnanajobitha et al. 2013a)

(Continued)

TABLE 3.2 (*Continued*) Characteristic Features of Green Synthesized Silver Nanoparticles Using Different Plant Extracts

Plant Name	Plant's Part	Precursor	Reported NPs Size	Shape	Responsible Biomolecules	References
Ipomoea indica	Flower	AgNO$_3$	10–50 nm	Spherical and cubic	—	(Pavani et al. 2013)
	Flower	AgNO$_3$	Average size of 14 nm	Spherical, tetrahedron, elongated decahedron, and fivefold twinned structure	Retinoic acid and proteins	(Bindhu et al. 2013)
Calotropis procera	Flower	AgNO$_3$	Average size of 45 nm	Cubic	—	(Babu and Prabu 2011)
Cassia auriculata	Flower	AgNO$_3$	10–40 nm	Spherical	—	(Velavan et al. 2012)
Millingtonia hortensis	Flower	AgNO$_3$	10–40 nm	Spherical	—	(Gnanajobitha et al. 2013b)
Alcea rosea	Flower	AgNO$_3$	Average size of 7.2 nm	Spherical	Oxygen-bearing functional groups	(Ebrahiminezhad et al. 2016b)
Matricaria chamomilla	Flower	AgNO$_3$	1.6–3.1 nm	Spherical	Carbohydrates and oxygen-bearing functional groups	(Ebrahiminezhad et al. 2016d)
Rhododendron dauricum	Flower	AgNO$_3$	25–40 nm	—	Phenolics, flavanones, or terpenoids	(Mittal et al. 2012)
Euphorbia nivulia	Latex	AgNO$_3$	5–10 nm	Spherical	Euphol and Proteins	(Valodkar et al. 2011)
Jatropha curcas	Latex	AgNO$_3$	20–30 nm with larger particles	Mostly spherical with uneven shapes for larger particles	Cyclic peptides	(Bar et al. 2009)
	Latex	AgNO$_3$	Multiple size	Round shape	Proteins, flavonoids, and terpenoids	(Patil et al. 2012)
Boswellia ovalifoliolata	Stem and Bark	AgNO$_3$	30–40 nm	Spherical	—	(Ankanna et al. 2010)
Breynia rhamnoides	Stem	AgNO$_3$	Average size of 64 nm	—	Phenolic glycosides	(Gangula et al. 2011)
Ephedra intermedia	Stem	AgNO$_3$	10–36 nm	Spherical	Carbohydrates	(Ebrahiminezhad et al. 2017b)
Annona squamosa	Peel	AgNO$_3$	20–60 nm	Irregular spherical	Water-soluble hydroxy functional group containing compounds	(Kumar et al. 2012)
Citrus sinensis	Peel	AgNO$_3$	35 ± 2 nm at 25°C & 10 ± 1 nm at 60°C	Spherical	Water-soluble fractions	(Kaviya et al. 2011)
Trachyspermum ammi	Seed	AgNO$_3$	Varied in size with majority having 87nm and few in 176, 320 and 988 nm	Triangular	Essential oils	(Vijayaraghavan et al. 2012)
Syzygium cumini	Seed	AgNO$_3$	3.2–7.6 μm	Spherical	Alkaloids	(Vijayaraghavan et al. 2012)
	Leaf, leaf water fraction, seed, and seed water fraction	AgNO$_3$	Average size of 30, 29, 92, and 73 nm	Spherical	Polyphenols	(Kumar et al. 2010)
Rumex hymenosepalus	Root	AgNO$_3$	2–40 nm	Face-centred cubic, and hexagonal	Polyphenols	(Rodríguez-León et al. 2013)
Trianthema decandra	Root	AgNO$_3$	36–74 nm	Spherical	Saponins	(Geethalakshmi and Sarada 2012)
Boswellia serrata	Gum	AgNO$_3$	7.5 ± 3.8 nm	Spherical	Hydroxyl and carbonyl groups of gum	(Kora et al. 2012)

2017, Vivekanandhan et al. 2014). In addition to leaves, the extract of plant other parts was also employed such as *Alcea rosea* and *Matricaria chamomilla* flower extract (Ebrahiminezhad et al. 2016b, c), *Ephedra intermedia* stem extract (Ebrahiminezhad et al. 2016f), *Cinnamon zeylanicum* bark extract (Sathishkumar et al. 2009), and *Curcuma longa* tuber extract (Sathishkumar et al. 2010).

Very little work has been performed to understand the exact mechanism behind the plant-mediated synthesis of silver nanoparticles. Various biomolecules including enzymes, proteins, carbohydrates, vitamins, alkaloids, phenolic acids, terpenoids, and polyphenols are the main components in plant extracts that play an important role in the bioreduction of silver ions to silver nanoparticles (Ebrahimi et al. 2016a, b, c, f, g, 2017c). Silver nanoparticles have a high surface energy state that makes them less stable and prone to aggregation in colloidal systems. Functional groups of the plant metabolites such as carbonyl groups have the potential to bind to the silver nanoparticles and prevent agglomeration. Thus, it is suggested that the biological components of plant extracts could be involved in reduction, formation, and stabilization of silver nanoparticles (Ahmed et al. 2016, Makarov et al. 2014a).

It has been shown that some factors including plant source, concentration of initial silver ions, and reaction conditions such as temperature, pH, and time are the key factors in the biosynthesis of silver nanoparticles using plant extracts. These factors have significant impacts on physicochemical characteristics of the produced particles (Chung et al. 2016).

3.8 Gold Nanoparticles

The simplicity of using plant extract for reducing a metal salt has also led to a massive investigation for the fabrication of gold nanoparticles. Apart from the applications in DNA labelling, drug delivery, and biosensors, gold nanoparticles have been known for the antimicrobial activity against human and animal pathogens (Mittal et al. 2013). As shown in Figure 3.3, gold nanoparticles can also serve as catalyst for the reduction of nitro compounds such as 4-nitrophenol (4-NP) to 4-aminophenol (4-AP) (Ghosh et al. 2012). A single-step procedure for plant-mediated gold nanoparticles preparation has attracted much attention as it is a rapid, cost-efficient, environment-friendly, and safe for clinical practices. In this approach, the plant extract is responsible for the bioreduction of Au ions, their growth, and stabilization (Mittal et al. 2013).

Many phytochemical compounds have been reported to be present in different parts of plants which can act as a reducing agent for the reduction of Au^{3+} ions to gold nanoparticles. Gold nanoparticles can be synthesized into various shapes such as nano-sphere, nano-rod, nano-star, nano-triangular, nano-cage, nao-prism, nano-plate, and nano-belt. Variation in the shape and size of the prepared particles results in the production of nanoparticles with

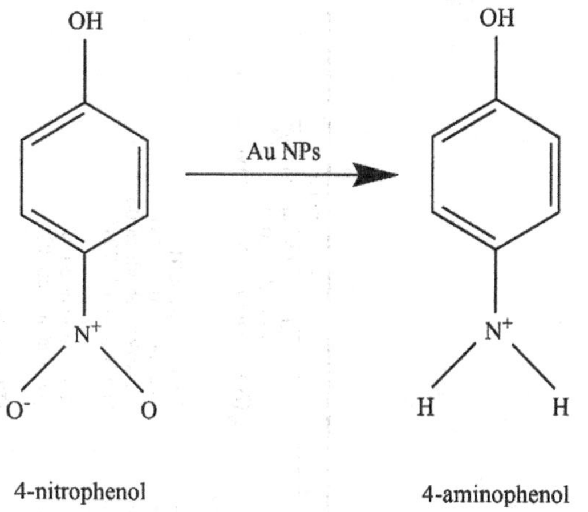

FIGURE 3.3 Catalytic activity of gold nanoparticles for reduction of 4-nitrophenol (4-NP) into 4-aminophenol (4-AP).

different physicochemical properties (Ganeshkumar et al. 2012). Shape and size are among the most important features for gold nanoparticles to modulate the optical properties in electronic and sensor applications and to minimize any potential toxic side effects in biomedical and biological purposes (Shankar et al. 2005). Huang et al. (2007) reported that Au ions can be reduced to triangular or spherical shaped nanoparticles using leaf extract of *Cinnamomum camphora* at ambient temperature. It was noted that the formation of nanoparticles by *C. camphora* strongly relies on the dosage of the plant extract dried biomass. Biochemical compounds such as terpenoids, flavones, and polysaccharides in the *C. camphora* biomass can act as capping agent for the fabricated nanoparticles. Once the biomass concentration increased from 0.5 to 1 g, the interaction between biomolecules present in plant extract and the surface of nanoparticles led to a size reduction. Since now, various parts of different plants such as root, pod, peel, leaf, tuber, flower, fruit, and seed have been used for the green synthesis of gold nanoparticles. These data are provided in detail in Table 3.3 along with the resulted particles size and shape.

3.9 Zinc Oxide Nanoparticles

Among metallic nanoparticles, zinc oxide (ZnO) nanoparticles have attracted much attention because they are non-toxic and hygroscopic. These properties make them a potent material for applications as a catalyst in many organic transformations and in optics and sensors (Madhumitha et al. 2016, Roopan and Khan 2010). In this section, the plant-mediated synthesis of ZnO nanoparticles is discussed. Zinc salts such as zinc nitrate, zinc acetate, and zinc sulphate are the most applied zinc precursor. Biochemical reduction of zinc ions starts when the plant extract is mixed with zinc salt aqueous solution and a change in the colour of the reaction mixture is an indication of nanoparticles formation. Once the nanoparticles formed, they are harvested by using centrifugation and kept in an oven for drying.

TABLE 3.3 Characteristic Features of Plant-Mediated Synthesized Gold Nanoparticles

Plant Name	Plant Part Used for Extract	Precursor	Reported NPs Size	Shape	Reference
Chenopodium album	Leaf	HAuCl$_4$	10–30 nm	Quasi-spherical	(Dwivedi and Gopal 2010)
Cinnamomum camphora	Leaf	HAuCl$_4$	10–40 nm	Triangular or spherical	(Huang et al. 2007)
Coriander	Leaf	HAuCl$_4$	6.75–57.91 nm	Spherical, triangle, truncated triangle, and decahedral	(Narayanan and Sakthivel 2008)
Pogostemon benghalensis	Leaf	HAuCl$_4$	10–50 nm	Spherical and triangular	(Paul et al. 2015)
Nerium oleander	Leaf	HAuCl$_4$	2–10 nm	Spherical	(Tahir et al. 2015)
Lemongrass	Leaf	HAuCl$_4$	–	Triangular and hexagonal	(Shankar et al. 2005)
Rosa rugosa	Leaf	HAuCl$_4$	11 nm	Spherical, triangular, and hexagonal	(Dubey et al. 2010b)
Terminalia Catappa	Leaf	HAuCl$_4$	10–35 nm	Spherical	(Ankamwar 2010)
Centella asiatica	Leaf	HAuCl$_4$	2–24 nm	Spherical, triangular, and hexagonal	(Das et al. 2010)
Mangifera indica	Leaf	HAuCl$_4$	17–20 nm	Spherical	(Philip 2010)
Memecylon umbellatum	Leaf	HAuCl$_4$	15–25 nm	Spherical, hexagonal, and triangular	(Arunachalam et al. 2013)
Memecylon edule	Leaf	HAuCl$_4$	10–45 nm	Triangular, spherical, and hexagonal	(Elavazhagan and Arunachalam 2011)
Cassia auriculata	Leaf	HAuCl$_4$	15–25 nm	Triangular and spherical	(Ganesh Kumar et al. 2011)
Salix alba	Leaf	HAuCl$_4$	50–80 nm	Spherical	(Islam et al. 2015)
Pear	Fruit	HAuCl$_4$	200–500 nm	Triangular and hexagonal	(Ghodake et al. 2010)
Gymnocladus assamicus	Pod	HAuCl$_4$	4.5 ± 0.23 to 22.5 ± 1.24 nm	Hexagonal, pentagonal, and triangular	(Tamuly et al. 2013)
Dioscorea bulbifera	Tuber	HAuCl$_4$	11–30 nm and 50–300 nm	Spherical and triangular	(Ghosh et al. 2011)
Mucuna pruriens	Seed	HAuCl$_4$	6–17.7 nm	Spherical	(Arulkumar and Sabesan 2010)
Abelmoschus esculentus	Seed	HAuCl$_4$	45–75 nm	Spherical	(Jayaseelan et al. 2013)
Mango (Mangifera indica Linn)	Peel	HAuCl$_4$	6.03 ± 2.77 to 18.01 ± 3.67 nm	Spherical	(Yang et al. 2014)
Panax ginseng	Root	HAuCl$_4$	10–40 nm	Spherical	(Singh et al. 2016)
Morinda citrifolia L.	Root	HAuCl$_4$	12.14–38.26 nm	Spherical and triangular	(Suman et al. 2014)
Achillea wilhelmsii	Flower	HAuCl$_4$	70 nm	Spherical	(Andeani et al. 2011)
Nyctanthes arbortristis	Flower	HAuCl$_4$	19.8 ± 5.0 nm	Triangular, pentagonal, rod shaped, and spherical	(Das et al. 2011)
Lonicera Japonica	Flower	HAuCl$_4$	8.02 nm	Triangular and tetrahedral	(Nagajyothi et al. 2012)
Curcuma mangga	–	HAuCl$_4$	2–35 nm	Spherical	(Foo et al. 2017)

The size and the morphology of fabricated nanoparticles are among the most important features that can influence their effectiveness in different applications. Table 3.4 illustrates the characteristics of ZnO nanoparticles fabricated using different plant parts. Similar to the chemical approaches for the nanoparticles synthesis, there are different factors that can affect the nanoparticles size. The morphological structure of nanoparticles has a significant effect in controlling the physical, chemical, electrical, and optical characteristics of them (Edison and Sethuraman 2012). It has been reported that the concentration of plant extract as well as the reaction temperature, pH, and time plays a key role in changing of the nanoparticles size, morphology, and quality. In this context, the scientists have attempted to modify the biosynthesis process and optimize the nanoparticle synthesis process by determining the optimal levels of effective variables. For instance, it was shown that the increase in the concentration of plant extract results in a reduction of nanoparticle size (Fu and Fu 2015). Anbuvannan et al. (2015a) found that the increase in the concentration of Anisochilus carnosus leaf extract from 30 to 50 mL leads to a 31% decrease in particle size while having no significant effect on the morphology of fabricated particles. The same observation was noticed by Elumalai et al. (2015) when the authors used various concentrations of leaf extract from Vitex trifolia. Due to the polarity and electrostatic attraction, the nanoparticles tend to agglomerate when a lower concentration of plant extract as a capping agent is used. However, the binding of smaller size nanoparticles which causes the formation of secondary bulk nanoparticles is responsible for agglomeration at the higher volume of leaf extract. Likewise, studies have shown that the changes in the pH of reaction solution result in the fabrication of nanoparticles with different shapes and sizes. It has been reported that a lower pH contributes to the commemoration of large particles (Dubey et al. 2010c). A higher pH has reported as a favourable condition for the formation of smaller metallic nanoparticles (Armendariz et al. 2004). The variation in particle size is due to the accessibility of functional groups at different pH levels during particles nucleation (Shah et al. 2015). The reaction time is another factor that influences both the particle size and the reaction yield. It has been noted that the increase in the reaction time results in a higher production of nanoparticles and this can increase the likelihood of agglomeration (Veerasamy et al. 2011). On the other hand, no significant reduction of precursor solution occurs in the early reaction time (Ahmad et al. 2016), while maintaining the reaction contributes to the further growth of particles and subsequently, a larger particle size is produced. The reaction temperature has also a significant effect on the nanoparticles size. It was found that the nanoparticles synthesis rate increases at elevated temperatures (Song et al. 2009). The authors noted that 100% of precursor converted to nanoparticles at 95°C in 5 min which this value was significantly higher than synthesis at room temperature at a given time.

3.10 Copper Nanoparticles

Copper nanoparticles show catalytic, antioxidant, antibacterial, and antifungal activities (Din et al. 2017). Leaves, barks, seeds, peels, fruits, roots, coir, and gum have been used as different parts of plants for the green synthesis of copper nanoparticles. Based on the literature, different aqueous solutions of precursors such as copper chloride, copper nitrate, copper sulphate, and copper acetate have been used for the preparation of copper nanoparticles. The reaction between plant extract and copper salt under a strong stirring condition at a controlled pH and temperature results in the formation of copper nanoparticles.

Table 3.5 shows the various sources of plant extract and reducing agents for the biosynthesis of copper nanoparticles. It has been reported that *Calotropis procera* L. latex is rich in protein including antioxidant enzymes (AOEs), cysteine protease with free thiol (–SH) group, and tryptophan (Freitas et al. 2007). Therefore, its extract can be used for the green fabrication of nanoparticles. The *Calotropis procera* L. latex-mediated synthesis of Cu NP was performed in the presence of copper acetate at room temperature and the particles subjected to characterization study (Harne et al. 2012). The characterization results showed monodisperse nanoparticles with an average diameter of 15 ± 1.7 nm. It was also found that the capping with latex proteins brings about a long-term stability of nanoparticles in an aqueous medium. In another investigation, the extract of Henna (Lawsonia inermis) leaves as a source of lawsone (2-hydroxy-1,4-naphthoquinone, $C10H6O3$), gallic acid, glucose, mannitol, fats, resin, mucilage, and alkaloids was used for copper nanoparticles synthesis (Cheirmadurai et al. 2014). Similarly, the extract obtained from dried leaves of *Ginkgo biloba* L. was identified as a source of polyphenolics and shown potential for the biosynthesis of copper nanoparticles (Nasrollahzadeh and Mohammad Sajadi 2015). The particles were synthesized upon the addition of *Ginkgo biloba* L. extract to copper chloride solution at a temperature of 80°C and pH of 9 for 30 min under vigorous shaking condition. Authors noted that the fabricated nanoparticles were stable for one month and had a narrow particle size distribution ranging from 15 to 20 nm. Copper nanoparticles were also synthesized using the aromatic dried flower buds of Myrtaceae tree (*Syzygium aromaticum*) (Subhankari and Nayak 2013). The copper precursor (5M copper sulphate) and plant extract were incubated for 1 h to allow the formation of nanoparticles nuclei. The characterization study was then performed using transmission electron microscopy (TEM) and scanning electron microscopy (SEM), and a particle size of 14–50 nm was noticed. Depending on the types of employed copper salts, different particle size and shape have been reported. For instance, Shanker and Rhim (Shankar and Rhim 2014) observed various particle shapes when they added different precursor salts into a solution containing ascorbic acid as a reducing agent. The presence of copper acetate resulted in the formation of rod shape nanoparticles, while the addition

TABLE 3.4   Characteristics of ZnO Nanoparticles Fabricated Using Different Plant Parts

Plant Name	Plant Part Used for Extract	Precursor	Reported NPs Size	Shape	Reference
Ocimum basilicum L. var. *purpurascens* Benth	Leaf extract	Zinc nitrate	50 nm	Hexagonal	(Abdul Salam et al. 2014)
Anisochilus carnosus	Leaf extract	Zinc nitrate	30–40 nm	Quasi-spherical	(Anbuvannan et al. 2015a)
Plectranthus amboinicus	Leaf extract	Zinc nitrate	88 nm	Rod shape	(Fu and Fu 2015)
Aloe barbadensis miller	Leaf extract	Zinc nitrate	25–40 nm	Spherical	(Sangeetha et al. 2011)
Aloe vera	Leaf extract	–	25–65 nm	Spherical and hexagonal	(Qian et al. 2015)
Aloe barbadensis Miller	Leaf extract	Zinc sulphate	8–18 nm	Spherical, oval and hexagonal	(Ali et al. 2016b)
Vitex negundo	Leaf extract	Zinc nitrate	75–80 nm	Spherical	(Ambika and Sundrarajan 2015a)
Azadirachta indica	Leaf extract	Zinc acetate	9.6–25.5 nm	Spherical	(Bhuyan et al. 2015)
Vitex trifolia	Leaf extract	Zinc nitrate	15–46 nm	Spherical	(Elumalai et al. 2015)
Parthenium hysterophorus L.	Leaf extract	Zinc nitrate	22–32 nm	Spherical and hexagonal	(Rajiv et al. 2013)
			82–86 nm		
Artocarpus gomezianus	Fruit extract	Zinc nitrate	<20 nm	Spherical	(Suresh et al. 2015)
Trifolium pratense	Flower extract	–	100–190 nm	–	(Dobrucka and Długaszewska 2016)
Eichhornia crassipes	Leaf extract	Zinc nitrate	28–36 nm	Spherical	(Vanathi et al. 2014)
Rosa canina	Fruit extract	Zinc nitrate	<50 nm	Spherical	(Jafarirad et al. 2016)
Solanum nigrum	Leaf extract	Zinc sulphate	29.79 nm	Quasi-spherical	(Ramesh et al. 2015)
Citrus paradisi	Peel extract	Zinc sulphate	12–72 nm	Spherical	(Kumar et al. 2014b)
Vitex negundo L.	Flower extract	Zinc nitrate	10–130 nm	–	(Ambika and Sundrarajan 2015b)
Artocarpus heterophyllus	Leaf extract	Zinc nitrate	15–25 nm	Hexagonal wurtzite	(Vidya et al. 2016)
P. trifoliate	Fruit extract	Zinc nitrate	8.48–32.51 nm	Spherical	(Nagajyothi et al. 2013)
Punica granatum	Peel extract	Zinc nitrate	50–100 nm	Spherical and square	(Mishra and Sharma 2015)
Agathosma betulina	Leaf extract	Zinc nitrate	15.8 nm	Quasi-spherical	(Thema et al. 2015)
Nephelium lappaceum L.	Peel extract	Zinc nitrate	50 nm	Needle like	(Yuvakumar et al. 2014)
Pongamia pinnata	Leaf extract	Zinc nitrate	100 nm	Spherical	(Sundrarajan et al. 2015)
Phyllanthus niruri	Leaf extract	Zinc nitrate	25.61 nm	Quasi-spherical	(Anbuvannan et al. 2015b)
Physalis alkekengi L.	Shoots extract	–	50–200 nm	Triangular	(Qu et al. 2011)
Coptidis rhizoma	Rhizome extract	Zinc nitrate	2.9–25.2 nm	Spherical and rod shaped	(Nagajyothi et al. 2014)

TABLE 3.5 Characteristics of Plant-Mediated Synthesized Copper and Copper Oxide Nanoparticles

NP	Plant Name	Plant Part Used for Extract	Precursor	Reported NPs Size	Shape	Reference
Cu	*Calotropis procera* L.	Latex extract	Copper acetate	15 ± 1.7 nm	Spherical	(Harne et al. 2012)
	Lawsonia inermis	Leaf extract	Copper sulphate	43 and 83 nm	Spherical	(Cheirmadurai et al. 2014)
	Ginkgo biloba L.	Leaf extract	Copper chloride	15–20 nm	Spherical	(Nasrollahzadeh and Mohammad Sajadi 2015)
	Syzygium aromaticum	Clove extract	Copper sulphate	<50 nm	Spherical	(Subhankari and Nayak 2013)
	Hibiscus rosasinensis	Leaf extract	Copper nitrate	–	Spherical	(Subbaiya and Selvam 2015)
	Magnolia kobus	Leaf extract	Copper sulphate	37–110 nm	Spherical	(Lee et al. 2013)
	Caesalpinia pulcherrima	Flower petal extract	Copper nitrate	18–20 nm	Spherical	(Kurkure et al. 2016)
	Euphorbia esula L.	Leaf extract	Copper chloride	20–110 nm	Spherical	(Nasrollahzadeh et al. 2014)
	Lemon	Fruit extract	Copper chloride	60–100 nm	Spherical	(Jayandran et al. 2015)
	Punica granatum	Peel extract	Copper sulphate	15–20 nm	Spherical	(Kaur et al. 2016)
	Ocimum sanctum	Leaf extract	Copper sulphate	25 nm	Rod, cylindrical and elliptical shape	(Shende et al. 2016)
CuO	Green tea	Leaf extract	Copper chloride	15–25 nm	Spherical	(Keihan et al. 2017)
	Cochlospermum Gossypium	Gum extract	Copper nitrate	19 nm	Spherical	(Suresh et al. 2016)
	Terminalia arjuna	Bark extract	Copper nitrate	20–30 nm	Spherical	(Yallappa et al. 2013)
	Olea Europaea	Leaf extract	Copper sulphate	20–50 nm	Spherical	(Sulaiman et al. 2018a)
	Drypetes sepiaria	Leaf extract	Copper nitrate	–	Spherical	(Narasaiah et al. 2017)
	Eichhornia crassipes	Leaf extract	Copper sulphate	28 ± 4 nm	Spherical	(Vanathi et al. 2016)
	Ferulago angulata	Plant extract	Copper acetate	44 nm	–	(Mehr et al. 2018)
	Pterospermum acerifolium	Leaf extract	Copper nitrate	266.4 ± 447.26 nm	Oval	(Saif et al. 2016)
	Abutilon indicum	Leaf extract	Copper nitrate	16.78 nm	Spherical	(Ijaz et al. 2017)
	Vitis vinifera cv.	Fruit extract	Copper chloride	25–50 nm	Spherical	(Gultekin et al. 2017)
	Saraca indica	Leaf extract	Copper chloride	40–70	Spherical	(Prasad et al. 2017)

of copper chloride and copper sulphate led to triangular and spherical shape, respectively. In another study, it was shown that the dropwise addition of olive tree (*Olea europaea*) leaf extract into an aqueous solution of copper sulphate at 100°C for 24 h results in the production of copper oxide (CuO) nanoparticles (Sulaiman et al. 2018a). The synthesized CuO nanoparticles showed a peak absorbance at 298 nm in UV-vis spectroscopy analysis, and the particles had a crystalline nature with a diameter of 20 nm.

3.11 Aluminium Oxide Nanoparticles

Aluminium oxide (Al_2O_3) nanoparticles are the other materials that have been used for diverse applications such as biomedical, antigen and drug delivery, biofiltration, and biosensors. These particles are biocompatible, and chemically inert and stable, and can be readily functionalized in the surface (Jalal et al. 2016, Saber 2012). The main attributes of plant-mediated synthesized Al_2O_3 nanoparticles are the exploitation of cost- and energy-efficient approach without using expensive equipment and toxic chemicals that are being used in chemical and physical methods.

Various sources of stabilizing agents and precursors for the green production of Al_2O_3 nanoparticles are given in Table 3.6. The reactions for the plant-mediated synthesis of Al_2O_3 nanoparticles with the help of heating are presented in the Eqs. 3.6 (Jalal et al. 2016). According to Eqs. 3.4, aluminium oxide-hydroxide (boehmite, AlOOH) can be synthesized by heating aluminium nitrate at different temperatures, and the generated AlOOH is used for the fabrication of stable Al_2O_3 nanoparticles. The lemongrass extract is a source of carboxylic acids (HO_2CR), and its reaction with AlOOH results in the production of carboxylate-functionalized Al_2O_3 nanoparticles (carboxylate-alumoxane, Eq. 3.5). As shown in Eq. 3.6, a stable form of Al_2O_3 (α-phase) is then produced upon the calcination of carboxylate-alumoxane at a temperature of 1100°C for 3 h.

$$Al(NO_3)_3 \cdot 9H_2O \rightarrow [Al(OH_2)_6]^{3+} \cdot 3NO_3^- + 3H_2O \ (85°C) \tag{3.1}$$

$$[Al(OH_2)_6]^{3+} \cdot 3NO_3^- \rightarrow [Al(OH)(OH_2)_5]^{2+} \cdot 2NO_3^- \\ + HNO_3 \ (150-180°C) \tag{3.2}$$

$$n\left\{[Al(OH)(OH_2)_5]^{2+} \cdot 2NO_3^-\right\} \rightarrow \left\{[AlO(OH)_2]^+ \cdot NO_3^-\right\}_n \\ + nHNO_3 + 4nH_2O \ (200-250°C) \tag{3.3}$$

$$[AlO(OH)_2]^+ \cdot NO_3^- \rightarrow [Al(O)(OH)] + HNO_3 \\ (300-400°C) \tag{3.4}$$

$$[Al(O)(OH)]_n + HO_2CR \rightarrow \left[Al(O)_x(OH)_y(O_2CR)_z\right]_n \tag{3.5}$$

$$\left[Al(O)_x(OH)_y(O_2CR)_z\right]_n \rightarrow Calcinations \ at \ 1100°C \\ \rightarrow \alpha - Al_2O_3 \ nanoparticles \tag{3.6}$$

3.12 Titanium Dioxide Nanoparticles

Due to their physicochemical and biological properties, titanium dioxide (TiO_2) nanoparticles have a wide range of applications. For instance, their potential oxidation strength, high photo stability, non-toxicity, and antibacterial activities make them useful in air and water purification as well as dye-sensitized solar cells (Mishra et al. 2014). In continuation of efforts for the synthesis of TiO_2 nanoparticles using chemical and physical methods, a greener approach using the plant extract has been tested. Table 3.7 shows the characteristics of TiO_2 nanoparticles fabricated using different plant parts. Sundrarajan and Gowri (2011) used Nyctanthes leaf extract due to its functional antioxidant, antifungal, and antimicrobial activities for the

TABLE 3.6 Plant-Mediated Synthesis of Aluminium Oxide (Al_2O_3) Nanoparticles

Plant Name	Plant Part Used for Extract	Precursor	Reported NPs Size	Shape	Reference
Lemongrass	Leaf extract	Aluminium nitrate	15–110 nm	Spherical	(Jalal et al. 2016)
Cymbopogon citratus	Leaf extract	Aluminium nitrate	9–180 nm	Spherical and spheroidal	(Ansari et al. 2015)
Tea	Leaf extract	Aluminium nitrate	50–100 nm	Spherical	(Sutradhar et al. 2013)
Coffee	Bean extract	Aluminium nitrate	<100 nm	Spherical	(Sutradhar et al. 2013)
Triphala	Seed extract	Aluminium nitrate	200–400 nm	Oval	(Sutradhar et al. 2013)

TABLE 3.7 Characteristics of TiO_2 Nanoparticles Fabricated Using Different Plant Parts

Plant Name	Plant Part Used for Extract	Precursor	Reported NPs Size	Shape	Reference
Nyctanthes	Leaf extract	Titanium tetraisopropoxide	100–150 nm	Spherical	(Sundrarajan and Gowri 2011)
Calotropis gigantea	Flower extract	$TiO(OH)_2$	160–220 nm	Spherical	(Marimuthu et al. 2013)
Eclipta prostrata	Leaf extract	$TiO(OH)_2$	<95 nm	Spherical	(Rajakumar et al. 2012)
Mangifera indica L.	Leaf extract	$TiO(OH)_2$	30 ± 5	Spherical	(Rajakumar et al. 2015)
Catharanthus roseus	Leaf extract	–	25–110 nm	Irregular	(Velayutham et al. 2012)
Psidium guajava	Leaf extract	$TiO(OH)_2$	32.58 nm	Spherical	(Santhoshkumar et al. 2014)
Euphorbia prostrata	Leaf extract	$TiO(OH)_2$	83.22 ± 1.5 nm	Spherical	(Zahir et al. 2015)

fabrication of TiO_2 nanoparticles. The leaf powder was obtained by grinding of shade-dried leaves. The plant extract was then achieved by mixing the leaf powder and ethanol under reflux condition at a temperature of 50°C for 5 h. The TiO_2 NPs were obtained upon the reaction of ethanolic leaf extract and titanium tetraisopropoxide at 50°C. After 4 h of vigorous stirring, the nanoparticles were formed and the separated particles were subjected to a calcination process at 500°C for 3 h. Their characterization data show that the resulted nanoparticles were crystalline with a spherical shape ranging from 100 to 150 nm. In another investigation, *Calotropis gigantea* flower extract and $TiO(OH)_2$ were employed as a stabilizing and precursor agent, respectively (Marimuthu et al. 2013). The mixture of plant extract and $TiO(OH)_2$ was kept stirring for 6 h and then subjected to ultrasonication for 30 min prior to nanoparticle separation. The results from the analyses show the fabricated nanoparticles were spherical with an average size of 160–220 nm. Rajakumar et al. (2012) proposed *Eclipta prostrata* for the biosynthesis of TiO_2 nanoparticles. The nanoparticles were synthesized upon the reaction of *E. prostrata* and $TiO(OH)_2$ under ambient temperature for 24 h. The SEM micrographs showed poorly dispersed spherical clusters with agglomeration size up to 95 nm. Few years later, the authors used the same protocol and precursor with a different plant extract "*Mangifera indica* L." for the biosynthesis of TiO_2 nanoparticles (Rajakumar et al. 2015). Their results show that the utilization of *Mangifera indica* L. significantly contributed to decreasing the particle size. The characterization data show that the prepared nanoparticles had a spherical shape with an average size of 30 ± 5 nm.

3.13 Conclusion and Perspective

Plant-mediated synthesis is now a growing field in the synthesis of nanostructures. Since now, there is no detailed knowledge about the exact mechanism in the green synthesis reactions. In the case of some nanoparticles such as silver and gold nanoparticles, it is revealed that the reaction is a reduction reaction. This reaction is mediated by oxygen-bearing functional groups in the phytochemicals such as phenolic compounds. Iron complex nanoparticles are formed by chelating ferrous ions in the polyphenol compounds from plant extract. But, still there are many undiscovered areas in this field that needs to be explored. Questions such as the effective molecules in the green synthesis reactions, shape directing functional groups and biomolecules, interactions of phytochemicals with plant-mediated synthesized nanostructures, and effects of plant extract preparation on the synthesis reaction need to be answered. Nevertheless, green synthesis is now one of the most economic, environment-friendly, and facile techniques in the fabrication of nanostructures. Soon, it is expected to be one of the main approaches for the large-scale production of nanostructures worldwide.

References

Abdul Salam, H., Sivaraj, R. and Venckatesh, R. 2014. Green synthesis and characterization of zinc oxide nanoparticles from Ocimum basilicum L. var. purpurascens Benth.-Lamiaceae leaf extract. *Materials Letters* 131: 16–18.

Ahmad, N., Sharma, S., Singh, V., Shamsi, S., Fatma, A. and Mehta, B.R. 2011. Biosynthesis of silver nanoparticles from Desmodium triflorum: A novel approach towards weed utilization. *Biotechnology Research International* 2011:8.

Ahmad, T., Irfan, M., Bustam, M.A. and Bhattacharjee, S. 2016. Effect of reaction time on green synthesis of gold nanoparticles by using aqueous extract of Elaeis guineensis (oil palm leaves). *Procedia Engineering* 148: 467–472.

Ahmed, S., Ahmad, M., Swami, B.L. and Ikram, S. 2016. A review on plants extract mediated synthesis of silver nanoparticles for antimicrobial applications: A green expertise. *Journal of Advanced Research* 7(1): 17–28.

Ajitha, B., Reddy, Y.A.K. and Reddy, P.S. 2015. Green synthesis and characterization of silver nanoparticles using Lantana camara leaf extract. *Materials Science and Engineering C* 49: 373–381.

Al-Ruqeishi, M.S., Mohiuddin, T. and Al-Saadi, L.K. 2016. Green synthesis of iron oxide nanorods from deciduous Omani mango tree leaves for heavy oil viscosity treatment. *Arabian Journal of Chemistry*. doi: 10.1016/j.arabjc.2016.04.003.

Ali, D.M., Sasikala, M., Gunasekaran, M. and Thajuddin, N. 2011. Biosynthesis and characterization of silver nanoparticles using marine cyanobacterium, Oscillatoria willei NTDM01. *Digest Journal of Nanomaterials and Biostructures* 6: 385–390.

Ali, I., Al-Othman, Z.A. and Alwarthan, A. 2016a. Green synthesis of functionalized iron nano particles and molecular liquid phase adsorption of ametryn from water. *Journal of Molecular Liquids* 221: 1168–1174.

Ali, K., Ahmed, B., Dwivedi, S., Saquib, Q., Al-Khedhairy, A.A. and Musarrat, J. 2015. Microwave accelerated green synthesis of stable silver nanoparticles with eucalyptus globulus leaf extract and their antibacterial and antibiofilm activity on clinical isolates. *PloS one* 10(7): e0131178.

Ali, K., Dwivedi, S., Azam, A., Saquib, Q., Al-Said, M.S., Alkhedhairy, A.A. and Musarrat, J. 2016b. Aloe vera extract functionalized zinc oxide nanoparticles as nanoantibiotics against multi-drug resistant clinical bacterial isolates. *Journal of Colloid and Interface Science* 472: 145–156.

Ambika, S. and Sundrarajan, M. 2015a. Antibacterial behaviour of Vitex negundo extract assisted ZnO nanoparticles against pathogenic bacteria. *Journal of Photochemistry and Photobiology B: Biology* 146: 52–57.

Ambika, S. and Sundrarajan, M. 2015b. Green biosynthesis of ZnO nanoparticles using Vitex negundo L. extract:

Spectroscopic investigation of interaction between ZnO nanoparticles and human serum albumin. *Journal of Photochemistry and Photobiology B: Biology* 149: 143–148.

Anbuvannan, M., Ramesh, M., Viruthagiri, G., Shanmugam, N. and Kannadasan, N. 2015a. Anisochilus carnosus leaf extract mediated synthesis of zinc oxide nanoparticles for antibacterial and photocatalytic activities. *Materials Science in Semiconductor Processing* 39: 621–628.

Anbuvannan, M., Ramesh, M., Viruthagiri, G., Shanmugam, N. and Kannadasan, N. 2015b. Synthesis, characterization and photocatalytic activity of ZnO nanoparticles prepared by biological method. *Spectrochimica Acta Part A: Molecular and Biomolecular Spectroscopy* 143: 304–308.

Andeani, J.K., Kazemi, H., Mohsenzadeh, S. and Safavi, A. 2011. Biosynthesis of gold nanoparticles using dried flowers extract of Achillea wilhelmsii plant. *Digest Journal of Nanomaterials and Biostructures* 6:1011–1017.

Ankamwar, B. 2010. Biosynthesis of gold nanoparticles (green-gold) using leaf extract of Terminalia catappa. *Journal of Chemistry* 7(4): 1334–1339.

Ankanna, S., Prasada, T.N.V.K.V., Elumalai, E. and Savithramma, N 2010. Production of biogenic silver nanoparticles using Boswellia ovalifoliolata stem bark. *Digest Journal of Nanomaterials and Biostructures* 5(2): 369–372.

Ansari, M.A., Khan, H.M., Alzohairy, M.A., Jalal, M., Ali, S.G., Pal, R. and Musarrat, J. 2015. Green synthesis of Al_2O_3 nanoparticles and their bactericidal potential against clinical isolates of multi-drug resistant Pseudomonas aeruginosa. *World Journal of Microbiology and Biotechnology* 31(1): 153–164.

Armendariz, V., Herrera, I., Jose-yacaman, M., Troiani, H., Santiago, P. and Gardea-Torresdey, J.L. 2004. Size controlled gold nanoparticle formation by Avena sativa biomass: Use of plants in nanobiotechnology. *Journal of Nanoparticle Research* 6(4): 377–382.

Arulkumar, S. and Sabesan, M. 2010. Biosynthesis and characterization of gold nanoparticle using antiparkinsonian drug Mucuna pruriens plant extract. *International Journal of Research in Pharmaceutical Sciences* 1(4): 417–420.

Arunachalam, K.D., Annamalai, S.K. and Hari, S. 2013. One-step green synthesis and characterization of leaf extract-mediated biocompatible silver and gold nanoparticles from Memecylon umbellatum. *International Journal of Nanomedicine* 8: 1307.

Arunachalam, K.D. and Annamalai, S.K. 2013. Chrysopogon zizanioides aqueous extract mediated synthesis, characterization of crystalline silver and gold nanoparticles for biomedical applications. *International Journal of Nanomedicine* 8: 2375.

Arunachalam, R., Dhanasingh, S., Kalimuthu, B., Uthirappan, M., Rose, C., Mandal, A.B. 2012.

Phytosynthesis of silver nanoparticles using Coccinia grandis leaf extract and its application in the photocatalytic degradation. *Colloids and Surfaces B: Biointerfaces* 94: 226–230.

Arunkumar, C., Astalakshmi, A., Nima, P. and Ganesan, V.J. 2013. Plant mediated synthesis of silver nanoparticles using leaves of odina wodier roxb. *International Journal of Advanced Research* 1: 265–272.

Asoufi, H.M., Al-Antary, T.M. and Awwad, A.M. 2018. Green route for synthesis hematite (α-Fe2O3) nanoparticles: Toxicity effect on the green peach aphid, Myzus persicae (Sulzer). *Environmental Nanotechnology, Monitoring & Management* 9: 107–111.

Awwad, A.M. and Salem, N.M. 2012a. A green and facile approach for synthesis of magnetite nanoparticles. *Nanoscience and Nanotechnology* 2(6): 208–213.

Awwad, A.M. and Salem, N.M. 2012b. A green and facile approach for synthesis of magnetite nanoparticles. *Journal of Nanoscience and Nanotechnology* 2(6): 208–213.

Awwad, A.M., Salem, N.M. and Abdeen, A.O. 2013. Green synthesis of silver nanoparticles using carob leaf extract and its antibacterial activity. *International Journal of Industrial Chemistry* 4(1): 29.

Azad, B. and Banerjee, A. 2014. Formulation of silver nanoparticles using methanolic extract of stem of plant Desmodium gangeticum, their characterization and antibacterial and anti-oxidant evaluation. *The Pharma Innovation* 3(7, Part B): 77.

Babu, P.J., Saranya, S., Sharma, P., Tamuli, R. and Bora, U. 2012. Gold nanoparticles: sonocatalytic synthesis using ethanolic extract of Andrographis paniculata and functionalization with polycaprolactone-gelatin composites. *Frontiers of Materials Science* 6(3): 236–249.

Babu, S.A. and Prabu, H.G. 2011. Synthesis of AgNPs using the extract of Calotropis procera flower at room temperature. *Materials Letters* 65(11): 1675–1677.

Bar, H., Bhui, D.K., Sahoo, G.P., Sarkar, P., De, S.P., Misra, A. 2009. Green synthesis of silver nanoparticles using latex of Jatropha curcas. *Colloids and Surfaces A: Physicochemical and Engineering Aspects* 339(1–3): 134–139.

Basavegowda, N., Idhayadhulla, A. and Lee, Y.R. 2014. Preparation of Au and Ag nanoparticles using *Artemisia annua* and their in vitro antibacterial and tyrosinase inhibitory activities. *Materials Science and Engineering C* 43(0): 58–64.

Basu, S., Maji, P. and Ganguly, J. 2016. Rapid green synthesis of silver nanoparticles by aqueous extract of seeds of Nyctanthes arbor-tristis. *Applied Nanoscience* 6(1): 1–5.

Belachew, N., Rama Devi, D. and Basavaiah, K. 2017. Green synthesis and characterisation of L-Serine capped magnetite nanoparticles for removal of Rhodamine B from contaminated water. *Journal of Experimental Nanoscience* 12(1): 114–128.

Benzie, I.F. and Szeto, Y. 1999. Total antioxidant capacity of teas by the ferric reducing/antioxidant power assay. *Journal of Agricultural and Food Chemistry* 47(2): 633–636.

Bharath, M.S., Lekshmi, N.P., Kumar, P.D., Brindha, J.R. and Jeeva, S. 2012. Synthesis of plant mediated silver nanoparticles using Solanum xanthocarpum fruit extract and evaluation of their anti microbial activities. *Journal of Pharmacy Research* 5(9): 4888–4892.

Bhuyan, T., Mishra, K., Khanuja, M., Prasad, R. and Varma, A. 2015. Biosynthesis of zinc oxide nanoparticles from Azadirachta indica for antibacterial and photocatalytic applications. *Materials Science in Semiconductor Processing* 32: 55–61.

Bindhu, M., Sathe, V., Umadevi, M. 2013. Synthesis, characterization and SERS activity of biosynthesized silver nanoparticles. *Spectrochimica Acta Part A: Molecular and Biomolecular Spectroscopy* 115: 409–415.

Bishnoi, S., Kumar, A. and Selvaraj, R. 2018. Facile synthesis of magnetic iron oxide nanoparticles using inedible Cynometra ramiflora fruit extract waste and their photocatalytic degradation of methylene blue dye. *Materials Research Bulletin* 97: 121–127.

Cai, Y., Shen, Y., Xie, A., Li, S. and Wang, X. 2010. Green synthesis of soya bean sprouts-mediated superparamagnetic Fe_3O_4 nanoparticles. *Journal of Magnetism and Magnetic Materials* 322(19): 2938–2943.

Cao, D., Jin, X., Gan, L., Wang, T. and Chen, Z. 2016. Removal of phosphate using iron oxide nanoparticles synthesized by eucalyptus leaf extract in the presence of CTAB surfactant. *Chemosphere* 159: 23–31.

Carrillo-López, L.M., Soto-Hernández, R.M., Zavaleta-Mancera, H.A. and Vilchis-Néstor, A.R. 2016. Study of the performance of the organic extracts of *Chenopodium ambrosioides* for Ag nanoparticle synthesis. *Journal of Nanomaterials* 2016: 1–13.

Cataldo, F. 2014. Green synthesis of silver nanoparticles by the action of black or green tea infusions on silver ions. *European Chemical Bulletin* 3(3): 280–289.

Cheirmadurai, K., Biswas, S., Murali, R. and Thanikaivelan, P. 2014. Green synthesis of copper nanoparticles and conducting nanobiocomposites using plant and animal sources. *RSC Advances* 4(37): 19507–19511.

Chitsazi, M.R., Korbekandi, H., Asghari, G., Bahri Najafi, R., Badii, A. and Iravani, S. 2016. Synthesis of silver nanoparticles using methanol and dichloromethane extracts of Pulicaria gnaphalodes (Vent.) Boiss. aerial parts. *Artificial Cells, Nanomedicine, and Biotechnology* 44(1): 328–333.

Chokshi, K., Pancha, I., Ghosh, T., Paliwal, C., Maurya, R., Ghosh, A. and Mishra, S. 2016. Green synthesis, characterization and antioxidant potential of silver nanoparticles biosynthesized from de-oiled biomass of thermotolerant oleaginous microalgae Acutodesmus dimorphus. *RSC Advances* 6(76): 72269–72274.

Chung, I.-M., Park, I., Seung-Hyun, K., Thiruvengadam, M. and Rajakumar, G. 2016. Plant-mediated synthesis of silver nanoparticles: their characteristic properties and therapeutic applications. *Nanoscale Research Letters* 11(1): 40.

Conde, E., Cara, C., Moure, A., Ruiz, E., Castro, E. and Domínguez, H. 2009. Antioxidant activity of the phenolic compounds released by hydrothermal treatments of olive tree pruning. *Food Chemistry* 114(3): 806–812.

Cruz, D., Falé, P.L., Mourato, A., Vaz, P.D., Luisa Serralheiro, M. and Lino, A.R.L. 2010. Preparation and physicochemical characterization of Ag nanoparticles biosynthesized by *Lippia citriodora* (*Lemon Verbena*). *Colloids and Surfaces B: Biointerfaces* 81(1): 67–73.

Das, R.K., Borthakur, B.B. and Bora, U. 2010. Green synthesis of gold nanoparticles using ethanolic leaf extract of Centella asiatica. *Materials Letters* 64(13): 1445–1447.

Das, R.K., Gogoi, N. and Bora, U. 2011. Green synthesis of gold nanoparticles using Nyctanthes arbortristis flower extract. *Bioprocess and Biosystems Engineering* 34(5): 615–619.

Din, M.I., Arshad, F., Hussain, Z. and Mukhtar, M. 2017. Green adeptness in the synthesis and stabilization of copper nanoparticles: Catalytic, antibacterial, cytotoxicity, and antioxidant activities. *Nanoscale Research Letters* 12: 638.

Dobrucka, R. and Długaszewska, J. 2016. Biosynthesis and antibacterial activity of ZnO nanoparticles using Trifolium pratense flower extract. *Saudi Journal of Biological Sciences* 23(4): 517–523.

Dubey, S.P., Lahtinen, M., Särkkä, H. and Sillanpää, M. 2010a. Bioprospective of *Sorbus aucuparia* leaf extract in development of silver and gold nanocolloids. *Colloids and Surfaces B: Biointerfaces* 80(1): 26–33.

Dubey, S.P., Lahtinen, M. and Sillanpää, M. 2010b. Green synthesis and characterizations of silver and gold nanoparticles using leaf extract of *Rosa rugosa*. *Colloids and Surfaces A: Physicochemical and Engineering Aspects* 364(1–3): 34–41.

Dubey, S.P., Lahtinen, M. and Sillanpää, M. 2010c. Tansy fruit mediated greener synthesis of silver and gold nanoparticles. *Process Biochemistry* 45(7): 1065–1071.

Dwivedi, A.D. and Gopal, K. 2010. Biosynthesis of silver and gold nanoparticles using Chenopodium album leaf extract. *Colloids and Surfaces A: Physicochemical and Engineering Aspects* 369(1–3): 27–33.

Ebrahimi, N., Rasoul-Amini, S., Ebrahiminezhad, A., Ghasemi, Y., Gholami, A. and Seradj, H. 2016a. Comparative study on characteristics and cytotoxicity of bifunctional magnetic-silver nanostructures: synthesized using three different reducing agents. *Acta Metallurgica Sinica (English Letters)* 29(4): 326–334.

Ebrahimi, N., Rasoul-Amini, S., Niazi, A., Erfani, N., Moghadam, A., Ebrahiminezhad, A. and Ghasemi, Y. 2016b. Cytotoxic and apoptotic effects of three types of silver-iron oxide binary hybrid nanoparticles. *Current Pharmaceutical Biotechnology* 17(12): 1049–1057.

Ebrahiminezhad, A., Bagheri, M., Taghizadeh, S., Berenjian, A. and Ghasemi, Y. 2016a. Biomimetic synthesis of silver nanoparticles using microalgal secretory carbohydrates as a novel anticancer and antimicrobial. *Advances in Natural Sciences: Nanoscience and Nanotechnology* 7, 1–8.

Ebrahiminezhad, A., Barzegar, Y., Ghasemi, Y. and Berenjian, A. 2016b. Green synthesis and characterization of silver nanoparticles using Alcea rosea flower extract as a new generation of antimicrobials. *Chemical Industry and Chemical Engineering Quarterly* 2016: 31–37.

Ebrahiminezhad, A., Berenjian, A. and Ghasemi, Y. 2016c. Template free synthesis of natural carbohydrates functionalised fluorescent silver nanoclusters. *IET Nanobiotechnology* 2016: 1–4.

Ebrahiminezhad, A., Berenjian, A. and Ghasemi, Y. 2016d. Template free synthesis of natural carbohydrates functionalised fluorescent silver nanoclusters. *IET Nanobiotechnology* 10(3): 120–123.

Ebrahiminezhad, A., Davaran, S., Rasoul-Amini, S., Barar, J., Moghadam, M. and Ghasemi, Y. 2012a. Synthesis, characterization and anti-*Listeria monocytogenes* effect of amino acid coated magnetite nanoparticles. *Current Nanoscience* 8(6): 868–874.

Ebrahiminezhad, A., Ghasemi, Y., Rasoul-Amini, S., Barar, J. and Davaran, S. 2012b. Impact of amino-acid coating on the synthesis and characteristics of iron-oxide nanoparticles (IONs). *Bulletin of the Korean Chemical Society* 33(12): 3957–3962.

Ebrahiminezhad, A., Ghasemi, Y., Rasoul-Amini, S., Barar, J. and Davaran, S. 2013. Preparation of novel magnetic fluorescent nanoparticles using amino acids. *Colloids and Surfaces B: Biointerfaces* 102: 534–539.

Ebrahiminezhad, A., Najafipour, S., Kouhpayeh, A., Berenjian, A., Rasoul-Amini, S. and Ghasemi, Y. 2014a. Facile fabrication of uniform hollow silica microspheres using a novel biological template. *Colloids and Surfaces B: Biointerfaces* 118: 249–253.

Ebrahiminezhad, A., Raee, M.J., Manafi, Z., Jahromi, A.S. and Ghasemi, Y. 2016e. Ancient and novel forms of silver in medicine and biomedicine. *Journal of Advanced Medical Sciences and Applied Technologies* 2(1): 122–128.

Ebrahiminezhad, A., Rasoul-Amini, S., Davaran, S., Barar, J. and Ghasemi, Y. 2014b. Impacts of iron oxide nanoparticles on the invasion power of *Listeria monocytogenes*. *Current Nanoscience* 10(3): 382–388.

Ebrahiminezhad, A., Rasoul-Amini, S., Kouhpayeh, A., Davaran, S., Barar, J. and Ghasemi, Y. 2015a. Impacts of amine functionalized iron oxide nanoparticles on HepG2 cell line. *Current Nanoscience* 11(1): 113–119.

Ebrahiminezhad, A., Taghizadeh, S.-M., Taghizadeh, S. and Ghasemi, Y. 2017a. Chemical and biological approaches for the synthesis of silver nanoparticles; A mini review. *Trends in Pharmaceutical Sciences* 3(2): 55–62.

Ebrahiminezhad, A., Taghizadeh, S., Berenjian, A., Heidaryan Naeini, F. and Ghasemi, Y. 2016f. Green synthesis of silver nanoparticles capped with natural carbohydrates using ephedra intermedia. *Nanoscience & Nanotechnology-Asia* 6: 1–9.

Ebrahiminezhad, A., Taghizadeh, S., Berenjian, A., Heidaryan Naeini, F., Ghasemi, Y. 2017b. Green synthesis of silver nanoparticles capped with natural carbohydrates using ephedra intermedia. *Nanoscience & Nanotechnology-Asia* 7(1): 104–112.

Ebrahiminezhad, A., Taghizadeh, S., Berenjiand, A., Rahi, A. and Ghasemi, Y. 2016g. Synthesis and characterization of silver nanoparticles with natural carbohydrate capping using *Zataria multiflora*. *Advanced Materials Letters* 7(6): 122–127.

Ebrahiminezhad, A., Taghizadeh, S., Berenjiand, A., Rahi, A. and Ghasemi, Y. 2016h. Synthesis and characterization of silver nanoparticles with natural carbohydrate capping using *Zataria multiflora*. *Advanced Materials Letters* 7(11): 939–944.

Ebrahiminezhad, A., Taghizadeh, S. and Ghasemi, Y. 2017c. Green synthesis of silver nanoparticles using Mediterranean Cypress (*Cupressus sempervirens*) leaf extract. *American Journal of Biochemistry and Biotechnology* 13(1): 1–6.

Ebrahiminezhad, A., Taghizadeh, S., Ghasemi, Y. and Berenjian, A. 2017d. Green synthesized nanoclusters of ultra-small zero valent iron nanoparticles as a novel dye removing material. *Science of the Total Environment* 621: 1527–1532.

Ebrahiminezhad, A., Taghizadeh, S., Ghasemi, Y. and Berenjian, A. 2017e. Green synthesized nanoclusters of ultra-small zero valent iron nanoparticles as a novel dye removing material. *Science of the Total Environment* 15(621): 1527–1532.

Ebrahiminezhad, A., Varma, V., Yang, S. and Berenjian, A. 2016i. Magnetic immobilization of *Bacillus subtilis* natto cells for menaquinone-7 fermentation. *Applied Microbiology and Biotechnology* 100(1): 173–180.

Ebrahiminezhad, A., Varma, V., Yang, S., Ghasemi, Y. and Berenjian, A. 2015b. Synthesis and application of amine functionalized iron oxide nanoparticles on menaquinone-7 fermentation: A step towards process intensification. *Nanomaterials* 6(1): 1–9.

Ebrahiminezhad, A., Zare-Hoseinabadi, A., Berenjian, A. and Ghasemi, Y. 2017f. Green synthesis and characterization of zero-valent iron nanoparticles using stinging nettle (Urtica dioica) leaf extract. *Green Processing and Synthesis* 6(5): 469–475.

Ebrahiminezhad, A., Zare-Hoseinabadi, A., Sarmah, A.K., Taghizadeh, S., Ghasemi, Y. and Berenjian, A. 2017g. Plant-mediated synthesis and applications of iron nanoparticles. *Molecular Biotechnology* 60: 1–15.

Ebrahiminezhad, A., Zare, M., Kiyanpour, S., Berenjian, A., Niknezhad, S.V. and Ghasemi, Y. 2017h. Biosynthesis of xanthan gum coated iron nanoparticles by using

Xanthomonas campestris. IET Nanobiotechnology 151: 684–691.

Edison, T.J.I. and Sethuraman, M.G. 2012. Instant green synthesis of silver nanoparticles using Terminalia chebula fruit extract and evaluation of their catalytic activity on reduction of methylene blue. *Process Biochemistry* 47(9): 1351–1357.

Ehrampoush, M.H., Miria, M., Salmani, M.H. and Mahvi, A.H. 2015. Cadmium removal from aqueous solution by green synthesis iron oxide nanoparticles with tangerine peel extract. *Journal of Environmental Health Science and Engineering* 13(1): 84.

Elavazhagan, T. and Arunachalam, K.D. 2011. Memecylon edule leaf extract mediated green synthesis of silver and gold nanoparticles. *International Journal of Nanomedicine* 6: 1265.

Elumalai, E., Prasad, T., Hemachandran, J., Therasa, S.V., Thirumalai, T. and David, E. 2010. Extracellular synthesis of silver nanoparticles using leaves of Euphorbia hirta and their antibacterial activities. *Journal of Pharmaceutical Sciences and Research* 2(9): 549–554.

Elumalai, K., Velmurugan, S., Ravi, S., Kathiravan, V. and Adaikala Raj, G. 2015. Bio-approach: Plant mediated synthesis of ZnO nanoparticles and their catalytic reduction of methylene blue and antimicrobial activity. *Advanced Powder Technology* 26(6): 1639–1651.

Fazlzadeh, M., Rahmani, K., Zarei, A., Abdoallahzadeh, H., Nasiri, F. and Khosravi, R. 2017. A novel green synthesis of zero valent iron nanoparticles (NZVI) using three plant extracts and their efficient application for removal of Cr (VI) from aqueous solutions. *Advanced Powder Technology* 28(1): 122–130.

Foo, Y.Y., Periasamy, V., Kiew, L.V., Kumar, G.G. and Malek, S.N.A. 2017. Curcuma mangga-mediated synthesis of gold nanoparticles: Characterization, stability, cytotoxicity, and blood compatibility. *Nanomaterials* 7(6): 123.

Francis, S., Joseph, S., Koshy, E.P. and Mathew, B. 2018. Microwave assisted green synthesis of silver nanoparticles using leaf extract of elephantopus scaber and its environmental and biological applications. *Artificial Cells, Nanomedicine, and Biotechnology* 46(4): 795–804.

Freitas, C.D.T., Oliveira, J.S., Miranda, M.R.A., Macedo, N.M.R., Sales, M.P., Villas-Boas, L.A. and Ramos, M.V. 2007. Enzymatic activities and protein profile of latex from Calotropis procera. *Plant Physiology and Biochemistry* 45(10–11): 781–789.

Fu, L. and Fu, Z. 2015. Plectranthus amboinicus leaf extract-assisted biosynthesis of ZnO nanoparticles and their photocatalytic activity. *Ceramics International* 41(2): 2492–2496.

Ganesan, V., Arunkumar, C., Nima, P. and Astalakshmi, A. 2013. Green synthesis of silver nanoparticles using leaves of Murraya paniculata (L.) Jack. *International Journal of Recent Scientific Research* 4: 1022–1026.

Ganesh Kumar, V., Dinesh Gokavarapu, S., Rajeswari, A., Stalin Dhas, T., Karthick, V., Kapadia, Z., Shrestha, T., Barathy, I.A., Roy, A. and Sinha, S. 2011. Facile green synthesis of gold nanoparticles using leaf extract of antidiabetic potent Cassia auriculata. *Colloids and Surfaces B: Biointerfaces* 87(1): 159–163.

Ganeshkumar, M., Sastry, T.P., Kumar, M.S., Dinesh, M.G., Kannappan, S. and Suguna, L. 2012. Sun light mediated synthesis of gold nanoparticles as carrier for 6-mercaptopurine: Preparation, characterization and toxicity studies in zebrafish embryo model. *Materials Research Bulletin* 47(9): 2113–2119.

Gangula, A., Podila, R., Karanam, L., Janardhana, C. and Rao, A.M. 2011. Catalytic reduction of 4-nitrophenol using biogenic gold and silver nanoparticles derived from Breynia rhamnoides. *Langmuir* 27(24): 15268–15274.

Geetha, N., Harini, K., Showmya, J.J. and Priya, K.S. 2012 Biofabrication of silver nanoparticles using leaf extract of Chromolaena odorata (L.) King and Robinson, pp. 56–59.

Geethalakshmi, R. and Sarada, D.V.L. 2012. Gold and silver nanoparticles from Trianthema decandra: Synthesis, characterization, and antimicrobial properties. *International Journal of Nanomedicine* 7: 5375.

Gengan, R., Anand, K., Phulukdaree, A., Chuturgoon, A. 2013. A549 lung cell line activity of biosynthesized silver nanoparticles using Albizia adianthifolia leaf. *Colloids and Surfaces B: Biointerfaces* 105: 87–91.

Ghaffari-Moghaddam, M. and Hadi-Dabanlou, R. 2014. Plant mediated green synthesis and antibacterial activity of silver nanoparticles using *Crataegus douglasii* fruit extract. *Journal of Industrial and Engineering Chemistry* 20(2): 739–744.

Ghodake, G.S., Deshpande, N.G., Lee, Y.P. and Jin, E.S. 2010. Pear fruit extract-assisted room-temperature biosynthesis of gold nanoplates. *Colloids and Surfaces B: Biointerfaces* 75(2): 584–589.

Gholami, A., Rasoul-Amini, S., Ebrahiminezhad, A., Abootalebi, N., Niroumand, U., Ebrahimi, N. and Ghasemi, Y. 2016. Magnetic properties and antimicrobial effect of amino and lipoamino acid coated iron oxide nanoparticles. *Minerva Biotecnologica* 28(4): 177–186.

Gholami, A., Rasoul-amini, S., Ebrahiminezhad, A., Seradj, S.H. and Ghasemi, Y. 2015. Lipoamino acid coated superparamagnetic iron oxide nanoparticles concentration and time dependently enhanced growth of human hepatocarcinoma cell line (Hep-G2). *Journal of Nanomaterials* 2015: 1–9.

Ghosh, S., Patil, S., Ahire, M., Kitture, R., Gurav, D.D., Jabgunde, A.M., Kale, S., Pardesi, K., Shinde, V. and Bellare, J. 2012. Gnidia glauca flower extract mediated synthesis of gold nanoparticles and evaluation of its chemocatalytic potential. *Journal of Nanobiotechnology* 10(1): 17.

Ghosh, S., Patil, S., Ahire, M., Kitture, R., Jabgunde, A., Kale, S., Pardesi, K., Bellare, J., Dhavale, D.D. and Chopade, B.A. 2011. Synthesis of gold nanoanisotrops using Dioscorea bulbifera tuber extract. *Journal of Nanomaterials* 2011: 45.

Gnanajobitha, G., Paulkumar, K., Vanaja, M., Rajeshkumar, S., Malarkodi, C., Annadurai, G. and Kannan, C. 2013a. Fruit-mediated synthesis of silver nanoparticles using Vitis vinifera and evaluation of their antimicrobial efficacy. *Journal of Nanostructure in Chemistry* 3(1): 67.

Gnanajobitha, G., Vanaja, M., Paulkumar, K., Rajeshkumar, S., Malarkodi, C., Annadurai, G., Kannan, C. 2013b. Green synthesis of silver nanoparticles using Millingtonia hortensis and evaluation of their antimicrobial efficacy. *International Journal of Nanomaterials and Biostructures* 3(1): 21–25.

Goodarzi, V., Zamani, H., Bajuli, L. and Moradshahi, A. 2014. Evaluation of antioxidant potential and reduction capacity of some plant extracts in silver nanoparticle synthesis. *Molecular Biology Research Communications* 3(3): 165–174.

Gultekin, D.D., Nadaroglu, H., Gungor, A.A. and Kishali, N.H. 2017. Biosynthesis and characterization of copper oxide nanoparticles using Cimin grape (Vitis vinifera cv.) extract. *International Journal of Secondary Metabolite* 4(3, Special Issue 1): 77–84.

Gupta, A.K. and Wells, S. 2004. Surface-modified superparamagnetic nanoparticles for drug delivery: Preparation, characterization, and cytotoxicity studies. *IEEE Transactions on NanoBioscience* 3(1): 66–73.

Haris, M., Kumar, A., Ahmad, A., Abuzinadah, M.F., Basheikh, M., Khan, S.A. and Mujeeb, M. 2017. Microwave-assisted green synthesis and antimicrobial activity of silver nanoparticles derived from a supercritical carbon dioxide extract of the fresh aerial parts of Phyllanthus niruri L. *Tropical Journal of Pharmaceutical Research* 16(12): 2967–2976.

Harne, S., Sharma, A., Dhaygude, M., Joglekar, S., Kodam, K. and Hudlikar, M. 2012. Novel route for rapid biosynthesis of copper nanoparticles using aqueous extract of Calotropis procera L. latex and their cytotoxicity on tumor cells. *Colloids and Surfaces B: Biointerfaces* 95: 284–288.

Harshiny, M., Iswarya, C.N. and Matheswaran, M. 2015. Biogenic synthesis of iron nanoparticles using Amaranthus dubius leaf extract as a reducing agent. *Powder Technology* 286: 744–749.

He, Y., Du, Z., Lv, H., Jia, Q., Tang, Z., Zheng, X., Zhang, K. and Zhao, F. 2013. Green synthesis of silver nanoparticles by *Chrysanthemum morifolium* Ramat. extract and their application in clinical ultrasound gel. *International Journal of Nanomedicine* 8: 1809–1815.

Hoag, G.E., Collins, J.B., Holcomb, J.L., Hoag, J.R., Nadagouda, M.N. and Varma, R.S. 2009. Degradation of bromothymol blue by 'greener'nano-scale zero-valent iron synthesized using tea polyphenols. *Journal of Materials Chemistry* 19(45): 8671–8677.

Huang, J., Li, Q., Sun, D., Lu, Y., Su, Y., Yang, X., Wang, H., Wang, Y., Shao, W. and He, N. 2007. Biosynthesis of silver and gold nanoparticles by novel sundried Cinnamomum camphora leaf. *Nanotechnology* 18(10): 105104.

Huang, J., Lin, L., Li, Q., Sun, D., Wang, Y., Lu, Y., He, N., Yang, K., Yang, X., Wang, H. 2008. Continuous-flow biosynthesis of silver nanoparticles by lixivium of sundried Cinnamomum camphora leaf in tubular microreactors. *Industrial & Engineering Chemistry Research* 47(16): 6081–6090.

Huang, L., Weng, X., Chen, Z., Megharaj, M. and Naidu, R. 2014a. Green synthesis of iron nanoparticles by various tea extracts: comparative study of the reactivity. *Spectrochimica Acta Part A: Molecular and Biomolecular Spectroscopy* 130: 295–301.

Huang, L., Weng, X., Chen, Z., Megharaj, M. and Naidu, R. 2014b. Synthesis of iron-based nanoparticles using oolong tea extract for the degradation of malachite green. *Spectrochimica Acta Part A: Molecular and Biomolecular Spectroscopy* 117: 801–804.

Iida, H., Osaka, T., Takayanagi, K. and Nakanishi, T. 2007. Synthesis of Fe_3O_4 nanoparticles with various sizes and magnetic properties by controlled hydrolysis. *Journal of Colloid and Interface Science* 314(1): 274–280.

Ijaz, F., Shahid, S., Khan, S.A., Ahmad, W. and Zaman, S. 2017. Green synthesis of copper oxide nanoparticles using Abutilon indicum leaf extract: Antimicrobial, antioxidant and photocatalytic dye degradation activities. *Tropical Journal of Pharmaceutical Research* 16(4): 743–753.

Islam, N.U., Jalil, K., Shahid, M., Rauf, A., Muhammad, N., Khan, A., Shah, M.R. and Khan, M.A. 2015. Green synthesis and biological activities of gold nanoparticles functionalized with Salix alba. *Arabian Journal of Chemistry*. DOI 10.1016/j.arabjc.2015.06.025, https://www.sciencedirect.com/science/article/pii/S187 8535215001975.

Jacob, S.J.P., Finub, J., Narayanan, A. 2012. Synthesis of silver nanoparticles using Piper longum leaf extracts and its cytotoxic activity against Hep-2 cell line. *Colloids and Surfaces B: Biointerfaces* 91: 212–214.

Jafarirad, S., Mehrabi, M., Divband, B. and Kosari-Nasab, M. 2016. Biofabrication of zinc oxide nanoparticles using fruit extract of Rosa canina and their toxic potential against bacteria: A mechanistic approach. *Materials Science and Engineering: C* 59: 296–302.

Jalal, M., Ansari, M.A., Shukla, A.K., Ali, S.G., Khan, H.M., Pal, R., Alam, J. and Cameotra, S.S. 2016. Green synthesis and antifungal activity of Al 2 O 3 NPs against fluconazole-resistant Candida spp isolated from a tertiary care hospital. *RSC Advances* 6(109): 107577–107590.

Jassal, V., Shanker, U. and Gahlot, S. 2016. Green synthesis of some iron oxide nanoparticles and their interaction with 2-Amino, 3-Amino and 4-Aminopyridines. *Materials Today: Proceedings* 3(6): 1874–1882.

Jayandran, M., Haneefa, M.M. and Balasubramanian, V. 2015. Green synthesis of copper nanoparticles using natural reducer and stabilizer and an evaluation of antimicrobial activity. *Journal of Chemical and Pharmaceutical Research* 7(2): 251–259.

Jayaseelan, C., Ramkumar, R., Rahuman, A.A. and Perumal, P. 2013. Green synthesis of gold nanoparticles using seed aqueous extract of Abelmoschus esculentus and its antifungal activity. *Industrial Crops and Products* 45: 423–429.

Jelle, A.A., Hmadeh, M., O'Brien, P.G., Perovic, D.D. and Ozin, G.A. 2016. Photocatalytic properties of all four polymorphs of nanostructured iron oxyhydroxides. *ChemNanoMat* 2(11): 1047–1054.

Jena, J., Pradhan, N., Dash, B.P., Panda, P.K. and Mishra, B.K. 2015. Pigment mediated biogenic synthesis of silver nanoparticles using diatom Amphora sp. and its antimicrobial activity. *Journal of Saudi Chemical Society* 19(6): 661–666.

Joseph, S. and Mathew, B. 2015. Microwave-assisted green synthesis of silver nanoparticles and the study on catalytic activity in the degradation of dyes. *Journal of Molecular Liquids* 204: 184–191.

Kahrilas, G.A., Wally, L.M., Fredrick, S.J., Hiskey, M., Prieto, A.L. and Owens, J.E. 2014. Microwave-assisted green synthesis of silver nanoparticles using orange peel extract. *ACS Sustainable Chemistry & Engineering* 2(3): 367–376.

Kannan, R.R.R., Arumugam, R., Ramya, D., Manivannan, K. and Anantharaman, P. 2013. Green synthesis of silver nanoparticles using marine macroalga Chaetomorpha linum. *Applied Nanoscience* 3(3): 229–233.

Kathiraven, T., Sundaramanickam, A., Shanmugam, N. and Balasubramanian, T. 2015. Green synthesis of silver nanoparticles using marine algae Caulerpa racemosa and their antibacterial activity against some human pathogens. *Applied Nanoscience* 5(4): 499–504.

Kaur, P., Thakur, R. and Chaudhury, A. 2016. Biogenesis of copper nanoparticles using peel extract of Punica granatum and their antimicrobial activity against opportunistic pathogens. *Green Chemistry Letters and Reviews* 9(1): 33–38.

Kaviya, S., Santhanalakshmi, J., Viswanathan, B., Muthumary, J., Srinivasan, K. 2011. Biosynthesis of silver nanoparticles using Citrus sinensis peel extract and its antibacterial activity. *Spectrochimica Acta Part A: Molecular and Biomolecular Spectroscopy* 79(3): 594–598.

Keihan, A.H., Veisi, H. and Veasi, H. 2017. Green synthesis and characterization of spherical copper nanoparticles as organometallic antibacterial agent. *Applied Organometallic Chemistry* 31(7): e3642.

Kesharwani, J., Yoon, K.Y., Hwang, J. and Rai, M. 2009. Phytofabrication of silver nanoparticles by leaf extract

of Datura metel: Hypothetical mechanism involved in synthesis. *Journal of Bionanoscience* 3(1): 39–44.

Khan, M., Mangrich, A., Schultz, J., Grasel, F., Mattoso, N. and Mosca, D. 2015. Green chemistry preparation of superparamagnetic nanoparticles containing Fe3O4 cores in biochar. *Journal of Analytical and Applied Pyrolysis* 116: 42–48.

Khan, Z., Bashir, O., Hussain, J.I., Kumar, S. and Ahmad, R. 2012a. Effects of ionic surfactants on the morphology of silver nanoparticles using Paan (Piper betel) leaf petiole extract. *Colloids and Surfaces B: Biointerfaces* 98: 85–90.

Khan, Z., Hussain, J.I. and Hashmi, A.A. 2012b. Shape-directing role of cetyltrimethylammonium bromide in the green synthesis of Ag-nanoparticles using Neem (Azadirachta indica) leaf extract. *Colloids and Surfaces B: Biointerfaces* 95: 229–234.

Kianpour, S., Ebrahiminezhad, A., Mohkam, M., Tamaddon, A.M., Dehshahri, A., Heidari, R. and Ghasemi, Y. 2016. Physicochemical and biological characteristics of the nanostructured polysaccharide-iron hydrogel produced by microorganism Klebsiella oxytoca. *Journal of Basic Microbiology* 2016(56): 132–140.

Kora, A.J., Sashidhar, R. and Arunachalam, J. 2012. Aqueous extract of gum olibanum (Boswellia serrata): A reductant and stabilizer for the biosynthesis of antibacterial silver nanoparticles. *Process Biochemistry* 47(10): 1516–1520.

Korbekandi, H., Asghari, G., Jalayer, S.S., Jalayer, M.S. and Bandegani, M. 2013. Nanosilver particle production using Juglans Regia L.(walnut) leaf extract. *Jundishapur Journal of Natural Pharmaceutical Products* 8(1): 20.

Kothai, S. and Jayanthi, B. 2014. Ultrasound Intensified green synthesis of silver nanoparticles using Camellia sinensis extract fortified with lemon and honey. *International Journal of ChemTech Research* 6: 248–253.

Kozma, G., Rónavári, A., Kónya, Z. and Kukovecz, A. 2015. Environmentally benign synthesis methods of zero-valent iron nanoparticles. *ACS Sustainable Chemistry & Engineering* 4(1): 291–297.

Krishnaraj, C., Jagan, E., Rajasekar, S., Selvakumar, P., Kalaichelvan, P., Mohan, N. 2010. Synthesis of silver nanoparticles using Acalypha indica leaf extracts and its antibacterial activity against water borne pathogens. *Colloids and Surfaces B: Biointerfaces* 76(1): 50–56.

Kuang, Y., Wang, Q., Chen, Z., Megharaj, M. and Naidu, R. 2013. Heterogeneous Fenton-like oxidation of monochlorobenzene using green synthesis of iron nanoparticles. *Journal of Colloid and Interface Science* 410: 67–73.

Kumar, B., Smita, K., Cumbal, L. and Debut, A. 2014a. Biogenic synthesis of iron oxide nanoparticles for 2-arylbenzimidazole fabrication. *Journal of Saudi Chemical Society* 18(4): 364–369.

Kumar, B., Smita, K., Cumbal, L. and Debut, A. 2014b. Green approach for fabrication and applications of zinc oxide nanoparticles. *Bioinorganic Chemistry and Applications* 2014: 1–7.

Kumar, B., Smita, K., Cumbal, L. and Debut, A. 2014c. Sacha inchi (Plukenetia volubilis L.) oil for one pot synthesis of silver nanocatalyst: An ecofriendly approach. *Industrial Crops and Products* 58: 238–243.

Kumar, B., Smita, K., Cumbal, L. and Debut, A. 2014d. Synthesis of silver nanoparticles using Sacha inchi (Plukenetia volubilis L.) leaf extracts. *Saudi Journal of Biological Sciences* 21(6): 605–609.

Kumar, B., Smita, K., Cumbal, L., Debut, A., Galeas, S. and Guerrero, V.H. 2016a. Phytosynthesis and photocatalytic activity of magnetite (Fe_3O_4) nanoparticles using the *Andean blackberry* leaf. *Materials Chemistry and Physics* 179: 310–315.

Kumar, R., Roopan, S.M., Prabhakarn, A., Khanna, V.G., Chakroborty, S. 2012. Agricultural waste Annona squamosa peel extract: biosynthesis of silver nanoparticles. *Spectrochimica Acta Part A: Molecular and Biomolecular Spectroscopy* 90: 173–176.

Kumar, V., Bano, D., Mohan, S., Singh, D.K. and Hasan, S.H. 2016b. Sunlight-induced green synthesis of silver nanoparticles using aqueous leaf extract of Polyalthia longifolia and its antioxidant activity. *Materials Letters* 181: 371–377.

Kumar, V., Yadav, S.C., Yadav, S.K. 2010. Syzygium cumini leaf and seed extract mediated biosynthesis of silver nanoparticles and their characterization. *Journal of Chemical Technology & Biotechnology* 85(10): 1301–1309.

Kurkure, R.V., Jaybhaye, S. and Sangle, A. 2016. Synthesis of Copper/Copper Oxide nanoparticles in ecofriendly and non-toxic manner from floral extract of Caesalpinia pulcherrima. *International Journal on Recent and Innovation Trends in Computing and Communication* 4: 363–366.

Lalitha, A., Subbaiya, R. and Ponmurugan, P. 2013. Green synthesis of silver nanoparticles from leaf extract Azhadirachta indica and to study its anti-bacterial and antioxidant property. *International Journal of Current Microbiology and Applied Sciences* 2(6): 228–235.

Latha, N. and Gowri, M. 2014. Bio synthesis and characterisation of Fe_3O_4 nanoparticles using Caricaya Papaya leaves extract. *Synthesis* 3: 1551–1556.

Lavanya, M., Veenavardhini, S.V., Gim, G.H., Kathiravan, M.N. and Kim, S.W. 2013. Synthesis, characterization and evaluation of antimicrobial efficacy of silver nanoparticles using Paederia foetida L. leaf extract. *International Research Journal of Biological Sciences* 2(3): 28–34.

Lee, H.J., Song, J.Y. and Kim, B.S. 2013. Biological synthesis of copper nanoparticles using Magnolia kobus leaf extract and their antibacterial activity. *Journal of Chemical Technology and Biotechnology* 88(11): 1971–1977.

Lee, S.Y., Krishnamurthy, S., Cho, C.-W. and Yun, Y.-S. 2016. Biosynthesis of gold nanoparticles using Ocimum sanctum extracts by solvents with different polarity.

ACS Sustainable Chemistry & Engineering 4(5): 2651–2659.

Li, S.Z., Ma, Y., Yue, X.L., Cao, Z. and Dai, Z.F. 2009. One-pot construction of doxorubicin conjugated magnetic silica nanoparticles. *New Journal of Chemistry* 33(12): 2414–2418.

Logeswari, P., Silambarasan, S. and Abraham, J. 2015. Synthesis of silver nanoparticles using plants extract and analysis of their antimicrobial property. *Journal of Saudi Chemical Society* 19(3): 311–317.

Lukman, A.I., Gong, B., Marjo, C.E., Roessner, U. and Harris, A.T. 2011. Facile synthesis, stabilization, and anti-bacterial performance of discrete Ag nanoparticles using *Medicago sativa* seed exudates. *Journal of Colloid and Interface Science* 353(2): 433–444.

Machado, S., Grosso, J.P., Nouws, H.P., Albergaria, J.T. and Delerue-Matos, C. 2014. Utilization of food industry wastes for the production of zero-valent iron nanoparticles. *Science of the Total Environment* 496: 233–240.

Machado, S., Pacheco, J., Nouws, H., Albergaria, J.T. and Delerue-Matos, C. 2015. Characterization of green zero-valent iron nanoparticles produced with tree leaf extracts. *Science of the Total Environment* 533: 76–81.

Machado, S., Pinto, S., Grosso, J., Nouws, H., Albergaria, J.T. and Delerue-Matos, C. 2013a. Green production of zero-valent iron nanoparticles using tree leaf extracts. *Science of the Total Environment* 445–446: 1–8.

Machado, S., Stawiński, W., Slonina, P., Pinto, A., Grosso, J., Nouws, H., Albergaria, J.T. and Delerue-Matos, C. 2013b. Application of green zero-valent iron nanoparticles to the remediation of soils contaminated with ibuprofen. *Science of the Total Environment* 461: 323–329.

Machala, L., Tucek, J. and Zboril, R. 2011. Polymorphous transformations of nanometric iron (III) oxide: a review. *Chemistry of Materials* 23(14): 3255–3272.

Madhumitha, G., Elango, G. and Roopan, S.M. 2016. Biotechnological aspects of ZnO nanoparticles: overview on synthesis and its applications. *Applied Microbiology and Biotechnology* 100(2): 571–581.

Makarov, V.V., Love, A.J., Sinitsyna, O.V., Makarova, S.S., Yaminsky, I.V., Taliansky, M.E. and Kalinina, N.O. 2014a. "Green" nanotechnologies: synthesis of metal nanoparticles using plants. *Acta Naturae* 6(1 (20)): 35–44.

Makarov, V.V., Makarova, S.S., Love, A.J., Sinitsyna, O.V., Dudnik, A.O., Yaminsky, I.V., Taliansky, M.E. and Kalinina, N.O. 2014b. Biosynthesis of stable iron oxide nanoparticles in aqueous extracts of Hordeum vulgare and Rumex acetosa plants. *Langmuir* 30(20): 5982–5988.

Mandal, M., Kundu, S., Ghosh, S.K., Panigrahi, S., Sau, T.K., Yusuf, S.M. and Pal, T. 2005. Magnetite nanoparticles with tunable gold or silver shell. *Journal of Colloid and Interface Science* 286(1): 187–194.

Manjamadha, V.P. and Muthukumar, K. 2016. Ultrasound assisted green synthesis of silver nanoparticles using weed plant. *Bioprocess and Biosystems Engineering* 39(3): 401–411.

Manquián-Cerda, K., Cruces, E., Rubio, M.A., Reyes, C. and Arancibia-Miranda, N. 2017. Preparation of nanoscale iron (oxide, oxyhydroxides and zero-valent) particles derived from blueberries: Reactivity, characterization and removal mechanism of arsenate. *Ecotoxicology and Environmental Safety* 145: 69–77.

Marimuthu, S., Rahuman, A.A., Jayaseelan, C., Kirthi, A.V., Santhoshkumar, T., Velayutham, K., Bagavan, A., Kamaraj, C., Elango, G., Iyappan, M., Siva, C., Karthik, L. and Rao, K.V.B. 2013. Acaricidal activity of synthesized titanium dioxide nanoparticles using Calotropis gigantea against Rhipicephalus microplus and Haemaphysalis bispinosa. *Asian Pacific Journal of Tropical Medicine* 6(9): 682–688.

Markova, Z., Novak, P., Kaslik, J., Plachtova, P., Brazdova, M., Jancula, D., Siskova, K.M., Machala, L., Marsalek, B. and Zboril, R. 2014. Iron (II, III)–polyphenol complex nanoparticles derived from green tea with remarkable ecotoxicological impact. *ACS Sustainable Chemistry & Engineering* 2(7): 1674–1680.

Martínez-Cabanas, M., López-García, M., Barriada, J.L., Herrero, R. and de Vicente, M.E.S. 2016. Green synthesis of iron oxide nanoparticles. Development of magnetic hybrid materials for efficient As (V) removal. *Chemical Engineering Journal* 301: 83–91.

Mehr, E.S., Sorbiun, M., Ramazani, A. and Fardood, S.T. 2018. Plant-mediated synthesis of zinc oxide and copper oxide nanoparticles by using Ferulago angulata (schlecht) boiss extract and comparison of their photocatalytic degradation of Rhodamine B (RhB) under visible light irradiation. *Journal of Materials Science: Materials in Electronics* 29(2): 1333–1340.

Mehr, F.P., Khanjani, M. and Vatani, P. 2015. Synthesis of Nano-Ag particles using sodium borohydride. *Oriental Journal of Chemistry* 31(3): 1831–1833.

Mishra, V. and Sharma, R. 2015. Green synthesis of zinc oxide nanoparticles using fresh peels extract of Punica granatum and its antimicrobial activities. *International Journal of Pharma Research and Health Sciences* 3(3): 694–699.

Mishra, V., Sharma, R., Jasuja, N. and Gupta, D. 2014. A review on green synthesis of nanoparticles and evaluation of antimicrobial activity. *International Journal of Green and Herbal Chemistry* 3: 081–094.

Mittal, A.K., Chisti, Y. and Banerjee, U.C. 2013. Synthesis of metallic nanoparticles using plant extracts. *Biotechnology Advances* 31(2): 346–356.

Mittal, A.K., Kaler, A., Banerjee, U.C. 2012. Free radical scavenging and antioxidant activity of silver nanoparticles synthesized from flower extract of Rhododendron dauricum. *Nano Biomedicine and Engineering* 4(3): 118–124.

Mo, Y.-Y., Tang, Y.-K., Wang, S.-Y., Ling, J.-M., Zhang, H.-B. and Luo, D.-Y. 2015. Green synthesis of silver nanoparticles using eucalyptus leaf extract. *Materials Letters* 144: 165–167.

Mohseniazar, M., Barin, M., Zarredar, H., Alizadeh, S. and Shanehbandi, D. 2011. Potential of microalgae and lactobacilli in biosynthesis of silver nanoparticles. *BioImpacts: BI* 1(3): 149.

Moosa, A.A., Ridha, A.M. and Al-Kaser, M. 2015. Process parameters for green synthesis of silver nanoparticles using leaves extract of Aloe vera plant. *International Journal of Current Research* 3: 966–975.

Muddineti, O.S., Kumari, P., Ajjarapu, S., Lakhani, P.M., Bahl, R., Ghosh, B. and Biswas, S. 2016. Xanthan gum stabilized PEGylated gold nanoparticles for improved delivery of curcumin in cancer. *Nanotechnology* 27(32): 325101.

Mukherjee, D., Ghosh, S., Majumdar, S. and Annapurna, K. 2016. Green synthesis of α-Fe2O3 nanoparticles for arsenic (V) remediation with a novel aspect for sludge management. *Journal of Environmental Chemical Engineering* 4(1): 639–650.

Muthukumar, H. and Matheswaran, M. 2015. Amaranthus spinosus leaf extract mediated FeO nanoparticles: Physicochemical traits, Photocatalytic and Antioxidant activity. *ACS Sustainable Chemistry & Engineering* 3(12): 3149–3156.

Mystrioti, C., Xanthopoulou, T., Tsakiridis, P., Papassiopi, N. and Xenidis, A. 2016. Comparative evaluation of five plant extracts and juices for nanoiron synthesis and application for hexavalent chromium reduction. *Science of the Total Environment* 539: 105–113.

Nadagouda, M.N. and Varma, R.S. 2008. Green synthesis of silver and palladium nanoparticles at room temperature using coffee and tea extract. *Green Chemistry* 10(8): 859–862.

Nagababu, P. and Rao, V.U. 2017. Pharmacological assessment, green synthesis and characterization of silver nanoparticles of sonneratia apetala Buch.-Ham. Leaves. *Journal of Applied Pharmaceutical Science* 7(08): 175–182.

Nagajyothi, P., Lee, S.-E., An, M. and Lee, K.-D. 2012. Green synthesis of silver and gold nanoparticles using Lonicera japonica flower extract. *Bulletin of the Korean Chemical Society* 33(8): 2609–2612.

Nagajyothi, P., Pandurangan, M., Kim, D.H., Sreekanth, T. and Shim, J. 2017. Green synthesis of iron oxide nanoparticles and their catalytic and in vitro anticancer activities. *Journal of Cluster Science* 28(1): 245–257.

Nagajyothi, P.C., Minh An, T.N., Sreekanth, T.V.M., Lee, J.I., Joo, D.L. and Lee, K.D. 2013. Green route biosynthesis: Characterization and catalytic activity of ZnO nanoparticles. *Materials Letters* 108: 160–163.

Nagajyothi, P.C., Sreekanth, T.V.M., Tettey, C.O., Jun, Y.I. and Mook, S.H. 2014. Characterization, antibacterial, antioxidant, and cytotoxic activities of ZnO nanoparticles using Coptidis Rhizoma. *Bioorganic & Medicinal Chemistry Letters* 24(17): 4298–4303.

Naik, L.S., Marx, K.P., Vennela, P.S., Devi, V. 2013. Green synthesis of silver nanoparticles using Strawberry leaf extract (Arbutus unedo) and evaluation of its antimicrobial activity-a Novel study. *International Journal of Nanomaterials and Biostructures* 3(3): 47–50.

Nakhjavani, M., Nikkhah, V., Sarafraz, M.M., Shoja, S. and Sarafraz, M. 2017. Green synthesis of silver nanoparticles using green tea leaves: Experimental study on the morphological, rheological and antibacterial behaviour. *Heat and Mass Transfer* 53(10): 3201–3209.

Narasaiah, P., Mandal, B.K. and Sarada, N. 2017. Biosynthesis of Copper Oxide nanoparticles from Drypetes sepiaria Leaf extract and their catalytic activity to dye degradation, *IOP Conference Series: Materials Science and Engineering* 263: 022012, IOP Publishing.

Narayanan, K.B. and Sakthivel, N. 2008. Coriander leaf mediated biosynthesis of gold nanoparticles. *Materials Letters* 62(30): 4588–4590.

Naseem, T. and Farrukh, M.A. 2015. Antibacterial activity of green synthesis of iron nanoparticles using Lawsonia inermis and *Gardenia jasminoides* leaves extract. *Journal of Chemistry* 2015: 1–7.

Nasiriboroumand, M., Montazer, M. and Barani, H. 2018. Preparation and characterization of biocompatible silver nanoparticles using pomegranate peel extract. *Journal of Photochemistry and Photobiology B: Biology* 179: 98–104.

Nasrollahzadeh, M. and Mohammad Sajadi, S. 2015. Green synthesis of copper nanoparticles using Ginkgo biloba L. leaf extract and their catalytic activity for the Huisgen [3+2] cycloaddition of azides and alkynes at room temperature. *Journal of Colloid and Interface Science* 457: 141–147.

Nasrollahzadeh, M., Sajadi, S.M. and Khalaj, M. 2014. Green synthesis of copper nanoparticles using aqueous extract of the leaves of Euphorbia esula L and their catalytic activity for ligand-free Ullmann-coupling reaction and reduction of 4-nitrophenol. *RSC Advances* 4(88): 47313–47318.

Natarajan, R.K., Nayagam, A.A.J., Gurunagarajan, S., Muthukumar, N.E. and Manimaran, A. 2013. Elaeagnus indica mediated green synthesis of silver nanoparticles and its potent toxicity against human pathogens. *World Applied Sciences Journal* 23(10): 1314–1321.

Ncube, B., Finnie, J.F. and Van Staden, J. 2011. Seasonal variation in antimicrobial and phytochemical properties of frequently used medicinal bulbous plants from South Africa. *South African Journal of Botany* 77(2): 387–396.

Niraimathee, V., Subha, V., Ravindran, R.E. and Renganathan, S. 2016. Green synthesis of iron oxide nanoparticles from Mimosa pudica root extract. *International Journal of Environment and Sustainable Development* 15(3): 227–240.

Njagi, E.C., Huang, H., Stafford, L., Genuino, H., Galindo, H.M., Collins, J.B., Hoag, G.E. and Suib, S.L. 2011. Biosynthesis of iron and silver nanoparticles at room temperature using aqueous sorghum bran extracts. *Langmuir* 27(1): 264–271.

Noruzi, M. and Mousivand, M. 2015. Instantaneous green synthesis of zerovalent iron nanoparticles by Thuja orientalis extract and investigation of their antibacterial properties. *Journal of Applied Chemical Research* 9(2): 37–50.

Ojha, S., Sett, A. and Bora, U. 2017. Green synthesis of silver nanoparticles by Ricinus communis var. carmencita leaf extract and its antibacterial study. *Advances in Natural Sciences: Nanoscience and Nanotechnology* 8(3): 035009.

Pal, A., Shah, S. and Devi, S. 2009. Microwave-assisted synthesis of silver nanoparticles using ethanol as a reducing agent. *Materials Chemistry and Physics* 114(2): 530–532.

Pandian, C.J. and Palanivel, R. 2016. Applications of L-arginine functionalised green synthesised nickel nanoparticles as gene transfer vector and catalyst. *Journal of Experimental Nanoscience* 11(15): 1193–1212.

Parashar, V., Parashar, R., Sharma, B., Pandey, A.C. 2009. Parthenium leaf extract mediated synthesis of silver nanoparticles: A novel approach towards weed utilization. *Digest Journal of Nanomaterials and Biostructures* 4(1).

Park, J., An, K.J., Hwang, Y.S., Park, J.G., Noh, H.J., Kim, J.Y., Park, J.H., Hwang, N.M. and Hyeon, T. 2004. Ultra-large-scale syntheses of monodisperse nanocrystals. *Nature Materials* 3(12): 891–895.

Parveen, A., Roy, A.S. and Rao, S. 2012. Biosynthesis and characterization of silver nanoparticles from Cassia auriculata leaf extract and in vitro evaluation of antimicrobial activity. *International Journal of Applied Biology and Pharmaceutical Technology* 3: 222–228.

Patil, S.V., Borase, H.P., Patil, C.D., Salunke, B.K. 2012. Biosynthesis of silver nanoparticles using latex from few Euphorbian plants and their antimicrobial potential. *Applied Biochemistry and Biotechnology* 167(4): 776–790.

Pattanayak, M. and Nayak, P. 2013. Green synthesis and characterization of zero valent iron nanoparticles from the leaf extract of Azadirachta indica (Neem). *World Journal of Nano Science Technology* 2(1): 06–09.

Paul, B., Bhuyan, B., Dhar Purkayastha, D., Dey, M. and Dhar, S.S. 2015. Green synthesis of gold nanoparticles using Pogestemon benghalensis (B) O. Ktz. leaf extract and studies of their photocatalytic activity in degradation of methylene blue. *Materials Letters* 148: 37–40.

Pavani, K., Gayathramma, K., Banerjee, A. and Suresh, S. 2013. Phyto-synthesis of silver nanoparticles using extracts of Ipomoea indica flowers. *American Journal of Nanomaterials* 1(1): 5–8.

Phanjom, P., Sultana, A., Sarma, H., Ramchiary, J., Goswami, K., Baishya, P. 2012. Plant-mediated synthesis of silver nanoparticles using Elaeagnus latifolia leaf extract. *Digest Journal of Nanomaterials and Biostructures* 7(3): 1117–1123.

Philip, D. 2010. Rapid green synthesis of spherical gold nanoparticles using Mangifera indica leaf. *Spectrochimica Acta Part A: Molecular and Biomolecular Spectroscopy* 77(4): 807–810.

Ponarulselvam, S., Panneerselvam, C., Murugan, K., Aarthi, N., Kalimuthu, K. and Thangamani, S. 2012. Synthesis of silver nanoparticles using leaves of Catharanthus roseus Linn. G. Don and their antiplasmodial activities. *Asian Pacific Journal of Tropical Biomedicine* 2(7): 574–580.

Prasad, A.S. 2016. Iron oxide nanoparticles synthesized by controlled bio-precipitation using leaf extract of Garlic Vine (*Mansoa alliacea*). *Materials Science in Semiconductor Processing* 53: 79–83.

Prasad, C., Gangadhara, S. and Venkateswarlu, P. 2016. Bio-inspired green synthesis of Fe_3O_4 magnetic nanoparticles using watermelon rinds and their catalytic activity. *Applied Nanoscience* 6(6): 797–802.

Prasad, K.S., Patra, A., Shruthi, G. and Chandan, S. 2017. Aqueous extract of saraca indica leaves in the synthesis of copper oxide nanoparticles: finding a way towards going green. *Journal of Nanotechnology* 2017: 1–6.

Prasad, T.N., Kambala, V.S.R. and Naidu, R. 2013. Phyconanotechnology: Synthesis of silver nanoparticles using brown marine algae Cystophora moniliformis and their characterisation. *Journal of Applied Phycology* 25(1): 177–182.

Prathna, T., Raichur, A.M., Chandrasekaran, N. and Mukherjee, A. 2014. Sunlight irradiation induced green synthesis of stable silver nanoparticles using citrus limon extract. *Proceedings of the National Academy of Sciences, India Section B: Biological Sciences* 84(1): 65–70.

Pulido, R., Bravo, L. and Saura-Calixto, F. 2000. Antioxidant activity of dietary polyphenols as determined by a modified ferric reducing/antioxidant power assay. *Journal of Agricultural and Food Chemistry* 48(8): 3396–3402.

Qian, Y., Yao, J., Russel, M., Chen, K. and Wang, X. 2015. Characterization of green synthesized nano-formulation (ZnO–A. vera) and their antibacterial activity against pathogens. *Environmental Toxicology and Pharmacology* 39(2): 736–746.

Qu, J., Yuan, X., Wang, X. and Shao, P. 2011. Zinc accumulation and synthesis of ZnO nanoparticles using Physalis alkekengi L. *Environmental Pollution* 159(7): 1783–1788.

Raee, M.J., Ebrahiminezhad, A., Gholami, A., Ghoshoon, M.B. and Ghasemi, Y. 2018. Magnetic immobilization of recombinant *E. coli* producing extracellular asparaginase: An effective way to intensify downstream process. *Separation Science and Technology* 53(9): 1–8.

Ragaseema, V., Unnikrishnan, S., Kalliyana Krishnan, V. and Krishnan, L.K. 2012. The antithrombotic and antimicrobial properties of PEG-protected silver nanoparticle coated surfaces. *Biomaterials* 33(11): 3083–3092.

Raĭkher, Y.L., Stepanov, V.I., Stolyar, S.V., Ladygina, V.P., Balaev, D.A., Ishchenko, L.A. and Balasoiu, M. 2010. Magnetic properties of biomineral particles produced by bacteria Klebsiella oxytoca. *Physics of the Solid State* 52(2): 298–305.

Rajakumar, G., Rahuman, A.A., Priyamvada, B., Khanna, V.G., Kumar, D.K. and Sujin, P.J. 2012. Eclipta prostrata leaf aqueous extract mediated synthesis of titanium dioxide nanoparticles. *Materials Letters* 68: 115–117.

Rajakumar, G., Rahuman, A.A., Roopan, S.M., Chung, I.-M., Anbarasan, K. and Karthikeyan, V. 2015. Efficacy of larvicidal activity of green synthesized titanium dioxide nanoparticles using Mangifera indica extract against blood-feeding parasites. *Parasitology Research* 114(2): 571–581.

Rajasekharreddy, P. and Rani, P.U. 2014. Biofabrication of Ag nanoparticles using Sterculia foetida L. seed extract and their toxic potential against mosquito vectors and HeLa cancer cells. *Materials Science and Engineering C* 39: 203–212.

Rajiv, P., Rajeshwari, S. and Venckatesh, R. 2013. Bio-Fabrication of zinc oxide nanoparticles using leaf extract of Parthenium hysterophorus L. and its size-dependent antifungal activity against plant fungal pathogens. *Spectrochimica Acta Part A: Molecular and Biomolecular Spectroscopy* 112: 384–387.

Ramesh, M., Anbuvannan, M. and Viruthagiri, G. 2015. Green synthesis of ZnO nanoparticles using Solanum nigrum leaf extract and their antibacterial activity. *Spectrochimica Acta - Part A: Molecular and Biomolecular Spectroscopy* 136(PB): 864–870.

Ranmadugala, D., Ebrahiminezhad, A., Manley-Harris, M., Ghasemi, Y. and Berenjian, A. 2017a. The effect of iron oxide nanoparticles on *Bacillus subtilis* biofilm, growth and viability. *Process Biochemistry* 62(2017): 231–240.

Ranmadugala, D., Ebrahiminezhad, A., Manley-Harris, M., Ghasemi, Y. and Berenjian, A. 2017b. Impact of 3–aminopropyltriethoxysilane-coated iron oxide nanoparticles on menaquinone-7 production using *B. subtilis*. *Nanomaterials* 7(11): 350.

Ranmadugala, D., Ebrahiminezhad, A., Manley-Harris, M., Ghasemi, Y. and Berenjian, A. 2017c. Iron oxide nanoparticles in modern microbiology and biotechnology. *Critical Reviews in Microbiology* 43(4): 493–507.

Ranmadugala, D., Ebrahiminezhad, A., Manley-Harris, M., Ghasemi, Y. and Berenjian, A. 2017d. Magnetic immobilization of bacteria using iron oxide nanoparticles. *Biotechnology Letters* 40(2): 237–248.

Ranmadugala, D., Ebrahiminezhad, A., Manley-Harris, M., Ghasemi, Y. and Berenjian, A. 2017e. Reduced biofilm formation in Menaquinone-7 production process by optimizing the composition of the cultivation medium. *Trends in Pharmaceutical Sciences* 3(4): 245–254.

Ranmadugala, D., Ebrahiminezhad, A., Manley-Harris, M., Ghasemi, Y. and Berenjian, A. 2018. High level production of menaquinone-7 by milking with biocompatible organic solvents. *Current Pharmaceutical Biotechnology* 19: 232–239.

Raut Rajesh, W., Lakkakula Jaya, R., Kolekar Niranjan, S., Mendhulkar Vijay, D. and Kashid Sahebrao, B. 2009. Phytosynthesis of silver nanoparticle using Gliricidia sepium (Jacq.). *Current Nanoscience* 5(1): 117–122.

Reddy, N.J., Nagoor Vali, D., Rani, M. and Rani, S.S. 2014. Evaluation of antioxidant, antibacterial and cytotoxic effects of green synthesized silver nanoparticles by Piper longum fruit. *Materials Science and Engineering C* 34(0): 115–122.

Rodríguez-León, E., Iñiguez-Palomares, R., Navarro, R.E., Herrera-Urbina, R., Tánori, J., Iñiguez-Palomares, C. and Maldonado, A. 2013. Synthesis of silver nanoparticles using reducing agents obtained from natural sources (Rumex hymenosepalus extracts). *Nanoscale Research Letters* 8(1): 318.

Roopan, S.M. and Khan, F.R.N. 2010. ZnO nanoparticles in the synthesis of AB ring core of camptothecin. *Chemical Papers* 64(6): 812–817.

Rufus, A., Sreeju, N., Vilas, V. and Philip, D. 2017. Biosynthesis of hematite (α-Fe2O3) nanostructures: Size effects on applications in thermal conductivity, catalysis, and antibacterial activity. *Journal of Molecular Liquids* 242: 537–549.

Saber, O. 2012. Novel self assembly behavior for γ-alumina nanoparticles. *Particuology* 10(6): 744–750.

Saif, S., Tahir, A., Asim, T. and Chen, Y. 2016. Plant mediated green synthesis of CuO nanoparticles: comparison of toxicity of engineered and plant mediated CuO nanoparticles towards Daphnia magna. *Nanomaterials* 6(11): 205.

Sangeetha, G., Rajeshwari, S. and Venckatesh, R. 2011. Green synthesis of zinc oxide nanoparticles by aloe barbadensis miller leaf extract: Structure and optical properties. *Materials Research Bulletin* 46(12): 2560–2566.

Santhoshkumar, T., Rahuman, A.A., Jayaseelan, C., Rajakumar, G., Marimuthu, S., Kirthi, A.V., Velayutham, K., Thomas, J., Venkatesan, J. and Kim, S.-K. 2014. Green synthesis of titanium dioxide nanoparticles using Psidium guajava extract and its antibacterial and antioxidant properties. *Asian Pacific Journal of Tropical Medicine* 7(12): 968–976.

Sathishkumar, G., Logeshwaran, V., Sarathbabu, S., Jha, P.K., Jeyaraj, M., Rajkuberan, C., Senthilkumar, N. and Sivaramakrishnan, S. 2018. Green synthesis of magnetic Fe3O4 nanoparticles using Couroupita guianensis Aubl. fruit extract for their antibacterial and cytotoxicity activities. *Artificial Cells, Nanomedicine, and Biotechnology* 46(3): 589–598.

Sathishkumar, M., Sneha, K., Won, S., Cho, C.-W., Kim, S. and Yun, Y.-S. 2009. *Cinnamon zeylanicum* bark extract and powder mediated green synthesis of nano-crystalline silver particles and its bactericidal activity. *Colloids and Surfaces B: Biointerfaces* 73(2): 332–338.

Sathishkumar, M., Sneha, K. and Yun, Y.-S. 2010. Immobilization of silver nanoparticles synthesized using *Curcuma longa* tuber powder and extract on cotton cloth for bactericidal activity. *Bioresource Technology* 101(20): 7958–7965.

Sathyavathi, R., Krishna, M.B., Rao, S.V., Saritha, R. and Rao, D.N. 2010. Biosynthesis of silver nanoparticles using Coriandrum sativum leaf extract and their application in nonlinear optics. *Advanced Science Letters* 3(2): 138–143.

Seifan, M., Ebrahiminezhad, A., Ghasemi, Y., Samani, A.K. and Berenjian, A. 2017. Amine-modified magnetic iron oxide nanoparticle as a promising carrier for application in bio self-healing concrete. *Applied Microbiology and Biotechnology* 102(1): 175–184.

Seifan, M., Ebrahiminezhad, A., Ghasemi, Y., Samani, A.K. and Berenjian, A. 2018a. The role of magnetic iron oxide nanoparticles in the bacterially induced calcium carbonate precipitation. *Applied Microbiology and Biotechnology* 102(8): 3595–3606.

Seifan, M., Sarmah, A.K., Ebrahiminezhad, A., Ghasemi, Y., Samani, A.K. and Berenjian, A. 2018b. Bio-reinforced self-healing concrete using magnetic iron oxide nanoparticles. *Applied Microbiology and Biotechnology* 102(5): 2167–2178.

Seifan, M., Sarmah, A.K., Samani, A.K., Ebrahiminezhad, A., Ghasemi, Y. and Berenjian, A. 2018c. Mechanical properties of bio self-healing concrete containing immobilized bacteria with iron oxide nanoparticles. *Applied Microbiology and Biotechnology* 102(10): 1–10.

Shah, M., Fawcett, D., Sharma, S., Tripathy, S.K. and Poinern, G.E.J. 2015. Green synthesis of metallic nanoparticles via biological entities. *Materials* 8(11): 7278–7308.

Shahwan, T., Abu Sirriah, S., Nairat, M., Boyacı, E., Eroğlu, A.E., Scott, T.B. and Hallam, K.R. 2011. Green synthesis of iron nanoparticles and their application as a Fenton-like catalyst for the degradation of aqueous cationic and anionic dyes. *Chemical Engineering Journal* 172(1): 258–266.

Shankar, S. and Rhim, J.-W. 2014. Effect of copper salts and reducing agents on characteristics and antimicrobial activity of copper nanoparticles. *Materials Letters* 132: 307–311.

Shankar, S.S., Rai, A., Ahmad, A. and Sastry, M. 2005. Controlling the optical properties of lemongrass extract synthesized gold nanotriangles and potential application in infrared-absorbing optical coatings. *Chemistry of Materials* 17(3): 566–572.

Shende, S., Gaikwad, N. and Bansod, S. 2016. Synthesis and evaluation of antimicrobial potential of copper nanoparticle against agriculturally important Phytopathogens. Synthesis 1(4): 41–47.

Sheny, D., Mathew, J., Philip, D. 2011. Phytosynthesis of Au, Ag and Au–Ag bimetallic nanoparticles using aqueous extract and dried leaf of Anacardium occidentale. *Spectrochimica Acta Part A: Molecular and Biomolecular Spectroscopy* 79(1): 254–262.

Si, S., Kotal, A., Mandal, T.K., Giri, S., Nakamura, H. and Kohara, T. 2004. Size-controlled synthesis of magnetite nanoparticles in the presence of polyelectrolytes. *Chemistry of Materials* 16(18): 3489–3496.

Singh, K., Panghal, M., Kadyan, S., Chaudhary, U. and Yadav, J.P. 2014a. Green silver nanoparticles of Phyllanthus amarus: as an antibacterial agent against multi drug resistant clinical isolates of Pseudomonas aeruginosa. *Journal of Nanobiotechnology* 12(1): 40.

Singh, P., Kim, Y.J., Wang, C., Mathiyalagan, R. and Yang, D.C. 2016. The development of a green approach for the biosynthesis of silver and gold nanoparticles by using Panax ginseng root extract, and their biological applications. *Artificial Cells, Nanomedicine, and Biotechnology* 44(4): 1150–1157.

Singh, P., Tiwary, D. and Sinha, I. 2014b. Improved removal of Cr (VI) by starch functionalized iron oxide nanoparticles. *Journal of Environmental Chemical Engineering* 2(4): 2252–2258.

Singh, P.P. and Bhakat, C. 2012. Green synthesis of gold nanoparticles and silver nanoparticles from leaves and bark of Ficus carica for nanotechnological applications. *International Journal of Scientific and Research Publications* 2(5): 1–4.

Sinha, S.N., Paul, D., Halder, N., Sengupta, D. and Patra, S.K. 2015. Green synthesis of silver nanoparticles using fresh water green alga Pithophora oedogonia (Mont.) Wittrock and evaluation of their antibacterial activity. *Applied Nanoscience* 5(6): 703–709.

Sirisha, N.G.D. and Asthana, S. 2018. Microwave mediated green synthesis of copper nanoparticles using aqueous extract of piper nigrum seeds and particles characterisation. *IAETSD Journal for Advanced Research in Applied Sciences,* 5(2): 859–870.

Sithara, R., Selvakumar, P., Arun, C., Anandan, S. and Sivashanmugam, P. 2017. Economical synthesis of silver nanoparticles using leaf extract of Acalypha hispida and its application in the detection of Mn(II) ions. *Journal of Advanced Research* 8(6): 561–568.

Sivaranjani, K. and Meenakshisundaram, M. 2013. Biological synthesis of silver nanoparticles using Ocimum basilicum leaf extract and their antimicrobial activity. *International Research Journal of Pharmacy* 4(1): 225–229.

Soliemanzadeh, A., Fekri, M., Bakhtiary, S. and Mehrizi, M.H. 2016. Biosynthesis of iron nanoparticles and their application in removing phosphorus from aqueous solutions. *Chemistry and Ecology* 32(3): 286–300.

Somchaidee, P. and Tedsree, K. 2018. Green synthesis of high dispersion and narrow size distribution of zerovalent iron nanoparticles using guava leaf (Psidium guajava L) extract. *Advances in Natural Sciences: Nanoscience and Nanotechnology* 9(3): 035006.

Song, J.Y., Jang, H.-K. and Kim, B.S. 2009. Biological synthesis of gold nanoparticles using Magnolia kobus and Diopyros kaki leaf extracts. *Process Biochemistry* 44(10): 1133–1138.

Subbaiya, R. and Selvam, M.M. 2015. Green Synthesis of Copper Nanoparticles from Hibiscus Rosasinensis and their antimicrobial, antioxidant activities. *Research Journal of Pharmaceutical Biological and Chemical Sciences* 6(2): 1183–1190.

Subhankari, I. and Nayak, P. 2013. Synthesis of copper nanoparticles using Syzygium aromaticum (Cloves) aqueous extract by using green chemistry. *World Journal of Nano Science and Technology* 2(1): 14–17.

Subramanian, V. and Suja, S. 2012. Green synthesis of silver nanoparticles using Coleus amboinicus lour, antioxidant activity and in vitro cytotoxicity against Ehrlich's ascite carcinoma. *Journal of Pharmaceutical Research* 5: 1268–1272.

Suganya, R., Priya, K. and Roxy, B. 2012. Phytochemical screening and antibacterial activity from Nerium oleander and evaluate their plant mediated nanoparticle synthesis. *International Research Journal of Pharmaceutical* 3: 285–288.

Sulaiman, G.M., Tawfeeq, A.T. and Jaaffer, M.D. 2018a. Biogenic synthesis of copper oxide nanoparticles using olea europaea leaf extract and evaluation of their toxicity activities: An in vivo and in vitro study. *Biotechnology Progress* 34(1): 218–230.

Sulaiman, G.M., Tawfeeq, A.T. and Naji, A.S. 2018b. Biosynthesis, characterization of magnetic iron oxide nanoparticles and evaluations of the cytotoxicity and DNA damage of human breast carcinoma cell lines. *Artificial Cells, Nanomedicine, and Biotechnology* 46(6): 1215–1229.

Suman, T.Y., Radhika Rajasree, S.R., Ramkumar, R., Rajthilak, C. and Perumal, P. 2014. The Green synthesis of gold nanoparticles using an aqueous root extract of Morinda citrifolia L. *Spectrochimica Acta Part A: Molecular and Biomolecular Spectroscopy* 118: 11–16.

Sun, J., Zhou, S., Hou, P., Yang, Y., Weng, J., Li, X. and Li, M. 2007. Synthesis and characterization of biocompatible Fe_3O_4 nanoparticles. *Journal of Biomedical Materials Research Part A* 80(2): 333–341.

Sun, Q., Cai, X., Li, J., Zheng, M., Chen, Z. and Yu, C.-P. 2014. Green synthesis of silver nanoparticles using tea leaf extract and evaluation of their stability and antibacterial activity. *Colloids and Surfaces A: Physicochemical and Engineering Aspects* 444: 226–231.

Sundrarajan, M., Ambika, S. and Bharathi, K. 2015. Plant-extract mediated synthesis of ZnO nanoparticles using Pongamia pinnata and their activity against pathogenic bacteria. *Advanced Powder Technology* 26(5): 1294–1299.

Sundrarajan, M. and Gowri, S. 2011. Green synthesis of titanium dioxide nanoparticles by Nyctanthes arbor-tristis leaves extract. *Chalcogenide Letters* 8(8): 447–451.

Suresh, D., Shobharani, R.M., Nethravathi, P.C., Pavan Kumar, M.A., Nagabhushana, H. and Sharma, S.C. 2015. Artocarpus gomezianus aided green synthesis of ZnO nanoparticles: Luminescence, photocatalytic and antioxidant properties. *Spectrochimica Acta—Part A: Molecular and Biomolecular Spectroscopy* 141: 128–134.

Suresh, Y., Annapurna, S., Bhikshamaiah, G. and Singh, A. 2016 Green luminescent copper nanoparticles. *IOP Conference Series: Materials Science and Engineering* 149: 012187. IOP Publishing.

Suresh, Y., Annapurna, S., Singh, A. and Bhikshamaiah, G. 2014. Green synthesis and characterization of tea decoction stabilized copper nanoparticles. *International Journal of Innovative Research in Science, Engineering and Technology* 3(4): 11265–11270.

Sutradhar, P., Debnath, N. and Saha, M. 2013. Microwave-assisted rapid synthesis of alumina nanoparticles using tea, coffee and triphala extracts. *Advances in Manufacturing* 1(4): 357–361.

Sutradhar, P., Saha, M. and Maiti, D. 2014. Microwave synthesis of copper oxide nanoparticles using tea leaf and coffee powder extracts and its antibacterial activity. *Journal of Nanostructure in Chemistry* 4(1): 86.

Taghizadeh, S.-M., Ghasemi, Y. and Ebrahiminezhad, A. 2017. Chemical and biological approaches for the synthesis of iron based nanoparticles. *Trends in Pharmaceutical Sciences* 3(4): 237–244.

Tahir, K., Nazir, S., Li, B., Khan, A.U., Khan, Z.U.H., Gong, P.Y., Khan, S.U. and Ahmad, A. 2015. Nerium oleander leaves extract mediated synthesis of gold nanoparticles and its antioxidant activity. *Materials Letters* 156: 198–201.

Tamuly, C., Hazarika, M. and Bordoloi, M. 2013. Biosynthesis of Au nanoparticles by Gymnocladus assamicus and its catalytic activity. *Materials Letters* 108: 276–279.

Tanghatari, M., Sarband, Z., Rezaee, S. and Larijani, K. 2017. Microwave assisted green synthesis of copper nanoparticles. *Bulgarian Chemical Communications*, Special Issue J, 347–352.

Thema, F.T., Manikandan, E., Dhlamini, M.S. and Maaza, M. 2015. Green synthesis of ZnO nanoparticles via Agathosma betulina natural extract. *Materials Letters* 161: 124–127.

Thilagam, M., Tamilselvi, A., Chandrasekeran, B. and Rose, C. 2013. Phytosynthesis of silver nanoparticles using medicinal and dye yielding plant of Bixa orellana L. leaf extract. *Journal of Pharmaceutical and Scientific Innovation* 2: 9–13.

Ullah, H., Wilfred, C.D. and Shaharun, M.S. 2018. Green synthesis of copper nanoparticle using ionic liquid-based extraction from *Polygonum minus* and their applications. *Environmental Technology* (just-accepted):118: 1–19.

Valodkar, M., Nagar, P.S., Jadeja, R.N., Thounaojam, M.C., Devkar, R.V., Thakore, S. 2011. Euphorbiaceae latex induced green synthesis of non-cytotoxic metallic nanoparticle solutions: A rational approach to antimicrobial applications. *Colloids and Surfaces A: Physicochemical and Engineering Aspects* 384(1–3): 337–344.

Vanaja, M. and Annadurai, G.J.A.N. 2013. Coleus aromaticus leaf extract mediated synthesis of silver nanoparticles and its bactericidal activity. *Applied Nanoscience* 3(3): 217–223.

Vanathi, P., Rajiv, P., Narendhran, S., Rajeshwari, S., Rahman, P.K.S.M. and Venckatesh, R. 2014. Biosynthesis and characterization of phyto mediated zinc oxide nanoparticles: A green chemistry approach. *Materials Letters* 134: 13–15.

Vanathi, P., Rajiv, P. and Sivaraj, R. 2016. Synthesis and characterization of Eichhornia-mediated copper oxide nanoparticles and assessing their antifungal activity against plant pathogens. *Bulletin of Materials Science* 39(5): 1165–1170.

Vanitha, V., Hemalatha, S., Pushpabharathi, N., Amudha, P. and Jayalakshmi, M. 2017 Fabrication of nanoparticles using Annona squamosa leaf and assessment of its effect on liver (Hep G2) cancer cell line. *IOP Conference Series: Materials Science and Engineering*: 191: 012010. IOP Publishing.

Vankar, P.S. and Shukla, D. 2012. Biosynthesis of silver nanoparticles using lemon leaves extract and its application for antimicrobial finish on fabric. *Applied Nanoscience* 2(2): 163–168.

Veerasamy, R., Xin, T.Z., Gunasagaran, S., Xiang, T.F.W., Yang, E.F.C., Jeyakumar, N. and Dhanaraj, S.A. 2011. Biosynthesis of silver nanoparticles using mangosteen leaf extract and evaluation of their antimicrobial activities. *Journal of Saudi Chemical Society* 15(2): 113–120.

Velavan, S., Arivoli, P. and Mahadevan, K. 2012. Biological reduction of silver nanoparticles using Cassia auriculata flower extract and evaluation of their in vitro antioxidant activities. *Nanoscience and Nanotechnology International Journal* 2(4): 30–35.

Velayutham, K., Rahuman, A.A., Rajakumar, G., Santhoshkumar, T., Marimuthu, S., Jayaseelan, C., Bagavan, A., Kirthi, A.V., Kamaraj, C. and Zahir, A.A. 2012. Evaluation of Catharanthus roseus leaf extract-mediated biosynthesis of titanium dioxide nanoparticles against Hippobosca maculata and Bovicola ovis. *Parasitology Research* 111(6): 2329–2337.

Vélez, E., Campillo, G., Morales, G., Hincapié, C., Osorio, J. and Arnache, O. 2018. Silver nanoparticles obtained by aqueous or ethanolic aloe Vera extracts: An assessment of the antibacterial activity and mercury removal capability. *Journal of Nanomaterials* 2018: 1–7.

Venkateswarlu, S., Rao, Y.S., Balaji, T., Prathima, B. and Jyothi, N. 2013. Biogenic synthesis of Fe_3O_4 magnetic nanoparticles using plantain peel extract. *Materials Letters* 100: 241–244.

Verma, A., Joshi, P., Arya, A. 2013. Synthesis of plant-mediated silver nanoparticles using plant extract of sonchus asper. *International Journal of Nanotechnology and Applications* 3(4): 11–18.

Vidya, C., Prabha, M.N.C. and Raj, M.A.L.A. 2016. Green mediated synthesis of zinc oxide nanoparticles for the photocatalytic degradation of Rose Bengal dye. *Environmental Nanotechnology, Monitoring & Management* 6: 134–138.

Vijayaraghavan, K., Nalini, S.K., Prakash, N.U., Madhankumar, D. 2012. One step green synthesis of silver nano/microparticles using extracts of Trachyspermum ammi and Papaver somniferum. *Colloids and Surfaces B: Biointerfaces* 94: 114–117.

Viola, A., Peron, J., Kazmierczak, K., Giraud, M., Michel, C., Sicard, L., Perret, N., Beaunier, P., Sicard, M. and Besson, M. 2018. Unsupported shaped cobalt nanoparticles as efficient and recyclable catalysts for the solvent-free acceptorless dehydrogenation of alcohols. *Catalysis Science & Technology* 8(2): 562–572.

Vivek, R., Thangam, R., Muthuchelian, K., Gunasekaran, P., Kaveri, K. and Kannan, S. 2012. Green biosynthesis of silver nanoparticles from Annona squamosa leaf extract and its in vitro cytotoxic effect on MCF-7 cells. *Process Biochemistry* 47(12): 2405–2410.

Vivekanandhan, S., Schreiber, M., Mason, C., Mohanty, A.K. and Misra, M. 2014. Maple leaf (Acer sp.) extract mediated green process for the functionalization of ZnO powders with silver nanoparticles. *Colloids and Surfaces B: Biointerfaces* 113(2014): 169–175.

Wang, C.-B. and Zhang, W.-X. 1997. Synthesizing nanoscale iron particles for rapid and complete dechlorination of TCE and PCBs. *Environmental Science & Technology* 31(7): 2154–2156.

Wang, T., Jin, X., Chen, Z., Megharaj, M. and Naidu, R. 2014a. Green synthesis of Fe nanoparticles using eucalyptus leaf extracts for treatment of eutrophic wastewater. *Science of the Total Environment* 466–467: 210–213.

Wang, T., Lin, J., Chen, Z., Megharaj, M. and Naidu, R. 2014b. Green synthesized iron nanoparticles by green tea and eucalyptus leaves extracts used for removal of nitrate in aqueous solution. *Journal of Cleaner Production* 83: 413–419.

Wang, Y.X., Hussain, S.M. and Krestin, G.P. 2001. Superparamagnetic iron oxide contrast agents: physicochemical characteristics and applications in MR imaging. *European Radiology* 11(11): 2319–2331.

Wang, Z. 2013. Iron complex nanoparticles synthesized by eucalyptus leaves. *ACS Sustainable Chemistry & Engineering* 1(12): 1551–1554.

Wang, Z., Fang, C. and Mallavarapu, M. 2015. Characterization of iron–polyphenol complex nanoparticles synthesized by Sage (*Salvia officinalis*) leaves. *Environmental Technology & Innovation* 4: 92–97.

Wang, Z., Fang, C. and Megharaj, M. 2014c. Characterization of iron–polyphenol nanoparticles synthesized by three plant extracts and their Fenton oxidation of azo dye. *ACS Sustainable Chemistry & Engineering* 2(4): 1022–1025.

Xiao, L., Mertens, M., Wortmann, L., Kremer, S., Valldor, M., Lammers, T., Kiessling, F. and Mathur, S. 2015. Enhanced in vitro and in vivo cellular imaging with green tea coated water-soluble iron oxide nanocrystals. *ACS Applied Materials & Interfaces* 7(12): 6530–6540.

Xie, J., Lee, J.Y., Wang, D.I. and Ting, Y.P. 2007a. Identification of active biomolecules in the high-yield synthesis of single-crystalline gold nanoplates in algal solutions. *Small* 3(4): 672–682.

Xie, J., Lee, J.Y., Wang, D.I. and Ting, Y.P. 2007b. Silver nanoplates: From biological to biomimetic synthesis. *ACS Nano* 1(5): 429–439.

Yallappa, S., Manjanna, J., Sindhe, M.A., Satyanarayan, N.D., Pramod, S.N. and Nagaraja, K. 2013. Microwave assisted rapid synthesis and biological evaluation of stable copper nanoparticles using T. arjuna bark extract. *Spectrochimica Acta Part A: Molecular and Biomolecular Spectroscopy* 110: 108–115.

Yang, L., Cao, Z., Sajja, H.K., Mao, H., Wang, L., Geng, H., Xu, H., Jiang, T., Wood, W.C. and Nie, S. 2008. Development of receptor targeted magnetic iron oxide nanoparticles for efficient drug delivery and tumor imaging. *Journal of Biomedical Nanotechnology* 4(4): 439–449.

Yang, N., WeiHong, L. and Hao, L. 2014. Biosynthesis of Au nanoparticles using agricultural waste mango peel extract and its in vitro cytotoxic effect on two normal cells. *Materials Letters* 134: 67–70.

Yilmaz, M., Turkdemir, H., Kilic, M.A., Bayram, E., Cicek, A., Mete, A. and Ulug, B. 2011. Biosynthesis of silver nanoparticles using leaves of *Stevia rebaudiana*. *Materials Chemistry and Physics* 130(3): 1195–1202.

Yu, W.W., Chang, E., Sayes, C.M., Drezek, R. and Colvin, V.L. 2006. Aqueous dispersion of monodisperse magnetic iron oxide nanocrystals through phase transfer. *Nanotechnology* 17(17): 4483–4487.

Yuvakkumar, R. and Hong, S. 2014. Green synthesis of spinel magnetite iron oxide nanoparticles. *Advanced Materials Research*, 1051: 39–42. Trans Tech Publ.

Yuvakkumar, R., Suresh, J., Nathanael, A.J., Sundrarajan, M. and Hong, S.I. 2014. Novel green synthetic strategy to prepare ZnO nanocrystals using rambutan (Nephelium lappaceum L.) peel extract and its antibacterial applications. *Materials Science and Engineering: C* 41: 17–27.

Zahir, A.A., Chauhan, I.S., Bagavan, A., Kamaraj, C., Elango, G., Shankar, J., Arjaria, N., Roopan, S.M., Rahuman, A.A. and Singh, N. 2015. Green synthesis of silver and titanium dioxide nanoparticles using Euphorbia prostrata extract showed shift from apoptosis to G0/G1 arrest followed by necrotic cell death in Leishmania donovani. *Antimicrobial Agents and Chemotherapy: AAC* 59: 4782–4799.

Zhuang, Z., Huang, L., Wang, F. and Chen, Z. 2015. Effects of cyclodextrin on the morphology and reactivity of iron-based nanoparticles using Eucalyptus leaf extract. *Industrial Crops and Products* 69: 308–313.

4

Cellulosic Nanomaterials

Subrata Mondal
National Institute of Technical Teachers'
Training and Research (NITTTR)

4.1 Introduction

In recent years, there has been significant research interest on the materials derived from bioresource for the diverse applications. Some of the well-known natural polymers include starch, alginate, chitosan, gelatin, and cellulose. Among these, cellulose is one of the most abundant natural polymers in the earth. Cellulose is derived from renewable and sustainable resources which are biodegradable and has low risks in health, safety, and environment. Cellulose is presented in plant-based materials viz. wood, cotton, hemp, flax, etc., and acts as a dominant reinforcing phase in the plant anatomy [1–4]. Cellulose is a natural homopolymer consisting of β-D-glucopyranose units which are linked together by $(1{\rightarrow}4)$-glycosidic bonds [5]. These units formed microfibrils of cellulose by side-by-side arrangement. The unit cells in a parallel conformation formed hydrogen bonding with adjacent chains. At present, there has been a growing research interest on cellulosic materials because of its sustainability, biodegradability, biosafety, and easy functionalization (to tailor surface properties) due to the presence of hydroxyl groups in its molecular structure [6].

Nanotechnology is the manipulation of matter at the dimension in the nanoscale, typically up to 100 nm, where unique properties provide a wide range of applications in novel areas [7]. Nanocellulose can be considered as a common name for the different type of cellulosic structures with at least one dimension $\leq$100 nm. Nanocellulose includes cellulosic nanofibers (CNF) and cellulosic nanocrystals (CNC) isolated from lignocellulosic biomass and also bacterial nanocellulose (BNC) synthesized by using certain strain of bacteria [8]. Generally, nanocellulose can be extracted from the cellulosic biomass due to its hierarchical and semicrystalline structure. Nanocellulose can be obtained from lignocellulosic materials by top-down approaches such as mechanical or chemical method or combination of mechanically and chemically induced deconstructing approaches [9]. Bacterial synthesis of nanocellulose can be considered as bottom-up approach of cellulosic nanomaterial synthesis.

The manufacturing of innovative nanomaterials from renewable bioresources is interesting and important, due to its lower impact on the environment [10]. Nanomaterials based on cellulose can be considered as sustainable nanosized materials because to its easy availability of source raw materials, nontoxicity, biodegradability, renewable in nature, etc. [11]. These novel forms of nanomaterials derived from abundant and renewable natural sources, therefore, has economic advantages in replacing synthetic nanomaterials [10]. In a top-down method of nanocellulose preparation, properties and yields of nanocellulose depend on the source of biomass and methods employed, whereas quality of bacteria-synthesized nanocellulose depends on cultural environment. This chapter presents an overview of various methods for the preparation of cellulosic nanomaterials, properties, and some of their major applications. The chapter concludes with some of the challenges of cellulosic nanomaterials.

4.2 Classifications of Cellulosic Nanomaterials

Cellulosic nanomaterials can be in various forms that have been given several names. Some of the common cellulosic nanomaterials include (a) nanofibrillated cellulose (NFC) from homogenized cellulose pulp, (b) cellulose nanocrystals (CNC) from acid hydrolyzed wood pulp, and (c) BNC from bacterial synthesis in a specific environment. NFC is the primary constituents of parent cellulose. CNC has lower aspect ratio as compared to NFC due to the partial hydrolysis of amorphous region of the cellulosic nanofibrils. On the other hand, BNC is synthesized by direct action of specific bacterial (such as Achromobacter, Alcaligenes, and Gluconacetobacter) strains in an artificial culture environment [11]. NFC and CNC can be prepared by top-down methods of nanomaterial preparation, whereas BNC preparation is by bottom-up approach of nanomaterial synthesis.

Cellulose nanocrystals derived from the lignocellulosic biomass by acid hydrolysis are elongated as rod-like or whisker-shaped particles. Typically, CNCs are crystals with diameter in the range of 3–20 nm and length can be up to 600 nm. In a top-down method, production of nanocellulose consists of two steps. In the first step, cellulosic raw materials can be purified by pretreatment which depends on the type of raw materials. The second step involved the individualization of cellulosic nanomaterials by using mechanical method, chemical method, or enzymatic hydrolysis [3].

4.3 Preparation of Cellulosic Nanomaterials

Nanocellulose can be prepared from the disintegration of plant cellulosic pulp or by specific bacterial synthesis in a controlled environment [11]. Nanocellulose can be obtained from various plant celluloses by using chemical, mechanical, and enzymatic methods or combination of more than one method. Whereas, in a bottom-up approach, nanocellulose can be synthesized in a controlled environment by using certain strain of bacteria.

4.3.1 Mechanical Methods

Cellulosic nanofibers (CNF) can be prepared by the disintegration of interfibrillar hydrogen bonding in the cellulose microfiber matrix under intense mechanical forces. By breaking amorphous domain or interfibrillar hydrogen bonding, the CNF can be obtained [12]. In a mechanical method, microfibrillated celluloses are prepared from cellulose fibers subjected to high mechanical shearing force. Afterwards, NFC can be prepared using microfibrillated cellulose by specific mechanical technique [3]. Wang et al. reported a process for the preparation of nanocellulose from eucalyptus (*Eucalyptus robusta* Smith) pulp by disintegrating the hydrogen bonded network structure of celluloses with high-pressure homogenization. The lateral dimension of extracted nanocellulose was 20–100 nm [13]. Prior to the mechanical process, certain pretreatments in the lignocellulosic biomass, it is possible to reduce

the energy requirement during nanofibrillation. Prior to mechanical processing, lignin from fibers can be removed by alkaline treatment and bleaching treatment, and this will help for separation of structural linkages between lignin and carbohydrates [2, 14, 15]. Energy consumption during mechanical disintegration can be reduced by enzymatic pretreatment due to the separation of wood fiber pulp into fibrillated nanofibers [16, 17]. Pretreatment with carboxymethylation and periodate oxidation of raw cellulose fibers before homogenization by using homogenizer produces highly nanofibrillated materials with significant concentration of aldehyde and carboxylic functionalities [18,19].

Ultrasound is widely used for accelerating various chemical reactions, including hydrolysis and depolymerization of cellulose-containing derivatives. Mishra et al. reported three different ways for the preparation of nanocellulose from native ultrasound-assisted 4-acetamido-TEMPO (2,2,6,6-tetramethyl-piperidin-1-oxyl)/NaBr/NaOCl system; the reported three approaches are (a) mechanical treatment in a blender, (b) with maximum intensity setting of an ultrasonic probe at 20 kHz, and (c) in various time span of an ultrasonic bath at 40 kHz, 600 W [20]. With similar treatment time, the subsequent ultrasound treatment yields higher nanocellulose than the subsequent mechanical treatment. Further, lateral dimension of cellulosic nanofibrils obtained by ultrasound-assisted treatment was lower compared to that produced by using mechanical treatment. The average diameter of nanocellulose, produced by ultrasound treatment, is 4.70 nm, whereas average diameter of nanocellulose produced by mechanical treatment is 5.51 nm. Transmission electron microscopic image revealed that nanocellulose produced by ultrasound method has more deformed structure than that obtained by mechanical treatment. Moreover, 10% more nanocellulose can be obtained from ultrasound-assisted TEMPO-oxidized pulp for the same duration of subsequent mechanical treatment compared to TEMPO-alone-oxidized pulp that was treated mechanically at the latter stage [20]. High-pressure homogenizer can be used for the fibrillation of cellulose in order to obtain cellulose nanofibers. Production of cellulosic nanofibrils (CNF) (Figure 4.1) in a high-pressure homogenizer is due to the combination of high shearing and impact forces which provides high degree of fibrillation for cellulose nanofibers [21]. Table 4.1 presents few methods, the source raw lignicellulosic biomass, and dimensions of nanocellulose produced by using various methods.

Steam explosion is a thermomechanical process of cellulosic nanomaterial preparation from the raw cellulosic materials. Steam penetrates into the matrix of cellulose fiber at high pressure through diffusion, and once the pressure is suddenly released, which creates shear force and subsequently hydrolyzes the glycosidic and hydrogen bonds, all these lead to formation of cellulose nanofibers [30,31]. Zhuo et al. reported an ecofriendly method with reduced energy consumption to obtained nanocellulose from shrub

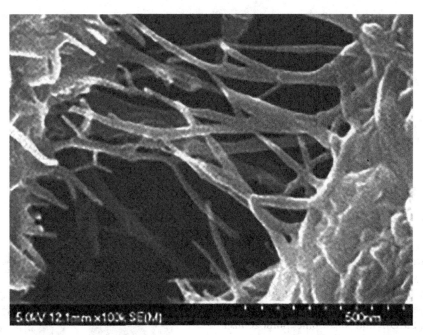

FIGURE 4.1 Scanning electron microscopic (SEM) image of cellulosic nanofibers (CNF) was obtained from alkali pretreated cellulose powder which processed at 1,400 bar pressure in high-pressure homogenizer of 12 passes. (Reproduced with permission from Ref. [21]. © 2012 The Korean Society of Industrial and Engineering Chemistry. Published by Elsevier B.V.)

TABLE 4.1 Characteristics of Cellulosic Nanomaterials Produced by Using Various Methods

Source	Method	Yield (%)	Width/diameter (nm)	Length (nm)	Ref.
Microcrystalline cellulose (MCC) with 20 μm mean diameter	TEMPO-mediated oxidation (TMO)	37.4 ± 8.3	40–80	200 nm to micrometers	[22]
MCC with 20 μm mean diameter	Acid hydrolysis	28.6 ± 8.0	30~40	200~400	[22]
"Spinifex", Triodia pungens	Optimized sulfuric acid hydrolysis	42	3.45 ± 1	497 ± 106	[23]
Cellulose powder	Planetary ball milling in the presence of ionic liquid	93.1	10–20	micrometer	[24]
MCC	Sulfuric acid hydrolysis	–	40–50	200–300	[25]
Pineapple leaf fiber	Steam explosion process coupled with acid treatment	–	5–60	200–300	[26]
Pinus radiata pulp fibers	Homogenization coupled with pretreatment by carboxymethylation	–	15 ± 0.8	–	[27]
Pinus radiata pulp fibers	Homogenization coupled with pretreatment by carboxymethylation and periodate	–	20 ± 0.8	–	[27]
MCC	Sulfuric acid hydrolysis	–	7	178	[28]
Cellulose powder	High-pressure homogenization	–	20–50	–	[21]
Licuri leaves	Sulfuric acid hydrolysis	33	5.7 ± 1.6	157 ± 24	[29]
Banana rachis	Acid hydrolysis (5% oxalic acid solution)	28.6	10–60	–	[10]
Sisal	Acid hydrolysis (5% oxalic acid solution)	38.8	20–80	–	[10]
Kapok	Acid hydrolysis (5% oxalic acid solution)	33.7	20–70	–	[10]
Pineapple leaf	Acid hydrolysis (5% oxalic acid solution)	40.1	50–150	–	[10]
Coir	Acid hydrolysis (5% oxalic acid solution)	23.5	40–90	–	[10]

plant, *Amorpha fruticosa* Linn., by a combined methods of grinding and homogenization. Extracted nanocellulose has a diameter of 10 nm with aspect ratio over 1,000, better thermal stability, and composed with a hydroxyl group and a crystal I structure [32].

4.3.2 Chemical Methods

Cellulosic fiber contains amorphous and crystalline regions which are oriented sequentially along the fiber direction, and the amorphous region separates nanofibrils in the cross section (Figure 4.2). Generally, crystalline part is hard to break due to the extensive hydrogen bondings between hydroxyl groups in the cellulose. However, it is relatively easy to break down the amorphous part of the raw cellulose (Figure 4.2). Therefore, nanocellulose can be obtained from the natural resources by pretreatment and subsequent breaking of amorphous part of cellulose. Acid hydrolysis is a well-known method to break down the amorphous region and extract the cellulosic nanocrystal (CNC) in the suspension. Thereafter, CNC can be separated by high-speed centrifugation [12,33]. Morphology of cellulose nanocrystals depends on the source of cellulosic raw materials and hydrolysis condition, such as concentration of acid, reaction time, temperature, and material-to-liquor ratio [34].

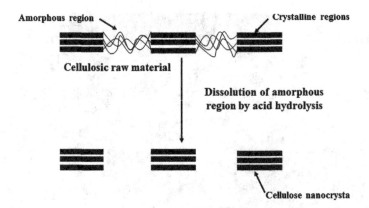

FIGURE 4.2   Schematic showing separation of cellulose nanocrystals from raw cellulose by acid hydrolysis.

Zhu et al. compared the properties of nanocellulose derived from microcrystalline cellulose (MCC) by using three approaches viz. acid hydrolysis using sulfuric acid, 2,2,6,6-tetramethylpiperidine-1-oxyl radical (TEMPO)-mediated oxidation (TMO), and ultrasonic homogenizer with 1,200 W power. Compared with ultrasonic and acid hydrolysis methods, good quality of nanocellulose can be obtained from the TMO approach. TMO method also provides higher yield than the other two approaches [22].

Niu et al. reported the influence of acid hydrolysis and ultrasonic treatment on dimension and morphology of nanocellulose. With increasing hydrolysis time, the radius of the nanocellulose decreases, whereas the crystallinity index of the nanocellulose increases. Furthermore, sulfate esters and sulfonate groups exist in the nanocellulose and formed more C-OH groups after hydrolysis [35].

Abraham et al. proposed a series of pretreatment methods prior to the acid hydrolysis in order to produce homogeneous dispersion of cellulosic nanofibers. Alkali pretreatment can remove some portion of lignin, hemicellulose, wax, and fat from the external surface of the fiber cell walls. Further, alkali treatment disintegrate native cellulose structure, separates the external cellulose microfibrils, and hence exposes short-length crystallites. Steam explosion can defibrillate the pretreated materials. Bleaching of steam-exploded fiber can be followed for the removal of remaining cementing materials. Then, the acid hydrolysis can be used to prepare cellulose nanofibrils. In this way, the nanofiber diameter can be reached to 5–50 nm when the raw lignocellulosic biomass came from banana, while it can be 15–25 nm when the raw lignocellulosic biomass is jute (Figure 4.3) [36].

4.3.3   Bacterial Synthesis of Nanocellulose

Nanocelluloses are generally extracted from lignicellulosic biomass by using mechanical, chemical, or combination of mechanical/chemical methods; however, several microorganisms such as bacteria, algae, and fungi can also synthesized nanocellulose naturally. Biosynthesized nanocellulose shows high purity with interesting physicochemical properties. A species of Gram-negative bacteria such as *Gluconacetobacter xylinus* can produce large amount of

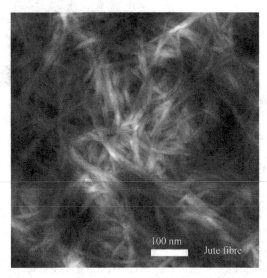

FIGURE 4.3   Scanning probe microscope of jute cellulosic nanofibrils dispersion. (Reproduced with permission from Ref. [36] © 2011 Elsevier Ltd.)

nanocellulose in a glucose solution [37,38]. In 1886, Brown observed that Acetobacter species bacteria can synthesized cellulose nanofibers by the fermentation of low-molecular-weight sugar [39]. The cellulose fibers synthesized by bacterial method are nanosized ribbon shape cellulosic fibrils with the largest lateral dimension from 25 to 86 nm and length up to few micrometers [40,41]. Figure 4.4 shows bacterial cellulose nanofibers and nanofibrillated cellulose [40].

BNC of highly hydrated fibrils (99% water) with improved mechanical strength can be synthesized by the bacterium *Gluconacetobacter xylinu* [42]. BNC film has great potentials for the biomedical application. The formation of uniform BNC films leads to a wide range of applications as biomaterials. Fu et al. presented BNC biosynthesis by *G. xylinus*. The microenvironment of culture medium affects the structure of BNC and ultimately surface properties of BNC. In a typical BNC synthesis, *G. xylinus* can be inoculated at 30 °C in a static culture medium with concentration (w/v%) of glucose, yeast extract powder, peptone, disodium phosphate, and citric acid, 2, 0.5, 0.5, 0.27, and 0.115, respectively [43].

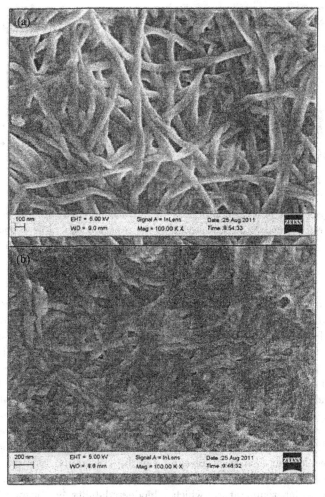

FIGURE 4.4 Scanning electron micrographs of (a) bacterial cellulosic nanofibers and (b) nanofibrillated cellulose manufactured by grinding of wet bleached birch kraft pulp (*Betula pendula*). (Reproduced with permission from Ref. [40]. © 2012 American Chemical Society.)

4.4 Surface Modification

Cellulose has a strong affinity to substances which contain hydroxyl groups or itself by hydrogen bonding. Based on the preponderance of hydroxyl groups, cellulose has high affinity with water [7]. Surface modification is an important method to tailor the surface morphology and properties of nanocelluloses. Modification of nanocellulose surface can be carried out with polar and non-polar molecules in order to accommodate it with wide varieties of matrices [30]. Abundancy of hydroxyl groups in the nanocellulose provides opportunities to the chemist to surface functionalize the nanocellulose surface by using different approaches such as esterification, etherification, oxidation, silylation, and polymer grafting. Apart from these covalent functionalization approaches, noncovalent surface modification such as the use of absorbing surfactants and polymer coatings are possible as well. One of the major challenges for the surface modification is to only change the surface characteristics while maintaining the original morphology in order to intact the integrity of the crystal [44,45]. Nanocellulose can also be modified by physical method. Some of the physical methods for the surface modification of nanocellulose are surface fibrillation, plasma treatment, ultrasonic, irradiation, electric currents, etc. Plasma treatment is an environment-friendly surface modification method of nanocellulose. Plasma treatment can introduce various functional groups on the nanocellulose surface depending on the types of plasma used [44].

Hydrophobicity of cellulose nanocrystals can be improved by the chemical modification of nanocellulose with hydrophobic molecules. Wei et al. proposed the functionalization of CNC by transesterification with canola oil fatty acid methyl ester (CME). Surface modification did not alter the crystal size and crystallinity of CNC, while hydrophobicity measured by water contact angle has been improved as compared to the control sample. Modified CNC with CME can be used as potential nanomaterials for hydrophobic coatings and reinforcing nanofillers to hydrophobic polymer to enhance the interaction between fillers and matrix during the fabrication of polymer nanocomposites [46].

Reactive vinyl groups can be functionalized on the nanocellulose surface by using several approaches such

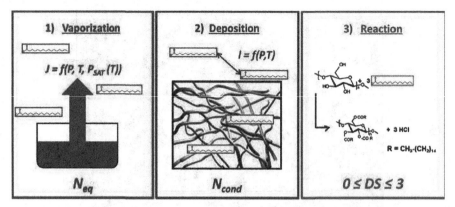

FIGURE 4.5 Various steps of gas-phase esterification process (J: vaporization flux, l: mean free path, P: reactor pressure, P_{SAT}: saturation pressure, T: temperature. N_{eq} and N_{cond}, correspond to the reagent molar ratio to anhydroglucose unit introduced and deposited. DS: number of hydroxyl reacted per anhydroglucose unit). (Reproduced with permission from Ref. [49]. © 2015, Springer Science Business Media Dordrecht.)

as direct esterification with oleic acid, and linseed or sunflower oil fatty acids, amidation of maleic acid/ethylene diamine with methyl ester of fatty acid. Surface-modified nanocellulose can be used to fabricate high-performance nanocomposite to increase the stress at break up to 90% [47]. Espino-Pérez et al. described an environment-friendly method for the carboxylic acid (CA) surface functionalization of nanocellulose. In the solvent-free esterification process, two nontoxic CAs, viz. phenylacetic acid and hydrocinnamic acid, were used. In this green process, CA acts not only as grafting molecules but also as a solvent media above their melting point. CA was added at a temperature higher than its melting point temperature which should be higher than the boiling point of water. This ensures the efficient esterification due to water distillation. Further, moderate temperature do not degrade reactant and the same can be easily recycled [48]. Fumagalli et al. proposed green gas-phase surface-restricted modification of nanocellulose-based aerogel by using fatty acid derivatives (Figure 4.5). In the gas-phase surface modification process, the reagents first required to be vaporized and afterwards deposited on the substrate surface, followed by esterification reaction [49].

4.5 Properties of Nanocellulose

Nanocellulose has many unique properties which include nanosized dimension (hence, very high surface area per volume/mass), better mechanical properties, biodegradability, transparency, and renewability. Physical properties of nanocellulose would depend on the source of raw materials and extraction process. Amiralian et al. reported the extraction of nanocellulose with high aspect ratio from Spinifex grass by using controlled sulfuric acid hydrolysis with crystallinity for the nanocellulose of 73%; further, high hemicellulose content (42%) provides lower traverse stiffness [23]. Crystallinity index and crystal size of the nanocellulose depend on the source of raw lignicellulosic biomass when the same method for the nanocellulose extraction would be

employed. Deepa et al. presented that crystallinity index varies from 81% to 92%, while crystal size varied from 3.4 to 3.7 nm when nanocellulose is extracted from various biomasses such as sisal, kapok, banana rachis, pineapple leaf, and coir by using the combination of chemical treatments which include alkaline treatment, bleaching, and acid hydrolysis (Figure 4.6). Further, isolated nanocellulose has higher thermal stability than corresponding raw lignocellulosic biomass [10]. Nanocellulose has Young's modulus of 100–130 GPa, and surface area of several 100 m^2/g can be considered as new promising reinforcement due to its reinforcing capability, abundance, low density, and biodegradability [9]. Suspension of nanocellulose extracted from cotton linters had a white gel-like appearance (Figure 4.7a), while nanocellulose bundles have agglomerated dispersed crystallites and individual crystallites (Figure 4.7b) [50].

4.6 Applications of Cellulosic Nanomaterials

Due to its multifunctional properties including inherent renewability, low density, easy availability, optical transparency, flexibility, and excellent mechanical, thermal, and physicochemical properties, nanocellulosic materials attract considerable attention for potential applications in various areas such as nano-biocomposite, biomaterials, electronic, energy, cosmetic, and food products [51,52].

4.6.1 Nanocellulose-Reinforced Polymer Nanocomposites

Composite materials consist of two or more phases; all phases maintain their physical identity while remaining bonded to each other by physical and/or chemical means. The major portion of the composite is known as matrix phase, whereas minor portion is referred to as reinforcement phase or dispersed phase. In the nanocomposite, the

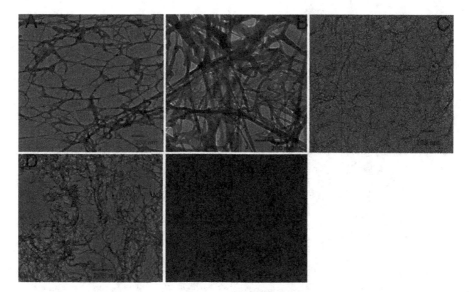

FIGURE 4.6 Transmission electron microscopic image of nanocellulose extracted from (a) banana rachis, (b) sisal, (c) kapok, (d) pineapple leaf, and (e) coir by using combined methods of chemical treatments which include alkaline treatment and bleaching followed by acid hydrolysis (5% oxalic acid solution). (Reproduced with permission from Ref. [10]. © 2015, Springer Science Business Media Dordrecht.)

reinforcement phase is typically nanomaterials with at least one dimension in the nanoscale range, i.e., ≤ 100 nm. In polymer nanocomposites, polymer is a matrix phase, while various nanosized organic or inorganic fillers with particular shape, size, and surface functional properties are nanoreinforcements [53]. Both degradable and nondegradable polymers can be used as matrix phase in the polymer nanocomposites. Apart from cellulosic nanofillers, some of the well-known nanoreinforcements for the fabrication of polymer nanocomposites are carbon nanotube, graphene, carbon nanofiber, clay nanomaterial, titanium dioxide nanoparticles, etc. Nanocellulose is renewable nanoreinforcing materials which can be used for the manufacturing of high-performance nano-biocomposite [40,54]. Nanocellulose is a promising renewable reinforcement with high specific surface area which provides interaction between filler and matrix, the most important parameter to control for the applications of nanocomposite [7]. Tensile strength and Young's modulus of polyvinyl alcohol (PVA) film first increase up to 4% (w/w) of nanocellulose fibrils reinforcement and then decreases (Figure 4.8). Improvement of mechanical properties of the PVA film is due to the strong reinforcing effect of nanocellulose and PVA matrix. Decline of mechanical properties after certain concentration of nanocellulose is possibly due to the agglomeration of nanofillers. The aggregation of NCFs in PVA caused local concentrations of stress, and tensile failure would occur readily during the testing stage [55]. Changes of mechanical properties for the various nanocellulose-based nanocomposites are tabulated in Table 4.2.

Solid plastic waste disposal is nowadays a global issue, and synthetic plastic derived from petroleum-based feedstock is widely used for the various consumer products and these are nondegradable. One such promising

application of plastic-based product is that it is used as packaging material. Global environmental impact due to the use of non-degradable packaging materials derived from petroleum-based raw materials has encouraged researchers in academic and industries to develop degradable or partially degradable packaging materials. Bio-based packaging materials for foods is gaining momentum due to its ecofriendly nature and its potential for the improvement of food quality and safety during packaging. Nanocellulose incorporated composites are promising for food packaging industries due to their improved mechanical and barrier properties together with their role as the carrier of bioactive substances [53]. Nanocellulose-reinforced polymer nanocomposite has excellent barrier properties due to the nanoscale reinforcements with its high crystalline nature, and abundant hydroxyl groups to form hydrogen bonds which formed strong network structure; hence, it is very hard for the various mass molecules to pass through it [63,64]. The mass molecules penetrate much more slowly through the CNF/CNC-reinforced film due to an increase in the tortuosity factor (Figure 4.9). CNCs with higher crystallinity together with their ability to form a network structure by hydrogen bonding which can improve gas barrier property [65].

Nanocellulose becomes important reinforcement due to the improvement of overall performance of nanocomposites. Apart from the food packaging, nanocellulose-reinforced nanocomposites with thermoplastic and thermosetting polymers can be used in packaging of consumer products, construction, automotive, electronic, furniture, etc. [30]. Xu et al. presented nanocellulose fiber (CNF) and graphene oxide (GO)-reinforced PVA film for the packaging application. Moisture permeability of PVA film decreased from 164.2% to 98.8% with the incorporation of 8% of CNF,

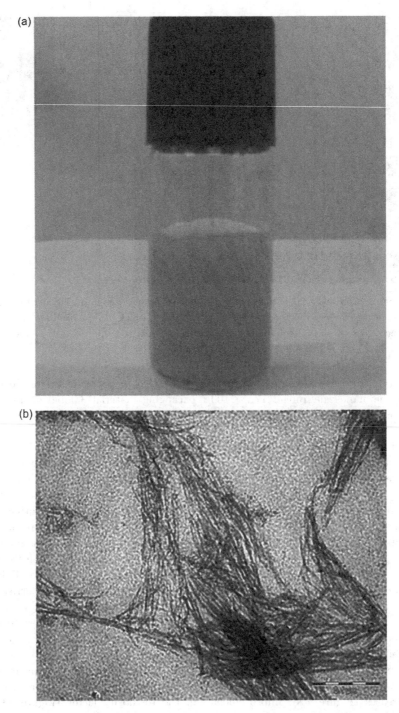

FIGURE 4.7 (a) Nanocellulose suspension and (b) transmission electron microscopic image of cotton linter cellulosic nanowhiskers. (Reproduced with permission from Ref. [50]. © 2012 Elsevier Ltd.)

and further, the moisture permeability decreased to 91.2% with the reinforcement of 0.6% of GO, while the oxygen permeability of PVA film decreased to ~12% and 76% with the incorporation of 8% CNF and 8% CNF with 0.6% GO, respectively [67]. Abraham et al. studied the biodegradation of cross-linked and non-cross-linked nanocellulose-reinforced natural rubber nanocomposites. Biodegradation of composites depends on the weight percentage of nanocellulose reinforcement. With an increasing nanocellulose percentage in the nanocomposite, the composite shows improved

biodegradation. Generally, biodegradation of composite begins with the accumulation of microbes on the periphery of nanocellulose in the composites. Hence, biodegradation starts from the nanocellulose and then goes to the matrix materials which usually needs more time for the degradation [68].

Ben Mabrouk et al. reported nanocellulose polymer nanocomposite, fabricated by miniemulsion polymerization of acrylic monomers in presence of nanocellulose and a coupling agent, methacryloxypropyl

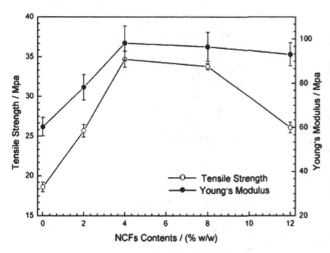

FIGURE 4.8 Tensile strength and Young's modulus graph of PVA/NCF nanocomposites. (Reproduced with permission from Ref. [55]. © 2013, Springer Science Business Media Dordrecht.)

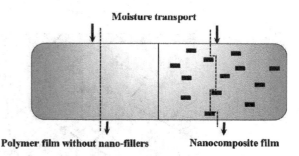

FIGURE 4.9 Schematic showing barrier property of cellulose nanoreinforcement polymer nanocomposite. (Reproduced with permission from Ref. [66]. © 2018 Taylor & Francis.)

trimethoxysilane (MPMS), in order to efficiently disperse nanofiller in the nanocomposite matrix. Silane in MPMS can react via methacrylic moiety with styrene (Sty) and 2-ethyl hexylacrylate (EHA), and as a result, alkoxysilane functional copolymer is formed. Alkoxysilane functional copolymer further hydrolyzed to form silanol-functionalized polymer particles. Cellulose nanocrystal (CNC) absorbed on the polymer particles through hydrogen bonding of silanol groups. Water evaporation at the time of film formation, physical interaction turned into chemical linkage by condensation reaction of silanol groups of MPMS and hydroxyl groups of CNC, and this ensures the homogeneous dispersion/distribution and proper interaction of nanofillers with the matrix polymer (Figure 4.10) [69].

4.6.2 Biomedical Applications

Nanocellulose materials have evolved as promising biomaterials in recent years because of their excellent physicochemical properties together with their biocompatibility, biodegradability, and low cytotoxicity. Nanocellulosic materials are widely applied for various medical applications including tissue engineering, drug delivery, wound-healing, cardiovascular applications, medical implants, 3D printing, etc. [6,70–72]. Three-dimensional bioprinting could revolutionize the tissue engineering and

regenerative medicine. Markstedt et al. reported cellulose nanofibers containing ink with the rapid cross-linking ability of alginate for the 3D bioprinting of living soft tissue with cells. Before cross-linking, the properties of NFC, e.g., shear thinning were dominating the bioink to obtained the desired shape with improved printing resolution as compared to 3D printing with pure alginate. After crosslinking, the property of alginate has been dominated, hence storage modulus has been increased due to the higher amount of alginate in the bioink [73]. Nanocellulose fibrils-alginate bioink can be suitable for bioprinting-induced pluripotent stem cells (iPSCs) to support cartilage production in co-cultured irradiated chondrocytes [74]. Saidi et al. presented non-cytotoxic and pH-sensitive nanocomposite membranes based on BNC and a polymer with amino acid pending moieties. The nanocomposite membrane shows pH-sensitive retention of selective drugs at pH 2.1 and released at pH 7.4, demonstrating their capability of controlled drug delivery in dermal as well as in oral applications [75]. Hydrogel and aerogel based on nanocellulose could be applicable in wide areas such as biomedical filed, 3D printing, energy storage, construction, separations, cosmetic and food related applications [76,77]. Leppiniemi et al. discussed nanocellulose-alginate hydrogel for 3D printing. Cross-linking of hydrogel by calcium ions improved mechanical properties and the hydrogel absorbed water in moist condition, suggesting potential for the wound dressing. Tissue compatible, bioactivated nanocellulose-alginate hydrogel could be used as biomedical device, wearable sensors, and materials for drug-release [77]. Nanocellulose-reinforced PVA hydrogel with excellent water content exhibits collagen-like

TABLE 4.2 Properties of Nanocellulose-Reinforced Polymer Nanocomposites

Matrix Material	Reinforcement Concentration	% change of Tensile Strength	% change of Modulus	% change of Elongation	Ref.
High-density polyethylene	3% (cellulose nanofibrils)	+40	+50	−20	[56]
Poly(vinyl alcohol)– Chitin	1 wt% Nanocellulose	+57.64	+50.66	–	[57]
Unsaturated polyester	0.5 wt% nanocellulose	+57%	–	–	[58]
Poly(vinyl alcohol)	5 wt% nanocellulose	+28%	+60%	–	[28]
Polyvinyl alcohol (PVA)/ starch blend	5% (v/v) nanocellulose	+48.7%	–	+30.54%	[59]
PVA	1% (w/w) nanocellulose (prepared by 1.5 M HBr)	+49%	+43.8%	–	[60]
PVA	4% (w/w) nanocellulose fibrils	+86%	+63%	–	[55]
PVA	6% (w/w) nanocellulose fibrils	+180%	+140%	–	[61]
Linear PVA	7.5 (w/w%) nanocellulose	+48%	–	+33%	[62]

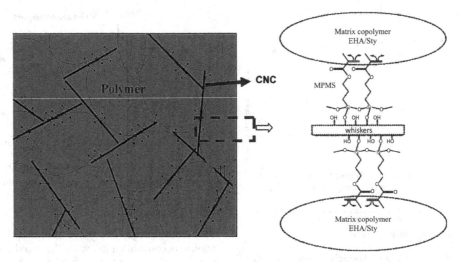

FIGURE 4.10 Schematic structure of nanocomposite by in situ miniemulsion polymerization [EHA: 2-ethyl hexylacrylate, Sty: styrene, MPMS: methacryloxypropyl trimethoxysilane]. (Reproduced with permission from Ref. [69]. © 2014 Elsevier B.V.)

mechanical property specifically required for soft tissues. Hydrogel shows strain-induced stiffening desired in stimulating the mechanical properties of collagenous soft tissues. Further, transparency (90%–95% at 550 nm) and excellent elastic properties (250%–350% strain) extend its applications in ophthalmology [78]. Nanocellulose as an excellent nanomaterial has potentiality for wide biomedical applications due to its (a) capability to self-assemble into 3D porous structure, (b) excellent capacity to maintain structural integrity in a moist environment, and (c) tailorable surface chemistry due to the abundancy of hydroxyl groups which provide opportunity for the modification with other suitable functionality. Chinga-Carrasco and Syverud reported pH-sensitive hydrogel based on nanocellulose (Figure 4.11) for potential application in wound dressing. Three-dimensional microporous structure of oxidized nanocellulose with poly-anionic surface of ionizable functional groups shows excellent swelling capacity and pH-responsive characteristics. On demand swelling degree with pH of the environment could find potential applications in chronic wound for controlled and smart release of active components into the biofilm. Nanocellulose gel has significantly higher swelling degree at neutral and alkaline condition than the acidic condition, and this will allow smart release of antibacterial component into the biofilm. Carboxylic functionality plays a prominent role in swelling of hydrogel. At higher pH, COO^- predominates which allows ionic repulsion between nanofibrils; hence, hydrogels swelled [27]. Cellulose nanocrystal-reinforced chitosan-based semi-interpenetrating polymer network hydrogel shows swelling which sensitive to pH. Hydrogel exhibited maximum swelling ratio in the acidic condition (pH 4.01). Improved mechanical properties with excellent pH sensitivity will broaden their application in several biomedical fields' viz. tissue engineering, pharmaceuticals, and drug delivery [25].

Nanocellulose can be used for wound-dressing applications due to its capability to form translucent film with excellent absorption capabilities. Adequate management of exudates is required for non-healing and chronic wounds. Moreover, the translucency will allow the development of wound without the requirement to remove the dressing from the wound surface [79].

Scaffold for tissue engineering should have tailorable biodegradability proportionate to the regeneration of new tissues, high porosity, interconnected pores, biocompatibility, and good mechanical properties. Ultrasonically suspended cellulose nanofibers (CNFs) with biodegradable poly(D, L-lactide-*co*-glycolide) (PLGA) can be used for the fabrication of 3D porous tissue engineering scaffold. The morphology, porosity, and mechanical properties of the scaffold can be engineered by the mass ratio of CNFs and PLGA. Mechanical properties of the scaffold can be improved by CNF reinforcement and comparable to that of cartilage tissue [80]. Bacterial nanocellulose (BNC) can be produced by *Gluconacetobacter xylinum*, and the BNC optimization can be achieved by optimum pH value and shaking rate. Nanostructure of BNC includes nanofibers up to 60 nm and mat with nanoporosity up to 265 nm. Appropriate mechanical strength and Young's modulus resembled BNC scaffold utilization for skin tissue [81]. BNC with certain concentration of cellulose is promising non-resorbable biomaterials for auricular cartilage tissue engineering due to the resemblance of auricular cartilage in terms of host tissue response and its mechanical property [42]. Composite scaffold with a dense nanocellulose layer on the surface of macroporous composite formed by of nanocellulose and alginate has been proposed for the tissue engineering of auricular cartilage. The bilayer BNC scaffolds offer favorable mechanical properties and structural integrity, and support cell ingrowth [82]. Nanocellulose-based biocomposites are nontoxic and biocompatible (Figure 4.12) and potential for biomedical applications which include soft tissue engineering and wound dressing [83].

Ferraz et al. presented polypyrrole (PPy)-Cladophora nanocellulosic membrane as potential candidate for the

FIGURE 4.11   Carboxymethylated pretreated structures after freeze-drying. (a) The images showing freeze-dried structures without (left) and with Ca^{2+} (right). (b) Field emission scanning electron microscopic (FESEM) image of the microporous structure of the pure carboxymethylated nanocellulose structure. (c) FESEM shows microporous structure of the ionic bonded with Ca^{2+} carboxymethylated nanocellulose (Reproduced from Ref. [27] (Open Access). © 2014 SAGE Publications).

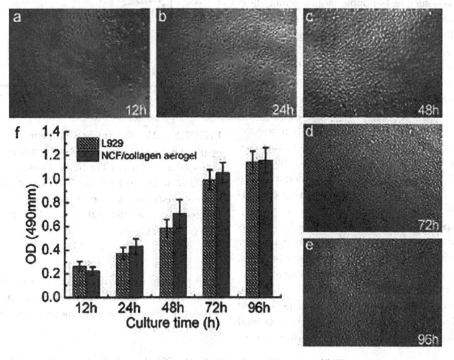

FIGURE 4.12   (a–e) Optical microscopic images of *in vitro* L929 cells proliferation. (f) The result of cytotoxicity test by MTT assays for nanocellulose composite aerogel for wound dressing and tissue engineering applications. (Reproduced with permission from Ref. [83]. © 2014 Elsevier Ltd.)

hemodialysis (HD). The benefits of using PPy-Cladophora nanocellulose HD membranes are as follows: (a) reduction of HD time by rapid, current-controlled removal of water-soluble ions, (b) tunable pore size distribution with highly porous structure that enables solute removal by combined ultrafiltration and electrochemically controlled process, and (c) potential of simultaneous electrochemically controlled release of drug [84].

4.6.3 Electronic- and Electrical-Related Applications

Nanocellulose in conducting polymer composite nanorod film can be synthesized by in situ polymerization. Electrochromic property of the composite film has been significantly improved because of the enhanced diffusion of electrolyte ions due to the space availability between nanocellu-lose/conducting polymer nanorods. Further, large surface area of the nanocomposite structure offered redox reactions. Nanocellulose-reinforced conductive composite films exhibited a higher color contrast, largest coloration efficiency, shortest response time, and best cycling stability required for chromic materials. All these properties of composite film make them very good candidate for the chromic material [85–87]. Rod-like core-shell structure of nanocellulose/polyaniline (PANI) nanocomposites can be synthesized via in situ polymerization. Nanocomposites show superior film-forming properties with superior electrochemical and electrochromic properties in electrolyte solution to be used as promising novel electrochromic materials [88].

Flexible supercapacitors can be used in various flexible electronic applications such as bendable mobile phone, flexible display, and wearable devices. Several excellent properties of BNC including excellent mechanical properties, low cost, and high porosity make BNC an ideal candidate for the manufacturing of flexible supercapacitors. Flexible supercapacitor can be prepared by interlocking GO flakes within nanocellulose network during BNC growth. The capacitors exhibit excellent energy storage performance with mechanical flexibility [89]. Wang et al. reported carbon aerogel fabricated by the pyrolysis of freeze-dried BNC in nitrogen atmosphere for potential as anode in lithium anode batteries. The carbon aerogel shows super capacity retention and rate performance (Figure 4.13) as compared to the other carbon-based materials [90].

Zheng et al. proposed fabrication of bio-based and low-cost PANI-based flexible electrodes with cellulose nanofiber (CNF) reinforcement for high-performance supercapacitors. Hybrid PANI electrode shows improved specific capacitance, good rate capability, and energy/power density balance. CNF binders in the composite substrate provide mechanical flexibility with enhanced electrochemical performance and capacitance retention in repeated bending conditions over 1,000 cycles [91].

Nanocellulose-reinforced polymeric ionic liquid, i.e., poly(methacroylcholine chloride) (PMACC), can be used to fabricate bio-based anion-exchange membranes.

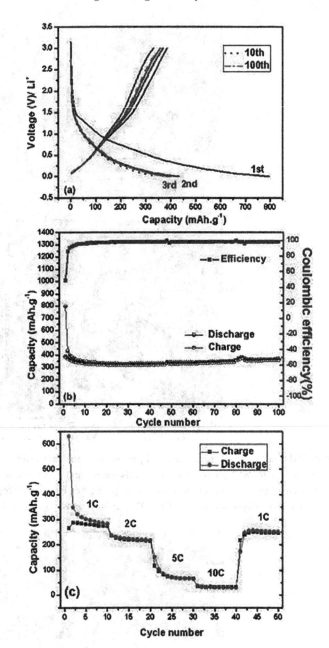

FIGURE 4.13 Electrochemical performance of carbon aerogel prepared from freeze-dried BNC and pyrolyzed at 900°C in nitrogen atmosphere for 2 h. (a) The trend of charge–discharge at a current rate of 75 mA/g (0.2 C). (b) Variation of capacity retention and coulombic efficiency with cycle number and (c) rate performance. (Reproduced from Ref. [90] (Open Access). © 2014 The Royal Society of Chemistry.)

Strongly humidity-dependent through-plane anionic conductivity of the membranes makes it potential as anion separator nanocomposite membranes for fuel cells and related areas [92].

4.6.4 Water Treatment

Nanocellulose with high specific surface area, high strength, chemical inertness, renewable character, and tailorable surface chemistry offered enormous opportunity for the water treatment [93]. Trimethyl-ammonium functionalized cellulose nanofibers (CNF) possess positive charge on the surface over a wide range of pH and ionic strength. Surface functionalized CNF shows excellent sorption and desorption capability for humic acid; therefore, CNF filter could be used for the separation of humic acid for multiple filter utilization [94]. Xu et al. presented black wattle (BW) tannin and nanocellulose based nanocomposite for the removal of heavy metal ions from aqueous solutions. In order, to prepare nanocomposite, at first nanocellulose has been oxidized with sodium periodate to obtain dialdehyde nanocellulose (DANC). Afterwards, BW tannin covalently immobilized onto DANC, which was used as both the matrix and cross-linker. Nanocomposite shows excellent adsorption efficiency of heavy metals, with a maximum adsorption capacity of nanocomposites of 51.846 mg/g, 53.371 mg/g, and 104.592 mg/g for Cu(II), Pb(II), and Cr (VI), respectively [95]. Zhu et al. reported crystalline nanocellulose/poly(vinyl alcohol-*co*-ethylene (PVA-*co*-PE)) nanocomposite absorptive membrane functionalized with 1, 2, 3, 4,-butanetetracarboxylic acid (BTCA) for the removal of heavy metal ions such as Pb (II), and the mixture of Pb (II), Cr (VI), Mn (II), and Cu (II). In order to improve

the absorptivity, the BTCA-modified membrane has been activated by NaHCO$_3$. The absorptive membrane exhibits excellent absorption capacity of single metal ion and metal ions mixture, and the equilibrium absorption could reach up to ~470 mg/g at 15°C. Further, the membranes are stable and reusable [96]. Nanocellulosic membranes can be fabricated by tailoring the support layer of sludge microfibers/cellulose nanofibers with a thin layer of cellulose nanocrystal by in situ TEMPO functionalization with high water permeability, mechanical stability, and improved the adsorption capacity of metal ions. In situ TEMPO functionalizations of membrane increase the adsorption capacity to approximately 1.3- and 1.2-fold for Cu(II) and Fe(II)/Fe(III), respectively. Membranes are stable in model wastewater, while completely degradable after 15 days in soil [97]. Nanocellulose can be used as potential biosorbents for the removal of insecticide from the wastewater. Moradeeya et al. reported nanocellulose-based biosorbents to remove an insecticide, chlorpyrifos (CP), from aqueous solutions, with an efficiency of 99.3% at 1.5 mg/l of CP [98]. Korhonen et al. proposed floating absorbent based on aerogel prepared from the cellulosic nanofibrils by using freeze-drying method (Figure 4.14). Native cellulose nanofibrils of aerogel can be functionalized with hydrophobic

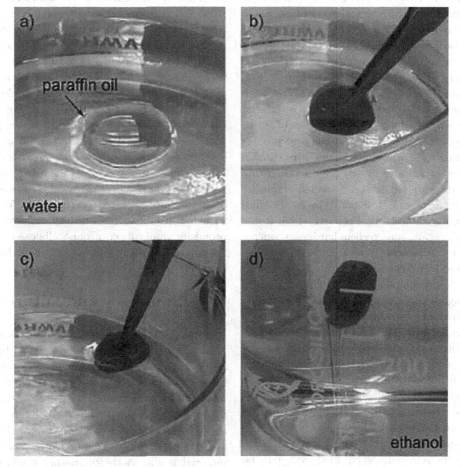

FIGURE 4.14 Oil spill removal from water by surface functionalized nanocellulosic aerogel: (a) floating of paraffin oil on water, (b) oil absorption by aerogel, and (c) complete floating oil absorbed aerogel. (d) The oil-filled aerogel can be recovered by immersing it in ethanol. (Reproduced with permission from Ref. [99]. © 2011, American Chemical Society.)

but oleophilic coating of titanium dioxide. Surface-modified aerogel allows to absorb organic contaminant from the water surface. The adsorbents are reusable, recyclable, or can be incinerated after oil adsorption, and environment-friendly as these are composed of nonhazardous titanium dioxide and cellulose nanofibrils [99].

Protein-based pharmaceutics are extensively used in healthcare applications. For the virus removal and bioprocessing, paper-based filtration is widely used, while fouling is a serious issue. Inherent inertness of nanocellulose exhibits low affinity of proteins on the nanocellulose surface, and presents nanocellulose-based paper filter an ideal candidate for the applications in advanced separation processes such as virus removal and bioprocessing [100].

4.6.5 Other Applications

Thermal insulation plays an important role to make energy-efficient building. Insulation properties of materials can be improved after the incorporation of nanocellulose. Wicklein et al. reported freeze casting suspensions of cellulose nanofibers (CNF), GO, and sepiolite (SEP) nanorods-based structure, which produces excellent insulating, fire retardant, and strong anisotropic foams, and these foams perform better than conventional polymer-based insulating materials for the building applications. Thermal conductivity for the radial orientation of CNF foam is as low as 18 mW/m K, and introduction of 10 wt% each of GO and SEP reduced the thermal conductivity to 15 mW/m K which is much lower than the thermal conductivity of air. Excellent thermal insulation property of the foam is due to much reduction of free path of gas molecules and also imparting significant interfacial thermal resistance which reduces the conduction of walls. Nanocomposite foam exhibits excellent mechanical properties and fire retardancy [101]. Cellulose nanofibers can act as heterogeneous nucleation sites for the cell growth during foaming of polymeric foam [102]. Transparent film can be fabricated by using NFC and titanium. The hybrid film with excellent optical and mechanical property, high transparency in the visible region, and excellent hardness property suggests potential use of nanocellulose-based hybrid transparent coating where high wear resistance and UV activity are required [103]. Nanocellulose can be used as coating to develop self-cleaning coating surface. Huang et al. proposed cellulose nanofibril monolayers coating solid surface by physical deposition of nanomaterials. Coated surface shows self-cleaning properties against various engine oils. Excellent self-cleaning properties of CNF are due to the uniformly, densely, and symmetrically arranged hydroxyl and carboxylic groups on the nanocellulosic surface [104]. Nanocellulose-reinforced epoxy resin can be used as coating application with anti-corrosion properties with 1 wt% of nanocellulose in composite coating showing pronounced anti-corrosion properties [105].

Food packaging is required, when food will not be consumed immediately after production. Intention of food packaging is to improve shelf life of food by providing protective barrier against physical and environmental hazards such as dirt, dust, oxygen, light, pathogenic microorganisms, moisture, and carbon dioxide. Due to its nontoxicity and excellent barrier properties against various hazards, nanocellulose-reinforced polymer nanocomposite film can be used for food packaging. Polymer nanocomposite can also be used in other fields where barrier property is desired. Bideau et al. proposed novel nano-biocomposite based on polypyrrole (PPy) and TEMPO-oxidized nanofibrillated cellulose (TOCN). Nanocomposite shows antioxidant property with high barrier properties against the mass molecules such as oxygen, carbon dioxide, and water vapor. Further, the nano-biocomposites are renewable and biodegradable in nature [106]. Nanocellulose precoating can improve the homogeneity and uniformity of the multi-layered alkyd resin/nanocellulose coating, and improve the water barrier properties. Excellent moisture barrier properties, renewability, and lightweight coatings could be useful for the packaging applications [107].

Some of the other applications of nanocellulose include applications in food processing and textile filed. BNC can be used to change the viscoelastic behavior of the masses which improves the baking quality of wheat flours. BNC can increase specific volume and moisture retention, decreased browning index and firmness of bread crumb, and provided larger average porous size of BNC crumb; hence, BNC as improver can enhance the bread-making performance [108]. Dai and Fan presents the nanocellulose modification of natural fiber viz. hemp fiber by using nanocellulose. The modification significantly increases the modulus (36.13%), tensile stress (72.80%), and tensile strain (67.69%) of hemp fibers. Microscopic images revealed that improvement of properties of hemp fibers is due to the following two reasons: (a) filling in the stria and (b) bonding the inter-fibriller gaps. Surface modification also increases crystallinity from 55.17% to 76.39% [109]. Interesting features of nanocellulose open up vast scope for possible applications in textile filed. Nanocellulose-treated fabric can clean themselves by similar principle of lotus effect. Few other promising applications of nanocellulose-treated textiles are UV-blocking textiles, wrinkle-resistant textile, antibacterial textiles, etc. One such promising application of nanocellulose-treated textile is in the area of medical textile such as in hospital and home care [34].

4.7 Summary and Future Outlook

Nanomaterials derived from natural resources, e.g., cellulosic nanomaterials, is an interesting topic for research in both industry and academia. Nanocellulose combines the chemical and surface characteristic properties of cellulose, which are the interesting features of nanostructured materials. Nanocellulose has many interesting features such as high mechanical strength, stiffness, large surface area, and biodegradability. Commercialization of cellulosic nanomaterial is still at the premature stage; however, there is huge market for cellulosic nanomaterials. Few technological breakthroughs for the application of nanocellulose, e.g., ultra strong, lightweight, and low-price nanocomposite comparable to the carbon fiber-reinforced composite, could trigger the nanocellulose market [12].

Cellulosic nanomaterial holds great promise in a wide verity of applications which includes biomedical, water treatment, energy, environment, and polymer nanocomposite. Nanocellulose is a renewable promising nanomaterials, which has abundant sources and is relatively cheap. Nanocellulose exhibits excellent optical, mechanical, thermal, and degradable properties, and it is nontoxic. Undoubtedly, nanomaterials based on cellulose has great potential for preparing novel class of biomaterials in terms of form and functionality. In order to reduce the cost of nanocellulose and increase the productivity, novel route for the preparation of nanocellulose should be explored, which must be technically and commercially viable. Being a nanosize particle, this is really challenging for the uniform dispersion and distribution of nanocellulose in the polymer matrix during the fabrication of polymer nanocomposites. Therefore, novel processing routes must be explored to control the nanocellulose dispersion and uniform distribution in the polymer nanocomposites.

Considering current state of the art of these renewable, abundant, nontoxic, and biocompatible nanomaterials, this author proposed that research and development on nanocellulosic material in the following areas will be interesting in both academia and industry:

i. Development of novel process which would enable increased yield of nanocellulose during manufacturing from lignocellulosic biomass by using appropriate method.

ii. Surface modification of nanocellulose with novel molecules for widening its application in biomedical, energy, environment, water, and polymer nanocomposite fields.

iii. Further, compatibility of nanocellulose with polar/non-polar matrix and hence to improve the interfacial interactions between nanocellulose and matrix material would be interesting to study.

iv. Due to concerns of its nondegradability and the increasing difficulty in managing solid plastic wastes, the research on nanocellulose-reinforced degradable or partially degradable polymer nanocomposites and its commercialization/scale-up will be encouraging.

v. Due to its ease of availability, non-cytotoxicity, and interesting properties of nanoscale materials, nanocellulose is interesting for the biomedical applications. Therefore, research of nanocellulose in diverse area of medical sciences will be interesting for the researches in both industry/academia.

References

1. A.K. Bledzki, J. Gassan, Composites reinforced with cellulose based fibres, *Progress in Polymer Science* 24(2) (1999) 221–274.

2. S.M. Yusop, I. Ahmad, N.F.M. Zain, Preparation and characterization of cellulose and nanocellulose from pomelo (Citrus grandis) albedo, *Journal of Nutrition and Food Science* 5(1) (2014) Article No 1000334.

3. S. Rebouillat, F. Pla, State of the art manufacturing and engineering of nanocellulose: A review of available data and industrial applications, *Journal of Biomaterials and Nanobiotechnology* 4 (2013) 165–188.

4. N. Soykeabkaew, N. Arimoto, T. Nishino, T. Peijs, All-cellulose composites by surface selective dissolution of aligned ligno-cellulosic fibres, *Composites Science and Technology* 68(10–11) (2008) 2201–2207.

5. E. Sjöström, *Wood Chemistry: Fundamentals and Applications*, Academic Press, San Diego, CA (1981).

6. M. Jorfi, E.J. Foster, Recent advances in nanocellulose for biomedical applications, *Journal of Applied Polymer Science* 132(14) (2015) Article No. 41719.

7. D.J. Gardner, G.S. Oporto, R. Mills, M. Samir, Adhesion and surface issues in cellulose and nanocellulose, *Journal of Adhesion Science and Technology* 22(5–6) (2008) 545–567.

8. Y. Aitomaki, K. Oksman, Reinforcing efficiency of nanocellulose in polymers, *Reactive & Functional Polymers* 85 (2014) 151–156.

9. A. Dufresne, Nanocellulose: A new ageless bionanomaterial, *Materials Today* 16(6) (2013) 220–227.

10. B. Deepa, E. Abraham, N. Cordeiro, M. Mozetic, A.P. Mathew, K. Oksman, M. Faria, S. Thomas, L.A. Pothan, Utilization of various lignocellulosic biomass for the production of nanocellulose: A comparative study, *Cellulose* 22(2) (2015) 1075–1090.

11. H. Wei, K. Rodriguez, S. Renneckar, P.J. Vikesland, Environmental science and engineering applications of nanocellulose-based nanocomposites, *Environmental Science- Nano* 1(4) (2014) 302–316.

12. J.H. Kim, B.S. Shim, H.S. Kim, Y.J. Lee, S.K. Min, D. Jang, Z. Abas, J. Kim, Review of nanocellulose for sustainable future materials, *International Journal of Precision Engineering and Manufacturing-Green Technology* 2(2) (2015) 197–213.

13. Y.H. Wang, X.Y. Wei, J.H. Li, F. Wang, Q.H. Wang, Y.D. Zhang, L.X. Kong, Homogeneous isolation of nanocellulose from eucalyptus pulp by high pressure homogenization, *Industrial Crops and Products* 104 (2017) 237–241.

14. C.J. Chirayil, L. Mathew, S. Thomas, Review of recent research in nano cellulose preparation from different lignocellulosic fibers, *Reviews on Advanced Materials Science* 37(1–2) (2014) 20–28.

15. B. Wang, M. Sain, Dispersion of soybean stock-based nanofiber in a plastic matrix, *Polymer International* 56(4) (2007) 538–546.

16. M. Paakko, M. Ankerfors, H. Kosonen, A. Nykanen, S. Ahola, M. Osterberg, J. Ruokolainen, J. Laine, P.T. Larsson, O. Ikkala, T. Lindstrom, Enzymatic

hydrolysis combined with mechanical shearing and high-pressure homogenization for nanoscale cellulose fibrils and strong gels, *Biomacromolecules* 8(6) (2007) 1934–1941.

17. M. Henriksson, G. Henriksson, L.A. Berglund, T. Lindstrom, An environmentally friendly method for enzyme-assisted preparation of microfibrillated cellulose (MFC) nanofibers, *European Polymer Journal* 43(8) (2007) 3434–3441.

18. M. Beaumont, T. Nypelo, J. Konig, R. Zirbs, M. Opietnik, A. Potthast, T. Rosenau, Synthesis of redispersible spherical cellulose II nanoparticles decorated with carboxylate groups, *Green Chemistry* 18(6) (2016) 1465–1468.

19. M. Bansal, G.S. Chauhan, A. Kaushik, A. Sharma, Extraction and functionalization of bagasse cellulose nanofibres to Schiff-base based antimicrobial membranes, *International Journal of Biological Macromolecules* 91 (2016) 887–894.

20. S.P. Mishra, A.S. Manent, B. Chabot, C. Daneault, Production of nanocellulose from native cellulose— Various options utilizing ultrasound, *Bioresources* 7(1) (2012) 422–436.

21. S.J. Chun, S.Y. Lee, G.Y. Jeong, J.H. Kim, Fabrication of hydrophobic self-assembled monolayers (SAM) on the surface of ultra-strength nanocellulose films, *Journal of Industrial and Engineering Chemistry* 18(3) (2012) 1122–1127.

22. Y.M. Zhou, S.Y. Fu, L.M. Zheng, H.Y. Zhan, Effect of nanocellulose isolation techniques on the formation of reinforced poly(vinyl alcohol) nanocomposite films, *Express Polymer Letters* 6(10) (2012) 794–804.

23. N. Amiralian, P.K. Annamalai, C.J. Garvey, E. Jiang, P. Memmott, D.J. Martin, High aspect ratio nanocellulose from an extremophile spinifex grass by controlled acid hydrolysis, *Cellulose* 24(9) (2017) 3753–3766.

24. P. Phanthong, S. Karnjanakom, P. Reubroycharoen, X.G. Hao, A. Abudula, G.Q. Guan, A facile one-step way for extraction of nanocellulose with high yield by ball milling with ionic liquid, *Cellulose* 24(5) (2017) 2083–2093.

25. U. Sampath, Y.C. Ching, C.H. Chuah, R. Singh, P.C. Lin, Preparation and characterization of nanocellulose reinforced semi-interpenetrating polymer network of chitosan hydrogel, *Cellulose* 24(5) (2017) 2215–2228.

26. B.M. Cherian, A.L. Leao, S.F. de Souza, S. Thomas, L.A. Pothan, M. Kottaisamy, Isolation of nanocellulose from pineapple leaf fibres by steam explosion, *Carbohydrate Polymers* 81(3) (2010) 720–725.

27. G. Chinga-Carrasco, K. Syverud, Pretreatment-dependent surface chemistry of wood nanocellulose for pH-sensitive hydrogels, *Journal of Biomaterials Applications* 29(3) (2014) 423–432.

28. M.J. Cho, B.D. Park, Tensile and thermal properties of nanocellulose-reinforced poly(vinyl alcohol) nanocomposites, *Journal of Industrial and Engineering Chemistry* 17(1) (2011) 36–40.

29. S.S. Costa, J.I. Druzian, B.A.S. Machado, C.O. de Souza, A.G. Guimaraes, Bi- functional biobased packing of the cassava starch, glycerol, licuri nanocellulose and red propolis, *Plos One* 9(11) (2014) Article No. e112554.

30. H.P.S.A. Khalil, Y. Davoudpour, N.A.S. Aprilia, A. Mustapha, S. Hossain, N. Islam, R. Dungani, Nanocellulose-based polymer nanocomposite: Isolation, characterization and applications. In: Thakur, V.K. (Ed.). Nanocellulose Polymer Nanocomposites: Fundamentals and Applications, John Wiley & Sons, Inc., Hoboken (2014), pp. 273–309.

31. J. Giri, R. Adhikari, A Brief review on extraction of nanocellulose and its application, BIBECHANA 9 (2013) 81–87.

32. X. Zhuo, C. Liu, R.T. Pan, X.Y. Dong, Y.F. Li, Nanocellulose mechanically isolated from amorpha fruticosa linn, *ACS Sustainable Chemistry & Engineering* 5(5) (2017) 4414–4420.

33. R.J. Moon, A. Martini, J. Nairn, J. Simonsen, J. Youngblood, Cellulose nanomaterials review: structure, properties and nanocomposites, *Chemical Society Reviews* 40(7) (2011) 3941–3994.

34. S.M. Salah, Application of nano-cellulose in textile, *Textile Science & Engineering* 3 (2013) Article No. 142.

35. F.G. Niu, M.Y. Li, Q. Huang, X.Z. Zhang, W.C. Pan, J.S. Yang, J.R. Li, The characteristic and dispersion stability of nanocellulose produced by mixed acid hydrolysis and ultrasonic assistance, *Carbohydrate Polymers* 165 (2017) 197–204.

36. E. Abraham, B. Deepa, L.A. Pothan, M. Jacob, S. Thomas, U. Cvelbar, R. Anandjiwala, Extraction of nanocellulose fibrils from lignocellulosic fibres: A novel approach, *Carbohydrate Polymers* 86(4) (2011) 1468–1475.

37. P. Aramwit, N. Bang, The characteristics of bacterial nanocellulose gel releasing silk sericin for facial treatment, *BMC Biotechnology* 14 (2014) Article No. 104.

38. A. Kurosumi, C. Sasaki, Y. Yamashita, Y. Nakamura, Utilization of various fruit juices as carbon source for production of bacterial cellulose by Acetobacter xylinum NBRC 13693, *Carbohydrate Polymers* 76(2) (2009) 333–335.

39. A.J. Brown, XIX.-The chemical action of pure cultivations of bacterium aceti, *Journal of the Chemical Society, Transactions* 49(0) (1886) 172–187.

40. K.Y. Lee, T. Tammelin, K. Schulfter, H. Kiiskinen, J. Samela, A. Bismarck, High performance cellulose nanocomposites: Comparing the reinforcing ability of bacterial cellulose and nanofibrillated cellulose, *ACS Applied Materials & Interfaces* 4(8) (2012) 4078–4086.

41. M. Iguchi, S. Yamanaka, A. Budhiono, Bacterial cellulose - a masterpiece of nature's arts, *Journal of Materials Science* 35(2) (2000) 261–270.

42. H.M. Avila, S. Schwarz, E.M. Feldmann, A. Mantas, A. von Bomhard, P. Gatenholm, N. Rotter, Biocompatibility evaluation of densified bacterial nanocellulose hydrogel as an implant material for auricular cartilage regeneration, *Applied Microbiology and Biotechnology* 98(17) (2014) 7423–7435.

43. L.N. Fu, P. Zhou, S.M. Zhang, G. Yang, Evaluation of bacterial nanocellulose-based uniform wound dressing for large area skin transplantation, *Materials Science & Engineering C-Materials for Biological Applications* 33(5) (2013) 2995–3000.

44. M. T. Islam, M.M. Alam, M. Zoccola, Review on modification of nanocellulose for application in composites, *International Journal of Innovative Research in Science, Engineering and Technology* 2(10) (2013) 5444–5451.

45. Y. Habibi, L.A. Lucia, O.J. Rojas, Cellulose nanocrystals: Chemistry, self-assembly, and applications, *Chemical Reviews* 110(6) (2010) 3479–3500.

46. L.Q. Wei, U.P. Agarvval, K.C. Hirth, L.M. Matuana, R.C. Sabo, N.M. Stark, Chemical modification of nanocellulose with canola oil fatty acid methyl ester, *Carbohydrate Polymers* 169 (2017) 108–116.

47. J.D. Rusmirovic, J.Z. Ivanovic, V.B. Pavlovic, V.M. Rakic, M.P. Rancic, V. Djokic, A.D. Marinkovic, Novel modified nanocellulose applicable as reinforcement in high- performance nanocomposites, *Carbohydrate Polymers* 164 (2017) 64–74.

48. E. Espino-Perez, S. Domenek, N. Belgacem, C. Sillard, J. Bras, Green process for chemical functionalization of nanocellulose with carboxylic acids, *Biomacromolecules* 15(12) (2014) 4551–4560.

49. M. Fumagalli, F. Sanchez, S. Molina-Boisseau, L. Heux, Surface-restricted modification of nanocellulose aerogels in gas-phase esterification by di-functional fatty acid reagents, *Cellulose* 22(3) (2015) 1451–1457.

50. J.P.S. Morais, M.D. Rosa, M.D.M. de Souza, L.D. Nascimento, D.M. do Nascimento, A.R. Cassales, Extraction and characterization of nanocellulose structures from raw cotton linter, *Carbohydrate Polymers* 91(1) (2013) 229–235.

51. H. Golmohammadi, E. Morales-Narvaez, T. Naghdi, A. Merkoci, Nanocellulose in sensing and biosensing, *Chemistry of Materials* 29(13) (2017) 5426–5446.

52. N. Lavoine, L. Bergstrom, Nanocellulose-based foams and aerogels: Processing, properties, and applications, *Journal of Materials Chemistry A* 5(31) (2017) 16105–16117.

53. A. Khan, T. Huq, R.A. Khan, B. Riedl, M. Lacroix, Nanocellulose-based composites and bioactive agents for food packaging, *Critical Reviews in Food Science and Nutrition* 54(2) (2014) 163–174.

54. A. Mandal, D. Chakrabarty, Synthesis and characterization of nanocellulose reinforced full-interpenetrating polymer network based on poly(vinyl alcohol) and polyacrylamide (both crosslinked) composite films, *Polymer Composites* 38(8) (2017) 1720–1731.

55. W. Li, X. Zhao, Z.H. Huang, S.X. Liu, Nanocellulose fibrils isolated from BHKP using ultrasonication and their reinforcing properties in transparent poly (vinyl alcohol) films, *Journal of Polymer Research* 20(8) (2013) Article No. 210.

56. T. Hees, F. Zhong, T. Rudolph, A. Walther, R. Mulhaupt, Nanocellulose aerogels for supporting iron catalysts and in situ formation of polyethylene nanocomposites, *Advanced Functional Materials* 27(11) (2017) Article No. 1605586.

57. C.F. Mok, Y.C. Ching, F. Muhamad, N.A. Abu Osman, R. Singh, Poly(vinyl alcohol)—chitin composites reinforced by oil palm empty fruit bunch fiber-derived nanocellulose, *International Journal of Polymer Analysis and Characterization* 22(4) (2017) 294–304.

58. C.J. Chirayil, L. Mathew, P.A. Hassan, M. Mozetic, S. Thomas, Rheological behaviour of nanocellulose reinforced unsaturated polyester nanocomposites, *International Journal of Biological Macromolecules* 69 (2014) 274–281.

59. N.S. Lani, N. Ngadi, A. Johari, M. Jusoh, Isolation, characterization, and application of nanocellulose from oil palm empty fruit bunch fiber as nanocomposites, *Journal of Nanomaterials* (2014) Article No. 702538.

60. S.Y. Lee, D.J. Mohan, I.A. Kang, G.H. Doh, S. Lee, S.O. Han, Nanocellulose reinforced PVA composite films: Effects of acid treatment and filler loading, *Fibers and Polymers* 10(1) (2009) 77–82.

61. W. Li, Q. Wu, X. Zhao, Z.H. Huang, J. Cao, J. Li, S.X. Liu, Enhanced thermal and mechanical properties of PVA composites formed with filamentous nanocellulose fibrils, *Carbohydrate Polymers* 113 (2014) 403–410.

62. A. Mandal, D. Chakrabarty, Studies on the mechanical, thermal, morphological and barrier properties of nanocomposites based on poly(vinyl alcohol) and nanocellulose from sugarcane bagasse, *Journal of Industrial and Engineering Chemistry* 20(2) (2014) 462–473.

63. S. S. Nair, J. Y. Zhu, Y. Deng, A. J. Ragauskas, High performance green barriers based on nanocellulose, *Sustainable Chemical Processes* 2 (2014) Article No. 23.

64. M. Chaichi, M. Hashemi, F. Badii, A. Mohammadi, Preparation and characterization of a novel bionanocomposite edible film based on pectin and crystalline nanocellulose, *Carbohydrate Polymers* 157 (2017) 167–175.

65. S. Belbekhouche, J. Bras, G. Siqueira, C. Chappey, L. Lebrun, B. Khelifi, S. Marais, A. Dufresne, Water

sorption behavior and gas barrier properties of cellulose whiskers and microfibrils films, *Carbohydrate Polymers* 83(4) (2011) 1740–1748.

66. S. Mondal, Review on nanocellulose polymer nanocomposites, *Polymer-Plastics Technology and Engineering* 57 (13) (2018) 1377–1391.

67. C.Y. Xu, L.N. Shi, L. Guo, X. Wang, X.Y. Wang, H.L. Lian, Fabrication and characteristics of graphene oxide/nanocellulose fiber/poly(vinyl alcohol) film, *Journal of Applied Polymer Science* 134(39) (2017) Article No. 45345.

68. E. Abraham, P.A. Elbi, B. Deepa, P. Jyotishkumar, L.A. Pothen, S.S. Narine, S. Thomas, X-ray diffraction and biodegradation analysis of green composites of natural rubber/nanocellulose, *Polymer Degradation and Stability* 97(11) (2012) 2378–2387.

69. A. Ben Mabrouk, M.C.B. Salon, A. Magnin, M.N. Belgacem, S. Boufi, Cellulose-based nanocomposites prepared via mini-emulsion polymerization: Understanding the chemistry of the nanocellulose/matrix interface, *Colloids and Surfaces a-Physicochemical and Engineering Aspects* 448 (2014) 1–8.

70. J.Y. Tang, X. Li, L.H. Bao, L. Chen, F.F. Hong, Comparison of two types of bioreactors for synthesis of bacterial nanocellulose tubes as potential medical prostheses including artificial blood vessels, *Journal of Chemical Technology and Biotechnology* 92(6) (2017) 1218–1228.

71. J. Henschen, P.A. Larsson, J. Illergard, M. Ek, L. Wagberg, Bacterial adhesion to polyvinylamine-modified nanocellulose films, *Colloids and Surfaces B-Biointerfaces* 151 (2017) 224–231.

72. A.G. Dumanli, Nanocellulose and its composites for biomedical applications, *Current Medicinal Chemistry* 24(5) (2017) 512–528.

73. K. Markstedt, A. Mantas, I. Tournier, H.M. Avila, D. Hagg, P. Gatenholm, 3D Bioprinting human chondrocytes with nanocellulose-alginate bioink for cartilage tissue engineering applications, *Biomacromolecules* 16(5) (2015) 1489–1496.

74. D. Nguyen, D.A. Hagg, A. Forsman, J. Ekholm, P. Nimkingratana, C. Brantsing, T. Kalogeropoulos, S. Zaunz, S. Concaro, M. Brittberg, A. Lindahl, P. Gatenholm, A. Enejder, S. Simonsson, Cartilage tissue engineering by the 3D bioprinting of iPS cells in a nanocellulose/alginate bioink, *Scientific Reports* 7 (2017) Article No. 658.

75. L. Saidi, C. Vilela, H. Oliveira, A.J.D. Silvestre, C.S.R. Freire, Poly(N-methacryloyl glycine)/nanocellulose composites as pH-sensitive systems for controlled release of diclofenac, *Carbohydrate Polymers* 169 (2017) 357–365.

76. K.J. De France, T. Hoare, E.D. Cranston, Review of hydrogels and aerogels containing nanocellulose, *Chemistry of Materials* 29(11) (2017) 4609–4631.

77. J. Leppiniemi, P. Lahtinen, A. Paajanen, R. Mahlberg, S. Metsa-Kortelainen, T. Pinornaa, H. Pajari, I. Vikholm-Lundin, P. Pursula, V.P. Hytonen, 3D-printable bioactivated nanocellulose-alginate hydrogels, *ACS Applied Materials & Interfaces* 9(26) (2017) 21959–21970.

78. G.K. Tummala, T. Joffre, R. Rojas, C. Persson, A. Mihranyan, Strain-induced stiffening of nanocellulose-reinforced poly(vinyl alcohol) hydrogels mimicking collagenous soft tissues, *Soft Matter* 13(21) (2017) 3936–3945.

79. F.Z. Sun, H.R. Nordli, B. Pukstad, E.K. Gamstedt, G. Chinga-Carrasco, Mechanical characteristics of nanocellulose-PEG bionanocomposite wound dressings in wet conditions, *Journal of the Mechanical Behavior of Biomedical Materials* 69 (2017) 377–384.

80. A.M. Tang, J. Li, S. Zhao, T.T. Liu, Q.W. Wang, J.F. Wang, Biodegradable tissue engineering scaffolds based on nanocellulose/PLGA nanocomposite for NIH 3T3 cell cultivation, *Journal of Nanoscience and Nanotechnology* 17(6) (2017) 3888–3895.

81. S. Khazeni, A. Hatamian-Zarmi, F. Yazdian, Z.B. Mokhtari-Hosseini, B. Ebrahimi- Hosseinzadeh, B. Noorani, G. Amoabedini, M.R. Soudi, Production of nanocellulose in miniature-bioreactor: Optimization and characterization, *Preparative Biochemistry & Biotechnology* 47(4) (2017) 371–378.

82. H.M. Avila, E.M. Feldmann, M.M. Pleumeekers, L. Nimeskern, W. Kuo, W.C. de Jong, S. Schwarz, R. Muller, J. Hendriks, N. Rotter, G. van Osch, K.S. Stok, P. Gatenholm, Novel bilayer bacterial nanocellulose scaffold supports neocartilage formation in vitro and in vivo, *Biomaterials* 44 (2015) 122–133.

83. T.H. Lu, Q. Li, W.S. Chen, H.P. Yu, Composite aerogels based on dialdehyde nanocellulose and collagen for potential applications as wound dressing and tissue engineering scaffold, *Composites Science and Technology* 94 (2014) 132–138.

84. N. Ferraz, A. Leschinskaya, F. Toomadj, B. Fellstrom, M. Stromme, A. Mihranyan, Membrane characterization and solute diffusion in porous composite nanocellulose membranes for hemodialysis, *Cellulose* 20(6) (2013) 2959–2970.

85. S.H. Zhang, R.F. Fu, S. Wang, Y.C. Gu, S. Chen, Novel nanocellulose/conducting polymer composite nanorod films with improved electrochromic performances, *Materials Letters* 202 (2017) 127–130.

86. S.H. Zhang, R.F. Fu, Z.F. Du, M.J. Jiang, M. Zhou, Y.C. Gu, S. Chen, High- performance electrochromic device based on nanocellulose/polyaniline and nanocellulose/poly(3,4-ethylenedioxythiophene) composite thin films, *Optical Engineering* 56(7) (2017) Article No. 077101.

87. S.H. Zhang, R.F. Fu, Y.C. Gu, L.Q. Dong, J.J. Li, S. Chen, Preparation of nanocellulose- based polyaniline composite film and its application in

electrochromic device, *Journal of Materials Science-Materials in Electronics* 28(14) (2017) 10158–10165.

88. S.H. Zhang, G. Sun, Y.F. He, R.F. Fu, Y.C. Gu, S. Chen, Preparation, characterization, and electrochromic properties of nanocellulose-based polyaniline nanocomposite films, *ACS Applied Materials & Interfaces* 9(19) (2017) 16426–16434.

89. Q.S. Jiang, C. Kacica, T. Soundappan, K.K. Liu, S. Tadepalli, P. Biswas, S. Singamaneni, An in situ grown bacterial nanocellulose/graphene oxide composite for flexible supercapacitors, *Journal of Materials Chemistry A* 5(27) (2017) 13976–13982.

90. L.P. Wang, C. Schutz, G. Salazar-Alvarez, M.M. Titirici, Carbon aerogels from bacterial nanocellulose as anodes for lithium ion batteries, *RSC Advances* 4(34) (2014) 17549–17554.

91. W.Z. Zheng, R.H. Lv, B. Na, H.S. Liu, T.X. Jin, D.Z. Yuan, Nanocellulose-mediated hybrid polyaniline electrodes for high performance flexible supercapacitors, *Journal of Materials Chemistry A* 5(25) (2017) 12969–12976.

92. C. Vilela, N. Sousa, R.J.B. Pinto, A.J.D. Silvestre, F.M.L. Figueiredo, C.S.R. Freire, Exploiting poly(ionic liquids) and nanocellulose for the development of bio-based anion- exchange membranes, *Biomass & Bioenergy* 100 (2017) 116–125.

93. H. Voisin, L. Bergstrom, P. Liu, A.P. Mathew, Nanocellulose-based materials for water purification, *Nanomaterials* 7(3) (2017) Article No. 57.

94. H. Sehaqui, L. Schaufelberger, B. Michen, T. Zimmermann, Humic acid desorption from a positively charged nanocellulose surface, *Journal of Colloid and Interface Science* 504 (2017) 500–506.

95. Q.H. Xu, Y.L. Wang, L.Q. Jin, Y. Wang, M.H. Qin, Adsorption of Cu (II), Pb (II) and Cr (VI) from aqueous solutions using black wattle tannin-immobilized nanocellulose, *Journal of Hazardous Materials* 339 (2017) 91–99.

96. Q. Zhu, Y. Wang, M.F. Li, K. Liu, C.Y. Hu, K.L. Yan, G. Sun, D. Wang, Activable carboxylic acid functionalized crystalline nanocellulose/PVA-co-PE composite nanofibrous membrane with enhanced adsorption for heavy metal ions, *Separation and Purification Technology* 186 (2017) 70–77.

97. Z. Karim, M. Hakalahti, T. Tammelin, A.P. Mathew, In situ TEMPO surface functionalization of nanocellulose membranes for enhanced adsorption of metal ions from aqueous medium, *RSC Advances* 7(9) (2017) 5232–5241.

98. P.G. Moradeeya, M.A. Kumar, R.B. Thorat, M. Rathod, Y. Khambhaty, S. Basha, Nanocellulose for biosorption of chlorpyrifos from water: Chemometric

optimization, kinetics and equilibrium, *Cellulose* 24(3) (2017) 1319–1332.

99. J.T. Korhonen, M. Kettunen, R.H.A. Ras, O. Ikkala, Hydrophobic nanocellulose aerogels as floating, sustainable, reusable, and recyclable oil absorbents, *ACS Applied Materials & Interfaces* 3(6) (2011) 1813–1816.

100. S. Gustafsson, L. Manukyan, A. Mihranyan, Protein-nanocellulose interactions in paper filters for advanced separation applications, *Langmuir* 33(19) (2017) 4729–4736.

101. B. Wicklein, A. Kocjan, G. Salazar-Alvarez, F. Carosio, G. Camino, M. Antonietti, L. Bergstrom, Thermally insulating and fire-retardant lightweight anisotropic foams based on nanocellulose and graphene oxide, *Nature Nanotechnology* 10(3) (2015) 277–283.

102. M.V.G. Zimmermann, M.P. da Silva, A.J. Zattera, R.M.C. Santana, Effect of nanocellulose fibers and acetylated nanocellulose fibers on properties of poly(ethylene-co-vinyl acetate) foams, *Journal of Applied Polymer Science* 134(17) (2017) Article No. 44760.

103. C. Schutz, J. Sort, Z. Bacsik, V. Oliynyk, E. Pellicer, A. Fall, L. Wagberg, L. Berglund, L. Bergstrom, G. Salazar-Alvarez, Hard and transparent films formed by nanocellulose-TiO$_2$ nanoparticle hybrids, *Plos One* 7(10) (2012) Article No. e45828.

104. S. Huang, D.Y. Wang, A simple nanocellulose coating for self-cleaning upon water action: Molecular design of stable surface hydrophilicity, *Angewandte Chemie-International Edition* 56(31) (2017) 9053–9057.

105. I.A.W. Ma, A. Shafaamri, R. Kasi, F.N. Zaini, V. Balakrishnan, R. Subramaniam, A.K. Arof, Anticorrosion properties of epoxy/nanocellulose nanocomposite coating, *Bioresources* 12(2) (2017) 2912–2929.

106. B. Bideau, J. Bras, N. Adoui, E. Loranger, C. Daneault, Polypyrrole/nanocellulose composite for food preservation: Barrier and antioxidant characterization, *Food Packaging and Shelf Life* 12 (2017) 1–8.

107. C. Aulin, G. Strom, Multilayered alkyd resin/nanocellulose coatings for use in renewable packaging solutions with a high level of moisture resistance, *Industrial & Engineering Chemistry Research* 52(7) (2013) 2582–2589.

108. M.L. Corral, P. Cerrutti, A. Vazquez, A. Califano, Bacterial nanocellulose as a potential additive for wheat bread, *Food Hydrocolloids* 67 (2017) 189–196.

109. D.S. Dai, M.Z. Fan, Green modification of natural fibres with nanocellulose, *RSC Advances* 3(14) (2013) 4659–4665.

Upconversion Nanoparticles: Design Strategies for Their Synthesis and Fabrication of Their Surface Chemistry

Karan Malhotra, Muhammad
Shahrukh, Hifza Najib, and
Ulrich J. Krull
University of Toronto Mississauga

5.1 Upconversion Nanoparticles—Functionalities by Design

"Nano" has become the term used to describe materials and structures with at least one dimension at the size scale of 1–100 nm. Further categorization can be based on the structure, composition, physical, chemical, or spectral properties of the material (Whitesides, 2003). One particular class of nanomaterials is upconversion nanoparticles (UCNPs). These nanoparticles are typically composed of lanthanide-doped crystals that absorb low-energy radiation and undergo anti-Stokes shift, emitting higher energy radiation. A specific example of this upconversion (UC) includes the absorption of low-energy infrared radiation ($\lambda = 980$ nm) and the subsequent luminescence at higher energies, i.e. ultraviolet and visible ($\lambda = 350$ nm – 700 nm). The luminescence profile of UCNPs can be tuned by modifications of the type and ratios of dopants giving rise to changes in the intensity of luminescence within specific colour channels. Spectroscopic properties can be further manipulated via growth of different shell materials on the core particles. The choice and combinations of dopants, shell materials, and particle size allow for application of this class of nanoparticles for biomedical device development, *in vivo* labels for biochemical interactions, and theranostic

agents for clinical treatment (Haase and Schäfer, 2011; DaCosta et al., 2014; Wang et al., 2010; Wang and Liu, 2009). In this section, the structure and composition of UCNPs will be discussed with considerations of the host lattice followed by a brief overview of the theory related to photon UC.

5.1.1 Structure and Composition of UCNP

Lanthanide-doped UCNPs display varying spectroscopic properties based on their structure and composition. UCNPs made of a doped core will have different spectroscopic properties than doped core particles that are encased with an inert shell layer (Wang and Li, 2007; Boyer et al., 2009). Control of spectroscopic properties of UCNP is further extended by the choice of materials for growing the shelling materials and by the implementation of multi-shell layer structures (Qian and Zhang, 2008). In core/shell structures, different dopants types and ratios may also be segregated into specific physical regions of the nanoparticle, further altering the spectroscopic profile of UCNPs (Qian and Zhang, 2008; Wang et al., 2011). This section will discuss the choice of dopants and host-lattice structure, and the core/core-shell architecture including the material choice and composition of core/shells.

Common Dopants used for UCNP Design

Traditional luminescence processes most commonly associated with molecular dyes and luminescence semiconductor-based nanomaterials (like quantum dots) function based on a ground state and an excited state system. Electrons in these systems are initially excited from the ground state to the excited state and subsequently transition back to the ground states undergoing radiative relaxation with a characteristic Stokes shift by having undergone some type of non-radiative relaxation processes, (see Figure 5.1a) (Algar, Tavares, and Krull, 2010; Chan et al., 2002; Medintz et al., 2005; Michalet et al., 2005; Algar et al., 2011; Gao et al., 2005; Petryayeva, Algar, Medintz, 2013). In contrast to these processes, UC operates with multiple excited states, where each energy level is organized in a ladder-based arrangement with respect to one another. UC is then accomplished via the sequential stacking of multiple, low-energy photons to reach higher energy levels and a subsequent radiative emission resulting in luminescence and a characteristic anti-Stokes shift (see Figure 5.1b). Unlike traditional luminescence, UC is a nonlinear process, and multiple factors can affect the photo-stability, quantum yield, and reproducibility of luminescence. The process of UC imposes a set of design criteria on the dopants driving UC including (a) the ladder-based arrangement of metastable energy levels, (b) buildup of multiple low-energy photons, and (c) long lifetimes for excited metastable states (Auzel, 2004; Auzel, 1966; Auzel, 1973; Auzel, 1990; Chen et al., 2014).

There are many different mechanisms of photon UC. However, the two most relevant mechanisms that impact dopant choice and UC nanoparticle design are ESA and ETU. ESA is based on the sequential absorption of multiple low-energy photons by a single activator dopant ion (emission source, see Figure 5.2a). The second mechanism of UC (ETU) is based on the energy transfer of an excited sensitizer dopant ion to an excited activator dopant ion resulting in radiative emission (see Figure 5.2b). For designing systems that can achieve the greatest UC efficiency, nanoparticles are often doped with activator and sensitizer ions to allow for both ESA- and ETU-based UC (Auzel, 2004; Auzel, 1966; Auzel, 1973; Auzel, 1990; Chen et al., 2014).

Activators

Several classes of elements meet the requirements for UC including transition metals (e.g., Ti^{2+}, Ni^{2+}, Mo^{3+}, Re^{4+}, and Os^{4+}), lanthanides (e.g., Er^{3+}, Tm^{3+}, Ho^{3+}), and actinides. Of these three groups, only lanthanides have shown promise for further development of luminescent methods suitable for biomedical applications. Transition metals and actinide-doped materials have been shown (in their respective host lattices) to undergo UC, however, these systems do not meet the optical requirements for an ideal

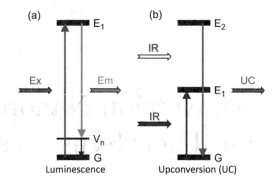

FIGURE 5.1 Schematic for (a) luminescence and (b) upconversion (UC) based processes. Here, G represents the ground state, E_1 and E_2 represent the first and second electronic states respectively, V_n represents a vibrational state, Ex represents the excitation photon, Em represents the emission photon, and IR represents the infrared photon.

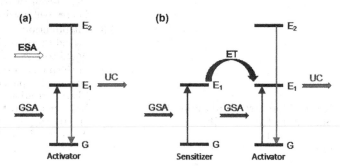

FIGURE 5.2 Energy level diagrams demonstrating the UC mechanism for (a) excited state absorption (ESA) and (b) energy transfer upconversion (ETU). Here, GSA represents ground state absorption, UC represents upconversion luminescence, ET represents energy transfer, G represents the ground state, E_1 represents the first excited state, and E_2 represents the second excited state. (Adapted with permission from DaCosta et al. Copyright 2014 Elsevier.)

nanocrystal system and mainly operate with weak luminescence and at low operating temperatures. Of the lanthanide series of elements, erbium (Er^{3+}), thulium (Tm^{3+}), and holmium (Ho^{3+}) have the highest UC efficiency when excited by low-energy photons, and these three lanthanides are also the most commonly investigated in literature (Haase and Schäfer, 2011; DaCosta et al.,2014; Auzel, 2004; Wenger and Güdel, 2001a; Wenger and Güdel, 2001b; Wenger et al., 2002; Gamelin and Güdel, 1998; Gamelin and Güdel, 2000a; Gamelin and Güdel, 1999; Gamelin and Güdel, 2000b; Wermuth and Güdel, 1999; Deren et al. 2000).

Lanthanide dopants in a UCNP can maintain an excited metastable state due to their long lifetime, and when excited, they can absorb a second photon of low energy to reach a higher excited state. With lanthanides such as Er^{3+} and Tm^{3+}, pairs of energy levels with similar energy gaps are present (see Figure 5.3) and electronic excitation within the ion can be achieved by a single monochromatic photon source. For erbium, the pairs of energy levels with similar energy gaps are the $^4I_{15/2}$ and $^4I_{11/2}$ energy levels, the

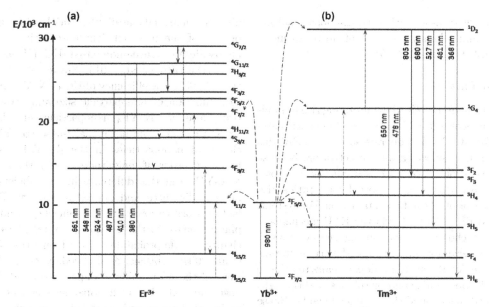

FIGURE 5.3 Energy level diagram representing the different mechanisms of UC for Yb^{3+} sensitizer coupled with Er^{3+} and Tm^{3+} activators. Solid lines represent radiative emission pathways, dotted lines represent non-radiative pathways, black lines represent vibrational relaxation, and curved dashed lines represent energy transfer between sensitizer and activator ions. (Adapted with permission from DaCosta et al. Copyright 2014 Elsevier.)

$^4I_{13/2}$ and $^4F_{9/2}$ energy levels, the $^4I_{11/2}$ and $^4F_{7/2}$ energy levels, and the $^4S_{3/2}$ and $^4G_{7/2}$ energy levels. These four electronic transitions can generate an excited-state population that would result in radiative emission in blue, and green, and red colour channels. With thulium, the pairs of energy levels with similar energy gaps are the $^3F_4$ and $^3F_2$ energy levels, the $^3H_4$ and $^1G_4$ energy levels, and the $^1G_4$ and $^1D_2$ energy levels. These three electronic transitions can yield an excited-state population that emits radiation in blue and red colour channels (Haase and Schäfer, 2011; DaCosta et al.,2014; Wang et al., 2010; Wang and Liu, 2009; Auzel, 2004; Liu et al., 2013; Gu et al., 2013; Guo et al., 2004; Xu et al., 2010; Kumar et al., 2007). Erbium and thulium have been doped into nanocrystals; however, there is a concentration limit to doping the ions before non-radiative processes quench luminescence resulting in decrease of emission efficiency of the nanocrystals. The dopant concentration limits are 3% for Er^{3+} ions and 0.5% for Tm^{3+} ions, which would make these single dopant systems inefficient for further UC applications (Wang and Li, 2007; Auzel, 2004; Auzel, 1966; Auzel, 1973; Auzel, 1990; Chen et al., 2014).

Sensitizers

To overcome the poor UC efficiency of lanthanide-doped nanoparticles, UCNPs can be co-doped with another type of ion (known as sensitizers) to take advantage of the greater UC efficiency from the ETU mechanism (Auzel, 2004; Auzel, 1966; Auzel, 1973; Auzel, 1990). Sensitizer ions chosen for this application are required to have a well-matched energy transition with the activator ions (activators are also called emitters). They should also have good cross section for absorption, and their positioning within the host lattice should facilitate the ETU (Haase and Schäfer,

2011; DaCosta et al., 2014; Wang and Liu, 2009; Liu et al., 2013; Lin et al., 2012; Chen and Zhao, 2012). For UCNP systems discussed thus far (using Er^{3+}, Tm^{3+}, or Ho^{3+} ions as activators), ytterbium (Yb^{3+}) is one of the most suitable sensitizer dopants. Yb^{3+} is an ideal lanthanide ion for facilitating the ETU mechanism of UC because it has spacing between energy levels (transition between the $^2F_7$ and $^2F_5$ energy levels) that match transitions in activator ions (Haase and Schäfer, 2011; DaCosta et al., 2014). Yb^{3+} is typically doped at higher concentrations than activators in co-doped systems (18–20% for systems with Er^{3+} and 30% for systems with Tm^{3+}) to maximize UC efficiency. UCNPs that contain Yb^{3+} and either Er^{3+}, Tm^{3+}, or Ho^{3+} absorb near-IR photons of 980 nm excitation wavelength to achieve the electronic transitions required for UC (Haase and Schäfer, 2011; DaCosta et al., 2014). However, these UCNPs pose a challenge for further application in biological systems because photons of 980 nm excitation are also weakly absorbed by water molecules. Use of 980 nm for excitation of UCNPs in cells could result in heating of tissue materials and cell death. One of the strategies to overcome this obstacle for *in vivo* application of UCNPs is to substitute Yb^{3+} with another sensitizer molecule that absorbs at a different energy band than biological samples. Neodymium (Nd^{3+}) is one of the more common substitutes of Yb^{3+} as sensitizer for UCNPs, absorbing excitation radiation at 808 nm. This absorption wavelength is more suitable for bioapplication. There is a compromise that the optimal dopant ratio for Nd^{3+} is around 2 mol%, making UC emissions with Nd^{3+} weak when compared with Yb^{3+} (Xie et al., 2013). To overcome this poor UC efficiency, nanostructures with Nd^{3+} ions are segregated into a shell surrounding the core particle (i.e. separated from activator ions). This reduces back-energy transfer with activator ions resulting in enhanced UC and allows for doping of up to 20 mol% (Wang et al., 2013; Wen et al., 2013).

Choice of Host Lattice

To achieve UC in a nanoparticle, dopants giving rise to luminescence must be incorporated into a crystalline host material. Typically, for UCNPs that operate solely via ESA, this can be done with low concentrations of activators because non-radiative relaxation processes can occur due to cross-relaxation pathways. However, the low dopant ratios also result in poor absorption of excitation radiation. For systems that are designed to achieve UC via the ETU mechanism, host materials are chosen for maximizing energy transfer efficiency. The criteria for suitable host-lattice materials include optimal ion–ion distance (for dopant pairs to achieve ETU), ideal spatial positioning and size of ions, and the right type of anions used to surround the cationic dopants (Haase and Schäfer, 2011; DaCosta et al., 2014; Wang et al., 2010; Wang and Liu, 2009; Auzel, 2004; Auzel, 1966; Auzel, 1973; Auzel, 1990; Chen et al., 2014; Liu et al., 2013; Diamente, Raudsepp, van Veggel, 2007). ETU relies on the transfer of excited state energy from a sensitizer to an activator dopant by dipole–dipole interactions with distance dependence, and host materials must provide for spatial positioning to maximize UC efficiencies. Crystals doped with ions that have a smaller or larger ionic radius than the host material cations have larger lattice stress and structural defects which increases non-radiative relaxation rates. Na^+, Ca^{2+}, and Y^{3+} are three cations of similar size to lanthanide dopants making them preferred cations in host lattices. Anions in host lattices can influence stability of the excited state because of the associated phonon energies (vibrational modes) of the crystal lattice that they compose. Fluoride is a particularly suitable anion for applications in UC host crystal because of the low phonon energies exhibited by its crystal lattice (Suyver et al., 2006; Wang and Liu, 2008; Chen, Ma, and Liu, 2007). A summary of phonon energies of common materials used for UC host lattices is presented in Table 5.1 (Diamente, Raudsepp, and van Veggel, 2007).

Many materials have been used to investigate the mechanism of UC including YAG, fluoride glass, LaF_3, and phosphate glass, which was made to investigate cooperative upconversion (CUC) mechanism. However, of the many different materials available, $NaYF_4$ has been found to be one of the most efficient host-lattice materials at the nanoscale for lanthanide-based UC (Haase and Schäfer, 2011; DaCosta et al., 2014; Suyver et al., 2006; Suyver et al., 2005). The efficiency of UC in $NaYF_4$ materials varies as a function of nanoparticle size and crystal phase. The two phases for $NaYF_4$ nanoparticles are the cubic phase (α-$NaYF_4$) and the hexagonal phase (β-$NaYF_4$). UC efficiency of nanoparticles in the β-$NaYF_4$ phase is found to be an order of magnitude greater than crystals in the α-$NaYF_4$ and this difference in UC efficiency arises from ion–ion distances and crystal symmetry. Together, these factors work to lower the UC efficiency from the α-$NaYF_4$ phase relative to the β-$NaYF_4$ crystal phase and result in changes of the probability of electronic transitions (Kramer et al., 2004; Aebischer et al., 2006; Klier and Kumke, 2015). Beyond the positioning of dopant ions and the selection of host materials, UC in nanoparticles can be modulated by design of the core and core/shell structure.

UCNP Core and Core/Shell Structures

Luminescence from UCNPs is in competition with non-radiative relaxation, with one important pathway being the interface between the UCNP and the surrounding environment. Typically, environments are a non-aqueous solvent after UCNP synthesis, or a capping ligand that assists the UCNP to dissolve in aqueous solution if there is interest in biomedical applications. Such capping ligands often contain amine and hydroxyl functional groups, which can quench UCNP luminescence via multiphonon relaxation processes associated with vibrational modes of the functional groups. Minimizing the surface quenching effects requires protection of the dopants in the host-lattice material, and this can be accomplished by coating the core with a shell layer to reduce relaxation at the outer surface (Haase and Schäfer, 2011; DaCosta et al., 2014; Wang and Liu, 2009). Many variations of core/shell structures have been explored to improve quantum yield, tune spectral properties, and impart further physiochemical properties to the core nanoparticles.

The Undoped Shell

Of the different types of core/shells, one of the simplest and most common structures involves the doped core material shelled with a passive, undoped host lattice. The shell in these structures is composed of the same host-lattice materials as the core but without any dopants, i.e. $NaYF_4$-based shell for $NaYF_4$:Yb^{3+}/Tm^{3+} or Er^{3+} type core. A requirement for these types of core/shell structures is that the interface between the doped core/undoped shell should be continuous and of high quality, and should not generate new quenching sites. Core/shell structures have been reported to yield fifteen- to thirtyfold increases in UC emission intensity when compared with bare cores (Haase and Schäfer, 2011; DaCosta et al., 2014; Wang et al., 2010; Wang and Liu, 2009; Yi and Chow, 2007). However, the extent of enhancement is reduced after ligand exchange using hydrophilic functional groups to allow the NPs to enter aqueous solution. The shell

TABLE 5.1 Common Host-Lattice Material and Their Associated Phonon Energies. (Adapted with permission from DaCosta et al., Copyright 2014 Elsevier.)

Material	Highest Phonon Energy [cm^{-1}]
Phosphate glass	1,200
Silica glass	1,100
$LaPO_4$	1,050
YAG (yttrium aluminium garnet)	860
YVO_4	600
Fluoride glass	550
Chalcogenide glass	400
$NaYF_4$	370
LaF_3	300
$LaCl_3$	240

thickness for these particles is one of the factors influencing the degree by which UC intensity is enhanced. UCNPs with thicker core/shells of >10 nm scale better passivate the cores from surface quenching in comparison with thinner shells. However, not all applications of UCNPs benefit from the addition of thick shells (Boyer et al., 2009).

Luminescence resonance energy transfer (LRET) is a near-field phenomenon operating in a distance-dependent manner. It is a dipole–dipole interaction taking place between an excited state donor and a ground state acceptor. Upon excitation of a donor particle such as a UCNP, dipole–dipole interactions with a ground state acceptor in the near field (1–10 nm) results in energy transfer to the acceptor with concomitant non-radiative relaxation of the donor. A requirement for close proximity of donor and acceptor presents a challenge for core/shell UCNP systems that are intended for LRET applications. A thick shell for UCNPs results in lowering of LRET efficiency, while thinner shells have lower intrinsic UC luminescence due to non-radiative relaxation pathways and thus lower LRET efficiency. The optimal shell thickness for an LRET system is obtained as a compromise of the two factors influencing the LRET efficiency (Wang et al., 2011).

The Doped Shell

UCNP systems that utilize core/undoped shell structures for UC are typically based around the ETU mechanism where the addition of an inert host-lattice layer to the core enhances luminescence efficiency (Haase and Schäfer, 2011; DaCosta et al., 2014; Wang et al., 2010; Wang and Liu, 2009). However, alternative UC mechanisms and luminescent properties are available via the application of doped shell materials. The core/shell model of UCNPs allows for segregation of different dopants into regions of the nanoparticle. One example of this is UCNPs that use energy migration-based upconversion (presented in Section 1.2.5). Energy migration upconversion (EMU) is achieved by separating sensitizers and activator into core and shell, respectively. Accumulator and migrator dopant ions are then added to shuttle energy from excited state sensitizer ions to activators in the shell region allowing for radiative luminescence. The advantage of this technique is greater control over the luminescence profile of the UCNP (Wang et al., 2011).

Doped shells are also being investigated to obtain UCNPs that luminescence in multiple colour channels. Systems co-doped with both Tm^{3+} and Er^{3+} luminesce primarily in the Er^{3+} colour channels because emission from Tm^{3+} emitters are quenched and energy transfer from Yb^{3+} to Er^{3+} may be preferred. However, if core particles are made with one type of activator and shell material is made from the second type of activator, then the UCNP can luminesce in both blue and green colour channels. To investigate this, core particles made of $NaYF_4$ doped with Yb^{3+} and Tm^{3+} (A) were synthesized with shell material of $NaYF_4$ doped with Yb^{3+} and Er^{3+} (B). This AB system showed the presence of both blue and green colour channels in one

nanoparticle, and synthesis of an ABA structure showed an even greater enhancement of intensity in the green colour channel. Core/shell structures composed of $NaYF_4$ form only one subcategory (Haase and Schäfer, 2011; DaCosta et al., 2014; Wang et al., 2010; Wang and Liu, 2009; Qian and Zhang, 2008). There are many other types of shell materials, each offering different spectral performance advantages to UCNPs.

Silica- and CaF₂-Based Shell Materials

One of the most commonly used and investigated materials for shelling nanoparticles is silicon dioxide (silica). Silica-based materials have many attractive features that make them suitable for integration with nanoparticles including ease of functionalization via application of well-known silica-based chemistry, chemical stability, and optical transparency (Haase and Schäfer, 2011; DaCosta et al., 2014; Wang et al., 2010; Wang and Liu, 2009; Muhr et al., 2014; Hu et al., 2009; Liu et al., 2014). Synthesis of silica shell thickness is also well investigated with thicknesses ranging from 2 to 10 nm. Silica shells have been reported to enhance UC luminescence and have also been used to facilitate LRET with acceptor dyes and quantum dots. Mesoporous silica shells have also been used to encapsulate photosensitizers and drug molecules, thereby facilitating therapeutics with simultaneous sensing of the *in vivo* environment (Hu et al., 2009; Liu et al., 2014).

There are many other shell materials including calcium fluoride (CaF_2)-based materials. CaF_2 has many of the same advantages as silica. Biological interaction with UCNPs coated with CaF_2 tends to be minimal because the calcium and fluoride components of the shell are native to calcified tissues in biosystems. CaF_2 is most commonly synthesized for cubic phase (α-$NaYF_4$) UCNPs because the crystal structure of the two materials has lower lattice stress than hexagonal-phase (β-$NaYF_4$) UCNPs. This hetero-core/shell structure may also offer an advantage over the previously mentioned materials. The mismatch of physiochemical properties of $NaYF_4$ material and CaF_2 may prevent leakage of dopant ions from the core nanoparticle to the environment. This may preserve the luminescent properties of UCNPs for extended periods and might be a significant factor for further biological applications of upconverting nanoparticles (Prorok et al., 2014; Shen et al., 2012; Chen et al., 2012).

5.1.2 Theories of Photon Upconversion

There are presently many distinct mechanisms proposed to explain photon UC for lanthanide-doped nanocrystals, operating both individually and in combination with one another, based on the sequential absorption of at least two low-energy (near-IR) photons. These mechanisms are broadly divided into classes of UC including ESA (Haase and Schäfer, 2011; DaCosta et al., 2014), addition de photon par transferts d'energie (APTE) (Haase and Schäfer, 2011; DaCosta et al., 2014; Auzel, 2004; Auzel, 1966; Auzel, 1973;

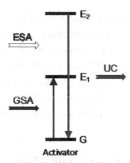

FIGURE 5.4 Energy level diagram representing the ESA-based mechanism of UC. Here, the energy levels are represented by G, E_1, and E_2 for ground, first excited, and second excited states, respectively. GSA represents the ground state absorption and UC represents upconversion luminescence. (Adapted with permission from DaCosta et al. Copyright 2014 Elsevier.)

Auzel, 1990), CUC (Auzel, 2004), photon avalanche (PA) (Haase and Schäfer, 2011; DaCosta et al., 2014), and EMU (Wang et al., 2011). APTE, also known as energy transfer upconversion, ESA, and PA are among the most efficient of the UC processes in nanocrystals. Much effort has gone into investigating synthetic methods for maximizing UC efficiency for the ESA and ETU processes. Each class of the mechanisms is represented pictorially via energy level diagrams (see Figure 5.4) with detailed explanations being presented in the following sections.

Excited State Absorption (ESA)

ESA is the only single dopant-based UC process consisting of the sequential absorption of multiple low-energy photons by an activator ion. In ESA, an electron in the ground electronic state (denoted as G) is excited to the first (metastable) excited electronic state (denoted as E1) via the absorption of a low-energy photon. This electronic transition, known as ground state absorption or GSA, is only achieved if the photon meets the resonance conditions between the G and E1 states. Metastable excited states in activator ions have long lifetimes (many micro- to milli-seconds) allowing for the electron in E1 to undergo further electronic transitions. These subsequent transitions are known as excited state absorption, and they result from absorption events where photons meet the resonance conditions between the different excited states. The UC process in ESA depends on the absorption of two photons meeting resonance conditions for two different electronic transitions. In specific dopant ions, the energetic spacing of these excited states is equal allowing for the use of a single-photon sources that meet the energy and resonance conditions of the electronic transitions. As mentioned previously, the dopant concentrations of activators are relatively low in host-lattice structures and the small cross sections of dopant ions mean that UC efficiency of ESA mechanisms is relatively poor. Co-doping of host-lattice structures with sensitizer ions improves UC efficiencies by the introduction of alternative mechanisms of UC (DaCosta et al., 2014; Auzel, 2004; Auzel, 1973; Gamelin and Güdel, 2000a).

Addition de Photon Par Transferts d'Energie (APTE) or Energy Transfer Upconversion (ETU)

ETU is one of the UC mechanisms related to ESA that can be found in nanocrystals doped with both sensitizer and activator ions. These dopant ions are present in proximity to one another in the host-lattice crystal and photon absorption by the sensitizer and activator ions in the nanoparticle result in electronic transitions (via GSA). After GSA of the dopant ions, both sensitizer and activator are in a metastable excited state. It is helpful to consider the case of one sensitizer and one activator ion in their respective excited states (see Figure 5.5). The sensitizer ion would undergo a non-radiative energy transfer with the neighbouring activator ions resulting in simultaneous electronic transition (via ESA) for the activator, and relaxation to the ground state for the sensitizer. Efficiency of energy transfer can be modulated by varying the distance between dopant ions that in turn varies the degree of overlap for the ion pair. In cases where the ion pair have energetic mismatches between electronic transitions, energy transfer must be assisted by phonon relaxation to achieving resonance. The advantages of co-doping nanoparticles are apparent when discussing UC efficiency which is reported to be two orders of magnitude greater for ETU than ESA. Design of UCNPs has been directed to take advantage of the enhanced UC efficiency of ETU, and most UCNPs reported in literature are based on co-doped particles (Haase and Schäfer, 2011; DaCosta et al., 2014; Wang et al., 2010; Wang and Liu, 2009; Auzel, 2004; Auzel, 1966; Auzel, 1973; Auzel, 1990).

Photon Avalanche (PA)

PA is another class of UC found to occur in nanoparticles that are co-doped with sensitizer and activator dopants. It is one of the more complex mechanisms for UC requiring a combination of both ESA and ETU to generate luminescence. PA can be described by three distinct nonlinear behaviours including transmission, emission, and rise time associated with pumping power intensity. To aid in the

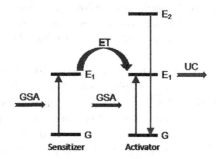

FIGURE 5.5 Energy level diagram representing the ETU-based mechanism of UC. Here, the energy levels are represented by G, E_1, and E_2 for ground, first excited, and second excited states, respectively. GSA represents the ground state absorption, ET represents energy transfer, and UC represents upconversion luminescence. (Adapted with permission from DaCosta et al. Copyright 2014 Elsevier.)

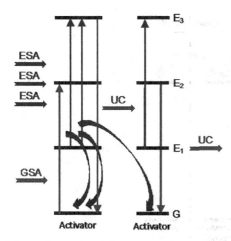

FIGURE 5.6 Energy level diagram representing the PA-based mechanism of UC. Here, the energy levels are represented by G, E_1, E_2, and E_3 for ground, first excited, second excited, and third excited states, respectively. GSA represents the ground state absorption, curved arrows represent energy transfer, ESA represents excited state absorption, and UC represents upconversion luminescence. (Adapted with permission from DaCosta et al. Copyright 2014 Elsevier.)

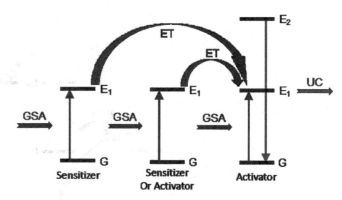

FIGURE 5.7 Energy level diagram representing the CUC-based mechanism of UC. Here, the energy levels are represented by G, E_1, and E_2 for ground, first excited, and second excited states, respectively. GSA represents the ground state absorption, curved arrows denoted with ET represent energy transfer, and UC represents upconversion luminescence. (Adapted with permission from DaCosta et al. Copyright 2014 Elsevier.)

description of PA-based UC, a schematic (see Figure 5.6) is illustrative. For PA, four distinct energy levels will be considered where three are meta-stable excited states (denoted as E1, E2, and E3, respectively), and the fourth is the ground state (G). An electron in G is initially excited and promoted by GSA to E2. This electron subsequently relaxes down to E1 and in the process can either a) promote another ground state electron in the **same** ion to E1 or b) promote a ground state electron in **another** ion (has to be **neighbouring**) to E1. The second process is also known as ion-pair relaxation energy transfer process. Next, the electrons in E1 are excited to E3 using the same photon source as the initial excitation event. These electrons in E3 relax back down to E1, undergoing ion-pair relaxation energy transfer with two electrons. The two electrons can be in the same ion or neighbouring ions or a combination of the two, resulting in a population of ions in E1. ESA excitation results in the transition of electrons from E1 to E3 generating a population of electrons in the E3 energy level. These processes repeat until a large group of electrons populate the E3 energy level, and radiative relaxation from E3 to G results in luminescence. PA when compared with other UC mechanisms is found to have higher UC efficiency than ESA and lower efficiency than ETU. There are further challenges with PA-based systems including delay in UC emission due to pumping of electrons to higher electronic levels and lower emission due to weak GSA (DaCosta et al., 2014; Wang et al., 2010; Auzel, 2004).

Cooperative Upconversion (CUC)

CUC is an additional class of UC based on nanoparticles with sensitizers and activators, using a combination of ESA and ETU mechanisms. CUC has only had minimal impact on nanoparticle design because it is not as efficient at the nanometer size scale as ETU. Studying CUC is also mechanistic complex because it is based on the simultaneous energy transfer from two dopant ions to a third dopant ion. This type of three-dopant system for UC (see Figure 5.7) typically involves two excited sensitizer/activators and a third activator (Wang et al., 2010; Auzel, 2004).

Energy Migration Upconversion (EMU)

EMU is a fairly complex UC mechanism utilizing the core/shell architecture of NPs and at least four different types of dopant ions including sensitizers, accumulators, migrators, and activations. The advantage of designing NPs with EMU is the enhanced tunability of UCNP luminescence. In previous mechanisms, activator dopants were required to have long lifetimes with respect to their excited states. This is no longer a critical factor for EMU-based activators because energy from photon interactions is stored in the other dopants of the host lattice before radiative relaxation by the activator. The mechanism for EMU (see Figure 5.8) begins with GSA by the sensitizer ion. Next, energy transfer and ESA processes occur with the neighbouring accumulator ions. This results in promotion of an electron in the accumulator to an excited state. These accumulator ions possess long-lived excited states to sustain the energy obtained from sensitizer ions. Accumulators then transfer the energy to migrator ions that then transfer energy to activators which undergo radiative relaxation. To shuttle energy within the crystal lattice of the nanoparticles effectively, UCNPs segregate dopants into different regions of the larger structure. Sensitizer and accumulators are sequestered to the core of the particle while migrator ions can be isolated in layers of shells. The final shell contains activator ions to complete the energy migration circuit (Wang et al., 2011).

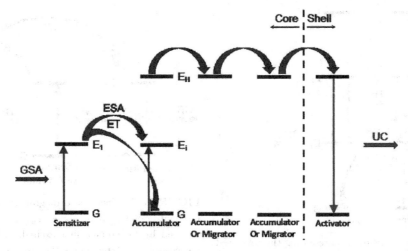

FIGURE 5.8 Energy level diagram representing the energy migration (EMU)-based mechanism of UC. Here, the energy levels are represented by G, E_i, and E_H for ground, initial excited, and high excited states, respectively. GSA represents the ground state absorption, curved arrows denoted with ET represent energy transfer, ESA represents excited state absorption, and UC represents upconversion luminescence. (Adapted with permission from DaCosta et al. Copyright 2014 Elsevier.)

5.2 Synthesis Methods

5.2.1 Overview of UCNP Synthesis Procedures

Synthesis of UCNPs has been extensively studied over the past two decades. In general, most UCNP applications require a narrow size distribution, pure crystal phase, and chemical compositions that display specific spectroscopic properties. Several synthetic approaches exist for creating UCNPs that satisfy the aforementioned criteria. The majority of synthetic methods can be divided into two categories: solvent-free synthesis and solution-based synthesis (Yan et al., 2016). Popular solvent-free synthetic approaches include vapour deposition (Swihart, 2003), combustion synthesis (Mansurov, 2013), and flaming synthesis (Qin, Yokomori, and Ju, 2007). Solution-based synthesis strategies include thermal decomposition (Mai et al. 2006), thermal coprecipitation (Stouwdam and van Veggel, 2002), micro-emulsion method (Ghosh and Patra, 2008), hydro-/solvothermal method (Tian et al., 2012), and other techniques. Fabricating UCNPs in solution is far more common than solvent-free synthetic routes. The preference for solution-based UCNP synthesis is due to reproducibility, versatility, and convenience (Chang et al., 2014). Reaction parameters in solution-based synthetic approaches can be adjusted to consistently yield nanoparticles of desired crystal morphology and composition. Solution-based approaches also produce nanoparticles that are conveniently dissolved in solvents that are required for further chemical modifications (Yan et al., 2016). The two most common solution-based UCNP syntheses are now explored in greater detail and include the thermal decomposition method and the thermal coprecipitation method.

The underlying mechanism of UCNP formation in solution-based synthesis procedures is analogous to the basic principles of crystallization. Work by Victor LaMer is often extended to describe the fundamentals of nucleation and growth processes of nanocrystal development (LaMer and Dinegar, 1950). Figure 5.9 depicts the various stages of development that take place in nanoparticle fabrication. Particle synthesis based on the LaMer mechanism is set up in such a way that the concentration of unstable, partially soluble reactants (monomers) rises rapidly in solution. As the concentration of monomer increases over time, it reaches a point of supersaturation. This is referred to as phase I in the LaMer plot. Nanocrystal nucleation and particle growth are not evident at this stage of development. In phase II, the monomer concentration reaches the critical

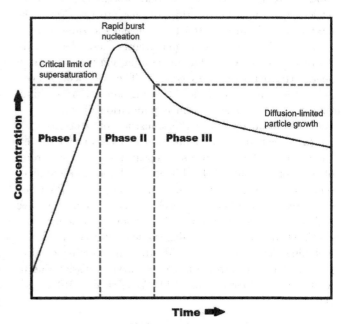

FIGURE 5.9 LaMer plot describing nucleation and growth processes in particle synthesis. (Adapted with permission from LaMer and Dingar. Copyright 1950 American Chemical Society.)

limit of supersaturation and exceeds this threshold. A rapid burst of nucleation occurs beyond the critical limit. The formation of nuclei depletes monomer concentration in solution, thereby relieving the stress of supersaturation. By the time phase III begins, no further nucleation events can take place because the monomer concentration is below the critical limit of supersaturation. During this stage, the newly formed nuclei undergo a slow diffusion-limited particle growth period until all monomers are consumed (Wang and Nann, 2010).

The LaMer mechanism plays a critical role in separating the nucleation stage from the particle growth stage in nanocrystal formation. However, the theoretical model fails to clarify the effects of particle growth that occur at sufficiently low monomer concentrations in phase III of the LaMer plot (Viswanatha and Sarma, 2007). In fact, a coarsening of the particle size distribution is observable even after the supposed depletion of monomer supply in solution. Work by Wilhelm Ostwald as well as the mathematical model based on Ostwald's work by Lifshitz and Slyozov, and Wagner helps explain the particle growth process in a theory now famously known as LSW (Ostwald, 1900; Lifshitz and Slyozov, 1961; Wagner, 1961). Based on this theory, the diffusion-limited growth of nuclei in phase III of the LaMer plot is dominated by energy associated with surfaces of these newly formed nanocrystals (Thanh, Maclean, and Mahiddine, 2014). Smaller particles tend to have a larger surface–to-volume ratio than large particles; thus, it can be deduced that surface free energy associated with smaller particles makes up a significant portion of the total energy of the particle. Driven by the need to achieve thermodynamic equilibrium, smaller particles with high surface energy must assimilate with larger particles to reduce the overall surface energy (Viswanatha and Sarma, 2007). Growth is dictated by the diffusion of monomers to the surface of a nanoparticle as well as the reactions that occur at this surface. Therefore, smaller particles dissolve and create a monomer population that will migrate to assemble onto larger particles, thereby reducing the surface free energy of the system (Wang and Nann, 2010; Viswanatha and Sarma, 2007). This diffusion-limited particle growth phenomenon that takes place in solution is referred to as Ostwald ripening.

5.2.2 Synthesis of Hydrophobic UCNP Core

Thermal Decomposition

Thermal decomposition is the most commonly practised method of UCNP synthesis. As the name suggests, the thermal decomposition method involves the breakdown of organometallic precursors under high heat conditions creating compounds that associate with host-lattice material in high boiling point organic solvents (Yan et al., 2016). This method originates from efforts by Mai and coworkers on the controlled synthesis of sodium rare-earth tetrafluoride nanocrystals (Mai et al., 2006). Since then,

this synthetic procedure has undergone various changes for creating nanoparticles with different host lattices, lanthanide dopants, and crystal structures (Naccache et al., 2009; Yang et al., 2011; Huang et al., 2014). Further discussion of thermal decomposition synthesis will be in reference to the well-known $NaYF_4$:Yb,Er system.

Numerous factors need to be taken into consideration when synthesizing UCNPs by thermal decomposition. A typical reaction setup for the synthesis of $NaYF_4$:Yb,Er UCNPs is illustrated in Figure 5.10. A three-necked round-bottom flask is equipped with a temperature probe, flow control adapter, and an adapter with a distillation apparatus in each respective flask opening. Organic solvents are introduced to the flask along with adequate amounts of lanthanides and host-lattice material to achieve a desired nanocrystal composition. The reaction vessel is fitted into a heating mantle connected to a voltage control unit. A magnetic stir bar is also introduced into the reaction vessel, and a stir plate underneath the vessel is used to mix all components creating a homogeneous solution. In an effort to create an inert environment, the vessel is evacuated, and the decomposition reaction is performed under a steady flow of argon. The reaction temperature is dependent on the types of lanthanide precursors that are used, and the crystal phase and morphology that are desired for further applications. Sufficient control over the reaction temperature is

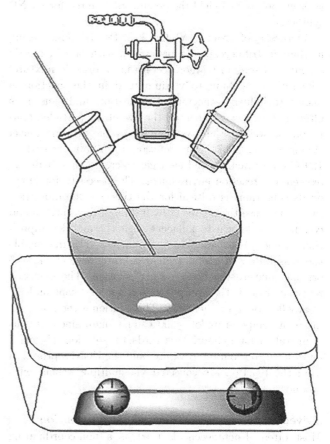

FIGURE 5.10 Reaction setup for thermal decomposition synthesis of UCNPs.

compulsory in this setup; typically, a temperature tolerance range of $\pm 1°C$ is required to produce monodisperse UCNPs.

There are many variations of the generalized reaction setup described above; however, for the sake of simplicity, this setup will be of primary focus as the thermal decomposition method is further described.

Lanthanide Precursors in Thermal Decomposition

UCNP synthesis via thermal decomposition is typically done with fluoride precursors of lanthanide ions. Zhang et al. were the first to demonstrate the thermal decomposition synthesis of LaF_3 triangular nanoplates using lanthanum trifluoroacetate precursors (Zhang et al., 2005). Work by Mai and coworkers as well as separate efforts by Boyer and coworkers also demonstrate the use of rare-earth trifluoroacetates $RE(CF_3COO)_3$ (Boyer et al., 2006; Mai et al., 2006). The widespread use of $RE(CF_3COO)_3$ precursors for UCNP synthesis is a result of the well-understood production and thermal decomposition mechanism of lanthanide trifluoroacetate complexes (Garner and Hughes, 1975; Logvinenko et al., 2003). Fluoride precursors are typically made by mixing lanthanide oxides in a 1:1 mixture of water and trifluoroacetic acid, and refluxing this mixture at $120°C$ for 12 h (Roberts, 1961). Initially, this mixture has a cloudy appearance. As the formation of lanthanide trifluoroacetates progresses, the solution becomes clear. The solvent is later evaporated to yield the precursors required for UCNP synthesis.

Synthesis of $NaYF_4$:Yb,Er UCNPs is done using lanthanide trifluoroacetate precursors with stoichiometric amounts of sodium trilfuoroacetate in an organic mixture. The ratio of sodium to yttrium used in the reaction is known to influence the rate of nucleation, and thus, it is often used as a way to control the size of the particle population, as well as the crystal phase of the growing nuclei (Liang et al., 2007). The solution is initially heated to $120°C$ under vacuum and inert gas to create an anhydrous, oxygen-free reaction environment. The necessity for anhydrous conditions is critical for the thermal decomposition regime because the degree of hydration of rare-earth metal complexes is known to influence their thermal decomposition profiles. Heat and vigorous magnetic stirring enable the dissolution of metal trifluoroacetate reagents in the organic mixture to create a clear solution. The degassed reaction vessel is then purged with argon and kept under a steady flow of argon thereafter. The solution is then brought up to a temperature of $\geq 300°C$ and maintained at this temperature for an hour. Afterwards, the solution is brought back down to room temperature, and the UCNPs produced from the reaction are collected via multiple washes with ethanol.

Solvent Conditions in Thermal Decomposition

Most often, 1-octadecene is used as a non-coordinating solvent to tolerate the necessary elevated temperature conditions of UCNP synthesis without evaporation. A capping ligand is incorporated in the 1-octadecene solution to regulate the growth of nanocrystals via coordination to the surface of growing nuclei, thus hindering particle growth reactions at the surface of the nanoparticle (Altavilla, 2016). In general, a capping ligand consists of a polar head group capable of coordinating to the nanoparticle surface and a long hydrocarbon chain that adds a hydrophobic character to nanoparticles rendering them soluble in organic media. Typically, oleic acid serves the function of a capping ligand in a thermal decomposition synthesis to yield hydrophobic UCNPs. The ratio of oleic acid to 1-octadecene has a great influence on the shape of the nanocrystal. Li and Zhang demonstrated that by simply adjusting the concentration of oleic acid in solution, one can create hexagonal nanoplates, spherical nanocrystals, and elliptical nanocrystals, respectively (Li and Zhang, 2008).

Aside from the nanocrystal shape, the crystal phase of the UCNP is also affected by the solvent conditions. Hexagonal-phase nanocrystals are more thermodynamically favourable than cubic phase, but in a thermal decomposition synthesis, the cubic phase nanocrystal forms first (Li and Zhang, 2008). Nevertheless, cubic phase nanoparticles grown in a mixture of oleic acid and 1-octadecene can transition into their hexagonal-phase counterparts through atomic rearrangement (Pin et al., 2018). One can facilitate this phase shift simply by elevating the reaction temperature providing sufficient energy to overcome the energy barrier separating the cubic phase from the hexagonal. However, working at elevated temperatures tends to yield larger nanoparticles; therefore, a trade-off between the more desirable hexagonal-phase nanocrystal and smaller biologically applicable nanocrystal size is imminent (Ostrowski et al., 2012). As such, producing sub-20 nm hexagonal-phase $NaYF_4$:Yb,Er nanoparticles in an oleic acid and 1-octadecene mixture is challenging.

Alternatively, addition of oleylamine (OLA) to the oleic acid and 1-octadecene mixture can also be used to promote the phase transition from cubic to hexagonal without increasing the reaction temperature (Mai et al., 2006). OLA coordination to the lanthanide ion occurs through an amine moiety which is significantly weaker than the carboxylic functionality of oleic acid. At temperatures greater than $250°C$, OLA can consume oleic acid to become N-oleyloleamide (Niu, Wu, and Zhang, 2011; Ostrowski et al., 2012). The neutral charge of OLA is more accommodating to the surface charge density of a hexagonal crystal phase than cubic crystal phase of $NaYF_4$ (Wang et al., 2010). The weak coordination of OLA to the lanthanide monomer allows greater reactivity, and subsequently, a greater number of nucleation events can take place to produce a large number of small nanoparticles (≤ 10 nm) (Ostrowski et al., 2012).

Thermal Coprecipitation

Although the trifluoroacetate precursors can be used to yield small, monodisperse nanoparticles with uniform crystal phase, thermal decomposition of trifluoroacetate reagents also produces toxic oxyfluorinated carbon by-products that must be handled with great caution (Chen, Li, and Shi, 2016; Liu et al., 2013). Therefore, more user-friendly synthetic approaches have emerged through further developments in UCNP synthesis. High-temperature thermal coprecipitation is one such method of creating UCNPs comparable in quality to particles produced via thermal decomposition without generating toxic side products (Yan et al., 2016). The thermal coprecipitation method is nearly identical to the thermal decomposition procedure; both methods utilize high heat to nucleate lanthanide and host-lattice reagents dissolved in degassed, dehydrated organic mixtures. However, the defining feature of the thermal coprecipitation technique is in its use of inorganic rare-earth salts instead of trifluoroacetate reagents. Use of lanthanide chlorides as well as lanthanide acetates in this form of UCNP synthesis has gained significant traction over the years due to their simplicity in preparation and low cost.

Lanthanide Precursors in Thermal Coprecipitation

Synthesis of lanthanide acetates is similar to lanthanide trifluoroacetates, except for the use of glacial acetic acid in a 1:1 mixture with water in place of trifluoroacetic acid. These lanthanide acetates can be made anhydrous through a further reaction with acetic anhydride (Edwards and Hayward, 1968). Lanthanide chlorides are synthesized by dissolving lanthanide oxides in hydrochloric acid at elevated temperatures. In efforts to produce lanthanide chlorides under milder conditions, chlorination of metals through thionyl chloride in water is performed at room temperature to yield lanthanide chloride products (Dell'Amico et al., 1995). The latter chlorination reaction is more commonly conducted than the former for UCNP synthesis because lanthanide chlorides produced via thionyl chloride are anhydrous. Much like the thermal decomposition method, the thermal coprecipitation method requires anhydrous solvent conditions. The hydroxyl group of water coordinating to growing nuclei surfaces can act as a capping ligand, thereby influencing the crystal size and shape in a manner that may not be intended. Use of anhydrous lanthanide reagents for particle synthesis also allows the accurate weighing of small amounts of lanthanide salts without the discrepancies associated with varying amounts of hydrate present.

UCNP fabrication using either lanthanide acetates or lanthanide chlorides requires the addition of an external sodium and fluorine source (Li, Zhang, and Jiang, 2008; Qian et al., 2009; Wang, Deng, and Liu, 2014). Stoichiometric amounts of ammonium fluoride and sodium hydroxide are dissolved in methanol via sonication. This methanol solution is injected into the degassed reaction solution at room temperature under a steady flow of argon. A cloudy solution is created upon mixing of all host-lattice materials in the organic mixture. At this point, methanol is distilled out of the reaction solution. The reaction vessel can then be brought up to temperatures suitable for nucleation and particle growth similar to the thermal decomposition procedure.

5.2.3 Synthesizing Core-Shell Nanoparticles

UCNPs are highly susceptible to surface quenching effects in aqueous environments. Phonons in the host matrix of the nanocrystal can provide for non-radiative relaxation through energy transfer interactions with high-energy vibrational modes of weakly bound hydroxyl and amine functionalities of capping ligands at the particle surface (DaCosta et al., 2014). This vibrational relaxation occurs through activators and sensitizers near the UCNP surface; therefore, the effect of quenching is more pronounced in smaller nanoparticle sizes as a result of the high surface area-to-volume ratios that these nanoparticles exhibit. In an effort to reduce such detrimental energy transfer processes from taking place, an inert shell is created around the nanoparticle core to separate the core dopant ions from capping ligands at the surface (Chang et al., 2014; Yan et al., 2016). There are several synthetic approaches for developing a core–shell UCNP. Two of the most commonly used shell synthesis procedures include the seed-mediated shell growth method and self-focussing by Ostwald Ripening method (Liu et al., 2013).

The seed-mediated shell growth is by far the most common method used in creating core-shell nanostructures (DaCosta et al., 2014). In this method, core-only UCNPs are first created using one of the many synthetic procedures previously described. The cores are separated from the reaction solution via multiple ethanol washes and are then dispersed in hexanes. Next, the hexane solution containing the core UCNPs is introduced to an organic mixture with shelling precursors. The hexanes are evaporated from the reaction vessel, and the core UCNPs are used as seeds for an epitaxial shell growth at high temperatures (Yi and Chow, 2007; Qian and Zhang, 2008). In order to grow a uniform shell around the core, the core UCNPs must have isotropic geometry, and a low lattice mismatch must exist among the core and shell (Liu et al., 2013). For the $NaYF_4$:Yb,Er model system, the $NaYF_4$ inert shell is commonly used to enhance the luminescence efficiency of the UCNP. Thickness of the shell is controlled by adjusting the concentration of shelling precursors added to the reaction solution (DaCosta et al., 2014). The isolation of the core synthesis reaction from the shelling reaction is critical in maintaining the integrity of the shell. Slight variations to the seed-mediated shell growth have been reported in literature in which the shelling precursor is added directly into the reaction solution used to create core nanoparticles. This results in mixing of unreacted core activators and sensitizers into the shell structure causing sub-optimal protection from surface quenching.

Although the seed-mediated epitaxial growth of shell on UCNP cores is commonly done, it is by no means an ideal synthetic approach. Seed-mediated shell growth is limited by the possibility that nucleation of shelling precursor can occur at high concentrations resulting in polydisperse, inhomogeneous nanoparticle distributions. For this reason, adequate control over the shell thickness is often a challenge with the seed-mediated growth method (Haase and Schäfer, 2011). Conversely, epitaxial shell growth using the method of self-focussing by Ostwald ripening does not suffer from the same limitations that effect the seed-mediated approach. This strategy of shell growth was first realized by van Veggal and coworkers who used small α-NaYF$_4$ nanocrystals as shell precursors for synthesizing β-NaYF$_4$:Yb,Er@NaYF$_4$ core–shell nanoparticles (Johnson et al., 2012). In this procedure, small α-NaYF$_4$ nanoparticles are first synthesized using thermal decomposition and nucleation of trifluoroacetate reagents at 290 °C for 30 minutes. These cubic phase cores are isolated and purified through washes with ethanol and then dispersed in 1-octadecene. After synthesizing β-NaYF$_4$:Yb,Er cores at 300 °C, the small α-NaYF$_4$ shelling precursors are injected into the reaction solution. By introducing smaller nanocrystals into a distribution of larger hexagonal-phase cores, the Ostwald ripening effect is exploited to grow an epitaxial shell layer on the larger β-NaYF$_4$:Yb,Er cores through rapid dissolution of the smaller α-NaYF$_4$ particles (Voss and Haase 2013). The shell thickness is adjusted by the amount of α-NaYF$_4$ particles injected into the reaction solution (Johnson et al., 2012).

The Ostwald ripening method of epitaxial shell growth has undergone significant changes since its inception. Injection of shelling precursor material at high temperatures can result in changes in temperature of the reaction solution due to mixing of the two solutions. Work by Haase and coworkers on the fabrication of sub-10 nm β-NaYF$_4$:Yb,Er@NaYF$_4$ UCNPs is representative of modification of the Ostwald ripening method through the manipulation of sodium to yttrium ratio (Rinkel et al., 2015). In their modified method, the α-NaYF$_4$ shelling precursors were prepared with a low Na/Y ratio, which results in a small number of β-phase seeds, and slower transition to the hexagonal phase at higher temperatures. The β-NaYF$_4$:Yb,Er cores were prepared with a high Na/Y ratio in order to promote a greater number of β-phase seeds. This results in formation of a large number of small hexagonal-phase nanoparticles in a short period of time at elevated temperatures. After isolated preparation and purification of these nanocrystals, the two are introduced to an organic mixture and heated to 300°C. The prolonged period of time required for the α-NaYF$_4$ to transition from α→β is crucial for this reaction because it prevents unwanted β-NaYF$_4$ particles in the shelling procedure. Only the β-NaYF$_4$:Yb,Er particles act as the β-phase seeds that consume monomers released from the dissolving α-NaYF$_4$ particles, resulting in highly uniform core–shell nanostructures.

5.3 Introduction to Surface Coatings for Biological Applications

UCNPs have garnered substantial interest for high-resolution imaging in biological systems due to their unique optical properties. An increasingly relevant application for fluorescence-based imaging is the ability to selectivity label sub-cellular structures (Kang and Kim 2010). Many molecular fluorescent dyes have been designed to selectively image biological systems. However, these conventional fluorescent dyes have drawbacks including photobleaching from prolonged irradiation with the excitation source and poor signal-to-noise ratio from autofluorescence of cellular material (DaCosta et al. 2014) UCNPs overcome many of these challenges including improved signal-to-noise ratio due to low background from near infrared (NIR) excitation sources and greater luminescent stability even after prolonged excitation (DaCosta et al. 2014).

Control of surface chemistry is vital in the design of interfacial properties of UCNPs for application in biological systems. Control and design of ligand chemistry on the surface of UCNPs can be used to endue specific bulk properties to nanoparticles in solution as required for biological analyses. The ideal surface chemistry of UCNPs must satisfy the following criteria for biological applications:

1. The surface coating must provide colloidal stability to nanoparticles in physiological conditions.
2. Surface coatings must maintain a small hydrodynamic diameter, particularly for intracellular analysis (average size <60 nm). (Jin et al. 2011)
3. UCNPs must retain their luminescence upon chemical modification and transfer to aqueous phase
4. The chemical coating must resist degradation *in vitro* and *in vivo*
5. The surface chemistry must be amenable for further bioconjugation.

As synthesized, UCNPs coated with organic ligands such as oleic acid and/or trioctylphosphine oxide do not meet these requirements, and the capping ligands must be modified or replaced. To achieve this task, several strategies have been developed including ligand exchange, ligand modification, addition of amphiphilic ligands, use of polymeric ligands via exchange and addition, addition of silica shells, and multistep exchanges of ligands using NOBF$_4$ reagent as an intermediate (Muhr et al. 2014). Table 5.2 illustrates common strategies to employ water soluble UCNPs. Some of these strategies are discussed in greater detail along with techniques used for characterizing NP surfaces.

5.3.1 Ligand-Based Surface Coatings

Many different types of small ligands have been used to successively alter the surface chemistry of oleic acid-coated UCNPs. Though these ligands may not provide colloidal stability in physiological conditions, they do provide an intermediate platform to add other types of chemistry required for water-soluble UCNPs. The functional groups commonly found on multidentate ligands

TABLE 5.2 Summary of the Most Common Strategies for Preparing Bio-compatible UCNPs.

Method	
Ligand Modification	The double bond of the OA UCNP is oxidized through various oxidation agents to yield water-soluble carboxylic-, epoxide-, and aldehyde-terminated UCNP (Chen et al. 2008; Zhou et al. 2009; Hu et al. 2008).
Ligand Exchange	Native ligand is replaced by a water-soluble ligand ideally with higher affinity to the nanoparticle surface. Ligand strength increases in the following order: sulphur, amine, carboxylic acids, and phosphates (Muhr et al. 2014).
Ligand Desorption	OA UCNPs are acidified to pH 4, the carboxylic group is protonated, the resultant neutral OA is no longer attracted to the surface and is stripped off to yield aqueous soluble bare UCNPs (Bogdan et al. 2011).
Bilayer coating	Hydrophobic backbone of the OA UCNP can undergo strong van der Waals interactions with amphiphilic moieties with hydrophobic chains. The hydrophilic moieties of the amphiphilic coating face outwards imparting water solubility to the UCNP (Muhr et al. 2014; Ren et al. 2012).
Inorganic Silica Shell	Silica shell can be grafted on both hydrophobic- and hydrophilic-coated UCNP surfaces through reverse microemulsion or the Stober method using various organosilanes, the latter method being more suitable for OA- or oleylamine-coated UCNPs (Muhr et al. 2014).

that successfully coordinate to the positively charged UCNP surface are carboxylic- and phosphorous-based moieties.

Citrate is an attractive ligand option as it meets two of the criteria mentioned above. The nanoparticles remain luminescent upon exchange of oleic acid to citrate ligand on the surface of the UCNPs. The carboxylic groups of citrate can be used to conjugate amine functionalized chemistries onto the surface of the nanoparticles. This ligand is commonly used for providing colloidal stability to gold nanoparticles in aqueous media. However, citrate has a relatively low affinity to UCNPs; thus, it is easily exchangeable for a higher affinity ligand. Upon conversion of OA-coated UCNPs to citrate-coated UCNPs, particles become water soluble, resist aggregation, and maintain their original size as illustrated in Figure 5.11. However, upon the removal of unbound, excess citrate from solution, the UCNPs tend to aggregate together as illustrated in Figure 5.11c.

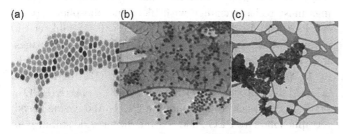

(a) (b) (c)

FIGURE 5.11 (a) Transmission electron microscope (TEM) oleic acid-coated UCNPs dispersed in hexanes, (b) citrate-coated UCNPs dispersed in excess citrate-based buffer, and (c) citrate-coated UCNPs in citrate-free solution (i.e. without excess ligand).

Conversely, phosphate ligands provide enhanced colloidal stability because the phosphate-lanthanide interaction on the UCNP surface is stronger than citrate (Zhang et al., 2018). Phosphoethanolamine (PEA) has been frequently used to obtain water-soluble UCNPs (Doughan et al. 2017). PEA is an attractive choice of ligand because its terminal amine is open to conjugation with chemical moieties to attach polymers and biorecognition elements. Doughan and coworkers utilized PEA UCNPs as LRET donors in nucleic acid hybridization paper-based assays (Doughan et al., 2017). As illustrated in Figure 5.12, PEA UCNPs were immobilized onto the surface of cellulose paper, and the paper-bound UCNPs were conjugated to N-hydroxysuccinimide-polyethylene glycol-streptavidin through amide coupling. The streptavidin on the PEG was further conjugated with biotin modified oligonucleotide which hybridized with complementary oligonucleotide targets (Doughan et al., 2017).

Layer-by-layer assembly of sequential chemical moieties to UCNPs immobilized on a solid surface do not suffer from aggregation but the same is not true for chemical modifications of UCNP surfaces in solution (Doughan et al., 2017). Functionalizing PEA-capped UCNPs in solution proves to be a challenge as upon removal of excess PEA ligand the particles aggregate. Removal of free ligands unbound to the nanoparticle is a prerequisite for surface further modification of surfaces; thus, an ideal ligand must provide stability in non-excess ligand conditions. As illustrated in Figure 5.13, OA UNCPs initially sized at around 80 nm aggregate to a few 100 nm after ligand exchange with PEA and removal of excess ligand.

5.3.2 Polymer Coatings

Due to the instability of ligand surface chemistries alone, often ligands with reactive groups are functionalized with hydrophilic polymers coatings to provide enhanced stability of UCNPs in solution. Polymer layers also shield UCNP cores from surface deactivation effects in aqueous environments to maintain optical integrity upon entering physiological conditions (Jin et al. 2011).

Multidentate Polymers

Multidentate polymers such as polyacrylic acid (PAA), polyethylenimine (PEI), and more recently multidentate derivatives of polyethylene glycol (PEG) have been utilized to yield UCNPs suitable for biological applications. PAA and PEI (see Figure 5.14) have charged carboxylic and amine groups that allow for coordination with lanthanide ions on the UCNP surface. These multidentate polymeric ligands are capable of instilling hydrophilic character to UCNPs while maintaining their surface morphology and size. Capobianco and colleagues have reported PAA-coated UCNPs that remained colloidal stable for over 3 months in aqueous conditions with no noticeable loss of emission intensity (Naccache et al., 2009). Carboxylic acids on the polymer can be used to graft a variety of functional groups

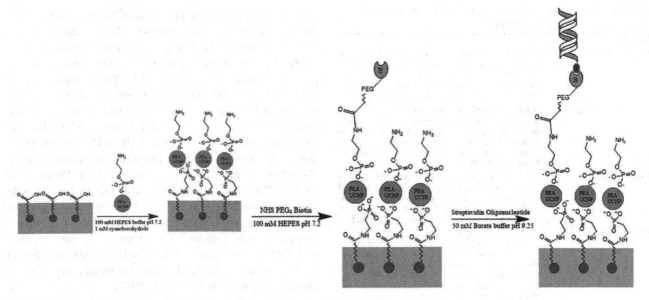

FIGURE 5.12 The scheme utilized to functionalize DNA-coated UCNPs onto cellulose paper.

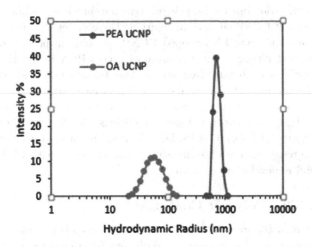

FIGURE 5.13 The hydrodynamic radius (nm) profile of oleic acid-coated UCNPs that underwent ligand exchange with PEA to yield PEA-coated UCNPs.

or other polymers. Cationic-branched PEI-polymer coated UCNPs have shown enhanced cellular uptake due to interaction between the positively charged UCNP surface and the negatively charged phospholipid bilayer (Verma et al., 2010; Bagheri et al., 2016). Such endocytotic processes can result in entrapment of nanoparticles in the cellular endosome. However, the abundance of free primary or secondary amines on the surface of a PEI-coated UCNP also provides a unique advantage in these highly acidic cellular environments. PEI can soak up protons creating a charge gradient between the endosome and cytosol causing a buildup of osmotic pressure that is relieved by bursting of the endosome allowing nanoparticles to perform further diagnostic or therapeutic work inside the cell (Bagheri et al. 2016). However, use of a PEI surface coating comes with the innate disadvantage of relatively high cellular toxicity in comparison with other polymers due to its highly cationic character. The cellular toxicity of PEI can be dramatically

reduced by creating copolymers of PEI and PEG (Neu et al. 2007).

Multidentate PEG coatings have been particularly popular for biological applications because neutral PEG coatings resist nonspecific adsorption of serum proteins (Chen et al., 2008). Neutral coatings are desirable for *in vivo* applications because nonspecific adsorption of serum proteins leads to the agglomeration of nanoparticles, as well as difficulty in clearance through the reticuloendothelial system (RES) (Verma et al., 2010). PEG coatings are synthetically advantageous because the terminal hydroxyls are available for further chemical modification. The viability of long PEG polymers functionalized with multidentate phosphorous ligands as surface coatings for biological applications has been extensively explored.

The Winnik Group has constructed many analogs of multidentate phosphorous PEGs to investigate stability of these coatings on UCNPs in biologically relevant media illustrated in Figure 5.15. Phosphate buffers are good screening buffers to examine the viability of UCNP coatings in *in vitro* and *in vivo* given that phosphate is a common electrolyte in blood serum and cellular growth media (Dwivedi et al., 2011). Inorganic phosphate acts as a competitive ligand for polymeric phosphate; therefore, successful polymeric coatings must resist exchange with inorganic phosphates (Cao et al. 2012). As illustrated in Figure 5.16, Winnik et al. reported that monophosphate PEG 2000 surface coatings were not sufficient in providing colloidal stability to UCNP in 1x phosphate-buffered saline (PBS) (137 mM NaCl, 2.7 mM KCl, 10 mM $Na_2HPO_4 \cdot H2O$, 2 mM KH_2PO_4, pH 7.4). However, diphosphonate PEG- and lysine-based tetraphosphonate PEG 2000 coating did provide stability in PBS (Cao et al. 2012).

Time-based studies are commonly conducted with *in vivo* and *in vitro* assays as it takes time for nanoparticles to reach their cellular target and perform their given function

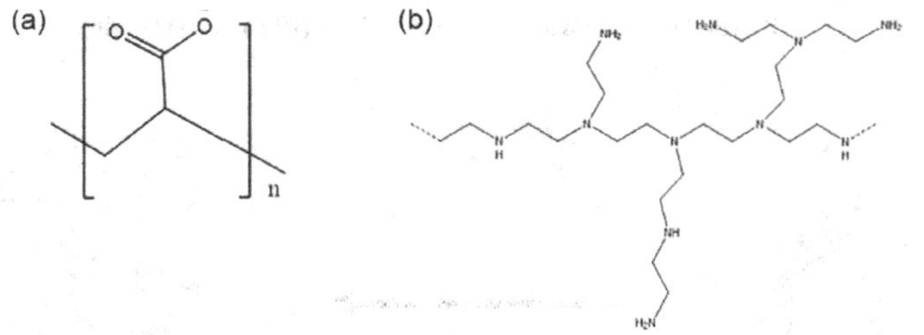

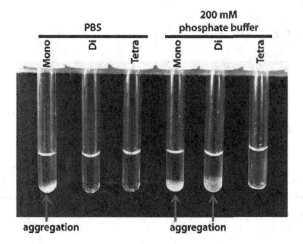

FIGURE 5.14 (a) PAA and (b) PEI polymer structures.

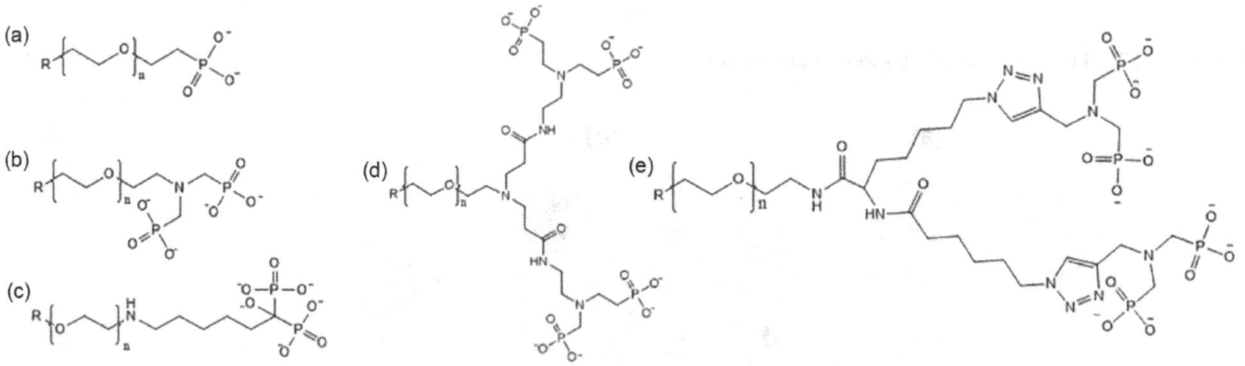

FIGURE 5.15 (a) Monophosphate-, (b) diphosphonate-, (c) neridronate-, (d) polyamidoamine- (PAMAM), and (e) lysine-based tetraphosphonate ligand conjugation schemes to PEG polymers reported in literature.

exhibits limited *in vitro* toxicity (Kostiv et al. 2017). The attractive feature of these relatively smaller bisphosphonate head moieties in comparison with larger tetraphosphate heads is that they take up less surface area on the UCNP surface, and this can allow for a higher density of ligands per unit area leading to improved stability.

Amphiphilic Polymers

Amphiphilic polymers are particularly advantageous classes of polymers as they are soluble in both hydrophobic and hydrophilic media. This type of surface coating route is far less laborious because it avoids the necessity for an intermediate ligand exchange or modification step as required to make OA UCNPs water soluble.

Polymeric anhydride (PMA) polymer is an amphiphilic polymer that has been used to impart water solubility in many classes of nanoparticles (Pellegrino et al., 2004). As illustrated by Figure 5.17, PMA has a hydrophobic alkyl chain and a maleic anhydride ring in each monomer unit. The alkyl chain can adsorb onto the oleic acid alkyl chain that is available from the nanoparticle synthesis procedure. The anhydride rings facing outwards can undergo nucleophilic ring opening reactions to impart water solubility and can target selective functionalities for conjugation making them attractive for biological applications.

5.3.3 Lipid Coatings

Liposomes provide an attractive surface chemistry for upconverting nanoparticles with the amphiphilic nature allowing them to act as hydrophilic and hydrophobic drug carriers (Pichaandi et al., 2017). The size and shape of

FIGURE 5.16 The suspension of lanthanide UCNPs coated with methoxy PEG phosphate, methoxy PEG diphosphonate, and methoxy PEG tetraphosphonate in different buffer condition. (Reprinted with permission from Cao et al. Copyright 2012 American Chemical Society.)

(Choi et al. 2014). Longevity of colloidal stability is an important parameter, and cellular studies tend to operate from a few minutes to 24 h. Thus, successful nanoparticle candidates for such work must be stable within such a time frame (Choi et al. 2014). Although the tetraphosphonate PEG 2,000 coatings can remain stable in 10 mM phosphate buffer, only the lysine analog maintains stability in 1x PBS for a prolonged period (Zhao et al., 2014). More recently, Horrak et al. synthesized a superior coating of neridronate conjugated to PEG 5,000 (see Figure 5.15c) which imparts colloidal stability for as long as 3 months in PBS and

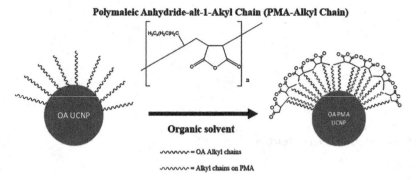

FIGURE 5.17 The scheme of PMA surface modification.

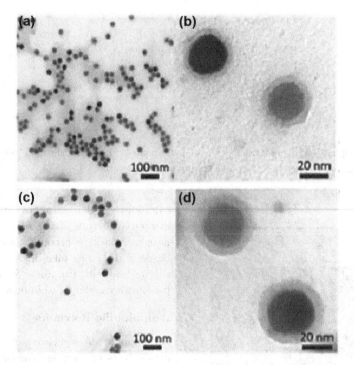

FIGURE 5.18 The TEM images of DEC221 coated UCNPs with 6 mol% mPEG2K-DSPE UCNPs. (a) and (b) are low- and high-resolution images of the UCNPs in water. (c) and (d) are low- and high-resolution images of the UCNPs in TRIS-buffered saline. (Reprinted with permission from Pichaandi et al. Copyright 2017 American Chemical Society.)

the liposomes can be tuned by changing the lipid composition, and they can be designed to carry small and large molecules (Pichaandi et al., 2017). Liposomes have not been commonly reported to encapsulate nanoparticles. It is a challenge to singly encapsulate a nanoparticle that is 30 nm or less because the radius of curvature for liposomes is too large and often multiple nanoparticles get trapped into one liposome (Pichaandi et al., 2017). Winnik et al. have demonstrated successful liposome encapsulation of individual 30 nm UCNPs (Pichaandi et al. 2017). The liposomes were composed of 2:2:1 mol ratio of dioleoylphosphatidyl choline (DOPC), egg sphingomyelin (ESM), and ovine cholesterol (Chol). The size of the liposome was reduced to fit 30 nm particles by incorporating 6 mol% 1,2-distearoyl-sn-glycero-3-phosphoethanolamine-N-[methoxy(polyethylene glycol)-2000 (PEG 2000 DSPE). Single encapsulation of the nanoparticles was confirmed

using TEM. As illustrated in Figure 5.18a and c, no clustering of nanoparticles was observed. Figure 5.18b and d illustrate close-up images of the UCNPs, which were observed to have a 5–6 nm grey coronas indicating successful liposome encapsulation (Pichaandi et al. 2017). Upon dispersion of the liposome-coated UCNPs into PBS, the size of the particles increased about 20%, and the particle size was maintained over a period of one month as illustrated by time-based dynamic light scattering (DLS) studies shown in Figure 5.19 (Pichaandi et al. 2017). For these UCNPs to be viable for selective detection in biological assays, they must not exhibit nonspecific binding. Nonspecific binding was tested in three different mammalian cell lines using flow cytometry. High dosages of 10,000–30,000 UCNP per cell resulted in very low amounts of nonspecific binding, and less than 0.1% was observed (Pichaandi et al. 2017).

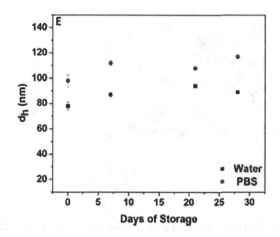

FIGURE 5.19 Hydrodynamic radius of liposome-coated UCNP over a period of 30 days in water and PBS. (Reprinted with permission from Pichaandi et al. Copyright 2017 American Chemical Society.)

5.3.4 Nucleic Acid Coatings

DNA is a very useful multifunctional coating in that DNA-coated UCNPs may be facilitated to cross the cell membrane without the use of transfection agents, and coatings with single-stranded oligonucleotides could interact with complementary sequences. Typically, a high cellular uptake is observed for positively charged coatings as these allow for interaction with negatively charged cellular membranes (Verma et al., 2010). Negatively charged oligonucleotides coated onto nanoparticles have shown entry into various different cell lines through interaction with serum proteins (Verma et al., 2010). Serum proteins adsorb onto oligonucleotides through electrostatic and hydrophobic interactions. The level of cellular uptake increases with greater density of oligonucleotide on the nanoparticle surface (Verma et al., 2010).

Lu et al. have demonstrated a one-step modification to obtain functional DNA-coated UCNPs that exhibit target selective imaging capability in cancer cells (Li et al. 2013). The negative charges on the phosphate backbone of DNA allow potent binding to the positively charged UCNP surface (Li et al., 2013). As illustrated in Figure 5.20a, OA UCNPs dispersed in hexanes were mixed with an aqueous solution of 30 base pair poly T DNA sequence to obtain water-soluble DNA-coated UCNPs (Li et al.

2013). Upon modification, these UCNPs were stable in physiological buffer and maintained their optical properties as illustrated in Figure 5.20b and c (Li et al. 2013).

To test their potential as specific hybridization probes, DNA-coated UCNPs were incubated with 5 nm gold nanoparticles that had been coated with complementary poly A DNA (Li et al. 2013). TEM images shown in Figure 5.21 show nanoassemblies of UCNP, and gold nanoparticles indicating the DNA on UCNP could selectively bind to complementary targets (Li et al. 2013). To further test the selective diagnostic ability in biological media, UCNPs were functionalized with aptamer specific for nucleolin, a protein overexpressed in the cellular membrane of breast cancer cells (Li et al. 2013). As illustrated in Figure 5.22, nucleolin aptamer-conjugated UCNPs were observed to bind selectively around MCF-7 breast cancer cells. Therefore, UCNPs coated with DNA can be used for selective molecular imaging (Li et al. 2013).

5.4 Surface Chemistry Characterization Techniques

5.4.1 Size and Morphology Analysis Using TEM and DLS

Quantitative and qualitative analysis upon surface modification is integral to assessing the viability of surface coatings for biological applications. Size characterization is important for surface modifications because it assesses the colloidal stability provided by the new surface coating. As previously mentioned, an ideal surface coating maintains the monodispersity of nanoparticles while resisting aggregation. Transmission electron microscopy (TEM) is an appropriate and direct method to view the changes in colloidal stability and morphology of nanoparticles upon surface modification. As illustrated in Figure 5.11b, the nanoparticles are monodispersed, and in contrast, Figure 5.11c shows an image of aggregated nanoparticles. TEM is a standard method used for nanoparticle characterization, capable of resolving structures to a scale of 0.2 nm (Shi et al. 2016). TEM is only useful for samples that are electron transparent and can withstand the electron bombardment.

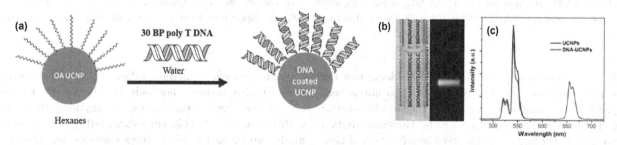

FIGURE 5.20 (a) Ligand exchange scheme to yield DNA-coated UCNPS, (b) DNA UCNP in water without 980 nm laser illumination (left) and DNA UCNP solution with 980 nm laser illumination, and (c) emission spectra of DNA UCNP under 980 nm laser illumination. (Adapted with permission from Li et al. Copyright 2013 American Chemical Society.)

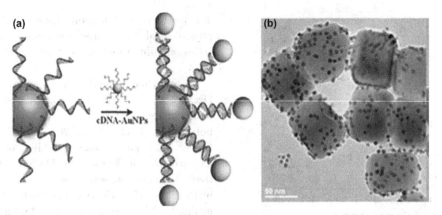

FIGURE 5.21   (**a**) Nanoassembly scheme between poly T-coated UCNPs and poly A-coated gold nanoparticles, and (**b**) the TEM image of the nanoassembly. (Reprinted with permission from Li et al. Copyright 2013 American Chemical Society.)

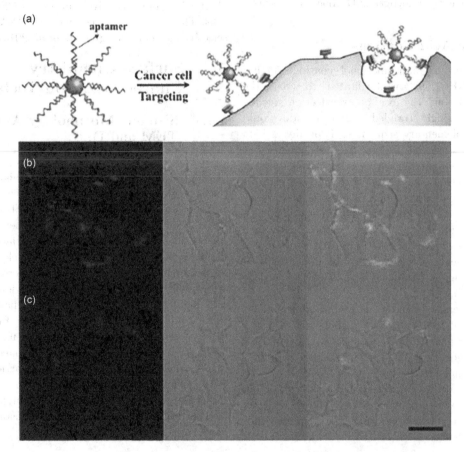

FIGURE 5.22   (**a**) Aptamer UCNP and MCF-7 targeting scheme, (**b**) confocal images of nucleolin aptamer UCNP-treated MCF-7 cells, and (**c**) random DNA UCNP-treated MCF-7 cells. (Reprinted with permission from Li et al. Copyright 2013 American Chemical Society.)

DLS is a particularly useful method to determine the presence of surface coatings. DLS is used to obtain size of particles by analysing how the nanoparticles scatter light as the scattering centres move by Brownian motion (Shaw 2018). The scattering observed as a function of time is related to rate of diffusion of the particles, which is described by the diffusion coefficient, D (Shaw, 2018). The diffusion coefficient can be related to hydrodynamic radius through Einstein's equation (Shaw, 2018). The hydrodynamic radius depends not only on the UCNP core but on any coatings on the particle or anything that moves with the particle in solution (Shaw, 2018). This is particularly advantageous when large coatings are placed on the surface, and a definitive increase in hydrodynamic radius can confirm the presence of the coating (Shaw, 2018).

Stoke Einstein's equation:

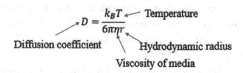

DLS outputs a size distribution profile along with an average hydrodynamic radius as illustrated in Figure 5.3. A caveat to DLS is that the size distribution profile produced can be skewed by a small population of larger particles as larger particles scatter more light (Shaw 2018).

5.4.2 Charge Characterization through Zeta Potential

Zeta potential measurements are used for characterizing charged surface coatings such as DNA, charged polymers, and ligands. Zeta potential is measured by principles of electrophoresis, which is the movement of charged particles under an applied electric field (Shaw, 2018). The sample solution is housed in a cuvette containing electrodes through which electric field is applied. The electrophoretic mobility, μ_E, of the particles is measured and can be related to zeta potential ζ through equation 5.2. ε is the dielectric constant of the solution, η is the viscosity of the solution, and $F(ka)$ is the Henry's function which describes the ratio between radius of the particle, r and the electric double layer thickness, known as the Debye length, K, as illustrated in Figure 5.23 (Shaw 2018).

$$\mu_E = \frac{2\zeta\ \varepsilon F(ka)}{3\eta} \qquad (5.2)$$

5.4.3 Surface Characterization through NMR and FTIR

Fourier transform infrared spectroscopy (FTIR) is commonly used as a qualitative technique to confirm the presence of many different functional groups (Movasaghi et al. 2008). The FTIR is an absorption experiment, and the instrument measures the absorption for a large

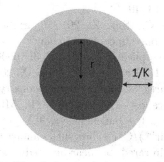

FIGURE 5.23 The thickness of electric double layer, K also known as Debye length in the units of K^{-1} and r, the radius of the nanoparticle.

range of infrared wavelengths ranging from 2.5 to 14 um (Thermofisher.com, 2018). Different functional groups commonly used as surface chemistry on UCNPs such as amines, amides, ethylene glycols, carboxylic acids, and phosphates have characteristic absorbances that can be determined by FTIR. Nuclear magnetic resonance spectroscopy (NMR) provides information about the presence of functional groups and can provide quantitative parameters such as density of surface chemistry (Tong et al. 2015). The NMR technique is insensitive, and often a few milligrams of nanoparticles are needed to see sufficient signal. Nonetheless, the technique is nondestructive, and sample can be recovered. A major caveat to NMR for the use of UCNPs is that only non-doped UCNPs can be utilized for analysis as paramagnetic lanthanides interfere with the magnetic field and produce broadened peaks (Satterlee 1990).

5.4.4 Composition Characterization through Atomic Emission Spectroscopy (AES)

AES is a common form of elemental analysis and can be used to determine the lanthanide composition, exact molecular weight of the UCNP, and surface chemistry density. Typically, UCNPs are dissolved in concentrated acid. The elements are detected through characteristic atomic emissions and can be quantified through calibration methods. AES analysis is very sensitive, and most metals can be quantified accurately in the parts per billion (ppb) range (Static.thermoscientific.com 2018).

5.4.5 Density Measurements through Thermogravimetric Analysis (TGA)

TGA analysis is routinely used for the quantification of nanoparticle coatings. TGA determines the quantity of surface chemistry by observing the changes in mass at elevated temperatures (Mansfield et al. 2014). TGA does not require much sample preparation aside from drying of the sample (Mansfield et al. 2014). Samples are dried under vacuum or heat, and are lyophilized prior to analysis (Tong et al. 2015). During analysis, initially the sample is further dried to ensure any trace solvents are evaporated prior to observing any mass change. After complete drying of the sample, the temperature is elevated to an oxidation temperature defined as the temperature at which bulk material decomposes (Mansfield et al. 2014). The analysis yields a decomposition curve which relates mass % as a function of temperature. This can quantify mass loss, which can be used to determine the quantity of the surface chemistry (Mansfield et al. 2014). TGA requires relatively greater quantities of UCNPs for analysis ($\sim$ few mg) in comparison with AES and is not sensitive to minor changes in mass (Mansfield et al. 2014). Microthermogravimetric analysis (u-TGA) may be more appropriate for nanoparticle work as it only requires 1 ug of sample and can measure nanogram changes in mass (Mansfield et al. 2014).

5.4.6   Crystal Structure Analysis through X-ray Diffraction

X-ray diffraction (XRD) is commonly used for the determination of crystal structure and phase. XRD utilizes photons with energies that typically are in the range of 100 eV–100 keV. Such photons can diffract from crystalline materials, and photons of $\geq$10 keV have sufficient energy to interact with core electrons through photoelectric absorption and scattering effects (Compton, Thomson/Rayleigh). Absorption and scattering of X-rays occurs to a greater extent for elements of larger atomic number due to the greater number of electrons. The diffraction and absorption of X-ray photons by nanomaterials provides information about elemental composition and spatial location at atomic resolution. In a UC nanocrystal, atoms are arranged in translational symmetry through identical unit cells repeating in space (Giannini et al., 2016). The incoming X-ray beam interacts with each atom present in the crystal unit cell via Rayleigh scattering, and secondary waves are generated as a result of this interaction. Interference (either constructively or destructively) created from the diffracted wavefield is representative of the atomic composition and relative positioning of atoms within the crystal lattice. The scattering of X-rays from atoms arranged in such periodic arrays generates diffraction patterns that contain information regarding the crystal atomic structure and the crystalline phase (Giannini et al., 2016). Each peak in an XRD profile of a nanocrystal is characteristic of an atomic plane within the unit cell, and distances between parallel atomic planes describe the relative position of diffraction peaks through Bragg's law:

$$\lambda = 2d_{HKL}\sin\theta$$

where λ, the wavelength, is related to the angle at which constructive interference of X-rays scattered by parallel atomic planes occurs. The geometric function d_{HKL} is a vector that describes the distance between parallel atomic planes through its magnitude and the shape of the unit cell through its orientation. Diffraction peak intensity and angular location are indicative of the type of atoms present in sample and their position within atomic planes of the crystal (Speakman, 2018).

References

Aebischer, A.; Hostettler, M.; Hauser, J.; et al. 2006. Structural and spectroscopic characterization of active sites in a family of light-emitting sodium lanthanide tetrafluorides. *Angew. Chem., Int. Ed.* 45: 2802–2806.

Algar, W.R.; Tavares, A.J.; Krull, U.J. 2010. Beyond labels: A review of the applications of quantum dots as integrated components of assays, bioprobes, and biosensors utilizing optical transduction. *Anal. Chim. Acta.* 673: 1–25.

Algar, W.R.; Susumu, K.; Delehanty, J.B.; Medintz, I.L. 2011. Semiconducting quantum dots in bioanalysis: Crossing the valley of death. *Anal. Chem.* 83: 8826–8837.

Altavilla, C. 2016. *Upconverting Nanomaterials.* CRC Press, Boca Raton, pp. 109–111.

Auzel, F. 2004. Upconversion and anti-stokes processes with f and d ions in solids. *Chem. Rev.* 104: 139–173.

Auzel, F. 1966. Quantum counter obtained by using energy transfer between two rare earth ions in a mixed tungstate and in a glass. *C. R. Acad. Sci. (Paris).* 262: 1016–1019.

Auzel, F. 1973. Materials and devices using double-pumped phosphors with energy transfer. *Proc. IEEE.* 61: 758–786.

Auzel, F. 1990. Upconversion processes in coupled ion systems. *J. Lumin,* 45: 341–345.

Bagheri, A.; Arandiyan, H.; Boyer, C.; Lim, M. 2016. Lanthanide-doped upconversion nanoparticles: Emerging intelligent light-activated drug delivery systems. *Advanced Science.* 3: 1500437.

Bogdan, N.; Vetrone, F.; Ozin, G.; Capobianco, J. 2011. Synthesis of ligand-free colloidally stable water dispersible brightly luminescent lanthanide-doped upconverting nanoparticles. *Nano Lett.* 11: 835–840.

Boyer, J.; Vetrone, F.; Cuccia, L.; Capobianco, J. 2006. Synthesis of colloidal upconverting $NaYF_4$ nanocrystals doped with Er^{3+}, Yb^{3+} and Tm^{3+}, Yb^{3+} via thermal decomposition of lanthanide trifluoroacetate precursors. *J. Am. Chem. Soc.* 128: 7444–7445.

Boyer, J.-C.; Manseau, M.-P.; Murray, J.I.; van Veggel, F.C.J.M. 2009. Surface modification of upconverting $NaYF_4$ nanoparticles with PEG-phosphate ligands for NIR (800 nm) biolabelling within the biological window. *Langmuir.* 26: 1157–1164.

Cao, P.; Tong, L.; Hou, Y.; et al. 2012. Improving lanthanide nanocrystal colloidal stability in competitive aqueous buffer solutions using multivalent PEG-phosphonate ligands. *Langmuir.* 28: 15425–15425.

Chang, H.; Xie, J.; Zhao, B.; et al. 2014. Rare earth ion-doped upconversion nanocrystals: Synthesis and surface modification. *Nanomaterials.* 5: 1–25.

Chan, W.C.W.; Maxwell, D.J.; Gao, X.; Bailey, R.E.; Han, M.; Nie, S. 2002. Luminescent quantum dots for multiplexed biological detection and imaging. *Curr. Opin. Biotechnol.* 13: 40–46.

Chen, C.; Li, C.; Shi, Z. 2016. Current advances in lanthanide-doped upconversion nanostructures for detection and bioapplication. *Adv. Sci.* 3: 1600029.

Chen, G.; Qiu, H.; Prasad, P.N.; Chen, X. 2014. Upconversion nanoparticles: Design, nanochemistry, and applications in theranostics. *Chem. Rev.* 114: 5161–5214.

Chen, G.; Shen, J.; Ohulchanskyy, T.Y.; et al. 2012. (α-$NaYbF_4$:Tm^{3+})/CaF_2 core/shell nanoparticles with efficient near-infrared to near-infrared upconversion for high-contrast deep tissue bioimaging. *ACS Nano.* 6: 8280–8287.

Chen, J.; Zhao, J.X. 2012. Upconversion nanomaterials: Synthesis, mechanism, and applications in sensing. *Sensors.* 12: 2414–2435.

Chen, X.; Ma, E.; Liu, G. 2007. Energy levels and optical spectroscopy of Er^{3+} in Gd_2O_3 nanocrystals. *J. Phys. Chem. C.* 111: 10404–10411.

Chen, Z.; Chen, H.; Hu, H.; et al. 2008. Versatile synthesis strategy for carboxylic acid functionalized upconverting nanophosphors as biological labels. *J. Am. Chem. Soc.*. 130: 3023–3029.

Choi, C. R.; Hao, L.; Wu, X.; Zhang, C.; Mirkin, C. 2014. Intracellular fate of spherical nucleic acid nanoparticle conjugates. *J. Am. Chem. Soc.* 136: 7726–7733.

DaCosta, M.V.; Doughan, S.; Han, Y.; Krull, U.J. 2014. Lanthanide upconversion nanoparticles and applications in bioassays and bioimaging: A review. *Anal. Chim. Acta.* 832: 1–33.

Deren, P.J.; Strek, W.; Zych, E.; Drozdzynski, J. 2000. Upconversion in elpasolite crystals doped with U^{3+}. *Chem. Phys. Lett.* 332: 308–312.

Dell'Amico, D.; Calderazzo, F.; della Porta, C.; et al. 1995. Improved preparation of anhydrous lanthanide chlorides under mild conditions. *Inorg. Chim. Acta.* 240: 1–3.

Doughan, S.; Uddayasankar, U.; Peri, A.; Krull, U. 2017. A paper-based multiplexed resonance energy transfer nucleic acid hybridization assay using a single form of upconversion nanoparticle as donor and three quantum dots as acceptors. *Anal. Chim. Acta.* 962: 88–96.

Diamente, R.; Raudsepp, M.; van Veggel, F.C.J.M. 2007. Dispersible Tm^{3+}-doped nanoparticles that exhibit strong 1.47 μm photoluminescence. *Adv. Funct. Mater.* 17: 363–368.

Dwivedi, A.; Saikia, G.; Iyer, P. 2011. Aqueous polyfluorene probe for the detection and estimation of Fe3+ and inorganic phosphate in blood serum. *J. Mater. Chem.* 21: 2502–2507.

Edwards, D.; Hayward, R. 1968. Transition metal acetates. *Can. J. Chem.* 46: 3443–3446.

Gamelin, D.R.; Güdel, H.U. 1998. Two-photon spectroscopy of d^3 transition metals: Near-IR-to-visible upconversion luminescence by Re^{4+} and Mo^{3+}. *J. Am. Chem. Soc.* 120: 12143–12144.

Gamelin, D.R.; Güdel, H.U. 1999. Spectroscopy and dynamics of Re^{4+} near-IR-to-visible luminescence upconversion. *Inorg. Chem.* 38: 5154–5164.

Gamelin, D.R.; Güdel, H.U. 2000a. Design of luminescent inorganic materials: New photophysical processes studied by optical spectroscopy. *Acc. Chem.Res.* 33: 235–244.

Gamelin, D.R.; Güdel, H.U. 2000b. Exited-state dynamics and sequential two-photon upconversion excitation of Mo^{3+}-doped chloro- and bromo-elpasolites. *J.Phys. Chem. B.* 104: 10222–10234.

Gao, X.; Yang, L.; Petros, J.A.; Marshall, F.F.; Simons, J.W.; Nie, S. 2005. *In vivo* molecular and cellular imaging with quantum dots. *Curr. Opin. Biotechnol.* 16: 63–72.

Garner, C.; Hughes, B. 1975. Inorganic compounds containing the trifluoroacetate group. *Adv. Inorg. Chem. Radiochem.* 17: 1–47.

Ghosh, P.; Patra, A. 2008. Tuning of crystal phase and luminescence properties of Eu^{3+} doped sodium yttrium fluoride nanocrystals. *J. Phys. Chem. C.* 112: 3223–3231.

Giannini, C., Ladisa, M., Altamura, D., Siliqi, D., Sibillano, T. and De Caro, L. (2016). X-ray diffraction: A powerful technique for the multiple-length-scale structural analysis of nanomaterials. *Crystals*, 6(8); 87.

Gu, Z.; Yan, L.; Tian, G.; Li, S.; Chai, Z.; Zhao, Y. 2013 Recent advances in design and fabrication of upconversion nanoparticles and their safe theranostic applications. *Adv. Mater.* 25: 3758–3779.

Guo, H.; Dong, N.; Zhang, W.; Lou, L.; Xia, S. 2004. Visible upconversion in Rare earth ion- doped Gd_2O_3 nanocrystals. *J. Phys. Chem. B.* 108: 19205–19209.

Haase, M.; Schäfer, H. 2011. Upconverting nanoparticles. *Angew. Chem. Int. Ed.* 50: 5808–5829.

Hu, H.; Xiong, L.; Zhou, J.; Li, F.; Cao, T. Huang, C. 2009. Multimodal-luminescence core-shell nanocomposites for targeted imaging of tumour cells. *Chem. Eur. J.* 15: 3577–3584.

Hu, H.; Yu, M.; Li, F.; et al. 2008. Facile epoxidation strategy for producing amphiphilic up-converting rare-earth nanophosphors as biological labels. *Chem. Mater.* 20: 7003–7009.

Huang, P.; Zheng, W.; Zhou, S.; et al. 2014. Lanthanide-doped $LiLuF_4$ upconversion nanoprobes for the detection of disease biomarkers. *Angew. Chem.* 53: 1252–1257.

Jin, J.; Gu, Y.; Man, C.; Cheng, J.; Xu, Z.; Zhang, Y. 2011. Polymer-coated NaYF4 :Yb3+, Er3+ upconversion nanoparticles for charge-dependent cellular imaging. *ACS Nano* 5(10): 7838–7847.

Johnson, N.; Korinek, A.; Dong, C.; van Veggel, F. 2012. Self-focusing by ostwald ripening: A strategy for layer-by-layer epitaxial growth on upconverting nanocrystals. *J. Am. Chem. Soc.* 134: 11068–11071.

Kang, D.; Kim, J. 2010. Review: Upconversion microscopy for biological applications. *Microscopy: Science, Technology, Applications and Education*, pp. 571–581.

Klier, D.T.; Kumke, M.U. 2015. Analysing the effect of the crystal structure on upconversion luminescence in Yb^{3+}, Er^{3+}-co-doped NaYF nanomaterials. *J. Mater. Chem. C.* 3: 11228–11232.

Kostiv, U.; Lobaz, V.; Kučka, J.; et al. 2017. A simple neridronate-based surface coating strategy for upconversion nanoparticles: Highly colloidally stable 125Iradiolabeled NaYF4:Yb3+/Er3+@PEG nanoparticles for multimodal in vivo tissue imaging. *Nanoscale.* 9: 16680–16688.

Kramer, K.W.; Biner, D.; Frei, G.; Güdel, H.U.; Hehlen, M.P.; Lithi, S.R. 2004. Hexagonal sodium yttrium fluoride based green and blue emitting upconversion phosphors. *Chem. Mater.* 16: 1244–1251.

Kumar, G.A.; Chen, C.W.; Ballato, J.; Riman, R.E. 2007. Optical characterization of infrared emitting rare-earth-doped fluoride nanocrystals and their transparent nanocomposites. *Chem. Mater.* 19: 1523–1528.

LaMer, V.; Dinegar, R. 1950. Theory, production and mechanism of formation Of monodispersed hydrosols. *J. Am. Chem. Soc.* 72: 4847–4854.

Li, L.; Wu, P.; Hwang, K.; Lu, Y. 2013. An exceptionally simple strategy for DNA-functionalized Up-conversion nanoparticles as biocompatible agents for nanoassembly, DNA delivery, and imaging. *J. Am. Chem. Soc.* 135: 2411–2414.

Li, Z.; Zhang, Y. 2008. An efficient and user-friendly method for the synthesis of hexagonal-phase $NaYF_4$:Yb, Er/Tm nanocrystals with controllable shape and upconversion fluorescence. *Nanotechnology.* 19: 345606.

Li, Z.; Zhang, Y.; Jiang, S. 2008. Multicolor core/shell-structured upconversion fluorescent nanoparticles. *Adv. Mater.* 20: 4765–4769.

Liang, X.; Wang, X.; Zhuang, J.; Peng, Q.; Li, Y. 2007. Synthesis of $NaYF_4$ nanocrystals with predictable phase and shape. *Adv. Funct. Mater.* 17: 2757–2765.

Lifshitz, I.; Slyozov, V. 1961. The kinetics of precipitation from supersaturated solid solutions. *J. Phys. Chem. Solids.* 19: 35–50.

Lin, M.; Zhao, Y.; Wang, S.; et al. 2012. Recent advances in synthesis and surface modification of lanthanide-doped upconversion nanoparticles for biomedical applications. *Biotechnol. Adv.* 30: 1551–1561.

Liu, Y.; Tu, D.; Zhu, H.; Chen, X. 2013. Lanthanide-doped luminescent nanoprobes: controlled synthesis, optical spectroscopy, and bioapplications. *Chem. Soc. Rev.* 41: 6924–6958.

Liu, B.; Li, C.; Yang, D.; et al. 2014. Upconversion-luminescent core/mesoporous-silica-shell-structures β-$NaYF_4$:Yb^{3+},Er^{3+}@SiO_2@$mSiO_2$ composite nanospheres: Fabrication and drug-storage/release properties. *Eur. J. Inorg. Chem.* 1906–1913

Logvinenko, V.; Chingina, T.; Sokolova, N.; Semyannikov, P. 2003. Thermal decomposition processes of lanthanide trifluoroacetates trihydrates. *J. Ther. Anal. Calorim.* 74: 401–405.

Mai, H.; Zhang, Y.; Si, R.; et al. 2006. High-quality sodium rare-earth fluoride nanocrystals: Controlled synthesis and optical properties. *J. Am. Chem, Soc.* 128: 6426–6436.

Mansfield, E.; Tyner, K.; Poling, C.; Blacklock, J. 2014. Determination of nanoparticle surface coatings and nanoparticle purity using microscale thermogravimetric analysis. *Anal. Chem.* 86, 1478–1484.

Mansurov, Z. 2013. Combustion synthesis of nanomaterials. *Adv. Mater. Res.* 699: 138–143.

Medintz, I.L.; Uyeda, H.T.; Goldman, E.R.; Mattoussi, H.; 2005. Quantum dot bioconjugates for imaging, labelling and sensing. *Nat. Mater.* 4: 435–446.

Michalet, X.; Pinaud, F.F.; Bentolila, L.A.; et al. 2005. Quantum dots for live cells, in Vivo imaging, and diagnostics. *Science.* 207: 538–544.

Movasaghi, Z.; Rehman, S.; Rehman, D. 2008. Fourier transform infrared (FTIR) spectroscopy of biological tissues. *Appl. Spectrosc. Rev.*, 43(2): 134–179.

Muhr, V.; Wilhelm, S.; Hirsch, T.; Wolfbeis, O.S. 2014. Upconversion nanoparticles: From hydrophobic to hydrophilic surfaces. *Acc. Chem. Res.* 47: 3481–3493.

Naccache, R.; Vetrone, F.; Mahalingam, V.; Cuccia, L.; Capobianco, J. 2009. Controlled synthesis and water dispersibility of hexagonal phase $NaGdF_4$:Ho^{3+}/Yb^{3+} nanoparticles. *Chem. Mater.* 21: 717–723.

Neu, M.; Germershaus, O.; Behe, M.; Kissel, T. 2007. Bioreversibly crosslinked polyplexes of PEI and high molecular weight PEG show extended circulation times in vivo. *J. Controlled Release.* 124: 69–80.

Niu, W.; Wu, S.; Zhang, S. 2011. Utilizing the amidation reaction to address the "Cooperative Effect" of carboxylic acid/amine on the size, shape, and multicolor output of fluoride upconversion nanoparticles. *J. Mater. Chem.* 21: 10894.

Ostrowski, A.; Chan, E.; Gargas, D.; et al. 2012. Controlled synthesis and single-particle imaging of bright, sub-10 Nm lanthanide-doped upconverting nanocrystals. *ACS Nano.* 6: 2686–2692.

Ostwald, W. 1900. Reviews-on the assumed isomerism between red and yellow mercuric oxide and on the surface-tension of solids. *J. Phys. Chem.* 5: 75–75.

Pellegrino, T.; Manna, L.; Kudera, S.; et al. 2004. Hydrophobic nanocrystals coated with an amphiphilic polymer shell: A general route to water sSoluble nanocrystals. *Nano Lett.* 4: 703–707.

Petryayeva, E.; Algar, W.R.; Medintz, I.L. 2013. Quantum dots in bioanalysis: A review of applications across various platforms for fluorescence spectroscopy and imaging. *Appl. Spectrosc.* 67: 215–252.

Pichaandi, J.; Tong, L.; Bouzekri, A.; et al. 2017. Liposome-encapsulated NaLnF4 nanoparticles for mass cytometry: Evaluating nonspecific binding to cells. *Chem. Mater.* 29: 4980–4990.

Pin, M.; Park, E.; Choi, S.; et al. 2018. Atomistic evolution during the phase transition on a metastable single $NaYF_4$:Yb,Er upconversion nanoparticle. *Sci. Rep.* 8: 2199.

Prorok, K.; Bednarkiewicz, A.; Cichy, B.; et al. 2014. The impact of shell host ($NaYF_4$/CaF_2) and shell deposition methods on the ip-conversion enhancement in Tb^{3+}, Yb^{3+} codoped colloidal α-$NaYF_4$ core-shell nanoparticles. *Nanoscale.* 6: 1855–1864.

Qian, H.-S.; Zhang, Y. 2008. Synthesis of hexagonal-phase core-shell $NaYF_4$ nanocrystals with tunable upconversion fluorescence. *Langmuir.* 24: 12123–12125.

Qian, H.; Guo, H.; Ho, P.; Mahendran, R.; Zhang, Y. 2009. Mesoporous-silica-coated up-conversion fluorescent nanoparticles for photodynamic therapy. *Small*. 5: 2285–2290.

Qin, X.; Yokomori, T.; Ju, Y. 2007. Flame synthesis and characterization of rare-earth (Er3+, Ho3+, And Tm3+) doped upconversion nanophosphors. *Appl. Phys. Lett.* 90: 073104.

Ren, W.; Tian, G.; Jian, S.; et al. 2012. Tween-coated NaYF4:Yb,Er/NaYF4 core/shell upconversion nanoparticles for bioimaging and drug delivery. *RSC Adv.* 2: 7037–7041.

Rinkel, T.; Raj, A.; Dühnen, S.; Haase, M. 2015. Synthesis of 10 Nm B-Nayf4:Yb,Er/Nayf4 core/shell upconversion nanocrystals with 5 Nm particle cores. *Angew. Chem.* 55: 1164–1167.

Roberts, J. 1961. Lanthanum and neodymium salts of trifluoroacetic acid. *J. Am. Chem. Soc.* 83: 1087–1088.

Satterlee, J. 1990. Fundamental concepts of NMR in paramagnetic systems. Part II: relaxation effects. *Concepts Magn. Reson.* 2(3): 119–129.

Shaw, R. 2018. Dynamic light scattering training achieving reliable nano particle sizing. (online) 149.171.168.221. http://149.171.168.221/partcat/wp-content/uploads/Malvern-Zetasizer-LS.pdf (Accessed 1 May 2018).

Shen, J.; Chen, G.; Ohulchanskyy, T.Y.; et al. 2012. Tunable near infrared to ultraviolet upconversion luminescence enhancement in (α-NaYbF$_4$:Tm)/CaF$_2$ core/shell nanoparticles for in situ real-time recorded biocompatible photoactivation. *Small*. 9: 3213–3217.

Shi, B. 2016, Transmission electron microscope, *Young Scientists Journal*. 18: 33–35.

Speakman, S. 2018. Basics of X-Ray Powder Diffraction. (online) Prism.mit.edu. http://prism.mit.edu/xray/oldsite/Basics%20of%20X-Ray%20Powder%20Diffraction.pdf (Accessed 9 May 2018).

Stouwdam, J.; van Veggel, F. 2002. Near-infrared emission of redispersible Er^{3+}, Nd^{3+}, and Ho^{3+} Doped LaF$_3$ nanoparticles. *Nano Lett.* 2: 733–737.

Suyver, J.F.; Grimm, J.; Kramer, K.W.; Güdel, H.U. 2005 Highly efficient near-infrared to visible up-conversion process in NaYF$_4$: Er^{3+}, Yb^{3+}. *J. Lumin.* 114: 53–59.

Suyver, J.F.; Grimm, J.; van Veen, M.K.; Biner, D.; Krämer, K.W.; Güdel, H.U. 2006. Upconversion spectroscopy and properties of NaYF$_4$ doped with Er^{3+}, Tm^{3+}, and/or Yb^{3+}. *J. Lumin.* 117: 1-12.

Swihart, M. 2003. Vapor-phase synthesis of nanoparticles. *Curr. Opin. Colloid Interface Sci.* 8: 127–133.

Thanh, N.; Maclean, N.; Mahiddine, S. 2014. Mechanisms of nucleation and growth of nanoparticles in solution. *Chem. Rev.* 114: 7610–7630.

Thermofisher.com. 2018. FTIR Technology—Thermo Fisher Scientific. (online) Available at: https://www.thermofisher.com/ca/en/home/industrial/spectroscopy-elemental-isotope-analysis/spectroscopy-elemental-isotope-analysis-learning-center/molecular-spectroscopy-information/ftir-information/ftir-technology.html (Accessed 1 May 2018).

Tian, G.; Gu, Z.; Zhou, L.; et al. 2012. Mn2+ Dopant-controlled synthesis of Nayf4:Yb/Er upconversion nanoparticles for in vivo imaging and drug delivery. *Adv. Mater.* 24: 1226–1231.

Tong, L.; Lu, E.; Pichaandi, J.; Cao, P.; Nitz, M.; Winnik, M. 2015. Quantification of surface ligands on NaYF4 nanoparticles by three independent analytical techniques. *Chem. Mater.* 27: 4899–4910.

Verma, A.; Stellacci, F. 2010. Effect of surface properties on nanoparticle–cell interactions. *Small*. 6, 12–21.

Viswanatha, R.; Sarma, D. 2007. Growth of nanocrystals in solution. *Nanomaterials Chemistry*. 139–170.

Voss, B.; Haase, M. 2013. Intrinsic focusing of the particle size distribution in colloids containing nanocrystals of two different crystal phases. *ACS Nano*. 7: 11242–11254.

Wagner, Z. 1961. Theory of precipitate change by redissolution. *Electrochemical* 65: 581–591.

Wang, F.; Banerjee, D.; Liu, Y.; Chen, X.; Liu, X. 2010. Upconversion nanoparticles in biological labelling, imaging, and therapy. *Analyst*. 135: 1839–1854.

Wang, F.; Deng, R.; Wang, J.; et al. 2011. Tuning upconversion through energy migration in core-shell nanoparticles. *Nat. Mater.* 10: 968–973.

Wang, F.; Deng, R.; Liu, X. 2014. Preparation of core-shell NaGdf4 nanoparticles doped with luminescent lanthanide ions to be used as upconversion-based probes. *Nat. Protoc.*. 9: 1634–1644.

Wang, F.; Han, Y.; Lim, C.; et al. 2010. Simultaneous phase and size control of upconversion nanocrystals through lanthanide doping. *Nature*. 463: 1061–1065.

Wang, F.; Liu, X. 2008. Upconversion multicolor fine-tuning: Visible to near-infrared emission from lanthanide-doped NaYF$_{4n}$ anoparticles. *J. Am. Chem. Soc.* 130: 5642–5643.

Wang, F.; Liu, X. 2009. Recent advances in the chemistry of lanthanide-doped upconversion nanocrystals. *Chem. Soc. Rev.* 38: 976–989.

Wang, H.; Nann, T. 2010. Upconverting nanoparticles. In: P. Hänninen and H. Härmä (eds.) *Lanthanide Luminescence*. Springer, Berlin, pp. 115–132.

Wang, L.; Li, Y. 2007. Controlled synthesis and luminescence of lanthanide doped NaYF$_4$ nanocrystals. *Chem. Mater.* 19: 727–734.

Wang, Y.-F.; Liu, G.-Y.; Sun, L.-D.; Xiao, J.-W.; Zhou, J.-C.; Yan, C.-H. 2013. Nd^{3+}-Sensitized upconversion nanophosphors: Efficient *In Vivo* bioimaging probes with minimized heating effect. *ACS. Nano*. 7: 7200–7206.

Wang, Y.; Liu, K.; Liu, X.; Dohnalová, K.; et al. 2011. Critical shell thickness of core/shell upconversion luminescence nanoplatform for FRET application. *J. Phys. Chem. Lett.* 2: 2083–2088.

Wen, H.; Zhu, H.; Chen, X.; et al. 2013. Upconverting near-infrared light through energy management in core-shell-shell nanoparticles. *Angew. Chem., Int. Ed.* 125, 13661–13665.

Wenger, O.S.; Bénard, S.; Güdel, H.U. 2002. Crystal field effects on the optical absorption and luminescence properties of Ni^{2+}-doped chlorides and bromides: Crossover in the emitting higher state. *Inorg. Chem.* 41: 5968–5977.

Wenger, O.S.; Güdel, H.U. 2001a. Chemical tuning of the photon upconversion properties in Ti^{2+}-doped chloride host lattices. *Inorg. Chem.* 40: 5747–5753.

Wenger, O.S.; Güdel, H.U. 2001b. Photon upconversion properties of Ni^{2+} in magnetic and nonmagnetic chloride host lattices. *Inorg. Chem.* 40: 157–164.

Wermuth, M.; Güdel, H.U. 1999. Photon avalanche in $Cs_2ZrBr_6:Os^{4+}$. *J. Am. Chem. Soc.* 121: 10102–10111.

Whitesides, G.M. 2003. The "right" size in nanobiotechnology. *Nat. Biotechnol.* 21 (10): 1161–1165.

Xie, X.; Gao, N.; Deng, R.; Sun, Q.; Xu, Q.H.; Liu, X. 2013. Mechanistic investigation of Photon Upconversion in Nd^{3+}-Sensitized Core-Shell Nanoparticles. *J. Am. Chem. Soc.* 135: 12608–12611.

Xu, Z.; Kang, X.; Li, C.; et al. 2010. Ln^{3+} (Ln = Eu, Dy, Sm, Er) Ion-doped YVO_4 nano/microcrystals with multiform morphologies: Hydrothermal synthesis, growing mechanism, and luminescent properties. *Inorg. Chem.* 49: 6706–6715.

Yan, C.; Zhao, H.; Perepichka, D.; Rosei, F. 2016. Lanthanide ion doped upconverting nanoparticles: Synthesis, structure and properties. *Small.* 12: 3888–3907.

Yang, D.; Li, C.; Li, G.; Shang, M.; Kang, X.; Lin, J. 2011. Colloidal synthesis and remarkable enhancement of the upconversion luminescence of $BaGdF_5:Yb^{3+}/Er^{3+}$ nanoparticles by active-shell modification. *J. Mater. Chem.* 21: 5923.

Yi, G.-S.; Chow, G.-M. 2007. Water-soluble $NaYF_4:Yb,Er(Tm)/NaYF_4$/Polymer Core/Shell/Shell nanoparticles with significant enhancement of upconversion fluorescence. *Chem. Mater.* 19: 341–343.

Zhao, G.; Tong, L.; Cao, P.; Nitz, M.; Winnik, M. 2014. Functional PEG–PAMAM-tetraphosphonate capped NaLnF4 nanoparticles and their colloidal stability in phosphate buffer. *Langmuir.* 30: 6980–6989.

Zhang, Y.; Sun, X.; Si, R.; You, L.; Yan, C. 2005. Single-crystalline and monodisperse LaF_3 triangular nanoplates from a single-source precursor. *J. Am. Chem. Soc.* 127: 3260–3261.

Zhang, Z.; Morishita, K.; Lin, W.; Huang, P.; Liu, J. 2018. Nucleotide coordination with 14 lanthanides studied by isothermal titration calorimetry. *Chin. Chem. Lett.* 29: 151–156.

Zhou, H.-P.; Xu, C.-H.; Sun, W.Yan. 2009. C-H Clean and flexible modification strategy for carboxyl/aldehyde-functionalized upconversion nanoparticles and their optical applications. *Adv. Funct. Mater.* 19: 3892–3900.

6

Microwave-Hydrothermal Synthesis of Perovskite Oxide Nanomaterials

Weiren Xia, Yao Lu, and
Xinhua Zhu
Nanjing University

6.1 Introduction

Microwaves belong to a portion of the electromagnetic spectrum, whose wavelengths are from 1 mm to 1 m, corresponding the frequencies between 300 MHz and 300 GHz. Their heating effect was first realized by an American engineer, Percy Spencer, at the Raytheon Company. In 1946, he accidentally discovered the microwave heating effect (a chocolate bar melted in his pocket) during his experiments on the microwave generation tube. Since then, microwave heating technique has been applied to various fields such as chemical reactions, drying, food processing, and ceramic processing (Clark and Sutton 1996; Katz 1992). During the microwave heating process, the materials couple with microwaves and absorb the electromagnetic energy volumetrically, transforming them into heat (Thostenson and Chou 1999; Yoshikawa 2010). Basically, microwave heating effect can be understood in terms of electric dipoles in materials which follow the alternating electric component of electromagnetic wave, i.e., water dipolar molecules rotate and its polarity orientation changes along with the external electric field, leading to the molecular friction motion effect and producing much amount of heat (Komarneni et al. 1992, 1993). Therefore, the heat is generated from inside the material, in contrast to the conventional heating where the heat is transferred from

outside to inside *via* heat conduction. As compared to the conventional heating manner, microwave heating technique exhibits several advantages such as rapid heating rate, reduced processing time, and energy efficiency, which has been used for many years to accelerate chemical reactions (Bilecka and Niederberger 2010; Rao et al. 1999; Tsuji et al. 2005; Vanetsev and Tret'yakov 2007; Yoshikawa 2010). A pioneering work on the microwave-assisted synthesis was performed in the middle of the 1980s. For example, Komarneni and Roy (1985) carried out the first microwave-assisted hydrothermal (M-H) synthesis of inorganic materials (e.g., titania gel spheres) in liquid phase, while Gedye et al. (1986) and Giguere et al. (1986) first reported the use of microwaves for the organic synthesis in 1986. Later in 1992, Komarneni et al. (1992) investigated the differences between the conventional hydrothermal syntheses (performed with conventional means of heating) and M-H syntheses (performed under heating special autoclaves with microwaves). He coined the term "the M-H process" for the reactions performed in a sealed vessel at temperatures above the boiling point of water and pressures higher than 1 atm under microwave heating. Nowadays, microwave heating technology is emerging as an alternative popular heating manner for rapid chemical reactions and materials synthesis in minutes, instead of hours or even days usually required by the conventional heating methods. Over the past two decades, the M-H method has been widely used to rapidly synthesize inorganic nanomaterials, and almost all classes of the functional materials have been synthesized, including metals, oxides, sulfides, phosphates, and halides (Balaji et al. 2009; Bilecka and Niederberger 2010; Chen et al. 2011; Hu and Yu 2008; Komarneni et al. 1995; Patzke et al. 2011). Therefore, microwave heating opens new synthetic pathways and allows the use of more environmentally friendly solvents, and it will also yield cleaner and pure products. The M-H synthesis of inorganic nanomaterials in liquid phase is currently a fast-growing area of research, especially in the field of the perovskite oxide nanomaterials (PONs) (Baghbanzadeh et al. 2011; Bilecka and Niederberger 2010; Zhu and Hang 2013), which display a wide spectrum of functional properties such as switchable polarization, piezoelectricity, pyroelectricity, and nonlinear dielectric behavior. These functional properties are indispensable for application in electronic nanodevices such as nonvolatile memories, sensors, micro-actuators, infrared detectors, and microwave phase filters (Zhu et al. 2010a, 2010b).

In the past two decades, large numbers of scientific publications are reported on this subject, and recently, the literatures are growing too rapidly to cite all the major contributions. In this chapter, we do not intend to provide an exhaustive overview; instead, we focus on overview of the recent advances on the structural characterization, physical properties, and applications of PONs synthesized by M-H process. This chapter is organized as follows: it begins with a short introduction of the basic principles, advantages, and limitations of the M-H process, and then the fundamentals of PONs are described. Next, an overview of the recent progress in the M-H synthesis of PONs is provided. The emphasis is focused on the physical properties, microstructural characterizations, and functional applications of the PONs obtained by this methodology. Finally, the future perspectives of this methodology in the synthesis of PONs are proposed.

6.2 Principles of Microwave-Hydrothermal (M-H) Synthesis

6.2.1 M-H Process: Basic Principles

The term of M-H process was coined by Komarneni in 1992, which refers to the chemical reactions performed in a sealed vessel (bomb, autoclave, etc.) at temperatures above the boiling point of water (as solvent) and pressures higher than 1 atm under microwave radiation. The critical point for water lies at 374°C and 218 atm. Above this temperature and pressure, water is said to be supercritical. Supercritical fluids exhibit characteristics of both a liquid and a gas: the interfaces of solids and supercritical fluids lack surface tension, yet supercritical fluids exhibit high viscosities and easily dissolve chemical compounds that would otherwise exhibit very low solubilities under ambient

conditions. In the M-H process, the microwave radiation is coupled with the material and converts the electromagnetic energy into thermal energy, which is absorbed by the material. The volumetric heating ability of microwaves allows for more rapid, uniform heating, decreased processing time, and reduction of thermal gradients inside the reaction mixture. The comparison between the temperature profiles in a reaction mixture undergoing both conventional hydrothermal and M-H treatments is schematically shown in Figure 6.1 (Li et al. 2013; Rosa et al. 2014), where the temperature gradient inside the reaction mixture is greatly reduced under microwave heating due to its volumetric heating ability. The M-H process allows many inorganic oxide nanomaterials to be prepared at temperatures substantially below those required by traditional solid-state reactions. Unlike the cases of co-precipitation and sol–gel methods, the M-H process also allows for substantially reduced reaction temperatures, and the products of M-H reactions are usually crystalline and do not require the additional post-annealing treatments. In addition, it also allows one to synthesize PONs with novel or metastable phases due to its fast crystallization kinetics (Komarneni and Katsuki 2002).

Microwave heating is particularly suitable for the synthesis of PONs because the absorption degrees of microwaves by them are much high due to their large dielectric constant and high dielectric loss. In the M-H process, the process parameters, including the microwave power and time, the pH value of the solution, and the reaction temperature and pressure, can be adjusted to optimize the final products. Therefore, more advanced M-H synthesis technology is required to control the pressure and

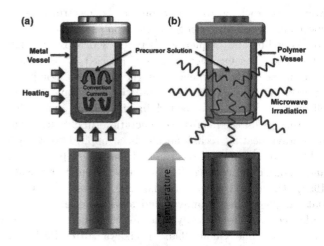

FIGURE 6.1 Schematic diagram showing the temperature profile comparisons occurring in a reaction mixture exposed to (a) conventional heating (inside a metal autoclave) and (b) microwave heating (inside a microwave transparent polymer reaction vessel). (Reprinted with permission from Rosa et al. 2014, Copyright 2014, John Wiley and Sons; Li et al. 2013, open access article distributed under the terms and conditions of the Creative Commons Attribution license, http://creativecommons.org/licenses/by/3.0/, © 2014 by the authors; licensee MDPI, Basel, Switzerland.)

temperature in the subcritical region of water. Nowadays, the M-H process becomes a promising approach towards the synthesis of PONs (Ma et al. 2014; Prado-Gonjal et al. 2015).

6.2.2 Advantages and Limitations of the M-H Process

The M-H process is emerging as a novel and innovative technology with many advantages over the conventional hydrothermal processing for the synthesis of oxide nanomaterials (Bilecka and Niederberger 2010; Gabriel et al. 1998; Oghbaei and Mirzaee 2010), such as (a) high heating rates achieved due to the internal heating manner within the material being processed rather than the heating from the surface, (b) no direct contact between the heating source and the reactants and/or solvents, (c) precisely controlling of the reaction parameters, (d) selective heating of the reaction mixtures containing compounds with different microwave absorbing properties, (e) higher yields of the crystalline products without the requirement of post-annealing treatments, and (f) both energy and time savings. The above advantages have been realized by the successful synthesis of wide ranges of PONs *via* the M-H process.

However, the M-H process also has some significant limitations. One of the major drawbacks is its high cost of the equipment (the need of expensive autoclaves). The prices of dedicated microwave reactors are normally in the range of ten thousands of US dollars, which is many times higher than that of the conventional heating equipment. The high price of the equipment has hindered from the wide use of M-H synthesis in academic laboratories around the world. This situation is expected to change over the next a few years when less-expensive equipment will likely become available. In addition, the short penetration depth of microwave irradiation into the liquid medium limits the size of the reactors, which is a serious problem for scale-up to the industry quantities. Currently, the largest microwave systems used on the pilot plant-scale comprise approximately 100 L (Ondruschka and Bonrath, 2006). Another problem is the impossibility of monitoring the M-H reaction process ("black box"), which means that the nucleation process and growth kinetics of oxide nanomaterials (e.g., nanoparticles, nanowires, or nanotubes) in the M-H process are difficult to be monitored *in situ* under microwave irradiation due to much complex interactions between the microwave and the reaction mixtures. However, in past few years, several techniques with advanced radiation sources for in situ monitoring the nucleation process and growth dynamics of nanoscale oxides have been developed, such as X-ray techniques, neutron scattering, NMR, and Raman spectroscopy, which have been used to monitor crystallization kinetics and growth mechanisms of oxide nanomaterials under the respective experimental conditions (Patzke et al. 2011). However, the controllable synthesis of complex systems has so far been proven to be a difficult task *via* the M-H technique, especially for the design and development

of periodical perovskite oxide nanostructures. While it is evident that further development of advanced microwave radiation heating route will push the exploration forward.

As for microwave radiations, the safe problem should be considered when working near microwave heating systems. It is well known that microwave radiations can cause serious health problems such as loss of appetite, irritation, discomfort, fatigue, and headaches. Effect of the microwave radiation on the human biochemistry and physiology depends upon the frequency, intensity, and duration of exposure of microwave radiation. It is better to use a microwave leakage meter monitoring the level of microwave radiations around the microwave heating system to ensure it below the prescribed limit. Otherwise, arrangements must be made to shield the system, or proper absorbing materials should be put to reduce the level below tolerable limit. It is good practice to always wear microwave safe gloves, apron, and goggles which are available in the market while working near microwave heating systems. Therefore, the hazards of microwave radiation can be well avoided by using safety protection methods.

6.3 Fundamentals of Perovskite Oxide Nanomaterials (PONs)

6.3.1 Perovskite Structures

Perovskite structure is coined from the mineral perovskite, $CaTiO_3$; now, it is a general name for oxides with the structural formula ABO_3. Many ABO_3 perovskites are cubic or nearly cubic in structure in their ideal form; however, some structural distortions may exist in the ideal cubic form of perovskite, leading to an orthorhombic, rhombohedral, hexagonal, or tetragonal form (Johnsson and Lemmens 2006; Peña and Fierro 2001). The general perovskite crystal structure is a primitive cube, with the A-larger cation in the corner, the B-smaller cation in the middle of the cube, and the oxygen anions, commonly in the center of the face edges (see Figure 6.2a)

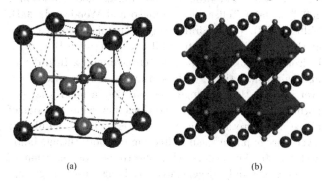

(a)   (b)

FIGURE 6.2 (a) Cubic perovskite-type structure ABO_3 (Reprinted with permission from http://www.3dchem.com/inorganicmolecule.asp?id=96, © Karl Harrison 3DChem.com), (b) corner-shared oxygen octahedral extending in three dimensions in an ABO_3 perovskite structure (Reprinted with permission from http://www.3dchem.com/inorganicmolecule.asp?id=96, © Karl Harrison 3DChem.com).

(http://www.3dchem.com/inorganicmolecule.asp?id=96, © Karl Harrison 3DChem.com), where A is a monovalent, divalent, or trivalent metal and B a pentavalent, tetravalent, or trivalent element, respectively. In the ABO_3 structure, the A-site cation is 12-fold coordinated, and the B-site cation is sixfold coordinated with oxygen anions. Alternatively, the perovskite structure is also regarded as a three-dimensional framework of BO_6 octahedra, as schematically shown in Figure 6.2b (http://www.3dchem.com/inorganicmolecule.asp?id=96, © Karl Harrison 3DChem.com).

In general, different sizes and charges of metal ions can be incorporated into the A- and/or B-sites of the perovskite structure. That demonstrates the great flexibility of composition and also leads to the distorted perovskite structure. However, the chemical substitutions at the A- and/or B-site cation are limited to a certain degree; otherwise, the three-dimensional framework of perovskite structure will be destroyed. All the cations in the perovskite structure should meet the restriction of $0.75 < t < 1.0$, where $t = (r_A + r_O)/\{\sqrt{2}(r_B + r_O)\}$ is the tolerance factor of perovskite oxides (Goldschmidt 1927), and r_A, r_B, and r_O are the radii of the respective ions, respectively. Ideal cubic perovskite structure only exists in some limited cases where t is very close to 1 and often at high temperatures. While $t < 1$ indicates that A is too small, $t > 1$ means A is too large to fit in the cavity between BO_6 octahedra. Although the t value, determined by the ionic size, is an important index for the stability of perovskite structures, the octahedral factor ($u = r_B/r_O$) and the contribution of the chemical nature of A and B atoms, such as the coordinating number of the constituent elements, are also needed to be considered to maintain the three-dimensional framework of perovskite structure (Li et al., 2008b).

Structural diversity arises from the distortions of ideal cubic structure, and the distorted perovskite structures maintaining the A- and the B-site oxygen coordination were achieved by the tilting of the BO_6 octahedrons (Glazer, 1972; Howard and Stokes, 1998; Lufaso and Woodward, 2001) or Jahn–Teller effect (Lufaso and Woodward, 2004; Millis, 1998). Octahedral tilting takes place when cation at the A-site is too small to fill cavity between BO_6 octahedrons, leading to a lower system symmetry. In order to maintain connectivity, each octahedron must rotate whenever one of them changes its position around the axis normal to the layer. These rotations may result in "out of phase" (when rotation direction is altering between different layers) or "in phase" (same direction rotation) tilting. Jahn–Teller distortion occurs when electronic configuration of cation is orbitally degenerated, yielding unstable ground states. This effect happens as a result of linear decrease in energy following displacement within the crystal, while at the same time, elastic energy of system increases. Distortional displacement can manifest through either elongation or shortening of the octahedral bonds, causing a minimum in total energy (Wolfram and Ellialtioglu, 2006). Therefore, the perovskite family includes not only compounds with an

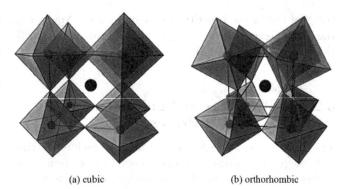

(a) cubic   (b) orthorhombic

FIGURE 6.3 Perovskite distortion from (a) cubic to (b) orthorhombic phase structure. (Reprinted with permission from Atta et al. 2016, open access book chapter, © 2016 The Author(s). Licensee In Tech. This chapter is distributed under the terms of the Creative Commons Attribution License (http://creativecommons.org/licenses/by/3.0), which permits unrestricted use, distribution, and reproduction in any medium, provided the original work is properly cited.)

ideal cubic perovskite lattice but also all compounds with distorted structures derived from the ideal one by small lattice distortions. A schematic diagram illustrating the structural distortion from cubic perovskite to orthorhombic one is shown in Figure 6.3 (Atta et al. 2016).

6.3.2 Fundamentals of PONs

PONs display a wide range of properties from colossal magnetoresistance (CMR) and high-temperature superconductivity to ferroelectricity and multiferroicity. Consequently, nanostructured PONs are central to the evolution of future electronics and information technologies, which provide a way to tackle large number of technological challenges including devices with lower energy consumption and transition to renewable sources of energy (Varghese 2013; Zhu and Hang 2013; Zhu et al. 2010a, 2010b). For example, memory storage based on perovskite ferroelectric polarization reversal is emerging as one of the emerging memory technologies, owing to their high read-write speed, low power consumption, and long rewriting endurance (Kinam et al. 2005).

Size and dimensionality play a critical role in determining the physical behaviors of PONs. It has been demonstrated that at nanoscale, PONs exhibit pronounced size effects, manifesting themselves in a significant deviation of the physical properties of low-dimensional structures from their bulk and film counterparts (Zhu 2009; Zhu and Liu 2011). For example, the ferroelectric properties of perovskite $PbTiO_3$ nanoparticles are strongly dependent upon the particle size (Akdogan et al. 2005; Chattopadhyay et al. 1995; Ishikawa et al. 1988; Zhong et al. 1993), and the ferroelectric properties of perovskite oxide films are strongly affected by the dead layers, grain boundaries, structural defects (such as oxygen vacancies), interfacial dislocations, and strain effects (Chu et al. 2004; Gregg 2009). Since the ferroelectricity is a cooperative phenomenon resulting from the arrangement of charge dipoles within a crystal structure, an increase of the surface area in perovskite oxide nanostructures could lead to immense changes in the long- and short-range ordering of dipoles such as depolarization field inside the perovskite oxide material that could change some features of the ferroelectric functionality, such as phase transition temperature or Curie temperature (T_C), domain dynamics, dielectric constant, coercive field, or spontaneous polarization, and piezoelectric response at the nanoscale (Ahn et al. 2004; Fong et al. 2004; Gregg 2009; Naumov et al. 2004; Shaw et al. 2000). Therefore, besides the intrinsic scientific value, understanding the size effects of physical properties in perovskite ferroelectric oxide nanostructures (e.g., nanoparticles, nanowires, and nanotubes) is also of importance for developing new generation of revolutionary electronic nanodevices (Zhu and Liu 2011; Zhu et al. 2010a, 2010b).

6.3.3 Quantum Confinement and Size Effects in PONs

Quantum confinement is referred to the changes of electronic and optical properties when the material sample is of sufficiently small size – typically 10 nm or less. The bandgap is increased as the size of the nanostructure is decreased. Specifically, the phenomenon is resulted from electrons and holes being squeezed into a dimension that approaches a critical quantum measurement, called the exciton Bohr radius. Quantum size effects, as illustrated in Figure 6.4 (Donegá 2010), have been extensively investigated in conventional semiconductor nanocrystals such as metal chalcogenides and utilized for a wide range of applications during the last two decades. As the sizes of the semiconductor nanocrystals approach to the exciton Bohr radius of the material, quantum confinement effects start to influence the excitonic wave function and the energetic states of the exciton, leading to blue-shifted photoluminescence. By precisely controlling the sizes of the semiconductor nanocrystals, the photoluminescence emission over wide wavelength ranges can be adjusted

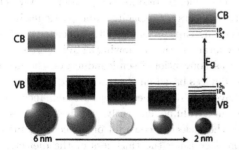

FIGURE 6.4 Schematic illustration of the quantum confinement effect on the energy levels of semiconductor CdSe nanocrystals. The emission of semiconductor nanocrystals can be tuned across the visible range by controlling their size. (Reprinted with permission from Donegá 2010, Copyright 2010, Royal Society of Chemistry.)

(Hintermayr et al. 2016; Polavarapu et al. 2017; Sichert et al. 2015). Similarly, quantum confinement effects in the PONs with reduced dimensionality (e.g., nanocrystals, nanowires, nanotubes, epitaxial perovskite oxide heterostructures) are also reported (Caviglia et al. 2008; Monkman et al. 2012; Ohtomo and Hwang 2004; Yoshimatsu et al. 2011). For example, in the perovskite oxide epitaxial heterostructures grown and controlled at atomic scales, the d electrons can be confined within a region of a few unit cells ($\sim$1 nm) in the epitaxial growth direction. Due to the quantum confinement, many novel physical phenomena occur (Mannhart and Schlom 2010; Zubko et al. 2011). Recently, quantum well states were also observed experimentally in two distinct oxide heterostructures: (a) $SrVO_3$ (SVO) ultrathin films (where electrons are geometrically confined inside the film) (Yoshimatsu et al. 2011) and (b) $SrTiO_3$ (STO) surfaces (where electrons are confined by a surface potential well) by angle-resolved photoemission spectroscopy (Santander-Syro et al. 2011). In both cases, very similar orbital-selective quantum well states were observed: d electrons with yz/xz orbital characters exhibited a large quantization of the energy levels, whereas xy electrons exhibited a much smaller level spacing (Zhong et al. 2013).

Historically, the size effects on the physical properties of PONs have been investigated since the 1950s, leading to a rich literature of both experimental and theoretical studies (Gregg 2009; Lichtensteiger et al. 2007; Rüdiger and Waser 2008; Zhu and Liu 2011). The early study of size effects in perovskite oxide ferroelectrics was from the 1950s. Jaccard et al. (1953) carried out the first systematic study on ferroelectric behavior of colloid KH_2PO_4 (KDP) particles embedded in an insulated medium with a low dielectric constant. It was found that the spontaneous polarization was reduced as decreasing the particle size, and below a critical particle size (the order of tens of nanometers), the ferroelectricity disappeared entirely (Rüdiger and Waser 2008). In recent years, advances in the characterized techniques allow ones to experimentally probe the fundamental size effects in perovskite ferroelectric oxide nanostructures at nanoscale, even at sub-Å level rather than those simply imposed by processing limitations. The experimental critical thickness for ferroelectricity in perovskite oxide thin films has been decreased by orders of magnitude over the years. For example, the critical thickness for BT films in the $SrRuO_3$/BT/$SrRuO_3$ sandwiched heterostructure grown epitaxially on ST(001) substrates is down to 5 nm (Kim et al. 2005) and 1.2 nm for $PbTiO_3$ (PT) films epitaxially grown on ST(001) substrates (Fong et al. 2004). The size effect in perovskite ferroelectric nanowires is also reported. A polarization perpendicular to perovskite BT oxide nanowire axis was induced and manipulated *via* an external electric field.

In parallel with experimentally investigating the size effects in perovskite ferroelectrics, theoretical computational models have also been developed, allowing ones to investigate the size effects from new ways. So far, theoretical models such as the Landau–Ginzburg–Devonshire (LGD) phenomenological theory (Li et al. 1996; Wang et al. 1995, 2008), first-principles-based computations (Ghosez and Rabe 2000; Junquera and Ghosez 2003; Meyer and Vanderbilt 2001), and the transverse Ising models (Cottam et al. 1984; Wang et al. 2000) have been developed to describe the size effects in perovskite oxide ferroelectrics. Within the LGD phenomenological theory, Kretschmer and Binder (1979) introduced a framework for considering size effects through two lengths, the correlation length, and the extrapolation length. This framework was then expanded and applied to numerous situations by Tilley and Zeks (1984). In the past several years, the most significant advance in theoretical studies of ferroelectricity is the development of powerful first-principles calculations, which has been widely applied to various perovskite ferroelectric nanostructures (Ghosez and Rabe 2000; Junquera and Ghosez 2003; Meyer and Vanderbilt 2001; Stengel and Spaldin 2006; Stengel et al. 2009), yielding several breakthroughs in understanding the nature of size effects in perovskite oxide ferroelectrics. For more details, the readers are referred to the recent reviews contributed by Kornev et al. (2006), Gregg (2009), Eliseev and Morozovska (2009), and book chapter from Ghosez and Junquera (2006). A review on recent progress of first-principles studies of perovskite magnetoelectric multiferroics is also available contributed by Ederer and Spaldin (2005).

6.4 Physical Properties of PONs Synthesized by M-H Process

6.4.1 Introduction

The applications of PONs synthesized by M-H process are dependent upon their multifunctional properties, such as dielectricity, ferroelectricity, multiferroicity, ferromagnetism, CMR, and optoelectronics, and these properties are indispensable for applications in novel electronic devices. The down-scaling of the material dimension not only implies a shrinkage of the active device which leads to higher packing density and lower power consumption, but also can significantly improve the device performance. In addition, when the dimension is reduced to a few nanometers, quantum mechanical effects start to play an important role. Therefore, a thorough understanding of the fundamental properties of the PONs is indisputably the prerequisite of research and development towards practical applications. The nanoscale multifunctional properties of perovskite oxide nanostructures (such as nanoparticles, nanowires/nanotubes), especially at the individual structure level, are of prime importance as far as the nanomaterial integration into modern nanotechnology is concerned. However, the nanoscale multifunctional properties cannot be measured *via* conventional techniques due to their small sizes of the perovskite oxide nanostructures. For this purpose, scanning probe techniques, in particular

piezoresponse force microscopy (PFM), have been developed, which play a critical role in the recent advances in the science and technology of nanoscale characterization of perovskite oxide nanostructures synthesized by the M-H process. Unfortunately, such PFM measurements with no access to the material internal structure have significantly limited the relevance of the data since any particular structural features of the perovskite oxide nanostructures modified during testing (determining the properties) have generally been hidden. Recently, *in situ* transmission electron microscopy (TEM) systems with precise structural and chemical analyses of a nanotube sample before, during, and after manipulation/deformation/electrical probing are possible through the parallel high-resolution imaging, electron diffraction, and spatially resolved spectroscopic methods (Golberg et al. 2009). In order to carry out *in situ* TEM physical property probing, special types of dedicated TEM holders with either scanning tunneling microscope (STM) or atomic force microscope (AFM) capabilities have been designed. Such holders have been commercialized, e.g., by "Nanofactory Instruments AB", Göteborg, Sweden (http://www.nanofactory.com). The usefulness of the advanced *in situ* TEM techniques for property analysis and delicate nanomanipulation of individual inorganic nanotubes under the ultimate spatial resolution typical of high-resolution transmission electron microscopy (HRTEM) has been demonstrated (Golberg et al. 2009). It is believed that a rapid progress will be made in this direction in the next several years. In this section, we will provide a collection of the physical properties of some representative PONs synthesized by the M-H process and discuss some multifunctional properties of perovskite oxide nanostructures that have been measured at nanoscale or the individual structure level.

6.4.2 Electrical Properties

PONs exhibit an array of electrical properties and a variety of solid-state phenomena from insulating, semiconducting, metallic, and superconducting characters; therefore, they are very fascinating to be studied and applied in a large scale. For example, perovskite BT powders are widely used as dielectrics in ceramic capacitors and particularly in multilayered ceramic capacitors (MLCCs) because of their high dielectric characteristics. Following a similar trend to miniaturization as conventional semiconductors, the down-sized electronic devices based on electroceramic materials such as MLCCs have been developed. In order to enhance the capacitance per unit volume, currently, much effort is focused on increasing the dielectric constant of the ceramic material as well as on decreasing the thickness of the dielectric layers in MLCCs. To achieve a thinner dielectric layer, dielectric BT nanomaterials with ultra-fine grains and high dielectric constant are highly required. Recently, Swaminathan et al. (2010) reported the ferroelectric and dielectric properties of tetragonal BT nanocubed powders synthesized by M-H process. Their remanent polarization (P_r) and saturation polarization (P_s) were measured to be 15.5 ± 1.6 and 19.3 ± 1.2 μC/cm^2, respectively. The dielectric constant of the BT pellets irradiated by microwave at 800 W for 15 min was measured to be 1068 at room temperature, while it became as high as 7903 after post-annealing at 120°C. The average leakage current density of the Pt/BT/Pt ceramic capacitors fabricated from BT truncated nanocubes was 5.78 ± 0.46 × 10^{-8} A/cm^2 measured @ 2 V. The above experimental results demonstrate that the BT nanocubes synthesized by M-H process have superior ferroelectric properties, which make them suitable for incorporation in charge storage devices.

The ferroelectric and piezoelectric properties of one-dimensional perovskite ferroelectric nanostructures (e.g., nanorods, nanowires, nanotubes) have recently been investigated with increasing intensity due to their potential applications in nonvolatile ferroelectric random-access memory, nano-electromechanical systems, energy harvesting devices, advanced sensors, and photonics. By using scanning probe microscopy (SPM) technique, the nanoscale ferroelectricity of an individual BT nanowire with a diameter of 15 nm grown by solution-based method was confirmed. It was demonstrated that a polarization perpendicular to the nanowire axis could be induced and manipulated *via* an external electric field. Such ferroelectricity could be retained in nanowires as thin as 0.8 nm, which was ascribed to surface absorbers such as OH$^-$ and carboxylates. Later, the axial polarization switching of a suspended ferroelectric BT nanowire prepared by molten salt technique was also realized by PFM with lateral mode (Wang et al., 2006a, 2006b). Besides the perovskite ferroelectric nanowires, the ferroelectric and piezoelectric properties of the perovskite ferroelectric nanotubes (e.g., BT, PZT, and BiFeO$_3$ (BFO)) fabricated by using a sol–gel template-based approach were also reported (Alexe et al., 2006; Luo et al., 2003; Zhang et al., 2005). However, the ferroelectric and piezoelectric properties of single-crystalline one-dimensional perovskite ferroelectric nanostructures synthesized by M-H process have not been reported yet. It is still a challenge for synthesizing high-quality one-dimensional perovskite ferroelectric oxides with controlled composition and morphology by the M-H process. It is expected that in this subject, a rapid progress will be seen in the near future.

6.4.3 Magnetic Properties

Perovskite manganites exhibit very interesting nanoscale physical properties due to their intrinsic coupling between charge, spin, and lattice degrees of freedom, giving rise to renowned phenomena such as CMR, charge ordering, and multiferroic behavior (Salamon and Jaime, 2001; Tokura, 2006). At nanoscale, perovskite manganites in the form of nanotubes or nanowires exhibit challenging electronic and magnetic properties because of their tendency to present coexistence of different phases in a wide range of scale lengths due to their small sizes and unconventional geometries. Understanding the magnetic properties of perovskite manganite nanowires/nanotubes could facilitate their sensory and memory applications, but is still in its infancy, demonstrating few studies in the literature. However, there are a few examples that are worth mentioning. The magnetic behavior of La$_{0.325}$Pr$_{0.300}$Ca$_{0.375}$MnO$_3$ (LPCMO) manganite nanotubes synthesized by microwave-assisted process using plastic templates was reported by Levy et al. (2003). Figure 6.5 shows the magnetization (M) as a function of temperature, while heating after field cooling (FC) and zero field cooling (ZFC) using a magnetic field of $H = 0.5$ T. Below 200 K, M increases notably, and it tends to saturate at lower temperatures as a FM-like material. However, a bifurcation between the M_{FC} and M_{ZFC} curves was observed at 70 K. The bottom inset of Figure 6.5 shows the inverse of the magnetic susceptibility (χ^{-1}), taken as H/M and showing a linear behavior for temperatures above 220 K. This paramagnetic behavior can be related by the Curie–Weiss relation ($\chi = C/(T - \Theta)$) with a positive Curie temperature ($\Theta = 170$ K), confirming that the dominant interactions are the ferromagnetic contributions. The top inset of Figure 6.5 displays a curve of M versus H at 5 K. The saturation magnetization reaches 55% of the bulk material value, which is $ns = 0.625n_S$ (Mn^{3+}) + 0.375n_S (Mn^{4+}) = 3.635 μ$_B$. This difference may be ascribed to both a mass overestimation (i.e., non-reacted raw materials and organic residue from the template) and to the extremely disordered nature of the material (high surface-to-volume ratio), affecting the bulk phase coexistence relation. The present magnetization results show that these La$_{0.325}$Pr$_{0.300}$Ca$_{0.375}$MnO$_3$ nanotubes are mainly ferromagnetic below 200 K and paramagnetic above this temperature. At low temperature, some degree of frustration was observed, and low-field MR around 40% for $H = 1$ T was achieved. Such structures

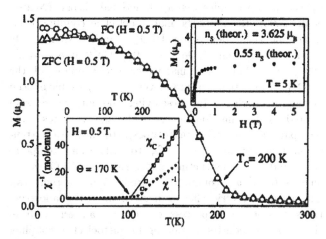

FIGURE 6.5 Magnetization versus the temperature for La$_{0.325}$Pr$_{0.300}$Ca$_{0.375}$MnO$_3$(LPCMO) tubes while warming after ZFC (triangles) and FC (circles). Bottom inset: raw and corrected inverse of the magnetic susceptibility as a function of temperature. Top inset: M versus H at 5 K after FC. (Reprinted with permission from Levy et al. 2003, Copyright 2003, AIP Publishing.)

constitute a tool both for spin-polarized injection and for their use as magnetic storage media.

6.4.4 Multiferroic Properties

Magnetoelectric multiferroic materials exhibit magnetic and ferroelectric order in the same temperature range (Spaldin and Fiebig 2005). In this class of compounds, perovskite BFO as a unique example of an intrinsic multiferroics shows simultaneously magnetic (antiferromagnetic), ferroelastic, and ferroelectric properties at room temperature, which has been widely investigated over the last few years in the forms of nanostructures (e.g., nanoparticles, nanotubes, nanowires, and nanostructured thin films) (Catalan and Scott 2009; Wu et al. 2016). Recently, Velasco-Davalos et al. (2016) reported the synthesis of epitaxial multiferroic BFO thin films on single-terminated Nb:SrTiO$_3$(111) substrates by intermittent M-H method, where the Bi(NO$_3$)$_3$·8H$_2$O and Fe(NO$_3$)$_3$·8H$_2$O along with KOH served as the precursors solution. Local phase hysteresis of the BFO thin films reveals the ferroelectric behavior of the BFO films. Single-phase BFO nanoparticles were also synthesized by M-H process using nitrates as the metallic source (Prado-Gonjal et al. 2011b). The observed magnetic behavior for the as-prepared M-H samples is similar to that reported for hydrothermally synthesized samples by Basu et al. (2008). Magnetization data collected at 5 K confirms that these samples show spontaneous magnetization not reaching saturation state, indicating a weak ferromagnetic nature (Azough et al. 2010). Zhu et al. (2011) also reported the synthesis of spherical BFO nanoparticles with diameters of 10–50 nm by M-H process, where Bi(NO$_3$)$_3$·5H$_2$O and Fe(NO$_3$)$_3$·9H$_2$O were used as the starting materials together with KOH as the mineralizer. The ferroelectric polarization and weak magnetization is coupled in single-phase BFO, which is confirmed by the first-principles density functional theory (Neaton et al. 2005). However, the magnetoelectric coupling coefficient of the single-phase BFO is rather small, only about 0.067 mV cm^{-1} Oe^{-1} at ~9.5 kOe (Suryanarayana 1994). To enhance the magnetoelectric coupling of the BFO at room temperature, several strategies have been developed, such as (a) destroying the long-range spin cycloid by forming low-dimensional nanostructures (e.g., nanoparticles, nanowires/nanotubes), (b) shifting the transition temperature T_C or T_N towards room temperature, and (c) constructing nanostructured BFO multiferroics (e.g., nanocomposites, core–shell nanostructures). The magnetoelectric effect generated in these nanostructures is due to a process of strain transferring from the phase boundaries, exhibiting magnetoelectric coupling coefficients three to five orders of magnitude larger than those of single-phase intrinsic multiferroic systems. Over the past few years, much work has been done to demonstrate the electric field control of magnetism in BFO/ferromagnet heterostructures, where the ferromagnet materials are either transitional metals (e.g., Co$_{0.9}$Fe$_{0.1}$

(Chu et al. 2008; Heron et al. 2011) and Permalloy (Lebeugle et al. 2009)) or oxides (e.g., La$_{0.7}$Sr$_{0.3}$MnO$_3$ (Wu et al. 2013). These heterostructures make full use of advantages of the intrinsic magnetoelectric coupling in BFO and the interface exchange coupling between BFO and the ferromagnetic layer. Recent advances in this topic are reviewed by Chu et al. (2007), Heron et al. (2014), Wu et al. (2016), and book chapter from Wu and Zhu (2018).

6.4.5 Photocatalytic Properties

Photocatalyses such as hydrogen generation from water splitting or degrading organic contaminant on photocatalysts under solar light are promising in solving current energy and environmental issues (Zhang et al. 2016; Zhu et al. 2014). PONs exhibit excellent promise as efficient photocatalysts under visible-light irradiation due to their unique crystal structures and electronic properties. The perovskite crystal structure provides a good framework in which to tune the bandgap values to enable visible-light absorption and band edge potentials to suit the needs of specific photocatalytic reactions. Among many photocatalyst materials, PONs have received much attention due to their controllable physicochemical properties, which enable them as energetic materials for catalysis. In recent years, BFO materials with a narrow optical bandgap (2.1–2.7 eV), a newly emerging photocatalytic material in parallel with the widely used TiO$_2$-based materials, have been widely investigated as a heterogeneous catalyst in treatment for the removal of organic compounds from wastewater (Gao et al. 2015). For example, Joshi et al. (2008) synthesized the single-crystalline BFO nanocubes with a size of 50–200 nm by M-H process and found that they exhibited photo-induced water oxidation activity in both photoelectrochemical and photocatalytic modes. Zhu et al. (2011) also synthesized BFO nanocrystals with diameters of 10–50 nm by M-H process, which exhibit high visible-light photocatalytic activity and can be used as promising photocatalysts for the degradation of organic compounds under visible light. These results indicate that the BFO nanoparticles are highly promising for visible-light-driven photochemistry. The introduce of BFO nanoparticles into TiO$_2$ nanofibers/nanotubes was reported to significantly enhance the photodegradation efficiency of TiO$_2$ photocatalyst under visible light as well as hinder the recombination of photogenerated electrons and holes, due to the p–n junction formed at the interface of p-type BFO and n-type TiO$_2$ (Yang et al. 2014). An enhanced photocatalytic performance was also observed in the BFO–graphene nanohybrids (Li et al. 2012). Besides the BFO nanoparticles, monodispersed BaTiO$_3$ nanocuboids synthesized by M-H process also exhibit high photocatalytic activity for RhB degradation under UV irradiation (Chen et al. 2016). Nearly complete degradation can be achieved in 180 min for 5 mol/L RhB solution. Shi et al. (2012) reported that perovskite NaTaO$_3$ nanocrystals synthesized by the M-H method exhibited quite a high activity in overall water splitting under UV irradiation, more than two times higher than those prepared under a conventional hydrothermal process that was ascribed to the good crystallinity and large surface area of $NaTaO_3$ nanocrystals achieved in a short reaction time in the M-H process. To further improve the photocatalytic properties, the development of new PONs that are sensitive to UV, visible, and near-infrared lights at the same time becomes a crucial issue (Kudo and Miseki 2009). It is also noticed that the small absorption coefficients of PONs further weaken their light-harvesting capabilities. In addition, to realize efficient charge separation and migration, PONs with both small size and high crystallinity are highly required because the small particle size provides shorter diffusion length and electrons or holes can be easily transported to the catalyst surface for reactions. The high degree of crystalline structure also avoids electron's trapping inside the particles. Clearly, this field is currently attracting much interest and provides an opportunity for further development of the PONs as promising photocatalyst materials under visible light.

6.4.6 Photovoltaic Properties

Photovoltaics (PV) is a term which covers the conversion of light into electricity by using semiconducting materials that exhibit the photovoltaic effect, a phenomenon studied in physics, photochemistry, and electrochemistry. The conversion process of light energy to electrical energy in photovoltaic devices relies on some form of built-in asymmetry that leads to the separation of electrons and holes. The fundamental physics behind this effect (e.g., in silicon-based cells) is charge separation using the potential developed at a p–n junction or heterojunction (Gur et al. 2005; O'Regan and Gratzel 1991). However, in most commercial solar cells based on the p–n junction or Schottky junction, the maximum photovoltage generated in these devices is limited by the energy barrier height and usually smaller than the semiconductor bandgap (Deng et al. 2010). On the contrary, the ferroelectric-photovoltaic (FE-PV) effect is distinctly different from the typical PV effect generated in silicon-based cells. In the FE-PV device, the polarization electric field is the driving force for the produced photocurrent, and the voltage output along the polarization direction can be significantly larger than the bandgap of the ferroelectric materials (Yuan et al. 2014). Recently, Yang et al. (2010) reported the above-bandgap voltages from FE-PV devices based on BFO thin films. In contrast to the semiconductor-based PV, the photovoltaic charge separation and photovoltage generation in BFO thin-film-based photovoltaic devices occurs exclusively at nanoscale (1–2 nm) ferroelectric domain walls in BFO thin films under white-light illumination (Yang et al. 2010). In addition, the photovoltaic effect can be reversed in polarity or turned off by electric field controlling the domain structures in the BFO thin films. Figure 6.6 demonstrates the

current–voltage (I–V) characteristics of the BFO thin films (with ordered arrays of 71° domain walls) measured in the two geometries of electrodes. One is the electrodes for electric transport measurements being perpendicular to the domain walls ($DW_\perp$) and the other one being parallel to the domain walls ($DW_{||}$) (Yang et al. 2010). Different photovoltaic behaviors were clearly observed. In the $DW_\perp$ direction (Figure 6.6a), a large photo-induced open-circuit voltage (V_{OC}) of 16 V (much higher than the bandgap of the BFO thin film, 2.3 ~ 2.7 eV) was measured, and the plane short-circuit current density (J_{sc}) was 1.2×10^{-4} A/cm^2. In contrast, dark and light I–V curves measured in the $DW_{||}$ direction (Figure 6.6b) exhibited a significant photoconductivity, but no photo-induced V_{OC}. It was also found that in the case of the electrodes perpendicular to the domain walls, the photo-induced voltages in the BFO film with different thickness were increased linearly with increasing the distance between the measured electrodes, as shown in Figure 6.6c. Most importantly, a single domain sample (i.e., with no domain walls between the platinum contacts) exhibits negligible level of photovoltage, which rules out a "bulk" photovoltaic effect arising from non-centrosymmetry (Fridkin 2001). In turn, this strongly suggests the prominent role of domain walls in creating the anomalous photovoltages. To interpret the strong photovoltaic response observed in the BFO thin-film-based FE-PV devices, the following factors should be considered. First, the internal electric field built in the BFO thin films is much high. The potential difference between the domains (with a typical domain-wall thickness of 2 nm) was 10 mV; thus, the corresponding internal electric field was 5 kV/mm, which was much higher than that (only 0.7 kV/mm) built in a conventional silicon p–n junctions (Yang et al. 2010). Second, in the ferroelectric BFO thin films, there are many 71° domains, and the ferroelectric polarizations within the adjunct ferroelectric domains form an angle of 71°. Therefore, the electric fields produced by these 71° domains are additive, forming a series of circuits. Thus, the generated photovoltaic voltage is higher than the bandgap. Bhatnagar et al. (2013) investigated the role of domain walls in the abnormal photovoltaic effect in BFO thin films, and they found that the bulk photovoltaic effect was the origin of the anomalous photovoltaic effect observed in BFO films. In addition, they also showed that irrespective of the measurement geometry (whether the electrodes running parallel or perpendicular to the domain walls), V_{OC} as high as 50 V could be achieved by controlling the conductivity of domain walls. It is also found that the photoconductivity of the domain wall is markedly higher than in the bulk of BFO. Numerous investigations (Bhatnagar et al. 2013; Choi et al. 2009; Ji et al. 2010; Yang et al. 2010) have shown that the photovoltaic effect in multiferroic BFO thin films is affected by many factors, such as the ferroelectric domains, interfaces, thickness, depolarization field, defect, and polarization direction. For example, Ji et al. (2010) found that the as-deposited BFO films were completely self-polarized

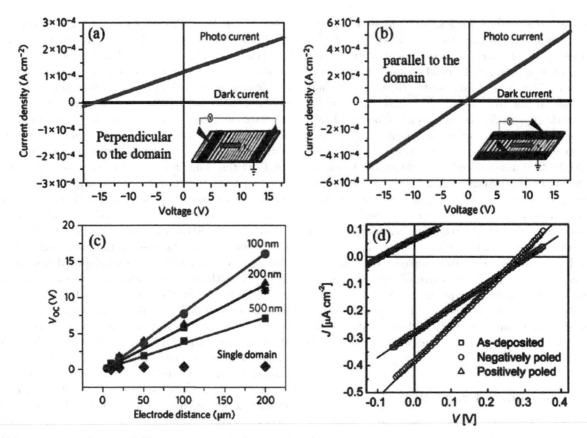

FIGURE 6.6 Light and dark $I-V$ curves of the BFO thin films measured with the two geometries: the electrodes for electric transport measurements (a) perpendicular to the domain walls ($DW_\perp$) and (b) parallel to the domain walls ($DW_{||}$). (c) The V_{OC} varies as a function of the distance between the measured electrodes for four different samples: 71° domain-wall samples with film thicknesses of 100 nm, 200 nm, and 500 nm as well as a monodomain BFO film having no domain walls. (Reprinted with permission from Yang et al. 2010, Copyright 2010, Nature Publishing Group). (d) $J-V$ characteristics of the BFO films (before and after polarization) were measured with incident light of 435 nm at 750 μW/cm² (Reprinted with permission from Ji et al. 2010, Copyright 2010, John Wiley and Sons.)

and produced large open-circuit photovoltage up to 0.3 V (18 kV/cm). The large portion of the photovoltage and photocurrent (about two thirds) is switchable in response to the switching of the ferroelectric polarization, with the direction of the photocurrent opposite to that of the polarization vector. Figure 6.6d demonstrates the $J-V$ curves of the BFO films with different states (as-deposited, positively polarized, and negatively polarized), which reveals that for the positively poled samples, the photocurrent is positive (i.e., it flows out of the top electrode) (Ji et al. 2010). In contrast, after the negative poling, the photocurrent direction is reversed. The magnitudes of both the photocurrent and photovoltage are smaller in positively poled samples than in negatively poled ones, indicating the photovoltaic effect in BFO thin films is closely related to the polarization directions of the BFO films. To understand the high photovoltaic efficiency in ferroelectric thin-film-based devices, Yang et al. (2010) and Seidel et al. (2011) have proposed a new photovoltage generated model as schematically shown in Figure 6.7, where the charge separation and photovoltage generation occur exclusively at nanometer-scale ferroelectric domain walls in ferroelectric

BFO thin films under white-light illumination. The domain walls act as nanoscale generators of the photovoltaic current. The steps in the electrostatic potential function accumulate electrons and holes on opposite sides of the walls while locally reducing the concentration of the oppositely charged carriers. Thus, the local reduction of the recombination rate leads to a net diffusion current. As a consequence, in an open circuit, the photovoltages generated from the periodically ordered domain walls are additive, and thus, the whole generated voltages are much larger than the bandgap of the ferroelectric thin film. Besides the BFO films, BFO nanofibers/nanowires also exhibit FE-PV properties. The $I-V$ curves of BiFeO₃ nanofibers and nanowires under dark and illumination condition are shown in Figure 6.8a–b, respectively (Fei et al. 2015; Prashanthi et al. 2015). It is found that much higher currents are generated under the illumination as compared to the dark condition, demonstrating the excellent photovoltaic behavior of 1D BFO nanostructures. It is reported that the current density of BFO nanofibers (~200 nm) in response to the applied voltage is about 1 mA/cm², which is two to ten times larger than that of BFO thin films (Fei et al. 2015).

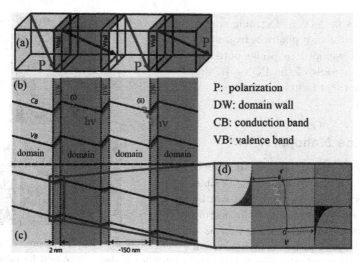

FIGURE 6.7 Schematic diagram of the new photovoltaic mechanism proposed by Ramesh's group in ferroelectric thin-film-based photovoltaic devices, which is based on the band structures of ferroelectric thin films in the dark and under illumination. (a) Schematic of four domains (three 71° domain walls). (b) Corresponding band structure in the dark and (c) band structure upon illumination, showing the valence band (VB) and conduction band (CB) across these domains and domain walls. Note that there is no net voltage across the sample in the dark. Section (i) illustrates a photon hitting in the bulk of a domain and section (ii) a photon hitting at a domain wall. (d) Detailed picture of the dynamics of photo-excited charges at a domain wall. (Reprinted with permission from Yang et al. 2010, Copyright 2010, Nature Publishing Group.)

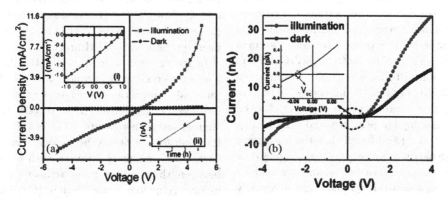

FIGURE 6.8 (a) I–V curve for BiFeO$_3$ nanofibers in dark and under illumination. Inset (i) shows expanded view of current density behavior around zero-bias. Inset (ii) shows averaged photocurrent after several measurements for different deposition time from 1 to 4 h. Reproduced with permission. (Reprinted with permission from Fei et al. 2015, Copyright 2015, American Chemical Society. (b) I–V curve for BiFeO$_3$ nanowires in dark and under illumination. Inset shows enlarged I–V curve in the portion of dotted circle. Reproduced with permission (Reprinted with permission from Prashanthi et al. 2015, Copyright 2015, Elsevier Ltd.)

The enhanced photovoltaic properties of BFO nanofibers are ascribed to the free-standing nanofibers with ferroelectric domains to be switched easily and more photons trapped by nanofibers due to their geometric confinement. In addition, the depolarization field may be helpful for driving electrons and holes in opposite directions, hampering their combination and thus enhancing the photovoltaic performance (Prashanthi et al. 2015). To date, PONs exhibit much better FE-PV activities; however, a clear physical picture about the formation mechanism of ferroelectric PV, especially the mechanisms related to ferroelectric domains, is not yet available. Much work remains to be done in understanding the mechanisms of perovskite ferroelectric-based photovoltaic devices, to develop powerful solar cells based on perovskite ferroelectric nanomaterials.

6.5 Microstructural Characterizations of PONs Synthesized by M-H Process

6.5.1 Introduction

Microstructural characterizations are essential in understanding the structures and properties of the PONs synthesized by M-H process. The differences in crystal and/or textural structure would certainly lead to variation in the properties, and for better understanding, the samples should be synthesized with pure perovskite phase and defined crystalline. To date, various approaches including X-ray powder diffraction (XRD), scanning electron microscopy (SEM) and TEM, SPM, HRTEM, and spectroscopy analyses such as energy-dispersive X-ray detector (EDX), electronic energy loss spectroscopy (EELS), and X-ray photoelectron spectroscopy (XPS), have been developed to probe both the macroscopic and microscopic details of the PONs. In this review, we want to introduce these techniques and provide general information on them.

6.5.2 Perovskite Oxide Nanoparticles

Perovskite oxide nanoparticles stand out as an important class of advanced nanomaterials due to their flexibilities with which their physical-chemical properties can be controlled *via* size, shape, and compositional engineering at their synthesis stage. Now microwave irradiation as a non-classical energy source is introduced to hydrothermal process, which has become increasingly popular in the preparation of perovskite oxide nanoparticles. Komarneni's research group in the Pennsylvania State University (USA) performed the pioneering works of M-H synthesis of perovskite oxide nanoparticles. As early as 1992, Komarneni et al. (1992) reported the M-H synthesis of the submicron-sized oxide powders (e.g., $KNbO_3$ and $BaTiO_3$) by controlling the M-H processing parameters (e.g., the concentration of chemical species, pH value, processing time, and temperature). In the next following years, they further applied the M-H process to synthesize the perovskite oxide powders such as ST, $Ba_{0.5}Sr_{0.5}TiO_3$, PT, $BaZrO_3$, $SrZrO_3$, PZT, $Pb(Mg_{1/3}Nb_{2/3})O_3$, and $Pb(Zn_{1/3}Nb_{2/3})O_3$ (Komarneni et al., 1993, 1994, 1995). Besides the influences of process parameters of precursor ratios such as H_2O/Ti, Ba/Ti, and OH^-/Ti ratios, heating temperature, and time (Jhung et al. 2004), Nyutu et al. (2008) found that the microwave frequency, microwave bandwidth sweep time, and aging time also affected the particle size, phase purity, and morphology of tetragonal BT nanoparticles. High microwave frequency (5.5 GHz) and variable frequency (3–5.5 GHz to 1 s) led to spherical BT nanoparticles with narrow and more uniform particle size distributions, whereas cubic BT nanoparticles were obtained by using the standard 2.45 GHz. Their surface areas were decreased with increasing the aging time using 4.0 and 5.5 GHz, but increased gradually with extending the aging time in variable frequency (3–5.5 GHz to 1 s) processing. The dependence of properties of barium titanate on microwave frequency could be ascribed to different transverse magnetic modes at different frequencies. Besides the spherical BT nanoparticles, Swaminathan et al. (2010) also reported the tetragonal truncated BT nanocubes synthesized by a facile M-H process. Figures 6.9a–d demonstrate the morphology and narrow size distribution of the truncated BT nanocubes. The selected area electron diffraction (SAED) pattern shown in Figure 6.9d clearly reveals the BT nanocube with a tetragonal structure with a space group $P4/mmm$ (lattice parameter $a = b = 4.01$ Å and $c = 4.03$ Å). In addition to BT nanoparticles, other perovskite oxide nanoparticles such as lanthanide-doped $CaTiO_3$ (Pereira et al. 2015), $LaMO_3$ (M = Al, Cr, Mn, Fe, Co) (Prado-Gonjal et al. 2011a), $NaTaO_3$ (Shi et al.

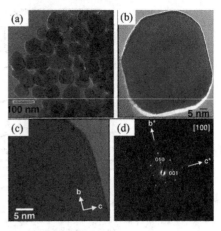

FIGURE 6.9 (a) TEM image of the truncated nanocubes of barium titanate synthesized by microwave-hydrothermal process. (b) Bright-field image of a single truncated nanocube of barium titanate, (c) HRTEM image of a single crystal taken from the [100] direction, and (d) corresponding SAED of c. (Reprinted with permission from Swaminathan et al. 2010, Copyright 2010, American Chemical Society.)

2012), sodium and potassium-sodium niobates ($NaNbO_3$ and $K_{0.5}Na_{0.5}NbO_3$) (López-Juárez et al. 2014), and $MSnO_3$ (M = Ca, Sr, and Ba) (Wang et al. 2014) nanoparticles have been synthesized by M-H process.

In recent years, scientific attention has been focused on magnetoelectric multiferroic materials that exhibit simultaneously magnetic (antiferro-magnetic), ferroelastic, and ferroelectric properties in a certain temperature range. Among them, perovskite BFO is the most interesting in the multiferroic family because it exhibits simultaneously charge and magnetic ordering with some mutual coupling between them above room temperature, thus leading to unique technological applications (Catalan and Scott, 2009). In the past few years, much work has been done to synthesize pure-phase perovskite BFO nanoparticles *via* the M-H process (Joshi et al., 2008; Zhu et al. 2011). In the early 1990s, Komarneni et al. (1996) first reported on the M-H synthesis of perovskite BFO nanoparticles at 194°C; however, their products exhibited highly crystalline agglomerated particles. Recently, Joshi et al. (2008) synthesized single-crystalline BFO nanocubes with size of 50–200 nm by M-H process at 180°C, whereas their products still had some degree of agglomeration. Therefore, an effort is needed to be focused on the M-H synthesis of single-crystalline perovskite BFO nanoparticles with a high degree of homogeneity and uniform particle size at fast rate and low temperatures. Zhu et al. (2011) first demonstrated the successful M-H synthesis of spherical perovskite BFO nanoparticles with diameters of 10–50 nm. Figure 6.10a displays the XRD pattern of the as-prepared perovskite BFO nanoparticles. As shown in Figure 6.10a, all the reflection peaks can be readily indexed as a rhombohedrally distorted perovskite BFO (JCPDS card No. 86-1518) with space group $R3c$ and lattice parameters of $a = 5.582$ Å and $c = 13.876$ Å. No peaks from other phase were detected. In addition,

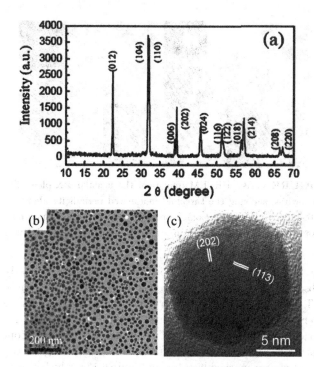

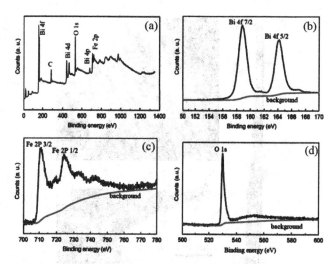

FIGURE 6.11 (a) Wide-scan XPS spectrum of the BFO nanoparticles and (b)–(d) the corresponding narrow-scan XPS spectra for Bi 4f, Fe 2p, and O 1s, respectively. (Reprinted with permission from Zhu et al. 2011, Copyright 2011, John Wiley and Sons.)

FIGURE 6.10 (a) XRD pattern of the as-synthesized perovskite $BiFeO_3$ nanoparticles. (b) TEM image of the perovskite $BiFeO_3$ nanoparticles, and (c) HRTEM image of a single $BiFeO_3$ nanoparticle with a diameter of $\sim$12 nm. (Reprinted with permission from Zhu et al. 2011, Copyright 2011, John Wiley and Sons.)

the sharp diffraction peaks indicated that highly crystallized and phase pure BFO nanoparticles were obtained. The TEM image of the perovskite BFO nanoparticles is shown in Figure 6.10b, which reveals that the perovskite BFO nanoparticles exhibit a spherical morphology with particle sizes of 10–50 nm. No agglomerated particles were observed, and nearly monodispersive behavior was observed in these nanoparticles. Figure 6.10c reveals an HRTEM image taken from a single BFO nanoparticle with a particle size of $\sim$12 nm, where the lattice fringes of $(20\underline{2})$ and (113) crystal planes are clearly resolved. To obtain further insights on the structural features of the as-obtained perovskite BFO nanoparticles, XPS measurements were also carried out. The wide-scan XPS spectrum of the BFO nanoparticles and the corresponding narrow-scan XPS spectra for Bi 4f, Fe 2p, and O 1s are shown in Figure 6.11a–d, respectively (Zhu et al. 2011). The XPS spectra clearly demonstrate that Fe element exists as the Fe^{3+} valence state, as well as Bi element as Bi^{3+} in the as-prepared BFO nanoparticles. Therefore, the M-H synthesis of single-phase crystalline BFO nanoparticles with a high degree of homogeneity and uniformity in particle size was first realized at low temperature of 180°C and fast reaction rate.

Recently, Ponzoni et al. (2013) investigated the influence of process parameters (i.e., precursor ratio, mineralizer concentration, temperature, time, and the use of inorganic chelating agents) on the phase formation, particle size distribution, and morphology of BFO powders. They found that the mineralizer concentration played a dominant role

with a synergic effect with the reaction temperature and mineralizer concentration required for the formation of single-phase BFO phase. The presence of Na_2CO_3 (acting as electron donor) allowed the M-H reaction to be performed at lower temperatures and KOH concentrations, e.g., 6 and 8 M at 180°C in 30 min in comparison with 8 M at 200°C in 30 min (without Na_2CO_3). With the presence of Na_2CO_3, the morphology of the particles was evolved in the following sequence: lamellar semi-cubic, cubic, and prismatic truncated octahedron shape, whereas without the presence of Na_2CO_3, the evolution followed the sequence: spherical, lamellar, lamellar semi-cubic, and cubic. At lower KOH concentration (<8 M), the presence of Na_2CO_3 did not affect the morphology of the BFO particles, but only their size.

6.5.3 1D Perovskite Oxide Nanostructures

Recently, one-dimensional (1D) perovskite oxide nanostructures such as nanowires, nanorods, and nanotubes have attracted a great attention due to their distinctive geometries, novel physical and chemical properties, and potential applications in the fields such as electronics, photonics, sensors, catalysts, energy harvesting, and information storage (Rørvik et al. 2011; Zhai et al. 2011; Zhu et al. 2010a, 2010b). They are expected to play an important role as both interconnects and functional units in the fabrication of nanodevices (Hu et al., 1999). Consequently, numerous studies on the M-H synthesis of 1D perovskite oxide nanostructures have been carried out over the past years (Paula et al. 2008; Zhu et al. 2008, 2010c). For example, the PX-phase single-crystalline PT nanowires synthesized *via* the M-H process were reported by Zhu et al. (2010c, 2008). Their microstructures were examined by TEM and HRTEM. Figure 6.12 shows the TEM images of the PX-phase single-crystalline PT nanowires.

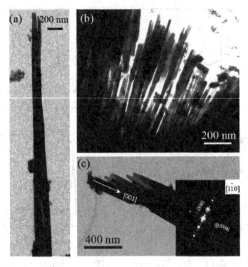

FIGURE 6.12 TEM images of the acicular PX-phase PbTiO₃ nanowires synthesized by microwave-hydrothermal method with different diameter sizes: (a) 40 nm, (b) 50 nm, and (c) 60 nm. (Reprinted with permission from Zhu et al. 2008, Copyright 2008, John Wiley and Sons.)

In Figure 6.12a, it can be seen that two to three acicular PT nanowires with a diameter of 40 nm, growing in parallel arrangement along their long axes, and the lengths of the wires are over 3 μm. As a consequence, their aspect ratio is close to 90. Such nanowires with a high aspect ratio, used as initial building blocks for nanodevices, will have potential applications in fabricating 1D ferroelectric nanostructures due to their unique geometry and anisotropic ferroelectric properties. As shown in Figure 6.12b, bundles of acicular PT nanowires are also observed, they tend to arrange side by side along their long axes, due to the van der Waals and/or electrostatic attraction forces. The diameter of an individual nanowire was about 50 nm and length usually up to 1 μm. Figure 6.12c displays several 1D acicular PT nanowires growing into a regular structure with parallel arrangement, and the average diameter of the wires is about 60 nm with length up to 2 μm. The SAED pattern (inset of Figure 6.12c) shows clear diffraction spots of the crystalline PX-phase PT, which is taken from the [1$\bar{1}$0] direction. Based on the TEM image and SAED pattern, it can be determined that the long axis of the nanowires is parallel to the [001] direction, and the nanowires grew along the [001] direction, as marked by a white arrow in Figure 6.12c. In addition, the superlattice electron diffraction spots with a threefold periodicity along the [110] direction are also clearly observed in the SAED pattern, which suggests that in the (001) plane, the PX-phase PT has a modulation with a periodicity of $3d_{600}$ (= 0.6165 nm) along the a and b axes, respectively. This result is confirmed by the following HRTEM images. Furthermore, in the [001] direction, the appearance of (008) diffraction spot with strong intensity and the forbidden (004) one are also observed, indicating that the PX-phase PT has a modulation with a periodicity of $4d_{008}$ (= 0.7245 nm) along the [001] direction. Figure 6.13 reveals the HRTEM images of the PX-phase PT nanowires.

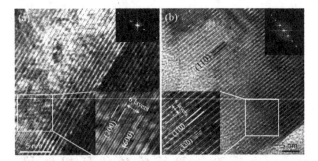

FIGURE 6.13 HRTEM images of the acicular PX-phase PT nanowires, showing the threefold modulated periodicity along (a) [100] and (b) [110] directions, respectively. (Reprinted with permission from Zhu et al. 2008, Copyright 2008, John Wiley and Sons.)

In Figure 6.13a, well-developed lattice fringes are clearly observed, indicating a good crystallinity of the PX-phase PT nanowires. The lattice fringes of the HRTEM image were examined to be 0.617 nm, close to the (200) lattice spacing of the PX-phase PT. The inset at right-bottom corner in Figure 6.13a is an enlarged HRTEM image of the selected area marked by a box in Figure 6.13a, which clearly demonstrates a modulated structure along the [100] direction. Three sub-layers are observed between two adjacent (200) lattice fringes, and each sub-layer corresponds well to the interplanar distance of (600) plane. Therefore, the PX-phase PbTiO₃ has threefold modulated periodicity along the [100] direction. The fast Fourier transform (FFT) pattern of the PX-phase PbTiO₃ nanowires is shown as an inset in Figure 6.13a (top-right corner), which demonstrates the periodically modulated structure of PX-phase PT. Similarly, the PX-phase PT also has threefold modulated periodicity along the [110] direction, as revealed by the HRTEM image shown in Figure 6.13b. The (110) lattice fringes of the PX-phase PT are clearly observed, and in the enlarged HRTEM image of the selected area, the (330) lattice fringes are also clearly observed, as marked by dot lines. The inset in Figure 6.13b at top-right corner is the corresponding FFT pattern of the HRTEM image, which reflects the modulated structure of the PX-phase PbTiO₃ along the [110] direction. That agrees well with the results obtained from the above SAED pattern. Based on the above structural data of PX-phase PT nanowires, a structural model of PX-phase PT unit cell ($3a_o \times 3a_o \times 4c_o$) is proposed, as shown in Figure 6.14 (Zhu et al. 2010c). In addition, unexpected Raman modes were also observed in the Raman spectra of the PX-phase acicular nanowires, which were ascribed to the breakage of the translation lattice symmetry across the above contiguous unit cells (Zhu et al. 2008).

6.5.4 Perovskite Oxide Nanoarchitectures

In addition to perovskite oxide nanowires (or nanorods), perovskite oxides nanoflowers of flower-like architectures are also synthesized by M-H process. From a viewpoint of the future device engineering, the architectures in the form of flowers are important because they exhibit

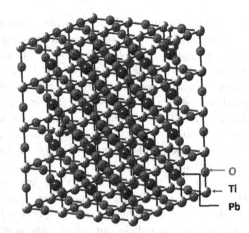

FIGURE 6.14 Proposed structural model for PX-phase PT unit cell-based HRTEM images and selected area electron diffraction pattern. (Reprinted with permission from Zhu et al. 2010c, Copyright 2010, Elsevier.)

excellent electrocatalytic activity (Jena and Raj 2007), high-dielectric permittivity (Fang et al. 2006), and photocatalytic activity (Yao et al. 2006), and these properties make them attractive in the applications as excellent field emitters (Tang et al. 2006), effective solar cells, or amperometric biosensors (Li et al. 2008a). Recently, Chybczyńska et al. (2014) reported the synthesis of perovskite BFO multiferroic flowers by M-H process at 200°C and 20 min, where $Bi(NO_3)_3·5H_2O$ and $Fe(NO_3)_3·9H_2O$ were used as the starting materials together with KOH and Na_2CO_3 as the mineralizer. It was found that the morphology of flowers was sensitive to KOH content and the time of microwave processing; therefore, homogeneous powders composed of flowers only are obtained for KOH concentration of 6 M and 20 min synthesis. Their SEM image is shown in Figure 6.15a, where the product contains only agglomerates in the form of flowers (Chybczyńska et al. 2014). The mean size of the flowers is about 15 μm, and the flowers are not perfectly spherical, exhibiting kind of hollows where the petals are packed less dense. The thickness of petals of BFO flowers is in the range of 30–200 nm, as presented in the histogram in Figure 6.15b. Most of the petals have a thickness in the range of 50–100 nm. TEM images reveal that the petals of BFO flowers petals are composed of BFO blocks with dimensions exceeding 100 nm. The finest petals with thickness of about 40 nm are composed of single layer of BFO nanoparticles and also some amount of the amorphous phase. Different KOH concentrations lead to the formation of other BFO structures, whereas the long-time synthesis (60 min) leads to powders containing large flowers but also irregular BFO agglomerates, as shown in Figure 6.15c. As shown in Figure 6.15c, the large flower is almost spherical with a diameter of 20 μm, which is composed of hundreds of closely packed petals. Based on the SEM images, the main stages of the formation of BFO flower-like architectures can be assumed as follows. At early stages of the growth process, a central part composed of only a few petals (or even four petals, shown in Figure 6.15d) was formed. The petals were

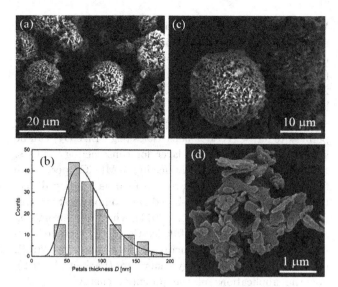

FIGURE 6.15 (a) SEM image of the product obtained at 6 M KOH concentration and 20 min synthesis, where the product contains only agglomerates in the form of flowers. (b) Histogram of the thickness of petals in the BFO flowers. The solid line is the best theoretical fit of the log-normal distribution to the data. (c) SEM image of a spherical flower from the sample produced at 6 M KOH concentration and 60 min synthesis. (d) SEM image of the initial stage of formation of a flower observed in the product obtained at 6 M KOH concentration and 20 min synthesis. Four petals in the center of the flower form a cage. (Reprinted with permission from Chybczyńska et al. 2014, Copyright 2014, Springer.)

connected at the edges and formed a kind of rectangular cage. Next, the following petals are attached successively to this structure as the reaction continues. Since the sizes of the nanoparticles forming the petals are smaller than the period of spin cycloid in BFO; therefore, uncompensated spins near the surface of the flowers appear, leading to an enhanced magnetization of the BFO flowers as compared to their bulk counterpart. The enhanced magnetic properties of the BFO flowers enable them to be used as active field emitters, effective solar cells, or catalyst supports.

6.6 Functional Applications of PONs Synthesized by M-H Process

PONs are of great attraction and show potential applications in different fields such as data memory storage, and solar energy conversion. The ferroelectric random-access memories (FeRAM) are one of the main applications of PONs in data memory storage, which are superior to the electrically erasable and programmable read-only memories (EEPROMs) and Flash memories in terms of write-access time and overall power consumption. Their operational principles are based on the reversible polarization of a perovskite ferroelectric materials used in nonvolatile memory, and the direction of polarization switching representing the binary "1" and "0" in data storage (Setter et al. 2006;

Vorotilov and Sigov 2012). Since the density of FeRAM nonvolatile memory can be increased thousands-fold by reading and writing in nanoparticle-based FeRAM, PONs (e.g., nanoparticles (Nuraje et al. 2006; O'Brien et al. 2001) and nanowires (Wouters et al. 2006; Yun et al. 2002) have been widely investigated for potential application in storage memory. Owing to the high dielectric permittivity, perovskite ferroelectric nanoparticles (e.g., $BaTiO_3$) can be also used as the dielectric layer for constructing MLCCs. Besides the practical applications in the MLCCs, perovskite ferroelectric nanoparticles are also used as a filling dielectric material in the polymer-based nanocomposite structures (Huang et al. 2010; Zhou et al. 2011), which have promising applications in the fields of dielectric energy storage, flexible thin-film dielectric capacitors (Almadhoun et al. 2012), and flexible electrical energy generators (Park et al. 2012). For the applications of the potential energy harvesting from collective mechanical movements, piezoelectric oxide nanowires are investigated extensively (Cha et al. 2011; Qi et al. 2011; Wang 2012a). For example, electrical energy collected from piezoelectric nanostructures has already been utilized effectively to power nanoelectronic devices (Xu et al. 2010) and sensors (Wang 2012b). One-dimensional perovskite piezoelectric nanostructures, especially piezoelectric oxide nanowires and rods, have mostly been used for piezotronics applications due to their large mechanical strain tolerance (Wang 2012a). Particularly, vertically or horizontally aligned arrays of piezoelectric nanowires are in an ideal configuration for energy harvesting applications due to their enhanced anisotropic piezoresponse (Xu et al. 2010; Wang 2012b).

PONs (e.g., BFO) exhibit multiferroic properties at room temperature and therefore are good candidates for potential multiferroic application in information technology. An emerging application seeks to exploit the multiferroic properties of PONs to develop novel multifunctional devices controlled by magnetic and electric fields. In recent years, applications of the magnetoelectric coupling and multiferroics in the fields of spintronics are increasing rapidly, and numerous possible device architectures have been proposed and fabricated (Bea et al. 2006; Kimura et al. 2003; Zutic et al. 2004). For example, in the ferroelectric antiferromagnets, such as the multiferroic perovskites, the magnetic structure could be modulated or controlled by the application of an electric field (Bea et al. 2006; Zutic et al. 2004). And also the nonvolatile control of the orientation of antiferromagnetic axis can be achieved by using the coupling between ferroelectricity and antiferromagnetism that can provide permanent (Kimura et al. 2003). Similarly, the coupling between the antiferromagnetic component and an adjacent ferromagnet also allows to switch the ferromagnetism by the application of an electric field. Therefore, potential new types of multiple-state memories and spintronic devices controlled by electric and magnetic fields can be developed (Gajek et al. 2007). It is expected that this field should be a hot topic in near future.

PONs also exhibit photovoltaic behavior, which has important applications in photovoltaic devices such as solar cells. The typical example is perovskite 1D $BiFeO_3$ nanofiber-based photovoltaic devices. The $I–V$ curves of BFO nanofibers and nanowires under dark and illumination condition shown in Figure 6.8 (Fei et al. 2015; Prashanthi et al. 2015) demonstrate that much higher currents are generated under the illumination as compared to the dark condition. The current density of $BiFeO_3$ nanofibers ($\sim$200 nm) in response to the applied voltage was reported to $\sim$1 mA/cm^2, which was two to ten times higher than that of $BiFeO_3$ thin films (Fei et al. 2015). In addition to the photovoltaic properties, PONs also exhibit photocatalytic behavior, which have promising applications in water splitting and environmental remediation. The representative example is perovskite BFO nanoparticles, which are promising materials for hydrogen generation under visible light due to their narrow bandgap and high efficiency of charge separation. The photocatalytic activities of BFO nanoparticles are related to their particle sizes and morphologies, e.g., $BiFeO_3$ nanoparticles with {100}c dominant cube exhibit good photocatalytic performance. In addition, the visible-light response of BFO nanoparticles can be improved by suitable elements doping or forming heterojunction (Gao et al. 2015). Although PONs have not yet been applied in industry up to date and there is still a long way to walk on before their commercialization, their easy availability, wide applicability, and especially controllable physicochemical properties enable them energetic materials for solar energy utilization. It is believed that the applications of PONs in photocatalytic and photovoltaic devices have a bright future.

6.7 Conclusions and Perspectives

PONs with formula of ABO_3 are a very important class of functional materials that exhibit a wide range of physical properties and have potential applications in the next generation of nanoelectronic devices. In the past decade, significant progresses have been achieved in the field of M-H synthesis of PONs, and their physical and microstructural characterizations. This chapter summarizes the recent advances on the microstructural and physical characterizations of PONs synthesized by M-H process. The potential applications of PONs in the fields of information storage memory, energy harvesting, and solar energy conversion are also discussed. Although the M-H process has been well developed to synthesize PONs such as perovskite oxide nanoparticles and nanowires, the influence of microwaves on the nucleation and growth of PONs is not well understood, and the fundamental principles have not yet been fully established. Up to now, the controllable synthesis of periodical perovskite oxide nanostructures has been proven to be a difficult task *via* the M-H technique. To solve the above problems, *in situ* measured techniques are highly required to be exploited for monitoring the nucleation process and growth dynamics of PONs during the M-H process. It is expected in the next few years, the methods for measuring the local temperatures with high resolution can be developed to better understand the selectively heating effect of microwaves. As a consequence, the advantages of microwave irradiation can be fully used in a rational way. Thus, a wide range of perovskite oxide nanostructures with complex multicomponents can be controllably synthesized.

Acknowledgments

The authors acknowledge the financial supports from National Natural Science Foundation of China (grant nos. 11674161, 11174122, and 11134004), six big talent peak project from Jiangsu Province (grant no. XCL-004), and open project of National Laboratory of Solid State Microstructures, Nanjing University (grant no. M28026).

References

Ahn, C. H., Rabe, K. M., Triscone, J. M. 2004. Ferroelectricity at the nanoscale: local polarization in oxide thin films and heterostructures. *Science* 303: 488–491.

Akdogan, E. K., Rawn, C. J., Porter, W. D., Payzant, E. A., Safari, A. 2005. Size effects in $PbTiO_3$ nanocrystals: effect of particle size on spontaneous polarization and strains. *J. Appl. Phys.* 97(8): 084305-1–8.

Alexe, M., Hesse, D., Schmidt, V. et al. 2006. Ferroelectric nanotubes fabricated using nanowires as positive templates. *Appl. Phys. Lett.* 89(17): 172907-1–3.

Almadhoun, M. N., Bhansali, U. S., Alshareef, H. N. 2012. Nanocomposites of ferroelectric polymers with surface-hydroxylated $BaTiO_3$ nanoparticles for energy storage applications. *J. Mater. Chem.* 22: 11196–11200.

Atta, N. F., Galal, A., El-Ads, E. H. 2016. Perovskite nanomaterials – synthesis, characterization, and applications. In: *Perovskite Materials – Synthesis, Characterisation, Properties, and Applications*, ed. L. K. Pan and G. A. Zhu, 107–151, Rijeka: In Tech.

Azough, F., Freer, R., Thrall, M., Cernik, R., Tuna, F., Collison, D. 2010. Microstructure and properties of Co-, Ni-, Zn-, Nb and W-modified multiferroic $BiFeO_3$ ceramics. *J. Eur. Ceram. Soc.* 30: 727–736.

Baghbanzadeh, M., Carbone, L., Cozzoli, P. D., Kappe, C. O. 2011. Microwave-assisted synthesis of colloidal inorganic nanocrystals. *Angew. Chem. Int. Ed.* 50(48): 11312–11359.

Balaji, S., Mutharasu, D., Subramanian, N. S., Ramanathan, K. 2009. A review on microwave synthesis of electrode materials for lithium-ion batteries. *Ionics* 15(6): 765–777.

Basu, S., Pal, M., Chakravorty, D. 2008. Magnetic properties of hydrothermally synthesized $BiFeO_3$ nanoparticles. *J. Magn. Magn. Mater.* 320: 3361–3365.

Bea, H., Bibes, M., Cherifi, S. et al. 2006. Tunnel magnetoresistance and robust room temperature exchange bias with multiferroic $BiFeO_3$ epitaxial thin films. *Appl. Phys. Lett.* 89: 242114.

Bhatnagar, A., Chaudhuri, A. R., Kim, Y. H., Alexe, M., Hesse, D. 2013. Role of domain walls in the abnormal photovoltaic effect in $BiFeO_3$. *Nat. Commun.* 4(4): 2835-1–8.

Bilecka, I., Niederberger, M. 2010. Microwave chemistry for inorganic nanomaterial synthesis. *Nanoscale* 2: 1358–1374.

Catalan, G., Scott, J. F. 2009. Physics and applications of bismuth ferrite. *Adv. Mater.* 21: 2463–2485.

Caviglia, A. D., Gariglio, S., Reyren, N. et al. 2008. Electric field control of the $LaAlO_3/SrTiO_3$ interface ground state. *Nature* 456: 624–627.

Cha, S. N., Kim, S. M., Kim, H. J. et al. 2011. Porous PVDF as effective sonic wave driven nanogenerators. *Nano. Lett.* 11: 5142–5147.

Chattopadhyay, S., Ayyub, P., Palkar, V. R. et al. 1995. Size-induced diffuse phase transition in the nanocrystalline ferroelectric $PbTiO_3$. *Phys. Rev. B* 52(18): 13177–13183.

Chen, P., Zhang, Y., Zhao, F., Gao, H., Chen, X., An, Z. 2016. Facile microwave synthesis and photocatalytic activity of monodispersed $BaTiO_3$ nanocuboids. *Mater. Characterization* 114: 243–253.

Chen, Y. C., Lo, S. L., Ou, H. H., Chen, C. H. 2011. Photocatalytic oxidation of ammonia by cadmium sulfide/titanate nanotubes synthesised by microwave hydrothermal method. *Water Sci. Technol.* 63(3): 550–557.

Choi, T., Lee, S., Choi, Y. J., Kiryukhin, V., Cheong, S. W. 2009. Switchable ferroelectric diode and photovoltaic effect in $BiFeO_3$. *Science* 324: 63–66.

Chu, W. M., Szafraniak, I., Scholz, R. et al. 2004. Impact of misfit dislocations on the polarization instability of epitaxial nanostructured ferroelectric perovskites. *Nat. Mater.* 3: 87–90.

Chu, Y. H., Martin, L. W., Holcomb, M. B., Ramesh, R. 2007. Controlling magnetism with multiferroics. *Mater. Today* 10(10): 16–23.

Chu, Y. H., Martin, L. W., Holcomb, M. B. et al. 2008. Electric-field control of local ferromagnetism using a magnetoelectric multiferroic. *Nat. Mater.* 7(6): 478–482.

Chybczyńska, K., Ławniczak, P., Hilczer, B. et al. 2014. Synthesis and properties of bismuth ferrite multiferroic flowers. *J. Mater. Sci.* 49(6): 2596–2604.

Clark, D. E., Sutton, W. H. 1996. Microwave processing of materials. *Annu. Rev. Mater. Sci.* 26: 299–331.

Cottam, M. G., Tilley, D. R., Zeks, B. 1984. Theory of surface modes in ferroelectrics. *J. Phys. C* 17: 1793–1823.

Deng, Q., Wang, X., Xiao, H. et al. 2010. Theoretical investigation of efficiency of a p-a-SiC:H/i-a-Si:H/n-μc-Si solar cell. *J. Semiconductors* 31: 103003-1–5.

Donegá, C. D. M. 2010. Synthesis and properties of colloidal heteronanocrystals. *Chem. Soc. Rev.* 40: 1512–1546.

Ederer, C., Spaldin, N. A. 2005. Recent progress in first-principles studies of magnetoelectric multiferroics. *Curr. Opin. Solid State Mater. Sci.* 9: 128–139.

Eliseev, E. A., Morozovska, A. N. 2009. General approach for the description of size effects in ferroelectric nanosystems. *J. Mater. Sci.* 44: 5149–5160.

Fang, X. S., Ye, C. H., Xie, T., Wang, Z. Y., Zhao, J. W., Zhang, L. D. 2006. Regular MgO nanoflowers and their enhanced dielectric responses. *Appl. Phys. Lett.* 88: 013101–013103.

Fei, L., Hu, Y., Li, X., et al. 2015. Electrospun bismuth ferrite nanofibers for potential applications in ferroelectric photovoltaic devices. *ACS Appl. Mater. Interfaces* 7(6): 3665–3670.

Fong, D. D., Stephenson, G. B., Streiffer, S. K. et al. 2004. Ferroelectricity in ultrathin perovskite films. *Science* 304: 1650–1653.

Fridkin, V. M. 2001. Bulk photovoltaic effect in noncentrosymmetric crystals. *Crystallogr. Rep.* 46: 654–658.

Gabriel, C., Gabriel, S., Grant, E. H., Halstead, B. S. J., Mingos, D. M. P. 1998. Dielectric parameters relevant to microwave dielectric heating. *Chem. Soc. Rev.* 27(3): 213–223.

Gajek, M., Bibes, M., Fusil, S. et al. 2007. Tunnel Junctions with multiferroic barriers. *Nat. Mater.* 6: 296–302.

Gao, T., Chen, Z., Huang, Q. et al. 2015. A review: preparation of bismuth ferrite nanoparticles and its applications in visible-light induced photocatalyses. *Rev. Adv. Mater. Sci.* 40(2): 97–109.

Gedye, R., Smith, F., Westaway, K., Ali, H., Baldisera, L., Laberge, L., Rousell, J. 1986. The use of microwave ovens for rapid organic synthesis. *Tetrahedron Lett.* 27: 279–282.

Ghosez, P., Junquera, J. 2006. First-principles modeling of ferroelectric oxides nanostructures. In: *Handbook of Theoretical and Computational Nanotechnology*, ed. M. Rieth and W. Schommers, 1–149. Stevenson Ranch: American Science Publishers.

Ghosez, P., Rabe, K. M. 2000. Microscopic model of ferroelectricity in stress-free $PbTiO_3$ ultrathin films. *Appl. Phys. Lett.* 76: 2767–2769.

Giguere, R. J., Bray, T. L., Duncan, S. M. 1986. Application of commercial microwave ovens to organic synthesis. *Tetrahedron Lett.* 27: 4945–4948.

Glazer, A. M. 1972. The classification of tilted octahedra in perovskites. *Acta Cryst B* 28(11): 3384–3392.

Golberg, D., Costa, P. M. F. J., Mitome, M., Bando, Y. 2009. Properties and engineering of individual inorganic nanotubes in a transmission electron microscope. *J. Mater. Chem.* 19(7): 909–920.

Goldschmidt, V. M. 1927. Crystal structure and chemical correlation. *Ber. Dtsch. Chem. Ges.* 60: 1263–1268.

Gregg, J. M. 2009. Ferroelectrics at the nanoscale. *Phys. Status Solidi A* 206: 577–587.

Gur, I., Fromer, N. A., Geier, M. L., Alivisatos, A. P. 2005. Air-stable all-inorganic nanocrystal solar cells processed from solution. *Science* 310: 462–465.

Heron, J. T., Schlom, D. G., Ramesh, R. 2014. Electric field control of magnetism using $BiFeO_3$-based heterostructures. *Appl. Phys. Rev.* 1: 021303.

Heron, J. T., Trassin, M., Ashraf, K. et al. 2011. Electric-field-induced magnetization reversal in a ferromagnet multiferroic heterostructure. *Phys. Rev. Lett.* 107(21): 217202.

Hintermayr, V. A., Richter, A. F., Ehrat, F. et al. 2016. Tuning the optical properties of perovskite nanoplatelets through composition and thickness by ligand-assisted exfoliation. *Adv. Mater.* 28: 9478–9485.

Howard, C. J., Stokes, H. T. 1998. Group theoretical analysis of octahedral tilting in perovskites. *Acta Crystallogr. Sect. B* 54(6): 782–789.

Hu, J., Odom, T. W., Lieber, C. M. 1999. Chemistry and physics in one dimension: synthesis and properties of nanowires and nanotubes. *Acc. Chem. Res.* 32(5): 435–445.

Hu, X. L., Yu, J. C. 2008. Continuous aspect-ratio tuning and fine shape control of monodisperse alpha-Fe_2O_3 nanocrystals by a programmed microwave hydrothermal method. *Adv. Funct. Mater.* 18(6): 880–887.

Huang, L., Jia, Z., Kymissis, I., O'Brien, S. 2010. High K capacitors and OFET gate dielectrics from self-assembled $BaTiO_3$ and $(Ba,Sr)TiO_3$ nanocrystals in the superparaelectric limit. *Adv. Funct. Mater.* 20: 554–560.

Ishikawa, K., Yoshikawa, K., Okada, N. 1998. Size effect on the ferroelectric phase transition in $PbTiO_3$ ultrafine particles. *Phys. Rev. B* 37(10): 5852–5855.

Jaccard, C., Känzig, W., Peter, M. 1953. Behavior of colloid seignette-electrics: I, potassium phosphate, KH_2PO_4. *Helv. Phys. Acta* 26(5): 521–544.

Jena, B. K., Raj, C. R. 2007. Synthesis of flower-like gold nanoparticles and their electrocatalytic activity towards the oxidation of methanol and the reduction of oxygen. *Langmuir* 23: 4064–4070.

Jhung, S. H., Lee, J. H., Ji, W. Y. et al. 2004. Effects of reaction conditions in microwave synthesis of nanocrystalline barium titanate. *Mater. Lett.* 58(25): 3161–3165.

Ji, W., Yao, K., Liang, Y. C. 2010. Bulk photovoltaic effect at visible wavelength in epitaxial ferroelectric $BiFeO_3$ thin films. *Adv. Mater.* 22: 1763–1766.

Johnsson, M., Lemmens, P. 2006. Crystallography and chemistry of perovskites. In: *Handbook of Magnetism and Advanced Magnetic Materials*, ed. H. Kronmueller and S. Parkin, 1–11. New York: John Wiley & Sons, Ltd.

Joshi, U. A., Jang, J. S., Borse, P. H., Lee, J. S. 2008. Microwave synthesis of single-crystalline perovskite $BiFeO_3$ nanocubes for photoelectrode and photocatalytic applications. *Appl. Phys. Lett.* 92(24): 242106-1–3.

Junquera, J., Ghosez, P. 2003. Critical thickness for ferroelectricity in perovskite ultrathin films. *Nature* 422: 506–509.

Katz, J. D. 1992. Microwave sintering of ceramics. *Annu. Rev. Mater. Sci.* 22: 153–170.

Kim, D. J., Jo, J. Y., Kim, Y. S. et al. 2005. Polarization relaxation induced by a depolarization field in ultrathin ferroelectric $BaTiO_3$ capacitors. *Phys. Rev. Lett.* 95: 237602.

Kimura, T., Kawamoto, S., Yamada, I., Azuma, M., Takano, M., Tokura, Y. 2003. Magnetocapacitance effect in multiferroic $BiMnO_3$. *Phys. Rev. B* 67: 180401.

Kinam, K., Gitae, J., Hongsik, J., Sungyung, L. 2005. Emerging memory technologies. In: *Proceedings of the IEEE 2005 Custom Integrated Circuits Conference*, 423–426. San Jose, CA, USA.

Komarneni, S., D'Arrigo, M. C., Leonelli, C., Pellacani, G. C., Katsuki, H. 1998. Microwave-hydrothermal synthesis of nanophase ferrites. *J. Am. Ceram. Soc.* 81: 3041–3043.

Komarneni, S., Katsuki, H. 2002. Nanophase materials by an novel microwave-hydrothermal process. *Pure App. Chem.* 74(9): 1537–1543.

Komarneni, S., Li, Q. H., Roy, R. 1994. Microwave-hydrothermal processing for layered and network phosphates. *J. Mater. Chem.* 4(12): 1903–1906.

Komarneni, S., Li, Q. H., Stefansson, K. M., Roy, R. 1993. Microwave-hydrothermal processing for synthesis of electroceramic powders. *J. Mater. Res.* 8: 3176–3183.

Komarneni, S., Pidugu, R., Li, Q. H., Roy, R. 1995. Microwave-hydrothermal processing of metal powders. *J. Mater. Res.* 10(7): 1687–1692.

Komarneni, S., Roy, R. 1985. Titania gel spheres by a new sol-gel process. *Mater. Lett.* 3: 165–167.

Komarneni, S., Roy, R., Li, Q. H. 1992. Microwave-hydrothermal synthesis of ceramic powders. *Mater. Res. Bull.* 27: 1393–1405.

Kornev, I. A., Fu, H. X., Bellaiche, L. 2006. Properties of ferroelectric ultrathin films from first principles. *J. Mater. Sci.* 41: 137–145.

Kretschmer, R., Binder, K. 1979. Surface effects on phase transitions in ferroelectrics and dipolar magnets. *Phys. Rev. B* 20: 1065–1076.

Kudo, A., Miseki, Y. 2009. Heterogeneous photocatalyst materials for water splitting. *Chem. Soc. Rev.* 38: 253–278.

Lebeugle, D., Mougin, A., Viret, M., Colson, D., Ranno, L. 2009. Electric field switching of the magnetic anisotropy of a ferromagnetic layer exchange coupled to the multiferroic compound $BiFeO_3$. *Phys. Rev. Lett.* 103: 257601.

Levy, P., Leyva, A. G., Troiani, H. E., Sa'nchez, R. D. 2003. Nanotubes of rare-earth manganese oxide. *Appl. Phys. Lett.* 83(25): 5247–5249.

Li, C., Fang, G., Liu, N., Ren, Y., Huang, H., Zhao, X. 2008a. Snowflake-like ZnO structures: self-assembled growth and characterization. *Mater. Lett.* 62: 1761–1764.

Li, C., Lu, X. G., Ding, W. Z., Feng, L. M., Gao, Y. H., Guo, Z. M. 2008b. Formability of ABX_3 (X = F, Cl, Br, I) halide perovskites. *Acta Crystallogr. B* 64: 702–707.

Li, S., Eastman, J. A., Li, Z., Foster, C. M., Newnham, R. E., Cross, L. E. 1996. Size effects in nanostructured ferroelectrics. *Phys. Lett. A* 212: 341–346.

Li, S., Nechache, R., Davalos, I. A. V. et al. 2013. Ultrafast microwave hydrothermal synthesis of $BiFeO_3$ nanoplates. *J. Am. Ceram. Soc.* 96(10): 3155–3162.

Li, Z., Shen, Y., Yang, C. et al. 2012. Significant enhancement in the visible light photocatalytic properties of $BiFeO_3$-graphene nanohybrids. *J. Mater. Chem. A* 1(3): 823–829.

Lichtensteiger, C., Dawber, M., Triscone, J. M. 2007. Ferroelectric size effects. In: *Physics of Ferroelectrics: A Modern Perspective, Topics Applied Physics*, ed. K. Rabe, C. H. Ahn and J. M. Triscone, 305–338. Berlin: Springer-Verlag.

López-Juárez, R., Castañeda-Guzmán, R., Villafuerte-Castrejón, M. E. 2014. Fast synthesis of $NaNbO_3$ and $K_{0.5}Na_{0.5}NbO_3$ by microwave hydrothermal method. *Ceram. Int.* 40(9): 14757–14764.

Lufaso, M. W., Woodward, P. M. 2001. Prediction of the crystal structures of perovskites using the software program SPuDS. *Acta Cryst. B* 57(6): 725–738.

Lufaso, M. W., Woodward, P. M. 2004. Jahn-Teller distortions, cation ordering and octahedral tilting in perovskites. *Acta Cryst. B* 60(1): 10–20.

Luo, L., Szafraniak, I., Zakharov, N. D. et al. 2003. Nanoshell tubes of ferroelectric lead zirconate titanate and barium titanate. *Appl. Phys. Lett.* 83(3): 440–442.

Ma, M. G., Zhu, J. F., Zhu, Y. J., Sun, R. C. 2014. The microwave-assisted ionic-liquid method: a promising methodology in nanomaterials. *Chem. Asian J.* 9: 2378–2391.

Mannhart, J., Schlom, D. G. 2010. Oxide interfaces – an opportunity for electronics. *Science* 327: 1607–1611.

Meyer, B., Vanderbilt, D. 2001. *Ab* initio study of $BaTiO_3$ and $PbTiO_3$ surfaces in external electric fields. *Phys. Rev. B* 63(20): 797–801.

Millis, A. J. 1998. Lattice effects in magnetoresistive manganese perovskites. *Nature* 392(6672): 147–150.

Monkman, E. J., Adamo, C., Mundy, J. A. et al. 2012. Quantum many body interactions in digital oxide superlattices. *Nat. Mater.* 11: 855–859.

Naumov, I. I., Bellaiche, L., Fu, H. 2004. Unusual phase transitions in ferroelectric nanodisks and nanorods. *Nature* 432: 737–740.

Neaton, J. B., Ederer, C., Waghmare, U. V., Spaldin, N. A., Rabe, K. M. 2005. First-principles study of spontaneous polarization in multiferroic $BiFeO_3$. *Phys. Rev. B* 71: 014113.

Nuraje, N., Su, K., Haboosheh, A. et al. 2006. Room temperature-synthesis of ferroelectric barium titanate nanoparticles using peptide nano-rings as templates. *Adv. Mater.* 18: 807–811.

Nyutu, E. K., Chen, C. H., Dutta, P. K., Suib, S. L. 2008. Effect of microwave frequency on hydrothermal synthesis of nanocrystalline tetragonal barium titanate. *J. Phys. Chem. C* 112(12): 1521–1528.

O'Brien, S., Brus, L., Murray, C. B. 2001. Synthesis of Monodisperse Nanoparticles of barium titanate: toward a generalized strategy of oxide nanoparticle synthesis. *J. Am. Chem. Soc.* 123: 12085–12086.

Oghbaei, M., Mirzaee, O. 2010. Microwave versus conventional sintering: a review of fundamentals, advantages and applications. *J. Alloys Compd.* 494(1–2): 175–189.

Ohtomo, A., Hwang, H. Y. 2004. A high-mobility electron gas at the LaAlO$_3$/SrTiO$_3$ heterointerface. *Nature* 427: 423–426.

Ondruschka, B., Bonrath, W. 2006. Microwave-assisted chemistry – a stock taking. *Chimia Intel. J. Chem.* 60(6): 326–329.

O'Regan, B., Gratzel, M. 1991. Low-cost high-efficiency solar cell based on dye-sensitized colloidal TiO$_2$ films. *Nature* 353: 737–740.

Park, K. I., Lee, M., Liu, Y. et al. 2012. Nanocomposites of ferroelectric polymers with surface-hydroxylated BaTiO$_3$ nanoparticles for energy storage applications. *Adv. Mater.* 24: 2999–3004.

Patzke, G. R., Zhou, Y., Kontic, R., Conrad, F. 2011. Oxide nanomaterials: synthetic developments, mechanistic studies, and technological innovations. *Angew. Chem.-Int. Edit.* 50(4): 826–859.

Paula, A. J., Parra, R., Zaghete, M. A., Varela, J. A. 2008. Synthesis of KNbO$_3$ nanostructures by a microwave assisted hydrothermal method. *Mater. Lett.* 62(17–18): 2581–2584.

Peňa, M. A., Fierro, J. L. G. 2001. Chemical structures and performance of perovskite oxides. *Chem. Rev.* 101: 1981–2017.

Pereira, S. C., Figueiredo, A. T., Barrado, C. M. et al. 2015. Facile microwave-assisted synthesis of lanthanide doped CaTiO$_3$ nanocrystals. *J. Braz. Chem. Soc.* 26(11): 2339–2345.

Polavarapu, L., Nickel, B., Feldmann, J., Urban, A. S. 2017. Advances in quantum-confined perovskite nanocrystals for optoelectronics. *Adv. Energy Mater.* 7: 1700267-1–9.

Ponzoni, C., Rosa, R., Cannio, M. et al. 2013. Optimization of BFO microwave-hydrothermal synthesis: influence of process parameters. *J. Alloys Compd.* 558(9): 150–159.

Prado-Gonjal, J., Arevalo-Lopez, A. M., Morán, E. 2011a. Microwave-assisted synthesis: a fast and efficient route to produce LaMO$_3$ (M = Al, Cr, Mn, Fe, Co) perovskite materials. *Mater. Res. Bull.* 46(2): 222–230.

Prado-Gonjal, J., Ávila, D., Villafuerte-Castrejón, et al. 2011b. Structural, microstructural and mössbauer study of BiFeO$_3$ synthesized at low temperature by a microwave-hydrothermal method. *Solid State Sci.* 13(11): 2030–2036.

Prado-Gonjal, J., Schmidt, R., Morán, E. 2015. Microwave-assisted routes for the synthesis of complex functional oxides. *Inorganics* 3: 101–117.

Prashanthi, K., Dhandharia, P., Miriyala, N., Gaikwad, R., Barlage, D., Thundat, T. 2015. Enhanced photo-collection in single BiFeO$_3$ nanowire due to carrier separation from radial surface field. *Nano Energy* 13: 240–248.

Qi, Y., Kim, J., Nguyen, T. D., Lisko, B., Purohit, P. K., McAlpine, M. C. 2011. Enhanced piezoelectricity and stretch ability in energy harvesting devices fabricated from buckled PZT ribbons. *Nano. Lett.* 11: 1331–1336.

Rao, K. J., Vaidhyanathan, B., Ganguli, M., Ramakrishnan, P. A. 1999. Synthesis of inorganic solids using microwaves. *Chem. Mater.* 11: 882–895.

Rørvik, P. M., Grande, T., Einarsrud, M. A. 2011. One-dimensional nanostructures of ferroelectric perovskites. *Adv. Mater.* 23(35): 4007–4034.

Rosa, R., Ponzoni, C., Leonelli, C. 2014. Direct energy supply to the reaction mixture during microwave-assisted hydrothermal and combustion synthesis of inorganic materials. *Inorganics* 2: 191–210.

Rüdiger, A., Waser, R. 2008. Size effects in nanoscale ferroelectrics. *J. Alloys Compd.* 449: 2–6.

Salamon, M. B., Jaime, M. 2001. The physics of manganites: structure and transport. *Rev. Mod. Phys.* 73(3): 583–628.

Santander-Syro, A. F., Copie, O., Kondo, T. et al. 2011. Two-dimensional electron gas with universal sub-bands at the surface of SrTiO$_3$. *Nature* 469: 189–193.

Seidel, J., Fu, D., Yang, S. Y. et al. 2011. Efficient photovoltaic current generation at ferroelectric domain walls. *Phys. Rev. Lett.* 107: 126805.

Setter, N., Damjanovic, D., Eng, L. et al. 2006. Ferroelectric thin films: review of materials, properties, and applications. *J. Appl. Phys.* 100: 051606-1–46.

Shaw, T. M., Troliermckinstry, S., Mcintyre, P. C. 2000. The properties of ferroelectric films at small dimensions. *Annu. Rev. Mater. Sci.* 30(1): 263–298.

Shi, J., Liu, G., Wang, N., Li, C. 2012. Microwave-assisted hydrothermal synthesis of perovskite NaTaO$_3$ nanocrystals and their photocatalytic properties. *J. Mater. Chem.* 22(36): 18808–18813.

Sichert, J. A., Tong, Y., Mutz, N. et al. 2015. Quantum size effect in organometal halide perovskite nanoplatelets. *Nano Lett.* 15: 6521–6527.

Spaldin, N. A., Fiebig, M. 2005. The renaissance of magnetoelectric multiferroics. *Science* 309: 391–392.

Stengel, M., Spaldin, N. A. 2006. Origin of the dielectric dead layer in nanoscale capacitors. *Nature* 443: 679–682.

Stengel, M., Vanderbilt, D., Spaldin, N. A. 2009. Enhancement of ferroelectricity at metal-oxide interfaces. *Nat. Mater.* 8: 392–397.

Suryanarayana, S. V. 1994. Magnetoelectric interaction phenomena in materials. *Bull. Mater. Sci.* 17(7): 1259–1270.

Swaminathan, V., Pramana, S. S., White, T. J., Chen, L., Chukka, R., Ramanujan, R. V. 2010. Microwave synthesis of noncentrosymmetric BaTiO$_3$ truncated nanocubes for charge storage applications. *ACS Appl. Mater. Interfaces* 2(11): 3037–3042.

Tang, Y. B., Cong, H. T., Wang, Z. M., Cheng, H. M. 2006. Catalyst seeded synthesis and field emission properties of flowerlike Si-doped AlN nanoneedle array. *Appl. Phys. Lett.* 89(25): 253112-1–3.

Thostenson, E. T., Chou, T. W. 1999. Microwave processing: fundamentals and applications. *Composites Part A* 30(9): 1055–1071.

Tilley, D., Zeks, B. 1984. Landau theory of phase transitions in thick films. *Solid State Commun.* 49: 823–828.

Tokura, Y. 2006. Critical features of colossal magnetoresistive manganites. *Rep. Prog. Phys.* 69(3): 797–851.

Tsuji, M., Hashimoto, M., Nishizawa, Y., Kubokawa, M., Tsuji, T. 2005. Microwave-assisted synthesis of metallic nanostructures in solution. *Chem. Eur. J.* 11: 440–452.

Vanetsev, A. S., Tret'yakov, Y. D. 2007. Microwave-assisted synthesis of individual and multicomponent oxides. *Russ. Chem. Rev.* 76(5): 397–413.

Varghese, J., Whatmore, R. W., Holmes, J. D. 2013. Ferroelectric nanoparticles, wires and tubes: synthesis, characterization and applications. *J. Mater. Chem. C* 1: 2618–2638.

Velasco-Davalos, V., Ambriz-Vargas, F., Kolhatkar, G., Thomas, R., Ruediger, A. 2016. Synthesis of BiFeO$_3$ thin films on single-terminated Nb:SrTiO$_3$(111) substrates by intermittent microwave assisted hydrothermal method. *AIP Adv.* 6(6): 3069–3097.

Vorotilov, K. A., Sigov, A. S. 2012. Ferroelectric memory. *Phys. Solid State* 54(5): 894–899.

Wang, C. L., Xin, Y., Wang, X. S., Zhong, W. L. 2000. Size effects of ferroelectric particles described by the transverse Ising model. *Phys. Rev. B* 62: 11423–11427.

Wang, J., Kamlah, M., Zhang, T. Y., Li, Y., Chen, L. Q. 2008. Size-dependent polarization distribution in ferroelectric nanostructures: phase field simulations. *Appl. Phys. Lett.* 92(16): 162905-1–3.

Wang, W., Liang, S., Ding, K. et al. 2014. Microwave hydrothermal synthesis of MSnO$_3$ (M^{2+} = Ca^{2+}, Sr^{2+}, Ba^{2+}): effect of M^{2+} on crystal structure and photocatalytic properties. *J. Mater. Sci.* 49(4): 1893–1902.

Wang, Y. G., Zhong, W. L., Zhang, P. L. 1995. Surface and size effects on ferroelectric films with domain structures. *Phys. Rev. B* 51: 5311–5314.

Wang, Z. L. 2012a. Progress in piezotronics and piezo-phototronics. *Adv. Mater.* 24: 4632–4646.

Wang, Z. L. 2012b. Self-powered nanosensors and nanosystems. *Adv. Mater.* 24: 280–285.

Wang, Z. Y., Suryavanshi, A. P., Yu, M. F. 2006a. Ferroelectric and piezoelectric behaviors of individual single crystalline BaTiO$_3$ nanowire under direct axial electric biasing. *Appl. Phys. Lett.* 89(8): 082903-1–3.

Wang, Z. Y., Hu. J., Yu, M. F. 2006b. One-dimensional ferroelectric monodomain formation in single crystalline BaTiO$_3$ nanowire. *Appl. Phys. Lett.* 89(26): 263119-1–3.

Wolfram, T., Ellialtioglu, S. 2006. *Electronic and Optical Properties of d-Band Perovskites.* Cambridge: Cambridge University Press.

Wouters, D. J., Maes, D., Goux, L. et al. 2006. Integration of SrBi$_2$Ta$_2$O$_9$ thin films for high density ferroelectric random access memory. *J. Appl. Phys.* 100: 051603-1–16.

Wu, H., Zhu, X. H. 2018. Nanostructured multiferroics. In: *Nanostructured Multiferroics: Current Trends and Future Prospects*, ed. B. Raneesh and P. M. Visakh. Berlin: Wiley-VCH.

Wu, J. G., Fan, Z., Xiao, D. Q., Zhu, J. G., Wang, J. 2016. Multiferroic bismuth ferrite-based materials for multifunctional applications: ceramic bulks, thin films and nanostructures. *Prog. Mater. Sci.* 84: 335–402.

Wu, S. M., Cybart, S. A., Yi, D., Parker, J. M., Ramesh, R., Dynes, R. C. 2013. Full electric control of exchange bias. *Phys. Rev. Lett.* 110: 067202.

Xu, S., Hansen, B. J., Wang, Z. L. 2010. Piezoelectric-nanowire-enabled power source for driving wireless microelectronics. *Nat. Commun.* 1: 93-1–4.

Yang, S. Y., Seidel, J., Byrnes, S. J. et al. 2010. Above-bandgap voltages from ferroelectric photovoltaic devices. *Nat. Nanotech.* 5: 143–147.

Yang, Y. C., Liu, Y., Wei, J. H., Pan, C. X., Xiong, R., Shi, J. 2014. Electrospun nanofibers of p-type BiFeO$_3$/n-type TiO$_2$ hetero-junctions with enhanced visible-light photocatalytic activity. *RSC Adv.* 4: 31941–31947.

Yao, W. T., Yu, S. H., Liu, S. J., Chen, J. P., Liu, X. M., Li, F. Q. 2006. Architectural control syntheses of CdS and CdSe nanoflowers, branched nanowires, and nanotrees via a solvothermal approach in a mixed solution and their photocatalytic property. *J. Phys. Chem. B* 110: 11704–11710.

Yoshikawa, N. 2010. Fundamentals and applications of microwave heating of metals. *J. Microwave Power EE* 44(1): 4–11.

Yoshimatsu, K., Horiba, K., Kumigashira, H., Yoshida, T., Fujimori, A., Oshima, M. 2011. Metallic quantum well states in artificial structures of strongly correlated oxide. *Science* 333(6040): 319–322.

Yuan, Y., Xiao, Z., Yang, B., Huang, J. 2014. Arising applications of ferroelectric materials in photovoltaic devices. *J. Mater. Chem. A* 2(17): 6027–6041.

Yun, W. S., Urban, J. J., Gu, Q., Park, H. K. 2002. Ferroelectric properties of individual barium titanate nanowires investigated by scanned probe microscopy. *Nano. Lett.* 2: 447–450.

Zhang, G., Liu, G., Wang, L. Z., Irvine, J. T. S. 2016. Inorganic perovskite photocatalysts for solar energy utilization. *Chem. Soc. Rev.* 45: 5951–5984.

Zhang, X. Y., Zhao, X., Lai, C. W. et al. 2005. Synthesis and ferroelectric properties of multiferroic BiFeO$_3$ nanotube arrays. *Appl. Phys. Lett.* 87(14): 143102-1–3.

Zhong, W. L., Jiang, B., Zhang, P. L. et al. 1993. Phase transition in PbTiO$_3$ ultrafine particles of different sizes. *J. Phys. Condens. Matter.* 5: 2619–2624.

Zhong, Z. C., Zhang, Q. F., Held, K. 2013. Quantum confinement in perovskite oxide heterostructures: tight binding instead of a nearly free electron picture. *Phys. Rev. B* 88: 125401-1–8.

Zhou, T., Zha, J. W., Cui, R. Y., Fan, B. H., Yuan, J. K., Dang, Z. M. 2011. Improving dielectric properties of BaTiO$_3$/ferroelectric polymer composites by employing surface hydroxylated BaTiO$_3$ nanoparticles. *ACS Appl. Mater. Interfaces* 3: 2184–2188.

Zhu, J., Li, H., Zhong, L. et al. 2014. Perovskite oxides: preparation, characterizations, and applications in heterogeneous catalysis. *ACS Catal.* 4: 2917–2940.

Zhu, X. H. 2009. Recent patents on perovskite ferroelectric nanostructures. *Recent Patents on Nanotechnol.* 3(1): 42–52.

Zhu, X. H., Liu, Z. G., Ming, N. B. 2010a. Perovskite oxide nanotubes: synthesis, structural characterization, properties and applications. *J. Mater. Chem.* 20(20): 4015–4030.

Zhu, X. H., Liu, Z. G., Ming, N. B. 2010b. Perovskite oxide nanowires: synthesis, property and structural characterization. *J. Nanosci. Nanotechnol.* 10(7): 4109–4123.

Zhu, X. H., Xing, Z. B., Zhang, Z. H. et al. 2010c. Microwave-hydrothermal synthesis and structural characterization of PX-phase single-crystalline PbTiO$_3$ nanowires by electron microscopy. *Mater. Lett.* 64(3): 479–482.

Zhu, X. H., Liu, Z. G. 2011. Size effects in perovskite ferroelectric nanostructures: current progress and future perspectives. *J. Adv. Dielectrics* 1(3): 289–301.

Zhu, X. H., Hang, Q. M. 2013. Microscopical and physical characterization of microwave and microwave-hydrothermal synthesis products. *Micron* 44: 21–44.

Zhu, X. H., Hang, Q. M., Xing, Z. B. et al. 2011. Microwave hydrothermal synthesis, structural characterization, and visible-light photocatalytic activities of single-crystalline bismuth ferric nanocrystals. *J. Am. Ceram. Soc.* 94: 2688–2693.

Zhu, X. H., Wang, J. Y., Zhang, Z. H. et al. 2008. Perovskite nanoparticles and nanowires: microwave-hydrothermal synthesis and structural characterization by high-resolution transmission electron microscopy. *J. Am. Ceram. Soc.* 91(8): 2683–2689.

Zubko, P., Gariglio, S., Gabay, M., Ghosez, P., Triscone, J. M. 2011. Interface physics in complex oxide heterostructures. *Annu. Rev. Condens. Matter Phys.* 2(1): 141–165.

Zutic, I., Fabian, J., Sarma, S. D. 2004. Spintronics: fundamentals and applications. *Rev. Mod. Phys.* 76: 323–410.

Gram-Scale Synthesis of Graphene Quantum Dots

Zheyuan Ding and Xiluan Wang
Beijing Forestry University

7.1 Introduction

Quantum dots (QDs) are defined as nanoparticles of 1–10 nm diameter within three dimensions, presenting unique photoluminescence (PL) under UV light or electrical excitation (Abbas et al. 2018, Baker and Baker 2010, Das et al. 2018, Haque et al. 2018, Lim et al. 2015, Ponomarenko et al. 2008, Wang et al. 2016a, Wei et al. 2014, Zhang et al. 2018a). QDs were formerly extensively applied into the biofields, such as bioimaging (Cao et al. 2007, Ko et al. 2018, Medintz et al. 2005, Yang et al. 2009b, Zhang et al. 2012, Zhu et al. 2011), biosensing (Liu et al. 2012, Yang et al. 2014) and bioprobes (Suvarnaphaet et al. 2016). More interestingly, the tunable fluorescence (Abbas et al. 2018, Bacon et al. 2014, Sarkar et al. 2016, Sk et al. 2014, Zhu et al. 2017, Zhu 2015a) and up-conversion property (Bacon et al. 2014, Wang et al. 2017b) of unique physiochemical property enable its present great potential for multiple fields, such as photocatalysis (Fernando et al. 2015, Hutton et al. 2017, Yan et al. 2018), electrocatalysis (ORR, OER, HER and CO_2 reduction) (Cailotto et al. 2018, Favaro 2014, Fei et al. 2014, Li et al. 2018a, 2011b, Luo et al. 2014, Wu et al. 2016a, b, 2017, Yuan et al. 2018b), supercapacitors (Kumar et al. 2016, Li et al. 2018c, Zhao et al. 2017), battery (Chao et al. 2015, Yuan et al. 2018b), solar cells, LED (Luo et al. 2016, Sekiya et al. 2014, Wang et al. 2017c, Yuan et al. 2016), anti-counterfeit (Mahesh et al. 2016), fingerprints detection (Chen et al. 2017), antibacterial (Luo et al. 2018) and phototherapy (Ge et al. 2014). However, the toxicity of heavy metal QDs could not meet the need of biocompatibility for biofields, thus producing carbon quantum dots (CQDs) or graphene quantum dots (GQDs) with low toxicity (Abbas et al. 2018, Lim et al. 2015, Wang et al. 2016a, Zhu 2015a), good biocompatibility (Das et al. 2018, Haque et al. 2018) and chemical inertness (Lim et al. 2015, Liu et al. 2013, Sekiya et al. 2014, Wang et al. 2017a, Zhu 2015). CQDs are generally composed of carbon atom with many functional groups around or inside the carbon network, which always present amorphous without clear lattice structure. GQDs, as special CQDs (without obvious definitions between them), always with graphene lattice structure, are made up of single or few graphene lattices. CQDs and GQDs have gained tremendous attentions for their unique physicochemical properties and potential applications. They are promising candidates competing with conventional QDs owing to their excellent solubility, tunable PL, low toxicity, good biocompatibility and easy functionalization.

Since the first discovery of CQDs in 2004 by the researchers from Clemson University (Xu et al. 2004), Xu et al. occasionally isolated the fluorescent nanoparticles (CQDs) from arc-discharge carbon soot during the purifications of single-walled carbon nanotubes (SWCNTs), generating much attention for this novel fluorescent nanomaterial. Sun's group (Cao et al. 2007) later explored the laser ablation to break the carbon sources into nanoscale particles accompanying with a functionalization (surface passivation) post-treatment, opening the way for large-scale synthesis of CQDs with high PL. Especially, the fluorescence nanoparticles above first received the name "carbon dots" (CQDs) in this work. Until now, it's a wide acceptance that tuning multi-color PL and enhancing quantum yield (QY) are increasingly considered as the key topic for CQDs, thus emerging many promoting routines, such as heteroatom doping, surface passivation, controlling graphitization and regular conjugated lattice structure. In addition, various precursors and different formation mechanisms for CQDs formation may severely affect QY and PL.

Consequently, the synthesis methods may play a key role for performance enhancements, and it is urgently needed to choose suitable synthesis methods preparing desired CQDs with specific PL color and high QY, which directly decide future application. Because QDs only have several nanometers very close to exciton Bohr radius, the quantum effect, such as quantum confinement effect and quantum size effect, may dominate nanoparticles physiochemical property. The particle size and intrinsic molecular structure could severely influence CQDs and GQDs property. Noted by previous reports, lateral size and sp^2 domain in GQDs may directly decide its PL property, such as QY and luminescent color. The increased particle size and conjugated sp^2 domain could lower the energy gap, thus inducing red-shift PL behavior as a result of quantum confinement effect, which could provide strategies for the multi-color PL tuning and QY enhancement.

Notably, the gram-scale or even large-scale synthesis of CQDs could be easily achieved by various methods. According to the mechanism differences, the synthetic routes for CQDs are generally divided into two distinct methods: top-down and bottom-up. Top-down method involves the breakdown of macro-scale carbon molecule into nanoscale CQDs, and the carbon resources including carbon nanotube (CNT) (Minati et al. 2012, Shinde and Pillai 2012), SWNTs (Minati et al. 2012, Xu et al. 2004), C_{60}, (Chua et al. 2015, Lu et al. 2011), carbon black (Gao et al. 2017), graphite (Kwon et al. 2014, Sun et al. 2006), graphene oxides (GO) (Liu et al. 2013, Xue et al. 2013), coal (Ye et al. 2013) and carbon fibers (CBs) (Peng et al. 2012) were subjected to exfoliation, oxidation, pyrolysis or electrolysis to form CQDs or GQDs under an intense condition with the presence of concentrated acids or severe oxidants (Hutton et al. 2017). Although it's a facile, low-cost and simple way to operate, the convenience also may bring the uncontrollability for the CQDs quality and poor size distribution, causing the appearances of defects, which is immediately responsible for low QY. On the contrary, bottom-up strategy may utilize relatively milder conditions, concerning the conversion from small molecule to CQDs with several nanometers. In general, small organic molecules, such as glucose (Tang et al. 2012), citric acid (Dong et al. 2012a, Liu et al. 2018, Schneider et al. 2017, Song et al. 2015, Sun et al. 2016, Zhu et al. 2013), acetylacetone (Umrao et al. 2015), urine (Essner et al. 2016), phenol (Chen et al. 2016a, Wang et al. 2017c, Yuan et al. 2018a), phenylenediamine (Ding et al. 2015, Jiang et al. 2015a, Liu et al. 2017) and pyrene (Wang et al. 2014), were undergone dehydration and condensed under a thermal, microwave or radiation treatment, and integrated into larger molecules, and they finally form CQDs in a continuous process. It's well accepted that bottom-up strategy could exhibit superiority over top-down strategy because of its easy controllability. Besides, the molecular fluorophore features existing in the citric acid-based CQDs contributing to the ultrahigh QY could be relatively inspiring for further improving PL property as a guidance for high-quality and high-performance GQDs synthesis (Figure 7.1). However, there are also a lot of commonly existing problems that need to be solved, such as high expenses, weak stability, low QY and intense conditions, which severely limits its practical large-scale produces. It's suggested that the origin of

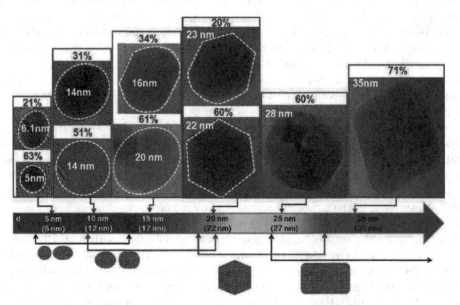

FIGURE 7.1 High-resolution transmission electron microscopy (HRTEM) images of GQDs for their major shapes and corresponding populations (p) with increasing average size of GQDs. Here, the dotted line indicates the region of a GQD, and p is defined as the ratio of number of GQDs with a major shape at each average size. Average sizes (d_a) of GQDs estimated from the HRTEM images at each d are indicated in the parentheses at the bottom of Figure 7.2. The connected arrows indicate the range of the average size in which GQDs with particular major shapes are found. (Reproduced with permission from Kim et al. 2012. Copyright 2012 American Chemical Society.)

starting materials could also severely effect the products; therefore, how to employ cheap precursors to produce high-quality and high-emission CQDs or GQDs with mild condition remains a facing challenge. In general, CQDs and GQDs always present much lower QY compared with semiconductor QDs or organic fluorescent dyes, which severely limits its potential applications. Accordingly, low-cost, simple and large-scale synthesis of CQDs and GQDs with high QY has been considered as current emphasis. As increasing researches paid into the origin of PL, diverse factors' effects on the QDs PL property could be noticed through size-dependent circular shape corresponding edge variations including zigzag and armchair types (see Figure 7.1) (Kim et al. 2012), and energy gap (see Figure 7.2) (Eda et al. 2010).

More recently, many reports have proposed a new idea: the utilization of natural biomass resources (especially those generally considered as biomass waste), to synthesize CQDs and GQDs based on two methods above, providing a new window for green, low-cost, large-scale synthesis of high-quality and high emission CQDs or GQDs. This chapter mainly focused on the synthesis of CQDs and GQDs on a large scale based on two approaches—top-down and bottom-up methods, and we also summarize various ongoing novel synthesis methods with a specific PL and high QY; most importantly, challenges are proposed for CQDs and GQDs' large-scale, green and high-quality synthesis with excellent PL from different visions. Thus, numerous natural materials, biomass resources and waste have intrigued large interests in recent days. The utilization of biomass raw materials has paved a wide way for summary of biomass-based CQDs (Table 7.1).

7.2 Synthetic Methods

7.2.1 Top-Down Method

The top-down method cut the macro-sized carbon sources with intrinsic carbon network structure into nano-sized CQDs with the help of severe chemicals (concentrated acid, alkali or strong oxidants in general). This process may utilize the oxygen functional groups as cutting sites to fragment the carbon structure, thus producing CQDs or GQDs. The oxidation-derived fragmentation may involve the intense condition and induce many severe chemicals. It is inevitable to produce many defects inside the GQDs' lattice plane, thus generally making top-down strategy-derived CQDs or GQDs low QY. Owing to its easy operation, former researches mainly focus on the utilization of top-down methods to make CQDs, and many reports have placed their recent efforts on how to improve the QY via a facile, green and low-cost strategy for a large-scale production.

Chemical Oxidation

The chemical oxidation strategy always requires severe oxidants, such as concentrated acid, strong oxidants or

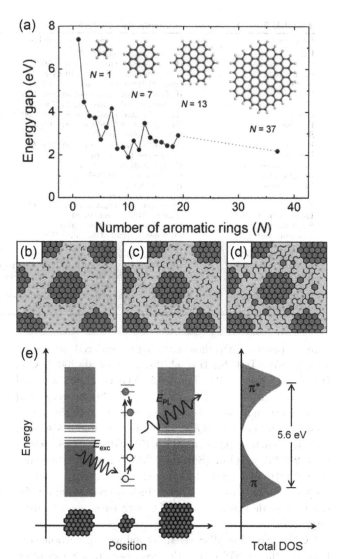

FIGURE 7.2 (a) Energy gap of π–π* transitions are calculated based on density functional theory (DFT) as a function of the number of fused aromatic rings (N). The inset shows the structures of the graphene molecules used for calculation Figure 7.4. (b–d) Structural models of GO at different stages of reduction. The larger sp² clusters of aromatic rings are not drawn to scale. The smaller sp² domains indicated by zigzag lines do not necessarily correspond to any specific structure (such as olefinic chains) but to small and localized sp² configurations that act as the luminescence centers. The PL intensity is relatively weak for (b) as-synthesized GO but increases with reduction due to (c) the formation of additional small sp² domains between the larger clusters because of the evolution of oxygen with reduction. After extensive reduction, the smaller sp² domains create (d) percolating pathways among the larger clusters. (e) Representative band structure of GO. The energy levels are quantized with large energy gap for small fragments due to confinement. A photogenerated electron–hole pair recombining radiatively is depicted. (Adapted and modified with permission from Eda et al. 2010. Copyright 2010 Wiley-VCH.)

peroxides, to oxide the carbon atom into epoxy or carbonyl group, which could act as cutting site to break the carbon resources into small pieces. From the very beginning, the

TABLE 7.1   Summary of Biomass-Based CQDs

Precursor	Synthetic Method	Quantum Yield (%)	Reference
Urine	Carbonization at 200°C	5.3	Essner et al. (2016)
Orange juice	Hydrothermal treatment at 120°C	26	Sahu et al. (2012)
Food waste	Ultrasound irradiation (40 kHz)	2.85	Park et al. (2014)
Garlic	Hydrothermal treatment at 200°C	17.5	Zhao et al. (2015)
Orange peel	Hydrothermal treatment at 180°C	36	Prasannan and Imae (2013)
Glucose	Hydrothermal treatment at 200°C	1.1–2.4	Yang et al. (2011a)
Grass	Hydrothermal treatment at 150–200°C	2.5–6.2	Liu et al. (2012)
Hemicellulose	Hydrothermal treatment at 200°C	16.18	Liang et al. (2014)
Soybean	Carbonization at 200°C	3.17	Xu et al. (2016)
Honey	Hydrothermal treatment at 180°C	6.2	Mahesh et al. (2016)
Rice husk	Ultrasonic treatment	8.1	Wang et al. (2016b)
Watermelon peel	Carbonization at 220°C	7.1	Zhou et al. (2012a)
Peanut shell	Carbonization at 250°C	9.91	Xue et al. (2016a)
Waste paper	Hydrothermal at 150–200°C	7.4–11.7	Wei et al. (2014)
Paper ash	Burning and dispersion	9.3	Wei et al. (2013)
Soy milk	Hydrothermal treatment at 180°C	2.6	Zhu et al. (2012)
Dead neem leaves	H_2SO_4, HNO_3 treatment	2	Suryawanshi et al. (2014)
Sago waste	Carbonization at 250–450°C	–	Tan et al. (2014)
Coconut water	Microwave irradiation (800 W)	2.8	Purbia and Paria (2016)
Coffee	Stirring at 90°C	5.5	Jiang et al. (2014)
Calcium alginate	Hydrothermal treatment	–	Sun (2015)
Durian	Hydrothermal treatment at 150°C with Pt	79	Wang et al. (2018)
Kraft lignin	Multi-step pyrolysis and with metal ion	12.4	Temerov et al. (2018)
Lignin	Hydrothermal treatment at 180°C	–	Chen et al. (2016b)
Cellulolytic enzyme lignin	Molecular aggregation, stirring at room temperature	2.47	Niu et al. (2017)
Alkali lignin	Two-step hydrothermal at 180°C	22	Ding et al. (2018) Our work

more expensive materials with highly ordered structure, such as SWNTs (Minati et al. 2012, Xu et al. 2004), C_{60}, (Lu et al. 2011), GO (Liu 2013, Xue et al. 2013), coal (Ye et al. 2013) and CBs (Peng et al. 2012), were tend to be used; however, the high cost may limit the large-scale production for GQDs. Low-cost raw materials, such as active carbon (Qiao et al. 2010), carbon black (Dong et al. 2012b), graphite (Sun et al. 2006) or even carbon soot (Ray 2009), were gradually utilized as starting materials to prepare highly emissive GQDs, which is greatly helpful for the large-scale synthesis of GQDs with excellent PL property.

CQDs were first discovered by subjecting carbon soot from arc-charge SWNTs as the precursors to the oxidation by nitric acid to break the carbon into fluorescent nanoparticles (see Figure 7.3), thus enabling oxidation cut or other strong condition possible to synthesize CQDs with a large diameter of 18.0 ± 0.4 nm (Xu et al. 2004). Sun et al. (2006) used laser ablation followed by a nitric acid reflux with surface passivation to produce highly luminescent CQDs from the mixture of graphite powder and cement. More interestingly, they found that the as-synthesized nanoparticles may be less luminescent without passivation. However, strong PL could be seen with organic molecules attached and present a perfect PL color control for the emission wavelength (see Figure 7.4). Sun and colleagues also give

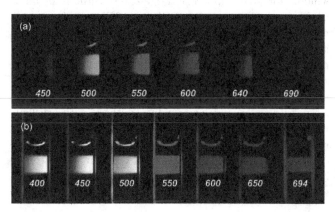

FIGURE 7.4 Aqueous solution of the PEG_{1500N}-attached carbon dots (a) excited at 400 nm and photographed through band-pass filters of different wavelengths as indicated, and (b) excited at the indicated wavelengths and photographed directly. (Reproduced with permission from Sun et al. 2006. Copyright 2006 American Chemical Society.)

a brief explanation for the PL's strong enhancement via surface passivation, first pointing out the quantum confinement effect of emissive energy traps that the passivation may contribute a lot to PL enhancement due to a large surface-to-volume ratio. Liu et al. (2007) exerted candle soot as precursor with a reflux of HNO_3 to synthesize small size CQDs within 2 nm, and they also acquire colorful PL (see Figure 7.5) after gel electrophoresis purification for nine CQDs separation; they make nine purified CQDs with emission wavelength from 415 to 615 nm under a single excitation wavelength UV light (315 nm). However, this oxidative acid treatment could only give a very low QY less than 2%. Ray and colleagues (Ray 2009) followed this cutting method in an intense condition, namely, concentrated acid oxidation, and the carbon soot was also refluxed for 12 h in the presence of nitric acid. It's noted that this work presents a simpler, easy-controlling way to produce CQDs with a diameter of

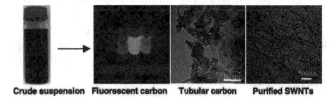

Crude suspension   **Fluorescent carbon**   **Tubular carbon**   **Purified SWNTs**

FIGURE 7.3 Fluorescent nanoparticles from arc-discharge carbon soot. (Reproduced with permission from Xu et al. 2004. Copyright 2004 American Chemical Society.)

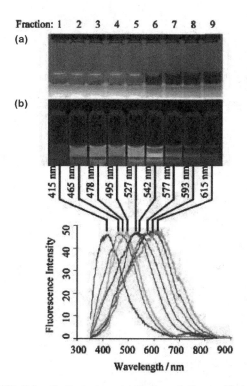

FIGURE 7.5 Optical characterization of the purified carbon nanoparticles (CNPs). Optical images illuminated under white (a) and UV light (312 nm; center). (b) Fluorescence emission spectra (excitation at 315 nm) of the corresponding CNP solutions. The maximum emission wavelengths are indicated above the spectra. (Reproduced with permission from Liu et al. 2007. Copyright 2007 Wiley-VCH.)

2–6 nm from cheaper materials with a yield of 20% owing great potential to scale up, but low QY (about 3%) still remains a big limit.

The pitch-based CBs were also utilized (Peng et al. 2012) trying to synthesize highly crystal GQDs, initiating a mixed acid, H_2SO_4 and HNO_3 treatment to etch and cut the CBs into massive GQDs with the size range of 1–4 nm (see Figure 7.6). Notably, edge structure, typically including zigzag and armchair edges in graphene, could be precisely controlled in this method, and graphene with zigzag edges structure could endow GQDs with specific electronic or magnetic property. Gao et al. (2017) prepared yellow GQDs with 2.37 ± 0.10 nm via a reflux of nitric acid from VCX-72 carbon black for 24 h at low temperature. Citing the edge state effects for PL property, they also developed a multi-color PL tuning strategy by simply coating polyethyleneimine (PEI) with different molecular weight. However, the QY for original GQDs, PEI_{1800} and PEI_{600} GQDs of 3.5%, 2.8% and 1.7% (see Figure 7.7), respectively, is still a very low value for further application. Acid oxidation always accompanied with a surface passivation, which could increase QY to some extent (Sun et al. 2006); however, the high cost for some passivation agents and difficulties of controlling passivation degree still limit the large-scale synthesis. Acid oxidation always accompanied with a surface passivation, which could increase QY

to some extent according to previous reports (from 4% to 10%) (Sun et al. 2006). The graphite was experienced tattering cutting with concentrated nitric acid following by amidative hydrothermal cutting to obtain range of the GQDs with certain size distributions from 2 to 10 nm through tuning the oleylamine (OAm) concentration (see Figure 7.8a and b). The highly crystalline structure could be clearly noticed (see Figure 7.8c) with specific edge structure including zigzag (blue line) and armchair (red line) (see Figure 7.8d), which specific electronic property, reflecting into the PL property with various energy gaps (see Figure 7.8e) in such GQDs, is narrowed down as size enlarged with a relatively low QY of ~10% defect sites. However, the high cost for some passivation agents and difficulties of controlling passivation degree still limit the large-scale synthesis, also contributing slightly to the QY increase (generally lower than 15%).

A simple approach towards GO implementing hydroxyl radicals cutting results in the production of small species GO and GQDs via an environmentally friendly method, which could self-assemble into GQDs/GO form through strong π–π^* interactions (see Figure 7.9). The small-sized GO resulted in the formation of GQDs/GO with red fluorescence by thiourea-mediated, blue-, green- and yellow GQDs with 40, 120 and 270 min under NH_4OH, with QY from 1% to 10% through hydroxyl radical-induced decomposition of GO. This approach makes effective strategy controllable for the preparation of multiple colored GQDs, also incorporated with GQ sheet.

The Fenton agent has been extensively applied into the aromatic organic pollutants degradation especially effective under the UV irradiation, exhibiting a great performance within short period. Actually, the GO could be generally considered as superb aromatic molecules, which should be actively reacted with the Fenton agent (noted as photo-Fenton reaction). Therefore, from the view of graphene chemistry property, the Fenton agent could pose as one type of cutting agent to break GO into smaller pieces, GQDs (see Figure 7.10). With this thought, photo-Fenton reaction developed in a mass production was conducted; micrometer-sized GO sheets could react with Fenton reagent ($Fe^{2+}/Fe^{3+}/H_2O_2$) efficiently under a UV irradiation, and as a result, the GQDs with periphery carboxylic groups could be generated with mass-scale production (Zhou et al. 2012b).

Hydrothermal and Solvothermal

Hydrothermal and solvothermal methods play an important role in making nanomaterials as the most universal approach, involving the decomposition and carbonization under a relatively high pressure and temperature with the presence of water or organic solvent. More specifically, this strategy may inevitably utilize concentrated acid or powerful oxidants as cutting agent. The utilization of hydrothermal route for the produce of GQDs was first achieved by using GO as the precursor (Pan et al. 2010). The GO sheets were pre-oxidized into graphene ribbons

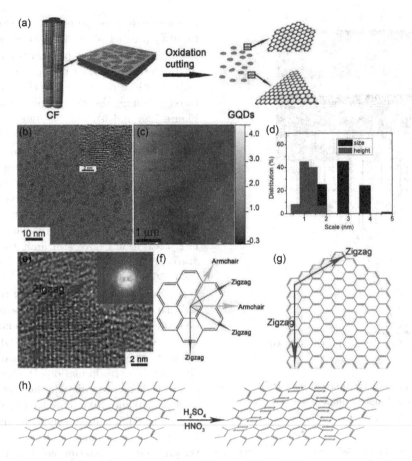

FIGURE 7.6 (a) Representation scheme of oxidation cutting of CB into GQDs. (b) Transmission electron microscopy (TEM) images of GQDs (synthesized reaction temperature at 120°C), inset of (b) is the HRTEM of GQDs. (c) Atomic force microscope (AFM) image of GQDs. (d) Size and height distribution of GQDs. (e) HRTEM image of the edge of GQD, inset is the 2D fast fourier transformation (FFT) of the edge in (e). (f) Schematic illustration showing the orientation of the hexagonal graphene network and the relative zigzag and armchair directions. (g) Schematic representation of the edge termination of the HRTEM image in (e). (h) Proposed mechanism for the chemical oxidation of CF into GQDs (Reproduced with permission from Peng et al. 2012. Copyright 2012 American Chemical Society.)

(GRs), and micrometer GRs were then cut into nano-sized GQDs (see Figure 7.11). In the whole process, the epoxy and carbonyl groups existed in GO are converted into many cutting sites under the hydrothermal treatment (unzipping mechanism), serving as a basis for the later researches, which broad the synthetic strategies for CQDs and GQDs. Specifically, in order to improve the QY, recommended by prior report about passivation could enhancing the QY to some extent, similar hydrothermal route by cutting hydrazine hydrate-reduced GO with polyethylene glycol (PEG) surface passivation was conducted (Shen et al. 2011), only achieving QY of 7.4%, which could be attributed to both existing many defects inside GQDs and lack of control for passivation degree. This route can be promoted by precise control of passivation, conducting pre-oxidized GO and PEG passivation one-step hydrothermal treatment at 200°C for 24 h without further reduction, and GQDs-PEG with diameter of 5–25 nm could reach a high QY of 28%. (Shen et al. 2012) PEG polymer chains were actually introduced to both the edge and basal plane of GQDs, which may be attributed to more emissive energy trapping to the GQDs surface due to

the quantum confinement effect, thus producing higher QY (Sun et al. 2006).

The GQDs easily attached on the CNT surface because of π–π stacking; when yielding acid cutting CNT into GQDs, π–π stacking may thus quenching the fluorescence emission to some extents. A novel procedure was advocated by PEI-assisted hydrothermal treatment based on previous works, finally achieving a QY of 21% (Xue et al. 2013). Considering the π–π stacking force, they introduce PEI as chemical both oxidizing agent to unzip the oxidized graphene sheet into one to three layers GQDs and stabilizing agent to prevent GQDs from layer stacking and lateral aggregation. It can be found that introducing PEI could efficiently prohibit the stacking of GQDs layers which was further identified by the statistical data of thickness. This observation reveals that the layer–layer coupling in multi-layer GQDs seems not optical features but quench PL to a certain extent.

There is a tendency that most publications may focus on the large-scale synthesis while neglecting the precise controlling of graphene-conjugated domains, thus causing many damages for the perfect lattice. Therefore, the

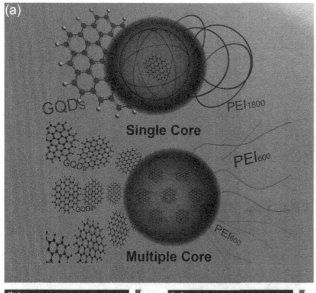

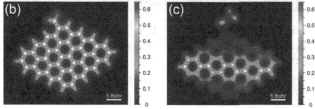

FIGURE 7.7 (a) G2 (the only carboxyl group linked with the nanographene core) and (b) G1-N (the nanographene core linked with the PEI branch chain). (c) Scheme of the formation mechanism of PEI$_{1800}$ GQDs (left) and red emitting PEI$_{600}$ GQDs (right). (Adapted and modified with permission from Gao et al. 2017. Copyright 2017 American Chemical Society.)

aromatic framework damages could introduce a low QY even with very high yield production of GQDs as high as over 40%, such as CX-72 carbon black hydrothermal prepared GQDs, solvothermal acid exfoliated graphite powder and H$_2$O$_2$ oxidized black carbon. It could be concluded that simple passivation strategy is much restricted to the increase of QY and intrinsic QDs structure may become the key point for the performance instead of massive surface modification strategies. Accordingly, most researchers have placed their effort on how to keep GQDs lattice structure complete while acquiring GQDs with passivation controlling. Zhu and co-workers implementing a one-step solvothermal strategy by GO tried to reduce defects inside GQDs as much as possible, but weak reduction agent could slightly reduce the formation of defects with a small increase achieving 11.4% (Zhu et al. 2011).

Therefore, novel strategies are urgently needed to reduce those defects aiming at improving QY. As depicted by Tetsuka et al. (2012), they used amino-hydrothermal to oxidize GO within ammonia solution with the highest QY of 29%, generally following Pan et al.'s unzipping mechanism and even achieve ~46% after PEG passivation. The edge-terminated amine was attached in GQDs' edge, and the amine-passivation around ordered graphene core would not disrupt sp^2 hybridization domain with less defects as much as possible (see Figure 7.12). This precise structural design dramatically lessens carboxylic and epoxide groups containing in as-prepared GQDs, acting as non-radiative electron–hole recombination centers, thus leading to higher QY. Ding's group (Zhu 2015) tried to use hydrogen peroxide (H$_2$O$_2$) as oxidant with the aid of tungsten oxide (W$_{18}$O$_{49}$) nanowires catalysis to oxidation-cut the GO sheet into GQDs with a high QY of 69%. The GO was undergone hydrothermal in the presence of H$_2$O$_2$ to produce blue GQDs ranged from 4 to 21 nm as the H$_2$O$_2$ dosages vary (see Figure 7.13). The ultra-long hydrothermal time within H$_2$O$_2$ solution could sufficiently break the carbon–carbon bond and thus in situ produce a periphery carboxylic groups around the GQDs, which could contribute to a high QY of 77.3%, giving a possibility for the large-scale preparation of GQDs with excellent PL property. This method is a mild, clean and efficient approach with ultra-high QY emission in the absence of by-products. However, the utilization of strong oxidants, specific catalysis and prolonged time with a predictable high cost may severely restrict its large-scale synthesis in a practical scope. In the same year, Ding's group (Sun 2015a) reported a nitrogen-doping strategy to efficiently improve GQDs performance to the maximum extent achieving 74% which is almost the highest value until now to the best of knowledge (see Figure 7.14). It systematically prepared the high-quality GQDs with ultrahigh QY and gave a reasonable mechanism for the origin of high QY, also achieving a yield of 25%, which have a great potential to be further applied into large scope. However, the GO origin could be responsible for the GQDs production because of the GO production with large scale remaining a big challenge, and thus may relatively both result in many derived troubles and increasing the expense.

The graphite nanoparticles were used to the synthetic preparation for GQDs using the chemical exfoliation of solvent and Hummers' method, respectively (Liu et al. 2013). Single-layered GQDs with 55% yield within 4 nm diameter GQDs were synthesized into numerously pure sp^2 carbon crystalline aromatic domains without exogenous defects inside the conjugated carbon network. Besides, it highlighted the higher energy mainly come from the intrinsic state of GQDs, while the lower energy PL emission originated from the defect states (see Figure 7.15).

The graphite as starting materials, yielding Hummers' method in making GO, typically by using a mixture of concentrated H$_2$SO$_4$ and HNO$_3$ with a ratio of 3:1, was oxidized at 120°C for 24 h to directly obtain GQDs, thus barely giving a with smooth sp^2 surface for the typical water-soluble GQD as expected (see Figure 7.16a). White PL light with edge functionalization of GQDs led to white emission was further prepared with GQD through the reaction of edge-oxidized with 4-propynyloxybenzylamine (bulky Frechet's dendritic wedges at the GQD-1 periphery). It's an uncommon feature that specific edge functionalization of GQDs could produce the white-light emission, which could hold great promises to engineer unique optoelectronic

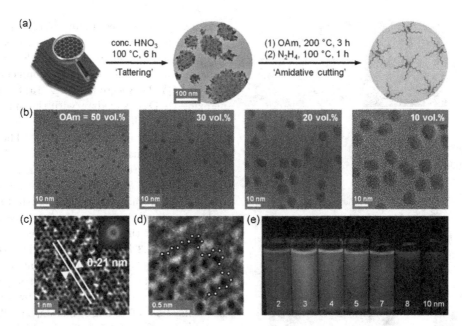

FIGURE 7.8 (a) Schematic of tattering graphite and amidative cutting of tattered graphite. (b) TEM images of the GQDs with varied sizes of 2, 4, 7 and 10 nm (left to right). The text insets represent the concentration of OAm. High-resolution TEM images showing (c) lattice spacing and (d) edge structure. (e) Photo of a series of the GQDs with various sizes under a 365 nm UV lamp. The captions represent the size of the GQDs. (Reproduced with permission from Kwon et al. 2014. Copyright 2014 American Chemical Society.)

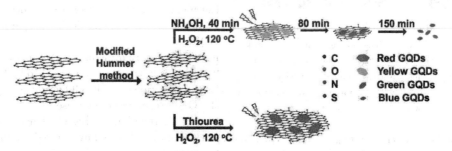

FIGURE 7.9 Preparation of fluorescent R-GQDs/GO, Y-GQDs/GO, G-GQDs/GO and B-GQDs through the hydroxyl radical-induced decomposition of GO. (Reproduced with permission from Ke et al. 2016. Copyright 2016 Wiley-VCH.)

devices in a large scale when QY could be further improved (Sekiya et al. 2014). It is worth highlighting that the conservation of graphene lattice structure and heteroatom doping could significantly improve QY, which is a key point in hydrothermal or solvothermal synthesis of GQDs with high QY. Besides, the large-scale synthesis of high emissive GQDs may require much cheaper resources, reliable origins and effective strategy to verify the whole produces.

Microwave Irradiation

Microwave irradiation could be regarded as a fast, green and ecological method compensating the prolonged period of hydrothermal method. This strategy exerts highly condensed energy to generate high temperature in short period, and with the presence of concentrated acid and other oxidants, different origins of precursors, such as GO, rGO and graphite, could be quickly cut into GQDs. Li et al. (2012a) synthesized GQDs with 2–7 nm diameter with negligible surface defects, achieving as high as 22.9% (see Figure 7.17a), which could serve as a foundation for

efficient enhancement for the QY. Mass production of GQDs was with a size distribution of 2–8 nm achieved by one-pot microwave irradiation from graphite in 1 hour with a high yield of ∼70% (Shin et al. 2014). Strong oxidant was used to exfoliate the graphite flakes into GO and simultaneously oxide the GO sheet into GQDs under a high-powered 600 W microwave irradiation (see Figure 7.17b). Actually, this technique requires massive KMnO$_4$, almost fivefold of graphite, inevitably resulting in high cost and great environmental burdens. More details, the as-prepared GQDs with many defects are rarely complete lattice structure, which may result in low PL emission.

In most case, directly cleaving GO into GQDs under microwave irradiation especially within severe oxidants could inevitably bring the edge of GQDs intrinsic defects, such as epoxy groups and carbonyl groups. An acid-free strategy was developed with GO as carbon resource and borax as boron resource via microwave treatment at 230°C in 30 min, presenting a facile, time-economical and environmentally friendly way (see Figure 7.17c). The

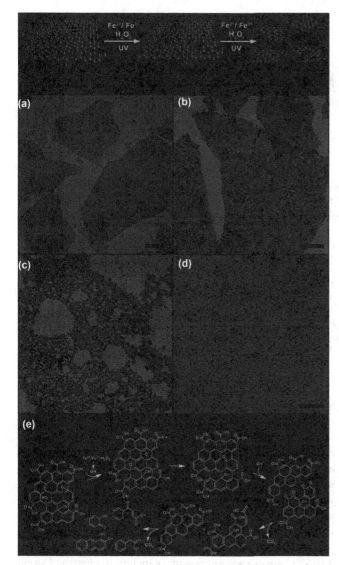

FIGURE 7.10 AFM images of GO sheets before (a) and after being reacted with the Fenton reagent under a UV irradiation (365 nm, 1,000 W) for 5 (b), 10 (c) and 15 (d) min, respectively. All images were acquired under the tapping mode. All scale bars equal 500 nm. (e) Schematic representation of a proposed mechanism for the photo-Fenton reaction of the GO sheets. (Reproduced with permission from Zhou et al. 2012b. Copyright 2012 American Chemical Society.)

obtained GQDs presented a complete graphene lattice with a relatively high QY of 21.1%, which could be attributed to the mild condition without oxidizing acid and suitable boron doping. More importantly, avoiding the defects' effects on the PL emission, they incorporated the boron atom doping for the restoration of defects with energy regulation, thus improving the QY (Hai et al. 2015). Wang's group also exert some efforts in the microwave synthesis of GQDs and its application (Luo et al. 2016, 2014, 2018). They proposed first work on the microwave-assisted solvothermal synthesis of S and N co-doping GQDs by heating mixture of GO and reduced glutathione. Actually, this work is mainly focused on the synthesis of N, S-doping GQDs

by microwave synthesis and the heteroatom-doping GQDs' electrochemical activity regardless of its massive procedures and prolonged period, illustrating GQDs good utilization in the electrochemical field (see Figure 7.18a). The white fluorescent GQDs microwave synthesis method (Luo et al. 2016) was extended with two-step strategy:: first, yellow-green fluorescent GQDs was prepared by microwave-assisted hydrothermal with mixed acid (H_2SO_4 and HNO_3 with a ratio of 3:1), then accompanying with microwave treatment for 12 h at 175°C under alkali condition (pH = 13) to generate white GQDs with 2–5 nm (see Figure 7.18b). The post-microwave treatment could facilitate fewer defects, which may give rise to GQDs white-light emission used as enhanced component for white LED devices. Owing to the need for the specific application, the as-prepared GQDs were further modified with adenine (A-GQDs) by two-step strategy with pretreatment at 80°C for 4 h following a microwave heating at 175°C for 12 h, which could serve as antibacterial agent (see Figure 7.18 c) (Luo et al. 2016, 2014, 2018).

Electrochemical Oxidation

Electrochemical oxidation is generally considered as a facile, green and ecological strategy. Typically, the carbon resources, such as graphite, MWNTs and graphene, were initiated with anodic potential to unzip the carbon network into GQDs. This method always generates free radicals, such as ·OH from water and other electrolyte-derived radicals to induce the oxidation of carbon resources into GQDs. However, due to the difference of starting materials and conducting conditions, the electrochemical oxidation is hard to control precisely compared with above strategies.

This method was first proposed using MWNTs covered in carbon paper as starting material, within an electrolyte of degassed acetonitrile solution, and they received a QY with the highest value of 6.7% (see Figure 7.19). Actually, the complexity and high cost for separation and purification arose massive modifications in order to optimize the severe condition. The occasional synthesis of GQDs was prepared from GNRs using a solid electrolyte, propylene carbonate and $LiClO_4$, which could provide a simpler way from MWCNTs to GQDs. It could be noticed that many defects were caused by uncontrollable oxidation, thus exhibiting low QY of 6.3%. The as-developed electrochemical methods to prepare the N-doped GQDs became another interest route for various electrocatalysis and photo-optical devices (see Figure 7.20a and b) (Li et al. 2011a, b). Ming et al. simultaneously used graphite rod as cathode and anode, and ultrapure water as the electrolyte to prepare CQDs via one-step electrochemical exfoliation, and this strategy has no need for specific electrolyte, which dramatically reduces the cost and complexity with high yield (Ming et al. 2012). In this process, similarly, Deng et al. (2015) reported a large production by one-step electrochemical tailoring with a yield of 65.5%, using similar mechanism to prepare multi-color GQDs with QY of ~7.8% (see

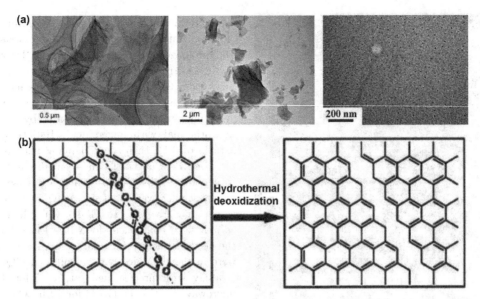

FIGURE 7.11 (a) TEM images of the pristine (left), oxidized graphene sheets (middle) and GQDs. (b) Mechanism for the hydrothermal cutting of oxidized graphene sheets into GQDs: a mixed epoxy chain composed of epoxy and carbonyl pair groups (left) is converted into a complete cut (right) under the hydrothermal treatment. (Reproduced with permission from Pan et al. 2010. Copyright 2010 Wiley-VCH.)

Figure 7.21). Yao et al. designed a novel electrolytic device combining ultrasonic and electrochemical method using via synergy effect of high-intensity ultrasonic field and electric field, which could prevent CQDs absorption in the graphite anode to improve electrolytic yield, preparing CQDs of 8.9% (Yao et al. 2014).

As concluded in hydrothermal and solvothermal strategy, keeping graphene lattice complete may dominate the QY enhancement. However, this strategy seems not suitable for the electrochemical method. The "molecular" sp^2 domains with a diameter about 3 nm small-sized red GQDs were observed with perfect graphene lattice, using $K_2S_2O_8$ instead of pure water as electrolyte to generate more free radicals for easily oxidizing and exfoliating commercial graphite rod into GQDs. (Tan et al. 2015) Highly active SO_4^- radicals from $S_2O_8^-$ may serve as electrochemical "scissors" to sharply cut the graphene sheet into small intact sp^2 structures. Unfortunately, its QY of 1.8% is much lower than previous reports, indicating the QY enhancement is still unclear in this method. It has been suggested that electrochemical strategy could provide a very facile, green and low-cost method for the preparation of GQDs with a large scale, but the PY generally presents a low value with unclear PL enhancing principle, which is urgently needed to be resolved.

Others Methods

In spite of conventional strategy, emerging novel technology has been proposed in recent years, such as plasma treatment (Gokus et al. 2009, Lee et al. 2012, Moon et al. 2014, Muramatsu et al. 2018, Yu et al. 2010, Zhi et al. 2002), nanolithography techniques (Wang et al. 2010), electron-beam lithography (Hu et al. 2011) and cage opening of fullerenes (Lu et al. 2011).

The plasma could be understood as etching with carbon atoms to generate oxide removing from surface, such as oxygen plasma etching yielding CO and CO_2 to successively remove the carbon atoms and functionalized carbon atoms at the same time. The plasma treatment is generally considered as an effective way to activate the carbon materials, endowing them with specific activity, which was used in SWCNT activation with good electrocatalytic activity and long-term stability (Yu et al. 2010), DWCNT's selective activation for outer CNT layer to increase wettability and ORR properties (Muramatsu et al. 2018). There are several resources including nitrogen (Moon et al. 2014, Yu et al. 2010), hydrogen (Zhi et al. 2002) and oxygen plasma treatment (Gokus et al. 2009) used as the activated source. As a result, counting on former reports of CNTs with plasma treatment (see Figure 7.22e, f and g) (Zhi et al. 2002), the possibility of applying into the GQDs was discussed to settle down conventional strategy weakness, such as disruptive defects, long term and uncontrollable functionalization. The oxygen plasma technique was formerly applied to prepare functionalized graphene as facile ways tuning the intrinsic properties of graphene (see Figure 7.22a and b) (Gokus et al. 2009). Uniform PL could be obtained by selective plasma oxidation especially in single-layer graphene while no PL in bi-layer graphene and multi-layer graphene. Inspired by such a phenomenon, N-doped GQDs preparation in a large-scale from as-grown monolayer graphene was achieved through a nitrogen plasma treatment (see Figure 7.23a) (Moon et al. 2014). The CVD-derived single-layer graphene as the precursors was irradiated under nitrogen plasma, and high-quality N-GQDs could be directly acquired with perfect lattice structure (see Figure 7.23b and c). Moreover, it could be dispersed into organic solvent

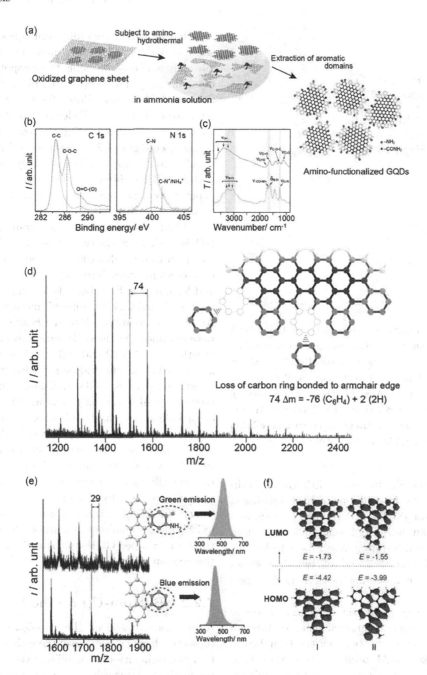

FIGURE 7.12 (a) Schematic illustration of the preparative strategy for af-GQDs. (b) C1s and N1s X-ray photoelectron spectra for af-GQDs prepared at 90°C and oxidized graphene sheet (OGS). (c) Matrix-assisted laser desorption ionization time-of-flight mass spectrometry (MALDI-TOFMS) mass spectrum of af-GQDs in aqueous solution prepared at 120°C. Note that peaks are detected as Na ion adducts ($[M + Na]^+$), the Na ion of which might have originated from residue in the reagents used for graphite oxidation to prepare the graphene oxide (GO). Inset, schematic representation of the edge structure of the af-GQDs. (d) Comparison of MALDI-TOFMS mass spectra for af-GQDs prepared at 120°C (bottom) and 90°C (upper). The inset shows the possible edge structures and their PL spectra excited at 350 nm. (e) Schematic illustrations of structures used for theoretical calculations with different functional groups. (f) The isosurface represents the highest occupied molecular orbital (HOMO) and the lowest unoccupied molecular orbital (LUMO). I: hydrogen-terminated edge structure, II: -NH₂/-CH₃ pair bonded to edge. The unit of energy is electron volt (eV). (Reproduced with permission from Tetsuka et al. 2012. Copyright 2012 Wiley-VCH.)

and then transferred onto arbitrarily shaped substrates, which facilitate the photocathode with excellent photoelectrochemical catalyst for hydrogen evolution reaction (HER). There is no doubt that this process could be rather complex and much more expensive, which could

not be possible to applying into synthesis of GQDs in practical large scale; however, this technique owns great potential for the modification, functionalization and doping with an easily controllable degree in an extremely short period, which is a great potential to be of assistance

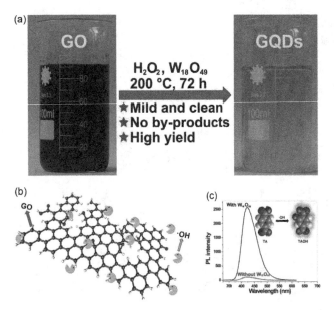

FIGURE 7.13 (a) Digital photos of GO aqueous solution (left) and GQD aqueous solution (right). This reaction is catalyzed by $W_{18}O_{49}$ under 200°C. (b) Schematic diagram of oxidize and cut progress. (c) PL spectra of the mixed solution of TA, $W_{18}O_{49}$ and H_2O_2 aqueous solution and only TA, and H_2O_2 after 12 h reaction. The inset is the reaction between OH and TA. The concentration of TA in all experiments is 50 mM. (Reproduced with permission from Zhu 2015. Copyright 2015 Royal Society of Chemistry.)

for conventional strategy. Electron-beam lithography technique has been utilized for the making of QD-based single-electron transistor (SET). The QD-based integrated charge sensor, with an integrated charge sensor via a standard electron-beam lithography and lift-off technique, acquired the isolated central island of twin-QD structure with diameter 90 nm, serving as SET devices (see Figure 7.22h and i).

The laser ablation was generally considered as an effective physical strategy preparing nanomaterial synthesis through laser irradiation. The interaction between the laser beams and the graphite flakes could produce an instant local high temperature and pressure, thus forming laser-induced bubble located at laser focus vapor and plasma pulse at the interface of the graphite flake and the surrounding liquid medium (see Figure 7.24a). One-step synthesis of CQDs was performed by laser irradiation of graphite flakes in polymer solution with increasing average sizes of ~3, 8 and 13 nm with fluorescent QY gradual variations of 12.2%, 6.2% and 1.2% under 0.3, 0.9 and 1.5 ms irradiative period, respectively (see Figure 7.24 b, c and d) (Hu et al. 2011).

The nanolithography techniques always related to extensively applications into semiconductor industry used to fabricate fine patterns. Recently, the nano-electronic fields drove great interests towards GQDs with precisely controllable size. However, conventional method is very difficult to present uniform size distribution, and based on that, the extremely uniform GQDs preparation was

reported by nanolithography techniques through templated silica nanodots. The silica nanodots based on the self-assembly of silica-containing block copolymers, polystyrene-*b*-polydimethylsiloxane (PS-PDMS) technique as previously developed (Jung and Ross 2009), were served as etching mask to single-layer CVD-derived graphene into GQDs with 10 and 20 nm corresponding to mono- and bi-layers via O_2 plasma (see Figure 7.25). It's noteworthy that this technique could provide an ultrafine strategy preparing extremely uniform GQDs; however, the weak fluorescence and high expense seem not suitable to the large-scale synthesis (Lee et al. 2012).

A highly promising route for gram-scale preparation of GQDs with a production yield of 26% was implemented via a bottom-up deflagration ignited reaction, and active carbon atoms grow into GQDs by breaking the C–F bond and forming the Si–F bond before ignition energetic materials combined with a Hummer's method exfoliation process. An alternative approach to synthesize active carbon atoms were obtained via the deflagration reaction of polytetrafluoroethylene (PTFE) and 100 nm Si as precursors to supply active carbon atoms and deflagration flame for the carbon growth process into GQDs. Considering the conventional strategy weakness easily leading to formation of small carbon clusters with variable size, cage opening of C_{60} with the assistance of Ru catalysis was implemented via GQDs with well-defined geometrical shapes could be produced from the C_{60}-derived carbon clusters, surface collisions or thermal effects is a complex process that typically mechanistic approach to the synthesis of a series of atomically defined GQDs. Series of scanning tunneling microscope (STM) images monitored the whole anamorphosis process of trapezium-shaped GQDs to triangular-shaped GQDs during the cage opening.

7.2.2 Bottom-Up Method

As mentioned above, the top-down strategy generally presents a low QY because of the utilization of severe condition and difficulty of controlling oxidation degree. Bottom-up method is regarded as most competitive strategy to prepare highly emissive CQDs and GQDs with a relatively milder condition for a large-scale production. Bottom-up, as the name implies, may involving the smaller molecules condensing into larger CQDs or GQDs molecules via dehydration and carbonization, which is easier to control the formation progress and molecular structure. Therefore, high QY is easier to be achieved by a bottom-up method, which may use various small organic molecules as precursors, such as pyrene, glucose, citric acid and phenolic molecules, facilitating an easier control for the GQDs structure and PL property. However, diversities of precursors derived various GQDs, following different mechanism, which could make the bottom-up strategy much more complicated for the whole reacting process involving different intermediates and uncertain enhancing PL mechanism.

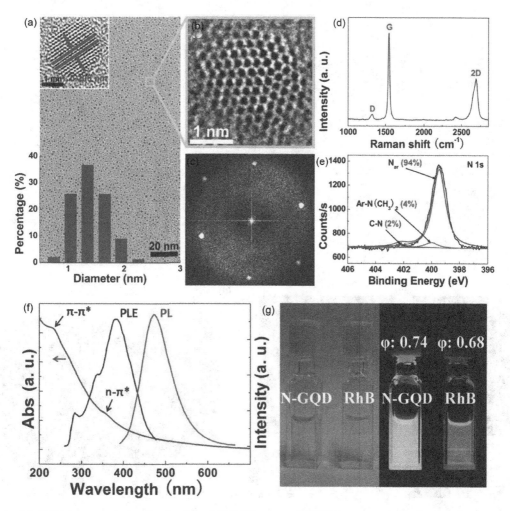

FIGURE 7.14 (a) TEM image and size distribution of N-GQDs, inset: HRTEM image with the lattice of 0.206 nm, (b) spherical aberration-corrected TEM image of a single dot, (c) FFT of a single dot in (a), (d) Raman spectra, (e) high-resolution N 1s spectrum of the N-GQDs, (f) UV-vis absorption, PL (λex = 400 nm) and PL excitation (PLE) (λex = 500 nm) spectra of N-GQDs aqueous solution, and (g) photographs of N-GQDs aqueous solution and RhB ethanol solution with the same mass concentration (10 μg m/L) under visible light (left) and 365 nm UV light (right). (Readapted with permission from Sun 2015. Copyright 2015 Wiley-VCH.)

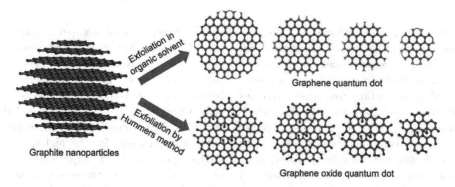

FIGURE 7.15 Synthesis and characterization of GQDs and GOQDs. Synthetic scheme for GQDs and GOQDs using chemical exfoliation of GNPs (oxygenous sites are shown as fringe dots). (Reproduced with permission from Liu et al. 2013. Copyright 2013 Wiley-VCH.)

Hydrothermal and Solvothermal

According to water solubility or polarity of starting materials, in particular, organic molecules as precursors always have poor water solubility. Hydrothermal and solvothermal methods being alternative for the bottom-up strategy were generally considered as the most universal strategy in the bottom-up methods for the synthesis of CQDs or

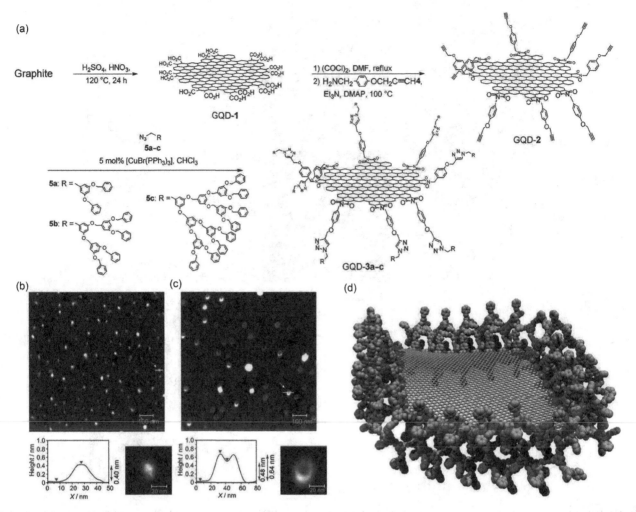

FIGURE 7.16   (a) Schematic preparation of GQD-2 and the introduction of dendritic wedges by copper-catalyzed Huisgen cycloaddition. DMF=N,N-dimethylformamide, DMAP=4-dimethylaminopyridine. AFM images (topography) of (b) GQD-2 (0.9 × 0.9 μm) and (c) GQD-3b (0.9 × 0.9 μm) on mica disks. Height profiles of the cross sections of the white lines in (b) and (c), and magnified images of GQD-2 and GQD-3b. (d) Calculated structure of GDQ-3c. (Reproduced with permission from Sekiya et al. 2014. Copyright 2014 Wiley-VCH.)

GQDs typically under Teflon-lined stainless-steel autoclave. Unlike top-down mechanism involving decomposition and carbonization from larger bulk molecules to nano-sized CQDs and GQDs, bottom-up hydrothermal always relates to the dehydration and aromatization from small molecules to CQDs and GQDs with continuing condensation, which allows for easy controlling of size and PL property under a prolonged period with a continuing heating and pressure. Many organic molecules were usually employed, such as many phenyl molecules, citric acid, carbohydrate and amino acids, which were generally considered as the most universal precursors. Because different molecules may holdi multiple reacting process, the PL emission enhancing mechanism could be significantly various. On the other hand, different resources-based quantum dots may inevitably present weakness due to the intrinsic character of precursor molecular structure and special reacting mechanism.

There's no doubt that graphene could be regarded as multi-phenyl aromatic macromolecules, and we could simply consider it as numerous graphene primitive unit, benzyl ring molecules oriented linking together. Therefore, emerging researchers are increasingly highlighting the possibility of phenyl molecules condensation into highly oriented GQDs, multi-phenyl carbon-rich materials. Considering the intrinsic phenyl structure's priority, the phenyl molecules could be regarded as the most potential precursors for the high-quality single-crystalline GQDs synthesis in a large scale. In a typical CQDs or GQDs synthesis procedure, heating the precursors via hydrothermal, solvothermal or pyrolysis could effectively facilitate the small aromatic molecules regenerating into larger carbon nanoparticles. However, the aromatic molecules could easily bring heavy toxicity needed to be solved, which could bring a great environmental burden and severely limit its bio-application. It also remains a big problem on how to balance the high quality and low costs in a large scale especially towards phenyl molecules (Figures 7.26 and 7.27).

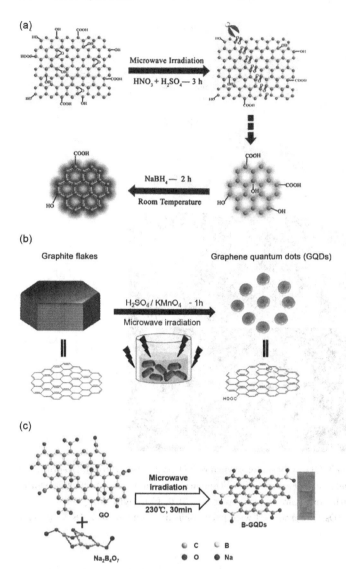

FIGURE 7.17 (a) Schematic representation of the preparation route for gGQDs and bGQDs (Li et al. 2012a). Copyright 2012 Wiley-VCH. (b) Schematic representation of the fabrication of GQDs from graphite powder by one-pot microwave irradiation under acidic conditions. (Reproduced with permission from Shin et al. 2014 Copyright 2014 Wiley-VCH.) (c) Preparation scheme for the B-GQDs by using GO and borax as carbon and boron sources. (Reproduced with permission from Hai et al. 2015. Copyright 2015 Royal Society of Chemistry.)

It's well accepted that the graphene could be broken into small pieces, GQDs in top-down route, via different oxidative, exfoliating or chemical cutting strategy. Through this modeling idea, the possibility from graphene primitive skeleton, organic aromatic molecules, to GQDs was gradually advocated in recent years. Gram-scale synthesis of single-crystalline GQDs was reported utilizing the pyrene molecule (Wang et al. 2014). The easy appearance of uncontrollable polycrystalline or greatly defective structures causes poor optical properties for most GQDs; moreover, they designed a controlled fusion procedure followed by the first nitration into 1,3,6-trinitropyrene by refluxing within

HNO_3 and then hydrothermal treatment in alkali solution to co-linked into larger conjugated molecules. After prenitration, three -NO_2 groups with positively charged sites were attached into the pyrene, thus significantly increasing the activity of nucleophilic substitution reactions with alkali, which facilitate the production of hydroxyl- and amine-functionalized GQDs in a large scale with maximum production yield over 63% (see Figure 7.28).

Small aromatic molecules intrinsically own great superiority for the synthesis of conjugated graphene structure when reactivity was improved with suitable method. The DCB molecules were yielded by photo-induced radicals coupling reaction under UV irradiation (see Figure 7.29a), obtaining perfectly hexagonal atomic arrangement lattice structure (see Figure 7.29d). The production yield could fulfil the theoretical values near 100% yield depending on the number of times the process carried out, which was never reported. Photochemical stitching as they depicted was developed for the preparation of GQDs based on the bottom-up route, only exhibiting a low QY of 2.56%, which still remains a challenge.

The ortho-phenylenediamine (OPD) (Vedamalai et al. 2014) as one alternative was used for obtaining CQDs with a mean size of 4 ± 0.3 nm under hydrothermal at 300°C for 2 h, likely experiencing radical coupling resulting in intramolecular cyclization, further condensation and surface amino/hydroxyl functional groups passivation. However, the QY is excessively low to 2% owing to simple thermal treatment without precisely controlling of ordered condensation. It's rather desirable to acquire multi-colored CQDs via a facile approach with a controllable synthesis. A facile solvothermal approach was developed to obtain three PL colored CQDs (green, blue and red, RGB typically) by three types of phenylenediamine isomers (ortho-, meta- and para-) under absolutely same condition at 180°C for 12 h (see Figure 7.30a) (Jiang et al. 2015b). Three CQDs noted as o-, m- and p-CQDs with average sizes of ~8.2, 6.0 and 10.0 nm, respectively, exhibited different PL QY of 10.4%, 4.8% and 20.6%, respectively (see Figure 7.30b).

The full visible band emissive CQDs with a high QY of up to 35% could be obtained via tuning the degree of surface oxidation. Typically, the co-mixed p-phenylenediamine and urea solution were undergone hydrothermal treatment at 160°C for 10 h following a critical separation by silica column chromatography to obtain eight types CQDs with various PL color emissions (see Figure 7.31a, b and c). A preferential theory was stated as the surface states dominantly controlling the various PL emission when considering total similar particle sizes with broad size distributions and only difference towards their surface states for overall samples (see Figure 7.31d). A higher oxidative degree on the surface could produce more surface defects, directly causing gradual PL red shift.

Recently, more strategies were continuingly developed about tuning multi-color PL emission. Similarly, the p-phenylenediamine was also yielded at 180°C for 12 h with under alkali solution (see Figure 7.32a), after careful

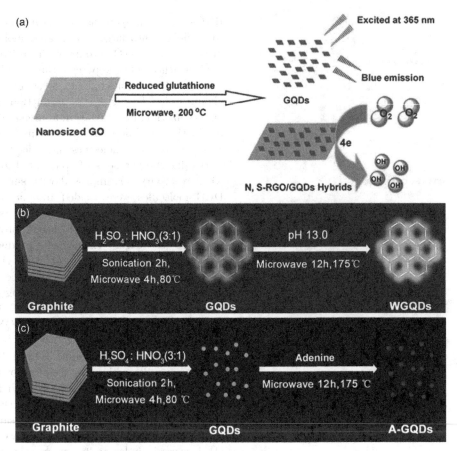

FIGURE 7.18   (a) Preparation of fluorescent GQDs and N, S-RGO/GQDs hybrids for oxygen reduction reaction (ORR). (Reproduced with permission from Luo et al. 2014. Copyright 2014 Royal Society of Chemistry.) (b) Schematic presentation of preparing white-light-emitting graphene quantum dots (WGQDs). (Reproduced with permission from Luo et al. 2016. Copyright 2016 Wiley-VCH.) (c) Schematic depicting the microwave-assisted preparation of A-GQDs. (Reproduced with permission from Luo et al. 2018. Copyright 2018 Elsevier.)

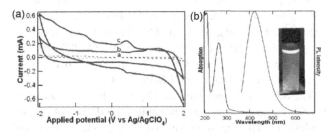

FIGURE 7.19   (a) Cyclic voltammetry of a: carbon paper, b: pristine CNTs and c: treated CNTs in acetonitrile with 0.1 M tetrabutylammonium perchlorate (TBAP) supporting electrolyte. (b) UV-vis absorption and PL spectra of carbon NCs in aqueous solution. PL spectrum was obtained under excitation at 365 nm. Inset is the solution illuminated by a UV lamp. (Reproduced with permission from Zhou et al. 2007. Copyright 2007 American Chemical Society.)

separation with a silica column chromatography, by simply adjusting the polarity of solvents, matrices or concentration of CQDs (see Figure 7.32b and c) recorded as polarity-dependent PL (Ren et al. 2018). The surface electronic structure could be varied with the surrounding solvent's polarity arising, directly leading to the declining of band

gaps and red shift in PL spectra, which agree well with the surface oxidative states strategy (Ding et al. 2015). It's suggested that the surface oxidative states or polarity tuning could be both ascribed into the energy band gaps, which could provide some universal strategy for efficiently guiding multiple PL-colored CQDs synthesis.

To obtain narrow-band emission CQDs meeting the need for new generation displaying technology, more recently, Yuan et al. (2018a) designed narrow bandwidth emission triangular CQDs (NBE-T-CQDs) via a solvothermal treatment at 200°C towards the phloroglucinol precursors with different reaction periods (see Figure 7.33a), following purification via silica column chromatography with high QY up to 54%–72%. Specifically, the highly ordered molecular refusion of threefold symmetric was easily achieved by three highly reactive meta-hydroxyl groups in phloroglucinol, thus endowing the CQDs with highly crystalline structure (see Figure 7.33b, c, d, e and j). Notably, as most publications report with full width at half maximum (FWHM) exceeding to 80 nm because of broad distribution of CQDs size, however, high color purity was achieved by the as-prepared NBE-T-CQDs with only 30 nm FWHM, owning great potentials into the CQDs-based LEDs displaying technology.

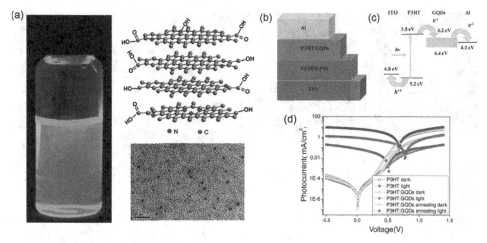

FIGURE 7.20 (a) The electrocatalytically active nitrogen-doped GQDs (N-GQDs). Schematic (b) and energy band (c) diagrams of the ITO/PEDOT:PSS/P3HT:GQDs/Al device. (d) J–V characteristic curves for the ITO/PEDOT:PSS/P3HT/Al, ITO/PEDOT:PSS/P3HT:GQDs/Al and ITO/PEDOT:PSS/P3HT:GQDs/Al devices after annealing at 140 °C for 10 min, single log scale. (Reproduced with permission from Li et al. Copyright 2011a American Chemical Society and 2011b Wiley-VCH.)

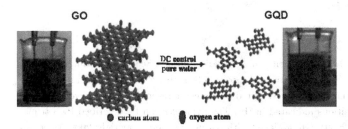

FIGURE 7.21 Illustration of the preparation of GQD from graphite oxide. (Reproduced with permission from Deng et al. 2015. Copyright 2015 Royal Society of Chemistry.)

Although the production yield is estimated to be relatively lower than previous reports (approximately 8%–13%), the high-quality and high-purity PL color still exerts significant potentials to scale up.

Similarly, the high reactivity of hydroxyl groups containing aromatic compounds could be superior to the generation of high-quality and high-performance CQDs with high QY. Fan's group implemented 1,3-dihydroxynaphthalene as precursors with solvothermal treatment at 180°C for 1 h aiming at enlarging the π-conjugated structure. To obtain high-quality red PL emissive CQDs, they innovatively yielded sequential dehydrative condensation and dehydrogenative planarization (DCDP) approach (see Figure 7.33 l) to prepare red PL CQDs with high QY of 53%. During the condensation, the 1,3-dihydroxynaphthalene with dual hydroxyl groups concomitantly served as a building block to cross-link into larger sized conjugated sp^2 cluster by controllable DCDP approach (Wang et al. 2017c).

Citric acid and citrates were frequently used for the synthesis of GQDs and CQDs because of their simple molecular structure and relatively clear process mainly including dehydration and condensation, which could easily control the morphology and sizes. As mentioned above, different carbon resource may exhibit various processes, citric acid

was easily co-fused with doping agent to produce in-situ heteroatom doping, yielding high QY and excellent performance. Most interestingly, when citric acid co-reacted with amine, most reports exhibited better performance with an ultrahigh QY, among which intrinsic mechanism was urgently needed to be explained. Many reports tried to clear the mechanism through heteroatom-doping theory; however, unlike top-down strategy generally considering the heteroatom as doping agent, citric acid-based CQDs always present unusual phenomenon when co-reacted with amine in most case according to recent publications, which make its PL origin and ultrahigh QY mechanism still uncertain.

Dong et al. (2012a) started to make some efforts into tuning the carbonization degree of citric acid through a pyrolysis method relatively low of 9.0% PL QY. Another promoting QY strategy by doping has been tried by using L-cysteine as N- and S-doping resources with citric acid via hydrothermal method, acquiring high QY with 73% (Dong et al. 2013). Considering the PL origin mechanism including emissive traps or electronic conjugate argued previously, the authors tend to attribute the PL into N- and S-doping effects resulting from irradiative recombination of the electrons and holes trapped on the CQDs surface (surface states). Actually, this chapter ascribes the high QY origin to N- and S-doping effect relating surface state, which may be relatively superficial for the deep understanding of citric acid CQDs' ultrahigh QY (Figure 7.34).

The full-color CQDs across the entire visible spectrum was first prepared through microwave-mediated hydrothermal heating typically using citric acid formamide mixed solution at 160°C for 1 h (see Figure 7.35a). When altering the excitation wavelengths from 330 to 600 nm, corresponding emissions nearly cover the entire visible spectrum; it's reported the highest QYs of the as-prepared CQDs are determined to be 26.2% (see Figure 7.35b). To fully understand the multi-color emission mechanism, lower temperature and shorter period synthetic condition was

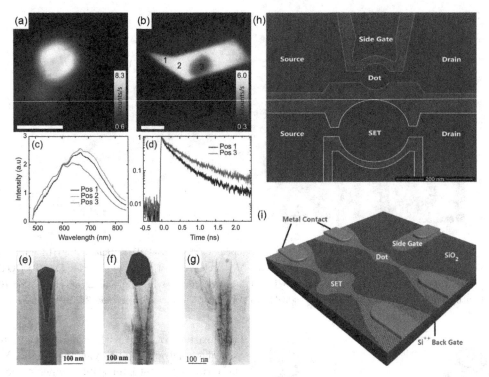

FIGURE 7.22 (a) Confocal PL image excited at 473 nm (2.62 eV) for a graphene sample oxidized for 3 s. Scale bar: 5 μm. The bright PL spots are spatially localized. (b) Uniform emission after 5 s. Scale bar: 10 μm. For position 3 in (b), PL is bleached intentionally by intense laser irradiation. (c) Spectra detected at the positions marked in (b) corrected for detector response. These have broad PL centered ∼700 nm (1.77 eV). (d) PL transients detected at the positions indicated in (b). The dynamics can be described by a triple-exponential with decay times ∼40, 200, and 1,000 ps (Gokus et al. 2009). (Reproduced with permission from Copyright 2009 American Chemical Society.) TEM images of CNTs with different plasma treatment duration: (e) is the nanotube with a catalyst particle on its top; (f) is a transitional etching process, where the particle is almost separated from the nanotube top; and (g) is a case where the catalyst particle has been etched off from the nanotube top. (Reproduced with permission from Zhi et al. 2002. Copyright 2002 American Institute of Physics Publishing.) (h) Scanning electron microscope image of the etched sample structure. The bar has a length of 200 nm. The upper small QD as the main device has a diameter of 90 nm, while the bottom SET as a charge sensor has a diameter of 180 nm. The bright lines define barriers and the graphene side gate. (i) Schematic of a representative device. (Reproduced with permission from Wang et al. 2010. Copyright 2010 American Institute of Physics Publishing.)

conducted to investigate the main emission region changes (mainly in ultraviolet region in this condition), but otherwise visible region corresponding for usual CQDs, directly reflecting its molecular- and polymer-like features ascribing to electronic absorption transitions, which is greatly beneficial for multidimensional sensing (Pan et al. 2015). As Dong et al. (2012a) yielded a pyrolysis to control the degree of carbonization of citric acid-based CQDs, the PL emission could be adjusted by alerting various molar ratios of citric acid (CA) to urea and different reaction temperatures with multiple PL color emission (see Figure 7.35c and d).

Liu et al. (2018) tried to synergistically control the electronic conjugate and surface state dual effects on the PL emission by different ion and solvent. The citric acid and urea were treated with solvothermal treatment at 180°C for 12 h, toluene as solvent, thus promoting the QY from 46% to 68.6% with the addition of Mn^{2+} ion under the same solvothermal condition (see Figure 7.36a and b).

There is one opinion holding that intermediated molecular fluorophores could be the key role in spite of photoluminescent emission (perhaps relates to surface/edge state

and quantum size effect in most case). More specifically, the co-reacted production may contain many unknown small molecules always owning high QY (Ehrat et al. 2017, Kasprzyk et al. 2013, Krysmann et al. 2011, Zhu et al. 2013). Giannelis's group (Krysmann et al. 2011) first elucidated the presence of highly emissive molecular fluorophores in citric acid-based CQDs. Both the molecular fluorophores and the carbogenic core contribute to the ultrahigh QY for the as-prepared citric acid-derived CQDs (dual PL emission) and the molecular fluorophores may play a very important role for the ultrahigh QY. Systematic investigation was conducted with ingredient temperature from low (180°C) to high (230°C, 300°C and 400°C) to detect corresponding products. The PL emission mainly originated from both the carbogenic core (CQDs) and the molecular fluorophores (molecular state emission), among which molecular fluorophores mainly contribute to the high QY (Krysmann et al. 2011). Although this pioneering work systematically studied the PL emission changes as temperature varied, which directly related to fluorophores. However, they haven't isolated the individual fluorophores molecules and further

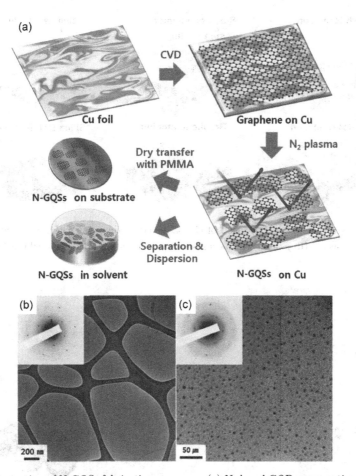

FIGURE 7.23 Schematic illustration of N-GQSs fabrication processes. (a) N-doped GQDs preparation in a large-scale from as-grown monolayer graphene achieved through a nitrogen plasma treatment. (b) and (c) The CVD-derived single layer graphene as the precursors was irradiated under nitrogen plasma, and high-quality N-GQDs directly acquired with perfect lattice structure. (Reproduced with permission from Moon et al. 2014. Copyright 2014 Wiley-VCH.)

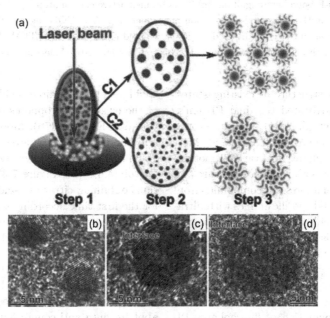

FIGURE 7.24 (a) Schematic of the mechanism of size control of CQDs obtained upon laser ablation. C1 and C2 corresponding to short and long laser pulse widths, respectively. (b–d) Typical HRTEM images of a sample produced from laser pulse widths of 0.3, 0.9 and 1.5 ms, respectively. (Reprinted with permission from Hu et al. 2011. Copyright 2011 Springer.)

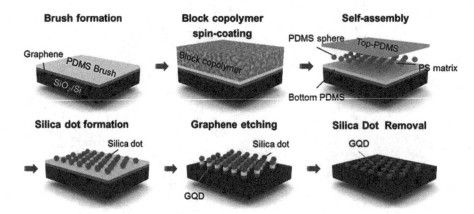

FIGURE 7.25 Schematic illustration of the fabrication of GQDs including the spin-coating of block copolymer, formation of silica dots and etching process by O₂ plasma. (Reprinted with permission from Lee et al. 2012. Copyright 2012 American Chemical Society.)

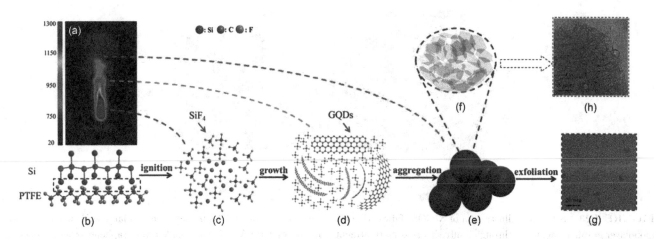

FIGURE 7.26 Schematic strategy proposed to synthesize GQDs. The 2 μm PTFE and 100 nm Si were used as precursors to supply active carbon atoms and deflagration flame for the carbon growth process into GQDs. (a) Temperature distribution image of the reaction flame measured via an IR thermometer. (b) The deflagration reaction between Si and PTFE was ready to take place through breaking the C–F bond and forming the Si–F bond before ignition. (c) The obtained active carbon atoms and SiF₄ molecules. (d) Active carbon atoms grow into GQDs. (e) The GQDs aggregated into carbon nanospheres. (f) The GQDs aggregation structure of the nanospheres. (g) TEM image of the GQDs exfoliated from the obtained GQDs-aggregated nanospheres. (h) TEM of the GQDs aggregation structure corresponding to (f). (Reprinted with permission from Liu et al. 2015b. Copyright 2015 American Chemical Society.)

proven the exact fluorophores structure or existing states. As Giannelis et al. exactly advocated the dual PL emission mechanism (Krysmann et al. 2011), various groups (Kasprzyk et al. 2013, Kasprzyk et al. 2018, Song et al. 2015, Zhu et al. 2013) aimed at the exploration of various photoluminescent molecular fluorophores about their structure and detailed states. A series of novel fluorescent compounds from condensation mixtures of CA and specific amines with ultrahigh QY. Taking the citric acid-based CQDs as example, they tried to clear the mechanism for the origin of high QY, thus reporting many novel fluorophores, 2-pyridone-based fluorophores in general (see Figure 7.37a).

Inspired by biodegradable photoluminescent polymers (BPLPs) nanoparticles (Yang et al. 2009a), much more effort has focused on the explanation of fluorophores PL from the co-condensed products of citric acid and amine compounds, yielding citric acid and L-cysteine by heating at 100°C for 40 min (Kasprzyk et al. 2018). This research

could also provide direct evidence supporting the inference of the existing fluorophores intermediates during the reaction (see Figure 7.37b). More recently, novel green molecular fluorophore was achieved through preparing CQDs with green fluoresce by the pyrolysis of citric acid and urea with QY of 14.6% (see Figure 7.37b) considering most publications obtaining citrazinic acid-based with blue fluorescence for the first time according to their depiction.

Hydrothermal strategy with a mixture of citric acid and ethylenediamine (EDA) at 150–300°C for 5 h to prepare CQDs with diameters of 2–6 nm has gained a production yield of ~58% with an ultrahigh QY of ~80%. (Zhu et al. 2013) The as-prepared CQDs were capable to serve as the printable photoluminescent ink for pattern on paper applying into anti-counterfeit and multi-color PVA/CQDs nanofibers (see Figure 7.38). At low temperature, the citric acid and EDA co-condensed into polymer-like CQDs where PL emission may be originated from amide-containing

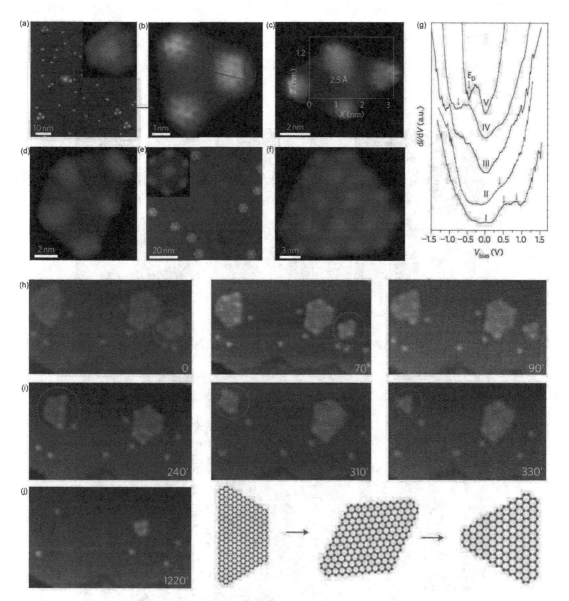

FIGURE 7.27 STM images of GQDs formed by decomposition of 0.08 ML C_{60} on Ru(0001). (a) A 0.08 ML C_{60}/Ru sample after annealing at 725 K for 2 min. Inset: magnified view of mushroom-shaped dots. (b–d) Magnified views of triangular (2.7 nm), (b), parallelogram-shaped (2.7 × 4.2 nm), (c), trapezoid-shaped (2.7 × 4.8 nm), (d) GQDs. Inset to c: line contour taken along the straight line in b. (e,f) Hexagon-shaped GQDs (5 nm and 10 nm) obtained after further annealing the sample at 825 K for 2 min. (g) Representative local STS data for differential conductances dI/dV of the GQDs in b (I), c (II), e (III), f (IV) and for giant monolayer graphene on Ru(0001) (V). Tunnelling parameters: V = 0.5 V, I = 0.1 nA; V = 0.3 V, I = 0.2 nA for the inset images in a and e; V = 0.3 V, I = 0.2 nA (b,c); V = 0.3 V, I = 0.1 nA (d,f). (h–j) Series of STM images monitoring the transformation of trapezium-shaped GQDs to triangular-shaped GQDs at 1,000 K. The numbers in the images indicate the time lapse in seconds. Tunneling parameters (h–j): V = 1.4 V, I = 0.3 nA; image size, 25 × 12 nm^2. (Reprinted with permission from Lu et al. 2011. Copyright 2011 Nature Publishing Group.)

fluorophores. As the process continued, at high temperature, further carbonization could produce carbogenic CQDs, thus generating many emissive carbogenic cores, whereupon the author attributing this phenomenon of ultrahigh QY could be the synergistic result of multi-carbogenic core and fluorophores. Even if they haven't paid efforts to prove the existence of the fluorophores and also explain what exactly it was, this method provides us more possibility for the PL enhancement mechanism towards the citric acid-based CQDs.

CQDs can be prepared from the citric acid and EDA with a simple hydrothermal method by Song et al. (2015). Considering the previous reports noting, temperature playing a key role in the whole synthesis, gradient hydrothermal temperature from 100°C to 200°C was designed to explore possible fluorophores molecules or even other intermediated state (see Figure 7.39a and b). It's notable that it further proves the existence of fluorophores with exact structure confirmation following a research about the photo and thermal stability about intermediated

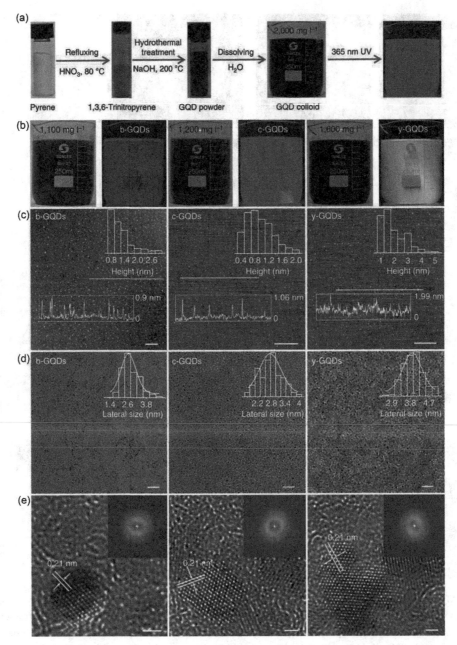

FIGURE 7.28 Preparation of OH-GQDs. (a) Synthetic procedure. Photographs of amine-GQDs. (b) Photographs of aqueous colloids of b-, c- and y-GQDs taken under UV (right) and visible (left) lights. (c) AFM images of b-, c- and y-GQDs (insets: height profiles along the white lines and height distributions). Scale bars, 2, 500 and 2 mm, respectively. (d) Their TEM images (insets: lateral size distributions). Scale bar, 20 nm. (e) Their HRTEM images (insets: FFT patterns). Scale bar, 1 nm. (Reprinted with permission from Wang et al. 2014. Copyright 2014 Nature Publishing Group.)

molecular fluorophores, which could play a leading role in the PL emission.

Stolarczyk's group (Ehrat et al. 2017) tried to track the origin of PL for CQDs; referring previous reports, the aromatic domains or molecular fluorophores may coherently contribute to the PL; and initial phase molecular fluorophores, likely 2-pyridone derivatives, account for the blue luminescence of the dots. As Zhu and co-workers concluded, many groups had studied different types of molecular fluorophores from various precursors. By far, it's extensively accepted that the molecular fluorophores-inducing PL was confirmed and makes great contribution to the high PL emission and QY in citric acid-based CQDs, gaining numerous attentions.

However, what's the detailed existing state? Whether molecular fluorophores covalently bonding onto the carbon cores above CQDs or just physically mixing with CQDs, the exact role of which played still puzzle us. Recently, Kumbhakar and colleagues (Sharma et al. 2017) typically synthesized CQDs with diameter around 6-10 nm by heating citric acid and urea. The edge state, mainly related to the boundary between sp^2 and sp^3-hybridized carbon and

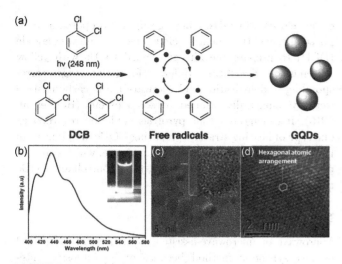

FIGURE 7.29 (a) Schematic representation of the mechanism, (b) PL spectrum (inset: fluorescence visible under UV lamp), (c) GQDs with magnified image of lattice fringes (inset) ($d = 0.36$ nm) and (d) hexagonal atomic arrangement visible at high resolution. (Reprinted with permission from Gokhale and Singh 2014. Copyright 2014 Wiley-VCH.)

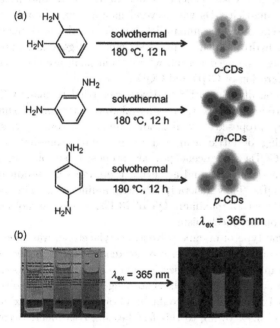

FIGURE 7.30 (a) Preparation of the RGB PL CDs from three different phenylenediamine isomers (i.e., oPD, mPD and pPD). (b) Photographs of m-CDs, o-CDs and p-CDs dispersed in ethanol in daylight (left), and under 365 nm UV irradiation (right). (Reprinted with permission from Jiang et al. 2015b. Copyright 2015 Wiley-VCH.)

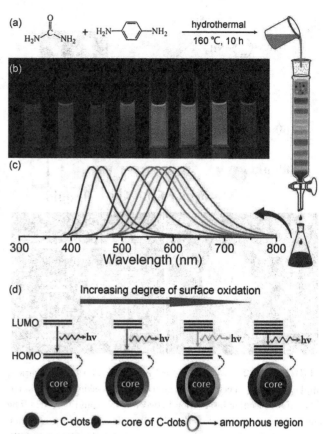

FIGURE 7.31 (a) One-pot synthesis and purification route for CDs with distinct fluorescence characteristics. (b) Eight CQD samples under 365 nm UV light. (c) Corresponding PL emission spectra of the eight samples, with maxima at 440, 458, 517, 553, 566, 580, 594 and 625 nm. (d) Model for the tunable PL of CQDs with different degrees of oxidation. (Reprinted with permission from Ding et al. 2015. Copyright 2015 American Chemical Society.)

surface functional groups, may be not responsible for CQDs PL emission origin. The exact effect of different nitrogen sources on the molecular fluorescence should be further studied. Different nitrogen sources, such as EDA, hexamethylenetetramine and triethanol-amine, could be employed to co-react with citric acid to gain three types CQDs under same hydrothermal condition at 200°C for 5 h, gaining three types CQDs with QY of 53%, 17% and 7%, respectively (see Figure 7.40) (Schneider et al. 2017).

There are increasing reports about the citric acid-based CQDs from multiple views, such as synthesis, mechanism for both PL origin and reacting process. From reported above, we could conclude that citric acid CQDs always present ultrahigh QY, which was urgently wanted in top-down methods; however, the fluorophores-derived QY enhancement may tend to be more unstable under photo or thermal irradiation, which was unprofitable for large-scale synthesis or practical application. Besides, because of the requirement of pure agents for controllable process and products, the expense of whole synthesis could be expected relatively high, severely limiting its large-scale synthesis.

Glucose was used as starting materials to prepare intrinsically fluorescent CQDs via a facile hydrothermal treatment at 200°C for 12 h in the presence of KH_2PO_4 (Yang et al. 2011b). KH_2PO_4 agent plays a key role in the formation of carbon dots and irregular carbon aggregates with average size of 15.7 nm generated when the absence of KH_2PO_4. For one thing, this prior research focused on the synthesis and characterization of CQDs, while lack

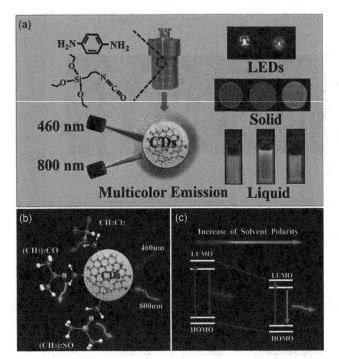

FIGURE 7.32 (a) Schematic illustration of the preparation, application, and mechanism of multi-color emission from CDs. (b) The intermolecular interaction between solvent and CDs. (c) The change of energy levels of the CDs in different polarity solvents. (Reproduced with permission from Ren et al. 2018. Copyright 2018 Wiley-VCH.)

of detailed study of the intrinsic mechanism on KH_2PO_4 prohibits the carbon aggregates growth. For another, the QY for C-blue and C-green was calculated as 2.4% and 1.1%, which is needed to be improved. Glucose-derived crystalline GQDs were also implemented with microwave-assisted hydrothermal treatment with a period from 1 to 10 min, thus causing a QYs variation from 7% to 11% (see Figure 7.41). By combining the superiority of both hydrothermal and microwave methods, this strategy could provide a fast, homogenous and efficient heating beneficial for the generation of GQDs with glucose self-passivation further improving its PL.

Pyrolysis

Different from macromolecules, pyrolysis method may be hard to control its carbonization degree and resultant products. In view of small molecules, direct pyrolysis could easily ensure the whole pyrolysis process transferring the organic precursors into nanoscale products. Mullen's group (Liu et al. 2011) delivered a prior research using hexa-peri-hexabenzocoronene (HBC) to synthesize large sized GQDs of ~60 nm diameter with a broad size distribution and 2–3nm thickness via two-step pyrolysis and oxidative exfoliation (bottom-up and top-down), and they infer that bright and colorful PL may originate from the chemical nature of graphene edges. As Dong et al. (2012a) yielded a pyrolysis to control the degree of carbonization of

citric acid-based CQDs, Ding's group has proposed a solid-phase thermal treatment of citric acid and dicyandiamide (DCD) to prepare the nitrogen-doped CQDs with yellow PL emission. Specifically, different formation stages corresponding to dehydration, polymerization and carbonization were systematically studied to some extent (Hou et al. 2016). It's suggested that pyrolysis could be regarded as one type of heating strategy preparing CQDs, different from hydrothermal treatment with high pressure, which could be relatively unprofitable to generate much controllable CQDs to some extent.

Microwave

Microwave or microwave-assisted method could be treated as fast synthetic method because of its intensive, integrated and concentrated energy for controlling elevated temperature within extremely short time, considered as a more promising fabrication technique for CQDs preparation. Typically, due to the fast heating treatment and consequent dramatic increase within such a short period, the starting materials could undergo extremely intensive reaction to easily generate a dehydration and condensation into ring-rich QDs with several nanometers. Notably, this strategy provides unparalleled rapid synthesis comparing with hydrothermal or pyrolysis methods generally experiencing prolonged period, which could facilitate the speedy preparation of GQD and CQDs.

Typically, the citric acid was ever tried to make CQDs due to its high reactivity under microwave treatment. Ding's group proposed several studies about microwave-assisted heating of citric acid and amine, and a one-pot route for CQDs fluorescent powder proposed, simplifying the synthetic process and achieving the large-scale production of CQDs. They generally fabricate amine-citric acid-based CQDs with a maximum QY of 73.1%, similar to molecular fluorophore mechanism.

One type of organic solvent, acetylacetone, was used for the controllable synthesis of switchable GQD under two-step microwave carbonization and aromatization processes with acetylacetone both as starting materials and solvent. Two PL color GQDs could be obtained with a maximum QY of 3.4% owing to this fast heating treatment. Because of the microwave usually incorporated with hydrothermal or pyrolysis mentioned above, the single microwave treatment could be relatively fewer reported, which could be generally considered as assisted methods for fast synthesis.

Others Methods

The typically established synthesis protocols including CVD were especially popular in recent years because of its extensive use in the synthesis of large-area high-quality graphene. Therefore, it could be of great interest if GQDs could be produced by CVD technique, the key to CVD is to restrain the nucleation during boosting growth by tuning the reaction parameters. Zhu's group (Fan et al. 2013) et al. first reported a strategy synthetic method towards GQDs by

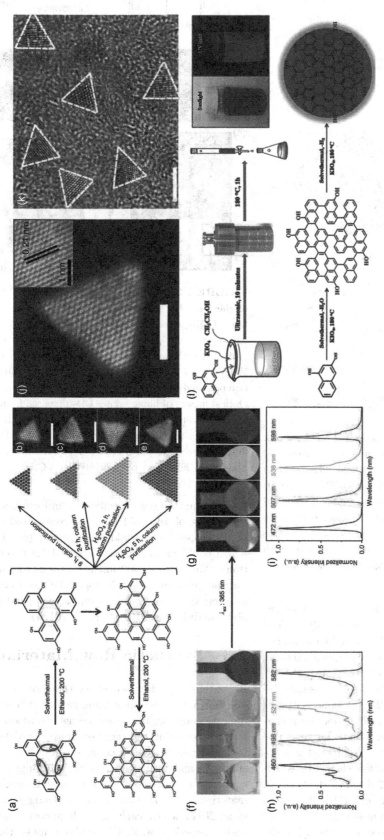

FIGURE 7.33 Design and synthesis of narrow bandwidth emission triangular CQDs. (a) Synthesis route of the NBE-T-CQDs by the solvothermal treatment of PG triangulogen. The typical aberration-corrected high-angle annular dark-field scanning transmission electron microscopy (HAADF-STEM) images of B- (b), G- (c), Y- (d) and R-NBE-T-CQDs (e), respectively. Scale bar, 2 nm. Photographs of the NBE-T-CQDs ethanol solution under daylight (f) and fluorescence images under UV light (excited at 365 nm) (g). The normalized UV-vis absorption (h) and PL (i) spectra of B-, G-, Y- and R-NBE-T-CQDs, respectively. Structural characterizations of the NBE-T-CQDs. (j) The typical aberration-corrected HAADF-STEM image of R-NBE-T-CQDs (the inset is the corresponding high-resolution image). (k) The wide-area TEM image of G-NBE-T-CQDs. Scale bar, 2 nm. (Reproduced with permission from Yuan et al. 2018a. Copyright 2018 Nature Publishing Group) (l) Schematic diagram showing the preparation and growth mechanism of R-CQDs. (Reproduced with permission from Wang et al. 2017c. Copyright 2017 Wiley-VCH.)

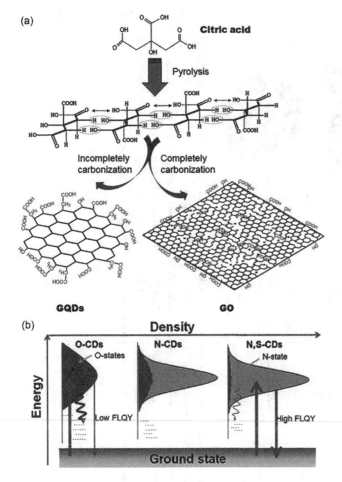

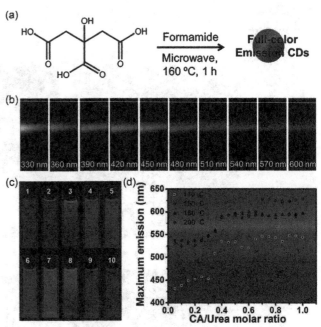

FIGURE 7.34 (a) Diagram for the synthesis of GQDs and GO. The black dots in the GO represent oxygen atoms. (Reproduced with permission from Dong et al. 2012a. Copyright 2012 Elsevier.) (b) Representation for the fluorescent mechanism of O-CDs, N-CDs and N,S-CDs. (Left) Electrons excited from the ground state and trapped by the surface states and excited electrons return to the ground state via a non-radiative route. (Right) excited electrons return to the ground state via a radiative route. (Reproduced with permission from Dong et al. 2013. Copyright 2013 Wiley-VCH.)

FIGURE 7.35 (a) A schematic illustration of the preparation of the full-color emission CDs. (b) FL emission photographs of the F-CQDs recorded from 330 to 600 nm in 30 nm increments. All spectra and photographs were obtained in deionized H_2O. (Adapted with permission from Pan et al. 2015. Copyright 2015 Wiley-VCH.) (c), (d) Optical property of the CQDs ethanol solution prepared from CA and urea at different reaction conditions. (c) Optical images of luminescence CDs prepared from different reaction conditions under different excitation light. (d) The maximum emission peaks of CQDs at different molar ratios of CA to urea and different reaction temperatures. (Adapted with permission from Miao et al. 2018. Copyright 2018 Wiley-VCH.)

CVD growth. Inspiring by their former deeply extensive studies of CVD grown graphene, they occasionally noticed the some carbon clusters during the CVD (see Figure 7.42a), which could be regarded as small pieces of graphene, serving as the growth core for further generating graphene. A precisely controlled parameter was set to obtain small size GQDs via a CVD growth (Figure 7.2b). Organic synthesis was considered as the most precise methods for the preparation of GQDs at a molecular synthesis scale. Large, stable and controllable synthesis of GQDs was conducted as stepwise refusion of benzyl through solution chemistry developed by Liang et al. (Mueller et al. 2010, Yan et al. 2012). The 3-(phenylethynyl) phenylboronic acid, Pd(PPh$_3$)$_4$, was served as the precursors, through complex multiple polymerization to generate into graphene moieties with the multiple $2',4',6'$-triakyl phenyl groups covalently attached to the edges of the graphene moieties primarily for stabilizing

the resultant GQDs. Specifically, uniform and tunable size composing of 168, 132 and 170 conjugated carbon atoms (1–3), respectively, could be prepared (see Figure 7.43a), which could provide great potential for extensive applications. (Yan et al. 2010) Similar synthesis was also conducted for the preparation of N-doped GQDs (see Figure 43b) as electrocatalysis with excellent size-dependent electrocatalytic activity for ORR (Li et al. 2012b).

7.3 Synthetic Raw Materials

It's extensively discussed above that both of the top-down and bottom-up methods using carbon-rich ordered macromolecules and organic molecules as precursors, respectively, have unavoidable weakness or limits, hard to balance the cost and quality, which could severely limit its large-scale synthesis and future application. Although the large-scale synthesis was possible to be reacted through some cheap materials, such as carbon black (Dong et al. 2012b, Lu et al. 2017), active carbon, the high cost, complex process, and low-quality, weak PL products make these strategies exhibit huge gap to be scaled up. Introducing much cheaper, more extensive and greener biomass materials are actually

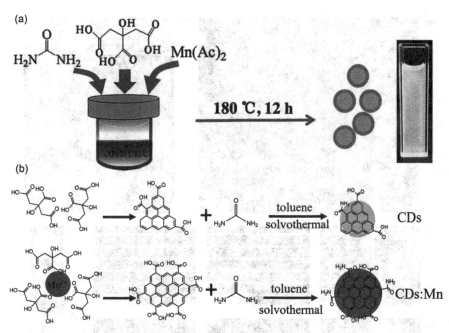

FIGURE 7.36 (a) Schematic diagram illustrates the preparation of fluorescence emissive carbon dots (CDs:Mn). A possible growth mechanism for (b) CDs and CDs: Mn. (Reproduced with permission from Liu et al. 2018. Copyright 2018 Elsevier.)

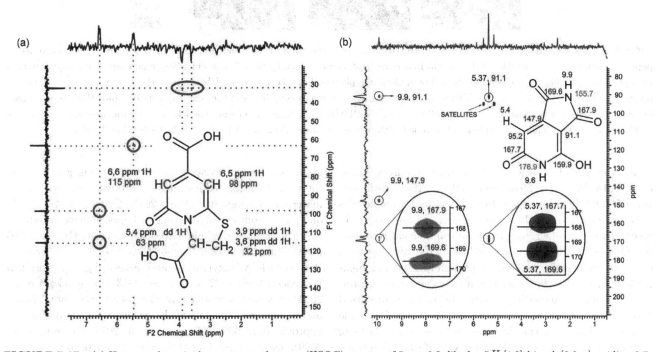

FIGURE 7.37 (a) Heteronuclear single quantum coherence (HSQC) spectra of 5-oxo-2,3-dihydro-5*H*-[1,3]thiazolo[3,2-*a*]pyridine-3,7-dicarboxylic acid (TPA). (Adapted with permission from Kasprzyk et al. 2013. Copyright 2013 Royal Society of Chemistry.) (b) HMBC[1]H-[13]C Nuclear magnetic resonance (NMR) spectrum of 4-hydroxy-1H-pyrrolo[3,4-*c*]pyridine-1,3,6(2*H*,5*H*)-trione (HPPT) and chemical structure of HPPT with NMR assignments (ppm). (Reproduced with permission from Kasprzyk et al. 2018. Copyright 2018 Royal Society of Chemistry.)

the future key point especially via simpler, more economic and environmentally friendly methods based on synthetic methods mentioned above. The starting materials with different resources and specific molecular structure could take a great effect on resultant products, which is responsible for choosing suitable methods and PL tuning strategy. Therefore, it's worth noting that low-cost biomass materials

could be meaningful to the preparation of CQDs in a large scale if high quality could be obtained.

7.3.1 Low-Cost Biomass-Based Resources

New perspectives about the synthesis of lignin-based nanomaterials have gained tremendous attentions for the

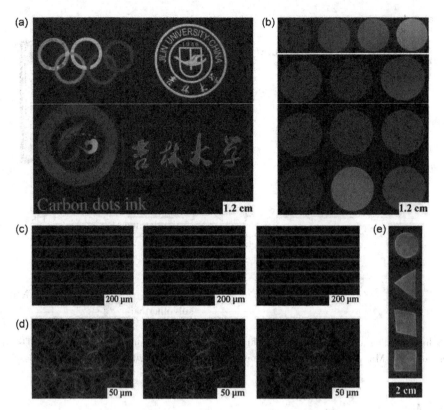

FIGURE 7.38 Printed patterns obtained by CD ink and the integration of CDs and polymers. (a) Different graphic patterns on paper (illuminated by a portable UV lamp). (b) Inks in multiple colors, tuned by the CD concentration in aqueous solution (illuminated by a portable UV lamp). (c) CD ink patterned on hydrophilic photoetching stripes (under UV excitation). (d) Fluorescence microscopy images of PVA/CD nanofibers with UV excitation. The fluorescence microscopy images in (c) and (d) were obtained through band-pass filters of different wavelengths: 450, 550 and 580 nm. (e) Bulk PDMAA/CD nanocomposites (The PL intensity was unchanged after 2000 W UV exposure for 30 min). (Adapted and modified with permission from Zhu et al. 2013. Copyright 2013 Wiley-VCH.)

high value-added utilization of biomass resources (Ding et al. 2018, Wu et al. 2015), especially for the transformation of GQDs in this chapter. In addition, natural molecules could provide greatly abundant resources as precursors. In recent years, the emergence from natural biomass resources to CQDs or GQDs was subsequently reported, aiming at not only the large-scale preparation of high-quality and high-performance QDs, but also the high value-added utilization of biomass. According to synthetic methods discussed above, the idea of using natural resources could provide a new window for green, low-cost, large-scale synthesis of high-quality and high-emission CQDs or GQDs. With those thoughts in mind, we systematically discussed recent publications about biomass-based QDs aiming at the large-scale preparation of high-quality GQDs.

Lignin

As mentioned above, the aromatic molecules could be intrinsically beneficial for the synthesis of high-quality GQDs (Gokhale and Singh 2014, Jiang et al. 2015b, Vedamalai et al. 2014, Wang et al. 2014). Natural aromatic molecules actually provide greatly abundant resources serving as precursors. It's well acknowledged that lignin is the largest

aromatic resource existing in nature, owning great potential to be transformed into myriad aromatic compound and materials (Georgakilas et al. 2015, Sun et al. 2018). As one major component existing in lignocellulosic resources, lignin is composed of phenylpropyl skeletons (H, G and S) mostly linked by carbon–carbon (C–C) bond and ether bond (C=O) with hydroxyl group- and methoxyl group-contained branches (Adler 1977, Chu et al. 2013, Nanayakkara et al. 2014). It's well accepted that the β-O-4 ether bond is the major type of linkage which accounts on total linkages approximately 60%–70% (Sun et al. 2018, Zakzeski et al. 2010). Different pulping techniques could produce various types of lignin raw materials, such as alkali lignin (AL), kraft lignin and lignosulfonate (LS) with different amount of monomeric units and linkages (Nanayakkara et al. 2014, Shi et al. 2016, Sun et al. 2018, Zakzeski et al. 2010).

In most cases, lignin has been viewed as one type of biomass waste or even pollutants from pulping and paper-making industry, typically treated by direct burning to generate energy, which is a huge resource waste urgently needed to be solved. To fully utilize this aromatic biomass resource, some researchers have exerted some efforts to synthesize lignin-derived GQDs and CQDs with good PL

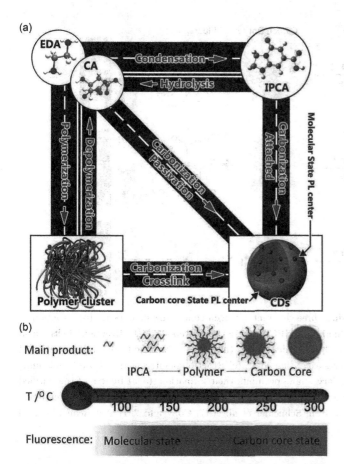

FIGURE 7.39 (a) Scheme 1: A schematic of the relationship between different products in the one-pot hydrothermal system of CA and EDA. (b) Scheme 2: A schematic of the CDs obtained at different hydrothermal temperatures. (Adapted and modified with permission from Song et al. 2015. Copyright 2015 American Chemical Society.)

property. Because of lignin's intrinsic multi-phenyl characters, the auto-fluorescence easily appeared in lignin (Donaldson and Radotic 2013, Donaldson et al. 2010, Albinsson et al. 1999). Previous researchers focusing on the wood cell wall fluorescence imaging have investigated the fluorescence lifetime changes in normal wood and compression wood, suggesting the existence of short-lifetime lignin fluorophore in compression wood, which could be ascribed to lignin special auto-fluorescence (Donaldson and Radotic 2013). However, the mechanism still remains a puzzle when designing lignin-based luminescent materials. Almost non-conjugated lignin could be PL emissive, actually drawing great interests. Therefore, more recently, the origins of lignin fluorescence were further studied referring to linkage between native lignin and hemicellulose (LCC) existing in natural wood cell wall, similar to aggregated state, which was believed to be appropriate for aggregation-induced emission (AIE) PL. Qiu's group (Xue et al. 2016b) first advocated that the lignin fluorescence could be related to AIE mechanism, and they carefully confirmed the AIE effect through critical aggregation concentration (CAC) to restrict intra-molecular aggregation rotation among lignin

and controlled sulfonation degree from agglomerated AL to expanded, stretched LS for demonstration of the stilbene (Ar-Cα=Cβ) and α-carbonyl (Ar-Cα=O) groups clustering effects (Xue et al. 2016b). The intrinsic fluorescence of lignin and similar aromatic structure could facilitate the generating of photoluminescent GQDs, which have drawn much more attentions in recent years to fully utilize the ultra-fine lignin aromatic resources and promote the large-scale synthesis of high-quality GQDs at the same time.

Counting on such AIE mechanism (Xue et al. 2016b), Liu's group (Niu et al. 2017) tried to utilize the molecular aggregation strategy of lignin benzyl unit for the preparation of CQDs, yielding the cellulolytic enzyme lignin (CEL) via very simple method with stirring for 6 h at room temperature. The π–π aggregation-induced molecular aggregation-derived CQDs with an average diameter of approximately 2.4 nm in the ethanol could emit bright blue-green fluorescence. They also proved that the as-prepared CQDs were mainly composed of CEL with low molecular weights, containing aromatic rings rich in oxygen-containing functional groups, which could be beneficial to the formation of π–π aggregation into CQDs. This work successfully prepares the fluorescent CQDs by using cheap biomass materials and cellulose enzymatic lignin (CEL), avoiding the harsh conditions and toxic chemical reagents. However, this mild strategy inevitably presents a non-crystalline structure for CQDs due to these single benzene rings' weak interaction forces between every single benzene units without any carbonization process to generate stable interaction, of which the stability and production yield still need to be discussed. Besides, the as proposed AIE mechanism from the lignin to CQDs still seems not persuasive enough for the exact formation mechanism. Most importantly, the lignin-based CQDs only exhibit a QY of 2.47%, extremely lower than previous reports, which could severely limit the practical usages in a large scale.

The first report about hydrothermal synthesis lignin-based CQDs was actually achieved by Liu's group (Chen et al. 2016b), and they typically yielded a fast, efficient and green hydrothermal strategy to synthesize fluorescent CQDs within the assistance of strong oxidants, H_2O_2, which could be considered as cutting agents to break the lignin molecular structure. The obtained CQDs ranged from 2 to 10 nm in diameter own blue fluorescence with maximum production yield of ~12%. More detailed structure information could be observed with a 0.21 nm lattice fringes. However, they haven't reported the QY, an important index for the PL property, and it's suggested that the as-synthesized CQDs generally present amorphous structure without any modification, thus directly causing relatively weak fluorescence because of reasonable awareness of lignin structural features. The cursory designing of lignin breaking and fusing process could straightforwardly diminish the PL property (Figures 7.44 and 7.45).

The LS (sulfite pulping-derived lignin) was first used for the preparation of fluorescent CQDs with the average diameter of ~4.6 nm via a microwave irradiation at 600 W

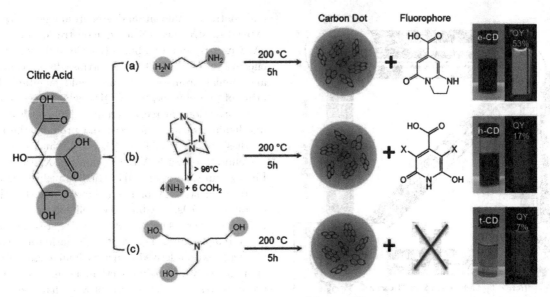

FIGURE 7.40 Synthesis conditions of citric acid-based CDs using three different nitrogen-containing precursors. (a) Reaction of citric acid and ethylenediamine (EDA), resulting in e-CDs and the fluorophore IPCA, as previously reported. (b) Reaction of citric acid with hexamethylenetetramine, producing h-CDs and citrazinic acid and/or 3,5 derivatives (marked by -X), due to the decomposition of hexamethylenetetramine to ammonia and formaldehyde at temperatures exceeding 96°C. (c) Reaction of citric acid and triethanolamine, resulting in t-CDs and no derivatives of citrazinic acid since the tertiary amine prohibits their formation. (a–c) Images of the purified reaction products under ambient light and corresponding diluted solutions under UV light excitation, which reveals fluorescent emission with PL QYs as labeled on the graph. (Reproduced with permission from Schneider et al. 2017. Copyright 2017 American Chemical Society.)

within 10 min (Rai et al. 2017). Because of good solubility, the LS could be uniformly dispersed in water and also sulfur-containing groups could be easily integrated under hydrothermal treatment, which could act as self-doping agent to greatly improve the PL performance. Furthermore, considering previous synthetic strategy (Li et al. 2012a, Tetsuka et al. 2012), the as-prepared CQDs with sulfur and oxygen-contained groups were further treated by reduction with NaBH$_4$, thus improving the QY from 31.27% to 47.3% due to the surface effect (see Figure 7.46b). Notably, a possible mechanism was proposed throughout whole process as follows: weak acid environment accelerated the subsequent decomposition of main linkage, such as β-O-4 ether, C–C and hydrogen linkages in lignin, and then, resultant small aromatic clumps could emerge a nucleation, which could be attributed to the carbocations site effect attacked by electron-rich phenyl rings, finally generating into CQDs (see Figure 7.46a) (Rai et al. 2017). However, it' a great pity that they haven't proved the exact existence of carbocations and also lack of specific proof to confirm the whole mechanism, which makes the whole progress remain controversial. For the other, it's recorded that this strategy could only accomplish very low production yield with a maximum yield of 5.02%; besides, the whole process still remains relatively complex and much expensive to scale up.

It's a common acceptance that previous reports generally covered amorphous CQDs or carbon dots with poor crystalline structure and PL property, paying little efforts to unveil the detailed mechanism. To fully understand the

exact mechanism and obtain high-quality, high-performance and low-cost GQDs, we first reported the gram-scale synthesis of high-quality single-crystalline GQDs derived from AL, using a two-step method involving the first pre-oxidized cleavage step with ultrasonication in nitric acid, followed by an aromatic refusion step under hydrothermal treatment at 180°C for 12 h (see Figure 7.47) (Ding et al. 2018). The as-prepared GQDs present excellent PL property with a high PL QY of 21%, superior to previous biomass-based QDs (including CQDs and GQDs). Furthermore, we first investigated the formation mechanism from lignin to GQDs systematically, which was neglected by previous researchers.

Unlike previous publications, the single-crystalline GQDs instead of CQD or carbon dots were prepared for the first time. We integrated both the advantages of top-down and bottom-up methods, easy operation and easy controllability, respectively (Ding et al. 2018). A peroxidation step was innovatively induced to cut the lignin into aromatic fragments (top-down process) and then yields these aromatic fragments to refuse multi-benzyl precursors into highly conjugated graphene structure via hydrothermal (bottom-up process). Specifically, to clear the whole period, we studied the mechanism through three periods, AL, oxidized AL and GQDs, and the characteristic peak intensity corresponding to sp^2 carbon resonate between 100 and 135 ppm in ^{13}C CP-MAS NMR spectra (see Figure 7.47c). It could reflect the possible presence of condensed lignin intermediates under acid environment (see Figure 7.47c: f

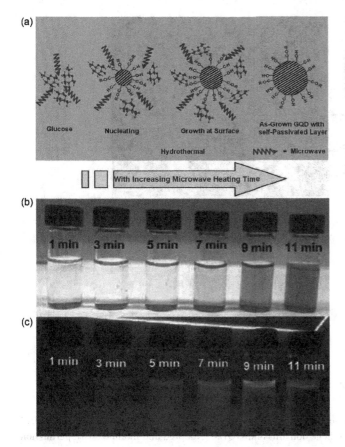

(a)

Glucose — Nucleating — Growth at Surface — As-Grown GQD with self-Passivated Layer

Hydrothermal ⟿ = Microwave

With Increasing Microwave Heating Time

(b)

1 min | 3 min | 5 min | 7 min | 9 min | 11 min

(c)

1 min | 3 min | 5 min | 7 min | 9 min | 11 min

FIGURE 7.41 (a) Schematic preparation of GQDs by microwave-assisted hydrothermal (MAH) method. (b), (c) the GQDs solutions irradiated by ambient light (top) and 365 nm UV lamp (bottom). The GQDs were prepared for 1–11 min with microwave heating at 595 W by 11.1 wt % glucose solution. (Adapted and modified with permission from Tang et al. 2012. Copyright 2012 American Chemical Society.)

to b), which could be beneficial for the further formation of highly ordered aromatic conjugated graphene network. The as-prepared GQDs distribute uniformly with a narrow diameter distribution of 2–6 nm (Figure 7.47a). Notably, more detailed atom-resolution structure could be clearly observed by HRTEM, indicative of highly crystalline structure with a lattice spacing of 0.21 nm, similar to the hexagonal pattern of graphene with (1,100) lattice fringes.[6,44] It could be clearly distinguished that approximately 340 benzene rings are arrayed in hexagonal honeycomb carbon network within single 3.7 nm GQD (Figure 7.47b). The selected area electron diffraction (SAED) pattern (see inset of Figure 7.47b) directly confirmed the actually high-quality graphene single-crystal feature of as-prepared GQDs (see Figure 7.47b), which has never been reported in previous publications about biomass-based GQDs (Ding et al. 2018). Considering the previous PL enhancement strategy, in-situ doping strategy by Tetsuka et al. (2012), the structure-performance correlation also could be demonstrated. The in-situ N-doping was successfully implemented during HNO₃ pretreatment and hydrothermal reduction process, demonstrating a single-crystalline graphitic network with N atoms

doped at the edges (Liu et al. 2017, Wang et al. 2014, Zhao et al. 2015, Tetsuka et al. 2012), which directly contribute to the high QY and excellent PL property. Most importantly, the overall production yield reached approximately 21%, which enables the high-quality, high-performance and low-cost preparation of GQDs in a large scale more practical to be reached.

Wang's group (Li et al. 2018b) typically yielded a hydrothermal treatment towards AL under 200°C for 12 h to generate CQDs with a diameter of 2.39 ± 0.38 nm. They also pretreated the lignin with ammonia solution aiming at N-doping and then conducted with hydrothermal treatment. By referring to previous energy trap theory (Yang et al. 2012), they successfully controlled the PL property by adjusting the concentration of ammonia, which attributed to electrons trapped by the induced N-doping or functionalized surface state, promoting a large amount of irradiative efficiency for the photo recombinations. The surface N-doping could be limited, thus inducing a maximum QY of approximately 14%, almost 10 times higher than lignin-based CQDs without N-doping (1.49%) (Li et al. 2018b). However, even if the N-doping strategy could greatly improve the QY from 1.4% to 14%, the whole method could barely rebuild the lignin molecules structure with controllable arraying of ordered benzyl to benzyl linkages (large conjugated domains), thus causing low QY and possibly still retaining amorphous features of lignin.

One type of novel approach for preparing GQDs was advocated based on top-down method. The preparation of GQDs in scheme was subsequently performed by the first synthesis of big carbon network via pyrolysis of kraft lignin with metallic salt, followed by a typical chemical cutting by top-down strategy (Temerov et al. 2018). As shown in the schematic presentation (see Figure 7.48), the kraft lignin together with three metallic salts (CoCl₃·06H₂O, NiCl₂·6H₂O, FeCl₃·6H₂O), respectively, was first performed at 500°C by a careful heating under argon gas (Ar), to prepare graphene-encapsulated metal nanoparticles (GEMNs). After HNO₃ dissolution to remove excessive salts, the GEMNs were further experienced 250°C hydrothermal for 12 h with the assisted cutting of KOH. By etching those amorphous carbons on the surface, the GQDs were successfully exfoliated via subsequent decomposition strategy. It's clearly observed that yellowish luminescence GQDs with large diameter of 20–25 nm could exhibit weak PL property (QY: 11.7%–12.4%) and poor structure. This work depicted a new approach based on top-down strategy, showing complex and high-cost process.

Similarly, one approach involving two steps converting the lignin into CQDs was also conducted by top-down strategy (Myint et al. 2018). The kraft lignin was first yielded to pre-carbonize into carbon nanoparticles using compressed liquid carbon dioxide (clCO₂) as an antisolvent (Myint et al. 2016) and then cut into CQDs species by chemical oxidation with mixed acid (H₂SO₄:HNO₃ = 3:1) via heating at 90°C for 8, 16 and 24 h (see Figure 7.49). According to their depiction, the as-prepared CQDs present a lattice parameter

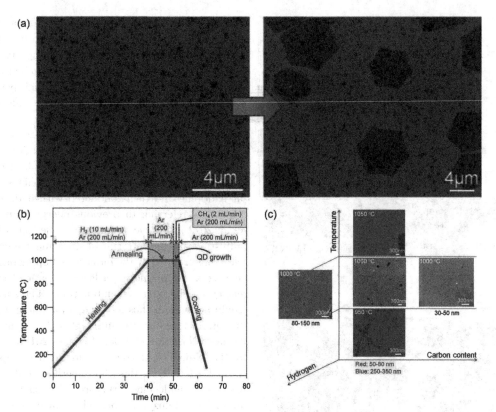

FIGURE 7.42   (a) Carbon clusters during the CVD grown graphene. (Adapted with permission from Fan et al. 2013. Copyright 2012 Wiley-VCH.) (b) CVD process for CGQDs growth. Size control of graphene domains and GQDs. (Adapted and modified with permission from Fan et al. 2013. Copyright 2013 Wiley-VCH.)

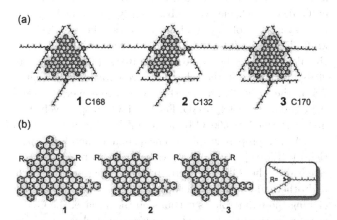

FIGURE 7.43   (a) Schematic presentation of synthesis of graphene quantum dots **1-3**. (Adapted with permission from Yan et al. 2010. Copyright 2010 American Chemical Society.) (b) Structures of N-doped graphene QDs 1, 2 and an undoped QD 3 for comparison studies. (Adapted and modified with permission from Li et al. 2012b. Copyright 2012 American Chemical Society.)

of ~0.29 nm, higher than 0.24 nm corresponding to the lattice fringes of graphene (1,120), indicative of CQDs noncrystalline characters with much lattice defects. The production yield generally varied from 21% to 23%. The defective structure could directly cause low QY of 13%. This work provides a facile, green and efficient pretreatment technique, the clCO$_2$ antisolvent method, by simply controlling the

pressure and temperature of CO$_2$ in a single stage, which could easily tune the carbon nanoparticles serving as precursors for the further generating of CQDs.

From these previous reports, we could draw some conclusions towards lignin-based QDs as following: (i) cursory pyrolysis or thermal treatments rarely produce high-quality GQDs or CQDs with ultrafine PL property, and heteroatom doping (basal plane or edge) including S and N atom could be regarded as effective strategy to improve the fluorescent QY to some extent. (ii) The experimental designing should exert more efforts into molecular level for the lignin, skillfully elaborating the superiority of aromatic skeleton by oriented conversion through stepwise controlling. The mechanism still requires much more detailed investigations throughout various approaches, which could provide guidance in large-scale preparation. (iii) The synthesis of low-cost and high-quality GQDs in a large scale could be achieved through our reported synthetic strategy combining the peroxidation and hydrothermal, which could provide a selective methods for the large-scale synthesis of GQDs.

Hemicellulose

Hemicellulose, as one amorphous polysaccharide component in lignocellulosic biomass, could be regarded as biocompatible molecules, owning great potentials into the biofield if hemicellulose-based nanomaterials could

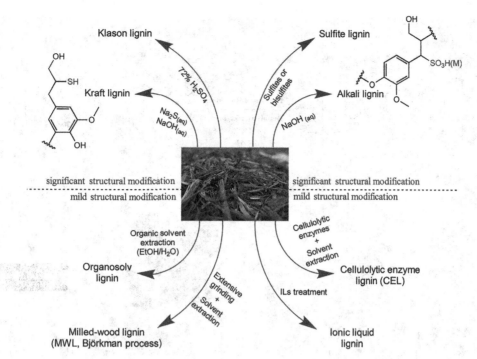

FIGURE 7.44 A summary of main types of lignin and procedures for the isolation of lignin from lignocellulose. (Reproduced with permission from Sun et al. 2018. Copyright 2018 American Chemical Society.)

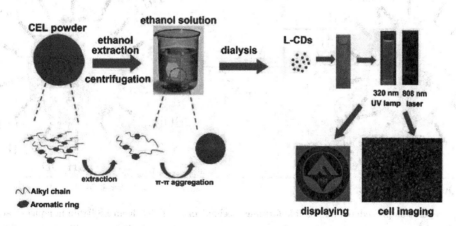

FIGURE 7.45 Diagram showing the preparation and luminescence of L-CDs. (Reproduced with permission from Niu et al. 2017. Copyright 2017 American Chemical Society.)

be obtained. The hemicellulose could be decomposed into oligosaccharide beneficial for human body as prebiotics (Pastell et al. 2009), such as xylan oligosaccharide, fructo-oligosaccharide and isomalto-oligosaccharides. However, there are fewer reports about the hemicellulose-based CQDs. The carbohydrate-based CQDs have been reported extensively (Wang et al. 2011), involving a simple dehydration and condensation process. Therefore, it could be expected that the polysaccharide, hemicellulose-based CQDs is of great possibility to be prepared. The possibility from hemicellulose to CQDs was extensively discussed because of the serious requirement of biocompatibility and low toxicity for bio-related fields.

Considering the low-cost preparation of CQDs and high value-added utilization of forestry and agricultural biomass waste, Wang's group (Liang et al. 2014) has tried the synthesis of hemicellulose-based N-doping CQDs by using facile method and low-cost passivation agent. The uniform dispersion for staring materials could be a key factor for the further generation of CQDs. It's well known that hemicellulose owns good solubility in alkaline environment, which could be easily extracted by ingredient diluted alkali treatment. They yielded the cheap alkali NH_4OH solution dissolving the xylan (representative of hemicellulose) to gain good dispersion, followed by a hydrothermal at 200°C for 12 h (see Figure 7.50a). The hemicellulose-based N-doping CQDs could be easily realized through good dissolution in NH_4OH solution, which not only enables the N-doping occurring on CQDs surface, but also produces more in-situ doping inside CQDs molecule, thus resulting

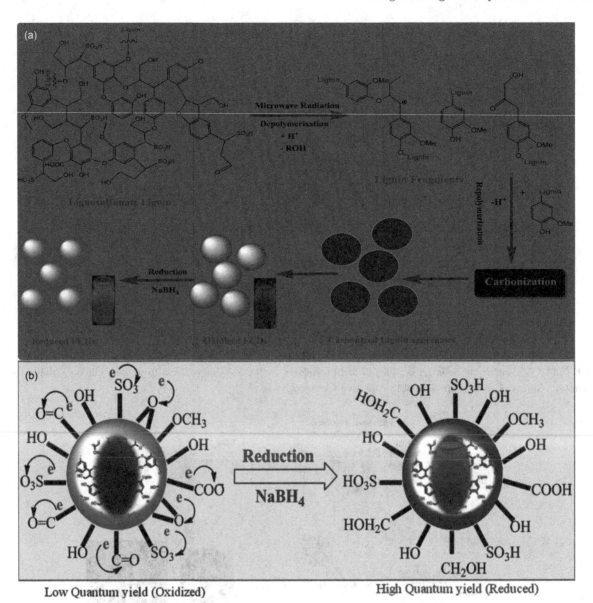

FIGURE 7.46 (a) Preparation procedure and possible forming mechanism of r-FCDs form LS lignin in aqueous acidic media, cleavage of β-O-4 (ether) bond and formation of carbocation and repolymerization reaction. (b) Schematic illustration of reduction mechanism of carbon dots. (Adapted and modified with permission from Rai et al. 2017. Copyright 2017 Elsevier.)

in good PL property with a maximum QY of 16% (see Figure 7.50b). They also investigated the effects of various nitrogen resources on the resultant CQDs' PL property. According to their depiction, the dilute NH_4OH solution was dispersed; hence, they stated that the dilute NH_4OH solution could exhibit much better results on the PL property because of much more stable surface defects instead of conventional organic amine compounds EDA and 4,7,10-trioxa-1,13-tridecanediamine (TTDDA) QY corresponding to QY of 7.48% and 7.26%, respectively. Considering the low-cost urea solution instead of more expensive amine agent, this work successfully transferred the biomass waste into highly missive CQDs, which could be scaled up when finely balancing the hemicellulose materials preparing cost.

The rapid one-pot microwave strategy towards synchronous preparation of CQDs and lignin nanoparticles was reported by using biomass rice straw as starting materials (Si et al. 2018). Under the acid assistance, it could certainly decompose the amorphous polysaccharide fraction (including cellulose amorphous region and hemicellulose) and partly separated the lignin from the heterogeneous natural lignocellulose, releasing soluble compounds holocellulose fraction (including xylose, arabinose, galactose and glucose), lignin fragments (including butanol, glycol, glyoxylic acid, 2-ethoxyethanol, 2,4-dimethylbenzaldehyde, benzoic acid, pentadecane and vinylguaiacol) under the acidolysis. It's suggested that these soluble molecules directly contributed to the easily generating of CQDs (see Figure 7.51). Besides, because of 2.3%–7.2% containment of nitrogen-containing protein substance in rice straw, the simultaneous N-doping by protein could be achieved during the preparation of CQDs

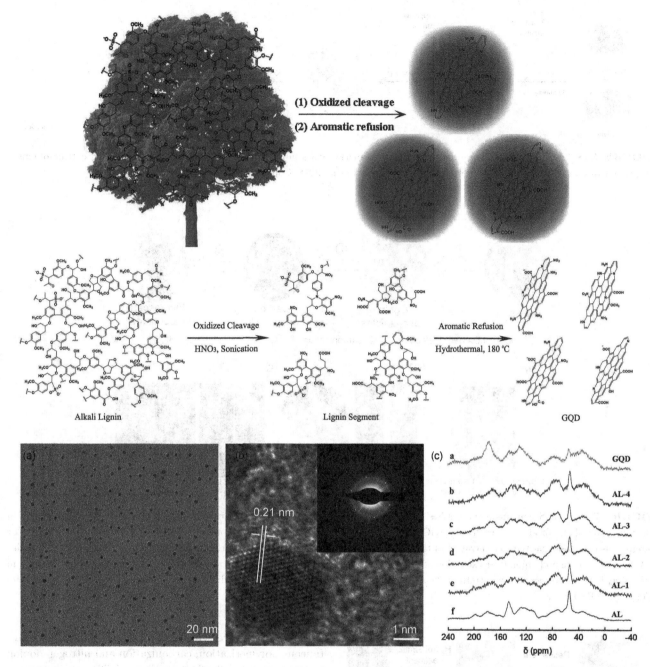

FIGURE 7.47 Scheme of two-step synthesis method of GQDs from AL. Morphology and size distribution of GQDs. (a) TEM image and (b) HRTEM image of the GQDs (inset: FFT pattern). (c) ^{13}C CP-MAS NMR spectra of a: GQDs, acid-treated AL for b: 12 h, c: 6 h, d: 1 h and e: 0.5 h and (f) AL. (Adapted and modified with permission from Ding et al. 2018. Copyright 2018 Royal Society of Chemistry.)

with a production yield of approximately 5.19%. We believed that the acidolysis carbohydrate could take main effect on the formation of CQDs with the assistance of soluble acidolysis-derived aromatic molecules. Actually, the majority of lignin component mainly existed in the formation of lignin nanoparticles, thus resulting in low yield of CQDs. It's noteworthy that this work presents a sparkling point of the direct utilization of lignocellulosic biomass to prepare CQDs via one-pot microwave irradiation without complex pretreatment, which provides detailed

explanations for the soluble molecules' benefits as guidance for the further synthesis. However, the non-neglected low production yield and poor PL property still need to be improved, which is unprofitable for the large-scale synthesis.

More recently, the microwave-assisted xylan-derived CQDs were conducted at a power of 200 W heating at 200°C within just 10 min using xylan as the precursor dissolving in ammonia solution, preparing relatively larger size CQDs with diameter of 4.44–13.34 nm (Yang et al. 2018).

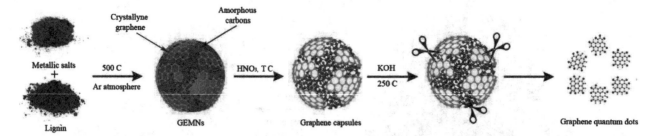

FIGURE 7.48 A schematic presentation of the GQDs formation via a fabrication of GEMNs and a hydrothermal cutting of empty graphene capsules. (Reproduced with permission from Temerov et al. 2018. Copyright 2018 Elsevier.)

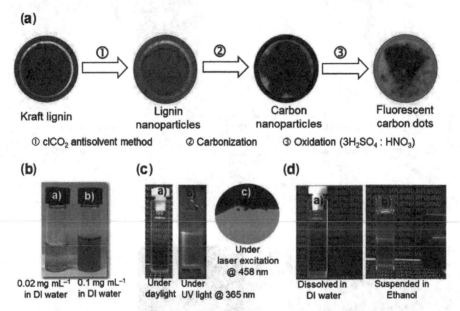

FIGURE 7.49 (A) Schematic of the synthesis of fluorescent carbon dots (CDs) using kraft lignin as a precursor, and photographs of the raw material and products at each stage (clCO$_2$ = compressed liquid CO$_2$). (B–D) Photographs of CQDs-24. B (a and b): The color intensity is dependent on the concentration of the CDs-24 solution in DI water; C (a–c): the fluorescent colors emitted under daylight, UV light at 365 nm and argon laser excitation at 458 nm observed using a confocal laser scanning microscope for the CDs-24 solution in DI water; D (a and b): solubility pattern of the as-synthesized CDs-24. (Adapted and modified with permission from Myint et al. 2018. Copyright 2018 Elsevier.)

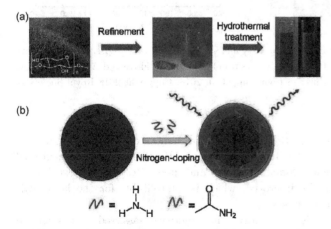

FIGURE 7.50 (a) A schematic diagram of the preparation route of the CDs. (b) Scheme of the passivation procedure of the CDs by nitrogen doping. (Reproduced with permission from Liang et al. 2014. Copyright 2014 American Chemical Society.)

The fast microwave synthesis of CQDs may involve simultaneous polymerization, carbonization and nitrogen-doping processes within 10 min (see Figure 7.52), which may be relatively milder than hydrothermal carbonization. When compared with Wang et al. reports with QY of ~16% (Liang et al. 2014), this work only obtained ~4% under same pretreatment condition with ammonia solution dissolution. It's suggested that the carbonization method could be directly responsible for the resultant low QY, and the mild microwave treatment could be severe enough to hydrolyze the polysaccharides. However, the potent hydrothermal treatment could efficiently depose the xylan into low-molecular-weight carbohydrate, facilitating the CQDs formation for fully utilizing bottom-up strategy's superiority. It could be noticed that there are fewer reports about hemicellulose-based CQDs, which still further dedicated into more efforts involving the high-quality quantum dots synthesis.

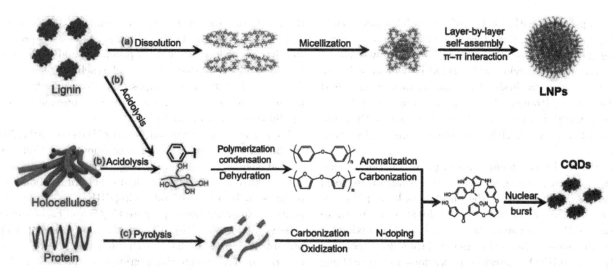

FIGURE 7.51 Scheme 1 Schematic mechanism for co-preparation of lignin nanoparticles (LNPs) and CQDs. (Reproduced with permission from Si et al. 2018. Copyright 2018 Royal Society of Chemistry.)

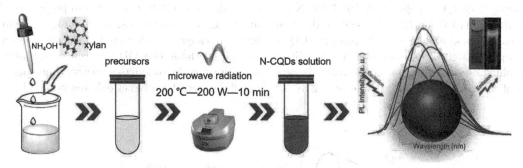

FIGURE 7.52 Synthesis of N-CQDs by microwave treatment. (Reproduced with permission from Yang et al. 2018. Copyright 2018 Elsevier.)

Cellulose

It's understandable that amorphous natural polymer, such as lignin and hemicellulose, could be easier to decompose and further regenerate into aromatization structure under carefully controlling of thermal process, which could be profitable for the synthesis of QDs with aromatic structure. However, in spite of amorphous components lignin and hemicellulose, the highly oriented lignocellulosic fibril resource, cellulose, the most abundant natural resources on earth, could also be yielded to prepare fluorescent CQDs. It's well known that the cellulose nanocrystal (CNC) have gained increasing numerous attentions in recent years, which induced much more interests about various nanomaterials preparation derived from CNC or cellulose derivatives.

The first reports about CNC-based CQDs were conducted with high-temperature pyrolysis from 300 to 1,000°C (see Figure 7.53). Different morphological carbon materials also could be prepared, such as CBs, carbon sheets and carbon sphere; furthermore, water-soluble CQDs with diameter of 4–8 nm could be obtained. These CNC fragments as precursors primarily in irregular rod-like particles with a length of 100–300 nm were prepared by using typical preparing procedure (de Rodriguez et al. 2006). They stated that the

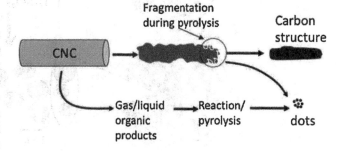

FIGURE 7.53 Schematic representation of the proposed mechanism of CNC pyrolysis to form CDs. (Reproduced with permission from da Silva Souza et al. 2016. Copyright 2016 Elsevier.)

formation of different carbon nanostructures using conventional structures remains unclear, which could be ascribed into the strong hydrogen bonding interactions between CNCs to aggregate into different nanostructure, and the thermal treatment may directly lead to welding of the carbon particles to fuse into CQDs (da Silva Souza et al. 2016). On the other hand, the schematic representation of the proposed mechanism from CNC to CQDs definitely indicated the low production yield by such a pyrolysis strategy because of uncontrollable structure and size, thus causing poor PL property with low QY of 1.64%. This prior study highlighted the generation of various carbon nanostructures during the pyrolysis of CNC relating to the self-assembling and aggregation behavior, which could provide an effective strategy for the synthesis of carbon nanomaterials based on the CNC. Although this work paid less efforts to the synthesis ultrafine CQDs and the improvement of PL property, it still paves the path for the cellulose-based QDs developments.

As the cellulose could be regarded as polysaccharide with intense linkages, similar to carbohydrate-based CQDs making mechanism, the eucalyptus kraft pulp (see Figure 7.54) (da Silva Souza et al. 2018) was yielded to prepare CQDs via a concentrated mixed acids refluxing for 12 h (subsequent dehydration and oxidation decomposition by H_2SO_4 and HNO_3) described in previous literature (Peng et al. 2009). The as-prepared CQDs with average diameter of 2.5 nm and a great amount of carboxylic groups on the surface could be further modified and passivated by thionyl chloride ($SOCl_2$) and oligomeric PEG1500 N via excessive refluxing at 110°C for 72 h. It could be deduced that this approach should obtain poor PL property CQDs with QY of approximately 3.2%. Because of the lack of precisely controlling of aromatization process and difficulty of breakdown of intense linkages in cellulose especially for

the crystalline region, the as-synthesized CQDs lacking of conjugated structure directly give rise to poor PL property. Besides, we believed that this approach could be relatively low in production yield because of reaching degree dehydration and oxidation decomposition, thus directly causing partial utilization of cellulose, similar to previous reports. (da Silva Souza et al. 2016)

Liu's group dedicated several works (Wu et al. 2014, 2015) to the simultaneous preparation of fluorescent CQDs and carbon sphere (see Figure 7.55a). Formerly, they generally yielded simple one-pot hydrothermal carbonization towards the bleach hard kraft pulp (BHKP) refinery-derived pentosan to synchronous prepare CQDs and carbon microsphere (Wu et al. 2015). The synthesized green PL CQDs present high QY of 24% with an average size of ∼30 nm, rich of oxygen-containing groups on the surface. Typically, they designed a similar experiment with co-mixture of carboxymethyl cellulose (CMC) and extra additions of urea as nitrogen sources, through similar one-pot hydrothermal carbonization at 210 °C for 12 h to prepare the CQDs with an average diameter of ∼32 nm (see Figure 7.55b) (Wu et al. 2014). It's suggested that the former researches mainly focus on the synthesis of CQDs; however, the later paid more attentions to the conductive microcarbon spheres, CQDs as one fluorescent by-product existing in liquid products.

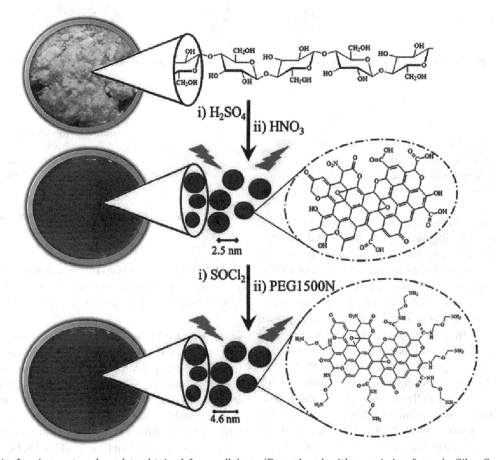

FIGURE 7.54 Luminescent carbon dots obtained from cellulose. (Reproduced with permission from da Silva Souza et al. 2018. Copyright 2018 Elsevier.)

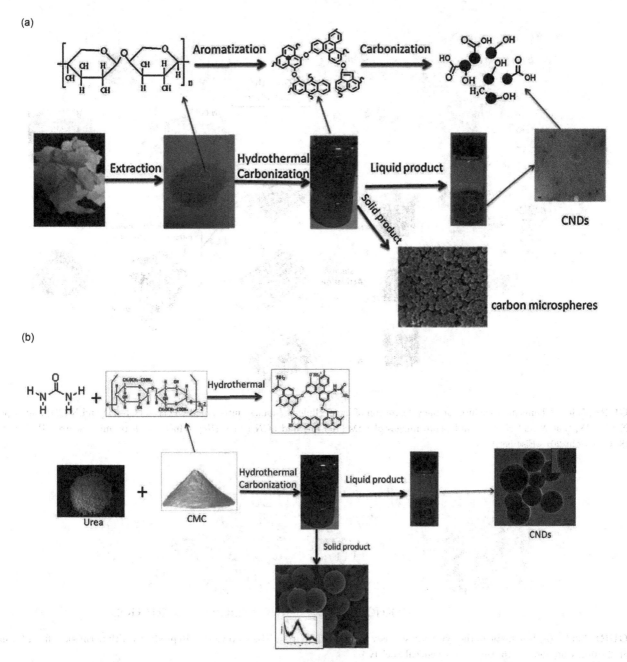

FIGURE 7.55 (a) Schematic illustration of carbon nano-dots (CNDs) synthesis. (Adapted and modified with permission from Wu et al. 2014. Copyright 2015 Elsevier.) (b) Schematic showing the one-pot hydrothermal carbonization of CMC for N-doped carbon spheres and the CNDs. (Adapted and modified with permission from Wu et al. 2015. Copyright 2015 Elsevier.)

A separation-free, straightforward, and controllable hydrothermal synthesis of heteroatom-doping CQDs was incorporated with phosphorus (P), sulfur (S) and nitrogen (N) for the synthesis of green fluorescence P-, S- and N-doped CQDs using wood waste, empty fruit bunch, exhibiting red, blue and yellow fluorescence, respectively (see Figure 7.56) (Park et al. 2018). More specific, hydrothermal was first implemented at 200°C for 12 h within the respective existence P, S and N containing compounds to P-, S- and N-doped CQDs. Then, the separation of liquid and a solid monolith was simultaneously accomplished during the hydrothermal period towards the starting materials, corresponding to cellulose component and lignin characters reflecting in ^{13}C CP-MAS NMR spectra.

7.3.2 Biomass Wastes Resources

It's widely accepted that the utilization of biomass waste, such as rice husk (Wang et al. 2016b), dead leave (Roy et al. 2014), waste peels (Narayanan et al. 2018, Prasannan and Imae 2013, Saxena and Sarkar 2012, Zhou et al. 2012a) or food waste (Park et al. 2014) and biomass derivatives (Si et al. 2018), has greatly contributed to the development of

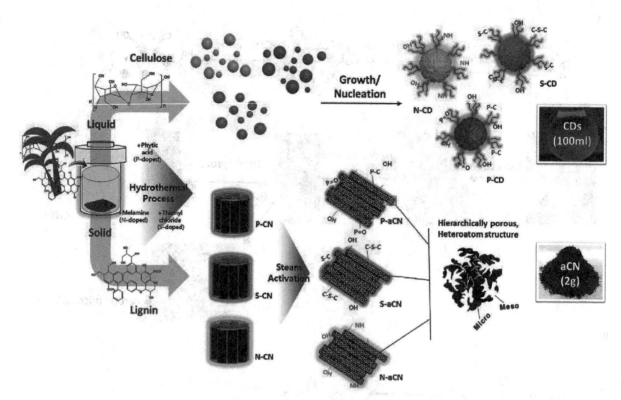

FIGURE 7.56 Chemical procedure for the valorization of lignocellulosic biomass into doped-CD (P-CD, S-CD and N-CD) and doped aCN (P-aCN, S-aCN and N-aCN) and photo images of CDs (100 ml) and aCN (2 g). (Reproduced with permission from Park et al. 2018. Copyright 2018 Elsevier.)

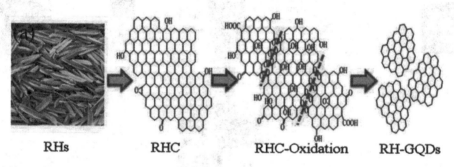

FIGURE 7.57 (a) Schematic of the synthetic strategy for RH-GQDs from RHs (top view). (Reproduced with permission from (Wang et al. 2016b). Copyright 2016 American Chemical Society.)

graphene and CQDs for non-negligible numerous emerging biomass waste-based QDs. The biomass waste could simultaneously consist of multiple components, which may greatly increase the difficulty in understanding the mechanism and tuning the PL property. However, there are still increasing number of researches relating to the biomass waste transforming into the CQDs, attracting by its characters of low cost, huge amounts and extensive origins. However, there are also numerous reports always present a low-quality and low QY products due to the lack of careful decomposition designing, thus disturbing its conjugated domains. The selection of suitable starting resources and corresponding methods could exert dramatic effects on the resultant CQDs products' structure, thus influencing fluorescent property. It's worth concluding that the weakness and advantages

towards various origins of biomass waste with tremendous differences in molecular structure and property are needed to be fully understood, which is practically meaningful for further large-scale synthesis.

Rice husk (Wang et al. 2016b), by-product of rice milling, composed of lignin, cellulose and hemicellulose (accounting on ~72%–85%), was utilized to prepare high-quality GQDs with subsequent carbonization, mixed concentrated acid treatment followed by hydrothermal treatment at 200°C for 10 h (see Figure 7.57). The reason for highlighting rice husk-based GQDs is mainly related to its representative features in components (main component of lignin, cellulose and hemicellulose in most biomass waste) and synthetic strategy. Top-down and bottom-up methods were combined with pre-carbonization into rice husk carbon flake (bottom-up), followed by acid oxidation to break the carbon into GQDs via hydrothermal treatment (top-down). The as-prepared blue fluorescent GQDs with a diameter of 3–6 nm present a lattice space of 0.24 nm, corresponding to the (1120) lattice fringes of graphene. The crystal structure seems not obvious, thus causing a relatively poor PL property with a QY of 8.1%. They highlight two types of PL emission (I and II, higher and lower energy PL) ascribing to the higher energy mainly corresponding to intrinsic state of GQDs, while lower energy PL emission originating from the defect states (Liu et al. 2013). Furthermore, in a typical procedure, the hydrothermal temperature effects were also investigated with the lateral diameter of GQDs gradually increasing from 4.5 to 9.0 nm as temperature decreasing from 200°C to 150°C corresponding to energy gap enlargement, thus causing red shifting, which could be attributed to quantum confinement effect. Besides, quantum confinement effect could also reflect into the temperature-dependent PL behavior, which was also observed with slight PL degradation trend (~20% decrease from 5°C to 65°C). They stated that the thermal activation of non-irradiative trapping excited by electrons–phonons interaction and in the crystal lattice (Li et al. 2014) could be responsible for the temperature-dependent PL behavior, thus decreasing energy-relaxation channel from phonon bottleneck effect. It's worth noting that rice husk-based GQDs were produced via two-step hydrothermal treatment, however, the relatively low production yield of 2 wt % based on starting materials, and rice husk was mainly originated from complex intermediated treatment, which could further increase the expense for the preparation of rice husk-derived GQDs in a large scale.

For another, the poor crystalline structure towards GQDs directly causes the low fluorescent QY, which needed further improvement.

The watermelon peel (Zhou et al. 2012a) has been used to prepare the blue luminescence CQDs with QY of 7.1% (see Figure 7.58a). The two steps involving low-temperature carbonization at 220°C for 2 h transforming the fresh watermelon peel into CQDs species with diameter of 2.0 ± 0.5 nm and the filtration separation under sonication to obtain CQDs solution carbonized under air atmosphere. This simple, green and economical synthetic method successfully transformed the fresh watermelon peel into CQDs, endowing added value for abandoned biomass waste. Furthermore, the orange waste peel pre-dried by sunlight was fully used to perform a green synthesis of CQDs without strong acid or an organic reagent (see Figure 7.58b). As shown in figure, the dried orange waste peel (Prasannan and Imae 2013) was oxidized by sodium hypochlorite solution (NaClO solution) for 4 h and then conducted by hydrothermal carbonization at 180°C for 12 h to generate CQDs of 2–7 nm in diameter with a production yield of 12.3% counting on the orange waste peels. The resultant CQDs products could present excellent PL property with QY of 36%, higher than previous biomass waste-based CQDs. We could insight the whole preparation process that the starting materials, orange waste peel subsequently treated by concentrated H_2SO_4 and NaClO solution, should have exhibited an amorphous structure, facilitating the hydrothermal carbonization transferring the precursors into aromatic. Besides, according to previous citric acid-based CQDs with ultrahigh QY fluorophore existence, whether this work with such a high QY is related to intermediated fluorophore or not still needs

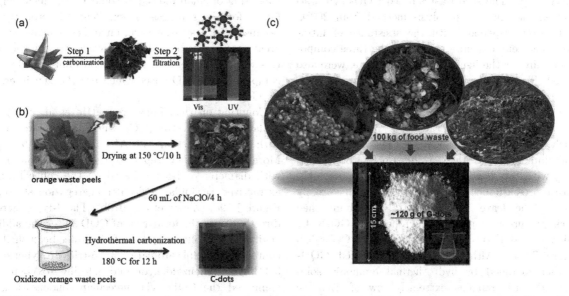

FIGURE 7.58 Schematic illustration of the synthesis of water-soluble fluorescent C-dots from watermelon peel. (Reproduced with permission from Zhou et al. 2012a. Copyright 2012 Elsevier.) (b) Scheme 1. Formation of C-dots from the hydrothermal treatment of orange waste peels. (Adapted with permission from Prasannan and Imae 2013. Copyright 2013 American Chemical Society.) (c) Schematic illustration of the synthesis of waste food-based CQDs. (Adapted and modified with permission from Park et al. 2014. Copyright 2014 American Chemical Society.)

further investigations. The food waste (Park et al. 2014), as one part of large waste resources, could also be generally considered as abundant carbon resources for the preparation of CQDs (see Figure 7.58c). The green fluorescence CQDs with 2.85% internal QY from food waste was first reported from food waste-derived sources via an ultrasound irradiation method at room temperature. This procedure could produce lots of food waste indeed (120 g of CQDs with a size 2–4 nm counting on per 100 kg food waste). In spite of the first synthesis of CQDs based on food waste in large quantities, this work also highlights the positive effects on plant growth of the by-products from food waste-derived sources in a huge quantity when neglecting the CQDs effects.

Coconut shell (Chunduri et al. 2017) was usually served as the starting materials for the preparation of activated carbon, which enables it possible to synthesis CQDs. Mechanical grounding into particle to get uniform precursor particles was first implemented following a hydrothermal treatment at 200°C for 3 h to prepare CQDs with diameter of 3–5 nm. Typically, the NaBH₄ causing reduction reaction could further control surface traps aiming at improving the PL property. The as-prepared CQDs own a lattice parameter of 0.262 nm corresponding to the (002) diffraction lattice of graphite with an illustration of polycrystalline nature by SAED pattern. The excitation-dependent property was shown with a blue fluorescence emission at excitation wavelengths within 300–380 nm and green luminescence light over excitation at 400–490 nm, corresponding to a QY of 35.74% at 360 nm and 11.57% at 490 nm, respectively. The dramatic reduction in QY demonstrated surface state emissive traps variation as the excitation wavelength changes, which corresponds to size effects and surface traps. The sago wastes-based CQDs synthesis was reported via one-step pyrolysis method from 200°C to 500 °C after dispersion with the assistance of ultrasonication. The temperature's effects on the three components, especially for the lignin and hemicellulose, were also investigated. Similarly, lack of precise molecular transformation tuning strategy, the PL property could be relatively weak under simple thermal pyrolysis (Tan et al. 2014).

Similarly, the dead neem leave (Suryawanshi et al. 2014) was implemented with bottom-up pre-carbonization combining with top-down oxidation, conducting as high-temperature pre-carbonization at 1,000°C for 5 h, directly transforming the leave into turbostratic carbon, then serving as precursors for further synthesis of GQDs by mixed H₂SO₄ and HNO₃ (3:1) refluxing at 90°C for 5 h (see Figure 7.59a). Although the as-synthesized GQDs were further modified by hydrothermal ammonia solution, the QY still remains extremely low of 2%. We believed that the pre-carbonization could be the key factor for further carbon precursors exfoliating into GQDs, because of amorphous structure intrinsically presenting adverse effects on the formation of destructive conjugated structure.

We could conclude that the typical procedure for the preparation of GQDs in several publications always involves a pre-carbonization following the chemical oxidation to exfoliate the biomass-derived carbon into GQDs. However, due to the structural complexity in bulky materials, simple thermal irradiation easily generates poor PL property products when lacking of careful designing.

Different from direct using complex bulky materials, Roy et al. (2014) tried to yield leaf extractives of neem (*Azadirachta indica*) and fenugreek (*Trigonella foenum-graecum*), generally composing of hydrocarbons and nitrogen-containing molecules (Ciftci et al. 2011), to synthesize the GQDs for first time (see Figure 7.59b), which was inspired by Tiwari's group (Kumar et al. 2011) implementing the neem oil to prepare the aligned CNT bundles for the first time. Subsequently, neem leaves were used as the starting materials, followed by the extraction of boiling water to obtain neem leaf extractives with a production yield of 25.2%. The GO sheets were first generated through a pre-pyrolysis, and undergone the leaf extractives fusion from small molecules into GO sheet at 300°C for 2 h and then a chemical cutting process by a facile hydrothermal treatment at 300°C for 8 h via an unzipping mechanism as discussed above (Pan et al. 2010). Both neem and fenugreek extractive-derived GQDs with average size of 5–7 nm (noted as N-GQDs and F-GQDs, respectively) own good PL property with high QY of 41.2% and 38.9%, respectively. They explained the temperature could be the key factor for the formation of GO sheet and further decomposition into GQDs, with a lower temperature (below 300°C) not enough for total carbonization into GO sheet and higher temperature (above 300°C) easily destroying the structure into amorphous carbon. Besides, the extractives generally consisting of small molecules could be the most important factor for the synthesis of excellent PL property GQDs because of its easy aromatization and controllability, owning great potential to be scaled up. However, it's suggested that this work lacks detailed information about the extractives component and GO sheet which are the key intermediate for GQDs synthesis.

Counting on this, Liu's group (He et al. 2018) yielded a flavonoid quercetin (QC) with flexible aromatic ring with a typical feature of AIE agent from *S. japonica*. More recently, they reported the preparation of CQDs with diameter of 1-5 nm based on the coffee bean shells extractive via similar molecular aggregation strategy (see Figure 7.59c) (Zhang et al. 2018b). The J-type aggregation directly drives the formation of CQDs after the addition of acid. The phenolic extractives from coffee bean shell mainly containing multiple aromatics, 3,4,5-trihydroxybenzoic acid, 3,4-dihydroxybenzaldehyde and 3,4-dihydroxybenzoic acid composed the CQDs via molecular condensation, which show great potentials into the application of antioxidation. However, owing to the good antioxidation capacity of as-prepared CQDs, the stability against external environment still remained a challenge which is not beneficial for CQDs storage and usages in a large scale.

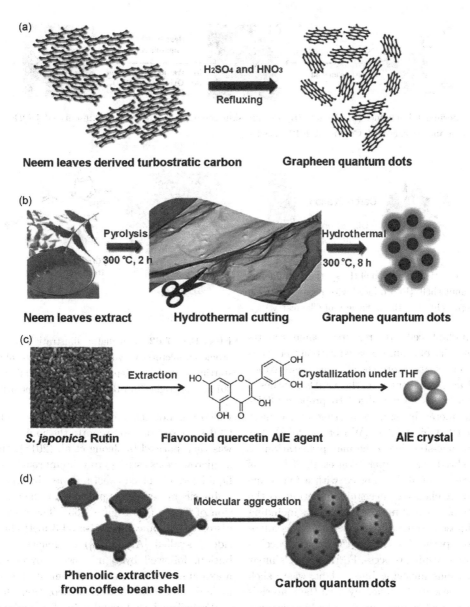

FIGURE 7.59 (a) Schematic for the fabrication of GQDs by the acidic oxidation of turbostratic carbon obtained from dead neem leaves. (Reproduced with permission from Suryawanshi et al. 2014. Copyright 2014 Royal Society of Chemistry.) (b) Schematic representation of the synthesis for GQDs from neem leaf extract via hydrothermal route (Roy et al. 2014, Copyright 2014 Royal Society of Chemistry.) (c) Schematic illustration of preparation of flavonoid quercetin AIE agent and AIE crystals with AIE fluorescence. (Adapted with permission from He et al. 2018. Copyright 2018 Wiley-VCH.) (d) Schematic illustration of the preparation of coffee shell beans extractive-based CDs. (Adapted and modified with permission from Zhang et al. 2018b. Copyright 2018 Royal Society of Chemistry.)

Peanut skin, composed of several compounds including tannins (Ciftci et al. 2011), long-chain saturated fatty acid, such as behenic and lignoceric, could be regarded as ultra-fine resource for materials making. The peanut skin (Saxena and Sarkar 2012) was first collected using roasting method to separate the skin from peanuts and then treated with 700°C heating under nitrogen atmosphere, finally followed by oxidation by nitric acid treatment to endow a good solubility (see Figure 7.60). However, considering the diameter of 10–40 nm, we'd rather place these products as carbon nanosphere, which exhibit very weak PL property ascribing to ultrahigh-temperature' s negative effects on the destructive structure. For another, inspired by above peanut

skin-based photoluminescent carbon nanosphere, the greatly abundant resources, peanut shell, were used for the synthesis of CQDs with lateral size of 0.4–2.4 nm through the typical pyrolysis at 250°C for 2 h (see Figure 7.61). Considering the higher temperature effect on neem leaf-based CQDs, this work chooses relatively lower temperature, which could avoid the carbon network collapse to some extent. Besides, the peanut shell is consisted of small-molecular-weight extractive, such as diverse chromone and terpenoid, which could serve as low-molecular-weight precursors for CQDs formation, thus causing a QY improvement of 9.91%.

As one typical lignocellulosic product, the waste paper's recycle and reutilization still remain challenging problems.

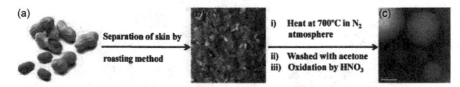

FIGURE 7.60 Experimental scheme. (a) Peanut. (b) Peanut skin separation. (c) Peanut skin-derived CQDs. (Reproduced with permission from Saxena and Sarkar 2012. Copyright 2012 Elsevier.)

FIGURE 7.61 Schematic illustration of the synthesis of fluorescent C-dots from peanut shell. (Reproduced with permission from Xue et al. 2016a. Copyright 2016 Royal Society of Chemistry.)

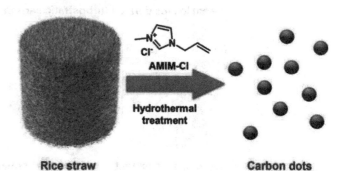

FIGURE 7.62 Schematic illustration for the ionic liquid promoted microwave-hydrothermal synthesis of CDs with straw as starting materials. (Reproduced with permission from Liu et al. 2015a. Copyright 2015 Royal Society of Chemistry.)

Owing to its abundant cellulose resource, some reports relating to its possibility of cellulose-based carbon materials were exerted in recent years. More specific, the waste paper has been utilized as precursors into the CQDs synthesis. The waste paper ash has been utilized to prepare CQDs within 2–5 nm diameter by simply heating under inert environment with PL QY of 9.3% (Wei et al. 2013). For another, a similar one-step hydrothermal preparation of soluble CQDs by shred waste paper (Wei et al. 2014) was implemented at 180°C for 10 h reported with a QY value of 11.7%. However, a plausible mechanism relating to the hydrothermal carbonization of major constituents including cellulose, hemicellulose and lignin in waste paper was advocated, which is too superficial to explain the detailed mechanism during the transforming process. Furthermore, similar to previous publications involving bulky plant, waste shell or biomass wastes, these works generally used the materials without any pretreatment to decomposing the biomass into lower molecular weight components, which directly result in the amorphous structure and poor PL property.

As Liu et al. (2015a) first inducing the ionic liquid pretreatment towards rice straw, the efficient dissolution of main components for the ionic liquid could own a good dissolving capacity which is feasible for the polysaccharide-based rice straw (see Figure 7.62). The 1-allyl-3-methylimidazolium chloride (AMIM-Cl) was used to uniformly dispersed raw materials, and then, microwave-assisted hydrothermal was at different temperature (160°C, 200°C, or 250°C) within short period to prepare three types of CQDs with a maximum QY of approximately 22.58%, much higher than previous reports. The emerging lignocellulosic pretreatment technique and further synthesis of CQDs were artfully combined to efficiently break the intense natural barrier within this unprecedented report, which could be directly responsible for good PL property with higher QY. However, to the best of our knowledge, the ionic liquid is always expensive, and the recyclability in practical industry still needs to be taken into consideration,

which was not economical to the large-scale preparation of CQDs. Based on this report, the ionic liquid recyclability was then noticed by Jeong et al. (2018), and they reported a microwave-assisted pyrolytic approach towards the ionic liquid pre-dissolved waste paper by the degradation of cellulosic materials in paper, thus facilitating the formation of CQDs (see Figure 7.63). The waste paper was first dissolved in ionic liquid 1-allyl-3-methylimidazolium chloride (so-called [Amim][Cl]) to generate a uniform distribution, followed by a microwave pyrolysis for 30 s at a maximum power degree of carbonization without consuming ionic liquid according to their depiction (Jeong et al. 2018).

We believed that good solubility and dispersion of raw materials via different pretreatment could be greatly responsible for the resultant CQDs quality and corresponding PL property, rather than simple thermal treatment without any modification, which could be considered as referable-worth strategy for the efficient overall utilization of biomass resource and further profitable for the large-scale synthesis of biomass-based CQDs.

7.3.3 Food-Based Resources

There are several types of food using as precursors to prepare the CQDs, such as coconut water (Purbia and Paria 2016), garlic (Zhao et al. 2015), honey (Mahesh et al. 2016) and bee pollens (Zhang et al. 2015). The coconut water (Purbia and Paria 2016) was used to synthesize the mono-dispersed CQDs with a diameter of 1–6 nm by microwave-assisted hydrothermal method within one minute by varying with microwave power and heating time. The as-prepared CQDs with QY of 2.8% were reported (Purbia and Paria 2016).

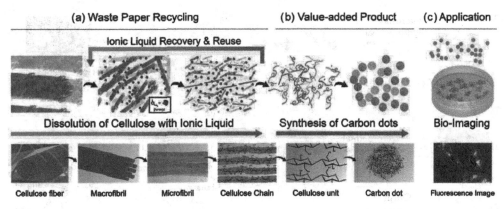

FIGURE 7.63 Schematic illustration. (a) Methods for recycling both waste paper and ILs; (b) the production of carbon dots from regenerated cellulose; (c) application for bioimaging. (Adapted and modified with permission from Jeong et al. 2018. Copyright 2018 American Chemical Society.)

Zhang's group previously conducted a typical top-down strategy, following a hydrothermal treatment of polythiophene (PT2) at 170°C for a period of 24 h to obtain GQDs in large quantities with ultrahigh singlet oxygen generation, which could be served as new generation of carbon-based nanomaterial photodynamic therapy agents (Ge et al. 2014). Considering the unsustainability, potential toxicity and low fluorescent QY of PT2-based GQDs, they further developed the N, S co-doping CQDs by using biocompatible resources garlic intrinsically rich in N and S elements, which could strongly take an effect on the resultant CQDs' photo-physical and chemical properties to some extent. In typical, a facile, green and low-cost hydrothermal method was conducted at 200°C for 3 h using garlic cloves as precursors to obtain fluorescent N,S-dual-doped CQDs with a QY of 17.5%, also showing good biocompatibility and favorable radical scavenging activity against the PT2-based GQDs (Zhao et al. 2015).

The hydrothermal treatment of grass for the preparation of N-doping photoluminescent polymer nanodots was reported for the first time. Typically, the grass was yielded using hydrothermal treatment at 180°C for period of 3 h to get the photoluminescent dots with the highest QY of 6.2%, also obtaining crystalline structure with 0.20 nm lattice spacing. More interestingly, they stated that this type of nanodots barely performs common points with conventional CQDs, but otherwise more like the polymerization of coumarin-like dye structure's formation into dot architecture, which was proved by the carbon-aromatic characteristic peaks' disappearance in the ^{13}C NMR spectrum (Liu et al. 2012). In addition, the honey has been yielded to prepare GQDs with good crystallinity via a cost-effective emulsion-templated carbonization method for the first time. The honey was mainly composed of reducing sugar and sucrose according to the analysis results of honey, which could be actually regarded as carbohydrate resource with good solubility in water. The hydrophilic honey solutions are first dispersed into stabilized emulsion in amphipathic 1-butanol via a form of water/oil (W/O) which could be

hydroxyl functional group that could stabilize the emulsions against coalescence, following a carbonization under inert environment. Considering the emulsion acting as numerous microreactors, it presents great priority for the synthesis of mono-dispersed emulsion, thus beneficial for generating GQDs with narrow size distribution. The as-prepared GQDs with well-resolved crystalline structure own weak PL property with a low value of 3.6% generally because of the aggregation effects (Mahesh et al. 2016).

Despite the increasing attempts of biomass-derived QDs using different type biomass in recent years, it's all the time highly pursued to prepare GQDs with high crystallinity and excellent fluorescent characters via a facile, environmentally friendly method from biomass. However, the easy occurrence of amorphous CQDs or carbon dots could be frequently reported in previous publications about biomass-based QDs. For another, the strategy for QY enhancement still needs much more precisely control, paying more insight of lattice doping rather simple surface or edge passivation (Figure 7.64).

Based on this, Ding's groups (Wang et al. 2018) implemented a hydrothermal treatment of durian intrinsically rich in sulfur element with the assistance of platinum catalyst, to obtain highly fluorescent S-doped CQDs with a ultrahigh QY of 79% higher than most reports (Figure 7.65c). The durian flesh was simply hydrothermal at 150°C for 12 h with the appearance of platinum sheets without other treatment, and the S-GQDs with an average size of approximately 4 nm could be acquired, showing high crystallinity with a lattice parameter of 0.24 nm (Figure 7.65a). Further analysis about the detailed structural features demonstrated that the characteristic thiophene structure (main bonding forms of S atom), corresponding to S-doping in graphic lattice rather than surface or edge passivation, could be proved by controllable sulfur doping in GQDs. Moreover, the S-doping concentrations could severely influence the PL emission, excitation wavelength and QY as shown in Figure 7.65 a (PL photograph from left to right corresponding to different S-doping concentration). To elucidate the detailed reaction mechanism and explore the importance

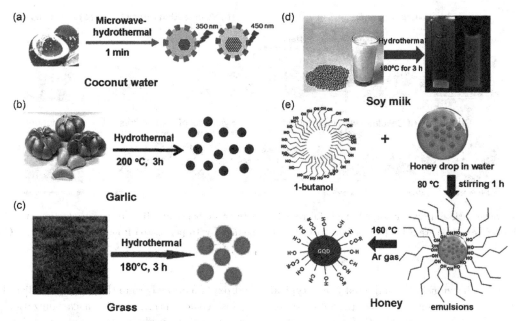

FIGURE 7.64 (a) Schemes of coconut water-based CQDs synthesis. (Reproduced with permission from Purbia and Paria 2016. Copyright 2016 Elsevier.) (b) Illustration of the formation process of CQDs from garlic. (Reproduced with permission from Zhao et al. 2015. Copyright 2015 American Chemical Society.) (c) Schematic presentation of the grass-based CQDs synthesis. (Reproduced with permission from Wang et al. 2016b. Copyright 2016 American Chemical Society.) (d) A schematic illustration of the preparation procedure of CQDs by the hydrothermal treatment of soy milk (photograph of the samples excited by daylight and a 365 UV lamp). (Reproduced with permission from Zhu et al. 2012. Copyright 2012 Royal Society of Chemistry.) (e) Schematic presentation of synthesis of GQDs from honey. (Adapted and modified with permission from Mahesh et al. 2016. Copyright 2016 Wiley-VCH.)

for the platinum, the MALDI-TOFMS was used to analyze the feature of the S-GQDs both with and without platinum. It's suggested that various sp^2-fused heterocyclic species were detected from the thiophene derivatives in the S-GQDs with Pt, directly conforming the occurrence of thiophene and platinum's positive effect on the refusion of heterocyclic structure, which was clearly reflected (Figure 7.65d).

This work successfully prepared the S-GQDs with ultrahigh fluorescent QY, and multiple-colored GQDs with high broadband light emission from 420 to 620 nm could be tuned by different doping concentration at the same time. More importantly, an investigation of mechanism was conducted to explore the actual as-fused structure in S-GQDs by MALDI-TOFMS, helpful for a deep understanding of the reaction for biomass-based GQDs synthesis. However, we could notice that this work could be relatively expensive owing to the utilization of noble metal, platinum and low production yield of 6.8 wt %. Besides, this work only presents a comparison experiment by platinum existence, proving the key factor of platinum for the high-quality GQDs preparation; however, the detailed reaction during the hydrothermal still remains a puzzle urgently needed to be explained.

Finally, from this work discussed above, we suggest that more efforts could be paid to the heterocyclic compounds, thiophene, which could be related to ultrahigh QY. Similar to citric acid-based fluorophore molecule producing ultrahigh QY, the thiophene structure existed in S-GQDs

perhaps owns possibility contributing to the ultrahigh QY, which needed to be further investigated.

7.4 Conclusions

This chapter extensively summarizes GQDs' synthesis methods based on the "top-down" and "bottom-up" approaches, highlight the ongoing biomass materials strategies with low expenses of large-scale GQDs synthesis and also review recent emerging novel synthetic methods beneficial for high-quality and high-performance GQDs preparation. GQDs synthesis in large scale was systematically discussed in this chapter, generally involving top-down and bottom-up strategies. Considering the inevitable weaknesses for both strategies, low quality, defects, uncontrollability for top-down routes, high expense, weak stability and complex synthetic process for bottom-up methods, respectively, the large-scale synthesis of GQDs utilizing a facile, green and low-cost method based on existed or integrated methods discussed above is highly desirable especially with high quality and high QY. It has been pointed that one of the most beneficial methods for the large-scale synthesis of GQDs, namely, the utilization of natural biomass resources (especially those generally considered as biomass waste), which could provide a new window for green, low-cost and large-scale synthesis of high-quality and high-emission CQDs or GQDs. Most importantly, challenges are proposed for CQDs and GQDs' large-scale, green and high-quality synthesis with excellent PL from different visions.

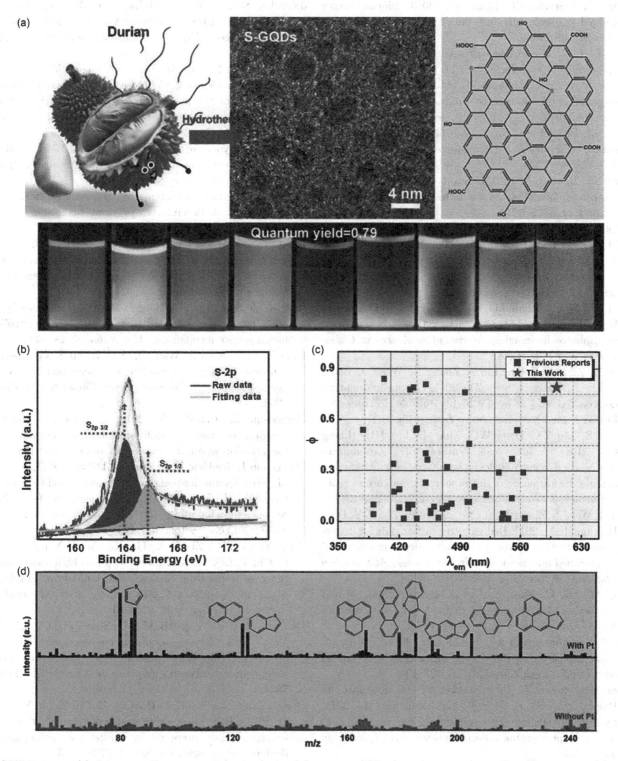

FIGURE 7.65 (a) Schematic illustration of the synthesis of fluorescent CQDs from durian with excellent PL property. (b) S-2p spectrum of the S-GQDs. (c) Summary of φ (QY) and emission wavelengths of S-GQDs obtained in this work and previous reports. (d) MALDI-TOFMS spectra of the reactant solution with (top) and without (bottom) the presence of catalyst (Pt). (Reproduced with permission from Wang et al. 2018. Copyright 2018 American Chemical Society.)

References

Abbas A, Mariana LT, Phan AN. 2018. Biomass-waste derived graphene quantum dots and their applications. *Carbon* 140: 77–99.

Adler E. 1977. Lignin chemistry-past, present and future. *Wood Science and Technology* 11: 169–218.

Albinsson B, Li S, Lundquist K, Stomberg R. 1999. The origin of lignin fluorescence. *Journal of Molecular Structure* 508: 19–27.

Bacon M, Bradley SJ, Nann T. 2014. Graphene quantum dots. *Particle & Particle Systems Characterization* 31: 415–428.

Baker SN, Baker GA. 2010. Luminescent carbon nanodots: emergent nanolights. *Angewandte Chemie International Edition* 49: 6726–6744.

Cailotto S, Mazzaro R, Enrichi F, Vomiero A, Selva M, Cattaruzza E, Cristofori D, Amadio E, Perosa A. 2018. Design of carbon dots for metal-free photoredox catalysis. *ACS Applied Materials & Interfaces*. doi:10.1021/acsami.8b14188

Cao L, Wang X, Meziani MJ, Lu F, Wang H, Luo PG, Lin Y, Harruff BA, Veca LM, Murray D. 2007. Carbon dots for multiphoton bioimaging. *Journal of the American Chemical Society* 129: 11318.

Chao D, Zhu C, Xia X, Liu J, Zhang X, Wang J, Liang P, Lin J, Zhang H, Shen ZX. 2015. Graphene quantum dots coated VO$_2$ arrays for highly durable electrodes for Li and Na ion batteries. *Nano Letters* 15: 565.

Chen BB, Liu ZX, Deng WC, Zhan L, Liu ML, Huang CZ. 2016a. A large-scale synthesis of photoluminescent carbon quantum dots: A self-exothermic reaction driving the formation of the nanocrystalline core at room temperature. *Green Chemistry* 18: 5127–5132.

Chen J, Wei J-S, Zhang P, Niu X-Q, Zhao W, Zhu Z-Y, Ding H, Xiong H-M. 2017. Red-emissive carbon dots for fingerprints detection by spray method: coffee ring effect and unquenched fluorescence in drying process. *ACS Applied Materials & Interfaces* 9: 18429–18433.

Chen W, Hu C, Yang Y, Cui J, Liu Y. 2016b. Rapid synthesis of carbon dots by hydrothermal treatment of lignin. *Materials* 9: 184.

Chu S, Subrahmanyam AV, Huber GW. 2013. The pyrolysis chemistry of a β-O-4 type oligomeric lignin model compound. *Green Chemistry* 15: 125–136.

Chua CK, Sofer Z, P Š, Jankovský O, Klímová K, Bakardjieva S, Hrdličková KŠ, Pumera M. 2015. Synthesis of strongly fluorescent graphene quantum dots by cage-opening buckminsterfullerene. *ACS Nano* 9: 2548–2555.

Chunduri LAA, Kurdekar A, Prathibha C, Kamisetti V. 2017. Single step synthesis of carbon quantum dots from coconut shell: evaluation for antioxidant efficacy and hemotoxicity. *Journal of Materials Sciences and Applications* 3: 83–93.

Ciftci ON, Przybylski R, Rudzinska M, Acharya S. 2011. Characterization of fenugreek (*Trigonella foenum-graecum*) seed lipids. *Journal of the American Oil Chemists Society* 88: 1603–1610.

da Silva Souza DR, Caminhas LD, de Mesquita JP, Pereira FV. 2018. Luminescent carbon dots obtained from cellulose. *Materials Chemistry and Physics* 203: 148–155.

da Silva Souza DR, de Mesquita JP, Lago RM, Caminhas LD, Pereira FV. 2016. Cellulose nanocrystals: A versatile precursor for the preparation of different carbon structures and luminescent carbon dots. *Industrial Crops and Products* 93: 121–128.

Das R, Bandyopadhyay R, Pramanik P. 2018. Carbon quantum dots from natural resource: A review. *Materials Today Chemistry* 8: 96–109.

de Rodriguez NLG, Thielemans W, Dufresne A. 2006. Sisal cellulose whiskers reinforced polyvinyl acetate nanocomposites. *Cellulose* 13: 261–270.

Deng J, Lu Q, Li H, Zhang Y, Yao S. 2015. Large scale preparation of graphene quantum dots from graphite oxide in pure water via one-step electrochemical tailoring. *RSC Advances* 5: 29704–29707.

Ding H, Yu S-B, Wei J-S, Xiong H-M. 2015. Full-color light-emitting carbon dots with a surface-state-controlled luminescence mechanism. *ACS Nano* 10: 484–491.

Ding Z, Li F, Wen J, Wang X, Sun R. 2018. Gram-scale synthesis of single-crystalline graphene quantum dots derived from lignin biomass. *Green Chemistry* 20: 1383–1390.

Donaldson L, Radotic K. 2013. Fluorescence lifetime imaging of lignin autofluorescence in normal and compression wood. *Journal of Microscopy* 251: 178–187.

Donaldson L, Radotic K, Kalauzi A, Djikanović D, Jeremić M. 2010. Quantification of compression wood severity in tracheids of Pinus radiata D. Don using confocal fluorescence imaging and spectral deconvolution. *Journal of Structural Biology* 169: 106–115.

Dong Y, Chen C, Zheng X, Gao L, Cui Z, Yang H, Guo C, Chi Y, Li CM. 2012b. One-step and high yield simultaneous preparation of single-and multi-layer graphene quantum dots from CX-72 carbon black. *Journal of Materials Chemistry* 22: 8764–8766.

Dong Y, Pang H, Yang HB, Guo C, Shao J, Chi Y, Li CM, Yu T. 2013. Carbon-based dots co-doped with nitrogen and sulfur for high quantum yield and excitation-independent emission. *Angewandte Chemie* 52: 7800–7804.

Dong Y, Shao J, Chen C, Li H, Wang R, Chi Y, Lin X, Chen G. 2012a. Blue luminescent graphene quantum dots and graphene oxide prepared by tuning the carbonization degree of citric acid. *Carbon* 50: 4738–4743.

Eda G, Lin YY, Mattevi C, Yamaguchi H, Chen HA, Chen IS, Chen CW, Chhowalla M. 2010. Blue photoluminescence from chemically derived graphene oxide. *Advanced Materials* 22: 505–509.

Ehrat F, Bhattacharyya S, Schneider J, Löf A, Wyrwich R, Rogach AL, Stolarczyk JK, Urban AS, Feldmann J. 2017. Tracking the source of carbon dot photoluminescence: Aromatic domains versus molecular fluorophores. *Nano Letters* 17: 7710–7716.

Essner JB, Laber CH, Ravula S, Polo-Parada L, Baker GA. 2016. Pee-dots: Biocompatible fluorescent carbon dots derived from the upcycling of urine. *Green Chemistry* 18: 243–250.

Fan L, Zhu M, Lee X, Zhang R, Wang K, Wei J, Zhong M, Wu D, Zhu H. 2013. Direct synthesis of graphene quantum dots by chemical vapor deposition. *Particle & Particle Systems Characterization* 30: 764–769.

Fei H, Ye R, Ye G, Gong Y, Peng Z, Fan X, Samuel ELG, Ajayan PM. Tour, JM. 2014. Boron-and nitrogen-doped graphene quantum dots/graphene hybrid nanoplatelets as efficient electrocatalysts for oxygen reduction. *ACS Nano* 8: 10837–10843.

Favaro M, Ferrighi L, Fazio G, Colazzo L, Di Valentin C, Durante C, Sedona F, Gennaro A, Agnoli S, Granozzi G. 2014. Single and multiple doping in graphene quantum dots: Unraveling the origin of selectivity in the oxygen reduction reaction. *ACS Catalysis* 5: 129–144.

Fernando KS, Sahu S, Liu Y, Lewis WK, Guliants EA, Jafariyan A, Wang P, Bunker CE, Sun Y-P. 2015. Carbon quantum dots and applications in photocatalytic energy conversion. *ACS Applied Materials & Interfaces* 7: 8363–8376.

Gao T, Wang X, Yang L-Y, He H, Ba X-X, Zhao J, Jiang F-L, Liu Y. 2017. Red, yellow, and blue luminescence by graphene quantum dots: Syntheses, mechanism, and cellular imaging. *ACS Applied Materials & Interfaces* 9: 24846–24856.

Ge J, Lan M, Zhou B, Liu W, Guo L, Wang H, Jia Q, Niu G, Huang X, Zhou H. 2014. A graphene quantum dot photodynamic therapy agent with high singlet oxygen generation. *Nature Communications* 5: 4596.

Georgakilas V, Perman JA, Tucek J, Zboril R. 2015. Broad family of carbon nanoallotropes: classification, chemistry, and applications of fullerenes, carbon dots, nanotubes, graphene, nanodiamonds, and combined superstructures. *Chemical Reviews* 115: 4744–4822.

Gokhale R, Singh P. 2014. Blue luminescent graphene quantum dots by photochemical stitching of small aromatic molecules: Fluorescent nanoprobes in cellular imaging. *Particle & Particle Systems Characterization* 31: 433–438.

Gokus T, Nair R, Bonetti A, Bohmler M, Lombardo A, Novoselov K, Geim A, Ferrari AC, Hartschuh A. 2009. Making graphene luminescent by oxygen plasma treatment. *ACS Nano* 3: 3963–3968.

Hai X, Mao QX, Wang WJ, Wang XF, Chen XW, Wang JH. 2015. An acid-free microwave approach to prepare highly luminescent boron-doped graphene quantum dots for cell imaging. *Journal of Materials Chemistry B* 3: 9109–9114.

Haque E, Kim J, Malgras V, Reddy KR, Ward AC, You J, Bando Y, Hossain MSA, Yamauchi Y. 2018. Recent advances in graphene quantum dots: Synthesis,

properties, and applications. *Small Methods*: 1800050. doi:10.1002/smtd.201800050.

He T, Niu N, Chen Z, Li S, Liu S, Li J. 2018. Novel quercetin aggregation-induced emission luminogen (AIEgen) with excited-state intramolecular proton transfer for in vivo bioimaging. *Advanced Functional Materials* 28: 1706196. doi:10.1002/adfm.201706196.

Hou J, Wang W, Zhou T, Wang B, Li H, Ding L. 2016. Synthesis and formation mechanistic investigation of nitrogen-doped carbon dots with high quantum yields and yellowish-green fluorescence. *Nanoscale* 8: 11185–11193.

Hu S, Liu J, Yang J, Wang Y, Cao S. 2011. Laser synthesis and size tailor of carbon quantum dots. *Journal of Nanoparticle Research* 13: 7247–7252.

Hutton GAM, Martindale BCM, Reisner E. 2017. Carbon dots as photosensitisers for solar-driven catalysis. *Chemical Society Reviews* 46: 6111–6123.

Jeong Y, Moon K, Jeong S, Koh W-G, Lee K. 2018. Converting waste papers to fluorescent carbon dots in the recycling process without loss of ionic liquids and bioimaging applications. *ACS Sustainable Chemistry & Engineering* 6: 4510–4515.

Jiang C, Wu H, Song X, Ma X, Wang J, Tan M. 2014. Presence of photoluminescent carbon dots in Nescafe® original instant coffee: Applications to bioimaging. *Talanta* 127: 68–74.

Jiang K, Sun S, Zhang L, Lu Y, Wu A, Cai C, Lin H. 2015b. Red, green, and blue luminescence by carbon dots: Full-color emission tuning and multicolor cellular imaging. *Angewandte Chemie International Edition* 54: 5360–5363.

Jiang K, Sun S, Zhang L, Wang Y, Cai C, Lin H. 2015a. Bright-yellow-emissive N-doped carbon dots: preparation, cellular imaging, and bifunctional sensing. *ACS Applied Materials & Interfaces* 7: 23231–23238.

Jung YS, Ross CA. 2009. Solvent-vapor-induced tunability of self-assembled block copolymer patterns. *Advanced Materials* 21: 2540–2545.

Kasprzyk W, Bednarz S, Bogda D. 2013. Luminescence phenomena of biodegradable photoluminescent poly (diol citrates). *Chemical Communications* 49: 6445–6447.

Kasprzyk W, Świergosz T, Bednarz S, Walas K, Bashmakova NV, Bogdał D. 2018. Luminescence phenomena of carbon dots derived from citric acid and urea – A molecular insight. *Nanoscale* 10: 13889–13894.

Ke CC, Yang YC, Tseng WL. 2016. Synthesis of blue-, green-, yellow-, and red-emitting graphene-quantum-dot-based nanomaterials with excitation-independent emission. *Particle & Particle Systems Characterization* 33: 132–139.

Kim S, Hwang SW, Kim M-K, Shin DY, Shin DH, Kim CO, Yang SB, Park JH, Hwang E, Choi S-H. 2012. Anomalous behaviors of visible luminescence from graphene quantum dots: Interplay between size and shape. *ACS Nano* 6: 8203–8208.

Ko NR, Nafiujjaman M, Cherukula K, Lee SJ, Hong SJ, Lim HN, Park CH, Park IK, Lee YK, Kwon IK. 2018. Microwave-assisted synthesis of biocompatible silk fibroin-based carbon quantum dots. *Particle & Particle Systems Characterization* 35: 1700300.

Krysmann MJ, Kelarakis A, Dallas P, Giannelis EP. 2011. Formation mechanism of carbogenic nanoparticles with dual photoluminescence emission. *Journal of the American Chemical Society* 134: 747–750.

Kumar R, Tiwari RS, Srivastava ON. 2011. Scalable synthesis of aligned carbon nanotubes bundles using green natural precursor: Neem oil. *Nanoscale Research Letters* 6: 92.

Kumar VB, Borenstein A, Markovsky B, Aurbach D, Gedanken A, Talianker M, Porat Z. 2016. Activated carbon modified with carbon nanodots as novel electrode material for supercapacitors. *The Journal of Physical Chemistry C* 120: 13406–13413.

Kwon W, Kim Y-H, Lee C-L, Lee M, Choi HC, Lee T-W, Rhee S-W. 2014. Electroluminescence from graphene quantum dots prepared by amidative cutting of tattered graphite. *Nano Letters* 14: 1306–1311.

Lee J, Kim K, Park WI, Kim BH, Park JH, Kim TH, Bong S, Kim CH, Chae G, Jun M. 2012. Uniform graphene quantum dots patterned from self-assembled silica nanodots. *Nano Letters* 12: 6078.

Li LL, Ji J, Fei R, Wang CZ, Lu Q, Zhang JR, Jiang LP, Zhu JJ. 2012a. A facile microwave avenue to electrochemiluminescent two-color graphene quantum dots. *Advanced Functional Materials* 22: 2971–2979.

Li Q, Zhang S, Dai L, Li LS. 2012b. Nitrogen-doped colloidal graphene quantum dots and their size-dependent electrocatalytic activity for the oxygen reduction reaction. *Journal of the American Chemical Society* 134: 18932–18935.

Li W, Liu Y, Wu M, Feng X, Redfern SA, Shang Y, Yong X, Feng T, Wu K, Liu Z. 2018a. Carbon-quantum-dots-loaded ruthenium nanoparticles as an efficient electrocatalyst for hydrogen production in alkaline media. *Advanced Materials*: 1800676. doi:10.1002/adma.201800676.

Li Y, Hu Y, Zhao Y, Shi G, Deng L, Hou Y, Qu L. 2011a. An electrochemical avenue to green-luminescent graphene quantum dots as potential electron-acceptors for photovoltaics. *Advanced Materials* 23: 776–780.

Li Y, Ren J, Sun R, Wang X. 2018b. Fluorescent lignin carbon dots for reversible responses to high-valence metal ions and its bioapplications. *Journal of Biomedical Nanotechnology* 14: 1543–1555.

Li Y, Shi Y, Zhu G, Wu Q, Li H, Wang X, Wang Q, Wang Y. 2014. A single-component white-emitting $CaSr_2Al_2O_6$: Ce^{3+}, Li^+, Mn^{2+} phosphor via energy transfer. *Inorganic Chemistry* 53:7668–7675.

Li Y, Zhao Y, Cheng H, Hu Y, Shi G, Dai L, Qu L. 2011b. Nitrogen-doped graphene quantum dots with oxygen-rich functional groups. *Journal of the American Chemical Society* 134: 15–18.

Li Z, Cao L, Qin P, Liu X, Chen Z, Wang L, Pan D, Wu M. 2018c. Nitrogen and oxygen co-doped graphene quantum dots with high capacitance performance for micro-supercapacitors. *Carbon* 139: 67–75.

Liang Z, Zeng L, Cao X, Wang Q, Wang X, Sun R. 2014. Sustainable carbon quantum dots from forestry and agricultural biomass with amplified photoluminescence by simple NH_4OH passivation. *Journal of Materials Chemistry C* 2: 9760–9766.

Lim SY, Shen W, Gao Z. 2015. Carbon quantum dots and their applications. *Chemical Society Reviews* 44: 362–381.

Liu F, Jang MH, Ha HD, Kim JH, Cho YH, Seo TS. 2013. Facile synthetic method for pristine graphene quantum dots and graphene oxide quantum dots: Origin of blue and green luminescence. *Advanced Materials* 25: 3657–3662.

Liu H, Ye T, Mao C. 2007. Fluorescent carbon nanoparticles derived from candle soot. *Angewandte Chemie International Edition* 119, 6593–6595.

Liu ML, Yang L, Li RS, Chen BB, Liu H, Huang CZ. 2017. Large-scale simultaneous synthesis of highly photoluminescent green amorphous carbon nanodots and yellow crystalline graphene quantum dots at room temperature. *Green Chemistry* 19: 3611–3617.

Liu R, Gao M, Zhang J, Li Z, Chen J, Liu P, Wu D. 2015a. An ionic liquid promoted microwave-hydrothermal route towards highly photoluminescent carbon dots for sensitive and selective detection of iron (III). *RSC Advances* 5: 24205–24209.

Liu R, Wu D, Feng X, Mllen K. 2011. Bottom-up fabrication of photoluminescent graphene quantum dots with uniform morphology. *Journal of the American Chemical Society* 133: 15221–15223.

Liu S, Tian J, Wang L, Zhang Y, Qin X, Luo Y, Asiri AM, Al-Youbi AO, Sun X. 2012. Hydrothermal treatment of grass: A low-cost, green route to nitrogen-doped, carbon-rich, photoluminescent polymer nanodots as an effective fluorescent sensing platform for label-free detection of Cu (II) Ions. *Advanced Materials* 24: 2037–2041.

Liu Y, Chao D, Zhou L, Li Y, Deng R, Zhang H. 2018. Yellow emissive carbon dots with quantum yield up to 68.6% from the solvent and manganese ions. *Carbon* 135: 253–259.

Liu Y, Gao B, Qiao Z, Hu Y, Zheng W, Zhang L, Zhou Y, Ji G, Yang G. 2015b. Gram-scale synthesis of graphene quantum dots from single carbon atoms growth via energetic material deflagration. *Chemistry of Materials* 27: 4319–4327.

Lu J, Yeo PSE, Gan CK, Wu P, Loh KP. 2011. Transforming C_{60} molecules into graphene quantum dots. *Nature Nanotechnology* 6: 247–252.

Lu Q, Wu C, Liu D, Wang H, Su W, Li H, Zhang Y, Yao S. 2017. A facile and simple method for synthesis of graphene oxide quantum dots from black carbon. *Green Chemistry* 19: 900–904.

Luo Z, Qi G, Chen K, Zou M, Yuwen L, Zhang X, Huang W, Wang L. 2016. Microwave-assisted preparation of white fluorescent graphene quantum dots as a novel phosphor for enhanced white-light-emitting diodes. *Advanced Functional Materials* 26: 2739–2744.

Luo Z, Yang D, Qi G, Shang J, Yang H, Wang Y, Yuwen L, Yu T, Huang W, Wang L. 2014. Microwave-assisted solvothermal preparation of nitrogen and sulfur co-doped reduced graphene oxide and graphene quantum dots hybrids for highly efficient oxygen reduction. *Journal of Materials Chemistry A* 2: 20605–20611.

Luo Z, Yang D, Yang C, Wu X, Hu Y, Zhang Y, Yuwen L, Yeow EKL, Weng L, Huang W. 2018. Graphene quantum dots modified with adenine for efficient two-photon bioimaging and white light-activated antibacteria. *Applied Surface Science* 434: 155–162.

Mahesh S, Lekshmi CL, Renuka KD, Joseph K. 2016. Simple and cost-effective synthesis of fluorescent graphene quantum dots from honey: Application as stable security ink and white-light emission. *Particle & Particle Systems Characterization* 33: 70–74.

Medintz IL, Uyeda HT, Goldman ER, Mattoussi H. 2005. Quantum dot bioconjugates for imaging, labelling and sensing. *Nature Materials* 4: 435.

Miao X, Qu D, Yang D, Nie B, Zhao Y, Fan H, Sun Z. 2018. Synthesis of carbon dots with multiple color emission by controlled graphitization and surface functionalization. *Advanced Materials*: 1704740. doi: 10.1002/adma.201704740.

Minati L, Torrengo S, Maniglio D, Migliaresi C, Speranza G. 2012. Luminescent graphene quantum dots from oxidized multi-walled carbon nanotubes. *Materials Chemistry & Physics* 137: 12–16.

Ming H, Ma Z, Liu Y, Pan K, Yu H, Wang F, Kang Z. 2012. Large scale electrochemical synthesis of high quality carbon nanodots and their photocatalytic property. *Dalton Transactions* 41: 9526–9531.

Moon J, An J, Sim U, Cho SP, Kang JH, Chung C, Seo JH, Lee J, Nam KT, Hong BH. 2014. One-step synthesis of N-doped graphene quantum sheets from monolayer graphene by nitrogen plasma. *Advanced Materials* 26: 3501–3505.

Mueller ML, Yan X, Dragnea B, Li L-S. 2010. Slow hot-carrier relaxation in colloidal graphene quantum dots. *Nano Letters* 11: 56–60.

Muramatsu H, Takahashi M, Kang C-S, Kim JH, Kim YA, Hayashi T. 2018. Synthesis of outer tube-selectively nitrogen-doped double-walled carbon nanotubes by nitrogen plasma treatment. *Nanoscale* 10: 15938–15942.

Myint AA, Lee H-W, Seo B, Son W-S, Yoon J, Yoon T-J, Park H-J, Yu J, Yoon J, Lee Y-Woo.2016. One pot synthesis of environmentally friendly lignin nanoparticles with compressed liquid carbon dioxide as an antisolvent. *Green Chemistry* 18: 2129–2146.

Myint AA, Rhim W-K, Nam J-M, Kim J, Lee Y-W. 2018. Water-soluble, lignin-derived carbon dots with high fluorescent emissions and their applications in bioimaging. *Journal of Industrial and Engineering Chemistry* 66: 387–395.

Nanayakkara S, Patti AF, Saito K. 2014. Chemical depolymerization of lignin involving the redistribution mechanism with phenols and repolymerization of depolymerized products. *Green Chemistry* 16: 1897–1903.

Narayanan DP, Cherikallinmel SK, Sankaran S, Narayanan BN. 2018. Functionalized carbon dot adorned coconut shell char derived green catalysts for the rapid synthesis of amidoalkyl naphthols. *Journal of Colloid and Interface Science* 520: 70–80.

Niu N, Ma Z, He F, Li S, Li J, Liu S-X, Yang P. 2017. Preparation of carbon dots for cellular imaging by molecular aggregation of cellulolytic enzyme lignin. *Langmuir* 33: 5786–5795.

Pan D, Zhang J, Li Z, Wu M. 2010. Hydrothermal route for cutting graphene sheets into blue-luminescent graphene quantum dots. *Advanced Materials* 22: 734–738.

Pan L, Sun S, Zhang A, Jiang K, Zhang L, Dong C, Huang Q, Wu A, Lin H. 2015. Truly fluorescent excitation-dependent carbon dots and their applications in multicolor cellular imaging and multidimensional sensing. *Advanced Materials* 27: 7782–7787.

Park SK, Lee H, Choi MS, Suh DH, Nakhanivej P, Park HS. 2018. Straightforward and controllable synthesis of heteroatom-doped carbon dots and nanoporous carbons for surface-confined energy and chemical storage. *Energy Storage Materials* 12: 331–340.

Park SY, Lee HU, Park ES, Lee SC, Lee J-W, Jeong SW, Kim CH, Lee Y-C, Huh YS, Lee J. 2014. Photoluminescent green carbon nanodots from food-waste-derived sources: Large-scale synthesis, properties, and biomedical applications. *ACS Applied Materials & Interfaces* 6: 3365–3370.

Pastell H, Westermann P, Meyer AS, Tuomainen P, Tenkanen M. 2009. In vitro fermentation of arabinoxylan-derived carbohydrates by bifidobacteria and mixed fecal microbiota. *Journal of Agricultural and Food Chemistry* 57: 8598–8606.

Peng H, Travas-Sejdic J. 2009. Simple aqueous solution route to luminescent carbogenic dots from carbohydrates. *Chemistry of Materials* 21: 5563–5565.

Peng J, Gao W, Gupta BK, Liu Z, Romero-Aburto R, Ge L, Song L, Alemany LB, Zhan X, Gao G. 2012. Graphene quantum dots derived from carbon fibers. *Nano Letters* 12: 844–849.

Ponomarenko L, Schedin F, Katsnelson M, Yang R, Hill E, Novoselov K, Geim A. 2008. Chaotic Dirac billiard in graphene quantum dots. *Science* 320: 356–358.

Prasannan A, Imae T. 2013. One-pot synthesis of fluorescent carbon dots from orange waste peels. *Industrial & Engineering Chemistry Research* 52: 15673–15678.

Purbia R, Paria S. 2016. A simple turn on fluorescent sensor for the selective detection of thiamine using coconut water derived luminescent carbon dots. *Biosensors and Bioelectronics* 79: 467–475.

Qiao Z, Wang Y, Gao Y, Li H, Dai T, Liu Y, Huo, Q. 2010. Commercially activated carbon as the source for producing multicolor photoluminescent carbon dots by chemical oxidation. *Chemical Communications* 46: 8812–8814.

Rai S, Singh BK, Bhartiya P, Singh A, Kumar H, Dutta P, Mehrotra G. 2017. Lignin derived reduced fluorescence carbon dots with theranostic approaches: Nano-drug-carrier and bioimaging. *Journal of Luminescence* 190: 492–503.

Ray SC. 2009. Fluorescent carbon nanoparticles: Synthesis, characterization, and bioimaging application. *Journal of Physical Chemistry C* 113: 18546–18551.

Ren J, Sun J, Sun X, Song R, Xie Z, Zhou S. 2018. Precisely controlled up/down-conversion liquid and solid state photoluminescence of carbon dots. *Advanced Optical Materials* 6: 1800115.

Roy P, Periasamy AP, Chuang C, Liou Y-R, Chen Y-F, Joly J, Liang C-T, Chang H-T. 2014. Plant leaf-derived graphene quantum dots and applications for white LEDs. *New Journal of Chemistry* 38: 4946–4951.

Sahu S, Behera B, Maiti TK, Mohapatra S. 2012. Simple one-step synthesis of highly luminescent carbon dots from orange juice: Application as excellent bio-imaging agents. *Chemical Communications* 48: 8835–8837.

Sarkar S, Sudolská M, Dubecký M, Reckmeier CJ, Rogach AL, Zboil R, Otyepka M. 2016. Graphitic nitrogen doping in carbon dots causes red-shifted absorption. *Journal of Physical Chemistry C* 120: 1303–1308.

Saxena M, Sarkar S. 2012. Synthesis of carbogenic nanosphere from peanut skin. *Diamond and Related Materials* 24: 11–14.

Schneider J, Reckmeier CJ, Xiong Y, von Seckendorff M, Susha AS, Kasák P, Rogach AL. 2017. Molecular fluorescence in citric acid-based carbon dots. *The Journal of Physical Chemistry C* 121: 2014–2022.

Sekiya R, Uemura Y, Murakami H, Haino T. 2014. White-light-emitting edge-functionalized graphene quantum dots. *Angewandte Chemie International Edition* 53: 5619–5623.

Sharma A, Gadly T, Neogy S, Ghosh SK, Kumbhakar M. 2017. Molecular origin and self-assembly of fluorescent carbon nanodots in polar solvents. *The Journal of Physical Chemistry Letters* 8: 1044–1052.

Shen J, Zhu Y, Chen C, Yang X, Li C. 2011. Facile preparation and upconversion luminescence of graphene quantum dots. *Chemical Communication* 47: 2580–2582.

Shen J, Zhu Y, Yang X, Zong J, Zhang J, Li C. 2012. One-pot hydrothermal synthesis of graphene quantum dots surface-passivated by polyethylene glycol and their photoelectric conversion under near-infrared light. *New Journal of Chemistry* 36: 97–101.

Shi J, Pattathil S, Parthasarathi R, Anderson NA, Im Kim J, Venketachalam S, Hahn MG, Chapple C, Simmons BA, Singh S. 2016. Impact of engineered lignin composition on biomass recalcitrance and ionic

liquid pretreatment efficiency. *Green Chemistry* 18: 4884–4895.

Shin Y, Lee J, Yang J, Park J, Lee K, Kim S, Park Y, Lee H. 2014. Mass production of graphene quantum dots by one-pot synthesis directly from graphite in high yield. *Small* 10: 866–870.

Shinde DB, Pillai VK. 2012. Electrochemical preparation of luminescent graphene quantum dots from multiwalled carbon nanotubes. *Chemistry-A European Journal* 18: 12522–12528.

Si M, Zhang J, He Y, Yang Z, Yan X, Liu M, Zhuo S, Wang S, Min X, Gao C. 2018. Synchronous and rapid preparation of lignin nanoparticles and carbon quantum dots from natural lignocellulose. *Green Chemistry* 20: 3414–3419.

Sk MA, Ananthanarayanan A, Huang L, Lim KH, Chen P. 2014. Revealing the tunable photoluminescence properties of graphene quantum dots. *Journal of Materials Chemistry C* 2: 6954–6960.

Song Y, Zhu S, Zhang S, Fu Y, Wang L, Zhao X, Yang B. 2015. Investigation from chemical structure to photoluminescent mechanism: A type of carbon dots from the pyrolysis of citric acid and an amine. *Journal of Materials Chemistry C* 3: 5976–5984.

Sun J, Yang S, Wang Z, Shen H, Xu T, Sun L, Li H, Chen W, Jiang X, Ding G. 2015a. Ultra-high quantum yield of graphene quantum dots: Aromatic-nitrogen doping and photoluminescence mechanism. *Particle & Particle Systems Characterization* 32: 434–440.

Sun S, Zhang L, Jiang K, Wu A, Lin H. 2016. Toward high-efficient red emissive carbon dots: Facile preparation, unique properties, and applications as multifunctional theranostic agents. *Chemistry of Materials* 28: 8659–8668.

Sun X, Tuo J, Yang W, Yang D. 2015b. Facile synthesis of photoluminescent carbon quantum dots from biomass. *Nano Reports* 2: 51–54.

Sun Y-P, Zhou B, Lin Y, Wang W, Fernando KS, Pathak P, Meziani MJ, Harruff BA, Wang X, Wang H. 2006. Quantum-sized carbon dots for bright and colorful photoluminescence. *Journal of the American Chemical Society* 128: 7756–7757.

Sun Z, Fridrich B, Santi AD, Elangovan S, Barta K. 2018. Bright side of lignin depolymerization: Toward new platform chemicals. *Chemical Reviews* 118: 614–678.

Suryawanshi A, Biswal M, Mhamane D, Gokhale R, Patil S, Guin D, Ogale S. 2014. Large scale synthesis of graphene quantum dots (GQDs) from waste biomass and their use as an efficient and selective photoluminescence on-off-on probe for Ag[+] ions. *Nanoscale* 6: 11664–11670.

Suvarnaphaet P, Tiwary CS, Wetcharungsri J, Porntheeraphat S, Hoonsawat R, Ajayan PM, Tang IM, Asanithi P. 2016. Blue photoluminescent carbon nanodots from limeade. *Materials Science & Engineering C* 69: 914–921.

Tan X, Li Y, Li X, Zhou S, Fan L, Yang S. 2015. Electrochemical synthesis of small-sized red fluorescent graphene quantum dots as a bioimaging platform. *Chemical Communications* 51: 2544–2546.

Tan XW, Romainor ANB, Chin SF, Ng SM. 2014. Carbon dots production via pyrolysis of sago waste as potential probe for metal ions sensing. *Journal of Analytical and Applied Pyrolysis* 105: 157–165.

Tang L, Ji R, Cao X, Lin J, Jiang H, Li X, Teng KS, Luk CM, Zeng S, Hao J. 2012. Deep ultraviolet photoluminescence of water-soluble self-passivated graphene quantum dots. *ACS Nano* 6: 5102–5110.

Temerov F, Beliaev A, Ankudze B, Pakkanen TT. 2018. Preparation and photoluminescence properties of graphene quantum dots by decomposition of graphene-encapsulated metal nanoparticles derived from kraft lignin and transition metal salts. *Journal of Luminescence* 206: 403–411.

Tetsuka H, Asahi R, Nagoya A, Okamoto K, Tajima I, Ohta R, Okamoto A. 2012. Optically tunable amino-functionalized graphene quantum dots. *Advanced Materials* 24: 5333–5338.

Umrao S, Jang MH, Oh JH, Kim G, Sahoo S, Cho YH, Srivastva A, Oh IK. 2015. Microwave bottom-up route for size-tunable and switchable photoluminescent graphene quantum dots using acetylacetone: New platform for enzyme-free detection of hydrogen peroxide. *Carbon* 81: 514–524.

Vedamalai M, Periasamy AP, Wang C-W, Tseng Y-T, Ho L-C, Shih C-C, Chang H-T. 2014. Carbon nanodots prepared from o-phenylenediamine for sensing of Cu^{2+} ions in cells. *Nanoscale* 6: 13119–13125.

Wang G, Guo Q, Chen D, Liu Z, Zheng X, Xu A, Yang S, Ding G. 2018. Facile and highly effective synthesis of controllable lattice sulfur-doped graphene quantum dots via hydrothermal treatment of durian. *ACS Applied Materials & Interfaces* 10: 5750–5759.

Wang L, Wang Y, Xu T, Liao H, Yao C, Liu Y, Li Z, Chen Z, Pan D, Sun L. 2014. Gram-scale synthesis of single-crystalline graphene quantum dots with superior optical properties. *Nature Communications* 5: 5357.

Wang L-J, Cao G, Tu T, Li H-O, Zhou C, Hao X-J, Su Z, Guo G-C, Jiang H-W, Guo G-P. 2010. A graphene quantum dot with a single electron transistor as an integrated charge sensor. *Applied Physics Letters* 97: 262113.

Wang R, Lu K-Q, Tang Z-R, Xu Y-J. 2017a. Recent progress in carbon quantum dots: Synthesis, properties and applications in photocatalysis. *Journal of Materials Chemistry A* 5: 3717–3734.

Wang X, Qu K, Xu B, Ren J, Qu X. 2011. Microwave assisted one-step green synthesis of cell-permeable multicolor photoluminescent carbon dots without surface passivation reagents. *Journal of Materials Chemistry* 21: 2445–2450.

Wang X, Sun G, Li N, Chen P. 2016a. ChemInform abstract: Quantum dots derived from two-dimensional materials and their applications for catalysis and energy. *Chemical Society Reviews* 45: 2239–2262.

Wang Y, Zhu Y, Yu S, Jiang C. 2017b. Fluorescent carbon dots: Rational synthesis, tunable optical properties and analytical applications. *RSC Advances* 7: 40973–40989.

Wang Z, Yu J, Zhang X, Li N, Liu B, Li Y, Wang Y, Wang W, Li Y, Zhang L. 2016b. Large-scale and controllable synthesis of graphene quantum dots from rice husk biomass: A comprehensive utilization strategy. *ACS Applied Materials & Interfaces* 8: 1434–1439.

Wang Z, Yuan F, Li X, Li Y, Zhong H, Fan L, Yang S. 2017c. 53% Efficient red emissive carbon quantum dots for high color rendering and stable warm white-light-emitting diodes. *Advanced Materials* 29: 1702910.

Wei J, Shen J, Zhang X, Guo S, Pan J, Hou X, Zhang H, Wang L, Feng B. 2013. Simple one-step synthesis of water-soluble fluorescent carbon dots derived from paper ash. *RSC Advances* 3: 13119–13122.

Wei J, Zhang X, Sheng Y, Shen J, Huang P, Guo S, Pan J, Liu B, Feng B. 2014. Simple one-step synthesis of water-soluble fluorescent carbon dots from waste paper. *New Journal of Chemistry* 38: 906–909.

Wu J, Ma S, Sun J, Gold JI, Tiwary CS, Kim B, Zhu L, Chopra N, Odeh IN, Vajtai R. 2016b. A metal-free electrocatalyst for carbon dioxide reduction to multi-carbon hydrocarbons and oxygenates. *Nature Communications* 7: 13869.

Wu J, Rodrigues MF, Vajtai R, Ajayan PM. 2016a. Tuning the electrochemical reactivity of boron- and nitrogen-substituted graphene. *Advanced Materials* 28: 6239–6246.

Wu J, Wen C, Zou X, Jimenez J, Sun J, Xia Y, Rodrigues MTF, Vinod S, Zhong J, Chopra N. 2017. Carbon dioxide hydrogenation over a metal-free carbon-based catalyst. *ACS Catalysis* 7: 4497–450.

Wu Q, Li W, Tan J, Wu Y, Liu S. 2015. Hydrothermal carbonization of carboxymethylcellulose: One-pot preparation of conductive carbon microspheres and water-soluble fluorescent carbon nanodots. *Chemical Engineering Journal* 266: 112–120.

Wu Q, Li W, Wu Y, Huang Z, Liu S. 2014. Pentosan-derived water-soluble carbon nano dots with substantial fluorescence: Properties and application as a photosensitizer. *Applied Surface Science* 315: 66–72.

Xu M, Huang Q, Sun R, Wang X. 2016. Simultaneously obtaining fluorescent carbon dots and porous active carbon for supercapacitors from biomass. *RSC Advances* 6: 88674–88682.

Xu X, Ray R, Gu Y, Ploehn HJ, Gearheart L, Raker K, Scrivens WA. 2004. Electrophoretic analysis and purification of fluorescent single-walled carbon nanotube fragments. *Journal of the American Chemical Society* 126: 12736–12737.

Xue M, Zhan Z, Zou M, Zhang L, Zhao S. 2016a. Green synthesis of stable and biocompatible fluorescent carbon dots from peanut shells for multicolor living cell imaging. *New Journal of Chemistry* 40: 1698–1703.

Xue Q, Huang H, Wang L, Chen Z, Wu M, Li Z, Pan D. 2013. Nearly monodisperse graphene quantum dots fabricated by amine-assisted cutting and ultrafiltration. *Nanoscale* 5: 12098–12103.

Xue Y, Qiu X, Wu Y, Qian Y, Zhou M, Deng Y, Li Y. 2016b. Aggregation-induced emission: the origin of lignin fluorescence. *Polymer Chemistry* 7: 3502–3508.

Yan X, Cui X, Li L-S. 2010. Synthesis of large, stable colloidal graphene quantum dots with tunable size. *Journal of the American Chemical Society* 132: 5944–5945.

Yan X, Li B, Li L-S. 2012. Colloidal graphene quantum dots with well-defined structures. *Accounts of Chemical Research* 46: 2254–2262.

Yan Y, Chen J, Li N, Tian J, Li K, Jiang J, Liu J, Tian Q, Chen P. 2018. Systematic bandgap engineering of graphene quantum dots and applications for photocatalytic water splitting and CO_2 reduction. *ACS Nano* 12: 3523–3532.

Yang J, Zhang Y, Gautam S, Liu L, Dey J, Chen W, Mason RP, Serrano CA, Schug KA, Tang L. 2009a. Development of aliphatic biodegradable photoluminescent polymers. *Proceedings of the National Academy of Sciences* 106: 10086–10091.

Yang P, Zhu Z, Chen M, Chen W, Zhou X. 2018. Microwave-assisted synthesis of xylan-derived carbon quantum dots for tetracycline sensing. *Optical Materials* 85: 329–336.

Yang S, Sun J, Li X, Zhou W, Wang Z, He P, Ding G, Xie X, Kang Z, Jiang M. 2014. Large-scale fabrication of heavy doped carbon quantum dots with tunable-photoluminescence and sensitive fluorescence detection. *Journal of Materials Chemistry A* 2: 8660–8667.

Yang S-T, Cao L, Luo PG, Lu F, Wang X, Wang H, Meziani MJ, Liu Y, Qi G, Sun Y-P. 2009b. Carbon dots for optical imaging in vivo. *Journal of the American Chemical Society* 131: 11308–11309.

Yang Y, Cui J, Zheng M, Hu C, Tan S, Xiao Y, Yang Q, Liu Y. 2012. One-step synthesis of amino-functionalized fluorescent carbon nanoparticles by hydrothermal carbonization of chitosan. *Chemical Communications* 48: 380–382.

Yang Z-C, Wang M, Yong AM, Wong SY, Zhang X-H, Tan H, Chang AY, Li X, Wang J. 2011a. Intrinsically fluorescent carbon dots with tunable emission derived from hydrothermal treatment of glucose in the presence of monopotassium phosphate. *Chemical Communications* 47: 11615–11617.

Yang Z-C, Wang M, Yong AM, Wong SY, Zhang XH, Tan H, Chang AY, Li X, Wang J. 2011b. Intrinsically fluorescent carbon dots with tunable emission derived from hydrothermal treatment of glucose in the presence of monopotassium phosphate. *Chemical Communications* 47: 11615–11617.

Yao S, Hu Y, Li G. 2014. A one-step sonoelectrochemical preparation method of pure blue fluorescent carbon nanoparticles under a high intensity electric field. *Carbon* 66: 77–83.

Ye R, Xiang C, Lin J, Peng Z, Huang K, Yan Z, Cook NP, Samuel EL, Hwang CC, Ruan G. 2013. Coal as an abundant source of graphene quantum dots. *Nature Communications* 4: 2943.

Yu D, Zhang Q, Dai L. 2010. Highly efficient metal-free growth of nitrogen-doped single-walled carbon nanotubes on plasma-etched substrates for oxygen reduction. *Journal of the American Chemical Society* 132: 15127–15129.

Yuan F, Wang Z, Li X, Li Y, Tan ZA, Fan L, Yang S. 2016. Bright multicolor bandgap fluorescent carbon quantum dots for electroluminescent light-emitting diodes. *Advanced Materials* 29: 1604436.

Yuan F, Yuan T, Sui L, Wang Z, Xi Z, Li Y, Li X, Fan L, Chen A, Jin M. 2018a. Engineering triangular carbon quantum dots with unprecedented narrow bandwidth emission for multicolored LEDs. *Nature Communications* 9: 2249.

Yuan G, Zhao X, Liang Y, Peng L, Dong H, Xiao Y, Hu C, Hu H, Liu Y, Zheng M. 2018b. Small nitrogen-doped carbon dots as efficient nanoenhancer for boosting the electrochemical performance of three-dimensional graphene. *Journal of Colloid and Interface Science.* doi:10.1016/j.jcis.2018.10.096.

Zakzeski J, Bruijnincx PC, Jongerius AL, Weckhuysen BM. 2010. The catalytic valorization of lignin for the production of renewable chemicals. *Chemical Reviews* 110: 3552–3599.

Zhang J, Yuan Y, Liang G, Yu SH. 2015. Scale-up synthesis of fragrant nitrogen-doped carbon dots from bee pollens for bioimaging and catalysis. *Advanced Science* 2: 1500002.

Zhang X, Jiang M, Niu N, Chen Z, Li S, Liu S, Li J. 2018a. Natural-product-derived carbon dots: From natural products to functional materials. *ChemSusChem* 11: 11–24.

Zhang X, Wang H, Ma C, Niu N, Chen Z, Liu S, Li J, Li S. 2018b. Seeking value from biomass materials: Preparation of coffee bean shell-derived fluorescent carbon dots via molecular aggregation for antioxidation and bioimaging applications. *Materials Chemistry Frontiers* 2: 1269–1275.

Zhang Y-Q, Ma D-K, Zhuang Y, Zhang X, Chen W, Hong L-L, Yan Q-X, Yu K, Huang S-M. 2012. One-pot synthesis of N-doped carbon dots with tunable luminescence properties. *Journal of Materials Chemistry* 22: 16714–16718.

Zhao S, Lan M, Zhu X, Xue H, Ng T-W, Meng X, Lee C-S, Wang P, Zhang W. 2015. Green synthesis of bifunctional fluorescent carbon dots from garlic for cellular imaging and free radical scavenging. *ACS Applied Materials & Interfaces* 7: 17054–17060.

Zhao X, Li M, Dong H, Liu Y, Hu H, Cai Y, Liang Y, Xiao Y, Zheng M. 2017. Interconnected 3 D network of graphene-oxide nanosheets decorated with carbon dots for high-performance supercapacitors. *ChemSusChem* 10: 2626–2634.

Zhi C, Bai X, Wang E. 2002. Enhanced field emission from carbon nanotubes by hydrogen plasma treatment. *Applied Physics Letters* 81: 1690–1692.

Zhou J, Sheng Z, Han H, Zou M, Li C. 2012a. Facile synthesis of fluorescent carbon dots using watermelon peel as a carbon source. *Materials Letters* 66: 222–224.

Zhou J, Booker C, Li R, Zhou X, Sham T-K, Sun X, Ding Z. 2007. An electrochemical avenue to blue luminescent nanocrystals from multiwalled carbon nanotubes (MWCNTs). *Journal of the American Chemical Society* 129: 744–745.

Zhou X, Zhang Y, Wang C, Wu X, Yang Y, Zheng B, Wu H, Guo S, Zhang J. 2012b. Photo-Fenton reaction of graphene oxide: A new strategy to prepare graphene quantum dots for DNA cleavage. *ACS Nano* 6: 6592–6599.

Zhu C, Zhai J, Dong S. 2012. Bifunctional fluorescent carbon nanodots: Green synthesis via soy milk and application as metal-free electrocatalysts for oxygen reduction. *Chemical Communications* 48: 9367–9369.

Zhu C, Yang S, Wang G, Mo R, He P, Sun J, Di Z, Kang Z, Yuan N, Ding J. 2015a. A new mild, clean and highly efficient method for the preparation of graphene quantum dots without by-products. *Journal of Materials Chemistry B* 3: 6871–6876.

Zhu S, Meng Q, Wang L, Zhang J, Song Y, Jin H, Zhang K, Sun H, Wang H, Yang B. 2013. Highly photoluminescent carbon dots for multicolor patterning, sensors, and bioimaging. *Angewandte Chemie* 125: 4045–4049.

Zhu S, Song Y, Wang J, Wan H, Zhang Y, Ning Y, Yang B. 2017. Photoluminescence mechanism in graphene quantum dots: Quantum confinement effect and surface/edge state. *Nano Today* 13: 10–14.

Zhu S, Song Y, Zhao X, Shao J, Zhang J, Yang B. 2015b. The photoluminescence mechanism in carbon dots (graphene quantum dots, carbon nanodots, and polymer dots): Current state and future perspective. *Nano Research*, 8: 355–381.

Zhu S, Zhang J, Qiao C, Tang S, Li Y, Yuan W, Li B, Tian L, Liu F, Hu R. 2011. Strongly green-photoluminescent graphene quantum dots for bioimaging applications. *Chemical Communications* 47: 6858–6860.

Electrocatalytic Optically Modulated Green Prepared Nanoparticles

Xolile Fuku and Mmalewane Modibedi

Council for Scientific and Industrial Research

8.1 Introduction

8.1.1 Nanotechnology and Nanoscience

Nanoscience has shown itself to be one of the most exciting areas in science, with experimental developments being driven by pressing demands for new technological applications. It is a highly multidisciplinary research field, and the experimental and theoretical challenges for researchers in different fields are substantial. Nowadays, scientists and research scholars have been developing new kinds of nanomaterials which could be used for forensic science, biology, electronic technology, environmental science, computer manufacturing, and sports facility production, as well as food industries. In 21 January 2000 California Institute of Technology, President Bill Clinton advocated nanotechnology development and raised it to the level of a federal initiative, officially referring to it as the National Nanotechnology Initiative (NNI). Nanoscience and nanotechnology are referred to as a type of applied science, studying the ability to observe, measure, manipulate, and manufacture materials at the nanometer scale (Rao et al. 2004). The prefix "nano" in the word nanometer (nm) is an SI unit of length, i.e., 10^{-9} or a distance of one-billionth of a meter (Sah et al. 2015, Qi and Warren 2002 and Rao et al. 2004). As a comparison, a head of a pin is about one million nanometers wide, or it would take about ten hydrogen atoms end-to-end to align in series in order to span the length of one nanometer (Singh and Nalwa 2011, Strappert and Fissan 2003 and Rao at al 2004). Because the matter it deals with is smaller than the macroscopic scale which couldn't be seen by our naked eye, but larger than the microscopic scale of the electrons and protons and that could only be sensed by cloud chambers, it dwells in a new realm called mesoscopic scale which contains the domain of 10^{-7}–10^{-9} nm. In other words, whenever a macroscopic device is scaled down to mesoscopic scale, it starts revealing quantum mechanical properties. While macroscopic scale could be studied by classical mechanics and microscopic scale could be expressed by quantum mechanics, mesoscopic scale is somewhere in between and our knowledge about this field is quite limited. This has stimulated the scientists to start a new territory dealing with the "bridge" which connects the macro and micro, this "bridge" being the so-called nanoscience. These nanotechnology materials are expected to bring about lighter, stronger, smarter, cheaper, cleaner, and more durable products. One of the main reasons why there are a lot more activities in producing nanotechnology products today than before is there are now many new kinds of facilities that can handle nanomaterials, including, but not limited to, transmission electron microscopy (TEM) which could directly see the atoms clusters; and atom force microscopy (AFM) which can measure, see, and manipulate nanometer-sized particles; and nanoimprint lithography (NIL) which is equipped with high-precision alignment system with accuracy within 500 nm and fine alignment up to 50 nm. Physical vapor disposition (PVD) and chemical vapor disposition (CVD) as well as molecular beam epitaxy (MBE) systems allow the scientists to accurately control the ingredients of the nanodevices during the manufacturing process. With more and more nanotechnologies emerging into our lives and the benefits it provides after manufacturing and becoming commercially available, it will also bring some ethical, legal, social, and moral issues. Most of them are not new problems, but because of nanotechnology, their importance and urgency have been emphasized to a new level.

8.1.2　Nanoparticles

Since nanotechnology was presented by Nobel Laureate Richard P. Feynman through his well famous 1959 lecture "There's Plenty of Room at the Bottom", there have been various ground-breaking developments in the field of nanotechnology. Nanotechnology produced materials of various types at nanoscale level, i.e., nanoparticles (NPs). NPs are microscopic and wide class materials that include particulate substances, which have one dimension less than 100 nm at least (Martin 2006 and Hasan 2015). Depending on the overall shape, these materials can be 0D, 1D, 2D, or 3D (Hasan 2015) as shown in Figure 8.1.

NPs are tiny materials having size ranges from 1 to 100 nm. They can be classified into different classes based on their properties, shapes, or sizes. The different groups include fullerenes, metal NPs, ceramic NPs, and polymeric NPs. NPs possess unique physical and chemical properties due to their high surface area and nanoscale size. Their optical properties are reported to be dependent on the size, which imparts different colors due to absorption in the visible region. Their reactivity, toughness, and other properties are also dependent on their unique size, shape, and structure (Martin 2006, Mohanraj and Chen 2006, Hasan 2015). Due to these characteristics,

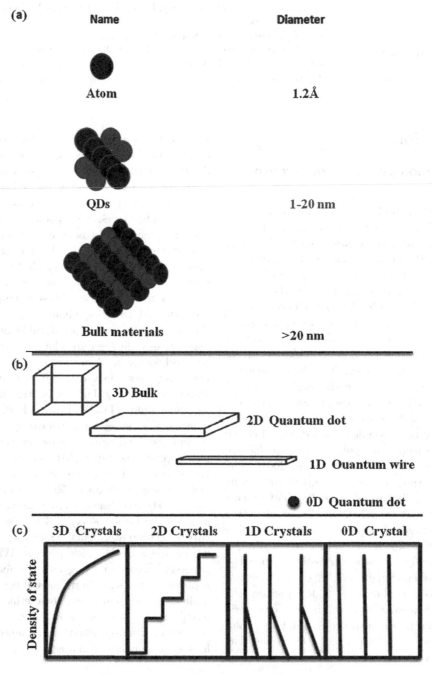

FIGURE 8.1　Schematic representation (a) quantum materials, (b) atoms, and (c) bulk materials.

they are suitable candidates for various commercial and domestic applications, which include catalysis, imaging, medical applications, energy-based research, and environmental applications. NPs are not simple molecules, and therefore, they're composed of three layers: (i) the surface layer, which may be functionalized with a variety of small molecules, metal ions, surfactants, and polymers; (ii) the shell layer, which is chemically different material from the core in all aspects; and (iii) the core, which is essentially the central portion of the NP. Various methods can be employed for the synthesis of NPs, but these methods are broadly divided into two main classes: (i) bottom-up approach (Mohanraj and Chen 2006 and Ranjit and Baquee 2013) and (ii) top-down approach (Roa et al. 2000, Bhadra et al. 2002, Mohanraj and Chen 2006 and Hasan 2015) as shown in Scheme 8.1. These approaches further divide into various subclasses based on the operation, reaction condition, and adopted protocols.

The successful synthesis and various physicochemical properties of NPs can be interrogated using different characterization techniques. These include techniques such as X-ray diffraction (XRD), X-ray photoelectron spectroscopy (XPS), infrared (IR) spectroscopy, scanning electron microscopy (SEM), TEM, UV-visible spectroscopy, photoluminescence, fluorescence, Brunauer–Emmett–Teller (BET), and particle size analysis. Figure 8.2 reveals synthesized NiO/MgO, Cu/Cu$_2$O/CuO/ZnO, and ZnO NPs using few of the latter techniques. Figure 8.2 shows (a) UV-vis, (b) energy diagram (c, e–f) SEM (EDX-insert), and (d) XRD.

UV-vis absorption spectroscopy is an effective technique used to study the particle growth of nanocrystals (Kang 2012) (Figure 8.2a). According to the effective mass model, the radius of the particle is related to the absorption band (Kang 2012). UV-vis characterization gives an idea about the size of the particles under investigation since the band edge depends on the particle size. Absorption peak occurring at lower wavelengths implies a blue-shifted absorption edge resulting from quantum confinement of the excitons present in the sample. The excitons present in the sample cause a more discrete energy spectrum of the individual NPs. If no visible absorption is seen in the visible region, the system can be said to be a non-linear optical (NLO) material. XRD (Figure 8.2c) on the other hand is a non-destructive technique which identifies crystalline phases and orientation, determines structural properties such as lattice parameters (10^{-4} Å), strain, grain size, epitaxy, phase composition, preferred orientation, order–disorder transformation, thermal expansion, to measure thickness of thin films and multi-layers, and determines atomic arrangement. Finally, the SEM is employed to determine particle morphology and shape (Figure 8.2c). Meanwhile, Figure 8.2b shows an energy diagram which distinguishes the effect of temperature and the energy required to move electrons. It is clear from the diagram that at low temperatures, p-type Cu/Cu$_2$O requires more energy for an electron to emit/move from low (LUMO) to higher levels (HOMO) due to high band gap energies 4.3 eV. However, the photon energy required to emit electrons decreased due to the more conducing Cu/Cu$_2$O (2.8 eV) and Cu/Cu$_2$O/CuO/ZnO (3.4 eV) NCs as we go higher in temperatures. Conclusively, it is evident that preparing the NCs at high temperature is a need due to lower photon energy required to shuttle an electron from ground state to the conduction state. Further, the nanostructured materials showed high conductance and absorptive properties, thus confirming their use in different fields such as solar absorber, optoelectronics, catalysis, and many others.

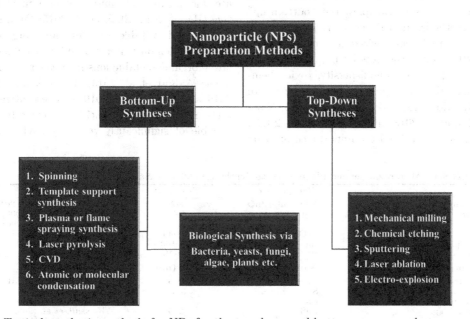

SCHEME 8.1 Typical synthetic methods for NPs for the top-down and bottom-up approaches.

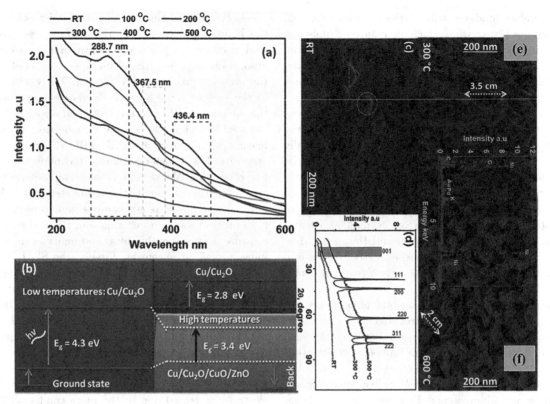

FIGURE 8.2 (a) UV-vis of the green-prepared metal oxides, (b) a Jablonski diagram representing the energy levels for a fluorescent molecule and several important transitions, and (c–f) SEM (EDX-insert) and XRD of the prepared oxide nanoparticles.

8.1.3 Methods of Synthesis—Green Chemistry, Physical, and Chemical Methods

Several synthesis routes to achieve shape, size, crystallinity, dispersity, and magnetic behavior have been developed. Most of the synthesis routes are discussed in Scheme 8.1. As previously explained, techniques for making different NPs can be categorized into two principles: the "bottom-up" and "top-down" methods. In this chapter, the bottom-up method will be largely emphasized (especially biologically prepared NPs) since the top-down methods have many drawbacks, (e.g., poor monodispersity, toxic chemicals, and contamination are quite high, and extensive waste of material) (Mohanraj and Chen et al. 2006; Kang 2012, Ranjit and Baquee 2013 and Sharma et al. 2017). In this account, scientists realized the importance of using

bottom-up technique in this case employing green chemistry as way forward. Green chemistry incorporates a new approach to the synthesis, processing, and application of chemical substances in such a manner as to reduce threats to health and the environment. Recent studies have shown that green biologically based methods using microorganisms and plants to synthesize materials (especially NPs) are safe, inexpensive, and an environment-friendly alternative (Haque 2014, Shah et al. 2015; Fuku et al. 2016b), as summarized in Table 8.1. Both microorganisms and plants have long demonstrated the ability to absorb and accumulate inorganic metallic ions from their surrounding environment (Akhtar et al. 2013, Rajan et al. 2015, Shah et al. 2015 and Fuku et al. 2016b). These attractive properties make many biological entities efficient biological factories capable of significantly reducing environmental pollution

TABLE 8.1 Microorganism and Plants Used to Synthesize Different Nanoparticulates

Plants	Nanoparticles	Size (nm)	References
Pomegranate	NiO, CuO/ZnO	5–20	Fuku et al. (2016b, 2017)
Aloe vera	In_2O_3	5–50	Chandran et al. (2006) and Maensiri et al. (2008)
Moringa oleifera	ZnO, CuO	10–40	Matinise et al. (2018)
Pyrus sp. (pear fruit extract)	Au	20–100	Ghodake et al. (2010)
Curcuma longa	Pd	10–15	Sathishkumar et al. (2009)
Microorganisms	**Nanoparticles**	**Size (nm)**	**References**
Yeast strain MKY3	Ag	2–5	Kowshik et al. (2003)
Saccharomyces cerevisiae	Sb_2O_3	3–10	Jha et al. (2009)
Bacteriophage	Ca	–	Wang et al. (2010)
Torulopsis sp.	PbS	2–5	Kowshik et al. (2002)
Tobacco mosaic	SiO_2, CdS	–	Shenton et al. (1999)

and reclaiming metals from industrial waste. Scheme 8.2 also reveals the proposed mechanism of interaction and different characterization techniques.

The advantage of using plants over other eco-friendly biologically based systems such as bacteria and fungi is that it avoids the use of specific, well-conditioned culture preparation and isolation techniques that are inclined to be expensive and intricate. Conversely, biosynthesis of NPs using plants or plant-based extracts tends to be safe, have relatively short production times, and have a lower cultivation cost. Further, plant extracts contain bioactive alkaloids, phenolic acids, polyphenols, proteins, sugars, and terpenoids which are believed to play a vital role in the reduction and stabilization of metal ions. Because of these interesting advantages and properties, plants have been considered a more environment-friendly route for biologically synthesizing metallic NPs and for detoxification applications. In this regard, Fuku and Matinise (Fuku et al. 2018, 2016a, 2017) have reported several papers on green synthesis of metallic NPs using different plants and fruit extracts. In their papers (Fuku et al. 2018, 2016a and 2016b, 2017), they prepared and eventually developed a mechanism of interaction (Schemes 8.2 and 8.3) with the nanoparticulates in coordination with extracted polyphenols.

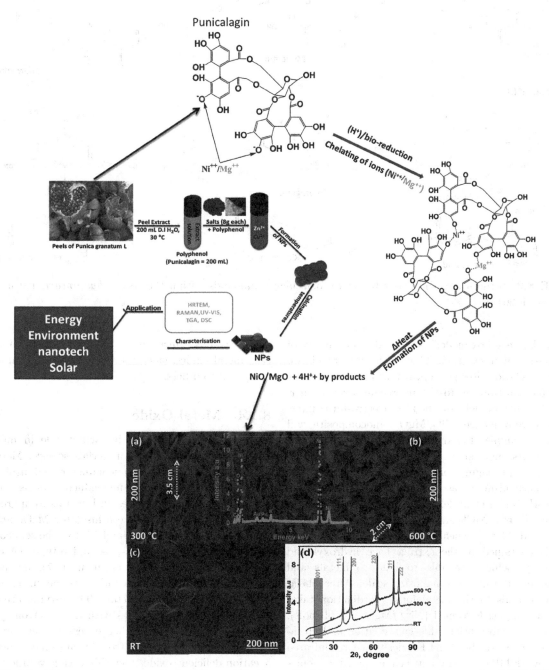

SCHEME 8.2 Schematic representation of the synthetic method/mechanism, (a-c) represents the SEM-EDX (insert) images of the prepared nanoparticles and (d) the corresponding XRD images (Fuku et al. 2016a and 2016b, 2017).

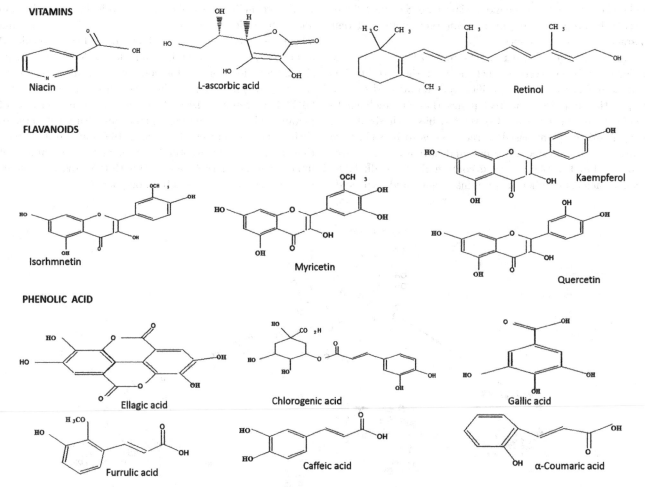

SCHEME 8.3 Structure of the major bioactive molecular compounds within the *Aspalathus linearis* natural extract (Matinise et al. 2018).

The popular "egg-box model" is referred to in the case of polyphenols-metal interaction. The unique feature of two consecutive chelation sites per repeat that provides a favorable entropic contribution to the interchain association is produced by this model governed by electrostatic interactions. For convenience, NiO/MgO nanocomposite will serve as an example. Based on the obtained SEM and HRTEM results, new structural models of Ni^{2+}/Mg^{2+}-polyphenols (punicalagin) complexes were proposed. The proposed mechanism is as explained in our previous research work (Fuku et al. 2018, 2016b, 2017) with minor adjustments. Briefly, $NiCl_2$ and $MgCl_2$ dissociate into metal ion (Ni^{2+} and Mg^{2+}) when in solution. Thus, phenolic compounds contained in the extract have hydroxyl and ketonic groups which are able to bind to metals and reduce the metal salt to a nanosize scale. After chelation of the ligands to the metal ions, coordination takes place and a complex is formed (Ni^{2+}/Mg^{2+}-polyphenols). Furthermore, to substantiate this coordination and chelation, a comparison study of the Fourier-transform infrared spectroscopy (FTIR) spectra between prepared nanomaterials, metal precursors, and the extract was carried out (Figure 8.3a–b).

The results affirm our choice of metal NPs in this case metal oxides, since they have superior characteristics compared to others.

8.1.4 Metal Oxides

Metal oxides play a very important role in many areas of chemistry, physics, and materials science. Metal oxides (MeOs) exhibit metallic, semiconductor, and insulator characteristics. Metallic and semiconductor MeOs are formed by the combination of oxides with metals from groups 3–12 of the periodic table, whereas insulator MeOs are formed from metals in groups 1, 2, and 13–18. The semiconductors are further classified into i-, n-, and p-types (Arora et al. 2016 and Raghunath and Perumal 2017). Intrinsic semiconductors are i-type semiconductors having properties of both insulators and conductors. The n-types are electron-excess semiconductor oxides with free electrons as charge carriers and will have either excess cations or deficient anions. The p-types are electron-deficit semiconductors with a cation-deficient oxide, and this cation vacancy provides the additional electrons for reactivity. At the nanoscale, these compounds can exhibit unique physical and chemical

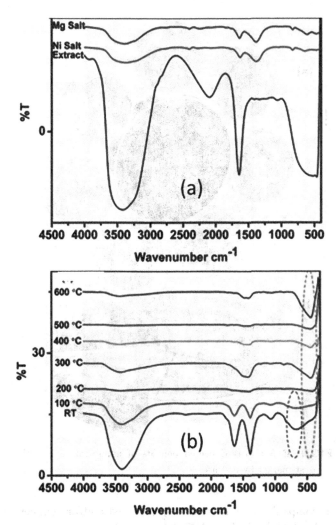

FIGURE 8.3 FTIR spectra of precursor salts (a) and (b) peel extract FTIR spectra of as-prepared NiO/MgO nanocomposite at different annealing temperatures (room temperatures (RT), 100°C, 200°C, 300°C, 400°C, 500°C, and 600°C).

properties due to their limited size and a high density of defect sites such as edges, corners, and point defects, thus offering versatility (Wells 1987, Noguera 1996 and Singh et al. 2016). MeO-NPs are key constituents in catalysis, diagnosis, drug delivery, semiconductors, sensing, and solid oxide fuel cells (Haddad and Seabra 2012 and Corr 2013). In the emerging field of nanotechnology, a goal is to make nanostructures or nanoarrays with special properties with respect to those of bulk or single particle species. In their bulk state, many oxides have wide band gaps and a low reactivity (Roy et al. 2013 and Raghunath and Perumal 2017). A decrease in the average size of an oxide particle does in fact change the magnitude of the band gap (Figure 1.7), with strong influence on the conductivity and (photo) chemical reactivity (Bandyopadhyay et al. 2011, Al-Hazmi et al. 2012 and Rakshit et al. 2013). Surface properties of oxides nanomaterials are of central importance in catalysis. Solid–gas or solid–liquid chemical reactions can be mostly confined to the surface and/or subsurface regions of the solid. As mentioned, the two-dimensional (2D) nature of surfaces has notable structural consequences, typically a rearrangement or reconstruction of bulk geometries, and electronic consequences, such as the presence of mid-gap states, which may act as trapping centers in photocatalysis, whose behavior depends on the relative position of their energy with respect to the valence and conduction band edge position (Rakshit et al. 2013). In the case of nanostructured oxides, surface properties are strongly modified with respect to 2D-infinite surfaces, producing solids with unprecedented sorption and acid/base characteristics, or metal–support interaction/epitaxy (Reddy et al. 2007, Huang et al. 2008, Rakshit et al. 2013, Jesline et al. 2014 and Verdier et al. 2014). Lastly, the presence of under-coordinated atoms or O vacancies in an oxide NP should produce specific geometrical arrangements as well as occupied electronic states located above the valence band (VB) of the corresponding bulk material, enhancing in this way the chemical reactivity of the system (Weller 2014 and Wang et al. 2005). These features confer unique chemical and physical properties to MeO-NPs through which they interact with biological, physical, and chemical systems (Weller 2014, Wang et al. 2005 and Singh et al. 2011). The different properties of MeO-NPs are the major contributors to antimicrobial activity. The alkalinity of the Cu/CuO/Cu$_2$O/ZnO and MgO NP surface is the significant component in conferring antimicrobial activity (Sawai et al. 2005 and Fuku et al. 2017). These alkali MeO-NPs are more soluble owing to their contribution to alkalinity in the medium, which cannot be found with MeO semiconductors such as neutral zinc oxide (ZnO) NPs. Nickel oxide/magnesium oxide (NiO/MgO) NPs are semiconductor photocatalysts that can inhibit the growth of even ultraviolet (UV) radiation-resistant and desiccation tolerant bacteria (Fuku et al. 2018). CuO/ZnO and CuO-CeO$_2$ mixed-metal oxides have important applications as electrodes in fuel cells and gas sensors (Wang et al. 2006) and as efficient catalysts for various reactions, such as the combustion of CO$_2$ reduction, synthesis, and the wet oxidation of phenol (Raghunath and Perumal 2017 and Fuku et al. 2016). Therefore, it is important to stress the need for a fundamental understanding of the properties of nanostructured oxides and their applications.

8.1.5 Catalysis

Catalysis occupies an important place in chemistry, where it develops in three directions, which still present very few overlaps: heterogeneous, homogeneous, and enzymatic. Since the end of the 1990s, and with the development of nanoscience, nanocatalysis has clearly emerged as a domain at the interface between homogeneous and heterogeneous catalysis, which offer unique solutions to answer the demanding conditions for catalyst improvement (Thomas 1988, 2010). The main focus is to develop well-defined catalysts, which may include both metal NPs and a nanomaterial as support. These nanocatalysts should be able to display the ensuing benefits of both homogenous and heterogeneous catalysts, namely, high efficiency and selectivity, stability, and easy recovery/recycling. Specific reactivity can be anticipated due to the nanodimension that can afford specific properties which cannot be achieved with regular, non-nano materials. In this approach, the environmental problems are also considered. Nanocatalysis is a rapidly growing field, which involves the use of nanomaterials as catalysts for a variety of catalytic reactions. Heterogeneous catalysis represents one of the oldest commercial applications of nanoscience; NPs of metals, semiconductors, oxides, and other compounds have been widely used for important chemical reactions. Although surface science studies have contributed significantly to our fundamental understanding of catalysis, most commercial catalysts are still produced by "mixing, shaking, and baking" mixtures of multi-components but their nanoscale structures are not well controlled and the synthesis–structure–performance relationships are poorly understood. Due to their complex physicochemical properties at the nanometer scale, even characterization of the various active sites of most commercial catalysts proves to be elusive. Key objectives of nanocatalysis research should include (i) increased selectivity and activity, low energy consumption, and long lifetime of catalysts by controlling pore size and particle characteristics; (ii) replacement of precious metal catalysts by catalysts tailored at the nanoscale with the use of base metals, thus improving chemical reactivity and reducing process costs; (iii) catalytic membranes by design that can remove unwanted molecules from gases or liquids by controlling the pore size and membrane characteristics. As discussed in the previous section, this can be achieved only by precisely controlling the size, shape, spatial distribution, surface composition and electronic structure, and thermal and chemical stability of the individual nanocomponents. In view of the numerous potential benefits (Figure 8.4a, b) that can grow through their use, nanostructured catalysts have been the subject of considerable academic and industrial research attention in recent times. Catalysts find application in a variety of industry segments spanning from refinery, petrochemical, pharmaceuticals, chemical, food processing sectors, and others (Figure 8.4a). An assessment of the global catalysts' market reveals the following: The global market for NPs used in catalytic applications increased from US $193.74 million in 2006 to an estimated US $200.695 million in 2007. It was expected to reach US $325 million by 2012, corresponding to a growth of 9.5% in the preceding five years, from 2007 onwards. Moreover, between 2013 and 2014, the global market for catalyst and catalyst regeneration reached nearly $23.2 billion and nearly $24.6 billion, respectively. This market is forecasted to grow at a compound annual growth rate (CAGR) of 4.0% to reach $29.9 billion in 2019. Commercially, well-established catalysts such as industrial enzymes, zeolites, and transition metal catalysts, till recently, accounted for about 98% of global sales. It is the smaller end-user segments whose consumption is growing the fastest. Not surprisingly, nanocatalysis is a growing business. The list

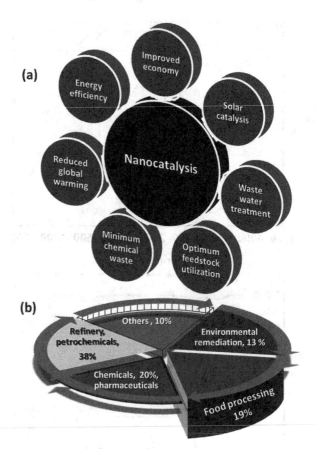

FIGURE 8.4 (a) Expected benefits of nanocatalysis and (b) catalyst market by end use

of companies that have already patented and/or commercialized technologies relating to nanocatalysts is already impressive. The dominant global players include Argonide Corporation, BASF Catalyst LLC, BASF SE, Bayer AG, Catalytic Solution, Inc., Evonik Degussa GmbH, Genencor International, Inc., Headwaters Nanokinetix, Inc., Hyperion Catalysis International, Johnson Matthey PLC, Mach I, Inc., Nanophase Technologies Corporation, NanoScale Corporation, NexTech Materials Ltd., Oxonica, PQ Corporation, Sachtleben Chemie Gmbh, S€ud-Chemie AG, Umicore NV, and Zeolyst International, among others.

In Section 8.2, the exciting opportunities of nanocatalysis in selected chemical processes and the challenges in developing nanostructured catalysts for research and possibly industrial applications are discussed.

8.2 Applications of Nanoparticles

Considering the unique properties discussed in Section 8.1.2, NPs can be used in an array of applications. The nanomaterials find application in a variety of industry segments spanning from petrochemical, pharmaceuticals, chemical, food processing sectors, and others (Serp 2013). An assessment of the global nanocatalysts' market was revealed in Section 8.1.5. Nanomaterials are at the leading edge of the rapidly developing field of nanotechnology. This brief chapter summarizes the most recent developments in the field of applied nanomaterials, in particular their application in biology, energy, and pollution. A decade ago, NPs were studied because of their size-dependent physical and chemical properties, and now, they have entered a commercial exploration period.

8.2.1   Electrocatalysis

The word "electrocatalysis" was first used by Grubb in 1963 in connection with the investigations of fuel cells. However, the first interpretation of electrocatalysis came before that in a famous paper by Horiuti and Polanyi (Minevski 1994). Electrocatalysis is a special field in electrochemistry that has gained a special growth after the late eighties due to the application of new hybrid techniques (Minevski 1994 and Korchagin 2013). The subject of electrocatalysis is the studying and realization of reaction rates underlying electrochemical methods of energy conversion, as well as searching for new electrode materials (Wells 1987, Minevski 1994, Wang 2010 and Korchagin 2013). The most significant trend is the studying of kinetics and mechanism of the key current producing reactions, in particular, the molecular oxygen electroreduction, hydrogen electrooxidation, and primary alcohols electrooxidation. The synthesized nanostructured

materials (NiO@C and Pd-NiO@C) were characterized by cyclic voltammetry (CV) to confirm the typical electrochemical response for Pd. Characterization and catalytic activity of the prepared NiO@C, Pd/C, and Pd-NiO@C nanocatalyst were scrutinized by XRD, SEM, and electrochemical techniques (Figure 8.5). The redox properties and formation of NiO@C, Pd/C, and Pd-NiO@C (Pd loading of 1.5 mg/cm^{-2} at ambient temperatures) were confirmed by CV in a N_2-saturated 0.1 M KOH over the wide range of voltage -1.0 V to $+0.5$ V (vs. Ag/AgCl) at 0.05 Vs^{-1} (Figure 8.5a). Figure 8.5a reveals the reversible peaks of Ni substrate at $(E_{p,(a,.c)}) = 0.25$ V which shifted to quartz-reversible couples (oxidation $(E_{p,a})$ − reduction peaks $(E_{p,c})$) due to both NiO NPs and C modification. The couples were found at $E_{p,a}= -0.2$ V (Figure 8.5A-a) and $E_{p,c}= -0.65$ V (Figure 8.5a-d) and the Ni-substrate peak potential moved to lower potentials of 0.015 V due to insertion of NiO nanocatalysts. Further, the CV profiles show the known characteristic peaks of palladium dispersed on Ni foam substrate recorded at lower potentials of -0.5 V to -0.1 V. The broad anodic peaks at the potentials of -0.5 V to -0.1 V are associated with the formation of Pd surface oxides, whereas the reduction of these oxides results in a well-defined cathodic peak between -0.9 V to -0.5 V. Meanwhile, the introduction of Pd catalyst on (NiO@C) gave high peak current density

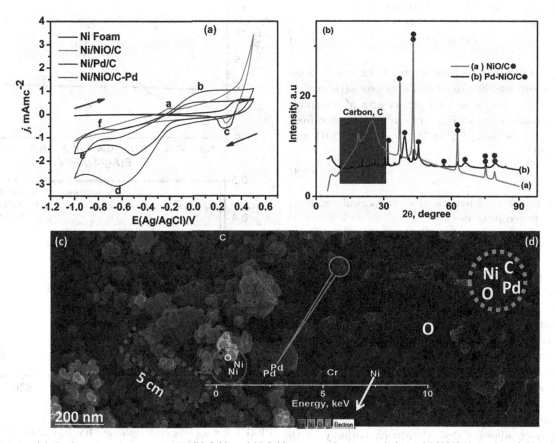

FIGURE 8.5   (a) Cyclic voltammograms of Pd/NiO/C and NiO/C nanocatalysts recorded in 1 M KOH electrolyte (pH 6.7) at a scan rate of 50 mV/s purged with either N_2, (b) XRD Powdered X-ray diffraction spectra of NiO/C and Pd-NiO/C and (c and d) SEM and mapping images of NiO/C and Pd-NiO/C.

of 1 mA/cm² (Pd-NiO/C) electrode vs. 0.5 mA/cm⁻² on NiO@C electrode. It is clear that the involvement of Pd catalyst significantly increased the NiO@C electrode activity. Clear oxidation–reduction waves are visible denoting the presence of Pd catalyst on NiO@C. An oxidation potential was observed at 0.1 V and the reduction peak negatively shifted to −0.65 V on Pd-NiO/C, proving the formation of the nanocatalyst Pd-NiO/C.

Parameters such as electroactive surface area, the active surface area, and the chemical or geometric surface area are of vital importance and can be estimated using the following equations: The electrochemical active surface area (S_{EL}) of the Pd-based electrocatalysts can be investigated by CV in 1M KOH saturated with nitrogen over the potential range −1.0 and +0.5 V vs. Ag/AgCl. The electroactive surface area of nanocatalyst can be calculated (at −0.5 Vs⁻¹ vs. Ag/AgCl), using Eq. (8.1) (Singh et al. 2011 and Dempsey 2013).

$$S_{EL} = \frac{Q_H}{Q_{ref+L_{Pd}}} \qquad (8.1)$$

where Q_{ref} is an average charge associated with hydrogen adsorption/desorption monolayer on the active surface of the palladium-supported NPs, and L_{Pd} is the palladium loading (mg). The active surface coverage (Θ) is an important parameter which gives an approximation of the catalytic activity and can be calculated by using the Eq. (8.2) (Singh et al. 2011 and Dempsey 2013).

$$\Theta = \frac{S_{EL}}{S} \qquad (8.2)$$

where S_{EL} is the electrochemical active surface area as explained above, and S is the chemical or geometric surface area (m²/g) also known as total surface area of the corresponding metallic nanocatalyst and can be estimated from the XRD analysis using Eq. (8.3) (Singh et al. 2011 and Dempsey 2013).

$$S = \frac{600}{\rho d} \qquad (8.3)$$

where d is the average particle size (nm) calculated from XRD analysis and ρ the density.

$$p = Wt\%_{pPd} + Wt\%_{pNiO/C} \qquad (8.4)$$

The enhancement of catalytic alcohol oxidation is generally interpreted in terms of the ability of the second metal to supply hydroxyl species (at lower potentials) necessary for the complete oxidation of the alcohol and for the removal of intermediates such as CO, which can poison the catalytic surface. On the other hand, electronic effects have also been attributed to the second metal, which can result in weakening of the bond between the adsorbed CO and the active catalytic surface. The bimetallic/alloy formation changes the structural and electronic properties of Pd-based electrocatalysts, allowing desired reaction pathways with lower activation energy. This is enabled via tuning energies of adsorption, surface reactions, and/or desorption reactions and facilitating production of oxygenated species from

water decomposition at low potentials for the oxidation of carbonaceous intermediates formed on the catalyst surface. The electrochemical activity of NiO/C, Pd-NiO/C, and Pd/C toward the oxidation of ethanol was investigated with an aqueous solution of 1 M KOH electrolyte (Figure 8.6a and b). The onset potential for both catalysts is around −0.5 V and −0.4 V (Ag/AgCl). The oxidizing current density increases linearly over the complete potential range (−0.6 to 0.01 V). With the same Pd loading, the C-NiO-supported catalyst reaches 23 mA/cm² at 0.01 V compared to 20 mA/cm² of the Pd/C and 2 mA/cm² of NiO catalysts. On the forward scan, Pd-NiO/C maintains a higher current density for all potentials above 0.01 V (Figure 8.6a). The enhanced activity for the ethanol oxidation reaction (EOR) with Pd-NiO/C with respect to Pd/C is due to the oxophilic nature of the NiO in the mixed C-NiO support and the three-phase (C, NiO and Pd) nanostructure. Electrochemical impedance (EI) spectroscopy was employed to confirm the kinetics of Pd-NiO@C and NiO@C at the electrode surface (Figure 8.6b). Compared to Ni/Pd/C and Ni/NiO/C, Ni/Pd-NiO@C revealed superior electrical

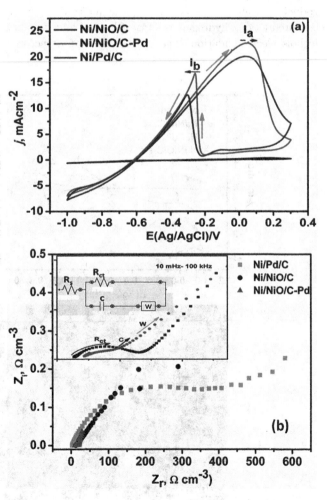

FIGURE 8.6 The cyclic voltammetric (a) and EIS (b) response for ethanol electro-oxidation by Pd-NiO/C-, NiO/C-, and Pd/C-modified electrode in alkaline conditions (1 M methanol + 1 M KOH) at a scan rate of 0.01 Vs⁻¹ vs. Ag/AgCl.

properties. The Randle Sevcik's equivalent circuit (inset, Figure 8.6b) was used to assess and simulate the specific parameters (R_s: solution resistance, R_{ct}: charge-transfer resistance, C: Faradaic pseudo-capacitance, and W: Warburg diffusion impedance) as presented by the behavior of our modified electrode. Remarkably, the Nyquist plots reveal characteristic circular curvature of Pd-NiO@C at a high-frequency region, signifying that the R_{ct} is prominent during charge–discharge (CD) process. Furthermore, the linearity or inclined line at the lower frequency regime denotes a capacitive behavior of the electrodes, whereas at high frequency, a diffusion-controlled system is apparent. The small R_{ct} proposes a faster electron shuttling, hence high conductivity at the electrode interface. The results are in agreement with fitted values and that of CV. Conclusively, the prepared electrochemical capacitive Pd-NiO@C possesses potential energy storage and other energy applications.

To further understand the electrocatalyst phenomena, we introduce other applications and methods. The electrocatalyst can be defined as an electrode material that interacts specifically with some species involved in the reaction and remains unaltered after the reaction or better defined by A. J. Appleby as an electrochemical reaction with an adsorbed species, either as reactant and/or product, which can change the kinetics of the reaction and in some cases also the mechanism (Minevski 1994, Thomas 2010 and Wang 2010). Scheme 8.4 illustrates the behavior of different materials for different applications and processes.

Nowadays, the application of newer concepts of electrocatalysis for industrial electrochemical processes has appeared as a requirement and attracting attention not only for chemists but for engineers. With a rising global population, increasing energy demands, and impending climate change, major concerns have been raised over the security of our energy. Developing sustainable, fossil-free pathways to produce fuels and chemicals of global importance could play a major role in reducing carbon dioxide emissions while providing the feedstocks needed to make the products we use on a daily basis (Scheme 8.5). One prospective goal is to develop electrochemical conversion processes that can convert molecules in the atmosphere (e.g., water, carbon dioxide, and nitrogen) into higher-value products (e.g., hydrogen, hydrocarbons, oxygenates, and ammonia) by coupling to renewable energy, Scheme 8.5b–c.

Electrocatalysts play a key role in these energy conversion technologies because they increase the rate, efficiency, and selectivity of the chemical transformations involved. There are various applications of electrocatalysis for technological electrochemical reactions, organic electrosynthesis, galvanoplasty, **supercapacitors**, electrode sensors, **fuel cells**, batteries preparations, and so on. Some of them are presented here in order to give an insight into the possibilities of this science. We further present and discuss different applications such as electrocatalyst, CO_2, and supercapacitors using different metal oxides. To ascertain the pseudo-capacitive nature of the electrochemically active binary and ternary nanocomposites (Cu/Cu_2O and $Cu/Cu_2O/CuO/ZnO$), different electrochemical methods such as CV, electrochemical impedance spectroscopy (EIS), and galvanostatic charge–discharge (GCD) were employed (Figure 8.7a–c). The GCD was employed to confirm the capacitive profiles of the cubes-like, rods-like, and platelets-like morphological structures at a constant current density of 0.1 mAg^{-1} (Figure 8.7a). The CD curves (Figure 8.7a) resulted in a nearly symmetrical behavior with a slight curvature. The distortion/shape of the CD curves indicates the pseudo-capacitance nature of the prepared nanostructures and was utilized to assess the C_s of individual nanostructures from Equation 8.4:

$$C_s = \frac{I \times t}{\Delta V \times m} \tag{8.5}$$

where t is the discharge time, m is the mass of the active material on the electrode, and ΔV is the potential range/window. The C_s value of the cube-like-Cu/Cu_2O

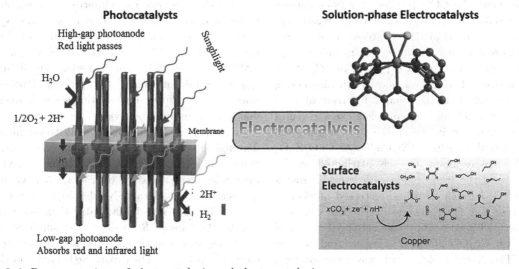

SCHEME 8.4 Representations of photocatalysis and electrocatalysis processes.

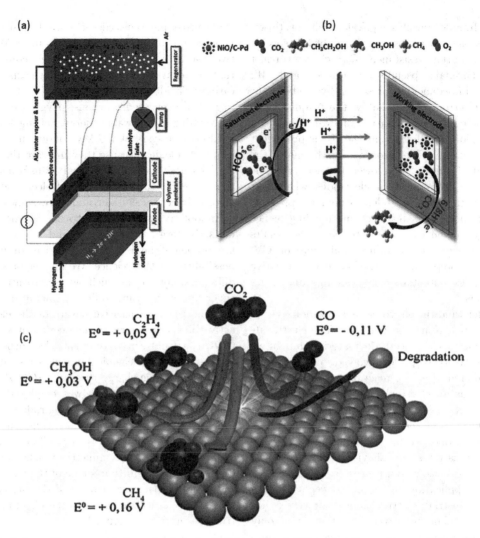

SCHEME 8.5 Schematic representation of (a) fuel cell setup, (b–c) CO$_2$ reduction with possible synfuels.

(at 200°C) was calculated to be 106 Fg^{-1}, of the rods-like Cu/Cu$_2$O/CuO/ZnO (at 400°C) to be 157 F/g, and of platelets-like Cu/Cu$_2$O/CuO/ZnO (at 600°C) to be 233 F/g at the same discharge current density.

Figure 8.7b shows the GCD profiles at various current densities (0.1–0.35 mA/cm^2) exhibiting reasonably symmetric curves. Potential (E) plateau in discharge curves was apparent at 0.3 V, which was in unison with the CV curves (not included). As expected, the C_s of the nanostructured electrode decreases linearly with increasing current densities, which is the typical behavior of electrochemical supercapacitors. The platelets-like structures showed a vital increment in the charge storage using the same electrochemical transducer (Ni/Cu/Cu$_2$O/CuO/ZnO) (Figure 8.7b). The resultant C_s was attributed to the unique structure of the metal oxide with irregular distributions of Cu, O, and Zn elements, increased number of oxidation states, and ionic penetration in the electrode surface. Further, EI spectroscopy was employed to reveal the kinetics of Cu/Cu$_2$O/CuO/ZnO at the electrode surface (Figure 8.7). The Nyquist plot which was recorded from 100 kHz to 25 mHz, revealed the electrical properties of

Ni/Cu/Cu$_2$O/CuO/ZnO (Figure 8.7c). The equivalent circuit (Figure 8.7) was used to evaluate and simulate the specific parameters (R_s: solution resistance, W: Warburg diffusion impedance, R_{ct}: charge-transfer resistance, and C: Faradaic pseudo-capacitance) as presented by the behavior of the transducer. Remarkably, the Nyquist plots reveal characteristic circular curvature of Ni/Cu/Cu$_2$O (at 200°C) and Ni/Cu/Cu$_2$O/CuO/ZnO (at 400°C and 600°C) at a high-frequency region, signifying that the R_{ct} is prominent during CD process. Furthermore, the linearity or inclined line at the lower frequency regime denotes a capacitive behavior of the electrodes, whereas at high frequency, a diffusion-controlled system is apparent. Through simulation, the charge-transfer resistance (R_{ct}) values at the Ni/Cu/Cu$_2$O, Ni/Cu/Cu$_2$O/CuO/ZnO electrodes were found to be 256 Ω, 342 Ω, and 184 Ω, respectively. Comparatively, the Ni/Cu/Cu$_2$O/CuO/ZnO (at 600°C) proves to be favorable than other transducers, due to their lower R_{ct} values. The small R_{ct} proposes a faster electron shuttling, hence high conductivity at the electrode interface. The capacitive behavior of the prepared transducer was determined by Eqs. 8.6–8.7:

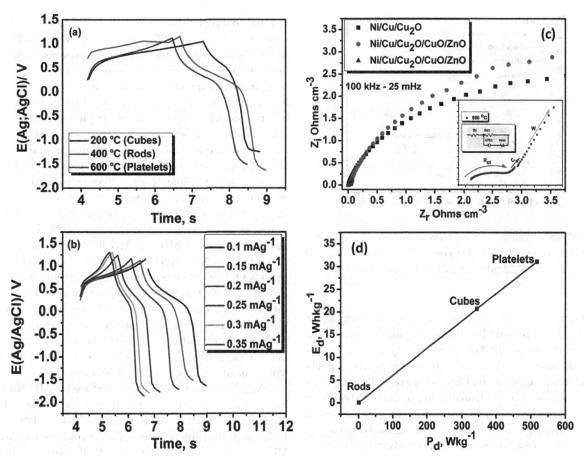

FIGURE 8.7 CD curves of Ni/Cu/Cu$_2$O/CuO/ZnO electrode in 2 M KOH at (a) constant current density (0.1mA/cm^2), (b) at various current densities (0.1–0.35 mA/cm^2), (c) Nyquist plots of Ni/Cu/Cu$_2$O/CuO/ZnO electrode in 2 M KOH at 50 mV and (d) the two main parameters of power applications.

$$Z'' = (2\pi f C)^{-1} \qquad (8.6\text{-}8.7)$$

$$C_{sp} = \frac{-2}{2(\pi m f / Z'' /)}$$

where $|Z''|$ is the imaginary impedance, f is the frequency, C is the capacitance, and C_{sp} is the specific capacitance. The capacitance and specific capacitance at 0.5 V calculated from the $Z''|$ values at the high frequency ($f = 100$ kHz) were found to be 85.2, 92.76, and 169–222 F g^{-1}, respectively. The results are in agreement with fitted values, GCD. Conclusively, the prepared electrochemical capacitive transducer (Ni/Cu/Cu$_2$O/CuO/ZnO) possesses potential energy storage and other energy applications. Figure 8.7d represents the two main parameters of power applications in electrochemical supercapacitors, i.e., energy density (Ed) and power density (Pd). The two applications including Cs are the main requisite of ultimate electrochemical supercapacitor at high charging–discharging rates. The Ragone plot (Ed vs. Pd) of Ni/Cu/Cu$_2$O/CuO/ZnO (nanorods), Ni/Cu/Cu$_2$O (nanocubes), and Ni/Cu/Cu$_2$O/CuO/ZnO (nanoplatelets) electrodes (Figure 8.7d)) was plotted and compared with the standard Ragone plot (Chen et al. 2015 and Li 2016). The results advocate the employment of nanoplatelets-structured electrode in the fabrication of

supercapacitor devices since they are within limits of the standard Ragone plot (Chen et al. 2015 and Li 2016). Furthermore, the prepared nanocatalysts (Pd-NiO@C and NiO@C) were applied and tested in the CO$_2$ reduction using CV. Blank cyclic voltammograms of the Pd-NiO@C and NiO@C were recorded in an alkaline medium of 1M NaHCO$_3$ solution that was purged with nitrogen (N$_2$) and compared with CVs of nanocatalysts purged with CO$_2$ (Figure 8.8).

Compared with N$_2$ purged system, the CO$_2$ purged system showed lessor current densities (4 mA/cm^2), indicating that the nanocatalysts have less interaction with CO$_2$. The particles with Pd insertion show high current densities (6 mA/cm^2) between the blank and CO$_2$ voltammograms, indicating that those particles have a stronger interaction with CO$_2$ than the NiO@C particles. Further, due to electrochemical reduction of CO$_2$, oxidation–reduction waves (Figure 8.8b, b' and c, c') emerged at peak potential of −0.2 V and −0.3 V, −0.1 V and −0.5 V vs. Ag/AgCl. The results elucidate catalytic activity of the Pd-NiO@C and NiO@C toward CO$_2$ electroreduction. Moreover, the observed reduction behavior could originate from the proton-induced processes, leading to the formation of formic acid/formate, alcohols, methane, and ethylene).

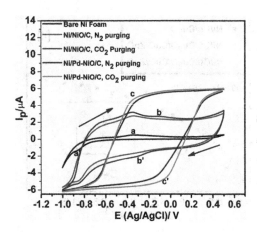

FIGURE 8.8 Cyclic voltammograms of Pd-NiO@C and NiO@C nanocatalysts recorded in a 1 M ethanol/1 M KOH electrolyte (pH 6.7) at a scan rate of 50 mV/s purged with either N_2 or CO_2.

In addition, since the pH near the electrode surface will increase as the potential is decreased, due to continuous hydrogen evolution, HCO_3 − will be generated in situ from CO_2 near the electrode surface due to the equilibrium between HCO_3 − and CO_2. Thus, regarding the reaction mechanism, the electroreduction of carbon dioxide at our nanostructured carbon catalyst does require proton discharge in spite of any possible activating interactions (e.g. the CO_2 adsorption at certain sites of carbon surfaces).

8.2.2 Photocatalysis

Photoinduced processes are studied in several industry-oriented applications. Despite the difference in character and use, all photoinduced processes are based on the excitation of a semiconductor by a light energy higher than the band gap, thus inducing the formation of energy-rich electron–hole pairs (Herrmann 2005 and Rodionov 2016). Photoinduced chemical processes have in common the so-called "initial" steps, which include the absorption of light, diffusion, and trapping of charge and fate (recombination or phase transfer to the gas/liquid media) of charge carriers. The term "photocatalysis" is commonly understood to be any chemical process catalyzed by a solid where the external energy source is an electromagnetic field with wavenumber in the UV-vis range (Herrmann 2005 and Rodionov 2016).

Photocatalysis is mainly involved in three areas: organic synthesis, degradation of pollutants (in both gas and liquid phases), and special reactions like H_2O reduction or N_2 fixation (Herrmann 2005 and Rodionov 2016). Photocatalysts are most of the time solid oxide semiconductors, among which TiO_2 in its anatase form is by far the most known and investigated photocatalyst, due to its chemical stability, nontoxicity, and well-positioned valence and conduction bands (Herrmann 2005 and Rodionov 2016). Besides the research concerning metal oxide materials, recent progress has expanded the chemical nature of photoactive systems by including (oxy) sulfides and (oxy) nitrides, doped

zeolites, and molecular entities embodied in zeolites or other composite materials. The interest of metal oxide nanomaterials as active phases in photocatalysis is discussed here by Fuku et al. (2018 and 2016a). Photocatalysts resulting from the combination of metallic metal oxides is a subject of intensive work because of the benefits induced by such association in terms of activity and selectivity. In this study, methylene orange (MO) and methylene blue (MB) were employed as a target pollutants to evaluate the real performance of the as-prepared catalyst (NiO/MgO) under visible-light illumination. Photocatalytic degradation of MO and MB was studied using UV-vis and scanned in the range of 190–800 nm, before and after UV light interaction (Figure 8.9). Figure 8.9 presents the maximum absorption plots of MB and MO before exposure, mixture of both MO+NiO/MgO (before and after exposure) and MB+NiO/MgO (before and after exposure). The absorption λ_{max} of NiO/MgO nanocomposite was at 452 nm while MO had two characteristic absorbance peaks at $\lambda_{max} = 270$ nm and $\lambda_{max} = 466$ nm and the mixture of MO + NiO/MgO with ratio 1:2 showed the same λ_{max} as MO but with slightly decrease in absorbance intensities. Meanwhile, the absorption maxima of MB was at $\lambda_{max} = 290$ nm and $\lambda_{max} = 630$ nm, while the mixture of MB+NiO/MgO had a $\lambda_{max} = 646$ nm. Figure 8.6 shows the effect of illumination time at selected λ_{max} (MO = 466 nm and MB = 646 nm), respectively. The degradation of MO and MB without the catalyst under UV illumination showed a slight/no change in the absorbance intensities confirming the need of a catalyst, and it attests that photocatalytic degradation is a feasible means of organic pollutants removal from wastewaters. After mixing the pollutants (MO and MB) with the catalyst (NiO/MgO), a slight decrease in absorption intensity was observed before irradiation. This was attributed to the good crystallinity, wide visible-light absorption range, and the efficient electron–hole pair separation properties of the NiO/MgO nanostructures (Zhang 2013). The behavior suggests or confirms the formation and adsorption of the materials (catalyst + pollutant). After 2 min of exposure, the catalyst degraded the pollutant by 34%, suggesting a good catalytic effect of the nanocomposite. However, after 10 (MO) and 12 (MB) min, saturation of the pollutants was observed, respectively. Thus, the optimum illumination time for the degradation of MO and MB was noted as 10 and 12 min. Many compounds were decomposed entirely after 3–5 h (Wang 2010, Zhou 2012 and Singh 2014). Therefore, the results suggest that the catalyst was efficient for the decomposition of MO and MB.

The percent degradation of MO and MB was calculated as follows:

$$\%D = \frac{(A_0 - A_t)}{A_0} \times 100 \tag{8.8}$$

where D is the percent degradation, A_0 is the absorbance of initial MO, and A_t is the absorbance of the solution after illumination. The catalyst degraded the pollutants by 73%–87% at 2 mg/mL concentration of NiO/MgO, 0.6 mg/mL concentration of pollutants at 10 min illumination

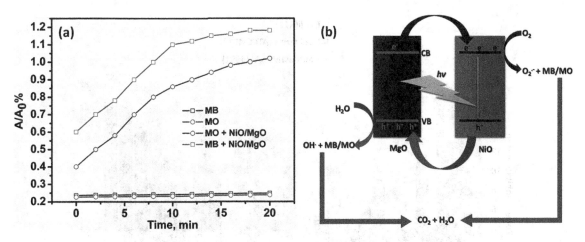

FIGURE 8.9 UV-vis spectra of MB and MO in the presence and absence of NiO/MgO nanocomposite at different UV-exposure time.

time (Figure 8.9a). The test demonstrates that no dye degradation in solution can be observed in the absence of the photocatalyst at the same conditions from a blank experiment. NiO/MgO material showed improved separation of photoinduced electron–hole pairs owing to the potential energy; therefore, NiO/MgO exhibited enhanced photocatalytic activity.

The schematic band structure and charge-transfer process of the NiO/MgO nanocatalyst is presented in Figure 8.9b. The photoinduced electron–hole pairs are separated from each other in NiO under UV irradiation. The electrons transit from the VB to the conduction band (CB) and leave positive holes (h^+) in VB. After the separation of electrons and holes, the dissolved oxygen (O_2) adsorbed on NiO surface will react with photoinduced electrons to form superoxide radical ($-\cdot O_2$). Considering the band structures of ZnO and CuO, direct transfer of photoinduced holes from MgO to NiO thermodynamically occurs in the NiO/MgO nanocatalyst, leading to low recombination rate of the photoinduced electron–hole pairs. The hydroxyl ions ($-OH$) will be oxidized into hydroxyl radicals ($\cdot OH$) by photoinduced holes (Mena 2012 and Santos and Schmickler 2010). Finally, the dye molecules are decomposed into simple organics by the continuously generated reactive oxidation species and further converted into CO_2 and H_2O. Therefore, the enhanced photocatalytic degradation of the NiO/MgO nanocatalyst should be attributed to the effective charge-transfer between NiO and MgO.

8.2.3 Antibacterial Agents

The term "antimicrobial" was derived from the Greek words anti (against), micro (little), and bios (life), and it refers to all agents that act against microorganisms (Galvez et al. 2008 and Weller 2014). Thus, antimicrobials include agents that act against bacteria (antibacterial), viruses (antiviral), fungi (antifungal), and protozoa (antiprotozoal) (Cui et al. 2012, Karlsson et al. 2013 and Chaterjee et al. 2014). Among these, antibacterial agents are by far the most widely known and studied class of antimicrobials. With

their unique physical, chemical, electrical, magnetic, optical, and biological properties, metal oxides (MeO-NPs) are of immense interest to scientists as antimicrobial agents (Cui et al. 2012, Karlsson et al. 2013 and Chaterjee et al. 2014). Several MeO-NPs have been reported to exhibit marked antibacterial activity allowing efficient eradication of various bacterial strains. This fact has attracted significant interest of environmental, agricultural, and health care industries that are searching for newer and better agents to control or prevent bacterial infections. Many studies have been undertaken to explain the efficacy and mechanisms of antibacterial action of MeO-NPs, but the existent literature is still divisive and partial. It was demonstrated, however, that when applied at well-defined sizes, crystal structure, and concentrations, these NPs are highly effective inhibitors against a wide range of bacteria. In their paper, Fuku et al. (2017) demonstrated the antimicrobial efficacy of green-synthesized $Cu/Cu_2O/CuO/ZnO$ against various bacterial pathogens such as Escherichia coli -K12, Proteus vulgaris-ATCC-49132, *Staphylococcus aureus*-MRSA-33591, and also when subjected to streptomycin and gentamicin antibiotics. Antimicrobial tests (zone of inhibition and minimum inhibitory concentration (MIC)) were performed on three different bacterial pathogens. The applications were performed by varying calcination temperatures but the same concentration of the prepared $Cu/Cu_2O/CuO/ZnO$ nanocomposite (Figure 8.10).

The antimicrobial activity of green-prepared $Cu/Cu_2O/CuO/ZnO$ nanostructures at different calcination temperatures (RT, 100°C, 200°C, 300°C, 400°C, 500°C, and 600°C but same concentration 10 mg/mL , see Table 8.2) toward various bacterial pathogens was tested by the well and disc diffusion agar methods (Figure 8.10). The presence of inhibition zone clearly indicates that the mechanism of the biocidal action of $Cu/Cu_2O/CuO/ZnO$ nanostructures involves disruption of the membrane while generating surface oxygen species to kill pathogens (Kaviyarasu et al. 2015) (Table 8.3). The study reveals that as we change the calcination temperatures and/or structure of the prepared nanocomposite, the growth of

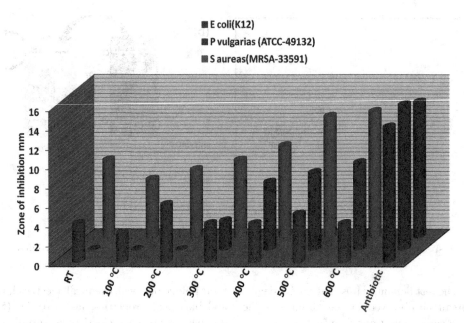

FIGURE 8.10 Zone of inhibition of Cu/Cu₂O/CuO/ZnO nanocomposite tested on various bacterial species (*E. coli*-K12, *P. vulgaris*-ATCC-49132, *S. aureus*-MRSA-33591).

TABLE 8.2 Antibacterial Activity of Different Size and Shapes of CuO NPs on Various Bacterial Species (*E. coli*-K12, *P. vulgaris*-ATCC-49132, *S. aureus*-MRSA-33591)

Conc. (mg/ml)	*E. coli* (K12)	*P. vulgaris* (ATCC-49132)	*S. aureus* (MRSA-3591)
10 mg/ml	4	0	8
	3	0	6
	6	0	7
	4	3	8
	4	7	9.5
	5	8	12.5
	4	9	13
Antibiotic	14	15	14
		Zone of inhibition (mm)	

TABLE 8.3 The MIC of Cu/Cu₂O/CuO/ZnO Nanocomposite on Various Bacterial Pathogens

Bacteria	MIC (Conc. mg mL)
E. coli K12	1.25
P. vulgaris-ATCC-49132	2.5
S. aureus-25923	5

inhibition increased as a function of temperature and structural change. Worth to note, temperatures at 500°C and 600°C show high zone of inhibition compared to low temperatures (RT, 100°C, 200°C, 300°C, 400°C). Among bacterial pathogens, maximum rate of activity was noted for *S. aureus*-MRSA-33591 followed by *P. vulgaris*-ATCC-49132 and then *E. coli*-K12 (Figure 8.10). Compared to the nanocomposites at low calcination temperatures, the antibiotics reveal slightly higher zone of inhibition as was expected. However, the results confirm that the nanocomposite at higher calcination temperature can be better inhibitors relative to the antibiotics.

To determine the MIC of the prepared Cu/Cu₂O/CuO/ZnO nanocomposite at 600°C (chosen for high inhibitory activity), different concentrations (10, 5, 2.5, 1.25, 0.625 mg/mL) of the nanocomposite were investigated. Table 8.2 reveals the MIC values (defined as

the lowest concentration of NPs that inhibits the growth of a microorganism) of Cu/Cu₂O/CuO/ZnO nanocomposite. From the MIC results (Table 8.2), the green-prepared Cu/Cu₂O/CuO/ZnO nanocomposite with hexagonal-like structures (at 600°C) showed enhanced inhibitory activity at lower concentration for *E. coli* (1.25 mg/mL) and *P. vulgaris* (2.5 mg/mL) but slightly higher for *S. aureus* (5 mg/mL). The calculated MIC values are in agreement with and slightly lower than those reported by other researchers (Mpenyana-Monyatsi et al. 2012, Cheng et al. 2013 and Mahmud 2013). The results show that a change in structure, size, concentration, and calcination temperatures of the nanocomposite plays a vital role in the inhibition studies of different bacterial pathogens. We further review the possible mechanisms of MeO-NPs antimicrobial activity revealed from recent studies (Scheme 8.6). Although the mechanisms are not exhaustive, research has revealed that retardation of bacterial growth is brought about by a combination of one or more mechanisms (i.e., reactive oxygen species (ROS) formation, metal ion release, particle internalization into bacteria, and direct mechanical destruction of bacterial cell wall and/or membrane) and it varies with the MeO-NPs nature and chemistry.

Briefly, metal ion homeostasis is essential for microbial survival as it regulates metabolic functions by assisting coenzymes, cofactors, and catalysts (Misra et al. 2003, Draper et al. 2005 and Ferré-D'Amaré and Winkler 2011). When bacteria have excess metal or metal ions, there will be a disorder in metabolic functions. Metal ions bind with DNA and disrupt the helical nature by cross-linking between and within DNA strands. This is evidenced on the exposure of microorganisms to CuO NPs. Metal ions released from the MeO-NPs carry a positive charge, bringing electrostatic interactions into play. The metal ions neutralize the charges

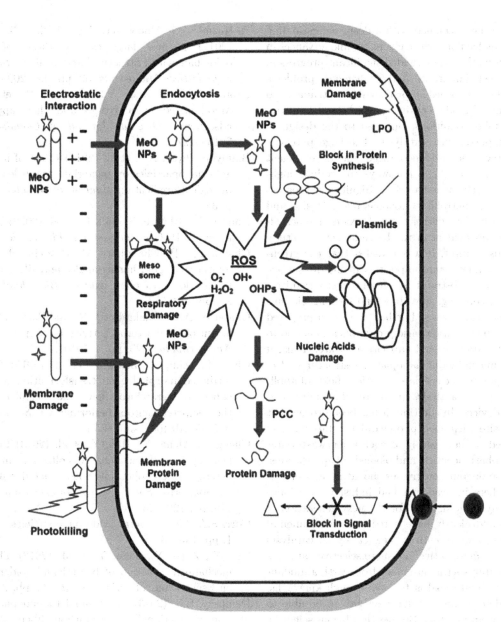

SCHEME 8.6 Overview of antimicrobial mechanisms by MeO-NPs. ROS, reactive oxygen species; LPO, lipid peroxidation.

on lipopolysaccharides (LPS) and increase permeabilization of the outer membrane. Bacterial growth is slowed down as the membranes become disorganized, with increased permeability contributing to the accumulation of MeO-NPs in the cells. Strong binding of MeO-NPs to the outer membrane inhibits active transport and the activities of dehydrogenase and periplasmic enzymes, as evidenced from the treatment of *E. coli* with TiO$_2$ and CdO NPs (Karthik et al. 2017 and Qin et al. 2017). Long-chain polycations coated onto cell surfaces are reported to efficiently kill both Gram-positive and Gram-negative bacteria (Yael et al. 2017). Released metal ions from MeO-NPs in cells are reduced to metal atoms by thiol (–SH) groups in enzymes and proteins, thus inactivating the essential metabolic proteins blocking respiration and leading to cell death (Yael et al. 2017). Metal

ions interact with –SH groups in the peptidoglycan layer and cause cell wall destruction. MeO-NPs slowly release metal ions through adsorption, dissolution, and hydrolysis; they are toxic and abrasive to bacteria and hence lyse the cells (Yael et al. 2017). The abrasive nature of MeO-NPs causes surface defects, thus bringing physical damage to the cell wall.

8.3 Conclusions and Remarks

The field of catalysis science is often criticized as being ad hoc and pragmatic. In that respect, the impact of nanoscience on catalysis, which has been already discussed by scientists from the heterogeneous catalysis community, is obvious. Creative use of the modern methods of nanomaterial synthesis coupled with a deep understanding of fundamental molecular surface chemistry and advances in characterization methods has yielded significant progress in some of the most important and challenging problems in the field. This knowledge principally contributes to the development of model catalysts, which can, however, operate under mild conditions, but also to the design of more robust-supported new catalysts. The huge potential of nanochemistry has also encouraged the development of new or improved synthetic pathways to produce high-performance catalytic nanomaterials. Significant progress has been made in controlling monometallic NP size and shape, particularly for different metal oxides nanocatalysis. The latter was evident in the bacterial activity study presented in this chapter. It was revealed that the green-prepared metal oxides nanocomposite can aid in surface coatings on various substrates to prevent microorganisms from attaching, colonizing, spreading, and forming biofilms in indwelling medical devices. Further, the green-prepared nanocomposite opens a wide variety of applications due to their structural, catalytic, and capacitive nature. Thus, if modified, the semiconducting nanoplatelets-structured electrode will be a potential candidate in various fields of application such as photocatalysis, antibacterial activity, and energy storage devices. In addition, it has been proven that the nanocomposite with good photocatalytic stability may be suitably used as new catalytic agents for wastewater treatment, and pharmaceutical and biomedical applications. In parallel, these ongoing progresses should rapidly enable catalyst researchers in academia and industry to achieve the goal of catalysis by design. For the next generation of catalysts, it should clearly have a reduced environmental impact. This will require entirely new catalysts: catalysts for new processes, more active and more selective catalysts and preferably catalysts that are made from earth-abundant elements. The ultimate goal is to have enough knowledge of the factors determining catalytic activity to be able to tailor catalysts atom-by-atom. Besides the obvious scientific impact of nanoscience and nanotechnology on catalysis, we would like to stress the fact that nanoscience and nanotechnology have also been a tremendous lever for two communities, the coordination chemists of homogeneous catalysis and the solid-state chemists of heterogeneous catalysis, to meet and collaborate efficiently on common objectives. This is perhaps the more important contribution that nanoscience and nanotechnology have brought today to catalysis. Nowadays, few scientists have a general background in all of these areas. It is thus important to open new academic programs in order to form the catalysis community of tomorrow.

References

Akhtar, M.S., Panwar, J., and Yun, Y.-S. (2013) Biogenic synthesis of metallic nanoparticles by plant extracts, *ACS Sustainable Chemistry and Engineering*, 1, pp. 591–602.

Al-Hazmi, F., Alnowaiser, F., Al-Ghamdi, A.A. et al. (2012) A new large-scale synthesis of magnesium oxide nanowires: Structural and antibacterial properties, *Superlattices Microstruct*, 52, pp. 200–209.

Arora, A.K., Jaswal, V.S., Singh, R. et al. (2016) Metal/mixed metal oxides and their applications as adsorbents, *Oriental Journal of Chemistry*, 14, pp. 3215–3227.

Bandyopadhyay, A.S. et al. (2011) Effect of iron oxide and gold nanoparticles on bacterial growth leading towards biological application, *Journal of Nanobiotechnology*, 9, p. 34.

Bhadra, D., Bhadra, S., Jain K. et.al. (2002) Pegnology: A review of PEG-ylated systems, *Pharmazie*, 57, pp. 5–29.

Chandran, S.P., Chaudhary, M., Pasricha, R. et al. (2006) Synthesis of gold nanotriangles and silver nanoparticles using Aloe vera plant extract, *Biotechnology Progress*, 22, pp. 577–583.

Chaterjee, A.K., Chakraborty, R., and Basu T. (2014) Mechanism of antibacterial activity of copper nanoparticles, *Nanotechnology*, 25, pp. 135–101.

Chen, L., Zhu, P., Zhou, F. et al. (2015) Copper salts mediated morphological transformation of Cu_2O from cubes to hierarchical flower-like or microspheres and their supercapacitors performances, *Scientific Reports*, p. 5. doi:10.1038/srep09672.

Cheng, J., Meng, X., Han, Y. et al. (2013) Interface engineering for efficient charge collection in Cu_2O/ZnO heterojunction solar cells with ordered ZnO cavity-like nanopatterns, *Solar Energy Materials and Solar Cells*, 116, pp. 120–125.

Corr, S.A. (2013) Metal oxide nanoparticles, *Nanoscience*, 1, pp. 180–234.

Cui, Y., Zhao, Y., Tian, Y. et al. (2012) The molecular mechanism of action of bactericidal gold nanoparticles on Escherichia coli, *Biomaterials*, 33, pp. 2327–2333.

Dempsey, B.S. (2013) Exceptional Pt nanoparticle decoration of functionalised carbon nanofibers: A strategy to improve the utility of Pt and support material for direct methanol fuel cell applications, *RSC Advances*, 3, pp. 2279–2287.

Draper, D.E., Grilley, D., and Soto, A.M. (2005) Ions and RNA folding, *Annual Review of Biophysics and Biomolecular Structure*, 34, pp. 221–243.

Ferré-D'Amaré, A. and Winkler, W.C. (2011) The roles of metal ions in regulation by riboswitches, *Metal Ions in Life Sciences*, 9, pp. 141–173.

Fuku, X., Matinise, N., Maaza, M. et al. (2016a) Punicalagin green functionalized Cu/Cu_2O/ZnO/CuO nanocomposite for potential electrochemical transducer and catalyst, *Nanoscale Research Letters*, 11, p. 386.

Fuku, X., Kasinathan, K., Kotsedi, L. et al. (2017) Size and concentration influence of Cu/Cu_2O/CuO/ZnO on selected environmental pathogenic bacteria: Escherichia coli, Staphylococcus aureus, and Proteus vulgaris, *Functional Nanostructures*. www.researchgate.net/publication/316741260-One-Central-Press.

Fuku, X., Matinise, N., Masikini, M. et al. (2018) An electrochemically active green synthesized polycrystalline NiO/MgO catalyst: Use in photo-catalytic applications, *Materials Research Bulletin*, 97, pp. 457–465.

Fuku, X., Diallo, A., Malik, M. et al. (2016b) Nanoscaled electrocatalytic optically modulated ZnO nanoparticles through green process of Punica granatum L. and their antibacterial activities, *International Journal of Electrochemistry*. doi:10.1155/2016/4682967.

Galvez, A., Lopez, R.L., Abriouel, H. et al. (2008) 'Application of bacteriocins in the control of foodborne pathogenic and spoilage bacteria, *Critical Reviews in Biotechnology*, 28, pp. 125–152.

Ghodake, G., Deshpande, N., Lee, Y. et al. (2010) Pear fruit extract-assisted room-temperature biosynthesis of gold nanoplates, *Colloids and Surfaces B: Biointerfaces*, 75, pp. 584–589.

Haddad, P. and Seabra, A.B. (2012) Biomedical applications of magnetic nanoparticles. In AI, M. (ed.) *Iron Oxides: Structure, Properties and Applications*. New York: Nova Science Publishers, Inc., pp. 165–188.

Haque, Z. (2014) Synthesis of ZnO/CuO nanocomposite and optical study of ammonia (NH_3) gas sensing, *International Journal of Scientific and Engineering Research*, 5, pp. 2229–5518.

Hasan, S. (2015) A review on nanoparticles: Their synthesis and types, *Research Journal of Recent Sciences*, 4, pp. 1–3.

Herrmann, J.M. (2005) Heterogeneous photocatalysis: State of the art and present applications, *Topics in Catalysis*, 34, pp. 1–4.

Huang, Z., Zheng, X., Yan, D. et al. (2008) Toxicological effect of ZnO nanoparticles based on bacteria, *Langmuir*, 24, pp. 4140–4144.

Jesline, A., John, N.P., Narayanan, P.M. et al. (2014) Antimicrobial activity of zinc and titanium dioxide nanoparticles against biofilm-producing methicillin-resistant Staphylococcus aureus, *Applied Nanoscience*, 5, pp. 157–62.

Jha, A.K., Prasad, K., Kulkarni, A. R. et al (2009) Plant system: Nature's nanofactory, *Colloids and Surfaces B: Bioinformatics*, 73, pp. 219–223.

Kang, Y.T. (2012) Immunosensor based on CdSe/ZnS quantum dots for detection of human serum albumin (HSA), *Biosensor and Bioelectronics*, 34, pp. 286–290.

Karlsson, H.L., Cronholm, P., Hedberg, Y. et al. (2013) Cell membrane damage and protein interaction induced by copper containing nanoparticles—importance of the metal release process, *Toxicology*, 313, pp. 59–69.

Karthik, S., Dhanuskodi, C., Gobinath, S. et al. (2017) Photocatalytic and antibacterial activities of hydrothermally prepared CdO nanoparticles, *Journal of Materials Science: Materials in Electronics*, doi:10.1007/s10854-017-6937-z.

Kaviyarasu, K., Fuku, X., Manikandan, E. et al (2015) Photoluminescence of well-aligned ZnO doped CeO_2 nanoplatelets by a solvothermal route, *Materials Letters*, 183, pp. 351–354.

Korchagin, M.R. (2013) Electrocatalysis and pH (A Review), *Russian Journal of Electrochemistry*, 49, pp. 600–618.

Kowshik, M., Arhtaputre, S., Kharrazi, S. et al. (2003) Extracellular synthesis of silver nanoparticles by a silver-tolerant yeast strain MKY3, *Nanotechnology*, 14, pp. 95–100.

Kowshik, M., Vogel, W., Urban, J. et al. (2002) Microbial synthesis of semiconductor PbS nanocrystallites, *Advance Materials,* 14, pp. 815–818.

Li, J.J. (2016) One-dimensional Cu-based catalysts with layered Cu–Cu_2O– CuO walls for the Rochow reaction, *Nano Research*, p. 2. doi:10.1007/s12274-12016-11033.

Maensiri, S., Laokul, P., Klinkaewnarong, J. et al. (2008) 'Indium oxide (In_2O_3) nanoparticles using Aloe vera plant extract: Synthesis and optical properties, *Journal of Optoelectronics and Advanced Materials*, 10, pp. 161–165.

Mahmud, S. (2013) Stress control in ZnO nanoparticle-based discs via high-oxygen thermal annealing at various temperatures, *Journal of Physical Science*, 24, pp. 1–15.

Martin, C.R. (2006) Welcome to nanomedicine, *Nanomedicine*, 1, p. 5.

Matinise, N., Fuku, X.G., Kaviyarasu, K. et al. (2018) ZnO nanoparticles via Moringa oleifera green synthesis: Physical properties & mechanism of formation, *Applied Surface Science*, 406, pp. 339–347.

Mena, P. (2012) Rapid and comprehensive evaluation of (poly) phenolic compounds in pomegranate (punica granatum l.) juice by UHPLC-MSN, *Molecules*, 17, pp. 14821–14840.

Minevski, J.C. (1994) Electrocatalysis: Past, present and future, *Ektrochlmlca Acta*, 39, pp. 1471–1479.

Misra, V.K., Shiman, R., and Draper, D.E. (2003) A thermodynamic framework for the magnesium-dependent folding of RNA, *Biopolymers*, 69, pp. 118–136.

Mohanraj, V.J. and Chen, Y. (2006) Nanoparticles – A review, *Tropical Journal of Pharmaceutical Research*, 5, pp. 561–573.

Mpenyana-Monyatsi, L., Mthombeni, N.H., Onyango, M.S., and Momba, M.N. (2012) Cost-effective filter materials coated with silver nanoparticles for the removal of pathogenic bacteria in groundwater, *International Journal of Environmental Research and Public Health*, 9, pp. 244–271.

Noguera, C. (1996) *Physics and Chemistry at Oxide Surfaces*. Cambridge, UK: Cambridge University Press.

Philippot, K. and Serp, P. (eds.) (2013) *Nanomaterials in Catalysis* Hoboken, NJ: Wiley-VCH Verlag GmbH & Co.

Qi, W.H. and Wang, M.P. (2002) Size effects on the cohesive energy of nanoparticles, *Journal of Material Science Letters*, 21, pp. 1743–1745.

Qin, Y., Feng, J., Xu, J. et al. (2017) Cytotoxicity of TiO_2 nanoparticles toward Escherichia coli in an aquatic environment: Effects of nanoparticle structural oxygen deficiency and aqueous salinity, *Environmental Science: Nano*, 4, pp. 1178–1188.

Raghunath, A., Perumal, E.(2017) Metal oxide nanoparticles as antimicrobial agents: A promise for the future, *International Journal of Antimicrobial Agents*, 49, pp. 137–152.

Rajan, R., Chandran, K., Harper, S.L. et al. (2015) Plant extract synthesized silver nanoparticles: An ongoing source of novel biocompatible materials, *Industrial Crops and Products*, 70, pp. 356–373.

Rakshit, S., Ghosh, S., Chall, S. et al. (2013) Controlled synthesis of spin glass nickel oxide nanoparticles and evaluation of their potential antimicrobial activity: A cost effective and eco friendly approach, *RSC Advances*, 3, p. 19348.

Ranjit, K. and Baquee, A.A. (2013) Nanoparticle: An overview of preparation, characterization and application, *International Research Journal of Pharmacy*, 4, pp. 2230–8407.

Rao, C.R., Muller, A., and Cheetham, A.K. (2004) *The Chemistry of Nanomaterials*. Weinheim: Wiley-VCH.

Reddy, M.P. and Venugopal, A.S. et al. (2007) Hydroxyapatite-supported $Ag–TiO_2$ as Escherichia coli disinfection photocatalyst, *Water Research*, 41, pp. 379–386.

Rodionov, I.A. (2016) Photocatalytic activity of layered perovskite-like oxides in practically valuable chemical reactions, *Russian Chemical Reviews*, 85, pp. 248–279.

Roy, A., Gauri, S.S., Bhattacharya, M.B. et al. (2013) Antimicrobial activity of CaO nanoparticles, *Journal of Biomedical Nanotechnology*, 9, pp. 1570–1578.

Sathishkumar, M., Sneha, K., Yun, Y. et al. (2009) Palladium nanocrystals synthesis using Curcuma longa tuber extract, *International Journal of Material Science*, 4, pp. 11–17.

Santos, E. and Schmickler, W. (2010) Recent advances in theoretical aspects of electrocatalysis. In P. Balbuena, V. Subramanian (eds.) *Theory and Experiment in Electrocatalysis*. Modern Aspects of Electrochemistry, vol. 50. New York: Springer Science+Business Media, pp. 25–88.

Sawai, J., Himizu, K. and Yamamoto, O. (2005) Kinetics of bacterial death by heated dolomite powder slurry, *Soil Biology and Biochemistry*, 37, pp. 1484–1489.

Shah, M., Sharma, S., Tripathy, S.K. et al. (2015) Green synthesis of metallic nanoparticles via biological entities, *Materials*, 8, pp. 7278–7308.

Sharma, G., Kumar, A., Sharma, S. et al. (2017) Novel development of nanoparticles to bimetallic nanoparticles and their composites: A review, *Journal of King Saud University – Science*, 31, pp. 257–269. doi:10.1016/j.jksus.2017.06.012.

Shenton, W., Douglas, T., Young, M. et al. (1999) Inorganic-organic nanotube composites from template mineralization of tobacco mosaic virus, *Advance. Material*, 11, pp. 253–256.

Singh, B., Murad, L., Laffir, F. et al. (2011) Pt based nanocomposites (mono/bi/tri-metallic) decorated using different carbon supports for methanol electro-oxidation in acidic and basic media, *Nanoscale*, 3, p. 3334.

Singh, D. (2014) Synthesis of copper oxide nanoparticles by a novel method and its application in the degradation of methyl orange, *Advanced Electronic and Electrical Engineering*, 4, pp. 83–88.

Singh, R. and Nalwa, H.S. (2011) Medical applications of nanoparticles in biological imaging, cell labeling, antimicrobial agents, and anticancer nanodrugs, *Journal of Biomedical Nanotechnology*, 7, pp. 489–503.

Strappert, S. and Fissan, H. (2003) Higher surface energy of free nanoparticles, *Physical Review Letters*, 91, pp. 106–102.

Thomas, J.M. (1988) Colloidal metals: Past, present and future, *Pure & Applied Chemistry*, 60, pp. 1517–1528.

Thomas, J.M. (2010) The advantages of exploring the interface between heterogeneous and homogeneous catalysis, *ChemCatChem*, 2, pp. 127–132.

Verdier, T., Countand, M., Bertron A. et al. (2014) Antibacterial activity of TiO_2 photocatalyst alone or in coatings on *E. coli:* The influence of methodological aspects, *Coatings*, 4, pp. 670–686.

Wang, W. (2010) Photocatalytic degradation of organic dye methyl orange with phosphotungstic acid, *Journal of Water Resources and Protection*, 2, pp. 979–983.

Wang, F., Cao, B., Mao, C. et al. (2010) Bacteriophage bundles with prealigned Ca^{2+} initiate the oriented nucleation and growth of hydroxylapatite, *Chemistry of Materials*, 22, pp. 3630–3636.

Wang, X., Jonathan,C., Hanson, C. et al. (2005) The structural and electronic properties of nanostructured $Ce1−x−yZrxTbyO_2$ ternary oxides: Unusual concentration of Tb^{3+} and metal↔oxygen↔metal interactions, *The Journal Of Chemical Physics*, 122, p. 154711-1.

Wang, X., Rodriguez, J.A., Hanson, J.C. et al. (2006) In situ studies of the active sites for the water gas shift reaction over $Cu-CeO_2$ catalysts: Complex interaction between metallic copper and oxygen vacancies of ceria, *Journal of Physical Chemistry*, 110, pp. 428–434.

Weller, M.T. (2014) *Solid State Chemistry and Its Applications*. Inorganic Materials Chemistry. Oxford, NY: Wiley.

Wells, A.F. (1987) *Structural Inorganic Chemistry*. 6th ed. New York: Oxford University Press.

Warren, C.W., Shuming N. et al. (2002) Luminescent quantum dots for multiplexed biological detection and imaging Current Opinion in Biotechnology, 13, pp. 40–46

Yael, N., Asnis, J., Urs, O., Bach, H et al. (2017) Metal nanoparticles: Understanding the mechanisms behind antibacterial activity, *Journal of Nanobiotechnology*, 15, p. 65.

Zhou, X. (2012) Photocatalytic degradation of methyl orange over metalloporphyrins supported on TiO_2 degussa P25, *Molecules*, 17, pp. 1149–1158.

Zhang, D. (2013) Synergetic effects of Cu_2O photocatalyst with titania and enhanced photoactivity under visible irradiation, *Acta Chimica Slovaca*, 6, pp. 141–149.

Carbon Nanotube Products from the Floating Catalyst Method

Hai M. Duong and
Thang Q. Tran
National University of Singapore

9.1 Introduction

Due to their outstanding mechanical and physical properties, carbon nanotubes (CNTs) are one of the most fascinating nanomaterials in the materials science since their discovery in 1991 by Iijima (1991). They possess cylindrical nanostructure built from sp^2-hybridized carbon bonded in a hexagonal lattice and can be categorized as single-walled nanotubes (SWNTs), double-walled nanotubes (DWNTs), and multi-walled nanotubes (MWNTs) (Bethune et al. 1993, Iijima 1991, Iijima and Ichihashi 1993). As CNTs have strong carbon–carbon covalent bonds and unique atomistic structures, they exhibit superior mechanical, electrical, and thermal properties. Specifically, their tensile strength is in the range of 11–63 GPa and their Young's modulus could reach 1 TPa (Yu et al. 2000). Furthermore, the thermal conductivity of SWNTs could approach 3,500 W/mK, which surpasses copper (Pop et al. 2006). Therefore, they have great potential in a wide range of applications such as nanotechnology (Tran et al. 2015a, b), advanced electronics (Cheng et al. 2015, Do and Visell 2017), biotechnology (Malarkey and Parpura 2007), nanomechanics (Li and Chou 2004), and other fields of materials science and technology (Do et al. 2016a, b).

The assembly of CNTs into macroscopic and engineered structures with controlled morphologies and properties is highly desirable to improve their real-world applications (Lu et al. 2012). Many methods have been developed to produce different high-performance macroscopic CNT structures such as CNT fibers (Zhang et al. 2004, Li et al. 2004, Ericson et al. 2004), CNT films (Zhang et al. 2005, Fraser et al. 2010, Mirri et al. 2012), and CNT aerogels (Mikhalchan et al. 2016a, Khoshnevis et al. 2018). Those CNT assemblies with high degree of CNT alignment have drawn great attention as their excellent anisotropic properties make them attractive candidates for many applications including biosensors (Wang et al. 2016a), high-performance composites (Cheng et al. 2015, Duong et al. 2016), artificial muscles (Foroughi et al. 2011), transmission wires (Kurzepa et al. 2014), and microelectrodes (Fairchild et al. 2014).

The CNT assemblies have been synthesized by different methods including wet spinning method (Ericson et al. 2004, Mirri et al. 2012), CNT array spinning method (Zhang et al. 2004, 2005), freeze-drying method (Luo et al. 2017), and floating catalyst method (Cheng et al. 2015, Duong et al. 2016, Khoshnevis et al. 2018, Li et al. 2004, Tran et al. 2015a, b). Among these methods, the floating catalyst method is the most widely used approach to synthesize the CNT assemblies thanks to its simple one-step process, good mass production potential, and high cost-efficiency (Janas and Koziol 2016, Tran et al. 2015b). In this method, a feedstock of a carbon source, catalyst precursor, promoter, and carrier gas is injected to a furnace at high temperature (up to 1,300°C) to form a CNT smoke during a chemical vapor deposition (CVD) process (Janas and Koziol 2016). Depending on the collecting methods, the CNT smoke can be pulled out of the furnace and forms different CNT assemblies such as CNT aerogels, CNT films, and CNT fibers (Janas and Koziol 2016, Tran et al. 2015b).

Compared to other fabrication methods, the floating catalyst method can address the issue of scalability in the manufacturing of the CNT products by its one-step process (Janas and Koziol 2016, Tran et al. 2015b). In fact, the high-quality and large-scale CNT aerogels, the CNT films, and the CNT fibers can be fabricated by this method with controlled morphology. Moreover, this method is more advantageous in terms of time consumption, energy and related costs. Natural gas or CO_2 can be used as a major carbon source in its process with high possibility to reduce financial costs and to enable the production to become more environmentally friendly by recycling both carrier gas and any unreacted hydrocarbon (Janas and Koziol 2016, Tran et al. 2015b). In fact, it is the first method to achieve commercial success (Nanocomp Technologies Inc.). In this chapter, we present an overview of the fabrication of the CNT products through the floating catalyst method together with its recent developments and applications.

9.2   Floating Catalyst Method

Figure 9.1a shows the formation process of a CNT smoke by the floating catalyst method using a CVD reactor. When the feedstock of carbon source (hydrocarbon), catalyst (ferrocene), and promoter (thiophene) is injected into the furnace heated at high temperature, these compounds pyrolyze gradually at the regions of non-uniform temperature distribution along the furnace (Gspann et al. 2014). Due to its least stability, ferrocene starts to break down first to provide iron atoms for catalytic sites of the CNT synthesis. When these particles travel through the furnace, they start to agglomerate and grow into larger clusters of iron with increasing temperature (Gspann et al. 2014,

Nasibulin et al. 2005). The decomposition of the hydrocarbon and thiophene then releases carbon fragments and sulfur for the CNT growth. The continuous buildup of CNTs forms a CNT smoke consisting of long CNTs that can be spun continuously into an aerogel, fiber, or film with good CNT alignment (Figure 9.1b). This continuous process can fabricate CNT aerogels with ultra-high deposition rate, or meter-long and thick CNT films, or kilometer-long CNT fibers and CNT thin films. Many process parameters such as the input flowrates of carbon source, catalyst, promoter, and carrier gas, and the synthesis temperature play crucial roles in the successful fabrication of high-quality CNT macroscopic products.

9.2.1   Carbon Source

CNT products with various morphologies have been successfully fabricated from many carbon sources such as methane (Liu et al. 2016a, b, Cheng et al. 2015), ethanol (Li et al. 2004, Motta et al. 2007b), toluene (Gspann et al. 2014, Khoshnevis et al. 2018, Tran et al. 2017), acetone (Song et al. 2014, Lee et al. 2015), and butanol (Alemán et al. 2016, 2015). The successful nucleation and growth of the CNTs require the pyrolysis of the carbon sources to be thermodynamically feasible under the synthesis conditions, and it must occur on the catalyst surface. Since each of these carbon sources has different pyrolysis temperatures, their selection has a significant influence on the availability of the carbon atoms during the synthesis and, therefore, affects the catalyst size and the CNT growth time in the process (Gspann et al. 2014, Sundaram 2012). Moreover, the supply rate of the carbon source in the process must be optimal and continuous for the nucleation and sustenance of the CNT growth (Sundaram 2012). A low carbon input may result in a low CNT yield while oversupply of the carbon source may lead to the deactivation and poisoning of the catalyst due to the formation of non-CNT products or defective

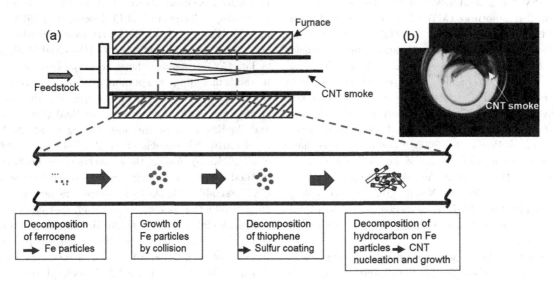

FIGURE 9.1   (a) Schematic of the floating catalyst process with a detailed schematic of the decomposition of the precursors resulting in the formation of an elastic CNT smoke, (b) a CNT smoke formed in the hot furnace.

CNTs and carbon encapsulation. For example, since toluene breaks down at lower temperatures than methane, the CNT fibers spun from the toluene source are stronger with low impurities compared to those spun from the methane source (Gspann et al. 2014).

9.2.2 Carrier Gas

Hydrogen is the most widely used carrier gas in the floating catalyst process, and therefore, most process is conducted in a vacuum or sealed system (Gspann et al. 2014, Motta et al. 2007b, Tran et al. 2015a, b, 2017). Nitrogen can also be used as a carrier gas to fabricate the CNT products in an open-air environment (Xu et al. 2016, Wang et al. 2014). In the CNT synthesis process, the carrier gas transfers the reactants along the CVD furnace and hence determines their retention time in the reaction zone for the reactant decomposition, CNT nucleation from iron particles, and CNT growth (Gspann et al. 2014, Liu et al. 2016b). In the process using hydrogen as a carrier gas, it displaces the equilibrium of the carbon source dehydrogenation to the reactant side by providing the reaction atmosphere reductive. This reduces the formation of excessive carbon that can lead to the formation of catalyst poisoning, defective CNTs, and non-CNT products (Gspann et al. 2014, Sundaram 2012).

Moreover, carrier gas also acts as a diluent to control the collision and growth dynamics of iron catalyst nanoparticles during the synthesis process. A low input of the carrier gas may result in high collision probability of the iron nanoparticles and therefore the formation of catalyst particles of larger diameter, while smaller particles are obtained with higher flow rates of carrier gas (Conroy et al. 2010, Motta et al. 2007a). Due to the strong correlation between catalyst size and CNT diameter during the CNT growth process, CNTs with small diameter can be fabricated by using high flow rates of carrier gas at the same carbon input (Motta et al. 2007a). However, excessive input of the carrier gas may result in the suppression of the thermolysis of carbon source and hydrocarbonation of CNTs. Besides, carrier gas also plays an important role in controlling the degree of CNT alignment and the fabrication rate of the CNT products (Tran et al. 2016b, Alemán et al. 2015). In fact, CNT assemblies synthesized at high input of carrier gas possess better CNT alignment and higher fabrication rate.

9.2.3 Catalyst

The floating catalyst process mostly uses ferrocene to provide iron nanoparticles as the standard catalyst for the CNT synthesis (Janas and Koziol 2016). While the catalyst sustains the thermolysis of the carbon source during the process, its size and state determine the type of carbon diffusion at a given temperature and therefore affect the nucleation and growth rate of CNTs (Sundaram 2012). Notably, the melting point of catalyst nanoparticles depends on their sizes, and it has been reported that the iron catalyst

particles are in liquid state during the CNT growth in the floating catalyst process due to the high synthesis temperature (1,000–1,300°C) (Alemán et al. 2016, Gspann et al. 2014). Furthermore, since the diameter of the growing CNTs is strongly correlated to the catalyst particle size at the point of nucleation, regulating the catalyst size is an important approach to control the CNT morphologies in the CVD process. For instance, small catalysts are reported to grow small few-walled CNTs, while large catalysts usually led to large-diameter CNTs with multiple walls (Conroy et al. 2010, Janas and Koziol 2016).

In the synthesis process, the catalyst nanoparticles are repelled away from the wall due to the "Ludwig–Soret" effect (Gspann et al. 2014). They grow larger by collision or coalescence, resulting in a catalyst-size distribution (Conroy et al. 2010, Gspann et al. 2014). The nonlinear velocity of the catalyst nanoparticles and furnace temperature distributions significantly affect the mechanics of the synthesis process (Conroy et al. 2010). Since the catalyst particles nearer the furnace wall move much slower than those in the furnace center, they have a longer residence time. Hence, their probability to collide and grow to larger sizes is higher, restricting their activity for the CNT growth. The catalyst particles in the furnace center, however, move faster and have an appropriate residence time for CNT growth with much lower collision probability (Conroy et al. 2010, Janas and Koziol 2016).

The size of the catalyst nanoparticles is significantly affected by their concentrations at the supply point and the reaction temperature. It is mainly determined by the decomposition temperature of the carbon source on the catalyst surface, ferrocene, and promoter compounds (Gspann et al. 2014). These temperatures correspond to different zones in the furnace due to the furnace temperature profile (Gspann et al. 2014, Lee et al. 2015). Additionally, the input flowrates of ferrocene and the carrier gas have a significant influence on the catalyst size by varying the catalyst concentration. Under the conditions of optimal carbon supply, appropriate catalyst size can prolong the CNT growth time and therefore increase the CNT length (Gspann et al. 2014, Janas and Koziol 2016). The CNT growth is usually terminated by the coarsening of the catalyst or their carbon poisoning. During the synthesis process, the catalyst nanoparticles are often incorporated into the CNT products as impurities, depending on the carbon supply, catalyst concentration, and particle-size distribution (Sundaram 2012).

9.2.4 Promoter

In the floating catalyst method, sulfur or sulfur compounds such as thiophene (Li et al. 2004, Motta et al. 2007a, Tran et al. 2016b, 2017), elemental sulfur (Lee et al. 2015), and carbon disulfide (Motta et al. 2008) have been used to fabricate high-quality CNT products. The role of sulfur in the synthesis process of the CNT products is manifold. First, it restricts the conglomeration of the catalyst nanoparticles and therefore controls their size. After being coated with sulfur, the iron catalyst particles possess strong iron–sulfur bonds and become more resistant to coalescence by collision (Gspann et al. 2014). Therefore, the temperature of atomic sulfur release should be controlled to arrest further growth of the catalyst particles and to preserve the appropriate catalyst size for the CNT growth.

Moreover, sulfur increases the nucleation rate of CNTs on the surface of the catalyst and hence improves the CNT growth reaction (Gspann et al. 2014, Motta et al. 2008, Sundaram 2012). It has been reported that growth rate of the CNTs in the floating catalyst process can be as high as 0.1–1 mm/s (Motta et al. 2008). Due to the low simultaneous solubility of carbon and sulfur in iron liquid at high temperature, the sub-surface diffusion of sulfur could modify the surface energies of the catalyst particles, leading to the surface diffusion of carbon with low solubility. Therefore, few-walled small-diameter CNTs can be formed by even very large iron catalyst particles (Motta et al. 2008).

9.2.5 Synthesis Temperature and Injector Tube Length

The synthesis temperature of CNT products must be sufficiently high (1,000–1,300°C) to pyrolyze the carbon sources, catalyst precursor, and promoter for the synthesis reaction (Janas and Koziol 2016). The process with excessively high synthesis temperatures could form amorphous carbon due to noncatalytic fragmentation of the carbon sources (Sundaram 2012). Different synthesis temperatures could influence the availability of the reactants for the synthesis reaction because of the strong correlation between the temperature gradient of the furnace and the synthesis temperature. Thus, it would determine the morphologies and quality of the CNT products. Additionally, injector tube length could alter the initial flow rate of the reactants and control their availability for the synthesis process (Paukner et al. 2014, Lee et al. 2015).

9.3 Aligned CNT Aerogels from Floating Catalyst Method

9.3.1 Fabrication of CNT Aerogels

A substrate is placed at the outlet of the CVD furnace to collect the continuous CNT smoke (Figure 9.2a) to fabricate the CNT aerogels. The most widely used substrates are sapphire, silicone oxide, or quartz due to their high working temperature, exceptional hardness, and scratch resistance (Mikhalchan et al. 2016a, Khoshnevis et al. 2018). During the synthesis process, the CNT smoke is deposited onto the substrate continuously to form a three-dimensional (3D) CNT network (CNT aerogel) via layer-by-layer stacking, as shown in Figure 9.2b.

The deposition rate along the thickness direction of the aerogels can be as high as 60 mm/h, while their

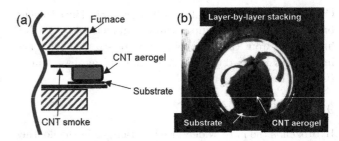

FIGURE 9.2 Fabrication of CNT aerogels from the floating catalyst process. (a) Schematic of CNT aerogel fabrication and (b) layer-by-layer stacking of CNT smoke.

other dimensions can be controlled by the substrate size (Mikhalchan et al. 2016a, Khoshnevis et al. 2018). By varying the size of the substrates and the deposition time, a wide range of CNT aerogels with different sizes and densities can be fabricated. For example, large but ultralight CNT aerogels with densities of 0.55–1.50 mg/cm^3 could be fabricated using a substrate of 5.0 × 8.5 cm at different deposition time (Figure 9.3a and b) (Mikhalchan et al. 2016b). As shown in Figure 9.3c, these CNT aerogels are noticeably lighter than those synthesized via critical point drying, freeze-drying, boron-doping CVD, and CVD methods, and are even comparable with the lightest graphene aerogels ever reported (Mikhalchan et al. 2016b).

9.3.2 Morphology and Advanced Multi-properties of CNT Aerogels

Figure 9.4 shows the structures and electrical properties of the CNT aerogels synthesized by the floating catalyst method. The aerogels consist of 3D pore networks, predominantly formed by mesopore (2–50 nm) (Mikhalchan et al. 2016b). Generally, the CNT assemblies fabricated by the floating catalyst method consist of highly oriented CNT bundles aligned in the synthesis direction. The CNT aerogel structure could be considered as an example of the reticulated and extremely open foam, where only the lineal boundaries remain (shown by CNT bundles). Meanwhile, all the classical "bubbles" or "cells" (attributes of the closed cells or semiclosed foams) are disappeared. The mesopores of several nanometers may be formed by the distance between the CNT bundles in the points of their contact and entanglement (Mikhalchan et al. 2016b).

The CNT aerogels also possess high surface area (up to 170.1 m^2/g), high pore volume (0.276 cm^3/g), and high porosity (>98%) (Mikhalchan et al. 2016b). Macropores might be formed as the spare space of hundreds of nanometers between the bundles, and they are difficult to be analyzed by conventional method such as N2 adsorption/desorption method. The specific surface area of the CNT aerogels is comparable to that of several aerogels synthesized by the freeze-drying methods or critical point drying (Mikhalchan et al. 2016b). Their porous structure and surface area could be enhanced by minimizing the cosynthesized impurities, controlling morphologies of the

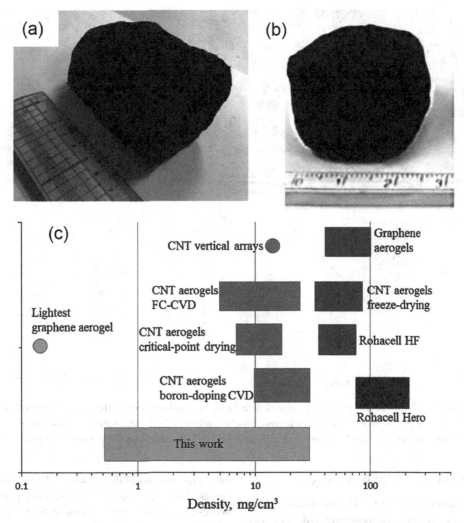

FIGURE 9.3 (a) The 1.50 mg/cm^3-density CNT aerogel, (b) the typical 20 mg/cm^3-density aerogel synthesized in 15–20 min, and (c) density range of the CNT aerogels fabricated by floating catalyst in comparison with other aerogels.

synthesized CNTs, or chemical post-treatments, whereas their density can be varied by using densification techniques (Mikhalchan et al. 2016b).

The CNT aerogels fabricated by the floating catalyst method are reported to have anisotropic behavior in their electrical conductivity with the in-plane conductivity three to five times higher than that in the through-thickness direction (Figure 9.4d) (Mikhalchan et al. 2016b). This may stem from the fact that some CNT bundles might have intrinsic preferences or more compact organization in one plane rather than another due to the self-deposition process. However, numerous strong and divergent connections are formed throughout the whole 3D structure of the CNT aerogels (Mikhalchan et al. 2016b). During the synthesis process, the elastic CNT smoke initially forms a planar network of randomly oriented CNT bundles, followed by a layer-by-layer stacking process to form the 3D aerogel structure. Their through-thickness conductivity (20–27 S/m) is comparable to that of the pure graphene, graphene/CNT aerogels, or the CNT aerogel coated with graphene, while the in-plane conductivity (90–106 S/m) is much better than the

CNT aerogels fabricated by the critical point drying, freeze-drying, or boron-doping CVD methods (Mikhalchan et al. 2016b).

The thermal conductivities of the CNT aerogels measured through-thickness are only 0.127–0.237 W/mK, comparable to that of other highly porous carbon materials such as commercial carbon fiber paper (0.210 W/mK) and graphene-based aerogels (0.120–0.360 W/mK) (Mikhalchan et al. 2016b). The low thermal conductivities of the CNT aerogels might stem from the possible structural defects and the coupling of CNTs into bundles, the low density, and extremely high porosity (>98%) of the aerogels with all tortuous pores occupied by air (thermal insulator, 0.026 W/mK). It is expected that the thermal conductivity of the CNT aerogels would be significantly affected by the complex morphology, structural density, and contact resistance between CNTs. Due to their low thermal conductivity and high surface area, the CNT aerogels have great potentials in a broad range of applications such as thermoacoustics, thermal insulation, and other applications for which this synergy is profitable (Mikhalchan et al. 2016b).

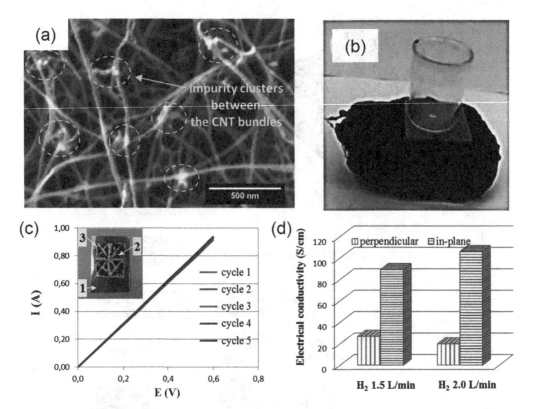

FIGURE 9.4 (a) Typical CNT network of the CNT aerogels fabricated by floating catalyst, (b) the 0.55 mg/cm^3-density CNT aerogel withstanding a weight of 8.5 g (~150 times higher than its own weight, and ~15,000 times higher than its density) without mechanical collapse, (c) representative current–voltage (I-V) curves (five cycles), and (d) anisotropic electrical behavior of the CNT aerogels.

9.3.3 Applications of CNT Aerogels

The high sorption properties of the CNT aerogels make them exceptional materials for environment-targeted functions such as sorption, filtration, and separation (Khoshnevis et al. 2018). Figure 9.5 shows the water contact angle and oil adsorption process of the CNT aerogels reported by Khoshnevis et al. (2018). As can be seen, the CNT aerogels are superhydrophobic with a water contact angle of 154°. Besides, they could float on the surface of water due to their

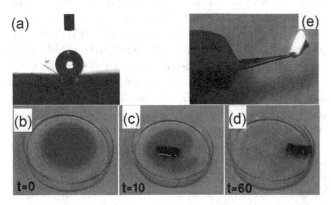

FIGURE 9.5 (a) Water contact angle on the surface of CNT aerogels, (b–d) snapshots of the adsorption of 5W50 motor oil staining with dye on the CNT aerogels, and (e) burning CNT aerogels after oil adsorption.

low densities and hydrophobic surfaces. Figure 9.5b–d illustrates the snapshots of oil adsorption process of the CNT aerogels from oil–water system at 25°C. The aerogels could adsorb oil from the water very fast, and the sorption capacities can be more than 100 g/g, depending on the oil types (Khoshnevis et al. 2018).

After the CNT aerogels become saturated with oil, they can be squeezed or burnt to eliminate the oil with low damages to their structure. In practice, the direct burning of the absorbed oil for heat conversion can be conducted at a power station before the aerogels are returned to the site of the environmental clean-up for reuse. Due to their high thermal and structure stability, the CNT aerogels exhibit good recyclability with a slight decrease in their sorption capacity after many adsorption-burning cycles, suggesting the great potential of the CNT aerogels as sorbent materials (Khoshnevis et al. 2018).

Additionally, the porous structure of the CNT aerogels remains easily accessible to host any other molecules of interest, thus broadening their practical utilization for advanced functional composite. For example, CNT aerogel-based composites are manufactured using polydimethylsiloxane (PDMS) as a matrix due to their inert, nontoxic, and non-flammable characteristics which are favorable for advanced flexible electronic composites and biointegrated applications (Mikhalchan et al. 2016b). Figure 9.6 shows the distinctively different surface patterns of the CNT aerogel/PDMS composite. The CNT bundles are fully

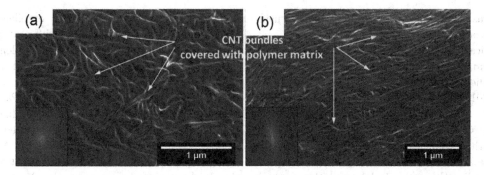

FIGURE 9.6 Surface morphology of the CNT aerogel/PDMS composite with (a) randomly and (b) preferentially aligned CNT bundles as a consequence of CNT arrangement and polymer infiltration. The inserts are derived from the two-dimensional (2D) fast Fourier transform (FFT) analysis of the corresponding images, showing (a) random and (b) anisotropic intensity distribution.

covered with the matrix and some of them are preferentially aligned and stacked together. The polymer infiltration process might induce densification of the CNT network into a more compact structure via merging and rearrangement of some CNT bundles, while still maintaining their 3D macro-structure and flexibility (Mikhalchan et al. 2016b).

The electrical conductivity of the CNT aerogels is about 33 S/m, higher than values reported for the PDMS composites with CNT vertical arrays, graphene, and CNT powder at the same weight fraction (Mikhalchan et al. 2016b). The significant improvement in their electrical conductivity compared to the pure PDMS matrix evidences that the CNT-percolated network is not disturbed by the polymer macromolecules upon infiltration and curing processes. Additionally, their thermal conductivity is much higher than that of the pure matrix and the CNT aerogel itself. This results may be attributed to the formation of a more compact CNT network during the polymer infiltration process, resulting in the increased CNT contacts and higher opportunities for phonon transferring between neighboring CNTs. Besides, air within the spaces between the CNT bundles is replaced by the polymer, leading to the relatively lower thermal resistance and therefore more effective thermal transport within the composite (Mikhalchan et al. 2016b).

9.4 Aligned CNT Films from Floating Catalyst Method

9.4.1 Fabrication and Morphologies of CNT Thin Films

To fabricate CNT films, a winding system is placed at the outlet of the CVD furnace to directly collect the continuous CNT smoke (Figure 9.7a and b) (Cheng et al. 2015, Liu et al. 2016b). The width and length of the films are determined by the size of the winders, while the film thickness is controlled by the winding time via layer-by-layer stacking (Figure 9.7b) (Liu et al. 2016b, Liu et al. 2016a, c). At high winding speed, the CNT films own high degree of CNT alignment and packing density due to the great tension applied to the

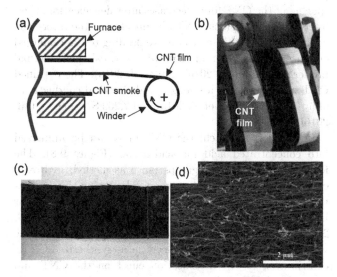

FIGURE 9.7 (a) Schematic of fabrication of CNT films from the floating catalyst process, (b) layer-by-layer stacking of CNT smoke in the fabrication process of the CNT films, (c) photograph of CNT films spun at 30 min, and (d) scanning electron microscope (SEM) images of surface morphologies of as-spun CNT films.

CNT smoke before its collection. As shown in Figure 9.7d, the as-synthesized CNT films consist of aligned CNTs and CNT bundles with a highly porous structure. They possess a low density, ranging from 0.30 to 1.39 g/cm³ (Liu et al. 2016b). The density and alignment of the CNT films can be improved by different post-treatments such as densification and purification.

9.4.2 Advanced Multi-properties and Post-treatments of CNT Thin Films

Due to the high porosity with many pores and spaces in their structures, the as-spun CNT films usually exhibit a low performance with mechanical strength, electrical conductivity, and thermal conductivity in a range of 0.1–0.2 GPa, 500–1,400 S/cm, and 100–120 W/mK, respectively (Liu et al. 2016b, Liu et al. 2016a, c). To improve the multi-properties of the CNT films for desired applications, many post-treatment methods have been developed such as liquid densification, mechanical densification, and acidization.

Liquid densification and mechanical densification are widely used to improve the performance of the CNT films due to its simplicity and effectiveness (Liu et al. 2016b, Gspann et al. 2017, Xu et al. 2016). In the liquid densification method, a liquid such as acetone or ethanol is absorbed by the CNT films before it is evaporated. The CNT films are then condensed into highly packed structures due to the surface tension of the solvents (Gspann et al. 2017). Regarding the mechanical densification method, a spatula (Liu et al. 2016b) or pressurized rolling system (Xu et al. 2016) can be used to apply much larger densification pressure onto the CNT film structure. Since the inter-tube spacing of the CNT films decreases after densification, their inter-tube interactions, CNT alignment, and the inter-tube contact areas significantly improve, leading to the remarked enhancement in their mechanical, electrical, and thermal properties (Liu et al. 2016b, Xu et al. 2016). The densified CNT films can reach a strength and electrical conductivity of up to 2.8 GPa (Xu et al. 2016) and 2,200 S/cm (Liu et al. 2016b), respectively.

In another approach, the CNT films can be immersed into concentrated acid for acidization (Figure 9.8). The formation of functional groups (such as methyl, hydroxyl, carbonyl, and methylene) on the CNTs during the acidization process results in the dipole–dipole interaction or hydrogen bonding between CNT bundles (Liu et al. 2016a, c). Such inter-tube interactions are stronger than the van der Waals interaction between the as-synthesized CNTs, contributing to the densification effect on the CNT film structure. Additionally, the acid treatment also purifies the CNT films by removing the amorphous and reactive carbon on the as-synthesized CNT films (Liu et al. 2016a, c). These carbonaceous impurities may serve as "defect" in the CNT film structure, limiting the inter-tube contact and inter-tube interaction and hence lowering the film performance. During the acid treatment, these amorphous carbons are oxidized and removed by the subsequent aqueous washing, leading to the remarkable enhancement in the CNT films' properties (Liu et al. 2016a, c). After the acidization, the CNT films could reach an electrical and thermal conductivity of up to 4,700 S/cm and 760 W/mK (Liu et al. 2016b).

The CNT films can achieve excellent performance by combining positive effects of different post-treatment methods. For example, both liquid densification and mechanical densification can be applied to significantly reduce the CNT film thickness and enhance their packing density. A tensile strength of up to 9.6 GPa and stiffness of up to 130 GPa are reported for the CNT films densified by a hybrid treatment of ethanol and pressurized rolling system, as presented in Figure 9.9 (Xu et al. 2016). Additionally, the hybrid treatment of mechanical densification and acidization is also reported to improve the electrical conductivity of the as-synthesized CNT films up to 5,000 S/cm (Liu et al. 2016b).

9.4.3 Applications of CNT Thin Films

Due to their excellent electrothermal properties, the conductive CNT thin films spun from the floating catalyst method exhibit great potentials to be the next generation of heating materials. The CNT films have been employed as resistive heaters with numerous advantages compared to classical resistive heaters (Figure 9.10a) (Janas and Koziol 2013). They are more lightweight, faster, and more efficient than classical resistive heaters made of nichrome or Kanthal. More importantly, their flexible and lightweight characteristics make them attractive to be used as flexible resistive heaters with the potential deposition onto any desired sites. Those unique properties are crucial in aerospace applications where lightweight requirements are critical. For example, CNT films spun from the floating catalyst method demonstrated high performance in a deicing system on a model of an aircraft wing (Janas and Koziol 2013), as shown in Figure 9.10b.

The CNT films spun from the floating catalyst method can be used to fabricate energy storage devices and sensors. A flexible supercapacitor with high strength and toughness can be fabricated by using the CNT film as a substrate for the efficient pulsed electrodeposition of polyaniline (PANI) (Benson et al. 2013). This advanced supercapacitor exhibits rapid ion adsorption with specific capacitances much better than that of the commercial activated carbon powders. Additionally, a composite of CNT thin films and nickel hydroxides can be synthesized by a facile electrochemical

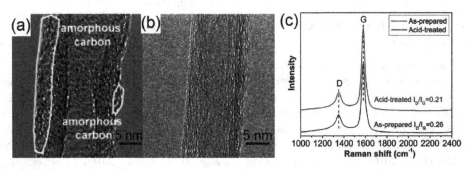

FIGURE 9.8 Transmission electron microscope (TEM) images of (a) as-synthesized CNT films and (b) acid-treated CNT films, (c) Raman spectra of the as-synthesized and acid-treated CNT films.

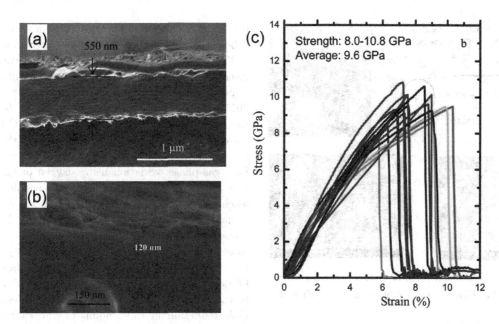

FIGURE 9.9 SEM images showing thin thickness of the CNT films densified by (a) ethanol and (b) both ethanol and pressurized rolling system, (c) stress versus strain curves of CNT thin films densified by the both ethanol and pressurized rolling system.

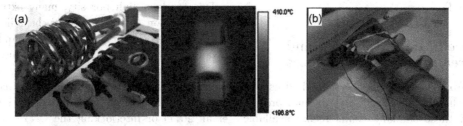

FIGURE 9.10 (a) A CNT film heater and (b) a model of an aircraft wing for deicing purposes using a CNT film heater.

deposition to produce an electric double-layer capacitor with good electrochemical performance and a high specific capacitance of over 1,200 F/g (Cheng et al. 2015). More importantly, with their good electrical conductivity and mechanical rigidity, these CNT-based supercapacitors are promising for the replacement of traditional supercapacitors made of metals that feature heavy weight and low energy density.

Additionally, the aligned CNT films spun from the floating catalyst method are excellent reinforcement for high-performance and multifunctional composites due to their high strength, electrical, and thermal properties. Many studies have reported the significant improvements in mechanical, electrical, and thermal properties of the CNT-based composite versus the baseline polymer. For example, CNT film/epoxy composite produced by a vacuum-assisted resin transfer molding method (VARTM) (Figure 9.11) exhibited an increase in strength and Young's modulus by a factor of 10 and 5, respectively, while their toughness could be as high as 6.39×10^3 kJ/m^3 (Liu et al. 2015a). More interestingly, their electrical conductivity could reach 252.8 S/cm, which is 20 times higher than that of the CNT/epoxy composites obtained by conventional dispersion methods.

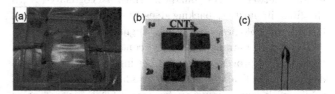

FIGURE 9.11 (a) AVRTM process for the fabrication of CNT/epoxy films, (b) CNT/epoxy films with 1, 5, 10, and 20 CNT plies, and (c) a flexible CNT/epoxy film bent by tweezers.

9.5 Aligned CNT Fibers from Floating Catalyst Method

9.5.1 Fabrication and Morphology of CNT Fibers

To fabricate CNT fibers, an ethanol/acetone spraying system is often used to densify the CNT smoke into fibers before their collection by the winder (Liu et al. 2016b, Motta et al. 2007a, Tran et al. 2016b, 2017), as shown in Figures 9.12a and b. Generally, acetone is the best solvent for the densification due to their high surface tension (Kai et al. 2010). Similar to the CNT films,

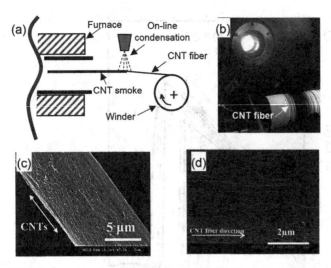

FIGURE 9.12 (a–b) Fabrication of CNT fibers from floating catalyst process, SEM images of CNT fibers (c) and their surface morphologies (d).

the CNT fibers also consist of aligned CNT bundles with many pores and spaces in their porous structure (Figures 9.12c and d).

9.5.2 Advanced Multi-properties and Post-treatments of CNT Fibers

CNT fibers spun from the floating catalyst method usually have excellent mechanical, electrical, and thermal properties which exceed those of many commercially available fibers presently (Lu et al. 2012, Koziol et al. 2007). Their performance is strongly determined by the inter-tube interactions between CNT bundles, which is correlated with the fiber impurities, packing density of the CNT bundles, CNT morphologies, and CNT alignment (Koziol et al. 2007). Those parameters can be controlled by optimizing the synthesis process to reduce fiber impurities, increase the degree of CNT alignment, and produce long, thin-walled CNTs with large diameters for larger CNT contact areas (Lu et al. 2012, Koziol et al. 2007) or by post-treatments (Tran et al. 2016b, 2017, Liu et al. 2016b). Notably, the mechanical properties of the direct-spun CNT fibers are highly length-dependent, with a strength of 1 GPa for 20 mm gauge length and 8.8 GPa for 1 mm gauge length (Koziol et al. 2007). The direct-spun CNT fibers with a dense structure and good alignment can exhibit a thermal conductivity of up to 770 ± 10 W/mK (Gspann et al. 2017), while the best reported electrical conductivity of the CNT fibers spun from the floating catalyst method in literature is 8,300 S/cm (Li et al. 2004).

Similar to the CNT thin films, the performance of the CNT fibers spun from the floating catalyst method can be significantly improved by different post-treatments including purification, densification, infiltration of extraneous materials, acidization, or their combination. Those post-treatments also help to improve the interactions between CNT bundles in the CNT fibers, resulting in better inter-tube load-transfer efficiency and lower inter-tube contact resistance, as discussed in Section 9.2.2.

The CNT fibers spun from the floating catalyst method often possess many iron catalyst particles or carbonaceous impurities which lower their performance (Tran et al. 2016b, 2017). These impurities can be removed by purification treatments. Most catalyst particles in the CNT assemblies are encapsulated in graphitic carbon cages that restrict the direct removal of the impurities by acid washing. Therefore, gas-phase oxidation using oxygen or air can be used to crack open these carbon layers to enable their efficient leaching by acids (Figure 9.13) (Tran et al. 2017). Additionally, the direct-spun CNT fibers can be purified by immersing the fibers in concentrated acid for a specific time (Liu et al. 2016b, Tran et al. 2016b). The acidization time should be optimum for the purification effect to be dominant on the CNT structural modification, leading to the removal of the amorphous carbon despite the negligible reduction in the amount of iron catalyst impurities (Tran et al. 2016b). Long acidization time is not favorable due to possible damage of the CNT structure (Tran et al. 2016b).

Due to their high porosity, many extraneous materials such as carbon (Lee et al. 2016), hexadiene (Boncel et al. 2011), copper (Hannula et al. 2016), and graphene oxide (Wang et al. 2016b) can be infiltrated into the structure of the CNT fibers to improve their performance. The treatment can improve the inter-tube interactions by forming cross-links between the CNT bundles (Figure 9.14) (Boncel et al. 2011) or interlocking the CNT bundles (Lee et al. 2016, Wang et al. 2016b) to enhance their shear interactions. Similar to the CNT films, the inter-tube interactions of the CNT fibers can also be enhanced by mechanical densification (Tran et al. 2016b, Wang et al. 2014). The CNT fibers densified by a spatula (Tran et al. 2016b) or pressurized rolling system (Wang et al. 2014) exhibit a denser structure with less pores and better CNT alignment. Since more van der Waals interactions between CNT bundles and higher inter-tube contact areas are obtained, the performance of the CNT fibers improves significantly (Wang et al. 2014, Tran et al. 2016b).

The direct-spun CNT fibers with excellent performance can also be fabricated by applying two different post-treatments. Some common hybrid treatments are the combination of mechanical densification and polymer infiltration (Tran et al. 2016b), or mechanical densification and acid treatment (Liu et al. 2016b), or purification and chemical cross-linking (Im et al. 2017), or purification and polymer infiltration (Tran et al. 2016b). Due to the combined effects of the hybrid treatments, the multi-properties of the CNT fibers improve significantly and are comparable to many commercial high-performance fibers. For example, CNT fibers treated with mechanical densification and epoxy infiltration could obtain an impressive enhancement factor of 13.5 times for strength and 63 times for stiffness, which are much better than many other single post-treatments (Tran et al. 2016b) (Figure 9.15). The strength and stiffness of

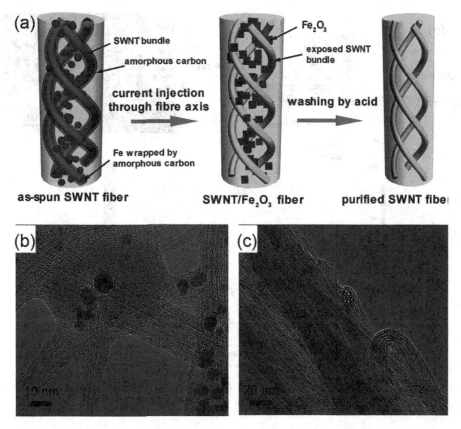

FIGURE 9.13 (a) Purification process of the direct-spun CNT fibers by instantaneous current injection and acid washing, TEM images of CNTs (b) before and (c) after purification.

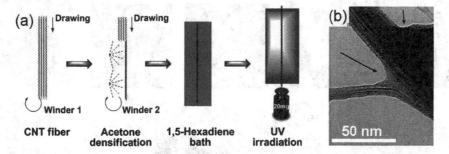

FIGURE 9.14 (a) Process for chemical treatment of the direct-spun CNT fibers by hexadiene and (b) TEM image of CNT bundles coated by extraneous materials.

the hybrid-treated CNT fibers could reach 3.6 and 266 GPa, respectively, comparable to those of commercial PAN carbon fibers (Tran et al. 2016b).

9.5.3 Applications of CNT Fibers

Due to their aligned structure, the CNT fibers spun from the floating catalyst method have been used as conductive wires to take advantage of the superior longitudinal properties of the individual CNTs (Janas and Koziol 2016, Lu et al. 2012). The continuous process of the floating catalyst method allows the production of conductive wires based on the CNT fibers with unlimited length and controlled morphologies for specific applications. For example, fully semiconductive CNTs are favorable for applications featuring transistors while metallic SWNTs with high electrical conductivity are beneficial for applications in power and data transmission (Janas and Koziol 2016).

Electrical insulation of CNT wires has been studied to ensure the proper current flows and prevent short-circuits (Lekawa-Raus et al. 2014a, Janas and Koziol 2016). Standard polymers such as polyethylene can be used successfully as insulation for the CNT-based wires (Lekawa-Raus et al. 2014a), while advanced polyimide aerogel can be a good insulation in lightweight and high-temperature applications (Liu et al. 2015b), as shown in Figure 9.16a. Additionally, the successful joining of the CNT wires with classical conductors by soldering (Figure 9.16b) leads to great potential for electrical/electronic and mechanical applications (Burda et al. 2015). Moreover, one important advantage of

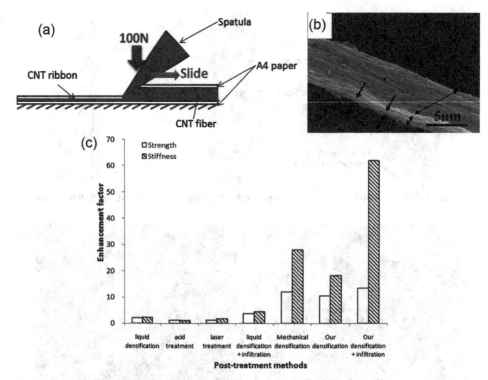

FIGURE 9.15 (a) Schematic diagram of mechanical densification of direct-spun CNT fibers by a spatula, (b) SEM images of densified CNT fibers, and (c) comparisons of enhancement factors of different post-treatments.

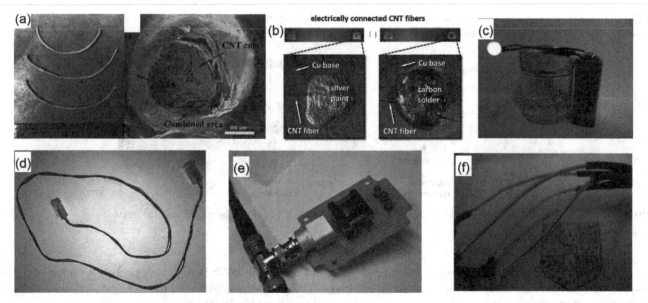

FIGURE 9.16 (a) CNT fibers coated with polyimide aerogel, (b) the solders on a Cu base to join CNT fibers using silver conductive paint and Sn3.6Ag0.7Cu2.5Cr alloy, (c) CNT wire submerged in concentrated sulfuric acid while it is used to drive electricity to power a green diode, (d) Ethernet cables, (e) electrical transformer, and (f) USB cables based on CNT wiring.

the CNT wires compared to traditional materials is their outstanding performance in applications that require harsh chemical conditions (Janas et al. 2013) or high temperatures (Janas et al. 2013), as shown in Figure 9.16c. Besides, many electronic devices based on CNT wiring such as electrical transformers (Kurzepa et al. 2014), USB (Janas et al. 2014), and Ethernet cables (Lekawa-Raus et al. 2014b) have been fabricated successfully with excellent performance and

are comparable to commercially available products (Figures 9.16 d–f).

Direct-spun CNT fibers with high thermal conductivity have also been used as cathodes of field emitters. They exhibit a stable emission in the current range of 1–2 mA and at cathode temperatures of up to 1,000°C (Fairchild et al. 2014). Moreover, the change in electrical properties of the CNT fibers and the epoxy-infused CNT fibers resulting from the densification effects caused by twisting suggests the great potential of the direct-spun CNT fibers as embedded torsional sensors in composite materials (Wu et al. 2012). Similar to the CNT films, the CNT fibers can be used as an excellent reinforcements for high-performance composite. The aligned CNT/epoxy composites employing direct-spun CNT fibers as reinforcement are reported to possess considerable pull-out of CNT bundles at the fracture surface (Mikhalchan et al. 2016b). This unique failure behavior suggests that subsidiary cracks are bridged by CNT bundles, resulting in structures reminiscent of crazes in polymer matrix, and higher fracture toughness compared to those of the conventional aligned carbon fiber composites (Mikhalchan et al. 2016b).

9.6 Conclusions and Suggestions

This chapter presents an overview of the floating catalyst method that uses a high-temperature horizontal CVD furnace to produce macroscopic CNT aerogels, films, and fibers. The method can fabricate high-quality and large-scale CNT products with controlled morphology in a single step while they possess many advantages in terms of energy, costs, time consumption, and waste materials. The CNT products with desired properties and controlled morphologies can be produced by optimizing different synthesis parameters such as the flow rate of reactants, synthesis temperature, and winding speed, whereas different post-treatments such as purification, acidization, and densification can be applied to significantly enhance their multi-properties for many practical applications. In fact, potential applications for the CNT aerogels, films, and fibers have been demonstrated in several multifunctional product prototypes such as oil absorbents, conductive wire, energy storage devices and sensors, composite materials, resistive heaters, and field emitters. Therefore, high-performance CNT products synthesized from floating catalyst method will potentially be incorporated into many different *commercial applications* in the upcoming future.

References

Alemán, B., Bernal, M. M., Mas, B., et al. 2016. Inherent predominance of high chiral angle metallic carbon nanotubes in continuous fibers grown from a molten catalyst. *Nanoscale* 8:4236–4244.

Alemán, B., Reguero, V., Mas, B. & Vilatela, J. J. 2015. Strong Carbon Nanotube Fibers by Drawing Inspiration from Polymer Fiber Spinning. *ACS Nano* 9:7392–7398.

Benson, J., Kovalenko, I., Boukhalfa, S., et al. 2013. Multifunctional CNT-polymer composites for ultra-tough structural supercapacitors and desalination devices. *Advanced Materials* 25:6625–6632.

Bethune, D. S., Klang, C. H., De Vries, M. S., et al. 1993. Cobalt-catalysed growth of carbon nanotubes with single-atomic-layer walls. *Nature* 363:605–607.

Boncel, S., Sundaram, R. M., Windle, A. H. & Koziol, K. K. K. 2011. Enhancement of the mechanical properties of directly spun CNT fibers by chemical treatment. *ACS Nano* 5:9339–9344.

Burda, M., Lekawa-Raus, A., Gruszczyk, A. & Koziol, K. K. K. 2015. Soldering of carbon materials using transition metal rich alloys. *ACS Nano* 9:8099–8107.

Cheng, H., Koh, K. L. P., Liu, P., Thang, T. Q. & Duong, H. M. 2015. Continuous self-assembly of carbon nanotube thin films and their composites for supercapacitors. *Colloids and Surfaces A: Physicochemical and Engineering Aspects* 481:626–632.

Conroy, D., Moisala, A., Cardoso, S., Windle, A. & Davidson, J. 2010. Carbon nanotube reactor: Ferrocene decomposition, iron particle growth, nanotube aggregation and scale-up. *Chemical Engineering Science* 65:2965–2977.

Do, T. N., Seah, T. E. T., Yu, H. & Phee, S. J. 2016a. Development and testing of a magnetically actuated capsule endoscopy for obesity treatment. *PLoS One* 11:e0148035.

Do, T. N., Seah, T. E. T. & Phee, S. J. 2016b. Design and control of a mechatronic tracheostomy tube for automated tracheal suctioning. *IEEE Transactions on Biomedical Engineering* 63:1229–1238.

Do, T. N. & Visell, Y. 2017. Stretchable, twisted conductive microtubules for wearable computing, robotics, electronics, and healthcare. *Scientific Reports* 7. doi:10.1038/s41598-017-01898-8

Duong, H. M., Gong, F., Liu, P. & Tran, T. Q. 2016. Advanced Fabrication and Properties of Aligned Carbon Nanotube Composites: Experiments and Modeling.

Ericson, L. M., Fan, H., Peng, H., et al. 2004. Macroscopic, neat, single-walled carbon nanotube fibers. *Science* 305:1447–1450.

Fairchild, S. B., Bulmer, J. S., Sparkes, M., et al. 2014. Field emission from laser cut CNT fibers and films. *Journal of Materials Research* 29:392–402.

Foroughi, J., Spinks, G. M., Wallace, G. G., et al. 2011. Torsional carbon nanotube artificial muscles. *Science* 334:494–497.

Fraser, I. S., Motta, M. S., Schmidt, R. K. & Windle, A. H. 2010. Continuous production of flexible carbon nanotube-based transparent conductive films. *Science and Technology of Advanced Materials* 11:1–6.

Gspann, T. S., Juckes, S. M., Niven, J. F., et al. 2017. High thermal conductivities of carbon nanotube films and micro-fibres and their dependence on morphology. *Carbon* 114:160–168.

Gspann, T. S., Smail, F. R. & Windle, A. H. 2014. Spinning of carbon nanotube fibres using the floating catalyst high temperature route: Purity issues and the critical role of sulphur. *Faraday Discussions* 173:47–65.

Hannula, P. M., Peltonen, A., Aromaa, J., et al. 2016. Carbon nanotube-copper composites by electrodeposition on carbon nanotube fibers. *Carbon* 107:281–287.

Iijima, S. 1991. Helical microtubules of graphitic carbon. *Nature* 354:56–58.

Iijima, S. & Ichihashi, T. 1993. Single-shell carbon nanotubes of 1-nm diameter. *Nature* 363:603–605.

Im, Y. O., Lee, S. H., Kim, T., et al. 2017. Utilization of carboxylic functional groups generated during purification of carbon nanotube fiber for its strength improvement. *Applied Surface Science* 392: 342–349.

Janas, D., Cabrero-Vilatela, A., Bulmer, J., Kurzepa, L. & Koziol, K. K. 2013a. Carbon nanotube wires for high-temperature performance. *Carbon* 64: 305–314.

Janas, D., Herman, A. P., Boncel, S. & Koziol, K. K. K. 2014. Iodine monochloride as a powerful enhancer of electrical conductivity of carbon nanotube wires. *Carbon* 73:225–233.

Janas, D. & Koziol, K. K. 2013. Rapid electrothermal response of high-temperature carbon nanotube film heaters. *Carbon* 59:457–463.

Janas, D. & Koziol, K. K. 2016. Carbon nanotube fibers and films: Synthesis, applications and perspectives of the direct-spinning method. *Nanoscale* 8:19475–19490.

Janas, D., Vilatela, A. C. & Koziol, K. K. K. 2013b. Performance of carbon nanotube wires in extreme conditions. *Carbon* 62:438–446.

Kai, L., Yinghui, S., Ruifeng, Z., et al. 2010. Carbon nanotube yarns with high tensile strength made by a twisting and shrinking method. *Nanotechnology* 21:045708.

Khoshnevis, H., Mint, S. M., Yedinak, E., et al. 2018. Super high-rate fabrication of high-purity carbon nanotube aerogels from floating catalyst method for oil spill cleaning. *Chemical Physics Letters* 693:146–151.

Koziol, K., Vilatela, J., Moisala, A., et al. 2007. High-performance carbon nanotube fiber. *Science* 318: 1892–1895.

Kurzepa, L., Lekawa-Raus, A., Patmore, J. & Koziol, K. 2014. Replacing copper wires with carbon nanotube wires in electrical transformers. *Advanced Functional Materials* 24:619–624.

Lee, J., Kim, T., Jung, Y., et al. 2016. High-strength carbon nanotube/carbon composite fibers: Via chemical vapor infiltration. *Nanoscale* 8:18972–18979.

Lee, S. H., Park, J., Kim, H. R., Lee, J. & Lee, K. H. 2015. Synthesis of high-quality carbon nanotube fibers by controlling the effects of sulfur on the catalyst agglomeration during the direct spinning process. *RSC Advances* 5:41894–41900.

Lekawa-Raus, A., Kurzepa, L., Peng, X. & Koziol, K. 2014a. Towards the development of carbon nanotube based wires. *Carbon* 68:597–609.

Lekawa-Raus, A., Patmore, J., Kurzepa, L., Bulmer, J. & Koziol, K. 2014b. Electrical properties of carbon nanotube based fibers and their future use in electrical wiring. *Advanced Functional Materials* 24:3661–3682.

Li, C. Y. & Chou, T. W. 2004. Strain and pressure sensing using single-walled carbon nanotubes. *Nanotechnology* 15:1493–1496.

Li, Y. L., Kinloch, I. A. & Windle, A. H. 2004. Direct spinning of carbon nanotube fibers from chemical vapor deposition synthesis. *Science* 304:276–278.

Liu, P., Fan, Z., Mikhalchan, A., et al. 2016a. Continuous carbon nanotube-based fibers and films for applications requiring enhanced heat dissipation. *ACS Applied Materials and Interfaces* 8:17461–17471.

Liu, P., Hu, D. C. M., Tran, T. Q., Jewell, D. & Duong, H. M. 2016b. Electrical property enhancement of carbon nanotube fibers from post treatments. *Colloids and Surfaces A: Physicochemical and Engineering Aspects* 509:384–389.

Liu, P., Lam, A., Fan, Z., Tran, T. Q. & Duong, H. M. 2015a. Advanced multifunctional properties of aligned carbon nanotube-epoxy thin film composites. *Materials and Design* 87:600–605.

Liu, P., Tan, Y. F., Hu, D. C. M., Jewell, D. & Duong, H. M. 2016c. Multi-property enhancement of aligned carbon nanotube thin films from floating catalyst method. *Materials and Design* 108:754–760.

Liu, P., Tran, T. Q., Fan, Z. & Duong, H. M. 2015b. Formation mechanisms and morphological effects on multi-properties of carbon nanotube fibers and their polyimide aerogel-coated composites. *Composites Science and Technology* 117:114–120.

Lu, W., Zu, M., Byun, J. H., Kim, B. S. & Chou, T. W. 2012. State of the art of carbon nanotube fibers: Opportunities and challenges. *Advanced Materials* 24:1805–1833.

Luo, S., Luo, Y., Wu, H., et al. 2017. Self-assembly of 3D carbon nanotube sponges: A simple and controllable way to build macroscopic and ultralight porous architectures. *Advanced Materials* 29. doi:10.1002/adma.2016 03549

Malarkey, E. B. & Parpura, V. 2007. Applications of carbon nanotubes in neurobiology. *Neurodegenerative Diseases* 4:292–299.

Mikhalchan, A., Fan, Z., Tran, T. Q., et al. 2016a. Continuous and scalable fabrication and multifunctional properties of carbon nanotube aerogels from the floating catalyst method. *Carbon* 102:409–418.

Mikhalchan, A., Gspann, T. & Windle, A. 2016b. Aligned carbon nanotube–epoxy composites: the effect of nanotube organization on strength, stiffness, and toughness. *Journal of Materials Science* 51:10005–10025.

Mirri, F., Ma, A. W. K., Hsu, T. T., et al. 2012. High-performance carbon nanotube transparent conductive films by scalable dip coating. *ACS Nano* 6:9737–9744.

Motta, M., Kinloch, I., Moisala, A., et al. 2007a. The parameter space for the direct spinning of fibres and films of carbon nanotubes. *Physica E: Low-Dimensional Systems and Nanostructures* 37:40–43.

Motta, M., Moisala, A., Kinloch, I. A. & Windle, A. H. 2007b. High performance fibres from 'dog bone' carbon nanotubes. *Advanced Materials* 19:3721–3726.

Motta, M. S., Moisala, A., Kinloch, I. A. & Windle, A. H. 2008. The role of sulphur in the synthesis of carbon nanotubes by chemical vapour deposition at high temperatures? *Journal of Nanoscience and Nanotechnology* 8:2442–2449.

Nasibulin, A. G., Pikhitsa, P. V., Jiang, H. & Kauppinen, E. I. 2005. Correlation between catalyst particle and single-walled carbon nanotube diameters. *Carbon* 43:2251–2257.

Paukner, C. & Koziol, K. K. K. 2014. Ultra-pure single wall carbon nanotube fibres continuously spun without promoter. *Scientific Reports* 4. doi:10.1038/srep03903.

Pop, E., Mann, D., Wang, Q., Goodson, K. & Dai, H. 2006. Thermal conductance of an individual single-wall carbon nanotube above room temperature. *Nano Letters* 6: 96–100.

Song, J., Kim, S., Yoon, S., Cho, D. & Jeong, Y. 2014. Enhanced spinnability of carbon nanotube fibers by surfactant addition. *Fibers and Polymers* 15:762–766.

Sundaram, R. M. 2012. *Production, Characterisation and Properties of Carbon Nanotube Fibres*, University of Cambridge, Cambridge.

Tran, T., Liu, P., Fan, Z., Ngern, N. & Duong, H. 2015a. Advanced multifunctional properties of aligned carbon nanotube-epoxy composites from carbon nanotube aerogel method. APS Meeting Abstracts, 17006.

Tran, T. Q., Fan, Z., Liu, P. & Duong, H. M. 2015b. Advanced morphology-controlled manufacturing of carbon nanotube fibers, thin films and aerogels from aerogel technique. Asia Pacific Confederation of Chemical Engineering Congress 2015: APCChE 2015, incorporating CHEMECA 2015, Engineers Australia, 2444.

Tran, T. Q., Fan, Z., Liu, P., Myint, S. M. & Duong, H. M. 2016a. Super-strong and highly conductive carbon nanotube ribbons from post-treatment methods. *Carbon* 99:407–415.

Tran, T. Q., Fan, Z., Mikhalchan, A., Liu, P. & Duong, H. M. 2016b. Post-treatments for multifunctional property enhancement of carbon nanotube fibers from the floating catalyst method. *ACS Applied Materials and Interfaces* 8:7948–7956.

Tran, T. Q., Headrick, R. J., Bengio, E. A., et al. 2017. Purification and dissolution of carbon nanotube fibers spun from the floating catalyst method. *ACS Applied Materials and Interfaces* 9:37112–37119.

Wang, H., Liu, Z., Ding, J., et al. 2016a. Conducting fibers: Downsized sheath–core conducting fibers for weavable superelastic wires, biosensors, supercapacitors, and strain sensors (Adv. Mater. 25/2016). *Advanced Materials* 28:4946.

Wang, J. N., Luo, X. G., Wu, T. & Chen, Y. 2014. High-strength carbon nanotube fibre-like ribbon with high ductility and high electrical conductivity. *Nature Communications* 5:3848.

Wang, Y., Colas, G. & Filleter, T. 2016b. Improvements in the mechanical properties of carbon nanotube fibers through graphene oxide interlocking. *Carbon* 98: 291–299.

Wu, A. S., Nie, X., Hudspeth, M. C., et al. 2012. Carbon nanotube fibers as torsion sensors. *Applied Physics Letters* 100. doi:10.1063/1.4719058

Xu, W., Chen, Y., Zhan, H. & Wang, J. N. 2016. High-strength carbon nanotube film from improving alignment and densification. *Nano Letters* 16:946–952.

Yu, M. F., Lourie, O., Dyer, M. J., et al. 2000. Strength and breaking mechanism of multiwalled carbon nanotubes under tensile load. *Science* 287:637–640.

Zhang, M., Atkinson, K. R. & Baughman, R. H. 2004. Multifunctional carbon nanotube yarns by downsizing an ancient technology. *Science* 306:1358–1361.

Zhang, M., Fang, S., Zakhidov, A. A., et al. 2005. Materials science: Strong, transparent, multifunctional, carbon nanotube sheets. *Science* 309:1215–1219.

Gold Nanoparticles by Green Chemistry

Rakesh Kumar Sharma,
Sneha Yadav, and Sriparna
Dutta
University of Delhi

10.1 Introduction

10.1.1 The Explosive Growth and Contribution of Nanotechnology to Mankind

The last few decades have experienced unprecedented transformations in political, economic and social spheres of society. These revolutions are stated to be kindled from various technological developments that have taken place in the nascent field of science and technology.[1] Nanotechnology has emerged as a noteworthy interdisciplinary technology in the 21st century. It has metamorphosed several technological and industrial sectors. Additionally, it is foretoken as a significant technology that smartly amalgamates varied disciplines of physics, chemistry, biology, medicine and engineering. It is a unique platform that provides ubiquitous exemplar of the association of scientists, engineers, entrepreneurs and industry representatives.[2,3] Richard Feynman, physicist and Nobel laureate, during his lecture in 1959 challenged the scientific community to think small by stating that **"There's Plenty of Room at the Bottom"**. He was the first person who ignited the spark of interest in the mind of researchers by revolutionizing the notion of "nano". The word "nanotechnology" was later on coined and popularized by Eric Drexler, who then stated

it to be a future technology for designing and engineering materials.[4,5] The structure-specific privilege and properties associated with nanotechnology stem from the fact that these materials offer the feasibility of tailoring the structures at even small scale and consequently allow automation of such tasks that were earlier found to be inaccessible due to physical restrictions. It is believed that the emergent field of nanotechnology offers the potential to permeate and improve all industrial sectors and influence their development towards sustainability. The major sections that have been privileged as a consequence of nanotechnology include the manufacturing sector, nanomedicine, energy production, electronic sector, better water purification and food protection methods (Figure 10.1).[6]

The pervasive developments have been highly appreciated by the scientific community, yet they usher various undesirable repercussions (such as ecological and environmental damage) along with it. Nanotechnology poses serious threat to mankind and environment, because the developed nanomaterial might possess toxicity. The toxicity associated with them is related to their increased mobility and reactivity. These potential hazards allied to the development of nanoparticles have been recognized widely. Moreover, the United States Environmental Protection Agency (USEPA) has started working in this direction. In fact, in Australia and UK, organic food sector has excluded the

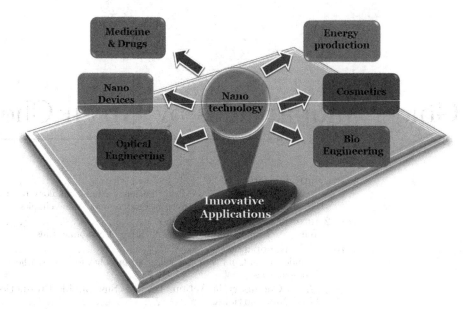

FIGURE 10.1 Applications of nanotechnology.

designed nanoparticles that are produced as a result of organic material.[7]

It has been a long-time misconception that nanotechnology could solve all the problems of sustainability by acting as "magic bullets". However, nanotechnology if wisely designed and applied could be an integral constituent of sustainable development.

10.1.2 Synergistic Integration of Green Chemistry and Nanotechnology

Very rightly said by John C. Warner, **"Nano technology and green chemistry have an intimate relationship and great potential to do good"**. Green nanotechnology that merges green chemistry and green engineering with nanotechnology offers the potential to address the problems related to health and environmental hazards as it provides a new vision on the peculiar and counter instinctive upshots that occur in nanoscale materials. The fusion of nanotechnology with green chemistry better termed as "green nanofactory" has provided a great impetus to the researchers as it efficiently fulfills indispensable criteria that strive to surpass natural processes. Dr. James Hutchis very well said that **"Green chemistry is a terrific way to do nanotechnology responsibly"**.[8,9]

The symbiotic integration of the principles of green chemistry with green engineering transcends sectoral boundaries by allowing the synthesis of benign novel products that promise profound upgradation in various spheres of life. It encompasses three complementary goals: (i) fosters the development of economically competitive and productive technology that incorporates nanotechnology, (ii) supports sustainable nanotechnology that minimizes the risks associated with the human health and environment, while designing and using nanotechnology and its products, and

(iii) endorses the renewal of current products with new lifecycle-based greener nanoproducts that are less burdensome to the environment and human health.[10] However, sustainability factors that restrain various strategies and approaches adopted by research engineers have always been under scrutinization. Although it is highly inappropriate to consider nanotechnology to be as green as Mother Nature, the developments or the advancements in the field of nanotechnology could be made green. This new paradigm shift towards greener nanotechnology is built on the pillars of interdisciplinary innovation that holds the potential to transform the nature of technology.

Nowadays, nanoproducts have grabbed an important position in the marketplace, and the number is expected to grow manifold in the coming years. Developing nanoproducts clean and green from the starting is a delightful strategy to outwit the hazards related to nanotechnology. Since nanotechnology is in its growing stage, it provides a wonderful prospect of including green methods and green designs for the product synthesis. Six nanotechnology sectors that are driven by greener and sustainable approach have been extensively explored by the research community as they are accompanied with biosynthetic routes having greater efficiency and reliability, which not only reduce the manufacturing costs but also reduce toxicity by eliminating hazardous substances. The sectors that have gained substantial momentum include nano-based solar cells, nano-generators, fuel catalysis, water treatment, energy storage and thermal energy.[11,12] It is proposed that in the coming years, a new arena of industries could evolve that incorporates green chemistry engineering principles in product design and synthesis. Indeed, it is splendid to say that the newly designed products via the aid of green technology may replace all existing products in the marketplace.

10.2 Introduction to Nanomaterials

10.2.1 What are Nanomaterials?

The word "nano" has been derived from the Greek word "nanos" which means dwarf. Nanomaterials, having dimension in the nanometer scale, belong to that class of materials in which the size of particles varies from 1 to 100 nm, and thus, they act as a bridge between bulk and atomic structures (Figure 10.2).[13] The growing significance of nanomaterials in the field of science and technology stems from the fact that they exhibit size- and shape-dependent physicochemical properties such as catalytic, magnetic, and optical properties.[14] For instance, when the macroscale materials are transformed into nanoscale, their properties vary. Many opaque materials like copper become transparent on being reduced to nanoscale. Similarly, chemically inert metal-gold becomes soluble and also acts as potent chemical catalyst on being reduced to nanoscale dimensions. These intriguing properties are attributed to the various quantum and surface phenomena that exist at nanoscale. In fact, the spherical particles also present additional advantages and applications due to an increase in surface area/volume ratio parameter. The surface area/volume ratio parameter is found to be inversely proportional to the radius, and increase in the surface area/volume ratio or significant reduction in the size of nanoparticles triggers the enhancement in various size-related physical properties. In addition, they are rich in chemically active sites, and therefore, their surface could be functionalized via various linking and surface protecting agents such as surfactants and polymers.[15,16]

Nanomaterials can be synthesized via two approaches: one is "TOP-DOWN" method, and another one is "BOTTOM-UP" method (Figure 10.3). The two methods differ from each other in their strategies as well as in the idea or discipline involved in its synthesis. TOP-DOWN approach is a destructive approach that entails breaking down of large bulk-sized material into nanosized material by milling, sputtering, and chemical etching technique. It is a quite fast method that requires high amount of energy, generates lot of waste, leads to imperfections in the surface structure of the synthesized nanoparticles, and hence is not suitable for large-scale synthesis of nanoparticles. BOTTOM-UP approach on the other hand is building up of small atoms and molecules for the generation of larger nanostructures. It broadly consists of gaseous-phase and liquid-phase methods. The former one, i.e., the gaseous phase, involves chemical vapor deposition (CVD) and physical vapor deposition methods (PVD), while the latter one includes spray pyrolysis, laser pyrolysis, sol–gel, solvothermal and co-precipitation techniques.[17]

10.2.2 Classification of Nanomaterials

In the past few decades, an enormous number of novel nanomaterials have been successfully synthesized. Nanomaterials can be broadly classified into four types based on their dimensions (Figure 10.4).[18]

1. **Zero-dimensional nanomaterials**: Zero-dimensional nanomaterials are those materials in which all the dimensions are measured within the nanoscale. It has been observed that nanoparticles belong to this class as they possess zero dimensions. Gold nanoparticles, silver nanoparticles and quantum dots are classic examples of zero dimension nanoparticles.

2. **One-dimensional nanomaterials**: In these materials, one dimension is found to be outside the nanoscale. One-dimensional nanoparticle includes nanorods, nanoribbons, nanowires, nanotubes and nanobelts. Although these materials are quite long, the diameter is found to be of only a few nanometers. One-dimensional nanoparticles have gained significant attention owing to their diverse applications in the field of nanoelectronics, nanodevices and nanocomposites.

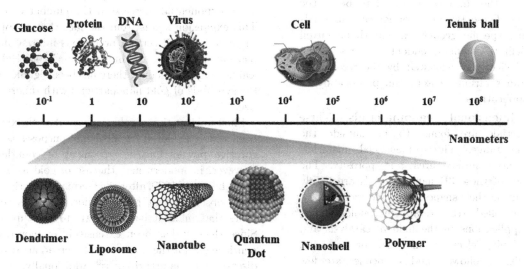

FIGURE 10.2 How small is small?

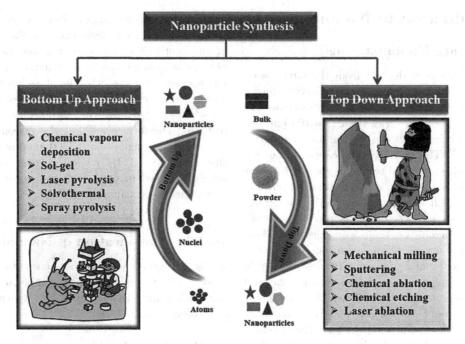

FIGURE 10.3   Synthetic approaches of nanomaterials.

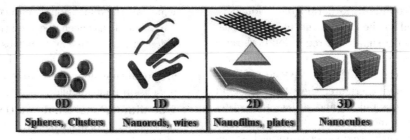

FIGURE 10.4   Types of nanomaterials based on their classification.

3. **Two-dimensional nanomaterials**: These nanostructures have two dimensions outside the nanoscale. Two-dimensional nanoparticles include nanoprisms, nanosheets, nanoplates, nanowalls and nanodisks. Although these materials possess large area, the thickness is found to be in the nanoscale. These two-dimensional nanomaterials possessing specific geometries and shape unveil some definite shape-dependent features which is being effectively exploited by the researchers in designing various sensors and photocatalysis-based applications.

4. **Three-dimensional nanomaterials**: These nanomaterials are found to be outside the nanoscale range. Three-dimensional materials possess large surface area and porosity. The properties of these 3D nanomaterials are highly dependent on their shapes, sizes and morphology. Their controlled structure and morphology offers varied applications in the field of catalysis and for developing electrode material for batteries. Nanoballs, nanoflowers and nanocones are few examples of this class of materials.

10.3   Gold Nanoparticles

10.3.1   Significance of Gold Nanoparticles

In recent years, the design and synthesis of nanoparticles has experienced rapid expansion/unprecedented boom resulting in an exponential increase in the number of publications. This expansion stems from the fact that nanoparticles lay the foundation of certain prodigious innovations that took place in last few decades in various technological and biological fields. In particular, there has been significant growth in the synthesis of gold nanoparticles with tailored shape and size.

Since ancient times, gold considered as "Elixir of Life" has been widely explored for medicinal purposes as it possesses curative powers for the treatment of countless diseases. However, in modern age, the use of gold as a therapeutic agent began in 1890, after the German bacteriologist Robert Koch observed that gold cyanide possesses certain properties that make it bacteriostatic to the tubercle bacillus. Since then, it has been extensively employed as a therapeutic agent for the treatment of a broad array of rheumatic diseases such as arthritis.[19,20] Additionally, the upsurge of interest in the properties of gold nanoparticles as well as the mechanisms underpinning its synthetic methods has created renaissance again in the emerging field of nanoscience. It was English scientist Michael Faraday who discovered that gold salts can be further reduced to gold nanoparticles. Indeed, he also delivered a lecture on the "Experimental Relations of Gold to Light" at the Royal Institute in London.[21]

Furthermore, numerous articles have been published in literature which clearly signifies the progress in the field of gold nanoparticles. Gold nanoparticles of various shapes such as spheres, rods, shells, cages, star, belt and prism can be synthesized.

Chemistry of Gold:

Gold is noble metal having atomic number 197 and exists in a number of oxidation states. Nevertheless, in aqueous solution, gold is found mainly in (0), (I) and (III) oxidation states which remains in equilibrium with each other. Metallic gold is inert, while gold in I and III oxidation states is unstable as gold III is easily reduced by thiol-based reductants to gold I and gold I readily reacts with S donor-based ligands. Gold I compounds have a strong tendency to react with thiol-based S donor atoms, rather than oxygen and nitrogen donor atoms. Moreover, these thiolate-based ligands also provide a perquisite for further ligand exchange-based reactions for desired applications.[19]

Properties of gold nanoparticles:

Gold nanoparticles possess certain unique physico-chemical attributes that make them excellent scaffolds as potent tools in various applications (Figure 10.5). Gold nanoparticles are highly stable and resistant to oxidation. In addition, they exhibit high surface area-to-volume ratio, low toxicity and outstanding biocompatibility. Additional enthusiasm for gold nanoparticles emanates from the fact that these gold nanoparticles also show excellent prospects of surface functionalization with various linking and protecting agents that render them appealing for use in various biomedical applications such as cancer therapy, sensing, cellular imaging and drug delivery.[22] Also, they offer tunable size- and shape-dependent optical properties.

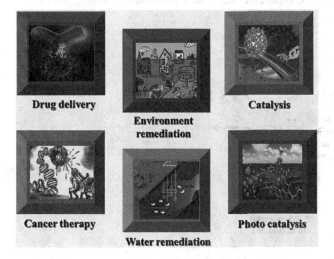

FIGURE 10.5 Gold nanoparticles for multifaceted intents.

For instance, spherical gold nanoparticles exhibit a wide range of colors from red wine color to brown, orange and purple. Similarly, gold nanorods are of blue or black, while metallic gold is of golden yellow color. In addition, gold nanoparticles also exhibit a distinctive phenomenon called "surface plasmon resonance" (SPR). These nanoparticles show absorption peak from 500 to 550 nm range. This absorption band is a result of the collective oscillations that is exhibited by the free conduction band electrons which are highly polarizable. Moreover, the surface of gold nanoparticles also supports surface plasmons, and when incident light is irradiated on the gold nanoparticles, these surface plasmons resonate with the wavelength of the incident light causing electrons to move under the influence of external field which then displays the optical phenomenon.[23,24]

10.3.2 Approaches for the Synthesis of Gold Nanoparticles

The synthesis of gold nanoparticles with precise control over shape, size and purity represents a huge milestone. Also, the large-scale industrial production of gold nanoparticles has not been achieved yet by the scientific community. These days, a large section of researchers have put their endeavors towards achieving the abovementioned goal. Literature reports clearly document that a number of methodologies have been reported and reviewed for the synthesis of gold nanoparticles. A few of them have been reported here.

Chemical Reduction Method

The very first synthetic approach that comes to our mind is the renowned Turkevich or citrate reduction, a conventional method for the synthesis of gold nanoparticles. In this classic approach, tetrachloroauric acid is reduced to gold nanoparticles via the aid of citric acid in boiling water. Here, citric acid plays the role of a reducing agent as well as that of a stabilizing agent, ultimately leading to the formation of water-soluble and citrate-capped gold nanoparticles having diameter of about 10–20 nm. The crucial factor that controls the size of resulting nanoparticles is the varying ratio of gold to citric acid.[25]

Another technique that created a breakthrough in the arena of gold nanoparticles is the well-known Brust–Schiffrin method for the synthesis of thiol-protected gold nanoparticles. This is a two-phase process, employing sodium borohydride as a reducing agent along with a phase transfer reagent, in which chemical reduction takes place at the oil–water interface. In this approach, gold chloride is transferred from the aqueous phase to toluene via the aid of tetraoctylammonium bromide (TOAB)-based surfactant. They are further reduced by sodium borohydride in the presence of thiol, for instance, dodecanethiol, which plays the role of a capping and stabilizing agent in the abovementioned scheme. Another remarkable attribute of the protocol is that these thiol-functionalized gold nanoparticles are found to be highly stable as a result of the strong gold thiol interactions.[26]

Additionally, numerous preparative methods have been reported where a diverse range of reducing agents such as hydrazine hydrate, hydroxylamine, ascorbic acid, glycerol, and oxalic acid have been efficiently and extensively used for the synthesis of negatively charged gold particles. Significant progress has been made towards the synthesis of negatively charged gold nanoparticles, wherein polyethyleneimine has been employed as a reducing agent.

Physical Methods

a. UV-assisted synthesis of gold nanoparticles

Photoinduced ultraviolet radiations-based synthesis is a promising approach for the synthesis of single-crystallite gold nanoparticles. Moreover, this approach allows the synthesis of gold nanoparticles with controlled morphology and dimensional features (Figure 10.6). These advantages offer prospects of making use of gold nanoparticles in various industrial applications such as photocatalysis, metallurgy, and for the fabrication of magnetic devices.[27]

Chen et al. reported the shape-controlled synthesis of gold nanoparticles at room temperature. They elucidated that the concentration of Au cations, polymer capping species and irradiated time play a significant role in the morphology of the resulting nanoparticles. They revealed that when solution containing Au cations was irradiated for around 48 h, peak at around 525 nm was observed in the UV-visible spectrum, thus clearly indicating the formation of gold nanoparticles.[28]

b. Ultrasonic-mediated synthesis of gold nanoparticles

Ultrasound-mediated synthesis is a relatively faster technique that occurs at ambient temperature. While synthesizing gold nanoparticles, use of ultrasonic radiations offers an opportunity to control the rate of reduction of gold nanoparticles. Water is generally employed in ultrasonic synthesis. At times, various stabilizers such as citrate and disulfide are added for the controlled and tunable synthesis of gold nanoparticles.

c. Microwave-assisted synthesis of gold nanoparticles

The use of microwave was first reported in 1986 by Gedye et al. for the synthesis of organic compounds. In fact, microwave has now been used for the synthesis of nanoparticles with controlled nucleation, growth and shape.[29] Kou et al. reported the microwave-irradiated synthesis of gold nanoparticles using glycerol as the reducing agent. In order to synthesize the desired gold nanoparticles with varying morphologies, authors reported the use of cetyltriethylammonium bromide (CTAB), polyvinylpyrrolidone (PVP) and sodium dodecyl sulfate (SDS)-based surfactants. It was observed that the use of CTAB resulted in the formation of hexagonal gold nanosheets, while the use of PVP resulted in the formation of monodispersed gold nanoparticles. In case of SDS, triangular gold nanosheets were obtained at high concentration of the surfactant, while lower concentration led to the formation of nanoparticles. In addition, microwave also offers adjustable heating rate for controlled formation of nanoparticles.[30]

d. Laser ablation-facilitated synthesis of gold nanoparticles

Laser ablation in liquid is a fast and efficient method for the synthesis of gold nanoparticles. Duan et al. reported the synthesis of gold nanoparticles via this process. In this technique, laser beam is focused onto a gold plate present in the deionized water. When the laser interacts with the gold target, high-temperature and high-pressure gold plasma is produced. Further, expansion of hot plasma takes place followed by cooling, which

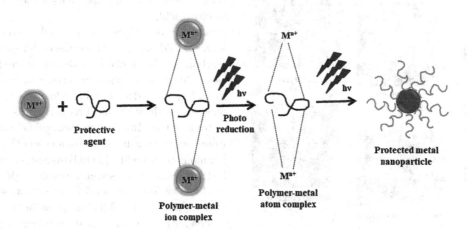

FIGURE 10.6 UV-irradiated reduction of gold ions to yield nanoparticles.

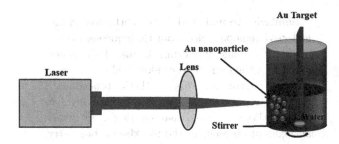

FIGURE 10.7 Laser ablation-assisted synthesis of gold nanoparticles.

then results in the fabrication of gold nanoparticles. Laser ablation method is considered to be an accurate approach in terms of dimensional and morphological features, in comparison with its counterparts (Figure 10.7). This approach entails concurrent evaporation and condensation mechanism and therefore allows tunable synthesis of gold nanoparticles via the reduction of gold (III) tetrachloroaurate salt.[31]

Microbial Method

A variety of biological sources present in nature such as microorganisms and enzymes could be efficiently accessed for the synthesis of gold nanoparticles. In addition to being cost-effective, these biological synthetic approaches are also eco-friendly and produce no harm to the environment. Moreover, they also provide an opportunity to design gold nanoparticles of desired shape, size and morphology which can be functionalized and further exploited for specific applications in clinical and biomedical domain. It is presumed that certain enzymes for instance ligninases, reductases and laccases play a key role in the sophisticated growth and crystal nucleation of the nanoparticles. In fact, the presence of certain amino acids and cysteine in the microbes also helps in the reduction and further stabilization of these nanoparticles. Ample of studies have been reported in literature wherein microbes have been employed for the synthesis of gold nanoparticles. Immanuel et al. reported Klebsiella pneumonia-based bacterium-mediated synthesis of gold nanoparticles.[32] Besides, several proteins like α-nicotinamide adenine dinucleotide phosphate hydrogen (NADPH)-based reductase, hydrolase and amylase present in fungi also help in the formation of metallic gold nanoparticles.

Drawbacks of the Conventional Strategies

It is worth mentioning here that the literature is flooded with a number of reports wherein several approaches including chemical reduction and physical methods have demonstrated the synthesis of gold nanoparticles. However, due to increasing societal and environmental concerns, these methods suffer from numerous drawbacks as they employ certain toxic chemicals and reagents along with hazardous equipment and technology. Indeed, several biogenic protocols using microorganisms such as bacteria,

fungi, algae and yeast have been reported in literature for the synthesis of gold nanoparticles. Nevertheless, the extensive sensitivity of these microorganisms with changes in the external conditions such as pH and temperature might inactivate the microbes. Also the requirement for specialized and expensive culture maintenance for the bioreduction process limits their practical applicability to various industrial uses.[33]

10.4 Green Chemistry in Action: Towards Sustainable Production of Gold Nanoparticles

In the 21st century, sustainable development has become a subject of paramount importance. The incorporation of green chemistry principles in nanotechnology has opened a door to the sustainable and environment-friendly synthesis of gold nanoparticles. However, the development of greener routes that can provide a ready access to the gold nanoparticles still remains to be a major challenge at the industrial level. Thus, immense attention has been paid towards exploring new sustainable and synthetic techniques that utilize non-toxic, environmentally benign chemicals and solvents.

10.4.1 Green Synthetic Approaches

A great deal of research efforts has been directed towards designing green chemistry approaches for the fabrication of gold nanoparticles. In this regard, ample of studies present in literature underpin the idea of utilizing green reducing and stabilizing agents for the biofabrication of gold nanoparticles. Plants contain a variety of antioxidant secondary metabolites. The fabrication of gold nanoparticles via plant-mediated approaches requires mainly three elements: the stabilizing agent, reducing agent and solvent medium. An exceptional attribute of plant-mediated synthesis is that the phytoconstituents present in plants play the key role of reducing and stabilizing agent. Moreover, green synthesis of nanoparticles using plants fulfills all the criteria of green chemistry as they generally occur at low temperature and pressure. Also, binding of certain functional molecules like phytomolecules to the surface of nanoparticles guides the shape and morphology of the synthesized nanoparticles. For instance, gold nanoprisms have been synthesized from the extract of lemon grass (Cymbopogon flexuosus). Two major approaches are known for the plant-mediated fabrication of gold nanoparticles.[34,35]

1. **Living plant/Plant biomass-mediated synthesis**

   This approach also known as intracellular synthesis takes into account the phenomenon of phytoextraction and phytoremediation. It was in 2002, when Gardea-Torresdey et al. for the first time reported the synthesis of gold nanoparticles inside the living plants of Alfalfa (Medicago sativa). For the accomplishment of the abovementioned goal, alfalfa plants were grown into gold-rich environment which then absorbed the gold metal. Authors further confirmed the controlled growth and nucleation of gold nanoparticles inside the plant.[34] Similarly, Sharma et al. reported the sesbania-mediated synthesis of gold nanoparticles in the plant tissues. Armendariz and co-workers described the synthesis of gold nanoparticles by oat (Avena sativa) biomass that efficiently converted Au (III) ions to Au (0) nanoparticles. HRTEM revealed that this methodology leads to the formation of the gold nanoparticles of face-centered cubic (FCC) tetrahedral, decahedral, hexagonal and rod shape.[36] A report present in literature by Gardea-Torresdey et al. clearly demonstrated the effect of different concentrations of thiocyanate for the synthesis of gold nanoparticles. Ammonium thiocyanate was utilized as a complexing agent to enhance the gold phytoextraction by *chilopsis linearis*. The concentration of ammonium thiocyanate was found to have pronounced effect on the uptake of gold by roots and stems X-ray absorption spectroscopy (XAS) studies disclosed the reduction of Au (III) to Au (0) by *chilopsis linearis* within the roots and shoots of the plants. In addition, extended X-ray absorption fine structure (EXAFS) studies revealed the formation of gold nanoparticles having an approximate radius of 0.55 nm. However, with this approach, scant reports are present in literature, owing to the involvement of difficult and expensive processing steps for recovering nanoparticles from the biomass.[37]

2. **Plant extract-mediated synthesis**

In this approach, phyto-constitutes present in the plants, which are responsible for reducing metal ions, are extracted, and therefore, it is also termed as an extracellular synthesis. Numerous reports present in literature clearly exemplify the green synthesis of gold nanoparticles either from the whole plant or through its components such as roots, leaves, barks, fruits, flowers and stems. For instance, Suman et al. reported the synthesis of gold nanoparticles from the aqueous root extract of Morinda citrifolia tree. Field emission scanning electron microscopy (FESEM) and transmission electron microscopy (TEM) studies divulged the formation of spherical and triangular shaped gold nanoparticles having an average diameter of 8–17 nm. Fourier-transform infrared spectroscopy (FTIR) results revealed that the protein present in the extract plays the role of a reducing agent in addition to being a stabilizing agent.[38]

Similarly Aromal et al. have synthesized spherical gold nanoparticles from the aqueous extract of Macrotyloma uniflorum beans. Polyphenols present in the extract were identified as the main reducing agent for this synthetic procedure.[39] Another report worth mentioning here involves the synthesis of gold nanoparticles, in which antioxidants present in the blackberry, blueberry, pomegranate and turmeric extracts worked as reducing and stabilizing agent. Turmeric extract is known to contain phenolic compounds like curcuminoids along with triterpenoids, alkaloid and sterols which are responsible for the reducing Au (III) ions to Au (0) nanoparticles. Similarly, blueberry and blackberry also contain polyphenol-based antioxidants, while pomegranate rich in anthocyanins, tannins, polyphenols and flavonoids are responsible for the reduction and shape of the resulting nanoparticles.[39] Similarly, Elia and co-workers reported the synthesis of gold nanoparticles from the plant extracts of four different plants: Salvia officinalis, *Lippia citriodora*, *Pelargonium graveolens* and *Punica granatum*. Authors also identified that seven antioxidants—geranyl acetate, geraniol, caffeic acid, vanillic acid, menthone, linalool and gallic acid present in these plant extracts acted as reducing and stabilizing agent.[40] Ingale et al. reported a simple, efficient and one-step procedure for the synthesis of gold nanoparticles using extract of banana fruit waste at room temperature. This method invariably resulted in the fabrication of gold nanoparticles possessing diverse shapes.[41]

Types of Phytoconstituents Present in Plant Extract

Till now, polyphenols (including phenolic acids, terpenoids, flavonoids), proteins and organic acids have been recognized as plausible phytoconstituents which act as bioreducing and stabilizing agent during the synthetic procedure of nanoparticles (Figure 10.8).

These phytoconstituents are reported below.

a. **Flavonoids**

Flavonoids are a diverse range of water-soluble secondary metabolites and one of the most essential bioreducing agents present in the plant extracts. They consist of two phenyl rings and a heterocyclic ring, and are majorly divided into six groups: flavanones, flavanonols, flavans, anthoxanthins, anthocyanidins and isoflavonoid. The reducing power of flavonoids is correlated with their tendency to lose electron or hydrogen atoms.

b. **Phenolic acids**

Phenolic acids are also plant metabolites containing a phenolic ring along with an organic carboxylic acid. The reducing properties of these

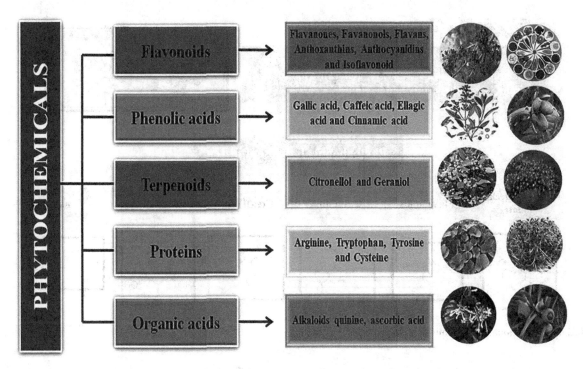

FIGURE 10.8   Types of phytoconstituents present in plant extract.

phenolic acids are associated with the tendency of the aromatic rings of phenolic acid to form a stable metal chelate. Many phenolic acid-based phytoconstituents such as gallic acid, caffeic acid, ellagic acid and protocatechuic acids are reported in literature which exhibit strong bioreducing properties. It was demonstrated by Aromal and coworkers that during the reduction of gold ions, caffeic acid gets converted into ferulic acid. During this course, the released hydrogen works as the reducing agent.[39]

c. **Terpenoids**

Terpenoids are low–molecular-weight organic compounds that belong to a large class of terpenes, diterpenes and sesquiterpenes. Terpenoids are the major cause of aroma, taste and color in various plants. It has been reported that certain hydroxyl functional groups present in various citronellol- and geraniol-based terpenoids are responsible for the reduction of gold ions to metallic gold nanoparticles.

d. **Proteins**

Proteins having complex structure possess antioxidant properties. It has been postulated that proteins contain certain carboxylic acid and amine groups that lead to the formation of nanoparticles via some sort of interactions or through complex formation. Numerous amino acids like arginine, tryptophan, tyrosine, cysteine, lysine and methionine have been known to interact with the metal ions, thus accounting for the production of nanoparticles.

e. **Organic acids**

At times, plants produce secondary metabolites, after coming in contact with metals. These secondary metabolites such as alkaloids and organic acids act as reducing agent for the synthesis of gold nanoparticles. For instance, ascorbic acid present in citrus sinensis peel extract has been reported to contain bioreducing properties for the synthesis of nanoparticles.

10.4.2   Synthesis of Gold Nanoparticles Using Tea Extract: A Green Chemistry Experiment

Since the discovery of tea by the second emperor of China, Shen Nung, in the bygone era of 2700 B.C., this beverage has become one of the most popular drinks in human history. Transitioning into the 21st century, consumption of tea has been directly attributed to a myriad of health benefits that include lowering the stroke, cancer and blood pressure risks, enhancing immune function, and preventing dental cavities. The evidences towards the increased health benefits of tea have aroused the research interest of the scientific fraternity to unravel the basis of the curing power of tea. A well-accepted scientific consensus stemming from all the scientific investigations is that tea has high levels of antioxidant polyphenols, including flavonoids and catechins, and all of which scavenge the dangerous free radicals in the body and thus prevent the progress of different diseases (Figure 10.9). While the enormous health benefits of these "chemical cocktails" present within tea are beyond doubt, the chemical reduction power of these phytoconstituents is also being

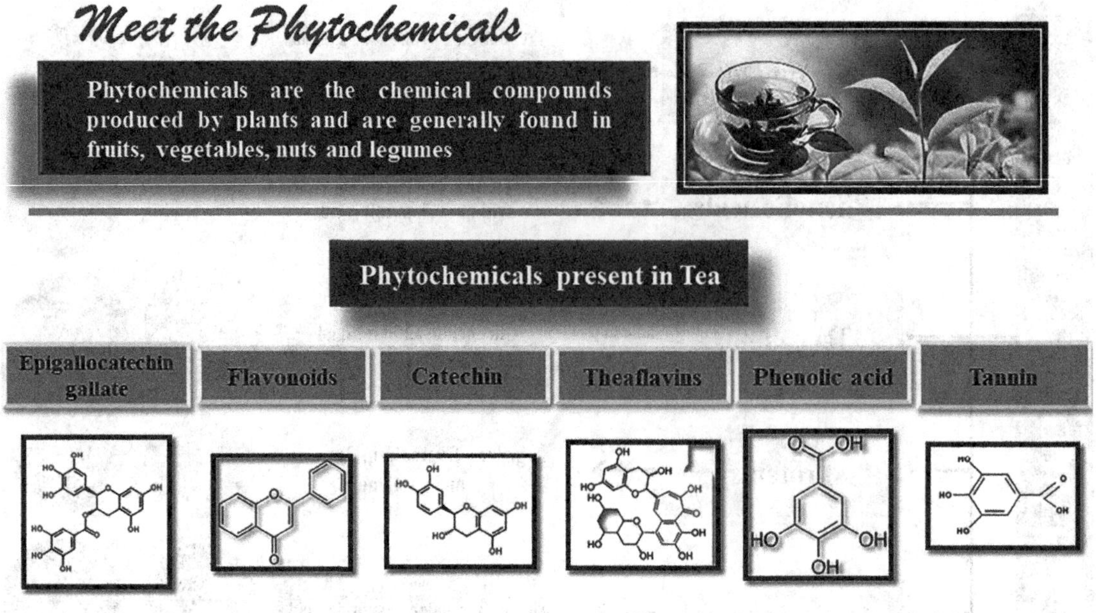

Meet the Phytochemicals

Phytochemicals are the chemical compounds produced by plants and are generally found in fruits, vegetables, nuts and legumes

Phytochemicals present in Tea

| Epigallocatechin gallate | Flavonoids | Catechin | Theaflavins | Phenolic acid | Tannin |

FIGURE 10.9   Various phytochemicals present in tea.

effectively exploited extensively. Within this context, it is worth mentioning that Sharma et al. have designed a simple, efficient and environmentally benign pathway for the one-step synthesis of gold nanoparticles using tea leaves in an aqueous media at room temperature.[42] It is hypothesized that the synergistic reduction potentials of polyphenols including flavonoids, catechins and various phytochemicals present in tea play the crucial role of reducing gold salts to gold nanoparticles.

The experiment has been designed with the goal to introduce chemistry students to the concept of green chemistry as well as nanotechnology. This single-step green protocol employs water as the environmentally benign solvent and does not require any surfactant, capping agent or template. This methodology reduces the use of hazardous toxic chemicals and thus is much more greener and superior in comparison with the conventional techniques that require external reducing and capping agents. The synthetic methodology for the production of gold nanoparticles involves mixing of the stock solution of Darjeeling tea with an aqueous solution of tetrachloroauric (III) acid. The moment yellow gold solution is added to the tea solution, the color of the resultant solution turns purple within 5 minutes, which implies the formation of gold nanoparticles. The overall reaction carried out at room temperature was completed within 30 minutes, signifying the essence of greener route (Figure 10.10).

Further, since the applications of AuNPs are highly dependent on their size, shape and morphology, the effect of different concentrations of tea (10%, 5% and 1% stock solution) on the size and dispersity of gold nanoparticles was examined. For this, authors prepared different concentrations of tea, i.e., 1%, 5% and 10% stock solution, while keeping the amount of gold solution that was being added to different tea concentrations fixed. It was found that as the concentration of tea was increased, the solution became

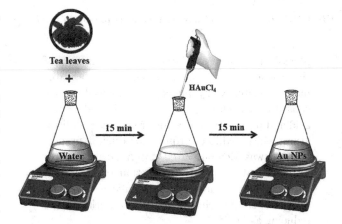

FIGURE 10.10   Synthesis of gold nanoparticles using tea extract.

more intensely purple red. This is because as the amount of tea was increased, the quantity of capping and reducing agents also increased, and this resulted in a decrease in the particle size of Au NPs.

The as-obtained gold nanoparticles were characterized by UV-vis absorption spectroscopy. When dispersed in liquid media, the nanoparticles exhibit a strong UV-vis absorption band that is not present in the spectrum of the bulk metal. This band results when the incident photon frequency is resonant with the collective excitation of the conduction electrons and is known as the SPR. Because the SPR depends on the size, shape and aggregation of AuNPs, UV-vis spectroscopy is a useful technique to estimate nanoparticles size, concentration and aggregation level. The UV-vis spectra of the dispersions of different-sized nanoparticles produced by using 1%, 5% and 10% tea stock solution, respectively, in this experiment are shown in Figure 10.11. As the concentration of tea leaves decreases, λmax changes from 530 to 563 nm. The observed shift could be due to an increase in the particle size. In other words, the wavelength shift observed in the plasmon bands is a consequence of different concentrations of tea added to the gold ions solution.

Further, in order to gain an insight into the size of the prepared nanoparticles, TEM was conducted and it was found that the size of the Au NPs lies between 20–25 nm as shown in Figure 10.12.

The advantages of this protocol are summarized in Figure 10.13.

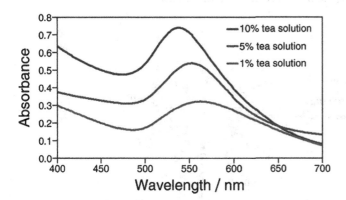

FIGURE 10.11 The UV-vis spectra of the dispersions of different-sized nanoparticles produced by using 1%, 5% and 10% tea.

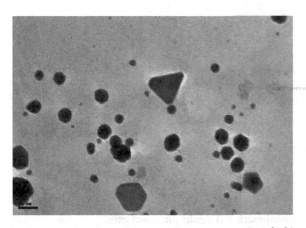

FIGURE 10.12 TEM of AuNPs. (Adopted from Ref. [42].)

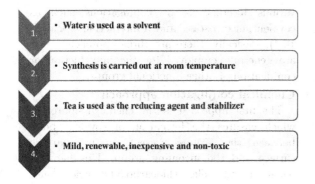

FIGURE 10.13 Special highlights of this tea-mediated synthesis of Au NPs.

10.5 Surface Modification of Gold Nanoparticles

10.5.1 Protective Surface Coatings

Gold nanoparticles at times show a strong tendency to undergo aggregation in solution under a wide range of biological, physical and environmental conditions, which ultimately causes a significant change in their properties. To impart stability to these nanoparticles, they are often coated with protecting agents like silica.

10.5.2 Surface Modification by Functionalization

Gold nanoparticles exhibit riveting electronic and optical properties that are found to be highly dependent on the functional molecules that are present on its surface. In this case, surface chemistry of the nanoparticles plays a very important role as it allows the introduction of varied functional groups onto the surface of nanoparticles which specifically interacts with certain cells and biological molecules, rendering them appealing for use in the field of sensing, delivery and imaging and therapies (Scheme 10.1). Three main strategies are known through which functional groups can be introduced onto the surface of gold nanoparticles.[23]

1. **Direct functionalization**

   In the first strategy, a functional ligand is introduced directly onto the surface of gold nanoparticles which then forms a dense protecting ligand layer. However, there are limited number of capping and functional groups. Polyethylene glycol (PEG) and mercaptocarboxylic acids belong to that class of ligands that imparts stability to the surface of gold nanoparticles, and alkane thiols are the most encountered surface capping ligands onto the gold nanoparticles (Scheme 10.2).[43,44]

2. **Ligand exchange**

   In the second strategy, the already existing ligand on the surface of gold nanoparticles gets exchanged with another incoming ligand possessing desired functionality for specific biological and medicinal applications. This strategy is commonly termed as ligand exchange (Scheme 10.3). For instance, gold nanoparticles

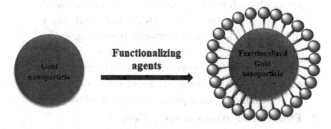

SCHEME 10.1 General schematic illustration for the functionalization of gold nanoparticles.

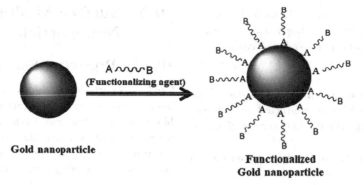

SCHEME 10.2 Schematic illustration showing direct functionalization of gold nanoparticles.

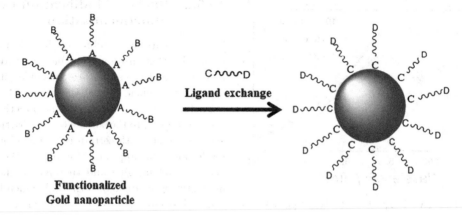

SCHEME 10.3 Schematic illustration demonstrating ligand exchange on the surface of gold nanoparticles.

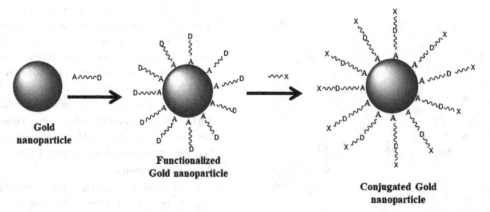

SCHEME 10.4 Schematic illustration showing functional group conjugation on gold nanoparticles.

synthesized by citrate reduction method contain adsorbed negatively charged citrate ions, which can be further replaced by other ligands such as mercaptocarboxylic acids (mercaptoacetic acid, mercaptopropionic acid, etc.) and sulfonated phosphines. Thiol-based ligands possessing sulfur atom have very strong affinity for the gold nanoparticles. The concept of ligand exchange is also valid for the nanoparticles that are needed to be transferred from aqueous to organic phase.[44,45]

3. Functional group conjugation

In the third strategy, bifunctional compounds are introduced on gold nanoparticles through various interactions or conjugations such as covalent, electrostatic and π interactions (Scheme 10.4). Basically, there are three approaches that are generally employed for the introduction or conjugation of other functional groups.

a. Chemical conjugation approach

The first approach is the chemical approach, which ensures the formation of covalent bond between the groups present on nanoparticle surface and the incoming group. This happens when ligands like thiocarboxylic acids once exchanged with the gold nanoparticles, covalently interacts with certain amino acids, proteins and peptides leading to the formation of amide bonds. Primary and secondary amine groups interact covalently with carboxylic-terminated gold nanoparticles via carbodiimide coupling. Similarly, ester derivatives can also be covalently linked with amine-functionalized gold nanoparticles. Azide click chemistry and esterifications reactions are other well-known approaches for the chemical functionalization of gold nanoparticles.[46]

b. **Physical conjugation approach**

This approach takes into consideration various hydrophobic and electrostatic interactions that are involved with the surface of gold nanoparticles. In this case, amphiphilic coating of the hydrophobic groups of the gold nanoparticles is carried out using hydrophilic groups. This kind of approach is also useful while transferring hydrophilic gold nanoparticles from aqueous to organic phase. Quaternary ammonium salts are the classic example of amphiphilic agents that get electrostatically adsorbed onto the surface of negatively charged gold nanoparticles.

c. **Bioconjugation approach**

Bioconjugation approach involves the functionalization of biomolecules such as peptides, sugar, proteins, enzymes, DNA and RNA, with gold nanoparticles. The approach exploits three types of intermolecular interactions for the conjugation of biomolecules to the nanoparticles.

  i. Electrostatic interaction between the positively charged biomolecules and negatively charged surface of gold nanoparticles

  ii. Chemisorption, i.e., this kind of interaction is similar to the one that occurs when thiol ligands get adsorbed onto the surface of gold nanoparticles

  iii. Covalent bonding between functional groups present on both biomolecules and nanoparticles.[47]

10.6 Catalysis by Gold: Exploitation of Catalytic Property of Surface Modified Gold Nanoparticles in Varied Industrially Significant Organic Reactions

10.6.1 Hybrid Au NPs (Supported Gold Nanoparticles Composites)

In today's era, when advancements in the field of nanoscience and technology have conquered new horizons, substantial efforts by the research community have been directed towards the synthesis of more complex gold-based

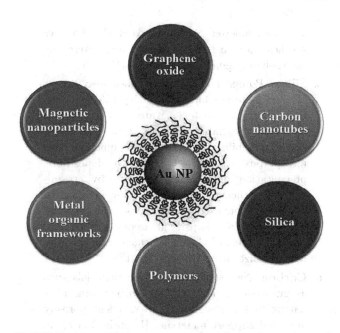

FIGURE 10.14 Various support materials for the synthesis of hybrid (supported) Au nanoparticles.

nanostructures. These complex nanostructured composites are amenable for the new technology-driven applications in the field of catalysis, sensing, optoelectronics, biomedicine and photocatalysis. Another remarkable attribute of these hybrid nanocomposites is the assimilation of distinctive physical and chemical properties that stimulate researchers to further employ these innovative advanced materials for noteworthy industrial applications. In this context, scientists have exploited the concept of "supported nanoparticles" wherein a diverse range of materials such as iron oxide, silica, carbon (graphene and carbon nanotubes), alumina, metal organic frameworks (MOF) and polymers have been employed as support materials for the stabilization of gold nanoparticles (Figure 10.14).[48] The crucial role of support material for gold nanoparticles besides providing large surface area and anchoring sites is the prevention of agglomeration. A variety of solid supports have been listed below.

a. **Magnetic metal oxide:** Significant progress has been made in the synthesis of multicomponent nanoparticles, among which iron-oxide-supported gold nanoparticles are of utmost importance. In these hybrid materials, first the nanoparticles of one kind of material are synthesized, followed by the nucleation of the surface functional groups, ultimately resulting in the deposition of other material. Impregnation, immobilization, deposition precipitation and hydrothermal depositions are some of the common routes for synthesizing supported gold nanoparticles. Ferrer et al. reported the synthesis of core–shell nanoparticles utilizing iron oxide as the core with gold nanoparticles being the shell.[49] Indeed, reverse strategy has also been followed by Alivisatos et al. who have synthesized gold nanoparticles as core covered by the shell of magnetic nanoparticles.[50]

b. **Silica**: Porous materials like silica also act as support materials for immobilizing gold nanoparticles. In this context, Kato and co-workers reported the fabrication of gold nanoparticles immobilized onto amine-functionalized mesoporous silica sheets using a deposition–precipitation phenomenon followed by crystallization (Scheme 10.5). First, they synthesized amine-functionalized mesoporous silica sheets. Then, the gold nanoparticles were dispersed onto the silica sheets followed by calcination in order to crystallize them on the support.[51]

c. **Carbon**: Nanostructured carbon materials such as graphene oxide (GO) and carbon nanotubes having high surface area and excellent porosity are good support materials. Recently, Liu et al. fabricated a hybrid composite of multiwalled carbon nanotubes-supported gold nanoparticles. First, the carbon nanotubes were coated with silica followed by amine functionalization using (3-Aminopropyl)trimethoxysilane (APTMS). Then, the presence of abundant functional groups on the surface of carbon nanotubes allows the immobilization of as-synthesized gold nanoparticles.[52]

Similarly, composites of gold nanoparticles with GO have also been reported. The Au-GO composite can be synthesized via two approaches: one is the dispersion of gold nanoparticles on the GO sheets, and the other is the encapsulation of gold nanoparticles inside the pores of GO. In the first approach, the precursor salts of gold are added into the suspension of GO. The oxygen functionalities present on the surface of GO nucleate the growth of gold ions, which are then subsequently reduced by certain reducing agents such as

sodium citrate, sodium borohydride and hydrazine hydrate to gold nanoparticles. While in the second approach, gold nanoparticles are synthesized separately and then embedded into the nanosheets of GO via several electrostatic, van der Waals and π–π interactions.[53]

d. **MOF**: Metal organic frameworks are an emerging class of porous crystalline materials that has gained immense interest as support material. The high thermal stability and the narrow pore size make them exceptional candidate for the immobilization of gold nanoparticles. Several techniques such as impregnation, CVD, deposition reduction, solid grinding and solution infiltration have been used in the past for immobilizing gold nanoparticles. A recent report by Wang and co-workers has been published in literature that documented the immobilization of dispersed gold nanoparticles onto an amine-functionalized metal organic framework via a solution-based synthesis approach involving absorption and reduction phenomenon. For this, authors first synthesized IRMOF-3-based MOF having amine functionalities. The functionalized MOF was then added into a batch solution containing gold precursor salt ($HAuCl_4$), which then coordinates to Au ions. The Au ions were further reduced by sodium borohydride which acts as the reducing agent and ultimately results in the formation of Au@MOF hybrid composite.[54]

e. **Polymers**: Polymeric materials are promising candidate for being used as support material. Within this context, Liu et al. have synthesized histamine-functionalized polymer-based monolithic capillaries as support material for immobilizing gold nanoparticles. For this, authors first silanized the silica capillaries using 3-(trimethoxysilyl)propyl methacrylate, followed by the photo-initiated free radical polymerization using poly(NAS-*co*-EDMA), (where NAS=N-acryloxysuccinimide and EDMA=ethylene glycol dimethacrylate). Then, they were subsequently functionalized using histamine followed by the immobilization of already synthesized gold nanoparticles. Electrostatic interactions between the positively charged histamine groups and negatively charged carboxylate moieties of the citrate groups present onto the surface of gold nanoparticles were the main driving force for immobilizing gold nanoparticles. Consequently, the designed material was utilized as a microreactor for the reduction of nitro compounds to amines.[55]

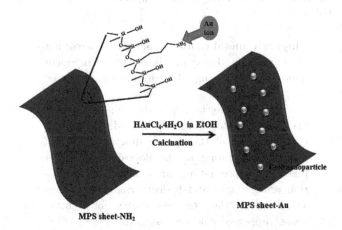

SCHEME 10.5 Schematic illustration for the synthesis of mesoporous silica sheet with immobilized gold nanoparticles.

It is believed that the integration of gold nanoparticles with the functional support materials will create great new prospects in the field of imaging, sensing, catalysis, drug delivery, optics and cancer therapy.

10.6.2 Catalysis by Hybrid Gold Nanoparticles

Ever since the pioneering works by Haruta and Hutchings in 1987, a breakthrough embarked the usage of gold nanoparticles in the field of catalysis. It was for the first time, the substance that was considered inert since ages remarkably proved its efficacy in the low-temperature CO oxidation and the hydrochlorination of ethyne to vinyl chloride. Since then, researchers have not stopped dreaming about developing valuable gold-based catalysts for the synthesis of industrially and pharmaceutically organic compounds. Driven by the constant demand of energy conservation and reduction, there has been an increasing trend towards the use of bare or hybrid-supported gold catalysts for various substantial organic reactions, including oxidation, reduction, coupling, esterification and condensation reactions (Figure 10.15).[56,57]

a. Oxidation reactions

Oxidation reactions that employ hydrogen peroxide, tert-butyl hydroperoxide (TBHP) and molecular oxygen as oxidants are highly significant in chemical industries for the production of numerous important compounds. In this regard, Yuan et al. have synthesized a mesoporous carbon-supported gold nanoparticle-based catalyst for the selective oxidation of glucose to gluconic acid under the base-free conditions (Scheme 10.6). TEM images revealed the uniform distribution of gold nanoparticles inside the pores of carbon. Also, the developed catalyst, in the presence of molecular oxygen, exhibited remarkable efficiency with 92% conversion percentage and 85% selectivity for the corresponding oxidation reaction.[58]

Another report by Lykasis and co-workers presented a titania-supported gold nanoparticle-based catalyst for the chemoselective oxidation of amines in the presence of hydrogen peroxide to nitroso compounds (Scheme 10.7). The mesoporous titania-supported gold catalyst has been synthesized via deposition–precipitation technique. Authors also demonstrated that the reaction between dienes and in situ formed nitroarene species resulted in the formation of Diels–Alder adducts. The protocol was highly efficient for the large-scale production of nitrosoarene derivatives having excellent yield and selectivity. The catalyst also showed good functional group compatibility as a wide range of amines were found to be efficiently oxidized to the nitrosoarenes.[59]

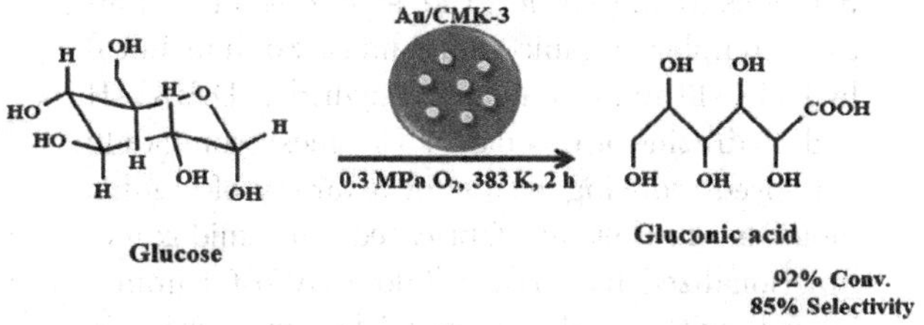

SCHEME 10.6 Aerobic oxidation of glucose to gluconic acid using mesoporous carbon (CMK-3)-supported gold catalyst.

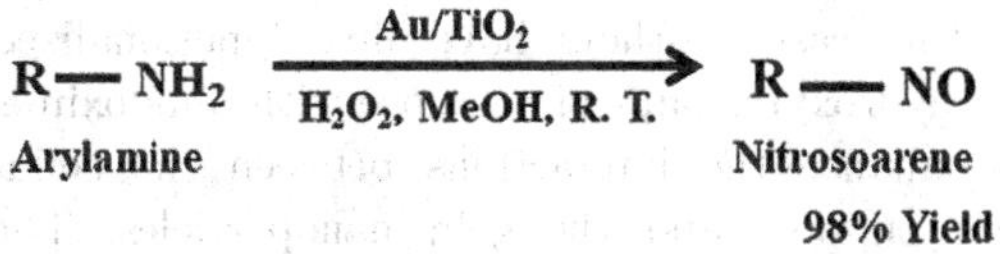

SCHEME 10.7 Selective oxidation of aryl amines into nitrosoarenes with H_2O_2 catalyzed by Au/TiO_2.

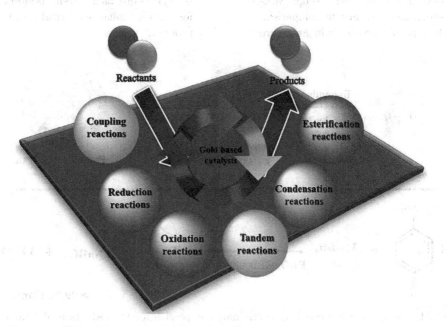

FIGURE 10.15 Applications of hybrid Au nanoparticle-based catalysts.

Martins et al. reported the fabrication of four catalysts synthesized by immobilizing gold nanoparticles on different oxides such as TiO_2, Fe_2O_3, ZnO and Al_2O_3. The developed catalysts were investigated in the cyclohexane oxidation using hydrogen peroxide. Au/Fe_2O_3 was found to be the best catalytic system as it showed exclusive cyclohexanol formation in a short span of time, i.e., 4 hrs. Also, the activities of the developed catalysts were also tested in the microwave-assisted oxidation of benzyl alcohol and methylbenzyl alcohol in the presence of TBHP. For this, Au/TiO_2 system was found to be the best catalytic system. In addition, it was found that the catalysts could be recycled and reused for subsequent multiple cycles.[60]

b. **Reduction reactions**

Reduction reactions are one of the most versatile reactions deployed for the synthesis of significant complex organic compounds. Sodium borohydride, lithium aluminum hydride, DIBAL-H and hydrazine are some of the most commonly employed reducing agents for a variety of organic moieties. Liu et al. fabricated an amidoxime-functionalized bacterial cellulose (AOBC) immobilized with gold nanoparticles via two-step assembly process for the reduction of 4-nitrophenol to 4-amino phenol. Bacterial cellulose possesses abundant hydroxyl groups on its surface which can be further functionalized with various groups. In this case, authors have first functionalized the hydroxyl groups of cellulose with amidoxime, to enhance the interactions between the cellulosic support and the gold nanoparticles. The AOBC plays the role of stabilizing as well as reducing agents for the reduction of $HAuCl_4$ to gold nanoparticles. The developed greener catalyst displayed remarkable efficacy in comparison with the already reported protocols in literature

for the reduction of 4-nitro phenol to 4-amino phenol.[61] For the similar reaction, Liu et al. fabricated gold nanoparticle-decorated magnetic nanocomposites via a direct reduction strategy (Scheme 10.8). Various studies reported in literature state that the lack of functional groups on Fe_3O_4 leads to aggregation of gold nanoparticles on its surface. Therefore, suitable coating agents such as polydopamine (PDA) can be used for forming a uniform and continuous layer on ferrite which directs controlled dispersion of gold nanoparticles due to various non-covalent and chelating interactions. In addition, catechol groups present in polydopamine possess redox property and directly reduce $HAuCl_4$ into gold nanoparticles. Thus, polydopamine acts as reducing agent in addition to being a coupling agent.[62]

c. **Esterification reactions**

Esterification reactions are important reactions in synthetic organic chemistry as they are generally employed for the conversion of alcohols to esters. Stellacci et al. fabricated gold nanoparticles decorated on magnetic support for the esterification of benzyl alcohol (Scheme 10.9). For the fabrication of final catalyst, first gold nanoparticles coated with 3-mercaptopropane-1-sulfonic acid (MPSA) and octanethiol (OT) were synthesized, surface-modified using N-hydroxysuccinimidyl 11-mercaptoundecanoate and finally added into (3-aminopropyl)triethoxysilane (APTES)-functionalized silica-coated magnetic nanoparticles. Subsequently, the catalytic efficiency of the catalyst $Fe_2O_3@SiO_2@Au$ was investigated in the acetylation of benzyl alcohol to benzyl acetate, owing to the diverse industrial applications of benzyl acetate as a solvent, odorant and precursor for many pharmaceutical and agrochemical product.[63]

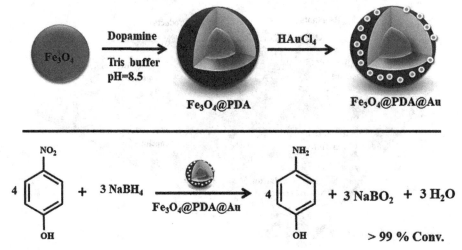

SCHEME 10.8 Gold nanoparticle-decorated magnetic nanocomposites for the reduction of 4-nitro phenol to 4-amino phenol.

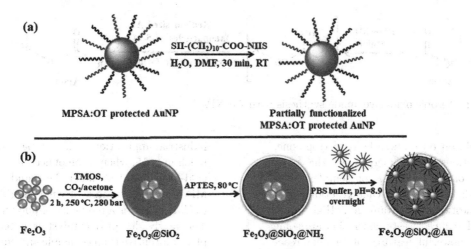

SCHEME 10.9 Schematic illustration of (a) the partial ligand exchange of MPSA:OT-covered AuNPs, (b) the synthesis of Fe_2O_3@SiO_2@Au hierarchical nanospheres.

d. **Coupling reactions**

Coupling reactions represent one of the most pivotal and versatile reactions that are generally employed in the synthetic organic chemistry. The coupling between two different starting materials via the aid of a metal catalyst leads to the formation of several industrial and pharmaceutical substantial compounds. A fascinating example of a gold-supported catalyst was reported by Karimi and co-workers. They fabricated gold nanoparticles supported on periodic mesoporous organosilica (PMO) with alkylimidazolium framework (Au@PMO-IL) having 2D hexagonal structure for the three-component coupling of alkyne, amine and aldehyde to form valuable synthetic product, propargylamine (Scheme 10.10). It was found that the Au (III) was the active component in the catalyst. Besides, TEM images revealed that the imidazolium units of ionic liquid embedded in the channels of mesoporous silica stabilized the gold nanoparticles. Moreover, the catalyst was stable, gave excellent yield and also reused for multiple cycles.[64]

e. **Condensation reactions**

A condensation reaction, by definition, is a chemical reaction in which reactants combine to form product with loss of smaller molecules such as water, ammonia, ethanol or acetic acid. The importance of amides in synthetic organic chemistry due to their versatile utility has led researchers to develop new synthetic methodologies over the last few decades. Recently, Kobayashi and co-workers synthesized a catalyst in which gold nanoparticles were immobilized in a copolymer having cross-linking groups, better termed as polymer-incarcerated gold nanoparticles.[65] The synthesized catalyst was then evaluated in the aerobic condensation of aldehydes and amines to yield amides (Scheme 10.11). Authors observed that size of nanoparticles directly influences the catalytic activity of the composite. Gold nanoparticles having 4.5–11 nm were found to exhibit good performance. In addition, a wide range of functional groups were tolerated with high yield. Also, the catalyst was recoverable and reused for multiple runs, without showing any discernible loss in its catalytic activity.

f. **Tandem Catalysis**

Tandem reactions also termed as cascade reactions combine different organocatalytic reaction within a single step, such that there is no need to isolate the intermediates and by just changing the reaction conditions. After the pioneering

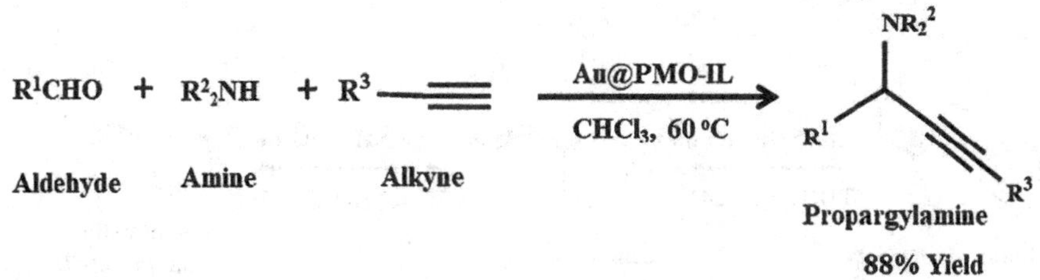

$$R^1CHO \ + \ R^2_2NH \ + \ R^3 \!\!-\!\!\!\equiv \ \xrightarrow[\text{CHCl}_3,\ 60\,°\text{C}]{\text{Au@PMO-IL}}$$

Aldehyde **Amine** **Alkyne**

Propargylamine

88% Yield

SCHEME 10.10 Three-component reaction of various aldehydes, secondary amines and terminal alkynes catalyzed by Au@PMO-IL.

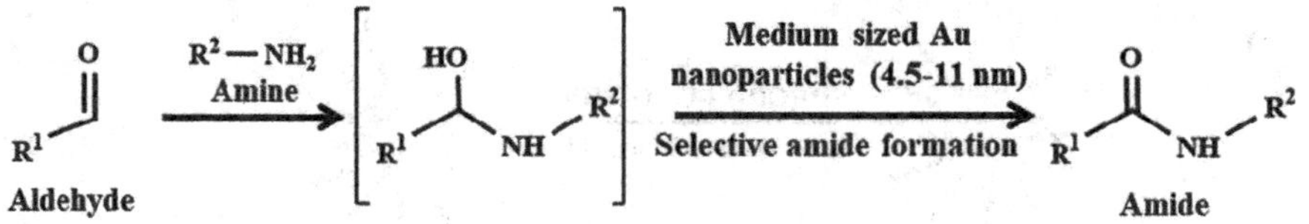

SCHEME 10.11 Aerobic oxidative amide synthesis with Au NPs.

work of Robinson on the synthesis of tropinone, tandem reactions have emerged as the most potent tool for the synthesis of complex heterocyclic synthons. These cascade reactions offer several unquestionable advantages. High atom economy and reduced waste generation are some of the most powerful features of tandem reactions that render them appealing for use in the synthesis of complex polycyclic ring moieties. Recently, a study has been reported wherein gold nanoparticles deposited onto NH$_2$-UiO-66 MOF served as a catalyst for one-pot selective oxidation of primary alcohols in tandem with the Knoevenagel condensation (Scheme 10.12). Authors demonstrated that the gold nanoparticles present on the surface of MOF were the actual active sites for the oxidation of cinnamyl alcohol to cinnamaldehyde, while the amine groups of MOF served as basic sites for the subsequent Knoevenagel condensation of the resulting cinnamaldehyde with malononitrile to form cinnamylidene malononitrile. The protocol was found to be efficient for the versatile coupling between α,β-unsaturated alcohols with active methylene compounds to form complex heterocyclic architectures.[66]

10.6.3 Industrial Case Studies

For decades, scientific research at both academia and industrial level has contributed immensely to advancing the frontiers of science. Driven by the constant demand towards miniaturization and high-throughput processes, great deals of research efforts have been directed towards developing new gold catalysts and gold-catalyzed processes.

a. **Pricat MFC**

Beginning its journey in the year 1817, Johnson Matthey, is a renowned British multinational company specialized in designing catalysts for industrial applications. The company rose to fame much in 2015, when it announced and marketed a gold-based catalyst under the brand name Pricat MFC, for the synthesis of vinyl chloride monomer (VCM), which is a monomer of polyvinyl chloride. The newly designed catalyst is a terrific exemplar of incentive from economic and environmental point of view, owing to its capability to replace all conventional mercury-based catalysts.[67,68] Pricat MFC, also termed as eggshell catalyst, employs gold as the active component prudently dispersed onto the surface of the carbon support material, making it more economic viable. The research team at Johnson Matthey, under the leadership of Johnson, set the development goals of the catalyst by limiting to a maximum gold loading of 0.25% and also eliminated the use of aqua regia during preparative steps. It is worth mentioning here that the idea of sustainability played a strategic role in the overall growth and development of company and was found to be a key driver for its competitiveness. Indeed, the pioneering work by Matthey has also been recognized in the Institution of Chemical Engineers (IChemE) Global Awards 2015, wherein the catalyst has been entitled as the Innovative Product of the Year.

b. **AUROlite**

A highly successful joint venture termed as **AuTEK** was formed between a mineral research organization, Mintek and three major gold mining houses—Anglogold Ashanti, Gold Fields and Harmony Gold Mining Company Ltd. whose main focus was research and development of novel industrial applications of gold. AuTEK was the first company that launched a plant for the reproducible synthesis of 1 wt% gold-based catalysts under the brand name of **AUROlite**. AUROlite includes supported gold catalysts of Au/TiO$_2$, Au/ZnO, Au/Al$_2$O$_3$ and

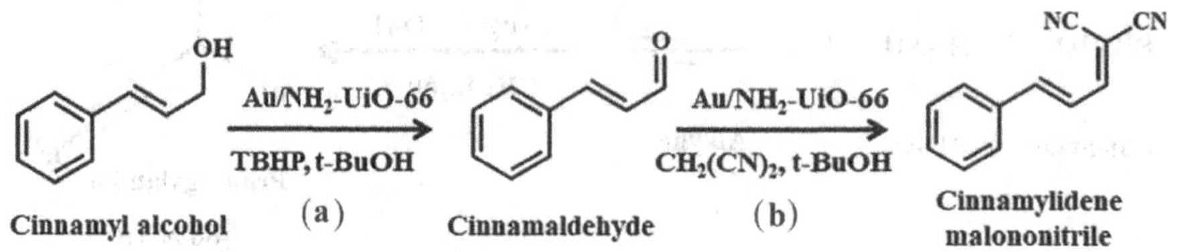

SCHEME 10.12 Oxidation of cinnamyl alcohol to cinnamaldehyde (a) and subsequent Knoevenagel condensation with malononitrile to form cinnamylidene malononitrile (b).

AUROlith (Au/Al_2O_3/cordierite) synthesized by the phenomenon of deposition and precipitation.[67] The designed catalysts were identified and then marketed as idealistically suited for the oxidation reactions in liquid and gas phase. In fact, till now the company has developed wide range of gold catalysts.

c. NanAucat

3M Corporation is a leading company in filtration technology known for developing world-class products for protecting workers from hazardous particulates. As part of their commitment towards reducing environmental impacts, in 2009, 3M Corporation have entered into a distribution partnership with Premier Chemicals. Under this commercial tie-up, 3M Corporation grants exclusive rights to Premier Chemicals to distribute its 3M's gold catalyst under the brand name of NanAucat.[67] NanAucat is marketed as a highly active and stable gold-based oxidation catalyst that oxidizes highly hazardous carbon monoxide to less hazardous carbon dioxide. The catalyst developed particularly for escape masks is synthesized via CVD technique, wherein Au nanoparticles are deposited onto the porous amorphous carbon. Significant remarkable attributes of NanAucat in comparison with other metal-based catalysts include higher catalytic activity and resistance to deactivation by moisture. NanAucat is a suitable catalyst of choice for the cost-effective removal of CO at high humidity even at ambient temperature.

10.7 Conclusion

Since the dawning of new millennium, the budding field of nanoscience and nanotechnology has completely transformed from material limited to application driven. This revolution has marked the beginning of a new era in which the scientific community has shifted from modus operandi research to a more sophisticated designing and engineering of materials and applications. Gold nanoparticles known since ages have always been under scrutiny owing to their promising potential in optical, catalytic, electronic, sensing, imaging and biomedical fields. They have been an immense subject of interest from academic and industrial point of view. Furthermore, an incredible hike in the exploration of various features and properties of gold nanoparticles has resulted into a more disciplined and systematic research field. Besides, the controlled synthesis and ease of functionalization make gold nanoparticle a tool box that never existed before. However, due to increasing societal and economic concerns, traditional chemical and physical techniques utilized for the synthesis of gold nanoparticles pose a serious threat to mankind and environment. In order to minimize the potential hazards associated with the conventional synthesis of gold nanoparticles, huge efforts are being made by the chemists to develop greener synthetic methodologies that incorporate use of natural resources and the principles of green chemistry. As a result, a great array of endeavors of researchers and chemists have focused on redesigning and replacing previously reported protocols via the aid of green chemistry and engineering. In this regard, the eco-friendly biogenic synthesis of gold nanoparticles using plants has evolved as an attractive alternative invincible route. It is worth mentioning here that significant achievement has been made in this context. Literature is flooded with numerous reports wherein biocompatible plant extracts have been directly utilized as reducing and stabilizing agent for the synthesis of gold nanoparticles. However, studies related to the identification of exact phytochemical present in plants are still in their nascent stage, and considerable research needs to be directed in this arena.

References

1. Mudhoo, A., & Sharma, S. K. (2010). *Green Chemistry for Environmental Sustainability.* CRC Press, Boca Raton, FL.
2. Fleischer, T., & Grunwald, A. (2008). Making nanotechnology developments sustainable. A role for technology assessment? *Journal of Cleaner Production, 16*(8–9), 889–898.
3. Matus, K. J., Hutchison, J. E., Peoples, R., Rung, S., & Tanguay, R. (2011). Green nanotechnology challenges and opportunities.
4. Maynard, A. D., Aitken, R. J., Butz, T., Colvin, V., Donaldson, K., Oberdörster, G., ... & Tinkle, S. S. (2006). Safe handling of nanotechnology. *Nature, 444*(7117), 267.
5. Zhu, W., Bartos, P. J., & Porro, A. (2004). Application of nanotechnology in construction. *Materials and Structures, 37*(9), 649–658.
6. Dowling, A. P. (2004). Development of nanotechnologies. *Materials Today, 7*(12), 30–35.
7. Morose, G. (2010). The 5 principles of "design for safer nanotechnology". *Journal of cleaner production, 18*(3), 285–289.
8. Hutchison, J. E. (2008). Greener nanoscience: A proactive approach to advancing applications and reducing implications of nanotechnology. *ACS Nano, 2*, 395–402.
9. Mulvihill, M. J., Beach, E. S., Zimmerman, J. B., & Anastas, P. T. (2011). Green chemistry and green engineering: a framework for sustainable technology development. *Annual Review of Environment and Resources, 36*, 271–293.
10. Schmidt, K. (2007). Green nanotechnology: it's easier than you think.
11. Dhingra, R., Naidu, S., Upreti, G., & Sawhney, R. (2010). Sustainable nanotechnology: through green methods and life-cycle thinking. *Sustainability, 2*(10), 3323–3338.

12. Iavicoli, I., Leso, V., Ricciardi, W., Hodson, L. L., & Hoover, M. D. (2014). Opportunities and challenges of nanotechnology in the green economy. *Environmental Health*, *13*(1), 78.

13. Kim, B. H., Hackett, M. J., Park, J., & Hyeon, T. (2013). Synthesis, characterization, and application of ultrasmall nanoparticles. *Chemistry of Materials*, *26*(1), 59–71.

14. Zhang, Q., Uchaker, E., Candelaria, S. L., & Cao, G. (2013). Nanomaterials for energy conversion and storage. *Chemical Society Reviews*, *42*(7), 3127–3171.

15. Mohanraj, V. J., & Chen, Y. (2006). Nanoparticles- a review. *Tropical Journal of Pharmaceutical Research*, *5*(1), 561–573.

16. Horikoshi, S., & Serpone, N. (2013). *Introduction to Nanoparticles: Microwaves in Nanoparticle Synthesis: Fundamentals and Applications*. Wiley-VCH, Weinheim, pp. 1–24.

17. Bhatia, S. (2016). Nanoparticles types, classification, characterization, fabrication methods and drug delivery applications. In *Natural Polymer Drug Delivery Systems* (pp. 33–93). Springer, Cham.

18. Tiwari, J. N., Tiwari, R. N., & Kim, K. S. (2012). Zero-dimensional, one-dimensional, two-dimensional and three-dimensional nanostructured materials for advanced electrochemical energy devices. *Progress in Materials Science*, *57*(4), 724–803.

19. Thakor, A. S., Jokerst, J., Zavaleta, C., Massoud, T. F., & Gambhir, S. S. (2011). Gold nanoparticles: A revival in precious metal administration to patients. *Nano letters*, *11*(10), 4029–4036.

20. Daniel, M. C., & Astruc, D. (2004). Gold nanoparticles: Assembly, supramolecular chemistry, quantum-size-related properties, and applications toward biology, catalysis, and nanotechnology. *Chemical reviews*, *104*(1), 293–346.

21. Abadeer, N. S., & Murphy, C. J. (2016). Recent progress in cancer thermal therapy using gold nanoparticles. *The Journal of Physical Chemistry C*, *120*(9), 4691–4716.

22. Yeh, Y. C., Creran, B., & Rotello, V. M. (2012). Gold nanoparticles: Preparation, properties, and applications in bionanotechnology. *Nanoscale*, *4*(6), 1871–1880.

23. Mieszawska, A. J., Mulder, W. J., Fayad, Z. A., & Cormode, D. P. (2013). Multifunctional gold nanoparticles for diagnosis and therapy of disease. *Molecular Pharmaceutics*, *10*(3), 831–847.

24. Grzelczak, M., Pérez-Juste, J., Mulvaney, P., & Liz-Marzán, L. M. (2008). Shape control in gold nanoparticle synthesis. *Chemical Society Reviews*, *37*(9), 1783–1791.

25. Herizchi, R., Abbasi, E., Milani, M., & Akbarzadeh, A. (2016). Current methods for synthesis of gold nanoparticles. *Artificial Cells, Nanomedicine, and Biotechnology*, *44*(2), 596–602.

26. Saha, K., Agasti, S. S., Kim, C., Li, X., & Rotello, V. M. (2012). Gold nanoparticles in chemical and biological sensing. *Chemical Reviews*, *112*(5), 2739–2779.

27. Sengani, M., Grumezescu, A. M., & Rajeswari, V. D. (2017). Recent trends and methodologies in gold nanoparticle synthesis–A prospective review on drug delivery aspect. *OpenNano*, *2*, 37–46.

28. Zhou, Y., Wang, C. Y., Zhu, Y. R., & Chen, Z. Y. (1999). A novel ultraviolet irradiation technique for shape-controlled synthesis of gold nanoparticles at room temperature. *Chemistry of materials*, *11*(9), 2310–2312.

29. Gedye, R., Smith, F., Westaway, K., Ali, H., Baldisera, L., Laberge, L., & Rousell, J. (1986). The use of microwave ovens for rapid organic synthesis. *Tetrahedron Letters*, *27*(3), 279–282.

30. Kou, J., Bennett-Stamper, C., & Varma, R. S. (2013). Green synthesis of noble nanometals (Au, Pt, Pd) using glycerol under microwave irradiation conditions. *ACS Sustainable Chemistry & Engineering*, *1*(7), 810–816.

31. Xu, X., Duan, G., Li, Y., Liu, G., Wang, J., Zhang, H., ... & Cai, W. (2013). Fabrication of gold nanoparticles by laser ablation in liquid and their application for simultaneous electrochemical detection of Cd^{2+}, Pb^{2+}, Cu^{2+}, Hg^{2+}. *ACS Applied Materials & Interfaces*, *6*(1), 65–71.

32. Prema, P., Iniya, P. A., & Immanuel, G. (2016). Microbial mediated synthesis, characterization, antibacterial and synergistic effect of gold nanoparticles using Klebsiella pneumoniae (MTCC-4030). *RSC Advances*, *6*(6), 4601–4607.

33. Mnisi, R. L., Ndibewu, P. P., & Mokgalaka, N. S. (2016). Green Chemistry in action: Towards sustainable production of Gold nanoparticles. *Pure and Applied Chemistry*, *88*(1–2), 83–93.

34. Gardea-Torresdey, J. L., Parsons, J. G., Gomez, E., Peralta-Videa, J., Troiani, H. E., Santiago, P., & Yacaman, M. J. (2002). Formation and growth of Au nanoparticles inside live alfalfa plants. *Nano Letters*, *2*(4), 397–401.

35. Sharma, N. C., Sahi, S. V., Nath, S., Parsons, J. G., Gardea-Torresde, J. L., & Pal, T. (2007). Synthesis of plant-mediated gold nanoparticles and catalytic role of biomatrix-embedded nanomaterials. *Environmental Science & Technology*, *41*(14), 5137–5142.

36. Armendariz, V., Herrera, I., Jose-yacaman, M., Troiani, H., Santiago, P., & Gardea-Torresdey, J. L. (2004). Size controlled gold nanoparticle formation by Avena sativa biomass: use of plants in nanobiotechnology. *Journal of Nanoparticle Research*, *6*(4), 377–382.

37. Gardea-Torresdey, J. L., Rodriguez, E., Parsons, J. G., Peralta-Videa, J. R., Meitzner, G., & Cruz-Jimenez, G. (2005). Use of ICP and XAS to

determine the enhancement of gold phytoextraction by Chilopsis linearis using thiocyanate as a complexing agent. *Analytical and Bioanalytical Chemistry, 382*(2), 347–352.

38. Suman, T. Y., Rajasree, S. R., Ramkumar, R., Rajthilak, C., & Perumal, P. (2014). The green synthesis of gold nanoparticles using an aqueous root extract of Morinda citrifolia L. *Spectrochimica Acta Part A: Molecular and Biomolecular Spectroscopy, 118*, 11–16.

39. Aromal, S. A., Vidhu, V. K., & Philip, D. (2012). Green synthesis of well-dispersed gold nanoparticles using Macrotyloma uniflorum. *Spectrochimica Acta Part A: Molecular and Biomolecular Spectroscopy, 85*(1), 99–104.

40. Elia, P., Zach, R., Hazan, S., Kolusheva, S., Porat, Z. E., & Zeiri, Y. (2014). Green synthesis of gold nanoparticles using plant extracts as reducing agents. *International Journal of Nanomedicine, 9*, 4007.

41. Deokar, G. K., & Ingale, A. G. (2016). Green synthesis of gold nanoparticles (Elixir of Life) from banana fruit waste extract–an efficient multifunctional agent. *RSC Advances, 6(78)*, 74620–74629.

42. Sharma, R. K., Gulati, S., & Mehta, S. (2012). Preparation of gold nanoparticles using tea: A green chemistry experiment. *Journal of Chemical Education, 89*(10), 1316–1318.

43. DeLong, R. K., Reynolds, C. M., Malcolm, Y., Schaeffer, A., Severs, T., & Wanekaya, A. (2010). Functionalized gold nanoparticles for the binding, stabilization, and delivery of therapeutic DNA, RNA, and other biological macromolecules. *Nanotechnology, Science and Applications, 3*, 53.

44. Neouze, M. A., & Schubert, U. (2008). Surface modification and functionalization of metal and metal oxide nanoparticles by organic ligands. *Monatshefte für Chemie-Chemical Monthly, 139*(3), 183–195.

45. Chen, Y., Xianyu, Y., & Jiang, X. (2017). Surface modification of gold nanoparticles with small molecules for biochemical analysis. *Accounts of Chemical Research, 50*(2), 310–319.

46. Biju, V. (2014). Chemical modifications and bioconjugate reactions of nanomaterials for sensing, imaging, drug delivery and therapy. *Chemical Society Reviews, 43*(3), 744–764.

47. Sperling, R. A., & Parak, W. J. (2010). Surface modification, functionalization and bioconjugation of colloidal inorganic nanoparticles. *Philosophical Transactions of the Royal Society of London A: Mathematical, Physical and Engineering Sciences, 368*(1915), 1333–1383.

48. Sharma, A. S., Kaur, H., & Shah, D. (2016). Selective oxidation of alcohols by supported gold nanoparticles: Recent advances. *RSC Advances, 6*(34), 28688–28727.

49. Canet-Ferrer, J., Albella, P., Ribera, A., Usagre, J. V., & Maier, S. A. (2017). Hybrid magnetite–gold nanoparticles as bifunctional magnetic–plasmonic systems: Three representative cases. *Nanoscale Horizons, 2*(4), 205–216.

50. Shevchenko, E. V., Bodnarchuk, M. I., Kovalenko, M. V., Talapin, D. V., Smith, R. K., Aloni, S., ... & Alivisatos, A. P. (2008). Gold/iron oxide core/hollow-shell nanoparticles. *Advanced Materials, 20*(22), 4323–4329.

51. Nakanishi, K., Tomita, M., Masuda, Y., & Kato, K. (2015). Gold nanoparticle–mesoporous silica sheet composites with enhanced antibody adsorption capacity. *New Journal of Chemistry, 39*(5), 4070–4077.

52. Zhang, K., Ji, J., Fang, X., Yan, L., & Liu, B. (2015). Carbon nanotube/gold nanoparticle composite-coated membrane as a facile plasmon-enhanced interface for sensitive SERS sensing. *Analyst, 140*(1), 134–139.

53. Turcheniuk, K., Boukherroub, R., & Szunerits, S. (2015). Gold–graphene nanocomposites for sensing and biomedical applications. *Journal of Materials Chemistry B, 3*(21), 4301–4324.

54. Luan, Y., Qi, Y., Gao, H., Zheng, N., & Wang, G. (2014). Synthesis of an amino-functionalized metal–organic framework at a nanoscale level for gold nanoparticle deposition and catalysis. *Journal of Materials Chemistry A, 2*(48), 20588–20596.

55. Liu, Y., Guerrouache, M., Kebe, S. I., Carbonnier, B., & Le Droumaguet, B. (2017). Gold nanoparticles-supported histamine-grafted monolithic capillaries as efficient microreactors for flow-through reduction of nitro-containing compounds. *Journal of Materials Chemistry A, 5*(23), 11805–11814.

56. Stratakis, M., & Garcia, H. (2012). Catalysis by supported gold nanoparticles: Beyond aerobic oxidative processes. *Chemical Reviews, 112*(8), 4469–4506.

57. Corma, A., & Garcia, H. (2008). Supported gold nanoparticles as catalysts for organic reactions. *Chemical Society Reviews, 37*(9), 2096–2126.

58. Qi, P., Chen, S., Chen, J., Zheng, J., Zheng, X., & Yuan, Y. (2015). Catalysis and reactivation of ordered mesoporous carbon-supported gold nanoparticles for the base-free oxidation of glucose to gluconic acid. *ACS Catalysis, 5*(4), 2659–2670.

59. Fountoulaki, S., Gkizis, P. L., Symeonidis, T. S., Kaminioti, E., Karina, A., Tamiolakis, I., ... & Lykakis, I. N. (2016). Titania-supported gold nanoparticles catalyze the selective oxidation of amines into nitroso compounds in the presence of hydrogen peroxide. *Advanced Synthesis & Catalysis, 358*(9), 1500–1508.

60. Martins, L. M. D., Carabineiro, S. A. C., Wang, J., Rocha, B. G. M., Maldonado-Hódar, F. J., &

Pombeiro, A. J. L. (2017). Supported gold nanoparticles as reusable catalysts for oxidation reactions of industrial significance. *ChemCatChem Catalysis*, *9*(7), 1211–1221.

61. Chen, M., Kang, H., Gong, Y., Guo, J., Zhang, H., & Liu, R. (2015). Bacterial cellulose supported gold nanoparticles with excellent catalytic properties. *ACS Applied Materials & Interfaces*, *7*(39), 21717–21726.

62. Liu, S., Qileng, A., Huang, J., Gao, Q., & Liu, Y. (2017). Polydopamine as a bridge to decorate monodisperse gold nanoparticles on Fe_3O_4 nanoclusters for the catalytic reduction of 4-nitrophenol. *RSC Advances*, *7*(72), 45545–45551.

63. Ertem, E., Murillo-Cremaes, N., Carney, R. P., Laromaine, A., Janeček, E. R., Roig, A., & Stellacci, F. (2016). A silica-based magnetic platform decorated with mixed ligand gold nanoparticles: A recyclable catalyst for esterification reactions. *Chemical Communications*, *52*(32), 5573–5576.

64. Karimi, B., Gholinejad, M., & Khorasani, M. (2012). Highly efficient three-component coupling reaction catalyzed by gold nanoparticles supported on periodic mesoporous organosilica with ionic liquid framework. *Chemical Communications*, *48*(71), 8961–8963.

65. Miyamura, H., Min, H., Soulé, J. F., & Kobayashi, S. (2015). Size of gold nanoparticles driving selective amide synthesis through aerobic condensation of aldehydes and amines. *Angewandte Chemie International Edition*, *54*(26), 7564–7567.

66. Hinde, C. S., Webb, W. R., Chew, B. K., Tan, H. R., Zhang, W. H., Hor, T. A., & Raja, R. (2016). Utilisation of gold nanoparticles on amine-functionalised UiO-66 (NH_2-UiO-66) nanocrystals for selective tandem catalytic reactions. *Chemical Communications*, *52*(39), 6557–6560.

67. Ciriminna, R., Falletta, E., Della Pina, C., Teles, J. H., & Pagliaro, M. (2016). Industrial applications of gold catalysis. *Angewandte Chemie International Edition*, *55*(46), 14210–14217.

68. Corti, C. W., Holliday, R. J., & Thompson, D. T. (2002). Developing new industrial applications for gold: Gold nanotechnology. *Gold Bulletin*, *35*(4), 111–117.

11

Self-assembly of Amphiphilic Molecules

Domenico Lombardo and Maria Teresa Caccamo
CNR–IPCF, Consiglio Nazionale delle Ricerche, Istituto per i Processi Chimico Fisici

Pietro Calandra
CNR-ISMN, Consiglio Nazionale delle Ricerche, Istituto Studio Materiali Nanostrutturati

11.1 Introduction

Amphiphilic molecules are natural (or synthetic) molecules having both hydrophilic and hydrophobic moieties and can spontaneously self-assemble into complex hierarchical nanostructures either in the bulk phases or at the systems interfaces. Self-assembly is one of the most important strategies for the fabrication of nanostructures over different length scales and has become one of the essential aspects for the development of advanced materials and novel technologies. Amphiphilic molecules, such as surfactants or block copolymers, undergo spontaneous self-assembly into several mesophases, in both aqueous and nonaqueous systems [1,2]. The self-assembled, complex, hierarchical geometries can be flexibly controlled or tuned by surfactant molecular architectures or solvent properties. Surfactants often exhibit remarkable phase behavior in water, forming various phases (micelles, vesicles, microemulsions, reverse micelles and liquid crystals), which depends on surfactant concentration and temperature [3]. The developments of multifunctional nanostructures using the recent concepts of nano-architectonics utilize advanced processes for the self-assembly. Finally, amphiphilic self-assemblies involving lipids or proteins are able to mimic the structure and functions of biological systems, thus highlighting the importance of a fundamental understanding of the amphiphilic self-assembly processes toward the realization of a wide range of complex processes operating in biological systems [4].

11.2 Basic Properties of Amphiphilic Molecules

Amphiphiles are (macro-) molecules possessing a water-loving (hydrophilic) component attached to a water-hating (hydrophobic) or a fat-loving (lipophilic) component. In conventional head/tail(s) amphiphiles, the hydrophilic head can be either nonionic or ionic, while the hydrophobic part consists generally of a long hydrocarbon chain. Due to their ability to reduce the interfacial tension, the amphiphilic molecules are often called surfactants (i.e., surface active agents). In aqueous solutions, the amphiphile (hydrophilic) polar headgroup interacts with the water, while the nonpolar (hydrophobic or lipophilic) chain will migrate above the interface (either in the air or in a nonpolar region of the liquid) [5]. In this case, the disruption of the cohesive energy at the interface favors a microphase separation between the selective solvent and the dispersed phase of the amphiphile with the formation of many smaller closed interfaces of micelles-like aggregates [6]. Due to this characteristic, many amphiphiles are employed as detergents, emulsifiers, dispersants, wetting and foaming agents in several applications [7].

Anionic amphiphiles (Figure 11.1a), which are widely used as soaps and detergents in cleaning processes, consist generally of negatively charged headgroups and positively charged counterions (such as sodium, potassium or ammonium ions). Commonly used polar groups are the carboxylate ($-COO^-$), sulfate ($-SO_4^-$), sulfonate and phosphate (which is the charged functionality mainly found in phospholipids).

Cationic amphiphiles consist of positively charged headgroups such as a quaternary ammonium and a halide ion as a counterion. As an example, the cetyltrimethylammonium bromide (CTAB) is one of the most employed cationic amphiphiles (Figure 11.1b).

Nonionic amphiphiles have either polyether or polyhydroxyl units as the hydrophilic group. Several conventional nonionic surfactants consist generally of a hydrophilic poly(ethylene oxide) chain, often called ethoxylates, connected with a hydrophobic alkyl chain, and are generally used in cleaning applications with anionic surfactants. For example, polyoxyethylene alkyl ethers, C_nE_m, are nonionic surfactants made of m hydrophilic oxyethylene units and an alkyl chain with n methylene groups (Figure 11.1c).

Finally, in *zwitterionic amphiphiles*, the headgroups possess both a positive charge and a negative charge. An example of zwitterionic amphiphiles is given by the vesicle-forming phospholipid phosphatidylcholine (Figure 11.1d).

Amphiphiles may present an architecture which is different from the conventional head-tail(s) topology found in traditional surfactants. For example, the *Gemini amphiphiles* consist of two hydrophilic headgroups linked (by a spacer) to two hydrophobic tails. This conformation confers a more compact molecular packing configuration that induces sensitively low surface tensions and good solubilization properties. Moreover, they favor the formation of sub-micellar thread-like aggregates with a critical micellar concentration (CMC.) which is up to two orders of magnitude lower than that of traditional single-chain surfactants [8].

Another example of non-conventional amphiphile is given by the *Bola amphiphiles* [9] that consist of two hydrophilic headgroups placed at both ends of a hydrophobic skeleton (generally one or more alkyl chains). This architecture yields a higher solubility in water, an increase in the CMC value and a decrease in aggregation number with respect to traditional single-headed amphiphiles. Together with the classical spheres-, cylinders-, disks- and vesicles-type aggregates, bolaamphiphiles are also known to form helical structures that can favor the formation of microtubular aggregates [9].

11.3 Amphiphiles Aggregates Formation

Various weak (noncovalent) forces acting in the amphiphiles are responsible for the amphiphiles self-assembly processes, including hydrogen bonding, hydrophobic effects, and electrostatic and van der Waals interactions. The stability of the amphiphiles in solution in their aggregate states is mainly driven by both the hydration of the hydrophilic (polar) headgroups (hydrogen bonding) and the insertion in the solvent of the hydrophobic tail(s) (hydrophobic effect) [1].

The *hydrogen bonding* is a crucial phenomenon for several important functions in biological systems as they are strong enough that bind biomolecules together but weak enough to be broken, when necessary, inside the living organisms and cells [10]. The polarity of water molecules results in the attraction of the negative (oxygen) portion and positive (hydrogen) partial charges, and is at the basis of the water hydrogen bonding, which has many important implications on the properties of water and its relevant functions in biological systems [11].

The *hydrophobic effect* is the second main driving force of amphiphile self-assembly [4]. When different nonpolar

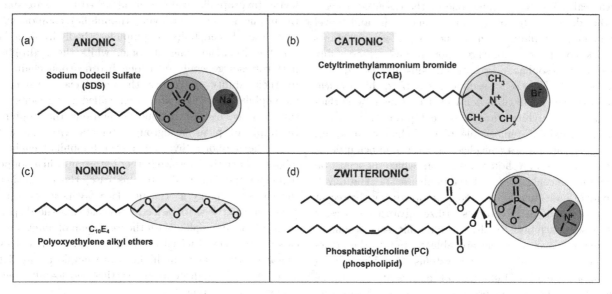

FIGURE 11.1   Example of common nonionic (a), anionic (b), cationic (c) and zwitterionic (d) amphiphilic molecules.

molecules are dissolved in water, the disruption of the H-bonding water network favors the creation of larger cavities (i.e. ice-like cage structures called *clathrates*) that accommodate an assembly of nonpolar (solute) molecules [12]. These entropically more favorable aggregated structures are generated to minimize the disruption of the water structure. This process corresponds to an effective mutual attraction among the nonpolar molecules in water (hydrophobic interaction). The synergistic effects of the hydrogen bond and hydrophobic interactions play a crucial role also in a wide range of biological processes such as the protein folding effects, which are important in keeping a protein alive and biologically active through the reduction of the undesired interactions with water [13].

Another important interaction for the self-assembly and colloidal stability of ionic amphiphilic molecules in solution is given by the so-called Derjaguin, Landau, Vervey, and Overbeek *(DLVO) interaction potential*, which was first introduced by a pioneering work on charged colloids in solution [14]. This theory represents the cornerstone to rationalize forces between charged amphiphiles (and charged colloids) at interfaces and to explain their aggregation behavior in solution. DLVO theory assumes that the interaction in solution between charged colloids can be approximated by two principal contributions, namely the *van der Waals* and *double-layer interactions* [15]. Despite the weakness of the involved interactions, the relevant number of these soft interactions will produce an overall effect that is strong enough to hold different amphiphile molecules together as well as to ensure their stability in solution [16,17]. Moreover, the weakness of the involved interactions makes the structure more flexible enabling the system to withstand minor perturbation while preserving the reversibility of the self-assembled structure.

11.3.1 Identification of the Amphiphile Critical Micellar Concentration

Amphiphilic molecules in solution solubilize as single molecules (unimers) only up to a certain limit concentration, called *critical micellar concentration*. Above the CMC, the surfactants in the bulk solution self-assemble into aggregates called *micelles*. From the thermodynamic point of view, the self-assembly of amphiphilic (macro-)molecules and their aggregate (micelles-like) formation are driven by the competition between the interfacial energy of the micelles core with solvent and the conformational distortion energy of the soluble chains emanating from the core region. According to the closed association model, the detected CMC can be used to obtain information on the thermodynamic parameters of the micellization process [1]. The standard Gibbs free energy change ΔG for the transfer of 1 mol of amphiphile from solution to the micellar phase (so-called *free energy of micellization*) can be expressed as a function of the absolute temperature T and the CMC molar fraction X_{CMC}, and can be approximated by $\Delta G \approx RT \ln(X_{cmc})$ (where R is the ideal gas constant) [6–9].

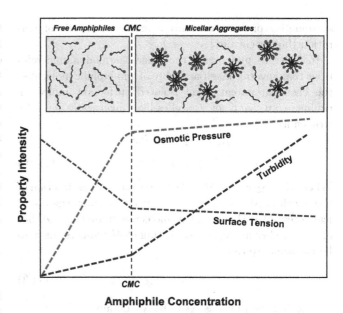

FIGURE 11.2 Discontinuity of physical properties associated with the formation of micelles at the CMC for amphiphilic molecules in a selective solvent.

Amphiphiles self-assembly can be demonstrated experimentally through the measurement of an inflection, usually associated with the formation of micelles, in some physical property at a given concentration (CMC) and temperature (critical micelle temperature—CMT). Together with theoretical modeling and simulation approaches, useful physical quantities for detecting the CMC are the turbidity, osmotic pressure and surface tension. The characteristic behavior of these quantities is shown in the Figure 11.2.

For commonly used surfactants, the CMC is typically less than about 0.01 M. Combining thermodynamic model with reaction rate theory, it has been demonstrated that, over most of the metastable concentration range, the presence of pre-micellar aggregates have been detected and are characterized by a macroscopic lifetime and a small polydispersity [18].

Finally, it is important to notice that the micellar aggregates should not be considered as "permanent" frozen structures, but they are rather thermodynamic equilibrium systems with a *dynamic structure* in which continuous exchange of unimers from the micelles and the bulk solution (and vice versa) takes place. Lifetime of molecule in a small micelle is between 10^{-5} and 10^{-3} s.

11.4 Thermodynamics of Amphiphiles Self-assembly

From the thermodynamic point of view, the main requirement to have a favorable spontaneous organization toward a self-assembled structure is the minimum energy configuration at equilibrium. More specifically, the generated self-assembled structure has a higher order (lower entropy) than the isolated components, while its surrounding presents generally a more disordered configuration (higher entropy).

In order to understand the basic thermodynamic relations that govern the self-assembly of amphiphilic molecules, the chemical potentials of isolated surfactant molecules (unimers) μ_1 and of (micelles-like) aggregates can be written as

$$\mu_1 = (\mu_1^0 + kT\ lnx_1) \tag{11.1a}$$

$$\overline{\mu}_M = (\overline{\mu}_M^0 + kT\ ln\overline{x}_M) \tag{11.1b}$$

Where the $\overline{x}_M = x_M/M$ is the molar (or volume) fraction of the micelles. At the thermodynamic equilibrium, the chemical potential of M amphiphilic molecules (unimers) and that of the micellar aggregate (consisting of M unimers) have to be the same, so that

$$M\mu_1 = \overline{\mu}_M \tag{11.2}$$

Comparing the previous relations, we obtain

$$\mu_1 = \frac{\overline{\mu}_M}{M} \tag{11.3a}$$

$$\mu_1^0 + kT\ logx_1 = \mu_M^0 + \frac{kT}{M}\ln\left(\frac{x_M}{M}\right) \tag{11.3b}$$

where the mean interaction energy per molecule in the aggregate is given by $\mu_M^0 = (\overline{\mu}_M^0/M)$.

In a system of amphiphilic molecules in solution forming aggregates (Figure 11.3), at thermodynamic equilibrium, the chemical potential of all identical molecules (in the different types of aggregates with different association number M) must be the same. We can express the equilibrium chemical potential per surfactant molecule as

$$\mu = (\mu_1^0 + kT\ lnX_1) = (\mu_2^0 + \frac{1}{2}kT\ ln\frac{X_2}{2}) = \ldots$$

$$= (\mu_M^0 + \frac{1}{M}kT\ ln\frac{X_M}{M}) = constant \tag{11.4a}$$

$$\mu = \mu_1 = \mu_2 = \ldots = \mu_M = constant \tag{11.4b}$$

In specific aggregates of aggregation number N (with $1 < N < M$), the energy per aggregate is $N{\cdot}\mu_N^0$, where μ_N

is the (mean) chemical potential of a molecule, $\mu^0 N$ is the mean interaction free energy per molecule (standard part of the chemical potential) and X_N is the concentration (activity) of amphiphilic molecules. These equations assume ideal mixing of diluted amphiphile systems, where inter-aggregate interactions are neglected.

Multiplying equation (3b) by M and rearranging the terms, we obtain

$$-\frac{M\left(\mu_M^0 - \mu_1^0\right)}{kT} = ln\left(\frac{x_M}{Mx_1^M}\right) \tag{11.5}$$

$$x_M = Mx_1^M\left[exp\left(-\frac{\left(\mu_M^0 - \mu_1^0\right)}{kT}\right)\right]^M = Mx_1^M K \tag{11.6}$$

where K is the ratio of the reaction rates of associating/dissociating unimers in the micellar aggregates. In fact, by using the law of mass action (Figure 11.3), we can introduce the equilibrium constant, K as the ratio of the two "reaction" rates (where $k_1 X_1^N$ is the rate of association, and $k_N(X_N/N)$ is the rate of dissociation).

$$K = k_1/k_N = \exp[-N(\mu_N^0 - \mu_1^0)/kT] \tag{11.7}$$

Assuming an arbitrary reference state of aggregates (or monomers) with aggregation number M, previous Eq. (11.7),can be expressed [1] in the equivalent form:

$$X_N = N\left\{(X_M/M)\exp[M(\mu_M^0 - \mu_N^0)/kT]\right\}^{N/M} \tag{11.8}$$

More specifically, for $M = 1$, previous Eq. (11.8) can be expressed as

$$X_N = N\left\{(X_1)\exp[(\mu_1^0 - \mu_N^0)/kT]\right\}^N \tag{11.9}$$

Moreover, the following conservation relation for the total solute concentration C should be considered:

$$C = X_1 + X_2 + \ldots + X_N = \sum_{N=1}^{\infty} X_N \tag{11.10}$$

These equations, together with the conservation relation for the total solute concentration C, completely define the system.

11.5 Amphiphiles Packing Factors and Aggregates Morphologies

The possibility to control the shapes and size of a micellar aggregate gives the possibility to manipulate topology and architectures of the self-assembled nanostructures, in order to obtain functional systems for advanced technological applications. The precise description of the aggregates morphology for amphiphiles having complex topology is difficult to rationalize, due to the different involved intermolecular interactions and complex synergistic effects present [19]. Moreover, the solution conditions such as surfactant concentration, temperature, pH

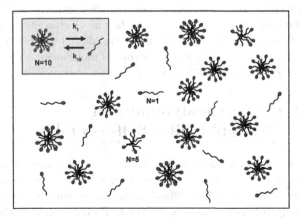

FIGURE 11.3 Self-association of N monomers (amphiphilic molecules) into a micellar aggregate.

and ionic strength play an important role in the final aggregate conformation at the thermodynamic equilibrium. However, for simple amphiphile topology, the shape and size of given micellar aggregates strongly depend on the molecular geometry of its component surfactant molecules [20,21].

An estimate of the size and shape of the amphiphile aggregate can be made by analyzing the so-called *critical packing parameter* $C_{pp} = V_0/(A_{mic} \cdot l_c)$ [1], where A_{mic} is the effective hydrophilic headgroup surface area at the aggregate–solution interface, V_0 is the effective volume of the hydrophobic chains in the aggregate core and l_c is the maximum effective length (critical chain length). In Figure 11.4, we report a summary of the amphiphile aggregate morphology that can be predicted starting from the value of the amphiphile critical packing parameter C_{pp}.

More specifically, by increasing the C_{pp} value, the aggregates structure passes from *spherical* ($C_{pp} \leq 1/3$), to *cylindrical* ($1/3 \leq C_{pp} \leq 1/2$) and *lamellar* ($C_{pp} = 1$) shape. At large values of the critical packing parameter (i.e., for $1/2 \leq C_{pp} \leq 1$), *vesicles (or microemulsions)* aggregates are generated, which correspond to closed spherical (or ellipsoidal) amphiphiles bilayer structures possessing an internal cavity containing the dispersion medium [2]. Bilayers vesicle structures formed by *phospholipids* (composed of one hydrophilic head and two hydrophobic tails) have been widely employed for research and industrial applications purposes. When phospholipids are dispersed in a selective solvent (like water), they aggregate spontaneously to form bilayers, which resemble the structures of the biological membranes [4].

For packing parameters larger than one ($C_{pp} > 1$), *inverse micelles* are favored, where the hydrophobic chains radiate away from the headgroups that surround the water solvent. Inverse aggregates also form in isotropic ternary systems such as *microemulsions*, where two immiscible phases ("water" and "oil") are present with a surfactant. The basic types of microemulsions are the "*direct*" oil-in-water (o/w), the "*reversed*" of water-in-oil (w/o) and the bicontinuous. These surfactant molecules (sometimes in combination with a cosurfactant) form a monolayer at the oil–water interface, with the hydrophilic headgroups located in the aqueous phase, while the surfactant hydrophobic tails are dissolved in the oil phase. This gives the formation of microemulsions where droplet radius is usually in the range between 5 and 50 nm [6].

With the advent of novel synthetized supramolecular systems containing multiple functional surface end groups, it is more difficult to predict the final aggregates architecture of assembling amphiphilic system. In this case, a wide range of different structures are available by modifying the balance between the hydrophilic/hydrophobic components. Beyond the more common micelles, vesicles and liquid crystalline phases, amphiphiles can self-assemble in a variety of hierarchically organized novel nanostructures including ribbons, fibers, (super-)helices and tubes, and distorted bilayer aggregates morphologies [22–25].

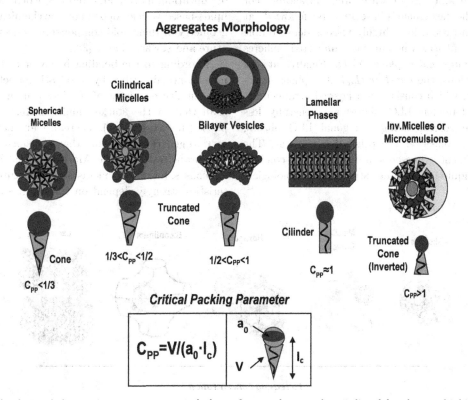

FIGURE 11.4 Analysis of the aggregate structures and shape factors that can be predicted by the amphiphile critical packing parameter C_{pp}.

11.5.1 Lyotropic Liquid Crystal Structures and Aggregation Phases of Amphiphiles

Complex structural and dynamic features as well as collective phenomena involving a large number of macromolecules are the main factors influencing the self-assembly processes of amphiphiles. The morphology of the self-assembled structures, especially at the higher concentrations, strongly depends on the solubility properties of the different ends of the amphiphilic molecules. While at very low concentrations, amphiphiles are randomly dispersed in solution without any order, at the higher concentrations, the balance of the inter- and intra-molecular forces may drive the amphiphiles molecules in selective solvents toward the formation of ordered lyotropic liquid crystals (LLC) structures [26–28]. Different classes of LLC phases and structures of several surfactant systems have been extensively investigated by means of neutron and X-ray diffraction and scattering experiments [29, 30] and nuclear magnetic resonance (NMR) measurements [31].

In Figure 11.5, a schematic representation of a typical progression of phases as a function of concentration for an amphiphile dissolved in a selective solvent is reported. At low concentration, the amphiphilic molecules self-assemble into a *cubic (I₁)* LLC phase that consists in a cubic arrangement of regular micelles in a solvent continuous. At the higher concentrations, the amphiphiles form micellar cylinders that arrange themselves into a very viscous *hexagonal* LLC phases, whose spacing depends upon the relative amounts of water and surfactant. For some systems, the hexagonal phase may be followed, at the higher concentration, by a highly viscous dense (cubic lattice) isotropic phase, wherein interconnected spheres form a *bicontinuous cubic phase (V1)*. Finally, at the higher concentrations, the *lamellar (Lα)* LCC phase structures are present, which consists of a layered arrangement of amphiphiles. Lamellar LLC phases are generally less viscous than the corresponding hexagonal LLC phases, as the layers slide one over the other during shear. The lamellar arrangement represents a fundamental configuration of the liquid crystalline structure of biological membranes.

At very high concentration, the amphiphilic molecules arrange themselves to maximize their polar–polar and apolar–apolar interactions while minimizing the steric hindrance. The low-content of water (or any other solvent) makes the structure more rigid, and a new region, characterized by the amphiphile *reversed (or inverted) liquid crystal phases*, is often observed [32]. In the extreme situation of pure amphiphiles (and liquid state), amphiphile generally has structures and dynamics typical of structured and glass-forming liquids. Opportune mixture of different amphiphiles can trigger therefore the arising of striking emerging properties that can be used for specific technological applications [33–35].

11.6 Lipid-Based Amphiphiles: Self-assembly of Nanocarriers for Drug Delivery Application

Self-assembly of lipid amphiphiles represents an important process for the construction of versatile nanostructured systems and prototypes for technological applications in the field of material science, nanomedicine and biotechnology [36]. (Phospho-)lipids amphiphilic molecules are generally composed of one hydrophilic head and two hydrophobic tails. When phospholipids are dispersed in water, they tend to aggregate spontaneously to form bilayers which can close into vesicles (liposomes), thus resembling the fundamental structures of biological membranes [37]. The final organization, morphology and physicochemical properties of lipids-based vesicles depend on the chemical nature, size and geometry of their lipid components, concentration, temperature and surface charge [38].

According to the bending behavior of lipid bilayers biomembranes, described by the elastic models of Helfrich [39], the curvature energy of a bilayer in a vesicle is higher than that in the stacked multilamellar, liquid crystalline phase (in excess water). Therefore, an energy cost is necessary to generate curvature that transforms an undisturbed membrane into a vesicle. Moreover, lipid bilayer vesicles in aqueous solution are metastable structures whose characteristics strongly depend on the conditions of preparation

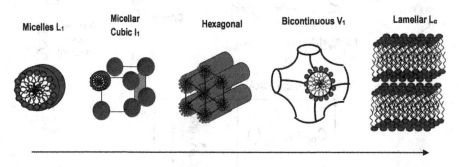

Increasing Concentration

FIGURE 11.5 Schematic representation of a typical progression of phases as a function of concentration for an amphiphile dissolved in a selective solvent.

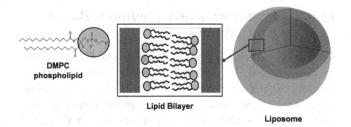

FIGURE 11.6 Schematic structure of a lipid bilayer and liposome composed of dimyristoylphosphatidylcholine (DMPC) phospholipids.

(i.e., stirring, sonication, extrusion, microfluidification or electroformation) [40] (Figure 11.6).

The size of liposomes nanocarriers employed for biomedical applications ranges mainly between 50 and 500 nm. Lipid vesicles systems may contain *multilamellar vesicles* (MLVs), which are composed of concentric bilayer surfaces in an onion-like structure (hydrated multilayers), *small unilamellar vesicles* (SUVs, diameter < 100 nm), *large unilamellar vesicles* (LUVs, diameter 100–1,000 nm) or *giant unilamellar vesicles* (GUVs, diameter > 1 μm).

The difference in the *drug release rate* strongly depends on the phospholipid bilayers that the active drug has to cross during the release process. In general, MLVs with large diameters have a greater entrapped volume than ULVs, which in turn have a much faster drug release rate. The study of the self-assembly of lipid bilayers is a crucial topic for understanding the physical and chemical bases of the structural organization and biological functions existing in bio-membranes [41–43]. It is also interesting for the creation of new materials, exploiting the self-assembly of supramolecular aggregates, and for the design of new drug delivery systems such as liposomes and mixed micelles [44–47].

11.7 Bio-inspired Amphiphiles: Peptide Amphiphiles

The introduction of biological motifs (such as peptides, carbohydrates or steroids) into the structure of an amphiphile allows to mimic specific functions of biological systems. This represents a formidable approach to the formation of *bio-inspired amphiphiles* for a wide range of potential applications in drug delivery and biotechnology. In this respect, *peptide amphiphiles* (PAs) represent one of the best examples of smart bio-functional nanomaterials for applications in biotechnology and nanomedicine [49]. PAs are synthesized through the conjugation of a short peptide sequence to a hydrophobic tail. When dissolved in water, PAs tend to self-assemble into cylindrical aggregates (nanofibers), while the selection of the amino acids determines, through folding and hydrogen bonding, the generated (secondary, tertiary and quaternary) structures.

As reported in Figure 11.7, the PA is composed of three main regions: the hydrophobic tail forming the PA nanofiber core (*region I*), a β-sheet forming amino acids to accomplish

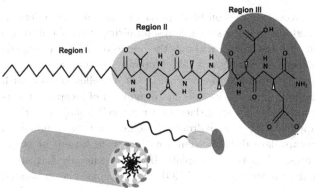

FIGURE 11.7 Structural characteristics of PA E2 (C16V2A2E2) and their self-assembled nanofiber [48].

specific functions (*region II*) and the hydrophilic moieties (*region III*) that facilitate water solubility of PAs and where a bioactive sequences can be attached at the water-exposed surface [48,49].

By exploiting their supramolecular structure, PAs represent good prototypes for the delivery of drugs and proteins as well as in regenerative medicine applications. For example, *in vivo* gelation can be obtained by injecting the PA solution within pre-formed nanostructures (at low ionic strength) into the high ionic strength environment of biological tissues. The drug travels from a dilute solution (outside of the cell) to a particular site in the cell, during their diffusion process into a bio-membrane.

11.8 Polymer-Based Amphiphiles

Polymer-based amphiphiles are obtained by linking hydrophilic and hydrophobic polymer blocks through a covalent bond. Depending on the molecular topology, polymer-based amphiphiles can be classified into the main category of linear (block) copolymers and hyperbranched polymers [50–53].

11.8.1 Linear Amphiphilic Block Copolymers

Self-assembly of linear amphiphilic block copolymers in selective solvents represents a versatile tool for the design and engineering of advanced materials as well as an important approach to investigate the complex processes in the field of soft matter and colloidal science [51–53]. These systems offer the possibility of molecular control by tuning the architecture and polymer composition. Moreover, by increasing the number of monomeric units, it is possible to facilitate a stronger interaction among the polymer chains, thus preventing a quick molecular exchange in comparison with traditional amphiphiles. Together with the linear AB diblock copolymers, thermoresponsive linear ABA triblock architectures have attracted enormous scientific interest and have already found broad application in material science, biotechnology and nanomedicine [53,54].

Recent studies on block polymeric micelles have demonstrated particular strength in solubilizing hydrophobic drugs in relevant doses without the inclusion of toxic organic solvents or surfactants [55]. Moreover, anticancer efficiency can be obtained by modifying the micelles surface with targeting ligands for specific recognition of receptors (over-expressed on the surface of tumor cells) [56]. For these applications, the hydrophobic block can be tailored to encapsulate drug molecules with a wide variety of structures, while the hydrophilic block is generally formed of poly(ethylene oxide) (PEO) that assures the requested biocompatibility and the desired "stealth" characteristic minimizing unwanted interactions with cellular components. In these systems, interpenetration of the external long hydrophilic PEO chains causes the depletion of the solvent in the outer layer of the nanocarrier, thus favoring the formation of larger clusters of entangled aggregates in some specific conditions of the amphiphilic system [57]. Temperature-dependent attractive interaction arising from the synergistic effects of different coexisting interactions in the system [57,58] can be exploited for triggering the assembly/disassembly process in these classes of nano-aggregates [59].

A special class of ABA triblock copolymers is represented by the commercially available *Pluronic-type* class of amphiphilic poly(ethyleneoxide)-poly(propyleneoxide)-poly(ethyleneoxide) PEO_m-PPO_n-PEO_m triblock copolymers (Figure 11.8). Apart from their applications in material science and industry, these systems represent a versatile component for applications in nanomedicine and biotechnology. A relevant number of investigations involving Pluronic block copolymers as drug delivery systems or bioformulations for (pre-) clinical use or trials are present in literature [53,56].

11.8.2 Hyperbranched Polymers-Based Amphiphiles

Another interesting class of polymer-based amphiphiles are constructed by employing highly branched three-dimensional polymer structures. These classes of macromolecules have attracted a great attention owing to their interesting properties resulting from the (hyper-)branched structure with a high number of functional groups generating high solubility and poor molecular entanglement [60]. They also received a great interest in both academic research and industrial fields because of their advantages in facile one-pot synthesis, low cost, excellent properties for a broad range of nanoscience and nanotechnology applications, including energy harvesting, catalysis, photonics, drug delivery and nanomedicine applications [60].

Hyperbranched linear–multiarm, linear–dendritic and dendritic–dendritic block copolymers reveal as promising materials for different advanced applications such as controlled drug delivery, environmental-specific nanoreactors, artificial enzymes and biosensors [61,62]. A schematic representation of some complex shape of amphiphiles and their self-assembly is reported in Figure 11.9.

Specific chemical synthesis protocols for the entanglement and linking of dendritic components have been employed to realize macromolecular amphiphiles with a large variety of different morphologies [63,64]. In this respect, highly monodispersed dendrimers with well-defined branched architectures represent fascinating building blocks for versatile application in host–guest, biotechnology and catalytic chemistry [65–67]. The dendrimer core region imparts to the central cavity the requested size, absorption ability and inclusion-release characteristics, while the topological and chemical variety of surface end-groups strongly influences the interaction, the solubility as well as the molecular recognition ability [68–70].

Synthesis and self-assembly processes of uniform aggregates of *jellyfish* [71]- and *cuttlefish* [72]-shaped *amphiphiles*, and fluorinated amphiphilic copolymers of different

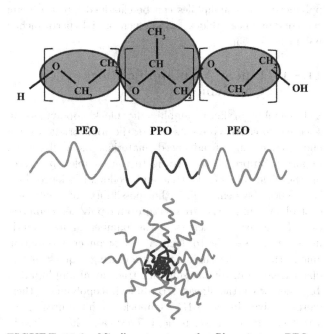

FIGURE 11.8 Micellar structure of a Pluronic-type PEOm-PPOn-PEOm amphiphilic triblock copolymer.

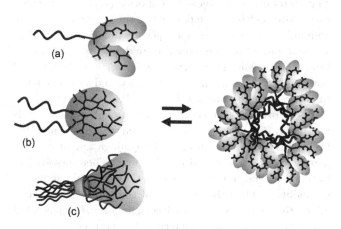

FIGURE 11.9 Sketch of the self-assembly process of hyperbranched amphiphiles: linear-hyperbranched (a) and linear-dendrimer (b) diblock copolymers, cuttlefish amphiphiles (c).

architectures (diblock, grafted, or palm tree [73,74]) have been recently reported. Finally, photo-responsive *Janus hyperbranched supramolecular polymers*, obtained on the basis of the host–guest molecular recognition of the cyclodextrin/azobenzene self-assembly (and disassembly) processes, have been recently synthesized [75].

11.9 Amphiphiles Self-assembly and Synthesis of Hybrid Nanostructures

The combination of molecular interactions together with the ability to control both length scale and structural morphologies makes amphiphiles particularly interesting as templates in the synthesis of porous materials with enhanced characteristics and properties. The use of amphiphiles self-assembly has been recognized as a promising bottom-up approach for the construction of organic–inorganic supramolecular nanostructures in which an amphiphilic (macro-)molecular template drives the formation of porous nanostructures with peculiar final properties [76,77].

When the silica source precursor (such as tetraethyl orthosilicate [TEOS] or other alkoxides with longer alkyl chains) is added to the amphiphiles solution system (a *sol*), it adsorbs on the micelles and hydrolyses. This causes an aggregation process between spherical micelles (that may become elongated and arranged in a hexagonal pattern), while a silica network is gradually formed (a *gel*) in which the liquid is enclosed. Finally, the gel is heated (calcinated) and the surfactants removed (i.e., decomposed or oxidized under oxygen or air atmospheres and evaporates), while the mesoporous silica network remains.

The whole process is also referred to as *soft templating*, for the use of a polymer-based, organic material as a (soft) template [78,79]. The specific soft interactions are responsible for nanoparticle stabilization processes, while both size and shape of the templating structures can be modulated by opportunely changing the composition and the reference amphiphiles. In a variant of this method (referred to as *liquid crystal templating*), a more concentrated (amphiphile) liquid crystalline phase solution is pre-formed before condensation of the inorganic framework upon addition of the precursors.

Amphiphiles templated mesoporous silica materials have attracted considerable interest in a wide range of applications including (heterogeneous) catalysis, adsorption and separation processes, electronics, optics, host–guest chemistry and biotechnology [80].

The first synthesis of silica mesoporous nanoparticles with a triblock copolymer as template was reported by Zhao et al. in 1998 [81] and named SBA-X (Santa Barbara Amorphous) where X is a number corresponding to a specific pore structure and surfactant. With the aim to solve the stability problem of synthesized zeolite-based porous nanostructures in solution, the successful formation of a mesoporous spherical complex, driven by the self-assembly of Linde-Type A zeolite synthesis in the presence of poly(dimethylsiloxane)–poly(ethylene oxide) [PDMS-PEO] block copolymers template has been recently reported [82]. The use of highly branched macromolecular structure as template furnishes an enhancement of template stability in solution during the zeolite synthesis [83,84]. Recently, Xu et al. reported an approach for designing a single (quaternary ammonium) head amphiphilic template by employing the noncovalent (aromatic–aromatic) π–π stacking interactions to stabilize the micellar structure. This has been achieved by introducing strong mutual interactions in the hydrophobic chain and giving an ordered orientation to crystal growth of MFI (Mobil-type five) zeolitic (nanosheet) framework [85].

In conclusion, it is worth mentioning that mesoporous silica nanoparticles (MSNs) are a highly promising platform for intracellular controlled release of drugs and biomolecules [86–89]. The ability to functionalize with organic groups the MSNs surface is used for the entanglement of functional molecules which are able to interact with nanosized intracellular structures. These systems represent then promising vehicles for drug targeting and drug delivery applications [98,90,91].

11.10 Supramolecular Self-assembly and Hierarchical Amphiphilic Nanostructures

The amphiphiles self-assembly can be considered a precursor of the development of the modern bottom-up approaches in nanoscience and nanotechnology. In this respect, a synergistic integration of approaches across different scientific disciplines promises to be very useful for advanced nanoscience applications. From a more general point of view, the self-assembly can be viewed as a process by which a set of disordered *building blocks* (undergoing mutual interactions) form *ordered structures* through a *spontaneous organization*. The building blocks are not necessarily amphiphiles molecules but can span a wide range of intermediate macromolecular nano- and mesoscopic structures with different chemical compositions, shapes and functionalities [92,93].

A hierarchical self-assembly based on the concept of supramolecular amphiphiles can be realized by employing suitable noncovalent driving forces including *hydrogen bonding, host–guest recognition, electrostatic forces, metal coordination and π–π stacking interaction.*

One advantage of this strategy is the wide selection of mono- or multifunctionalized building blocks (e.g., monomer, hyperbranched polymer, macrocycles, porous nanostructures) and a variety of synthetic methods using noncovalent interactions rather than the most expensive chemical synthesis [94–96].

The *π–π stacking interaction*, that involves direct attraction between aromatic rings, has been implicated widely in supramolecular chemistry and has great potential applications in biotechnology and material sciences [97,98], while metal–organic coordination interaction can be employed in construction of complex reversibly tunable supramolecular amphiphiles nanostructures that are able to coordinate with metal ions [99].

Among the self-assembled hierarchical amphiphiles, the construction of supramolecular nanostructures based on *host–guest interaction* provides a flexible platform for the development of smart nanomaterials and functional supramolecular devices [100,101]. Molecular recognition of a host–guest system consists of a receptor molecule (host) interacting with a ligand molecule (guest) through noncovalent interactions [102,103].

Finally, the reversible nature of noncovalent interactions may allow a dynamic switching of structure, morphology and functions in response to various external stimuli, such as stress, pH, temperature and electromagnetic radiation. For example, nanostructures self-assembly could disassemble upon the activation of an external stimulus such as pH or ultraviolet (UV) electromagnetic radiation, and this effect can be exploited for potential applications in the field of controlled drug delivery. Among numerous external stimuli, electromagnetic radiation (light) is of special interest. For example, azobenzene, which has two isomers (trans and cis), can be recognized by β-CD reversibly upon light irradiation [104]. Stimuli-responsive supramolecular amphiphiles self-assemblies have potential applications in a wide range of fields, including memory storage, smart supramolecular polymers nanodevices and drug delivery systems [105].

11.11   Conclusion

Amphiphiles self-assembly can be considered a precursor of the development of the bottom-up approaches in modern nanotechnology. Segregation of hydrophilic and hydrophobic regions represents the primary noncovalent driving force for the formation of organized nanostructures such as micelles, vesicles, bilayers, microemulsions and LLC. Design of more complex structures can be obtained by introducing structure-directing interactions that influence, in a fascinating way, the morphology of the final assemblies. In this respect, a large variety of synthetic systems are able to self-assemble in a highly ordered fashion capable of mimicking biological activity, thus playing an important role in several applications including drug release, tissue engineering and gene therapy. Moreover, the incorporation of functional moieties, such as simple peptides or carbohydrates, has enormously increased the design and fabrication of dedicated bio-inspired functional materials. Amphiphiles' self-assembly, which has been the central topic of intense research activity for colloid, polymer and materials science for several years, continues to stimulate the renewed interest of nanomaterial scientists, combining the processes of colloid self-assembly with the more advanced concept of supramolecular chemistry.

References

1. Israelachvili, J. N. 2011: *Intermolecular and Surface Forces.* Elsevier, Academic Press: Amsterdam.
2. Domb, C., Lebowitz, J. L., Gompper, G., and Schick, M. 1994: *Self-Assembling Amphiphilic Systems* (Phase Transitions and Critical Phenomena). Academic Press: London.
3. Glotzer S. C. and Solomon. M. J. 2007: Anisotropy of building blocks and their assembly into complex structures, *Nature Materials* 6: 557–562.
4. Tanford, C. 1991: *The Hydrophobic Effect: Formation of Micelles and Biological Membranes.* Krieger: Malabar, FL.
5. Holmberg, K., Jonsson, B., Kronberg, B., and Lindman, B. 2002: *Surfactants and Polymers in Aqueous Solution,* 2nd Edn. Wiley: Chichester.
6. Rosen, M. J. 1989: *Surfactants and Interfacial Phenomena,* 2nd Edn. Wiley, New York.
7. Cui, X., Mao, S., Liu, M., Yuan, H.. and Du Y. 2008: Mechanism of surfactant micelle formation, *Langmuir,* 24(19): 10771–10775.
8. Camesano T. A. and Nagarajan, R. 2000: Micelle formation and CMC of gemini surfactants: A thermodynamic model, *Colloids and Surfaces A: Physicochemical and Engineering Aspects,* 167: 165–177.
9. Fuhrhop, J. H. and Wang, T. 2004: Bolaamphiphiles, *Chemical Review,* 104: 2901–2937.
10. Ball. P. 2008: Water as an active constituent in cell biology, *Chemical Review,* 108(1):74–108.
11. Fenimore, P. W., Frauenfelder, H., Magazù, S., McMahon, B. H., Mezei, F., Migliardo, F., and Young, R. D. 2013: Concepts and problems in protein dynamics, *Chemical Physics,* 424: 2–6.
12. Finney, J. L. 1999: The Structural basis of the hydrophobic interaction. In Hydration Processes in Biology, Ed. M.-C. Bellissent-Funel, pp. 115–124. IOS Press: Amsterdam.
13. Angell, C. A., Poole, P. H., and Shao, J. 1994: Glass-forming liquids, anomalous liquids, and polyamorphism in liquids and biopolymers, *Il Nuovo Cimento D,* 16(8): 993–1025.
14. Verwey, E. J. W. and Overbeek, J. T. G. 1948: *Theory of the Stability of Lyophobic Colloids.* Elsevier: Amsterdam.
15. Hunter R. J. 1986: *Foundations of Colloid Science,* vol. I–I.I Oxford University Press: New York.
16. Lombardo, D., Calandra, P., Barreca, D., Magazù, S., and Kiselev, M. A. 2016: Soft interaction in liposome nanocarriers for therapeutic drug delivery, *Nanomaterials,* 6 (7): 125.
17. Lombardo, D., Kiselev, M. A., Magazù, S., Calandra, P. 2015: Amphiphiles self-assembly: Basic concepts and future perspectives of supramolecular approaches, *Advances in Condensed Matter Physics,* 2015: 151683.

18. Hadgiivanova, R. and Diamant, H. 2009: Premicellar aggregation of amphiphilic molecules: Aggregate lifetime and polydispersity, *Journal of Chemical Physics*, 130(1–5): 114901.

19. Calandra, P., Caschera, D., Liveri, V. T., and Lombardo, D., 2015. How self-assembly of amphiphilic molecules can generate complexity in the nanoscale, *Colloids and Surfaces A: Physicochemical and Engineering Aspects*, 484: 164–183.

20. Shimizu, T., Masuda, M., and Minamikawa, H. 2005: Supramolecular nanotube architectures based on amphiphilic molecules, *Chemical Review* 105(4):1401–1443.

21. Lombardo, D., Munaò, G., Calandra, P., Pasqua, L., and Caccamo, M. T. 2019: Evidence of pre-micellar aggregates in aqueous solution of amphiphilic PDMS–PEO block copolymer, *Physical Chemistry Chemical Physics*, 21(22): 11983–11991.

22. Luzzati, V. and Tardieu, A. 1974: Lipid phases: Structure and structural transitions, *Annual Review of Physical Chemistry*, 25: 79–94.

23. Kiselev, M. A., Lombardo, D., Lesieur, P., Kisselev, A. M., Borbely, S., Simonova, T. N., and Barsukov, L. I. 2008: Membrane self-assembly in mixed DMPC/NaC systems by SANS, *Chemical Physics*, 345: 173–180.

24. Breton, M., Berret, J.-F., Bourgaux, C., Kral, T., Hof, M., Pichon, C., Bessodes, M., Scherman, D., and Mignet, N. 2011: Protonation of lipids impacts the supramolecular and biological properties of their self-assembly, *Langmuir*, 27: 12336–12345.

25. Kluzek, M., Tyler, A. I. I., Wang, S., Chen, R., Marques, C. M., Thalmann, F., Seddon, J. M., and Schmutz, M. 2017: Influence of a pH-sensitive polymer on the structure of monoolein cubosomes, *Soft Matter*, 13: 7571–7577.

26. Demus, D., Goodby, J., Gray, G. W., Spiess, H. W., and Vill, V. 1998: *Handbook of Liquid Crystals*. Wiley: Weinheim.

27. Zana, R. 2005: *Dynamics of Surfactant Self-Assemblies: Micelles, Microemulsions, Vesicles and Lyotropic Phases*, Taylor & Francis, London.

28. Luzzati, V. and Tardieu, A. 1974. Lipid phases: Structure and structural transitions, *Annual Review of Physical Chemistry*, 25: 79–94.

29. Pabst, G., Kučerka, N., Nieh, M.-P., and Katsaras, J. 2014: *Liposomes, Lipid Bilayers and Model Membranes: From Basic Research to Application*. CRC Press: Boca Raton, FL.

30. Kiselev, M. A. and Lombardo, D. 2017: Structural characterization in mixed lipid membrane systems by neutron and X-ray scattering, *Biochimica et Biophysica Acta (BBA) - General Subjects*, 1861: 3700–3717.

31. Price, W. S. 2009: *NMR Studies of Translational Motion: Principles and Applications.* Cambridge University Press: New York.

32. Fong, C., Le, T., and Drummond, C. J. 2012: Lyotropic liquid crystal engineering-ordered nanostructured small molecule amphiphile self-assembly materials by design, *Chemical Society Reviews*, 41(3): 1297–1322.

33. Longo, A., Calandra, P., Casaletto, M. P., Giordano, C., Venezia, A., and Turco Liveri, V., 2006: Synthesis and physico-chemical characterization of gold nanoparticles softly coated by AOT, *Materials Chemistry and Physics*, 96, 66–72.

34. Calandra, P., Longo, A., Ruggirello, A., and Turco Liveri, V., 2004: Physico-chemical investigation of the state of cyanamide confined in AOT and lecithin reversed micelles, *Journal of Physical Chemistry B*, 108: 8260–8268.

35. Calandra, P., Turco Liveri, V., Ruggirello, A. M., Licciardi, M., Lombardo, D., and Mandanici, A. 2015: Anti-Arrhenian behaviour of conductivity in octanoic acid–bis (2-ethylhexyl) amine systems: A physico-chemical study, *Journal of Materials Chemistry C*, 3(13): 3198–3210.

36. Sackmann, E. 1995: Physical basis of self-organization and function of membranes: Physics of vesicles. In Handbook of Biological Physics, Eds. R. Lipowsky, E. Sackmann, vol. 1, pp. 213–303. Elsevier: Amsterdam.

37. Chang, H. I., Yeh, M. K. 2012: Clinical development of liposome-based drugs: Formulation, characterization, and therapeutic efficacy. *International Journal of Nanomedicine*, 7: 49–60.

38. Gennis, R. B. 1989: *Biomembranes: Molecular Structure and Function*, Springer: New York.

39. Helfrich, W. 1973: Elastic properties of lipid bilayers: Theory and possible experiments, *Z Naturforsch*, 28(11): 693–703.

40. Allen, T. M. and Cullis, P. R., 2013: Liposomal drug delivery systems: From concept to clinical applications, *Advanced Drug Delivery Reviews*, 65(1): 36–48.

41. Nagle, J. F. and Tristram-Nagle, S. 2000: Structure of lipid bilayers, *Biochimica et Biophysica Acta (BBA) – Reviews on Biomembranes*, 1469(3): 159–195.

42. Lesieur, P., Kiselev, M. A., Barsukov, L. I., and Lombardo, D. 2000: Temperature-induced micelle to vesicle transition: Kinetic effects in the DMPC/NaC system, *Journal of Applied Crystallography*, 33(31): 623–627.

43. Kiselev, M. A., Janich, M., Hildebrand, A., Strunz, P., Neubert, R. H. H., and Lombardo, D. 2013: Structural transition in aqueous lipid/bile salt [DPPC/NaDC] supramolecular aggregates: SANS and DLS study, *Chemical Physics*, 424: 93–99.

44. Dan, N., Pincus, P., and Safran, S. A., 1993: Membrane-induced interactions between inclusions, *Langmuir*, 9(11): 2768–2771.

45. Bourgaux, C. and Couvreur, P. 2014: Interactions of anticancer drugs with biomembranes: what can we learn from model membranes? *Journal of Controlled Release*, 190: 127–138.

46. Lombardo, D., Calandra, P., Bellocco, E., Laganà, G., Barreca, D., Magazù, S., Wanderlingh, U., and Kiselev, M. A. 2016: Effect of anionic and cationic polyamidoamine (PAMAM) dendrimers on a model lipid membrane, *Biochimica et Biophysica Acta (BBA) - Biomembranes*, 1858: 2769–2777.

47. Lombardo, D., Calandra, P., Magazù, S., Wanderlingh, U., Barreca, D., Pasqua, L., and Kiselev, M. A., 2018: Soft nanoparticles charge expression within lipid membranes: The case of amino terminated dendrimers in bilayers vesicles, *Colloids and Surfaces B: Biointerfaces*, 170: 609–616.

48. Hartgerink, J. D., Beniash, E., and Stupp S. I., 2001: Self-assembly and mineralization of peptide-amphiphile nanofibers, *Science*, 294(5547): 1684–1688.

49. Korevaar, P. A., Newcomb, C. J., Meijer, E. W., and Stupp, S. I. 2014: Pathway selection in peptide amphiphile assembly, *Journal of American Chemical Society*, 136(24): 8540–8543.

50. Yates, C.R., and Hayes, W. 2004: Synthesis and applications of hyperbranched polymers, *European Polymer Journal*, 40: 1257–1281.

51. Mallamace, F., Beneduci, R., Gambadauro, P., Lombardo, D., and Chen, S. H. 2001: Glass and percolation transitions in dense attractive micellar system. *Physica A: Statistical Mechanics and Its Application*, 302 (1–4): 202–219.

52. Mai, Y. and Eisenberg, A. 2012: Self-assembly of block copolymers, *Chemical Society Reviews*, 41(18): 5969–5985.

53. Feng, H., Lu, L., Wang, W., Kang, N.-G., and Mays, J. W. 2017: Block copolymers: Synthesis, self-assembly, and applications, *Polymers* 9(10): 494.

54. Tan, C., Wang Y., and Fan, W. 2013: Exploring polymeric micelles for improved delivery of anticancer agents: Recent developments in preclinical studies, *Pharmaceutics*, 5(1): 201–219.

55. Toughraï, S., Malinova, V., Masciadri, R., Menon, S., Tanner, P., Palivan, C., Bruns, N., and Meier, W. 2015: Reduction-sensitive amphiphilic triblock copolymers self-assemble into stimuli-responsive micelles for drug delivery, *Macromolecular Bioscience* 15(4): 481–489.

56. Babikova, D., Kalinova, R., Zhelezova, I., Momekova, D., Konstantinov, S., Momekov, G., and Dimitrov, I. Functional block copolymer nanocarriers for anticancer drug delivery, *RSC Advances*, 6: 84634–84644.

57. Lombardo, D., Micali, N., Villari, V., and Kiselev, M. A. 2004: Large structures in diblock copolymer micellar solution, *Physical Review E*, 70 (2): 021402.

58. Chen, S. H., Mallamace, F., Faraone, A., Gambadauro, P. et al. 2002. Observation of a re-entrant kinetic glass transition in a micellar system with temperature-dependent attractive interaction. *The European Physical Journal E*, 9 (3): 283–286.

59. Yin, J., Chen, Y., Zhang, Z.-H. and Han, X. 2016: Stimuli-responsive block copolymer-based assemblies for cargo delivery and theranostic applications, *Polymers*, 8: 268.

60. Chen, S., Zhang, X. Z., Cheng, S. X., Zhuo, R. X., and Gu, Z. W. 2008: Functionalized amphiphilic hyperbranched polymers for targeted drug delivery, *Biomacromolecules*, 9(10): 2578–2585.

61. Sun, F., Luo, X., Kang, L., Peng, X., and Lu, C. 2015: Synthesis of hyperbranched polymers and their applications in analytical chemistry, *Polymer Chemistry*, 6: 1214–1225.

62. Yan, D., Gao, C., and Frey, H. (eds.) 2011: *Hyperbranched Polymers Synthesis, Properties, and Applications*. John: Hoboken.

63. Fan, X., Zhao, Y., Xu, W., and Li, L. 2016: Linear-dendritic block copolymer for drug and gene delivery, *Materials Science and Engineering: C*, 62: 943–959.

64. Zhou, Z., Forbes, R. T., and D'Emanuele, A. 2017: Preparation of core-crosslinked linear-dendritic copolymer micelles with enhanced stability and their application for drug solubilisation, *International Journal of Pharmaceutics*, 523(1): 260–269

65. Singh, J., Jain, K., Mehra, N. K., and Jain, N. K. 2016: Dendrimers in anticancer drug delivery: Mechanism of interaction of drug and dendrimers, *Artificial Cells, Nanomedicine, and Biotechnology* 44(7):1626–1634.

66. Lombardo, D. 2009: Liquid-like ordering of negatively charged poly (amidoamine) (PAMAM) dendrimers in solution, *Langmuir*, 25(5): 3271–3275.

67. Sharma, A. K., Gothwal, A., Kesharwani, P., Alsaab, H., Iyer, A. K., and Gupta, U. 2017: Dendrimer nanoarchitectures for cancer diagnosis and anti-cancer drug delivery, *Drug Discovery Today*, 22(2), 314–326.

68. Abbasi, E., Aval, S. F., Akbarzadeh, A., Milani, M., Nasrabadi, H. T., Joo S. W., Hanifehpour, Y., Nejati-Koshki, and K., Pashaei-Asl, R. 2014: Dendrimers: Synthesis, applications, and properties, *Nanoscale Research Letters*, 9(1): 1–10.

69. Lombardo, D. 2014: Modeling dendrimers charge interaction in solution: Relevance in biosystems, *Biochemistry Research International*, 2014: 1–10.

70. Jain, K. 2017: Dendrimers: Smart nanoengineered polymers for bioinspired applications in drug delivery, *Biopolymer-Based Composites: Drug Delivery and Biomedical Applications*, 7: 169–220.

71. Shao, S., Si, J., Tang, J., Sui, M., and Shen. Y. 2014: Jellyfish-Shaped amphiphilic dendrimers: Synthesis

and formation of extremely uniform aggregates, *Macromolecules*, 47(3): 916–921.

72. Lombardo, D., Longo, A., Darcy, R., and Mazzaglia, A. 2004: Structural properties of nonionic cyclodextrin colloids in water, *Langmuir*, 20(4): 1057–1064.

73. Alaimo, D., Beigbeder, A., Dubois, P., Broze, G., Jérôme, C., and Grignard. B. 2014: Block, random and palm-tree amphiphilic fluorinated copolymers: Controlled synthesis, surface activity and use as dispersion polymerization stabilizers, *Polymer Chemistry*, 5: 5273–5282.

74. Wang, D., Jin, Y., Zhu, X., and Yan, D. 2017: Synthesis and applications of stimuli-responsive hyperbranched polymers, *Progress in Polymer Science*, 64: 114–153.

75. Liu, Y., Yu, C., Jin, H., Jiang, B., Zhu, X., Zhou, Y., Lu, Z., and Yan, D. A. 2013: Supramolecular janus hyperbranched polymer and its photoresponsive self-assembly of vesicles with narrow size distribution, *Journal of the American Chemical Society*, 135(12): 4765–4770.

76. Botella, P., Corma, A., and Quesada, M. 2012: Synthesis of ordered mesoporous silica templated with biocompatible surfactants and applications in controlled release of drugs, *Journal of Material Chemistry*, 22: 6394–6401.

77. Deng, Y., Wei, J., Sun Z., and Zhao, D. 2013: Large-pore ordered mesoporous materials templated from non-Pluronic amphiphilic block copolymers, *Chemical Society Reviews*, 42: 4054–4070.

78. Wan, Y. and Zhao, D. 2007: On the controllable soft-templating approach to mesoporous silicates, *Chemical Reviews*, 107(7): 2821–2860.

79. Bonaccorsi, L., Lombardo, D., Longo, A., Proverbio, E., and Triolo, A. 2009: Dendrimer template directed self-assembly during zeolite formation, *Macromolecules*, 42: 1239–1243.

80. Shibata, H. 2017: Fabrication and functionalization of inorganic materials using amphiphilic molecules, *Journal of Oleo Science*, 66(2): 103–111.

81. Zhao, D., Huo, Q., Feng, J., Chmelka, B. F., and Stucky, G. D. 1998: Nonionic triblock and star diblock copolymer and oligomeric surfactant synthesis of highly ordered, hydrothermally stable, mesoporous silica structures, *Journal of American Chemical Society*, 120: 6024–6036.

82. Bonaccorsi, L., Calandra, P., Kiselev. M. A., Amenitsch, H., Proverbio, E., and Lombardo, D. 2013: Self-assembly in poly(dimethylsiloxane)-poly(ethylene oxide) block copolymer template directed synthesis of Linde type A zeolite, *Langmuir*, 29(23): 7079–7086.

83. Pagis, C., Prates, A. R. M., Farrusseng, D., Bats, N., and Tuel, A. 2016: Hollow zeolite structures: An overview of synthesis methods, *Chemistry of Materials*, 28(15): 5205–5223.

84. Bonaccorsi, L., Calandra, P., Amenitsch, H. Proverbio, E., and Lombardo, D. 2013: Growth of fractal aggregates during template directed SAPO-34 zeolite formation, *Microporous and Mesoporous Materials*, 167: 3–9.

85. Xu, D., Ma, Y., Jing, Z., Han, L., Singh, B., Feng, J., Shen, X., Cao, F., Oleynikov, P., Sun, H., Terasaki, O., and Che, S. 2014: π–π interaction of aromatic groups in amphiphilic molecules directing for single-crystalline mesostructured zeolite nanosheets, *Nature Communications*, 5(1–9): 4262.

86. Lu, J., Liong, M., Li, Z., Zink, J. I., and Tamanoi, F. 2010: Biocompatibility, biodistribution, and drug-delivery efficiency of mesoporous silica nanoparticles for cancer therapy in animals, *Small*, 6(16): 1794–1805.

87. Morelli, C., Maris, P., Sisci, D., Perrotta, E., Brunelli, E., Perrotta, I., Panno, M. L., Tagarelli, A., Versace, C., Casula, M. F., Testa, F., Andò, S., Nagy, J. B., and Pasqua, L. 2011: PEG-templated mesoporous silica nanoparticles exclusively target cancer cells, *Nanoscale*, 3(8) 3198–3207.

88. Pasqua, L., Leggio, A., Sisci, D., Andò, S., and Morelli, C. 2016: Mesoporous silica nanoparticles in cancer therapy: Relevance of the targeting function, *Mini-Reviews in Medicinal Chemistry*, 16(9): 743–753.

89. Aiello, R., Cavallaro, G., Giammona, G., Pasqua, L., Pierro, P., and Testa, F. 2002: Mesoporous silicate as matrix for drug delivery systems of non-steroidal antinflammatory drugs, *Studies in Surface Science and Catalysis*, 142: 1165–1172.

90. Argyo, C., Weiss, V., and Bräuchle, C. 2014: Multifunctional mesoporous silica nanoparticles as a universal platform for drug delivery. *Chemistry of Materials*, 26: 435–451.

91. Watermann, A. and Brieger, J. 2017: Mesoporous silica nanoparticles as drug delivery vehicles in cancer, *Nanomaterials*, 7(7): 189.

92. Stoffelen, C. and Huskens J. 2016: Soft supramolecular nanoparticles by noncovalent and host-guest interactions, *Small* 12(1): 96–119.

93. Hrubý, M., Filippov, S. K., and Štěpánek P. Supramolecular structures and self-association processes in polymer systems, *Physiological Research*, 65(Suppl. 2): S165–S178.

94. Feng, Z., Zhang, T., Wang H., and Xu B. 2017: Supramolecular catalysis and dynamic assemblies for medicine, *Chemical Society Review*, 46(21): 6470–6479.

95. Tang, F., Li, L., and Chen, D. 2012: Mesoporous silica nanoparticles: Synthesis, biocompatibility and drug delivery. *Advanced Materials*, 24(12): 1504–1534.

96. Pasqua, L., Testa, F., Aiello, R., Cundari, S., and Nagy, J. B. 2007: Preparation of bifunctional hybrid mesoporous silica potentially useful for drug targeting,

Microporous and Mesoporous Materials, 103(1–3), 166–173.

97. Fu, H. L.-K., Po, C., Leung, S. Y.-L., and Yam V. W.-W. 2017: Self-assembled architectures of alkynylplatinum(II) amphiphiles and their structural optimization: A balance of the interplay among Pt⋯Pt, π–π stacking, and hydrophobic-hydrophobic interactions, *ACS Applied Materials & Interfaces*, 9 (3): 2786–2795.

98. Tiekink, E. R. T., 2017: Supramolecular assembly based on "emerging" intermolecular interactions of particular interest to coordination chemists, *Coordination Chemistry Reviews*, 345: 209–228.

99. Chakrabarty, R., Mukherjee, P. S., and Stang, P. J. 2011: Supramolecular coordination: Self-assembly of finite two- and three-dimensional ensembles, *Chemical Review*, 111(11): 6810–6918.

100. Yang, H., Yuan, B., Zhang, X., and Scherman, O. A., 2014: Supramolecular chemistry at interfaces: Host-guest interactions for fabricating multifunctional biointerfaces, *Accounts of Chemical Research*, 47(7): 2106–2115.

101. Wang, J., Wang, X., Yang, F., Shen, H., You, Y., and Wu, D. 2014: Self-assembly behavior of a linear-star supramolecular amphiphile based on host–guest complexation, *Langmuir*, 30(43): 13014–13020.

102. Dan, N., 2000: Synthesis of hierarchical materials, *Trends in Biotechnology*, 18(9): 370–374.

103. Lombardo, D., Kiselev, M. A., and Caccamo, M. T. 2019: Smart nanoparticles for drug delivery application: development of versatile nanocarrier platforms in biotechnology and nanomedicine, *Journal of Nanomaterials*, 2019: 3702518.

104. Guo, Y., Gong, Y., Gao, Y., Xiao, J., Wang, T., and Yu, L. 2016: Multi-stimuli responsive supramolecular structures based on azobenzene surfactant-encapsulated polyoxometalate, *Langmuir* 32(36): 9293–9300.

105. Zhuang, J., Gordon, M. R., Ventura, J., Li, L., and Thayumanavan, S. 2013: Multi-stimuli responsive macromolecules and their assemblies, *Chemical Society Reviews*, 42: 7421–7435.

Pre-programmed Self-assembly

Carlos I. Mendoza
Universidad Nacional Autónoma de México

Daniel Salgado-Blanco
Instituto Potosino de Investigación Científica y Tecnológica

12.1 Introduction

In the last decade, we have witnessed an avalanche of progress in the design of artificial nanostructures with mechanical, electronic, or magnetic properties essential for specific needs. Applications as important as the capture and storage of energy, electronics, photonics, diagnosis, and therapy, among others, have benefited from these advances.

Self-assembly, the spontaneous organization of disordered components to form ordered structures without external intervention, relies on the interactions of the building blocks, provided that appropriate conditions are met (Whitesides and Grzybowski 2002).

Successful self-assembly of a structure depends on a number of factors: the components that interact to each other, the interactions with their balance between attractive and repulsive contributions, their specificity and directionality, reversibility to allow the components to adjust their position within a structure, the environment where the components move, and the thermal motion usually required to provide movement to the components.

The programmed assembly of structures from their components requires instructions that direct the formation of a particular assemblage. The location and connectivity of the building blocks within the assembled structure, as well as the order in which they are added to it, are specified in such assembly instructions (Cademartiri and Bishop 2015). There are two main strategies to assemble a structure: in a top-down approach, assembly is directed by an external assembler (such as a robot), while in a bottom-up approach, the instructions are encoded in the components (e.g., in their interactions) and are executed spontaneously without external guidance. The bottom-up approach is an attractive route to the fabrication of nanostructures. Work in this area spans from DNA nanotechnology to colloidal assembly. Although we cannot yet program the self-assembly of nanoscopic components with arbitrary complexity as it is done with top-down techniques such as 3D printing, in the last decade, we have witnessed enormous advances in this area. For instance, with current experimental techniques, it has been possible to encode constraints, strength, range, specificity, and directionality in the interactions between the components and to precisely control the shape and size of nanoparticles. Thanks to these advances it is now possible to program the formation of DNA-folded structures (Jones, Seeman and Mirkin 2015 and Hong et al. 2017), well-defined clusters (Xu et al. 2012) as well as nanocrystal superlattices (Biancaniello, Kim and Crocker 2005, Jones et al. 2010, Macfarlane et al. 2011 and Miszta et al. 2011), and self-similar structures (Shang et al. 2015) (Figure 12.1).

From a theoretical and computational point of view, advances have come from inverse statistical-mechanics approaches, numerical simulations of systems of particles whose interactions have directionality and functionality (the so-called "patchy" particles), and the use of mixtures of a large number of species.

In what follows, we describe different strategies to study the programmed self-assembly of structures, starting from the simplest cases of infinite lattices to the more intricate finite constructs.

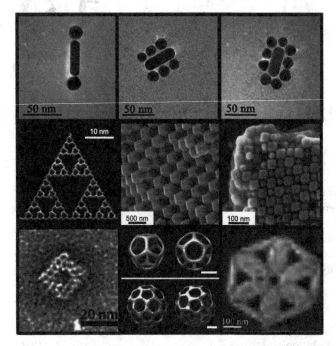

FIGURE 12.1 Examples of self-assembled structures. Thanks to advances in the design of interactions and shape of components, it is now possible to program the formation of well-defined clusters, nanocrystal superlattices, DNA-folded structures, and other complex designs. Image adapted with permission from Xu et al. 2012 (nanorod assemblies), American Chemical Society; Shang et al. 2015 (small-molecule Sierpinski triangle), Macmillan Publishers Limited; Henzie et al. 2012 (nanoparticle assembly), Nature Publishing Group; Lu et al. 2015 (cube-sphere binary assembly), Macmillan Publishers Limited; Douglas et al. 2009 (multilayered DNA origami structure), Macmillan Publishers Limited; He et al. 2008 (DNA wireframe polyhedra. Scale bars 20 nm), Nature Publishing Group; and Zhao, Liu and Yan 2011 (super DNA origami), American Chemical Society.

12.2 Inverse Optimization Techniques

Of all possible structures to assemble, probably infinite crystal lattices are the simplest ones to achieve, in particular single-component close-packed lattices. In two dimensions (2D), a triangular lattice is the closest packed crystal; analogously, in three dimensions (3D), face-centered cubic (fcc) and hexagonal close packed (hcp) lattices are the most compact lattices. According to the second law of thermodynamics, the final equilibrium assembly is the one that minimizes the free energy $F = U\text{-}TS$, in which U is the total effective potential (the sum of all interactions) and S is the entropy. At low temperatures, the term "TS" can be ignored, and U is the relevant quantity. Let us imagine a lattice made of spherical particles that interact through short-range attractions in addition to the ineludible excluded volume. Then, not surprisingly, the obtained lattice is the one that maximizes the number of (favorable)

contacts between spheres, because this minimizes U and thus F; therefore, close-packed lattices are formed.

At positive temperatures, the equilibrium state is determined not only by the energy U, but also by the entropy S that starts to play an important role. For example, the crystallization of monodisperse isotropic particles into colloidal crystals is an entropy-driven process in which configurational entropy and free volume entropy dominate at low and high concentrations, respectively. Concentrating the dispersions allows a disorder-to-order transition in which originally disordered configurations crystallize into fcc and hcp structures because they are thermodynamically favored, as they gain more free volume entropy than other packing symmetries.

Single-component open lattices are more difficult to obtain than closed packed ones. Nonetheless, powerful theoretical strategies have been designed to construct interaction potentials that favor the assembly of arbitrary crystalline lattices.

Traditionally, theoretical and computational materials science has relied on a "forward" approach of statistical-mechanics to study new materials. In this approach, a known material system of interest is identified, a convenient approximation of the interactions between the constituents is proposed, and finally, simulation and analytical methods are used to predict non-obvious details concerning structural, thermodynamic, and kinetic features of the system.

In recent years, several "inverse" statistical-mechanics methods have been devised that find optimized interactions that robustly and spontaneously lead to a targeted many-particle configuration of the system for a wide range of conditions (Rechtsman, Stillinger and Torquato 2005, Torquato 2009, Adorf et al. 2018 and Piñeros et al. 2018). Output from these optimization techniques could be applied to create particles with interactions that yield these structures at the nanoscopic length scale. Although these theoretical schemes are not restricted to pairwise additive or spherically symmetrical interactions, we are going to show the conceptual basis of the procedure using such simplified potentials (Torquato 2009).

A classic inverse design problem is the optimization of decision variables $\boldsymbol{a}$ for a system of identical particles interacting via a given isotropic pairwise interaction potential, $\phi(r;\,\boldsymbol{a})$, to stabilize a desired structure. Here, r is the distance between particle centers, and $\boldsymbol{a}$ are the parameters required for the pair potential. By minimizing either the ground state energy of the ideal configuration (relative to competing structures) or the free energy of an associated configurational ensemble at a higher temperature (relative to competing phases), researchers have successfully found isotropic interactions that stabilize a wide variety of structures and phases including, for example, two-dimensional honeycomb and Kagomé lattice assemblies as well as three-dimensional simple cubic and diamond crystals (Piñeros et al. 2018). Figure 12.2 (upper right panel) shows a 2D honeycomb lattice stabilized using the pair potential shown in the upper left panel.

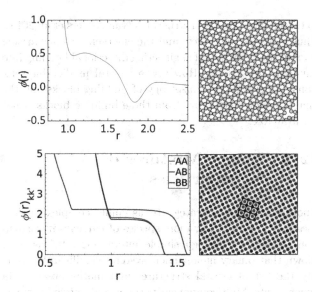

FIGURE 12.2 Inverse optimization techniques to self-assemble target structures. Upper panel: optimized pair potential and ground state configuration of a honeycomb crystal. Image reproduced with permission from Rechtsman, Stillinger and Torquato 2005, The American Physical Society. Lower panel: optimized pair potentials and ground state configuration of a binary square lattice. Image reproduced with permission from Piñeros et al. 2018, The American Institute of Physics.

Recently, a strategy called relative-entropy minimization (Shell 2008 and Chaimovich and Shell 2011), which enables "on-the-fly" optimization of the potential parameters directly within a molecular simulation, led to the discovery of isotropic interactions that promote self-assembly of a variety of two- and three-dimensional crystals, quasicrystals as well as clustered fluids and porous mesophases (Piñeros et al. 2018). Ground-state and relative-entropy-based inverse design strategies have been used (Piñeros, Jadrich and Truskett 2017) to infer a set of "design rules" that help to understand the properties of single-well pair potentials that can stabilize two-dimensional target lattices (isotropic pair interactions comprising a repulsive core and a single attractive well are models for effective interactions that are ubiquitous in colloidal fluids and interesting for material design applications due to their possible tunability). These rules can be summarized as follows: (i) the interaction range must span a minimum number of coordination shells to differentiate the target structure from its competitors, (ii) the well must be relatively narrow and located toward the end of the minimally required interaction range, and (iii) nontrivial repulsive features (e.g., shoulders) in the potential spanning specific coordination shells in the target and competing lattices are required to impose energetic advantages to the target structure.

One of the main challenges of the inverse optimization approaches consists in the development of robust potentials for target structures that can be realistic, in other words, that can be synthesized experimentally. However, standard optimization techniques tend to produce isotropic pair potentials with too many features and length scales that are not actually critical and are challenging to map to mechanistic models. Recently, a method has been developed for the optimization of the pair potentials that minimizes the relative entropy while considering only those length scales most relevant for the self-assembly of a target structure (Adorf et al. 2018). The resulting interaction potentials are smoother and simpler in real space, and therefore likely easier to make.

The relative-entropy optimization strategy has been extended to determine isotropic interactions that promote the assembly of targeted multicomponent phases (Piñeros et al. 2018). While this extension increases the complexity of the optimization problem, it might also lead to interparticle interactions that, for a given target, have simpler (and perhaps more easily realizable) functional forms than those discovered under the constraint of a single-component design. Figure 12.2 (lower panels) shows a binary assembly obtained using this scheme and the optimized potentials used to assemble it.

12.3 Directional Interactions

The pursuit of assembled structures using isotropic interactions may lead to involved potentials difficult to translate into a realistic system; thus, alternative strategies have been developed. An approach consists in trying to emulate the self-assembly of living systems. Since most biomolecular objects interact through directionally specific forces, a large amount of work has been done to mimic the anisotropic nature of these interactions (Glotzer and Solomon 2007), specifically, with the design and use of patchy particles (Duguet et al. 2016 and Gong et al. 2017) and shape-induced directional binding (Lu et al. 2015).

By properly designing the patch shape and symmetry, patchy particles can crystallize in a specified target morphology. The location and size of the patches assures that the minimum energy configuration at zero temperature is precisely the desired structure. Figure 12.3 shows patchy particles and their assembly into three different crystalline lattices (square, honeycomb, and Kagomé), with two possible designs of patchy particles to assemble the Kagomé lattice (Romano and Sciortino 2011). The second alternative for the Kagomé lattice has been experimentally validated at the microscale (Chen, Bae and Granick 2011).

Although the addition of directional binding can significantly broaden structural diversity, nanoscale implementation remains challenging. Another way to induce directionality in the assembly blocks is by means of shape anisotropy (Glotzer and Solomon 2007, van Anders et al. 2014, Lu et al. 2015 and Duguet et al. 2016) (Figure 12.4), also called entropically patchy particles to distinguish them from the traditional enthalpically patchy particles, where anisotropic interactions arise from patterned coatings, functionalized molecules, DNA, and other enthalpic means; they create the possibility for directional binding of particles into

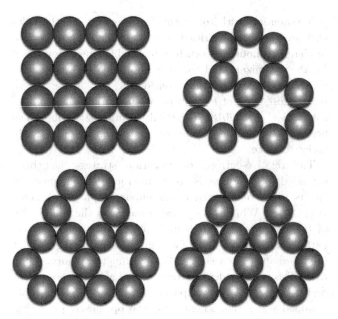

FIGURE 12.3 Self-assembly of patchy particles. Properly designed patchy particles to assemble square (upper left), honeycomb (upper right), and Kagomé lattices (lower).

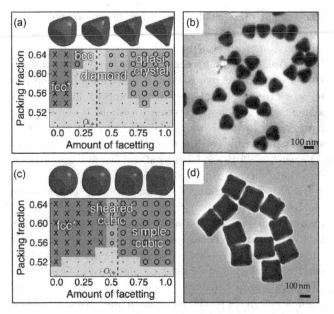

FIGURE 12.4 Entropically patchy particles. Different crystal structures formed by the assembly of tetrahedrally (a) and cubically (c) faceted spheres as the amount of faceting varies. Image reproduced with permission from van Anders et al. 2014. Copyright 2014 American Chemical Society. Transmission electron microscopy (TEM) views of dimpled silica particles with four (b) and six (d) concave entropic patches. Image reproduced with permission from Désert et al. 2013. Copyright 2013 John Wiley and Sons.

higher-ordered structures. The directional nature of entropic forces has recently been elucidated; it has been numerically shown how to modify entropic patchiness by introducing geometric features to the particles via shape operations so as

to target specific crystal structures (van Anders et al. 2014). Thus, entropic patchiness and the emergent valency arising from it provide a way of engineering directional bonding into nanoparticle systems. However, a general predictable relationship between the anisotropy of building blocks and the final assembled structure from these building blocks is not yet available.

12.4 Binary Mixtures of Nanoparticles

Another successful strategy to assemble complex lattices uses mixtures of two or more species of particles in suspension interacting through simple interactions. It has been shown that binary nanoparticle systems significantly diversify the list of crystal structures and motifs observed in experiments. Multicomponent periodic and aperiodic superlattices, including dodecagonal quasicrystals, have been constructed using colloidal inorganic nanoparticles interacting through simple isotropic potentials (Figure 12.5) (Podsiadlo et al. 2011). In general, the use of electrical charges, entropic, van der Waals, steric, and dipolar forces, contributes to stabilize the obtained lattice. The construction of a given structure will be the result of a combination of interparticle interactions and sphere packing phenomena governed by entropy.

12.5 Multicomponent Assemblies

At the nanoscale, co-crystallization of differently sized nanoparticles possessing different functionalities is expected to lead to the formation of homogeneous solids, which exhibit high degree of internal ordering and well-defined stoichiometry. The precise ordering across large dimensions offers a unique opportunity to control the electronic, optical, and magnetic coupling between the individual nanoparticles, which may result in new collective properties. It is also expected that the use of multicomponent systems would expand the available design space for assembly. New and complex structures, which are inaccessible to single-component systems, can be stabilized using multicomponent systems. Furthermore, the self-assembly of intricate finite-sized structures with low symmetry may be constructed using a large number of species of particles as long as precise control of the interactions is achieved.

The impressive progress in using the molecular recognition properties of DNA (see forthcoming sections below) to direct the self-assembly of nanoparticles suggests that DNA hybridization[1] would enable one to design arbitrary nanoparticle clusters. Thanks to the directionality

---

[1]Hybridization is a phenomenon in which single-stranded DNA or ribonucleic acid (RNA) molecules anneal to complementary DNA or RNA.

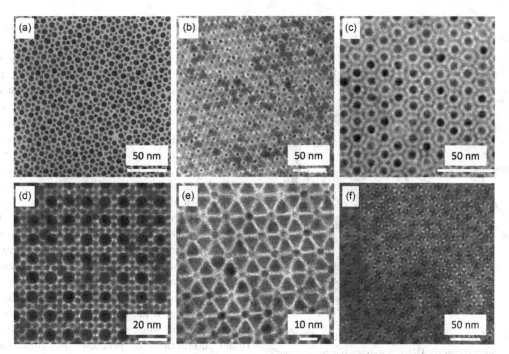

FIGURE 12.5 Binary mixtures. Diversity of binary superlattices (SL) prepared by slow evaporation method observed with transmission electron microscope. (a) Electron micrograph of an ordered raft comprising Au nanoparticles of two distinct sizes with size ratio of <0:58 (Reproduced with permission from Kiely et al. 1998, Macmillan Publishers Ltd). (b) AB-type SL composed of 13-nm Fe and 5-nm Au nanoparticles. Reproduced with permission from Saunders and Korgel 2005, Wiley-VCH Verlag GmbH& Co. KGaA, Weinheim. (c) AB_5-type SL composed of 8.1-nm CdTe and 4.4-nm CdSe nanoparticles. Reproduced with permission from Chen et al. 2007, American Chemical Society. (d) *cub*-AB_{13}-type SL from 10.3-nm PbSe and 5.8-nm CdSe nanoparticles. Reproduced with permission from Overgaag et al. 2008, American Chemical Society. (e) SL composed of triangular LaF_3 nanoplates (9.0-nm side) and 5.0-nm spherical Au nanoparticles. Reproduced with permission from Shevchenko et al. 2006, Nature Publishing Group. (f) Dodecagonal quasicrystalline SL composed of 13.4-nm Fe_2O_3 and 5-nm Au nanoparticles. Reproduced with permission from Talapin et al. 2009, Macmillan Publishers Limited.

and specificity of DNA functionalized nanoparticles, one may expect that the use of a large number of distinct component types would allow the assembly of equilibrium structures with prescribed arbitrary complexity, not only infinite lattices or small clusters (Ke et al. 2012, Halverson and Tkachenko 2013, Hedges, Mannige and Whitelam 2014 and Jacobs, Reinhardt and Frenkel 2015) (Figure 12.6).

Systems made of a large number of components ($\approx 10^3$) possessing "designed" interactions, chosen to stabilize an equilibrium target structure in which each component type has a defined spatial position, as well as "undesigned" interactions that allow components to bind in a compositionally random way, are shown to assemble with high fidelity in the equilibrium target structure even in the presence of substantial attractive undesigned interactions. However, for structures of macroscopic size ($\approx 10^{24}$), thermodynamic and dynamic factors would impose limitations; for example, interactions between distinct components would require increasingly strong interactions to counterbalance the loss of entropy upon assembly. Also, the timescale for the growth of structures would increase as the number of component type increases. It has been shown through simulations that structures with highly variable shapes made out of dozens of particles can form with high yield, as long as each particle in the structure binds only to the

particles in their local environment. For larger structures, made of hundreds of particles, high yield can still be significant. The reason is that the large number of competing states that appear for large structures is not as competitive as one may expect. However, robust self-assembly of even larger structures would require removing the assumption of equilibrium interactions and to include nonequilibrium design as the use of irreversible bonds, the possibility of using error correction, or the use of allosteric interactions, in which the binding energy of a particle depends on the set of particles that it binds to (Zeravcic, Manoharan and Brenner 2014). The study of such schemes in nanoparticle-mediated DNA interactions is an important topic for future research.

Although the final assembled structure in a multicomponent system does not necessarily coincide with the target structure due to kinetic traps, it has been experimentally (Ke et al. 2012) and numerically (Halverson and Tkachenko 2013, Reinhardt and Frenkel 2014 and Hedges, Mannige and Whitelam 2014) shown that the self-assembly of equilibrium structures is possible if the interactions between components are sufficiently selective. The combination of directionality and selectivity of interactions allows one to avoid unwanted metastable configurations, which lead to slow self-assembly kinetics.

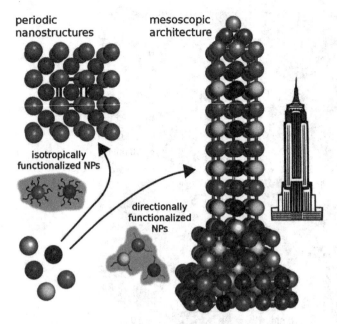

FIGURE 12.6 Nanoparticles isotropically functionalized with complementary single-stranded DNA self-assemble into periodic nanostructures like a body-centered cubic crystal. When the nanoparticles are directionally functionalized by attaching DNA strands at specific locations, the particles can be designed to self-assemble into finite-size mesoscopic architectures, such as the Empire State Building. Image reproduced with permission from Halverson and Tkachenko 2013, American Physical Society.

In constructing a model for the self-assembly of addressable structures, the designed interactions should be much stronger than any attractive interactions between subunits that are not adjacent in a correctly assembled structure. Strategies have been designed to minimize undesigned interactions such that the target structure has considerable lower energy than competing configurations (Madge and Miller 2017).

12.6 Template-Assisted Growth

In a general sense, a template is a pre-existing structural feature, such as a preorganized motif or a surface support, that is external to the system of interest but intimately related to it and is used to "seed" the growth of the structure without waiting for spontaneous nucleation. The basic principle behind the use of template-assisted growth is to increase the probability of the units meeting and therefore binding.

12.7 Hierarchical Self-assembly

Ensuring the thermodynamic stability of information-rich self-assembled nanostructures in which each component is of a distinct type and has a defined spatial position, requires inter-component interaction energies to increase (logarithmically) with structure size in order to counter the entropy gained upon mixing component types in solution.[2] However, self-assembly done in the presence of strong interactions results in general in kinetic trapping, because strong bonds can prevent the correction of mistakes that happen when components undergoing Brownian motion collide randomly. This is so even if the only interactions present are those designed to stabilize the intended structure, thus imposing limits to the size of a self-assembled structure built from distinguishable components. A possible way to overcome this limitation is to consider stage-by-stage hierarchical self-assembly. In this scheme, each stage of assembly involves the formation of distinct substructures that in a subsequent stage combine to form higher-order structures. In order to achieve this assembly pathway, one can either combine reactants in stages or "switch on" inter-component interactions in stages, as illustrated in Figure 12.7. The example shown in this figure represents the hierarchical assembly of an information-rich square. In the first stage of assembly, selected interactions between monomers promote the assembly of squares of size 4. One then turns on additional interactions to promote the formation of second-stage structures, squares of size 16, and so on. This process can continue as long as structures can be brought into contact, so permitting the generation of an arbitrarily large information-rich array (Whitelam 2015). Figure 12.7 shows a series of snapshots of a computational realization of such scheme.

The main drawback of this scheme is that it is particularly susceptible to kinetic traps caused by "undesigned" interactions, by which we mean interactions that are not required to stabilize the intended structure (Figure 12.7). Therefore, the scale on which this scheme can be realized depends upon how effectively this latter kind of interaction can be suppressed. The use of directional interaction, shape-complementarity or by confining the random Brownian motion of the components to a limited space, helping pieces to meet and bind (template-assisted growth) can help to restrict the number of isoenergetic states, eventually leading to a narrower energy profile with the target structure as uniquely defined minimum.

12.8 DNA Bonds

DNA provides one of the simplest yet most flexible platforms for the programmable assembly of molecular devices (Seeman 2003 and Turberfield 2011). Certainly, this is due to the specificity and high predictability of canonical Watson-Crick base-pairing interactions (Figure 12.8).

---

[2]Let us remind that the free energy $F = U - TS$ minimizes at thermodynamic equilibrium. An increase in the number of components of the system increases the entropy S associated with the mixing of the components. Thus, an increase in the inter-component interaction U is needed to make the specific target structure the one with the largest (negative) U so that it has a lower free energy than the large number of disordered mixed-component structures.

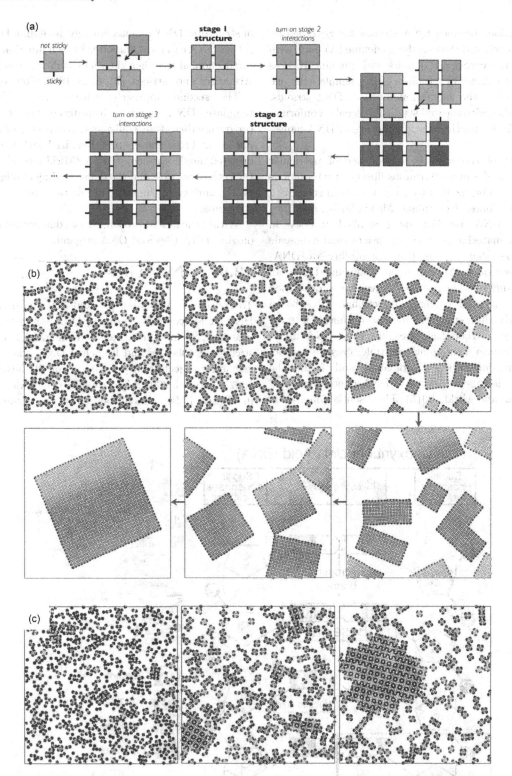

FIGURE 12.7 Hierarchical assembly. Self-assembly of large, information-rich structures in thermal equilibrium will likely require strong inter-component interactions. (a) The hierarchical assembly scheme is immune to the kinetic traps associated with strong native interactions, that is, interactions designed to stabilize the assembled structure. Each stage of assembly involves the formation of squares, mediated by four chemically specific internal bonds. No matter the strength of these bonds or the order in which they form, kinetic trapping in the form of mis-binding cannot occur. (b) A time-ordered series of snapshots from a single trajectory of the stage-by-stage hierarchical assembly procedure. The procedure results in error-free assembly of the target structure. Order of snapshots, clockwise from top left. (c) A time-ordered series of snapshots (time increases left to right) from a simulation run in the presence of attractive non-native contacts. The square-upon-square hierarchical dynamics is overcome by a second organizational process that sees 4-mers assemble into compositionally disordered structures. As a result, the hierarchical assembly procedure all but stops. Image reproduced with permission from Whitelam 2015, The Royal Society of Chemistry.

The interactions between DNA strands are governed by the pairing of nucleotide bases, where adenine (A) pairs with thymine (T) and cytosine (C) pairs with guanine (G), as shown in Figure 12.8. On the basis of this simple rule, one can easily program the interactions between DNA strands. This fact together with the well-studied molecular conformation of DNA facilitates the creation of designer DNA nanostructures.

Permutation of the nucleobase sequence of particular DNA strands results in an enormous library of interactions that can direct hybridization to occur with high selectivity and specificity (Jones, Seeman and Mirkin 2015).

Directional DNA bonding or controlled valency in programmable materials relies on the use of rigid nanoscale building blocks that retains the tailorability of DNA-mediated interactions (Mirkin, et al. 1996 and Li, et al. 1996). Two different approaches have been developed that achieve the goal of rigid, directional, DNA-based bonds through different fundamental chemical interactions. The first methodology uses branched DNA architectures (molecules containing multiple crossover junctions between double-helical domains) and forms the basis for what is called "structural DNA nanotechnology" (Jones, Seeman and Mirkin 2015). The essential foundation

of structural DNA nanotechnology, including DNA origami, is that DNA can form immobile branch junctions that can be further joined together via sticky ends to form higher-order structures and lattices. (Li, et al. 1996, Hong et al. 2017).

The second approach, referred to as "nanoparticle template DNA bonds", introduced the concept of a programmable atom equivalent, comprised of a rigid non-nucleic acid core densely functionalized with a layer of highly oriented single strands of DNA (Mirkin, et al. 1996). Both the central particle and the dense loading of oligonucleotides on the surface of the structure dictate the valency in these structures.

Within the first approach, we can distinguish two methodologies: DNA tiles and DNA origami.

12.9 DNA Tiles

To construct interesting 2D and 3D nanostructures from DNA, it is necessary to design DNA structures containing branches. Seeman first presented the idea of combining the sticky-end cohesion and branched DNA junctions to make geometric objects and periodic 2D or 3D lattices (Seeman 1982, Yan et al. 2003, Lin et al. 2006). This idea is illustrated in Figure 12.9a, which shows that a four-arm branched

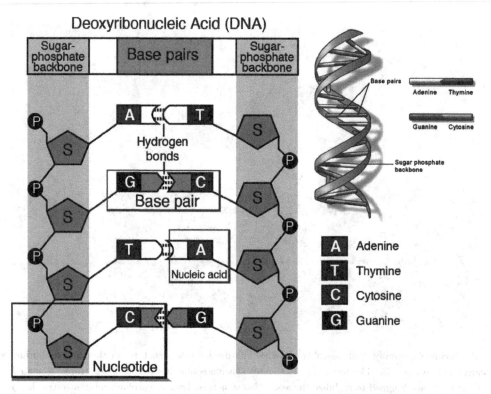

FIGURE 12.8 DNA structure and hybridization. A nucleotide is the basic structural unit and building block for DNA. These building blocks are hooked together to form a chain of DNA. A nucleotide is composed of three parts: a five-sided sugar (S), a phosphate group (p), and a nitrogenous base. The sugar and phosphate group make up the backbone of the DNA double helix, while the bases (A, T, C, G) are located in the middle. A chemical bond between the phosphate group of one nucleotide and the sugar of a neighboring nucleotide holds the backbone together. Chemical bonds (hydrogen bonds) between the bases that are across from one another hold the two strands of the double helix together. Short DNA strands are also called oligonucleotides. Image courtesy of the National Human Genome Research Institution. Inset: U. S. National Library of Medicine with permissions.

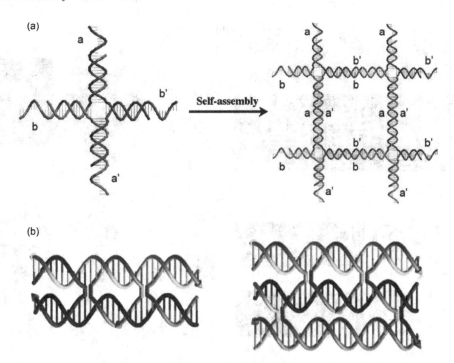

FIGURE 12.9 Key concept of DNA tile-based self-assembly. (a) Combining branched DNA junctions with sticky-end associations (e.g., a–a′ and b–b′ pairings) to self-assemble 2D lattices. In the example shown, a four-branch junction assembles a square lattice. (b) Left: Double-crossover motif (DX). Right: Triple crossover motif (TX).

DNA junction with complementary sticky ends could self-assemble into 2D square lattices.

By joining two double helices together through strand exchange, a group of branched complexes called double-crossover molecules (DX) can be constructed (Figure 12.9b). These molecules are roughly twice as stiff as linear DNA. With proper sticky-ends design, the DX molecules can self-assemble into periodic 2D lattices. Similarly, a variety of rigid, branched DNA tile molecules can be constructed to self-assemble into 2D lattices. Figure 12.10a illustrates some representative examples of the DNA tiles and their corresponding self-assembly into 2D periodic lattices.

This ability to rationally assemble synthetic DNA-based tiles has enabled a number of important advances in both the construction of scaffolds for the immobilization of other nanoobjects and for the development of dynamic nanoscale materials. Directional binding domains in three dimensions have been constructed using structures known as tensegrity triangles[3] (Figure 12.10b) (Liu et al. 2004). Unlike previous structures, the arms of this tile are not all coplanar and point in three separate directions in 3D space like the axes of a rhombohedral coordinate system. Tensegrity triangles have been assembled into macroscopic 3D crystalline materials with considerable long-range order (Figure 12.10b)

(Zheng et al. 2009). These 2D or 3D DNA crystals allow for the precise organization of periodic nanoparticles, quantum dots, and biomolecules (Hong et al. 2017).

Another method, based on DNA tile-based building blocks with intrinsic curvature, has been used to construct 3D wireframe polyhedral structures (Figure 12.10c) (He et al. 2008). The intrinsic curvature of the tiles together with their tailored flexibility dictates the morphology of the structure. More complex 3D objects can be made with a more diverse library of tiles from which to build.

A remarkable feature of DNA-mediated interactions is that the number of distinct species in the system can be as large as the total number of "building blocks". Furthermore, the sequences can be designed to control all the possible interactions between the various species. With such control, it is possible to program the self-assembly of a target structure by first breaking it down into subunits, then determining how many unique species are needed, and finally mixing various species of building blocks that bind only to their corresponding neighbors within the target structure. In this scheme, each building block is encoded with information about where it must go in the final structure.

This scheme is exploited in a method in which, in contrast to the previous methods that use rigid tiles, single-stranded DNA units have been employed to construct extremely complex discrete 2D and 3D objects (Figure 12.10d) (Wei, Dai and Yin 2012 and Ke et al. 2012). In this case, Lego-like building blocks based on polypeptides (chains of amino acids) form a simple, modular, and robust framework for assembling complex structures from short DNA strands. Each brick represents a single DNA strand composed of 32

---

[3]Normally, geometric structures or periodic arrays will result from rigidly joining rigid building blocks. Tensegrity, an alternative strategy, involves rigid struts and flexible tendons. Struts push outward, and tendons pull inward. The balance between these two forces leads to stable, rigid structures.

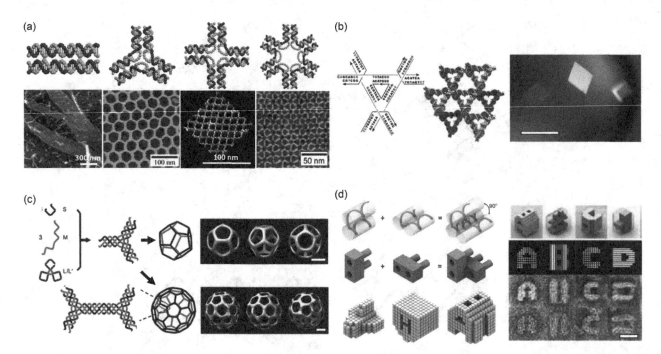

FIGURE 12.10 DNA hybridization-based tiles. (a) Multiarm DNA structural motifs with different numbers (2, 3, 4, and 6) junctions tethered together. Their 2D lattice assemblies were imaged by AFM and are shown. (Modified from Hong et al. 2017, with permission). The 4 by 4 tile assembly scale bars are 100 nm. (Modified from Yan et al. 2003 and Winfree et al. 1998, with permission). Three-dimensional DNA hybridization-based tiles. (b) The tensegrity triangle assembled into a macroscopic 3D rhombohedral lattice. Scale bar, 500 mm. (Modified from Seeman 2010 and Zheng et al. 2009, with permission) (c) The three-point star motif assembled into a variety of wireframe polyhedrons based on the intrinsic curvature of the tile and its concentration. Scale bars, 20 nm. (Modified from He et al. 2008, with permission) (d) Single-stranded DNA assembled into arbitrary 3D brickwork-like nanoscale objects. Scale bar, 20 nm. (Modified from Ke et al. 2012, with permission).

nucleotides that can be visualized as a Lego, with two pins representing the head domains and two holes representing the tail domains.

The pins and holes are in different shapes, such that one brick's pins will only plug into the holes of another if the DNA strands are complementary. Provided the base pairs of the two domains in question match up, the bricks attach at right angles. Every domain is one voxel (the 3D equivalent of a pixel), so a complex shape can be constructed on a 3D molecular canvas, and then an algorithm is used to determine which DNA bricks would self-assemble into the desired structure. With this "maximal specificity" technique, extremely complex aperiodic patterns can be constructed, including 2D symbols and images, and 3D block letters and shapes (Figure 12.10d) (Wei, Dai and Yin 2012, Ke et al. 2012 and Ke et al. 2014).

Experiments (Ke et al. 2012) with DNA bricks use complementary DNA sequences to encode an addressable structure with each distinct single-stranded brick belonging to a specific location within the target assembly.

12.10   DNA Origami

DNA origami has become a powerful tool for building discrete nanoscale materials, generating a variety of complex and dynamic structures. It is a technique that uses hundreds

of short DNA oligonucleotides, typically 15-60 nucleotide long, called staple strands to fold a long single-stranded DNA, which is called a scaffold strand, into a desired target shape (Rothemund 2006, Hong et al. 2017). The process of design and assembly of DNA origami is schematized in Figure 12.11a. First, a target planar shape is approximated with tightly aligned parallel cylinders, representing the DNA double helices. Then, periodic crossovers are introduced to hold all of the helices together. One large strand continuously routes through each adjacent helix and serve as the scaffold. Its complementary strand is cut into hundreds of shorter oligonucleotides (staple strands), which travel between neighboring helices via the crossovers. The sequences of the staples are determined by the routing pathway of the scaffold, the location of the crossovers used to maintain the overall shape, and the nick point positions (Hong et al. 2017).

The advantage of DNA origami over tile-based assembly is that hybridization occurs frequently via intramolecular interactions on the scaffold strand, resulting in a high local concentration of complementary oligonucleotides that drives the system toward the intended sequence-specific thermodynamic product (Pinheiro et al. 2011). This allows for relatively high yields of particularly complex objects and obviates the need for a precise stoichiometry of all components in the system (Jones, Seeman and Mirkin 2015).

Origami methods have found particular utility in constructing 2D scaffold materials because individual sticky ends and other DNA-based features can be placed in nearly any arrangement (Figure 12.11a). The principles that govern the building of 2D objects can be extended to those that fill 3D space (Douglas et al. 2009). This is achieved by imagining the scaffold strand being rasterized into an unrolled 2D schematic of the target shape. This extended sheet is then encouraged to fold back and forth onto itself via the interactions of staple strands to form pleated sheets of antiparallel double helices which ultimately pack into a honeycomb lattice (Figure 12.11b). Using this approach, a variety of complex 3D shapes were synthesized (Douglas et al. 2009, Jones, Seeman and Mirkin 2015).

Large DNA origami structures are highly desired because they can be used to create more complex structures. However, the length of the scaffold limits the size of a DNA origami structure. To scale up the DNA origami structure assembly, several methods have been designed; for example, hierarchical assembly has been implemented by using individual DNA origami structures as the tile units (Pfeifer and Saccà 2016). Other methods include surface-assisted assembly or DNA origami-templated tile assembly (Hong et al. 2017). Because each staple strand is different, some

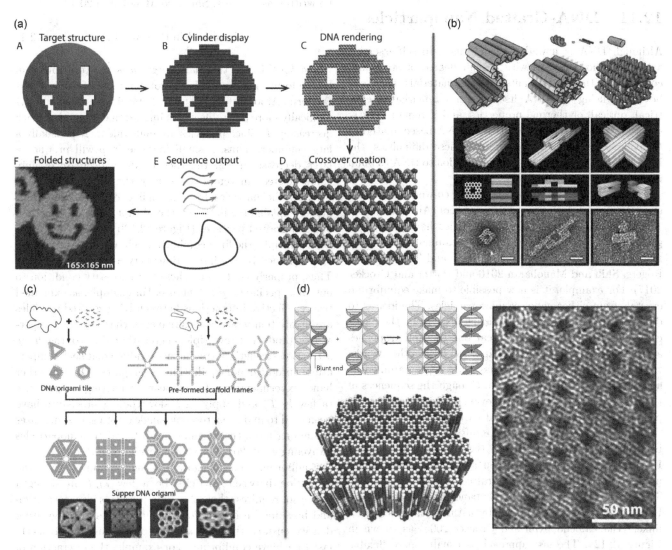

FIGURE 12.11 DNA, super DNA origami, and shape-complementarity design and assembly. (a) Illustration of the assembly of a smiley face pattern. A rasterized image forms the pattern for the scaffold strand folded by staple stands. The atomic force microscope (AFM) image of the representative assembled DNA origami structure. The size of the image is 165 × 165 nm. Image reproduced with permission from Rothemund 2006. Copyright 2006 Macmillan Publishing Ltd. (b) Extension of the origami principle to 3D by using staple strands which promote the formation of pleated sheets of duplexes, which ultimately pack into a honeycomb lattice. Scale bars, 20 nm. Modified from Douglas et al. 2009, Copyright 2009 Macmillan Publishing Ltd, with permission. (c) DNA origami was used as staples to interact with the prefolded scaffold frames to generate the desired structures. The scale bars are 100 nm. Images reproduced from Zhao, Liu and Yan 2011. Copyright 2011 American Chemical Society. (d) Shape-complementarity relies on base stacking and is used to create higher-order DNA origami structures. Images adapted from Gerling et al. 2015. Copyright 2015 AAAS.

can be designed to present DNA sticky ends at programmed locations on the periphery of the final assembled object, allowing these structures to act as scaffolds with spatially prescribed DNA bonds that capture and organize other nanoobjects (Figure 12.11c).

Base stacking is a major force that stabilizes DNA duplex structures and has been used to assemble higher-order discrete three-dimensional DNA components in solution on the basis of shape-complementarity instead of sequence complementarity, that is, without base pairing (Gerling et al. 2015). An example of stacking based on shape-complementarity in 3D is shown in Figure 12.11d.

12.11 DNA-Grafted Nanoparticles

Although DNA is nowadays widely used to self-assemble artificial nanostructures with a high degree of structural complexity, there are limits to the materials that can be made using it. DNA itself has no remarkable electrical, optical, or thermal properties, and it remains costly to produce large-scale quantities of DNA-based materials (Pinheiro et al. 2011). To overcome these difficulties, the integration of other components in addition to DNA to build materials has been proposed.

Pioneered by the work of Mirkin and co-workers (Mirkin et al. 1996) and Alivisatos and co-workers (Alivisatos et al. 1996), colloidal nano- and microparticles have been integrated with DNA to build fully programmable materials made of grafted particles (Boles, Engel and Talapin 2016, Rogers, Shih and Manoharan 2016 and Porter and Crocker 2017). For example, it is now possible to make equilibrium crystals from either nano- or microparticles. The idea is to direct the desired structure by using DNA while the particles confer unique electrical, optical, or structural properties at a lower cost than DNA-only approaches. In other words, the desired outcome of the self-assembly is programmed by adding information to the system through the sequences of the DNA strands while the relevant physical property is partially conferred by the grafted particles.

The net result of the DNA-grafting is an attraction between the grafted particles via the formation of transient DNA bridges, which forms when a strand grafted to one particle hybridizes to a strand grafted to a second particle. This process can be direct (Nykypanchuk et al. 2008 and Park et al. 2008) or mediated by a third strand called a linker (Biancaniello, Kim and Crocker 2005) as shown in Figure 12.12a. The last approach is usually more flexible because the interactions can be tuned without resynthesizing the particles, for example, by adjusting the concentration of linkers (Rogers, Shih and Manoharan 2016).

Understanding the interactions between DNA-grafted particles provides insight into how to design them to obtain a desired equilibrium self-assembled structure. A coarse-grained approach, in which the molecular level details of the transient DNA bridges can be incorporated into an effective interaction between the grafted particles, will depend on the DNA sequences and the temperature. The attraction between particles should be obtained considering the possibility to form more than one DNA bridge at a time (Figure 12.12b). This so-called "multivalency" depends on the density of grafted strands, whereas the strength of each DNA bridge depends on the hybridization free energy. Opposing the attraction between particles, due to the transient bridge formation, is found repulsion due to the compression of the grafted DNA molecules. The overall result of both attractive and repulsive contributions is a time-averaged effective potential that has a minimum at the equilibrium distance between the particles.

The attractive component of the effective potential can be written as (Rogers, Shih and Manoharan 2016)

$$F_a = k_B T \ln \left[(1 - p)^N \right], \qquad (12.1)$$

where $k_B T$ is the thermal energy, p is the probability of bridge formation, and N is the total number of bridges that can form. At lower temperatures, hybridization is thermodynamically more favorable which implies that p increases with decreasing T. Since particles are multivalent, N is usually a large number. Thus, a small increase in p will produce a large decrease in F_a. In other words, a decrease of only a few degrees in temperature may change the attraction between the particles from negligible to very large. This model explains qualitatively the abrupt melting transition of DNA-grafted particles (Figure 12.12b).

In principle, the final equilibrium self-assembled structure is determined by the interactions between the components. Thus, properly controlling them is a necessary condition to obtain a specific target structure. For example, as explained before, effective interactions between DNA-grafted particles are highly temperature sensitive such that too low temperatures produce irreversible aggregation due to the large attraction strength. Experiments conducted at such temperatures produce nonequilibrium aggregates. On the other hand, experiments conducted near the melting temperature (a few $k_B T$) and applying a slow cooling procedure have been used to make macroscopic single crystals, in some cases containing more than a million DNA-grafted nanoparticles (Auyeung et al. 2014) as shown in Figure 12.12c.

Equilibrium self-assembly not only requires that the attraction between particles be a few $k_B T$ in strength but also requires that the kinetics of bridge formation and breaking be fast at room temperatures. High grafting density ensures that the kinetics of binding is sufficiently fast to achieve equilibrium. For example, the interaction of particles with a small number of strands that bind strongly (corresponding to low N and large p in Eq. (12.1)) will be of a few $k_B T$ in strength. However, due to the strong binding of the strands, the lifetime of each DNA bond will be large, which hinders rotational diffusion and, hence, equilibration of the particles. On the other hand, interaction of particles with a large number of strands that bind weakly (corresponding to large N and low p in Eq. (12.1)) will be also of a few $k_B T$ in strength. However, in this case, the lifetime

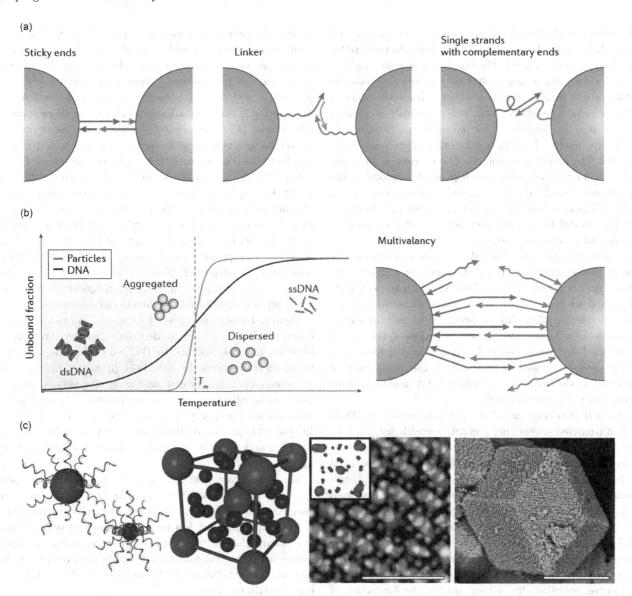

FIGURE 12.12 DNA hybridization induces an effective interaction potential between DNA-grafted particles. (a) DNA-grafted particles experience an effective attraction due to bridge formation, which can be induced by direct hybridization of grafted double-stranded
DNA (dsDNA) with dangling "sticky ends", binding of a partially complementary linker strand from solution or hybridization of grafted
single strands with complementary ends. (b) Unlike hybridization of DNA in solution, which is a gradual function of temperature,
suspension of DNA-grafted particles transitions from a dispersed state to an aggregated one over only a few degrees Celsius, owing to
multivalency in the interactions; ssDNA, single-stranded DNA; Tm, melting temperature. Images reproduced from Rogers, Shih and
Manoharan 2016. Copyright 2016 2016 Macmillan Publishers Limited. (c) Spherical nanoparticles functionalized with the appropriate
DNA strands can assemble into a variety of superlattices (AB6-type structure shown in the scheme and TEM image with tomographic
reconstruction, inset), some of which form large faceted single crystals. Scale bars, TEM image, 100 nm; scanning electron microscopy
image, 1 mm. Modified from Auyeung et al. 2014, Copyright 2014 Macmillan Publishers Limited., with permission.

of each DNA bond will be short, allowing rotational diffusion of the particles and thus equilibration. For this reason,
recent grafting methods aim to achieve high grafting density
(Wang et al. 2015).

Most self-assembled equilibrium configurations studied
using DNA-grafted particles consist of crystalline lattices.
Varying interaction strengths and particle sizes, it is possible
to obtain different crystal structures (Figure 12.12c). As

explained before, according to the second law of thermodynamics, the equilibrium structure is the one that minimizes
the free energy $F = U\text{-}TS$, in which U is the total effective
potential (the sum of all interactions) and S is the entropy.
At low temperatures, the term "TS" can be ignored, and
U is the relevant quantity. Not surprisingly, the obtained
lattice is the one that maximizes the number of (favorable) contacts between spheres with complementary strands

(Macfarlane et al. 2011, 2013) because this minimizes U and thus F. In the case of long-range interactions, the potential U is no longer proportional to the number of nearest-neighbor pairs since next-nearest-neighbor interactions are possible. Consequently, maximizing the number of favorable contacts between nearest-neighbor pairs not necessarily minimizes U. Similarly, for higher T, the entropic contribution TS to the free energy F plays a role and can even dominate over the interaction potential U. The entropy S increases with the number of collective degrees of freedom of the crystal. Accordingly, if a given crystal symmetry can sustain "soft" vibrational modes,[4] the entropic term can dominate and the crystal be favored over other crystal symmetries that do not sustain soft modes.

So far and thanks to statistical-mechanical models, we are able to relate experimental variables such as temperature, DNA sequences, and grafting densities, with the strength, range, and specificity of DNA-grafted particle interactions. This knowledge allows us to program to certain extent the resulting self-assembled lattice. Nonetheless, the variety of the final self-assembled architectures is still limited to simple periodic arrays, in contrast to the complexity of the structures obtained in the field of DNA nanotechnology (using DNA tiles and origami).

The next step is to extend the programmability of DNA-grafted particles so that they can self-assemble not just into crystals, but into any prescribed structure.

As previously discussed, particles with directional interactions can be used to program self-assembly. Directionality in DNA-grafted nanoparticles can be incorporated in a number of different ways, for example, by chemically patterning particles with patches of DNA or using DNA-encoded nanoparticles assembled on a solid support (Maye et al. 2009). Also, directionality arises naturally from shape anisotropy; specific orientations between particles can be favored enthalpically if they permit the formation of additional DNA bridges or entropically if they increase the configurational freedom of the system (Jones et al. 2010, Lu et al. 2015 and O'Brien et al. 2015). Recently, nanoparticles (about 100 nm in size) that combine chemical and geometrical directionality have been fabricated (Désert et al. 2013). Shape and sequence complementarity are promising strategies that are being currently investigated to maximize specificity of DNA-grafted particles; however, the number of different species that can be built is still far from what can be obtained using DNA tiles.

12.12   Conclusions and Outlook

Pre-programmed self-assembly of nanostructures has come a long way from a cutting-edge concept to a flourishing field that spans from DNA nanotechnology to nanoparticle science. Currently, although it is possible to program the final result of self-assembly and obtain a large variety of structures, we cannot yet program the self-assembly of arbitrary assemblages possessing desired physical and chemical properties, with the same freedom and versatility offered by 3D printing in the fabrication of centimeter-scale constructs. The use of DNA segments to program the assembly has proven to be a very versatile tool to impart a high degree of specificity to the elements that are to be assembled. Structures with high degree of complexity have been constructed using this approach. Nonetheless, such methodology shows limitations in terms of physical properties and is still expensive. A combination of nanoparticles with DNA appears to be an interesting way to surmount the limitations in terms of physical properties that materials obtained using DNA-only methods encounter. Nanoparticles impart the desired physical property (optical, electric, magnetic, etc.) while assembly is directed by DNA-functionalization.

From a theoretical point of view, statistical-mechanics-based models are able to describe effective interactions between the components to the point that design rules based on their knowledge have been proposed. Inverse optimization techniques have allowed proposing general interactions to assemble target lattices with nontrivial coordination numbers such as honeycomb, Kagomé, square, and others. In spite of this, the resulting interaction potentials are still complex and difficult to translate into a realizable experimental system.

We have shown that kinetic difficulties inherent to self-assembly process can be overcome by the use of seeds and templates and by suitable designed hierarchical strategies.

Enormous progress has been made to incorporate information into the building blocks in order to program a desired assembled structure; nonetheless, much road remains to be traveled and challenges to be overcome. This will be an exciting trip to make.

Acknowledgments

This work was supported in part by Grants DGAPA IN-110516 and IN-103419.

References

Adorf, C. S., Antonaglia, J., Dshemuchadse, J., and Glotzer, S. C. 2018. Inverse design of simple pair potentials for the self-assembly of complex structures. arXiv:1711.04153v1.

Alivisatos, A. P., Johnsson, K. P., Peng, X., et al. 1996. Organization of 'nanocrystal molecules' using DNA. *Nature* 382: 609–611.

Auyeung, E., Li, T. I., Senesi, A. J., et al. 2014. DNA-mediated nanoparticle crystallization into Wulff olyhedral. *Nature* 505: 73–77.

Biancaniello, P., Kim, A., and Crocker, J. 2005. Colloidal interactions and self-assembly using DNA hybridization. *Physical Review Letters* 94: 058302 (4pp).

---

[4]Soft modes are excitations above the ground state whose energy vanishes in the limit of long wavelengths.

Boles, M. A., Engel, M., and Talapin, D. V., 2016. Self-assembly of colloidal nanocrystals: From intricate structures to functional materials. *Chemical Reviews* 116: 11220–11289.

Cademartiri, L. and Bishop, K. J. M. 2015. Programmable self-assembly. *Nature Materials* 14: 2–9.

Chaimovich, A. and Shell, M. S. 2011. Coarse-graining errors and numerical optimization using a relative entropy framework. *The Journal of Chemical Physics* 134: 094112 (15 pp).

Chen, Q., Bae, S. C., and Granick, S. 2011. Directed self-assembly of a colloidal kagome lattice. *Nature* 469: 381–384.

Chen, Z. Y., Moore, J., Radtke, G., Sirringhaus, H., and O'Brien, S. 2007. Binary nanoparticle superlattices in the semiconductor–semiconductor system: CdTe and CdSe. *Journal of The American Chemical Society* 129: 15702–15709.

Désert, A., Hubert C., Fu, Z. 2013. Synthesis and site-specific functionalization of tetravalent, hexavalent, and dodecavalent silica particles. *Angewandte Chemie International Edition* 52: 11068–11072.

Douglas, S. M., Dietz, H., Liedl, T., Högberg, B., Graf, F., and Shih, W. M. 2009. Self-assembly of DNA into nanoscale three-dimensional shapes. *Nature* 459: 414–418.

Duguet, É., Hubert, C., Chomette, C., Perro, A., and Ravaine, S. 2016. Patchy colloidal particles for programmed self-assembly. *Comptes Rendus Chimie* 19: 173–182.

Gerling, T., Wagenbauer, K. F., Neuner, A. M., and Dietz, H. 2015. Dynamic DNA devices and assemblies formed by shape-complementary, non–base pairing 3D components. *Science* 347: 1446–1452.

Glotzer, S. C. and Solomon, M. J., 2007. Anisotropy of building blocks and their assembly into complex structures, *Nature Materials* 6: 557–562.

Gong, Z., Hueckel, T., Yi, G.-R., Sacanna, S. 2017. Patchy particles made by colloidal fusion. *Nature* 550: 234–237.

Halverson, J.D. and Tkachenko, A. V. 2013. DNA-programmed mesoscopic architecture. *Physical Review E* 87: 062310 (7pp).

He, Y., Ye, T., Su, M., et al. 2008. Hierarchical self-assembly of DNA into symmetric supramolecular olyhedral. *Nature* 452: 198–201.

Hedges, L. O., Mannige, R. V., and Whitelam, S. 2014. Growth of equilibrium structures built from a large number of distinct component types. *Soft Matter* 10: 6404–6416.

Henzie, J., Grunwald, M., Widmer-Cooper, A., Geissler, P. L. and Yang, P. D. 2012. Self-assembly of uniform polyhedral silver nanocrystals into densest packings and exotic superlattices. *Nature Materials* 11: 131–137.

Hong, F., Zhang, F., Liu, Y., and Yan, H. 2017. DNA origami: Scaffolds for creating higher order structures. *Chemical Reviews* 117: 12584–12640.

Jacobs, W. M., Reinhardt, A., and Frenkel, D. 2015. Rational design of self-assembly pathways for complex multicomponent structures. *Proceedings of the National Academy of Sciences USA* 112: 6313–6318.

Jones, M. R., Macfarlane, R. J., Lee, B., et al. 2010. DNA-nanoparticle superlattices formed from anisotropic building blocks. *Nature Materials* 9: 913–917.

Jones, M. R., Seeman, N. C., and Mirkin, C. A. 2015. Programmable materials and the nature of the DNA bond. *Science* 347:1260901 (11 pp).

Ke, Y., Ong, L. L., Shih, W. M., et al., P. 2012. Three-dimensional structures self-assembled from DNA bricks. *Science* 338: 1177–1183.

Ke, Y., Ong, L. L., Sunet, W., et al., 2014. DNA brick crystals with prescribed depths. *Nature Chemistry* 6: 994–1002.

Kiely, C. J., Fink, J., Brust, M., Bethell, D., and Schiffrin, D. J. 1998. Spontaneous ordering of bimodal ensembles of nanoscopic gold clusters. *Nature* 396: 444–446.

Li, X., Yang, X., Qi, J., and Seeman, N. C. 1996. Antiparallel DNA double crossover molecules as components for nanoconstruction. *Journal of the American Chemical Society* 118: 6131–6140.

Lin, C., Liu, Y., Rinker, S., and Yan, H. 2006. DNA tile based self-assembly: Building complex nanoarchitectures. *ChemPhysChem* 7: 1641–1647.

Liu, D., Wang, M., Deng, Z., Walulu, R., and Mao, C. 2004. Tensegrity: Construction of rigid DNA triangles with flexible four-arm DNA junctions. *Journal of the American Chemical Society* 126: 2324–2325.

Lu, F., Yager, K. G., Zhang, Y., Xin, H., and Gang, O. 2015. Superlattices assembled through shape-induced directional binding. *Nature Communications* 6: 6912 (10 pp).

Macfarlane, R. J., Lee, B., Jones, M. R., et al. 2011. Nanoparticle superlattice engineering with DNA. *Science* 334: 204–208.

Macfarlane, R. J., O'Brien, M. N., Petrosko, S. H., and Mirkin, C. A. 2013. Nucleic acid-modified nanostructures as programmable atom equivalents: forging a new "table of elements". *Angewandte Chemie International Edition* 52: 5688–5698.

Madge, J. and Miller, M. A. 2017. Optimizing minimal building blocks for addressable self-assembly. *Soft Matter* 13: 7780–7792.

Maye, M. M., Nykypanchuk, D., Cuisinier, M., et al. 2009. Stepwise surface encoding for high-throughput assembly of nanoclusters. *Nature Materials* 8: 388–391.

Mirkin, C. A., Letsinger, R. L., Mucic, R. C., and Storhoff, J. J., 1996. A DNA-based method for rationally assembling nanoparticles into macroscopic materials. *Nature* 382: 607–609.

Miszta, K., de Graaf, J., Bertoni, G., et al. 2011. Hierarchical self-assembly of suspended branched colloidal nanocrystals into superlattice structures. *Nature Materials* 10: 872–876.

Nykypanchuk, D., Maye, M. M., van der Lelie, D., et al. 2008. DNA-guided crystallization of colloidal nanoparticles. *Nature* 451: 549–552.

O'Brien, M. N., Jones, M. R., Lee, B., and Mirkin, C. A. 2015. Anisotropic nanoparticle complementarity in DNA-mediated co-crystallization. *Nature Materials* 14: 833–839.

Overgaag, K., Evers, W., de Nijs, B., Koole, R., Meeldijk, J., and Vanmaekelbergh, D. 2008. Binary superlattices of PbSe and CdSe nanocrystals. *Journal of The American Chemical Society* 130: 7833–7835.

Park, S. Y., Lytton-Jean, A. K. R., Lee, B., et al. 2008. DNA-programmable nanoparticle crystallization. *Nature* 451: 553–556.

Pfeifer, W. and Saccà, B. 2016. From Nano to Macro through Hierarchical Self-Assembly: The DNA Paradigm. *ChemBioChem* 17: 1063–1080.

Pinheiro, A. V., Han, D., Shih, W. M., et al. 2011. Challenges and opportunities for structural DNA nanotechnology. *Nature Nanotechnology* 6: 763–772.

Piñeros, W. D., Jadrich, R. B., and Truskett, T. M. 2017. Design of two-dimensional particle assemblies using isotropic pair interactions with an attractive well. *AIP Advances* 7: 115307 (10 pp).

Piñeros, W. D., Lindquist, B. A., Jadrich, R. B., and Truskett, T. M. 2018. Inverse design of multicomponent assemblies. *The Journal of Chemical Physics* 148: 104509 (8 pp).

Podsiadlo, P., Krylova, G. V., Demortière, A., and Shevchenko, E. V., 2011. Multicomponent periodic nanoparticle superlattices. *Journal of Nanoparticle Research* 13: 15–32.

Porter, C. L. and Crocker, J. C. 2017. Directed assembly of particles using directional DNA interactions. *Current Opinion in Colloid & Interface Science* 30: 34–44.

Rechtsman, M. C., Stillinger, F. H., and Torquato, S. 2005. Optimized interactions for targeted self-assembly: Application to a honeycomb lattice. *Physical Review Letters* 95: 228301 (4 pp).

Reinhardt, A. and Frenkel, D., 2014. Numerical evidence for nucleated self-assembly of DNA brick structures. *Physical Review Letters* 112: 238103 (4 pp).

Rogers, W. B., Shih, W. M., and Manoharan, V. N. 2016. Using DNA to program the self-assembly of colloidal nanoparticles and microparticles. *Nature Reviews* 1: 16008 (14 pp).

Romano, F. and Sciortino, F. 2011. Two-dimensional assembly of triblock Janus particles into crystal phases in the two bonds per patch limit. *Soft Matter* 7: 5799–5804.

Rothemund, P. W. K., 2006. Folding DNA to create nanoscale shapes and patterns. *Nature* 440: 297–302.

Saunders, A. E. and Korgel, B. A. 2005. Observation of an AB phase in bidisperse nanocrystal superlattices. *ChemPhysChem* 6: 61–65.

Seeman, N. C. 1982. Nucleic acid junctions and lattices. *Journal of Theoretical Biology* 99: 237–247.

Seeman, N. C. 2003. DNA in a material world. *Nature* 421: 427–431.

Seeman, N. C. 2010. Nanomaterials based on DNA. *Annual Review of Biochemistry* 79: 65–87.

Shang, J., Wang, Y., Chen, M., et al. 2015. Assembling molecular Sierpiński triangle fractals. *Nature Chemistry* 7: 389–393.

Shell, M. S. 2008. The relative entropy is fundamental to multiscale and inverse thermodynamic problems. *The Journal of Chemical Physics* 129: 144108 (7 pp).

Shevchenko, E. V., Talapin, D. V., Kotov, N. A., O'Brien, S., and Murray, C. B. 2006. Structural diversity in binary nanoparticle superlattices. *Nature* 439: 55–59.

Talapin, D. V., Shevchenko, E. V., Bodnarchuk, M. I., Ye, X. C., Chen, J., and Murray, C. B. 2009. Quasicrystalline order in self-assembled binary nanoparticle superlattices. *Nature* 461: 964–967.

Torquato, S. 2009. Inverse optimization techniques for targeted self-assembly. *Soft Matter* 5: 1157–1173.

Turberfield, A. J., 2011. Geometrical self-assembly. *Nature Chemistry* 3: 580–581.

van Anders, G., Ahmed, N.K., Smith, R., Engel, M., and Glotzer, S. C. 2014. Entropically patchy particles: Engineering valency through shape entropy. *ACS Nano* 8: 931–940.

Wang, Y., Wang, Y., Zheng, X. et al. 2015. Crystallization of DNA-coated colloids. *Nature Communications* 6: 7253.

Wei, B., Dai, M., Yin, P. 2012. Complex shapes self-assembled from single-stranded DNA tiles. *Nature* 485: 623–626.

Whitelam, S. 2015. Hierarchical assembly may be a way to make large information-rich structures. *Soft Matter* 11: 8225–8235.

Whitesides, G. M. and Grzybowski, B. 2002. Self-assembly at all scales. *Science* 295: 2418–2421.

Winfree, E., Liu, F., Wenzler, L. A., and Seeman, N. C. 1998. Design and self-assembly of two-dimensional DNA crystals. *Nature* 394: 539–544.

Xu, L. et al. 2012. Region specific plasmonic assemblies for *in situ* Raman spectroscopy in live cells. *Journal of the American Chemical Society* 134: 1699–1709.

Yan, H., Park, S. H., Finkelstein, G., Reif, J. H., LaBean, T. H. 2003. DNA-templated self-assembly of protein arrays and highly conductive nanowires. *Science* 301: 1882–1884.

Zeravcic, Z., Manoharan, V. N., and Brenner, M. P. 2014. Size limits of self-assembled colloidal structures made using specific interactions. *Proceedings of the National Academy of Sciences USA* 111: 15918–15923.

Zhao, Z., Liu, Y., and Yan, H. 2011. Organizing DNA origami tiles into larger structures using preformed scaffold frames. *Nano Letters* 11: 2997–3002.

Zheng, J., Birktoft, J. J., Chen, Y., et al., 2009. From molecular to macroscopic via the rational design of a self-assembled 3D DNA crystal. *Nature* 461: 74–77.

Survey of Nanomaterials Synthesis, Fabrication, and Growth

Xuefeng Song, Kai Tao,
Shuai Wang, Xiaofeng Zhao, and
Jijin Xu
Shanghai Jiao Tong University

13.1 Introduction

In the 21st century, the rapid development of new energy, information, biotechnology, environment, advanced manufacturing technology, and national defenses will inevitably put new demands on materials and materials properties, such as high energy density, ultrafast transmission of components, miniaturization, intelligence, and high integration. Recent years, nanotechnologies have made great progress, and nanomaterials have been found to have many important uses in research and application. Owing to the surface effect, quantum size effect, small size effect, and macroscopic quantum tunneling effect, nanomaterials exhibit remarkable specific properties different from that of the bulk, which show excellent prospects in every field as next-generation materials. There are a large number of approaches to fabricate nanomaterials. However, there is no specific standard of classification. According to original state of the material, the preparation method can be divided into solid-phase method, liquid-phase method, and gas-phase method. A more simple classification, which involves physical and chemical methods according to the main aspect of matter change of the preparation roads, is introduced in this chapter.

Along with the development of science and technology, especially after the appearance of the microscope, the understanding of crystallization associated with nanomaterial has been significantly expanded. The unique properties of nanomaterials are closely related to their size, shape, and structure. Therefore, understanding the formation mechanism of nanocrystals is necessary for synthesizing nanomaterials with desired properties. There are many similarities between the formation of nanocrystals and general crystals. The formation of crystals can be described that a new phase emerges from the old phase, and in this process, the whole system free energy becomes lower and lower until the system stays in a stable state. In a ward, crystallization process generally contains three basic steps: first, attainment of supersaturation or supercooling; second, formation of critical nuclei; and third, growth of these nuclei into crystals.[1]

In the initial stage, the new phase will not appear until the system reaches the thermodynamical phase boundary. Due to reaction kinetics requirements, it is essential for the system to achieve certain degree of supersaturation or supercooling. Then, system dynamic fluctuations bring a sufficient number of atoms or molecules together to form clusters that have a stable size. Only those clusters that grow up to the certain critical size can become stable nuclei. The critical size will be described later. This process is called nucleation of the new phase.[2]

There are predominantly two nucleation mechanisms. Generally, nucleation occurs at nucleation sites where energy barriers are lower. Nucleation sites usually locate at the surfaces contacting the liquid or vapor. In particular, suspended particles or minute bubbles also can provide nucleation sites sometimes. Nucleation without preferential nucleation sites is called homogeneous nucleation. Nucleation occurring at nucleation sites is called heterogeneous nucleation.

After stable nuclei formed, it will begin to grow into crystal with larger size along predetermined planes. For nanocrystals, the final grown crystal size is in nanosize. Researchers have been working on the kinetics of crystal growth. In the earlier time, Gibbs suggested that growth of perfect crystals must proceed by growth of monolayer nucleus to the extremities of the crystal face. Thus, energetically, there should be a critical supersaturation to provide driven force. This is the prototype of the layer-by-layer growth mechanism.[3] However, the layer-by-layer growth mechanism needs to overcome higher energy barrier at the initial stage of each layer formation. It only can take place at high supersaturation. In 1931, Volmer and Schultze discovered Gibbs's theory was not perfect. They found that a number of systems don't need such critical supersaturation, contradicting this growth theory of perfect crystals. They proposed that the surface layers play a very important role, which act as collectors of atoms or molecules. The rough surface made the crystal growth possible without surface nucleation. In the later work, Frank suggested that the discrepancy observed by Volmer and Schultze had a different origin and that the surfaces of real crystals were imperfect with enormous dislocations. A new crystal-growth mechanism based on screw dislocation theory was then established.[4]

In this chapter, two general methods, involving physical and chemical techniques, are first introduced, which can help the reader understand the synthesis and fabrication of nanomaterials. Second, the nucleation and growth mechanisms of nanocrystals will be elucidated.

13.2 Nanomaterials Fabricated by Chemical Techniques

13.2.1 Chemical Vapor Deposition Method

Chemical vapor deposition (CVD) is a chemical process to synthesize high-quality, high-performance solid materials. This technology is often used in the semiconductor industry to produce thin films. In a typical CVD process, three steps are involved: (i) the formation of volatile substances, (ii) transferring the above material to the depositional area, and (iii) chemical reaction on the solid substrate. CVD is commonly used to deposit conformal films and ameliorate surface of the substrate with ways that are not capable for traditional surface modification techniques. What's more, CVD has significant advantages for the deposition of extremely thin layers compared with other methods. But there are still some defects for CVD such as high cost for large-scale fabrication and the by-products which must be removed by a continuous gas flow. Owing to

various advantages, CVD has attracted tremendous attention for nanostructure construction and wide applications.

CVD growth has always been considered as one of the most promising methods for the growth of single-walled carbon nanotube (SWNT). A thermodynamically optimized SWNT structure with a specific chirality has the best catalytic property. But selecting a chirality-specific SWNT is still a drastic challenge. Under constant condition of the normal CVD growth, the SWNTs nucleation in all direction makes it hard to optimize the chirality. After continuous elongation process, the chirality of the SWNT is unlikely to be changed. If perturbations imposed on the reaction system during the growth process, the change of chirality of the SWNT and optimizing the SWNT–catalyst interface become possible.

Based on the perturbation theory, Zhao et al. designed a new CVD process "tandem plate CVD" which takes advantage of a consecutive reaction while making sure each step in equilibrium state.[5] By utilizing the periodically changed temperature during the elongation process, the energetically preferred SWNT–catalyst interface is built up by most smallest helix angle chirality (~72%) of SWNT. This novel CVD method makes it possible for high-quality synthesis of chirality-selective nanomaterials.

Besides, CVD method provides possibility for complicated structure designing and synthesis, which is widely used in energy field. Song et al. fabricated a hierarchical SnO_2@HCS@Sn architecture by combining the sol–gel and CVD methods.[6] As shown in Figure 13.1, this hierarchical architecture was composed of three building units, namely well-defined SnO_2 nanoplates, mesoporous hollow carbon spheres (HCSs), and tin nanoparticles (Sn).

This synthetic approach can successfully diminish uniform tin nanoparticles down to a range of less than 20 nm rather than one of over submicrometers or micrometers due to the confinement of rattle-type nanoreactors. When used as an anode material for Li-ion battery, the SnO_2@HCS@Sn hierarchical hybrids exhibit exceptional properties in terms of specific capacity, cycling performance, and rate capacity. The superior electrochemical performance can be attributed to a heterogeneous architecture which unifies voids in the form of nanoreactors with active anode material to compensate the mechanical strain causing capacity loss and to alleviate the pulverization of active anode material. This delicate hierarchical structure design provides a guideline for the fabrication of composite nanostructure.

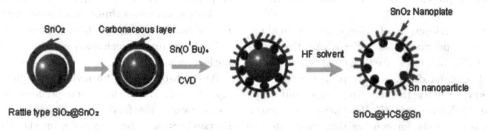

FIGURE 13.1 Synthesis strategies to obtain SnO_2@HCS@Sn composite structures.

Ma et al. put forward a novel design of hollow-structured SnO_2@Si nanospheres by a simple CVD method.[7] SiO_2 nanospheres were selected as template because of the easy removal and nontoxic. Then, SiO_2 was coated by Na_2SnO_3. The SnO_2 coating was formed by the decomposition of Na_2SnO_3. With the removal of SiO_2 and subsequent silicon coating by CVD reaction, the hollow-structured SnO_2@Si is formed (Figure 13.2). By controlling the deposition time of CVD, the thickness of silicon layer can be tailored. This hollow SnO_2 nanospheres (denoted as h-SnO_2@Si) show impressive volumetric specific capacity $(1,600\,\mathrm{mAh/cm^3})$ when used for lithium-ion batteries. At a current density of $300\,\mathrm{mA/g}$, the capacity retention of $1,030\,\mathrm{mAh/cm^3}$ over 500 cycles can be achieved. The excellent performance mainly attributes to the hollow structure in which silicon acts not only as a reactive layer of high capacity but also as a protective layer to suppress the aggregation of Sn nanoparticles.

For synthesis of two-dimensional (2D) thin film with high performance, CVD method shows dominant advantages. For example, a facile confinement CVD approach was put forward by Pang et al. for making graphene directly grow on Si wafers (Figure 13.3).[8] Comparing with the vast body of work concerning about the fabrication of large-area graphene using CVD, the mostly used catalytic substrates were Cu or Ni. But for electronic applications, removing the graphene sheets (GNS) from the metal means potentially damage and contamination of the graphene. By simply sandwiching two Si/SiOx wafers together, large-area uniform single-layer graphene films can be obtained. The as-synthesized graphene via this way shows highly competitive optical and electric performance when compared to monolayer graphene produced on other metal substrates. This technology promotes the production of large area graphene directly on the silicon wafer and may be extended to other 2D material generally.

13.2.2 Hydrothermal Method

Compared with all other synthetic methods, the hydrothermal method based on a water system has its unique advantages such as simple, easy control, homogeneous resultant, low energy cost, and lower pollution. Hydrothermal synthesis can be considered as a method of synthesis material by solubility changes or chemical reactions of the reactant under a high temperature and pressure atmosphere. By precisely controlling the reaction condition of the liquid or the environment, such as pH value, pressure, temperature, time, concentration, templates, and additives, the chemical composition, size, structure and morphology of the nanostructures can be modulated. Based on the presence or absence of additives and templates in the reaction conditions, the hydrothermal method can be mostly divided into three categories: template-assisted, organic additive-assisted, organic additive- and template-free.[9] We will have a brief discussion based on the categories in the following text.

Template-assisted hydrothermal method has a unique advantage in controllable shape and size of the target products. In principle, as long as the template is various enough, the corresponding nanostructure can be generally synthesized by the template-assisted method. The fabrication of hollow nanoparticles with asymmetric structure has been considered as a challenge. Most of the methods such as bowl-shaped, cage-structured and other nonspherical particles are limited to several metal oxides, silica or polymer latex mainly involving the participation of caustic reagents or toxic organic polymers. Chen et al. developed a sustainable route based on cheap and naturally abundant sugars, using a simple, soft-template, hydrothermal synthesis method to fabricate asymmetric flask-like hollow carbonaceous material (Figure 13.4).[10] During the hydrothermal synthesis, an acidic environment formed because of the decomposition of precursors (ribose) and sodium oleate was converted into oleic acid nanoemulsions. At the same time, the

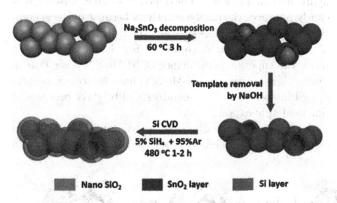

FIGURE 13.2 Schematic illustration of the fabrication process of SnO_2@Si hollow nanospheres.

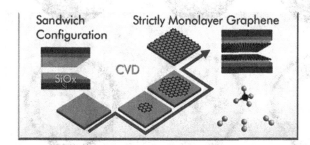

FIGURE 13.3 Schematic representation of direct synthesis of graphene over Si/SiOx using thermal CVD.

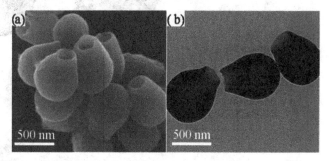

FIGURE 13.4 Morphologies of hollow and open carbonaceous nanoflasks (HOCF): (a) SEM and (b) TEM images of the HOCF.

co-surfactant P123 makes the dispersion of nanoemulsions better, and acts as a template and a benign solvent for amphiphilic derivatives of precursors. The synergetic interaction between template and biomass could be the possible explanation for the asymmetric flask-like structure. Furthermore, by observing the microstructure of resultant product, it shows a high uniform morphology with narrow size distribution. The hollow flask-like structure has a high surface area of 2,335 m²/g, displaying a good performance of supercapacitors. These findings provide a new thought for the fabrication of asymmetric nanostructure in a simple template-assisted method.

Similarly template-assisted hydrothermal method could design complex structure. Core-shell structure with high porosity and large surface area show excellent performance in the field of Li-ion battery. The core-shell structure has improved active material utilization and fast electrochemical kinetics because of promoting the electrolyte penetration and reducing Li-ion diffusion distance compared to simple composite material. In addition, compared to single-shelled structure, multishelled hollow structures have other advantages such as shell permeability, high volumetric energy density, and homogenous dispersed nanosized subunits. However, synthesizing a multishelled hollow structure is usually time-consuming and need to control the structures and components of the products repeatedly.

Luo et al. put forward a simple strategy for the synthesis of multishelled transition metal oxides (TMO) microspheres used for lithium-ion battery by using a one-pot hydrothermal synthesis method (Figure 13.5).[11] Using metal acetate polysaccharide microspheres as the template, metal ions bond with polysaccharides construct a metal−polysaccharide network. After subsequently thermal treatment, a porous multishelled hollow micrometer structure with nanosized subunits were obtained by the carbon removing and the metal ion oxidizing. By varying concentration and proportion of metal ions, controllable and universal hollow microspheres with multiple morphology and composition were attained through this simple synthetic and economic method. Furthermore, the multishelled TMO

microspheres have abundant empty interiors to relax the strain and inhibit the volume expansion during the charge−discharge processes. Benefiting from abundant electrode/electrolyte interfaces, which shorten the lithium-ion diffusion path and provides numerous Li⁺ conversion reaction sites, the multishelled hollow microspheres exhibit high volumetric and gravimetric energy density as well as high cycle stability of 1,470 mAh/g at 0.2 A/g for over 250 cycles.

As long as the templates are rich enough, the shape and size of the target products can be directly determined by the templates. So template-assisted method is very versatile for printing nanostructures with various complex shapes. While, there are still some problem need to be concerned, such as the design and selection of templates, feasible way of the templates removal, the compatibility between the host and template.

Compared with other hydrothermal routes, the organic additive-assisted hydrothermal method means the introduction of various organic additives, which may help further control of the morphology, dispersibility and uniformity of the nanostructure. The commonly recognized MoS₂-based catalysts with well-defined structures have high-activity catalytically sites and are considered as ideal catalysts for hydrogen generation. However, most of MoS₂−based electrocatalysts synthesized under lab conditions have lots of limitation (such as low output, strict reaction conditions and high cost), which can't satisfy the large-scale industrial appliances. Miao et al. developed a unique functional Ni-Mo-S/C electrode through a facile organic-assisted hydrothermal method.[12] The doped Ni atom in Mo-S significantly improves the catalytic activity because of the generation of substantial defect sites and the enhanced morphology of Ni-Mo-S network. A compared test is also designed to prove the superior performance of Ni-Mo-S/C over that of MoS₂/C. It shows that Ni-Mo-S/C have easier access integrated into existing H₂ generators and brighter prospect of practical application.

Wang et al. synthesized polyaniline (PANi)/GNS composites with different nanoscale morphologies through

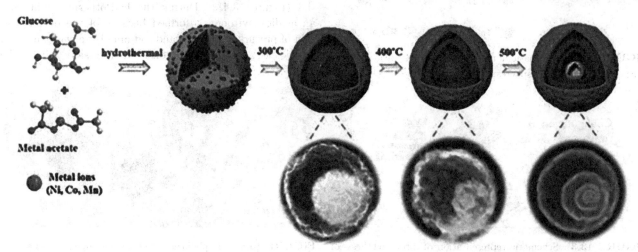

FIGURE 13.5   Formation process of TMO quadruple-shelled hollow spheres.

an organic additive-assisted hydrothermal method.[13] By adding the organic additive of aniline and ammonium persulfate, ultrathin PANi layers can react with GNS by a two-step hydrothermal-assistant chemical oxidation polymerization process. The GNS/PANi generated with precise control of the reaction conditions has ultrathin layer and porous structure, with the PANi homogeneously dispersed onto the surface and tightly connected. The GNS/PANi composites electrode exhibits a specific capacity ability of 532.3–304.9 F/g at scan rates of 2–50 mV/s. It also showed excellent stability at the scanning rate of 50 mV/s, and the capacitance retention rate was 99.6%, indicating that there is a great potential in supercapacitors.

Due to the advantages of convenient manipulation and the avoidance of extraneous impurities, organic additive- and template-free method becomes the most frequently used strategy for the synthesis of inorganic nanostructures. Traditional methods for fabricating electrocatalysts nanoparticles mainly involved the introduction of toxic additive (such as stabilizer, reducing agents, and surfactant), which is complicated and environmentally unfriendly. Therefore, an eco-friendly way without any organic surfactants, where the electrochemically active bacteria is utilized to reduce the metal ions into nanoparticles, is proposed. Liu et al. in situ synthesized alloyed bio-Pd/Au nanoparticles with the participation of electrochemically active bacteria through a hydrothermal reaction,[14] in which organic additive and template are not used. Compared with commercial Pd/C under the same condition, it exhibits better electrocatalytic activity and durability.

Undeniably, there are still several issues need to be totally considered, which drastically hinder the hydrothermal synthesis of nanomaterials, listed as follows: (i) the growth and formation mechanisms of some nanostructures are still unknown or remain in the hypothetical stage; (ii) although variations in nanostructures with different structure are accessed, it still can't cover the whole periodic table elements; (iii) in compositionally systems, it is hard to control and predict the structure in terms of (dimensionality, size, morphology, and crystal type); and (iv) the widely used periodically heterogeneous nanostructures in semiconductor industry have proven hard to achieve via hydrothermal method.

13.2.3 Sol–Gel Method

The sol–gel method is a wet-chemical technique for producing solid materials from small-molecule precursor, which is mainly used for the fabrication of metal oxides and metal chlorides. In a typical sol–gel process, three major steps are involved: (i) the conversion of monomers into a colloidal solution (sol), (ii) the formation of a gel-like diphasic system containing both liquid phase and solid phase, and (iii) post-processing for phase separation (Figure 13.6). The precursor sol can also be used for the deposition of film, the synthesis of powders, and casting of desired shape. The sol–gel approach allows for cheap and low-temperature synthesis of materials with a controllable composition. So a variety of materials fabricated by sol–gel

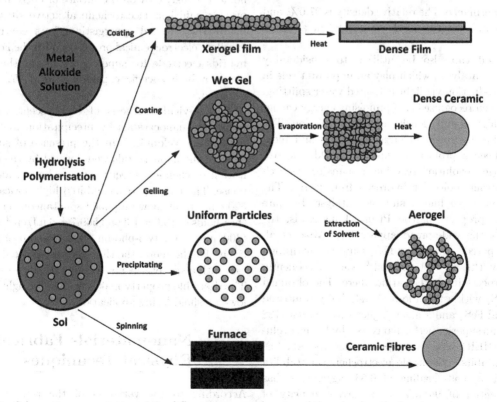

FIGURE 13.6 Schematic representation of the different stages and routes of the sol–gel method.

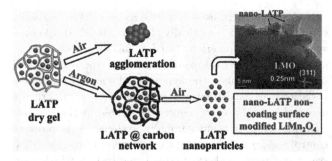

FIGURE 13.7 Schematic illustration of sol–gel methods for the preparation of $Li_{1.4}Al_{0.4}Ti_{1.6}(PO_4)_3$ (LATP) by different calcination conditions.

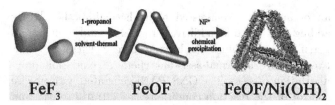

FIGURE 13.8 Schematic illustration for the synthesis of FeOF/Ni(OH)$_2$ core–shell structure.

process have wide applications in energy, electronic, catalyst sensors, medicine, optics fields, etc.

Liu et al. prepared LATP nanoparticles by a modified citric-acid-assisted sol–gel method which is simple and reproducible (Figure 13.7).[15] LATP has high lithium-ion conductivity, high electrochemical oxidative potential ($\sim$6 V vs Li/Li$^+$) and excellent stability in air and water, which is a potential material for all-solid-state energy storage devices. The conventional synthetic methods such as solid-state reaction and melting–quenching mainly utilize ball milling to refine the LATP, and the size is larger than 100 nm. Without the adoption of ball milling, no impurities were introduced, and the method merely involves a two-step heat treatment in different atmospheres. Compared with conventional methods, the obtained LATP has smaller size (<40 nm) and a narrower size distribution showing less aggregation performance. The relative density is 97.0%, and it has a high conductivity of 5.9×10^{-4} S/cm at room temperature.

Sol–gel method can also be utilized to considerably synthesize electrocatalysts, which play an important role in the hydrogen production via light-activated water splitting. Hydrogen has been recognized as a promising energy carrier because of its high energy density and zero-emission. The way of producing hydrogen in large quantities and inexpensively has been a problem around the world. Electrocatalytic hydrogen evolution reaction is a major research direction to extract molecular hydrogen from water. The key of this issue is to find a suitable catalyst. In spite of high-efficient property of the Pt-based materials, the high cost limits the widespread applications. Miao et al. successfully prepared mesoporous FeS$_2$ materials with high surface area by the sol–gel method[16] and subsequently sulfurization process in an H$_2$S atmosphere. The obtained mesoporous FeS$_2$ with a much lower crystallinity compared with commercial FeS$_2$ and a much higher surface area (128 m^2/g). It exhibits superior performance as a hydrogen evolution reaction (HER) electrocatalyst. Under the alkaline condition, it exhibits remarkable electrochemical stability for 24 h. With a mass loading of 0.53 mg/cm^2, it has a low overpotential of 96 mV at a current density of 10 mA/cm^2.

13.2.4 Precipitation Method

Precipitation is a process of forming solid phase from solution. First, the ions homogeneously dispersed in the solution. Then, a uniform precipitate of various components can be obtained by precipitation reaction involving the nucleation and growth processes. The advantages for chemical precipitation are listed as follows: (i) uniform chemical composition, (ii) small particle size and uniform distribution of particle, (iii) controllable and time-effective preparation, and (iv) a simple and low-cost method.

Wang et al. designed and synthesized well-controlled core–shell hierarchical nanostructures by using a chemical precipitation method combining with solvothermal method (Figure 13.8).[17] On the basis of oxyfluoride and hydroxide, the composite material was first obtained with FeOF nanorod acting as core and porous Ni(OH)$_2$ nanosheets as shell. The fluorine in FeOF nanorods is partially substituted by oxygen, which reduces the intrinsic electrical resistance and facilitates the charge transfer process. At the same time, the external Ni(OH)$_2$ nanosheets with large surface area homogeneously coat on the surface of FeOF nanorod, which provide sufficient electrochemical active sites. The unique microstructure and synergistic effects contribute to the excellent electrochemical property. With the FeOF/Ni(OH)$_2$ hybrids electrode for supercapacitors, at the current of 1 Ag^{-1}, a high specific capacitance of 1,452 Fg^{-1} can be achieved.

A high-yield synthesis of luminescent CH$_3$NH$_3$PbBr$_3$ perovskite nanocrystals by precipitation method was put forward by Veldhuis.[18] In the presence of surface-binding ligands, the nanocrystals were formed by the direct precipitation of chemical precursors in a benzyl alcohol/toluene phase. The nanocrystals exhibited high optical gain (about 520 cm^{-1}), ultralow amplified spontaneous emission (ASE) thresholds of 13.9 ± 1.3 and 569.7 ± 6 µJ/cm^2 at one-photon (400 nm), and two-photon (800 nm) absorption, respectively. Furthermore, the thin film constructed by nanocrystals with a size of $\sim$11 nm remains stable >4 months in air. The desirable property makes it ideal for light-emitting and low-threshold lasing application.

13.3 Nanomaterials Fabricated by Physical Techniques

According to the variation of the raw material, physical techniques can be generally classified into "top-down" approaches and "bottom-up" approaches. "Top-down" approaches mean departing the large bulk materials into small part which has a desired size and morphology by introducing some externally controlled tools (such as milling, vibration) or energy, whereas "bottom-up" approaches are similar to the behavior of crystal growth. Small units or components, such as atoms, ions, or molecules orderly aggregate, self-organize and self-assemble into the designed structures. It is commonly companied with preferential growth which is confined by environmental conditions and inherent characteristics of material.

13.3.1 Top-Down Approaches

Mechanical Attrition

Mechanical attrition process is a novel and cost-effective method for the synthesis of nanomaterials. By using mechanical energy to activate chemical reactions and structural changes, the particles suffer the deformation, fracturing, and cold welding during repeated collisions, and finally the size reduced to the nanosize. It is a simple process and can be scaled up for industrial production. However, a large number of defects, grain boundaries, and phase boundaries are simultaneously generated by the stress and strain. As a result, the nanoparticles obtained have a nonuniform size distribution and enormous defects.

Heintz et al. successfully synthesize stable alkyl passivated silicon nanoparticles through a "green mechanical" approach by using high-energy ball milling (HEBM).[19] The impacts and collisions during the milling provide abundant mechanical energy to the system, causing fracture of the large silicon pieces, create fresh surface, and decrease the particle to nanosize. By using this top-down approach, it simultaneously makes production of silicon nanoparticles and passivation of silicon surface possible. At the same time, it avoids the introduction of unstable hydrogen-terminated intermediates and any dangerous chemicals. The experimental results demonstrate that the nanoparticles less

than 4 nm in diameter are produced and they are thermally stable over periods of months. It proves that mechanical attrition is a simple and effective way to produce nanomaterials.

An HEBM method was utilized to fabricate BP nanosheets, combining the expansion effect of plane distance caused by the LiOH (Figure 13.9). And the application of BP nanosheets for photocatalytic H_2 evolution was first studied by Zhu.[20] The results show BP nanosheets involving no noble metal cocatalyst exhibit high visible-light photocatalytic hydrogen evolution rate of 512 μmol/hg, which is comparable or even higher than that of g-C_3N_4. The newly discovered property of BP extends the applications of metal-free elemental photocatalysts.

Lithography

Lithography also refers to as photoengraving that is the process of transferring a pattern into a photoresist-like reactive polymer film, which will subsequently be used to replicate that pattern into an underlying thin film or substrate. Through this method, nanomaterials (tiny metals or semiconductor crystals) with feature size less than 100 nm can be produced in a large scale. With continuous improvement of this technology, it is expected that the feature size will keep decreasing further. The limitation of this method is the possible damage on the surface of fabricated nanostructures, which may have drastic impact on the properties and performance.

Nanomaterials have unique properties that make them ideal candidates for making electronic devices. Lithography exhibits advantages for fabrication of arbitrary structures in the nanometer scale, which is an ideal method for proof-of-concept and prototyping nanotechnologies. In particular, lithography is commonly used in electronic industry, which is an important manufacturing process. Wang et al. reported a general method for direct lithography of functional inorganic nanomaterials without the participation of photoresists.[21,22] The materials used

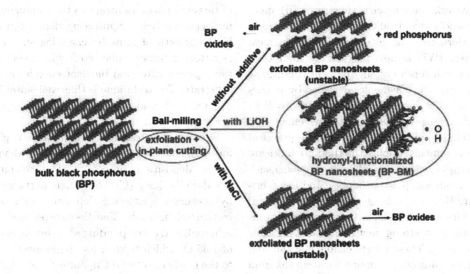

FIGURE 13.9 Schematic depiction of ball-milling treatment of black phosphorus (BP) with or without additive.

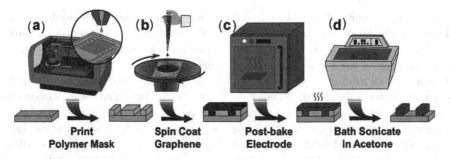

FIGURE 13.10   Description of the four-step manufacturing process for IML graphene patterning.

for etching can be metals, semiconductor, magnetic, oxides, or rare-earth mixtures. The absence of organic impurities in the graphic layer helps to achieve good electronic and optical properties. The conductivity, carrier mobility, and dielectric and luminescent properties of the optically etched layer are consistent with the properties of state-of-the-art solution processing materials.

Hondred et al. invented a photolithography-free, high-resolution solution-phase graphene patterning technique, coined inkjet maskless lithography (IML) (Figure 13.10).[23] Compared with current inkjet printing resolution of graphene (line widths ~60 μm), this new lithography method was able to form high-resolution, flexible graphene films(line widths down to 20 μm). The markedly improved resolution could potentially promise high-resolution patterns for concentrated, conductive nanoparticle inks which are compatible with a variety of substrates.

13.3.2   Bottom-Up Approaches

Physical Vapor Deposition

Physical vapor deposition (PVD) is characterized by atoms and molecules transferring from the source to the surface of substrate. And PVD is generally used for the deposition of thin films with superior mechanical, optical, chemical, or electronic functions. The advantages of PVD are listed as follows: (i) available for any type of inorganic and some organic coating materials on diverse substrate, (ii) more environmentally friendly than traditional coating processes, and (iii) high performance for mechanical and chemical properties. However, PVD is also plagued by high cost, energy waste, and harsh control of environment condition.

Zhou et al. fabricated a plasmonic absorber by a one-step PVD process.[24] The ideal absorber can absorb light effectively in a wide range of wavelengths, which is very important for fundamental research as well as practical applications. The commonly applied method to fabricate absorbers is lithography, which has inherent problems such as limited spatial resolution, poor scalability, and low fabrication throughput. However, during the PVD possessing, metallic nanoparticles are self-assembled onto a nanoporous template contributing to strong multiple scattering, low effective refractive index. These effects enable efficient and broadband plasmonic absorption. Among wavelengths from 400 nm to 10 μm, the absorbers have an average measured

absorbance of ~99%, which was recognized as the most efficient and broadband up to date. The high-throughput deposition process to synthesize absorbers with pronounced optical absorption may lead to large-scale production of high-performance nanomaterials and devices.

Qin et al. synthesized large-size, high-quality 2D selenium nanosheets by using the PVD method.[25] Selenium has been considered as a promising material for the application of optoelectronics. But due to the anisotropic atomic structure, there is strong tendency for selenium to grow into the nanowire shape. According to Hooge's rule, the 1D structure means decreased size and a high ratio of edge to bulk, which corresponds to poor electrical transport and electrical noise. These phenomena hinder the exploration of its potential applications. By using the PVD method, high-quality 2D selenium nanosheets could be massively synthesized. The surface of the obtained 2D selenium nanosheets is atomically flat showing that the Se nanosheets are well crystallized with good quality. It has a large lateral size up to 30 μm, and the minimum thickness could be as thin as 5 nm (Figure 13.11). What's more, it exhibits a strong in-plane anisotropic property and an excellent photo-responsivity of 263 A/W in a low response time. These results provide evidence of the applied possibility for 2D selenium nanosheets device.

Sputtering Deposition Method

Sputtering deposition can be defined as the ejection process of targeted particles owing to the impingement of energetic projectile particles. Sputtering deposition offers versatility in the growth of films by changing sputtering parameters (sputtering power, time, etc.) which control the composition, growth rate, and incident energies of atoms upon the substrate. The technique is thus well suitable for the growth of compounds and alloys with desired composition or spatial variation in composition.

Shin et al. presented a high-quality pinhole-free films for high-efficiency photovoltaic (PV) devices by a sputtering deposition process.[26] The selenium-incorporated $Cu_2BaSnS_{4-x}Se_x$ (CBTSSe) nanoparticles were fabricated by precursor sputtering deposition and successive sulfurization/selenization. The Se-incorporated sulfide film was achieved, and corresponding absorber layers have a bandgap of 1.55 eV which is ideal for single-junction PV. Compared to the newly developed $Cu_2ZnSn(S,Se)_4$ (CZTSSe), it avoids the fundamentally cationic disorder and related band tailing

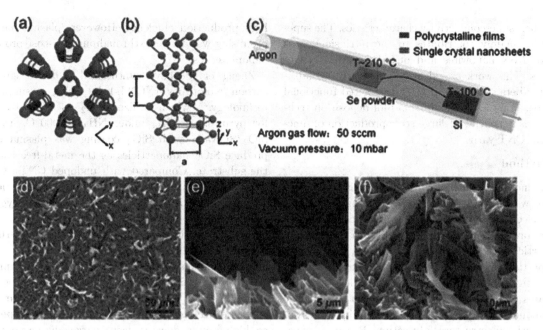

FIGURE 13.11 Growth of Se nanosheets and characterization by PVD. (a,b) Atomic structure of selenium. (c) Schematic diagram of the PVD method. (d) Low-magnification SEM image for synthesized selenium nanosheets on a Si (111) substrate. (e) Enlarged view of typical Se nanosheets with saw-like structure and (f) feather-like twin structure.

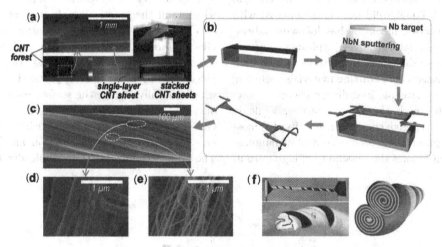

FIGURE 13.12 Fabrication of NbN–CNT yarns by sputtering method. (a) A free-standing CNT-sheet substrate prepared by direct mechanical drawing of CNT sheets from a vertically grown CNT forest, followed by stacking the drawn sheets on a U-shaped jig using a stepper motor. (b) Illustration of the NbN sputtering process on a substrate of CNT sheets, followed by twisting, to obtain a yarn. (c) A SEM image of the fabricated NbN-CNT yarn twisted by 10-L CNT sheets with the nominal NbN layer thickness of 250 nm. (d) An amplified SEM image showing the NbN-deposited region. (e) An amplified SEM image showing the surface region without NbN. (f) Images of model system and schematics showing the internal and surface structures of the twisted NbN-CNT yarn, explaining the dual Archimedean scrolls.

of CZTSSe. The CBTSSe-based PV device shows 5.2% power conversion efficiency (PCE) with a total area of 0.425 cm^2, which is over 2.5 times better than the pure sulfide device.

Except for individual phase film, the sputtering deposition can also be applied to fabricate hybrid species. Since the discovery of carbon nanotube (CNT), much attention has been paid to it. Based on the superior mechanical properties of CNT, CNT sheets always combine with other materials and use as host material. After subsequently twisting and spinning, functional yarns can be obtained. Recently,

Kim et al. reported that a superconducting composite fiber was formed by sputtering deposition of NbN nanowires on free-standing CNT sheets followed by post-twisting process (Figure 13.12).[27] By evaluating the superconductivity of the composite system, the quality of deposited layers can be sensitively valued. Compared with typical NbN thin film on a normal solid substrate, the NbN–CNT yarns show superior mechanical flexibility and keep a clear superconductivity transition behavior without any substantial degradation of quality. Furthermore, even under severe mechanical deformations, these NbN–CNT composite yarns still retain their superconductivity characteristics. The superior flexibility mainly contributes to the intrinsic porous structure of NbN nanowires and quasi-superconductivity of the CNTs. This work provides a possible way to fabricate flexible fibers consisting of CNT-templated functional nanowires based on a commonly thin-film deposition technique, which is beneficial for large-scale production of functional hybrid-CNT yarns.

Plasma Method

Plasma commonly considered as either partially or fully ionized gas was first put forward by Erving Langmuir. Plasma is electrically neutral in a macro-scale perspective, while on the molecular level, plasma consists of charged particles such as electrons, cations, and anions (in some particular cases). The variety of excited species leads to a high chemical reactivity, which lead to high reaction rates compared to ordinary reaction media. The "plasma enhancement" property provides a unique environment for novel nanostructure fabrication. Recent plasma-assisted techniques can be briefly categorized into four parts: plasma-enhanced CVD, thermal plasma evaporation, plasma treatment of solids and condensation, and thermal plasma sintering with liquid/solid feeding. Compared with ordinary chemical reaction media, it has following advantages for nanostructure fabrication: (i) "plasma enhancement" feature improves the reaction reactivity, which indicates much broader selection of starting materials including inert materials. Inert materials in ordinary chemical reactions cannot able to reach the energy barrier, while in plasma states, it can keep high chemical reactivity; (ii) most current research emphasizes on precise control of nanomaterials fabrication but ignores the possibility of high output,

high production efficiency. However, plasma method is a promising way for high-throughput, low-cost production of nanomaterials.

Zheng et al. have synthesized N-doped single-walled carbon nanotubes (NCNTs) by using a simple plasma technology,[28] which utilizes SiO_2 as nonmetal catalysts for pyrolysis of CH_4 under NH_3 at 900°C. Typically, a SiO_2/Si wafer with SiO_2 coating was plasma etched to produce SiO_2 nanoparticles as the metal-free catalysts on the substrate. Compared with undoped CNTs, the metal-free NCNTs show relatively good electrocatalytic activity and maintain considerably stable during the oxygen reduction reaction in an acidic medium.

Plasma can tune the intrinsic properties of carbon materials, involving surface hydrophobicity, surface roughness, etc., which substantially effects the electrochemical properties. For example, graphene fiber-based supercapacitors (GFSCs) have attracted considerable attention to power wearable electronics because of the superior properties such as high power density, fast charge–discharge rate, long cycling life, and safe operation conditions. However, the complex preparation process, severe stacking of GNS, and hydrophobicity of graphene fibers are the main factors, which limit the practical application of GFSCs. Meng et al. developed a facile and green plasma method to fabricate GFs (Figure 13.13).[29] By rapidly and uniformly modifying the surface energy and morphology of graphene layers, the plasma can tune the intrinsic properties of carbon materials, which not only change the surface roughness but also introduce additional function groups. As a result, the fabricated GFs possess unique properties such as high specific surface area, tunable pore size distribution, and high hydrophilicity. The plasma-treated GFSCs exhibit ultrahigh rate capability

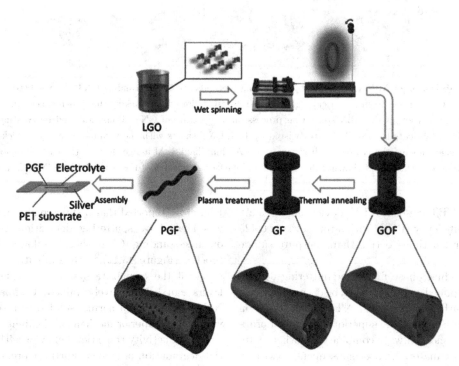

FIGURE 13.13   Schematic diagram showing the preparation method of graphene fibers (GFs) and plasma treatment.

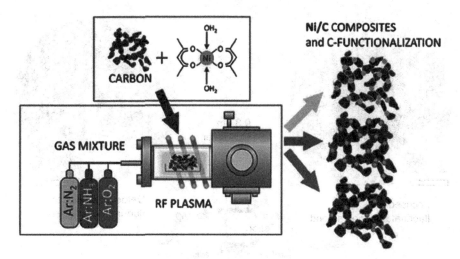

FIGURE 13.14 A sketch illustrating the plasma synthesis process of the Ni/C nanocatalysts.

(69.13% for 40 s plasma treated) and superior cycle stability (96.14% capacitance retention after 20,000 cycles for 1 min plasma treated).

Haye et al. investigated a new method for the synthesis of nickel/carbon nanocatalyst by low-pressure plasma radio-frequency treatment based on solid (powder) precursor (Figure 13.14).[30] During the plasma discharge, the organometallic nickel precursor simultaneously decomposed into nickel NPs and a mesoporous carbon xerogel, which can be used as supporting material for nickel NPs. By varying the plasma power and the composition of gas mixture (Ar, N_2, O_2, and NH_3), different plasma and chemical atmosphere can be realized, which leads to a wide range of size distribution and chemical reactivity of the homogeneously dispersed nickel NPs. Moreover, it proves that the functionalization of carbon substrate is achieved by using the plasma treatment.

Laser Method

With simple experimental setup and varieties of source materials, laser method has been proved to be an essentially simple and clean technique to synthesize desired nanoparticles. The laser with high energy density irradiates on a solid target, producing high density of plasma species at high temperature and high pressure. This highly unequilibrium process (ultrafast chemical reaction and thermal quenching) with partial disorder of atom arrangement and abundant defect states provides the possibility of thermodynamic nucleation and kinetic growth, phase transition, and chemical reaction of the nanomaterials. As a result, the obtained products possess defect-related properties that can hardly be achieved in an ambient environment.

Powell et al. synthesize a hybrid crystalline/amorphous Si nanospheroids with high surface-enhanced Raman scattering (SERS) activity by using an ultrafast femtosecond laser method.[31] Other than individual crystalline and amorphous nanoparticles, with a laser pulse depositing onto the silicon substrate and forming a nanoweb layer, the nanospheroids have both regimes of crystalline and

amorphous silicon. By changing the ionization mechanism and temperature in the laser plume, the content ratio of crystalline Si to amorphous Si in the hybrid silicon nanomaterials and the size distribution can be precisely controlled. These hybrid silicon nanomaterials are considered as an effective chemical sensing substrate for nano-Raman sensing applications.

Ye et al. reported a laser-assisted method for direct synthesis of FND or FG.[32] The functionalized carbon allotropes were obtained just by treating polytetrafloroethylene (Teflon or PTFE) with a 9.3 μm pulsed CO_2 laser under argon atmosphere. After the absorption of 9.3 μm laser, due to the photothermal and photochemical effects, the C−F bonds break and the carbon skeleton reform. As a result, the FND or FG can be obtained (Figure 13.15). Compared with conventional approaches for the fluorination of carbon materials, it avoids the commonly used fluorine sources such as F_2 and XeF_2, which is toxic and expensive.[33] The laser used in this system is commercially accessible, and the method is easy to implement.

The techniques mentioned above are the most common methods for the synthesis and fabrication of nanomaterials, which are effective for controlling the size, morphology, microstructure, and chemical composition of nanomaterials. The properties of nanomaterials obtained by various techniques are closely related to their size, shape, and structure, which is drastically relevant to nucleation and growth of nanocrystals. Therefore, understanding the formation mechanism of nanocrystals is necessary for synthesizing nanomaterials with desired properties. In the next section, the nucleation and growth mechanism of nanomaterials will be systematically elaborated.

13.4 Nucleation and Growth of Nanomaterials

This section includes two parts. The first part is about the formation of nanocrystals nuclei, mainly including

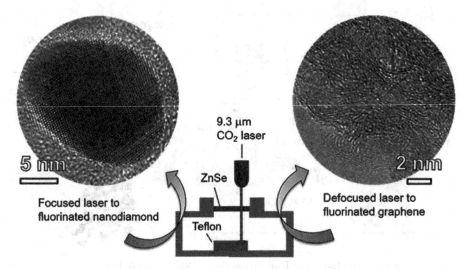

FIGURE 13.15   Reaction scheme for the synthesis of fluorinated nanodiamonds (FND) and fluorinated graphene (FG) from Teflon.

homogeneous nucleation and heterogeneous nucleation. The second part elaborates the growth mechanism of nanocrystals, mostly including layer-by-layer growth mechanism and screw dislocation growth mechanism. The layer-by-layer growth mechanism is applicable in high supercooling or high supersaturation conditions, and the screw dislocation growth mechanism could explain the growth in low supercooling or supersaturation conditions.

13.4.1   Homogeneous Nucleation

Homogeneous nucleation usually occurs spontaneously and randomly. And homogeneous nucleation is much more difficult than heterogeneous nucleation in real thermodynamic systems. When reaction circumstance has met the condition of supersaturation or supercooling, the nuclei are formed by spontaneous random aggregation of atoms or molecules. However, because of the existence of large surface tension, the initial aggregation of molecules or atoms is energetically unfavorable. So it is in unstable state, and only those clusters that grow up to the certain critical size can become stable nuclei. This is a dynamic process. In the process of nucleation, the surface free energy is increased, while the volume free energy is reduced. Therefore, for a spherical drop with a radius r, the total energy change ΔG can be expressed:

$$\Delta G = 4\pi r^2 \sigma - \frac{4}{3}\pi r^3 G_{\mathrm{v}}$$

where ΔG, the total energy change; r, the radius of a spherical nuclei; σ, surface free energy per unit; G_{v}, the volume free energy per unit.

Then, we can obtain the size of a critical nucleus by setting $(d\Delta G/dr) = 0$ to determine the maximum,

$$0 = 8\pi r\sigma - 4\pi r^2 G_{\mathrm{v}} \longrightarrow G_{\mathrm{v}} = \frac{8\pi r\sigma}{4\pi r^2} = \frac{2\sigma}{r}$$

Therefore,

$$\Delta G_c = 4\pi r^2 \sigma - \frac{4}{3}\pi r^3 \frac{2\sigma}{r} = \frac{4}{3}\pi\sigma r_c^2$$

where ΔG_c, the total energy change when the radius is the critical nucleus radius; r_c, the critical radius.

As shown in Figure 13.16, it clearly exhibits us those clusters with radiuses less than the critical radius r_c will tend to lower their energy by moving to the left of curve. This means they fail to be mature nuclei and re-dissolve into the old phase. When the clusters radiuses are beyond r_c, they can naturally grow to reduce their free energy. For those clusters with an initial radius r_c, it can go either to the left or to the right to lower their system free energy.

In the last century, examples of homogeneous nucleation have been rarely discovered. However, with the help of electron microscopy, the phenomenon of homogeneous

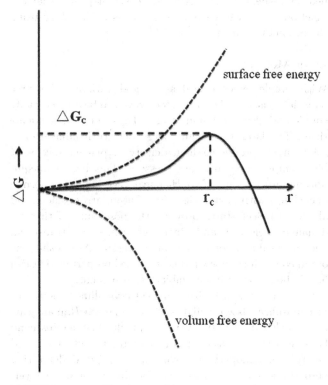

FIGURE 13.16   The change of free energy in nucleation process.

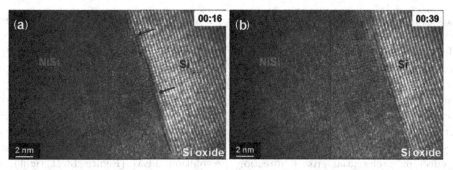

FIGURE 13.17 High-resolution TEM images of motion of steps on epitaxial NiSi/Si interface at 450°C. (a) One NiSi atomic layer grows from the middle region of the interface, and two steps are formed. (b) The motion of the two steps is towards both ends of oxide. After two steps reaches to the oxide ends, the layer growth ends and the interface becomes flat without steps.

nucleation can be observed directly by controlling examination condition carefully. Chou et al. reported homogeneous nucleation in epitaxial growth of CoSi$_2$ and NiSi silicides in the axial direction of Si nanowire.[34] They observed that the growth of every single atomic layer required nucleation. Figure 13.17a,b shows that one NiSi atomic layer nucleates and grows from the middle region of the interface, and spreads towards the both edges of the Si nanowire. This behavior is repeated for every atomic layer growth, which is a direct evidence of homogeneous nucleation of silicide in the epitaxial interface of Si nanowires. The model of heterogeneous nucleation at the edge is depicted in Figure 13.18. If the heterogeneous nucleation occurred, as shown in Figure 13.18a, the low energy oxide/Si interface would be replaced by the high energy oxide/silicide interface. It makes no sense in thermodynamics, so the heterogeneous nucleation is suppressed. The epitaxial interface between Si and silicide is a low energy interface. Homogeneous nucleation of a circular disk in the center of nanowire becomes possible, and the cross section is depicted in Figure 13.18c.

Another classical example is the homogeneous nucleation of ice. Water is the source of life; meantime, it is the most important and common liquid. In general, water freezes due to the impurities through heterogeneous nucleation. Homogeneous nucleation of ice under atmospheric conditions remains a challenge. Nowadays, by confining water to micrometer-to-nanometer scales, researchers can limit heterogeneous nucleation. Suzuki et al. reported that the nucleation mechanism of water could be precisely regulated by confinement in nanoporous alumina.[35] Their results show that the nucleation mechanism could be precisely regulated by confining water within self-ordered nanoporous aluminum oxide (AAO) with pore diameters ranging from 400 to 25 nm. Homogeneous nucleation is the dominant mechanism at lower temperatures in the smaller pores. Yao et al. also found similar results.[36] In their research, the multiple length scales involved in hollow silica (HS) spheres affect the ice nucleation mechanism. They found homogeneous nucleation inside the water-filled capsules, whereas heterogeneous nucleation prevailed in the surrounding dispersion medium.

Outer conditions are always the most important factors that limit homogeneous nucleation to happen. It can be seen that the outer conditions required for homogeneous nucleation are too strict. The external factors, such as electric and magnetic field, ultraviolet light, X-rays, and sonic and ultrasonic radiation, could have a huge impact on homogeneous nucleation. Therefore, in real practice, homogeneous nucleation is not the first choice. The best condition to prepare nanocrystals derived from homogeneous nucleation is to seed the small particles of nanomaterial to be crystallized in the supersaturated solution.

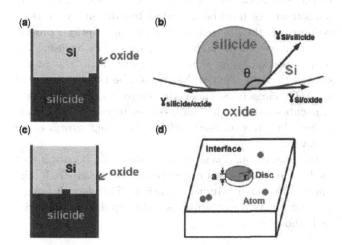

FIGURE 13.18 Schematic diagrams of the cross section of the epitaxial interface between silicide and Si. (a) Cross section of heterogeneous nucleation of a step at the right-hand side corner. To form the step, the oxide/Si interface will be replaced by the oxide/silicide interface, which is energetically unfavorable. (b) Triple point configuration of a heterogeneous nucleus. (c) Cross section of homogeneous nucleation of a disk in the center of the silicide/Si interface. (d) Schematic diagram of the nucleation of a circular disk on the interface.

13.4.2 Heterogeneous Nucleation

In above section, we have known that the formation of a nucleus requires overcoming the high-energy barrier under the thermodynamically stable condition. But, the high-energy barrier can be bypassed by heterogeneous nucleation because crystal seeds can serve as basement for molecular or atoms assemble. In essence, this is similar to the catalyst of crystal nucleation. Therefore, heterogeneous nucleation can happen more easily than homogeneous nucleation (Figure 13.19).

It is worthy to note that heterogeneous nucleation not only promotes the crystallization process, but also controls morphology and structures of crystals. This is due to the fact that seed crystals serve as the basis of the nucleus, which can greatly effect on preferential growth direction of nanocrystal, lattice parameters, and exposed crystal planes. These factors also determine the crystal structure, morphology, and related properties of nanomaterials. Many works have been made to prove this viewpoint.

Song et al. used a solvothermal method to synthesize NiO hierarchical architectures with controllable morphologies and sizes by controlling the nucleation rate of product in the solution.[37] They found that heterogeneous nucleation occurred in the low-level degree of supersaturation, while the homogeneous nucleation occurred in the high-level degree of supersaturation. And the resultant products showed extremely different morphologies and high photocatalytical performance, as shown in Figure 13.20.

WO_3 is an important n-type semiconductor. Miyauchi et al. reported that WO_3 nanoparticles were segregated on

the surface of p-type $CaFe_2O_4$(CFO) particles by a heterogeneous nucleation process under controlled hydrothermal conditions.[38] They prepared CFO/WO_3 by hydrothermal and mechanical ball milling, respectively, and named CFO/WO_3(H) and CFO/WO_3(M). In Figure 13.21, the WO_3 particles are abundantly attached and distributed on the surface of CFO particles in CFO/WO_3(H), while most WO_3 particles formed aggregates in CFO/WO_3(M). Furthermore, crystal lattice fringes were observed by high-resolution TEM (Figure 13.21 right), and the distances matched well with the (101) and (100) planes of WO_3 and the (210) plane of CFO. These results indicate that highly crystallized WO_3 particles are attached on the surface of CFO crystals.

Many researches on nanocrystals synthesis have shown how the kinetics and thermodynamics of nucleation and growth of nanocrystals in solution can be affected by the presence of outer seeds.[39] CdSe/CdS nanocrystals are important colloidal nanocrystals which have enormous application in many fields, such as optics, PVs, nanoelectronics, and biosensing. Kim et al. reported the influence of chloride ions (Cl^-) on the seeded growth synthesis of colloidal branched CdSe(core)/CdS(pods) nanocrystals.[40] First, they hoped to set an optimal concentration of $CdCl_2$ to keep stable growth of octapods. However, they found that at higher concentration of $CdCl_2$ added, octapods were initially formed. Afterwards, many of them subsequently evolved into tetrapods over time. This transformation demonstrates an additional role of Cl^- species in regulating the growth rate and stability of various crystal facets of the CdS pods, as shown in Figure 13.22.

Another classical example for heterogeneous nucleation is the formation of Ag nanowires. One-dimensional Ag nanostructures have been studied broadly in recent years because of their excellent electric and optical properties. Schuette et al. prepared Ag nanowires by using the polyol method.[41] As shown in Figure 13.23, the results showed that the additive NaCl could be rapidly converted to AgCl nanocubes under the polyol media. The AgCl nanocubes subsequently induced heterogeneous nucleation of metallic silver on their surfaces, initiating growth of Ag nanowires.

By comparison, it is generally evident that heterogeneous nucleation happens easily in real events of nucleation. It does not require harsh external conditions. Therefore, heterogeneous nucleation mechanism is widely applied in industries and laboratories.

13.4.3　Growth of Nuclei

It is well known that one-dimensional nanomaterials (nanowires, nanorods, and nanotubes, etc.) and two-dimensional nanomaterials (nanoplates, nanosheets, and nanoribbon, etc.) possess interesting physical and chemical properties due to their unique structures. Controlling the growth of nanomaterials is important for their synthesis and practical application.[42]

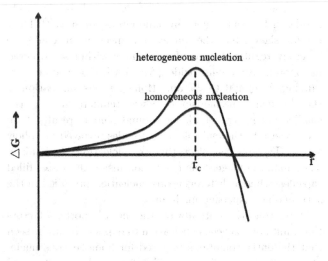

FIGURE 13.19　Comparison of energy barriers for homogeneous and heterogeneous nucleation.

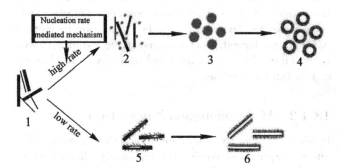

FIGURE 13.20　One-pot preparation of hollow hierarchical NiO superstructures.

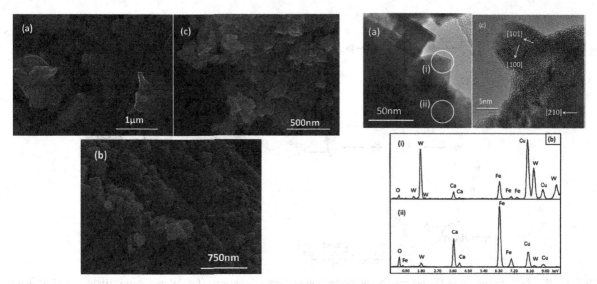

FIGURE 13.21 Left: SEM images of CFO (a), CFO/WO$_3$(H) (b), and CFO/WO$_3$(M) (c). The weight percentage of CFO in CFO/WO$_3$(H) and CFO/WO$_3$(M) was 10%. Right: TEM images for CFO/WO$_3$(H) (a), EDX point analysis for points i and ii in panel a (b), and high-resolution TEM image of CFO/WO$_3$(H) (c). The weight percentage of CFO in CFO/WO$_3$(H) was 10%.

FIGURE 13.22 Typical evolution process of the CdSe/CdS multipod shape, when large molar fractions of CdCl$_2$ are introduced in the synthesis of core-shell CdSe/CdS nanocrystals.

Then, here comes a question: How do nanocrystals form and grow from nuclei? An understanding about this question is important for the development of synthetic methods. For nanocrystals, the growth processes are mainly controlled by the properties of the precursor and reaction conditions. Unfortunately, although many researchers and scientists have made great efforts, the real mechanism involved in these processes is still not entirely clear. In the following section, we will lend concepts in classical crystallography to illustrate the growth mechanism of nanocrystals by several specific examples.

In classical crystal-growth theory, two basic growth modes are distinguished in terms of the supersaturation conditions.[43] At high supersaturation, the layer-by-layer growth and dendritic growth are able to happen. In contrast, at very low supersaturation, screw dislocation-driven (SDD) growth is disclosed, and the resultant products contain novel properties due to their anisotropic growth, which has attracted a lot of attention recently.[44]

Growth at High Supersaturation

Layer-by-layer growth mechanism is an earlier crystal-growth theory. In this way, a nucleus forms on the perfect crystal faces, and it absorbs fresh materials and expands to a new layer. Such as the growth of the one-dimensional Si nanowire is based on the layer-by-layer growth mechanism. However, this growth is difficult to occur at the initial stage of each layer formation. It needs to overcome higher energy barrier, so high degree of supersaturation is needed to provide the driving force of reaction.

Zhang et al. investigated the formation procedure of γ-Na$_x$Co$_2$O$_4$ crystals by XRD, field-emitted scanning electron microscopy (FESEM), and high-resolution transmission microscopy (HREM). The flaky hexagonal crystals (Figure 13.24a) are observed, where they are stacked layer-by-layer along the perpendicular direction to the flaky hexagonal plane. The layer thickness (Figure 13.24c) is about 15–30 nm.[45]

The specific growth process is described in Figure 13.25. The network structure colloidal precursor is formed by ultrasound irradiation and composed of bovine serum albumin (BSA), cobalt, and sodium sources due to the linking of BSA molecules. Through a freeze-drying process, the oriented layered structure of precursor is obtained. At 800°C, with the help of an oriented layered structure, the γ-Na$_x$Co$_2$O$_4$ hexagonal crystals are assembled by hexagonal nanosheets along the perpendicular direction to flaky hexagonal plane, which exhibits the layer-by-layer growth mechanism is predominant during the crystals' growth.

Liao et al. used a liquid cell reactor which can be placed in a special TEM sample holder to observe colloidal nanocrystal growth in real time.[46] In fact, there have been two main mechanisms to explain the formation of the five-twinned nanoparticles. The layer-by-layer mechanism suggests that the small nanocrystals initially nucleate by a non-crystallographic atom packing process, and afterwards, they are transformed to a five-twinned structure through layer-by-layer growth. This

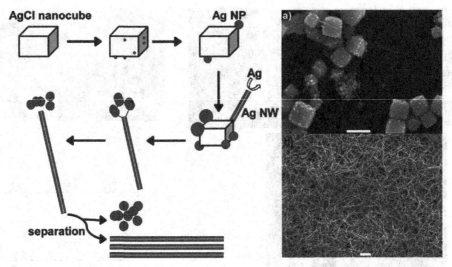

FIGURE 13.23   Left: Depiction of the heterogeneous nucleation and growth pathway for Ag nanowires. Right: (a) SEM image taken of an aliquot from a reaction mixture employing presynthesized AgCl nanocubes. The aliquot was removed 23 min after the start of dropwise addition of AgNO$_3$ solution. The scale bar is 1 µm. (b) SEM image of purified AgNWs deriving from AgCl nanocubes.

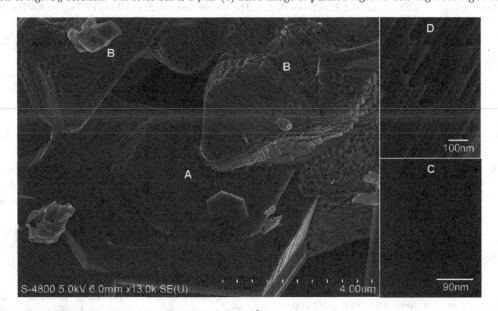

FIGURE 13.24   (A–D) FESEM images of sample obtained at 800°C.

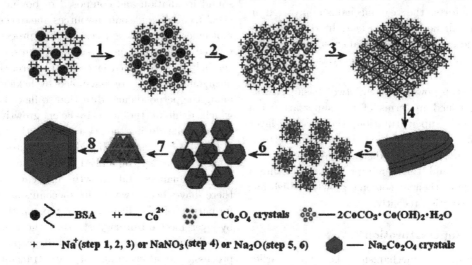

FIGURE 13.25   Growth schematic of layer-by-layer nanostructured γ-Na$_x$Co$_2$O$_4$ hexagonal crystals.

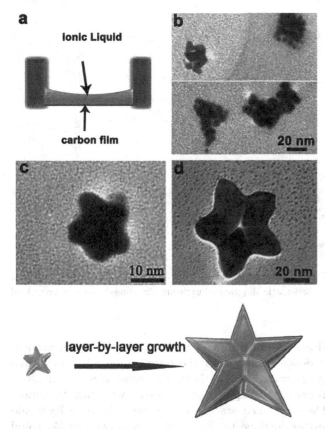

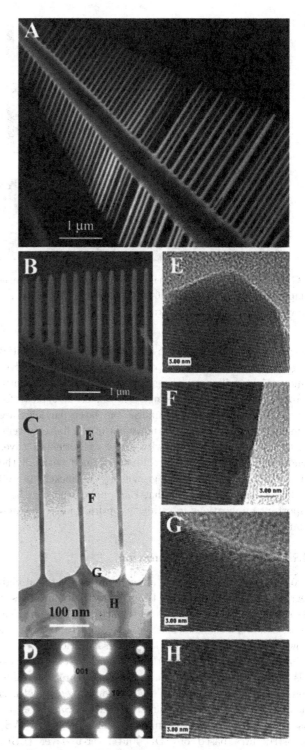

FIGURE 13.26 (a) Schematic of mini-pool and thin layer of ionic liquid, (b, c) small nanoparticles aggregates and fivefold twinned gold seed in solution after the reaction start for 20 min, and (d) bigger particle found at 1.5 h after the reaction started.

inference is approved firmly by their observation. After 20 min of the reaction, many aggregated small particles are evidently found (Figure 13.26b), and small five-fold twinned gold seeds (Figure 13.26c), which already have a preliminary star shape, also are observed. Figure 13.26d provides a typical particle found in the solution after 1.5h. The results clearly confirm that layer-by-layer growth contributes to the growth of star-shaped crystals.

Another common growth mode at high supersaturation degree is called as dendritic growth. The products have a tree-like branching structure. For the unique structure, the formation of branches usually depends on the supersaturation of solution.[47] Yan et al. synthesized comb-like structures made of periodic arrays of single-crystalline ZnO nanowires, as shown in Figure 13.27. The samples were synthesized within a chemical vapor transport and condensation system to ensure that the supersaturation degree meets the requirements. The evaporation of zinc metal induced high supersaturation level and fast condensation in an oxygen environment. This work is a representative example of dendritic growth mechanism.[48]

Mandke et al. reported that silver dendritic nanostructures could be prepared by electrochemical deposition from an aqueous solution of $AgNO_3$ in the presence of citric acid.[49] The left part of Figure 13.28 showed the

FIGURE 13.27 (A,B) Scanning electron microscope (SEM) images of comb structures made of ZnO nanowires. (C) Transmission electron microscope (TEM) image of an electron-transparent comb structure. (D) Electron diffraction pattern recorded on the entire comb structure in (C). (E–H) High-resolution TEM images recorded in different regions of the comb structure in (C).

high magnification of silver dendritic structures with time. It could be easily seen that there is a gradual increase in the number of dendritic structures with deposition time. And the right part of Figure 13.28 explained the growth

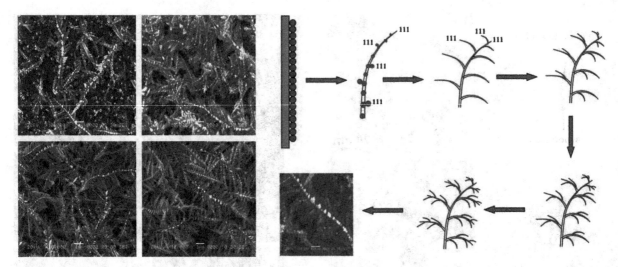

FIGURE 13.28   Left: High-magnification ($\times 10{,}000$) SEM images of as-prepared silver dendritic nanostructures for sample A – 100 s, sample B – 300 s, sample C – 600 s, and sample D – 1,200 s. Right: Schematic diagram illustrating the time-dependent growth of individual silver dendritic nanostructures.

process of the silver dendritic nanostructures. At the initial stage, Ag nanoparticles form on the substrate. Then, silver nanoparticles will grow along the (111) direction to form a rod-like silver trunk because the silver is typical fcc metal. As the reaction continues, the later particles are launched on a silver trunk or move out of the finite system. As the reaction proceeded with time, well-oriented silver dendritic nanostructures were finally formed with all the trunks, branches, and leaves grown bigger, thicker, as well as denser. It is remarkable that these nanostructures have important applications in catalysis, optics, bioscience, and solar cells.

Growth at Low Supersaturation

Two growth modes mentioned above can be activated at high degree of supersaturation. But the growth of nanomaterials at low supersaturation also plays an important role in practice. Thus, a new mechanism must be exploited to explain the growth of crystals under the low supersaturation.

In most cases, the growth environment of nanocrystals is difficult to satisfy the high supersaturation condition. A new mechanism, i.e., dislocation-driven growth, is then proposed under low supersaturation. Sarma et al. discussed the growth mechanism of these pyramidal WS_2 nanostructures based on SDD and layer-by-layer (LBL) growth models.[50] Figure 13.29a,b shows the AFM image and the corresponding height profile for a thick WS_2 flake with an LBL stacked configuration, respectively. WS_2 monolayer thickness is ~ 0.86 nm. Figure 13.29c and d shows a zoomed AFM image of a WS_2 flake with a spiral stacking morphology and its corresponding height profile with a step height of ~ 0.95 nm, respectively. Both spiral stacking structures resulted from a dislocation-driven growth of WS_2 during CVD process. This is obtained by controlling the initial nucleation rates, which generates defects and dislocation in the basal

plane, leading to pyramid-shaped stacked configuration. A detailed analysis of the growth mechanism was given by Figure 13.30. Initially, the basal plane is uplifted. Towards the exposed edges, atoms will start to diffuse. The layer then starts to grow over the base layer and further continue to grow ending up in pyramid-like spiral structure.

Similarly, Liu et al. also demonstrated that the SDD growth mode could be tuned to the LBL growth mode by increasing the precursor concentration.[51] In Figure 13.31, the spirals are composed of a few connected SnSe layers with decreasing areas; each basal plane has a rectangular shape and spirals up by layer by layer to the summit center, just like pyramids.

Others

In addition to the mechanisms mentioned above, there are still other broadly acceptable mechanisms. These mechanisms are also important to the synthesis of nanomaterials. Therefore, they need to be listed and explained by examples.

For the growth of 1D nanomaterials, the catalyst-driven mechanisms have been substantially established, including vapor–liquid–solid (VLS), solution–liquid–solid (SLS), and vapor–solid–solid (VSS) growth. They are commonly used for the synthesis of 1D nanomaterials. However, some scholars propose that the catalyst-driven mechanisms still are LBL growth, but the nucleation barrier for LBL growth is lowered due to the presence of catalyst. Therefore, the crystal growth is much faster at this interface. Sang et al. significantly observed a different behavior for self-catalytic VLS growth of sapphire nanowires by in situ transmission electron microscopy. The growth occurs in a layer-by-layer fashion, which is similar to the layer-by-layer growth.[52] Apart from mechanisms mentioned above, there are still some uncommon mechanisms on the growth of

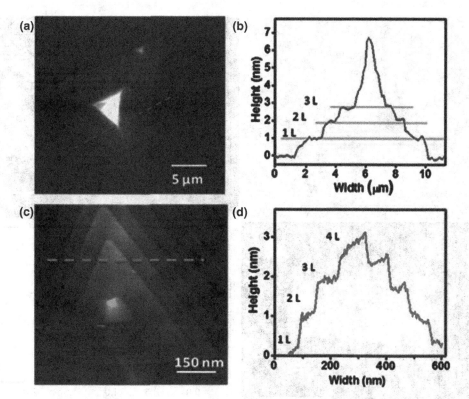

FIGURE 13.29 AFM analysis of WS$_2$ structures with LBL and spiral stacking morphologies. (a) AFM image and (b) AFM height profile of a LBL stacked pyramidal WS$_2$. Multiple layers stack one over other forming a pyramid with a step height of $\sim$ 1 nm and a terrace width of $\sim$ 64 nm. The pyramid-like WS$_2$ has about seven to eight layers. (c) AFM image of a spiral few-layer WS$_2$ structure and the corresponding (d) height profile along the marked line. AFM height profile clearly shows step heights of $\sim$ 1 nm and a terrace width of $\sim$ 50 nm.

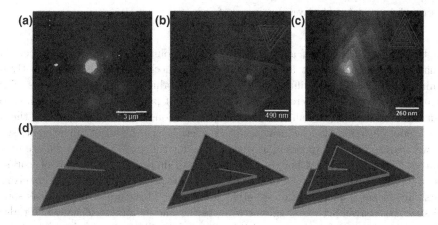

FIGURE 13.30 Growth mechanism of spiral WS$_2$. (a) AFM image of a large WS$_2$ domain with multi-centered dislocation-driven growth. (b) AFM image of a multi-centered stacked WS$_2$ showing both left- and right-handed spirals. Inset shows the schematic of the multi-centered growth. (c) AFM image of WS$_2$ domains with alternately stacked spirals having the same plateau region. Inset shows the growth model. (d) Schematic illustration of SDD growth.

nanomaterials. Taking the work reported by Song et al., for instance, they synthesized size engineerable NiS$_2$ hollow spheres using a facile hydrothermal process. Their results represent a morphology transform from rod-like species into hollow sphere, involving a synergic effect of the Rayleigh instability and the Kirkendall effect. These novel theories also pave a new way for the synthesis of nanocrystals with special configuration.[53]

13.5 Conclusion

In the field of nanotechnology, synthesis and fabrication of nanomaterials and nanostructures are the essential aspects. Controlling synthesis of nanomaterials with desired size, morphology, microstructure, and chemical composition is the basis for the studies on properties and applications of nanomaterials. In the last decade, considerable effort

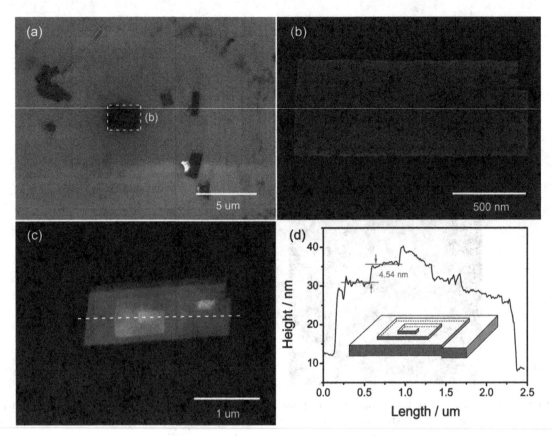

FIGURE 13.31 (a) Low-magnification SEM image of SnSe nanoplates obtained from a standard synthetic procedure. (b) High-magnification SEM image of a representative layered spiral-type SnSe nanoplate shown in (a) with a dashed square. (c) AFM amplitude image of the nanoplate. (d) AFM height profile of the nanoplate along the dashed white line in (c). The inset is a schematic representation of the incorporation of screw dislocation into the layered structure of SnSe.

has been focused on the design, fabrication, and growth mechanism of nanomaterials using abundant approaches, publishing overwhelming literatures across different disciplines. In the first part of this chapter, different fabrication methods for the nanostructures are generally discussed and briefly categorized into physical and chemical techniques. Furthermore, we analyze the corresponding advantages and disadvantages for each one. Although great progresses for the synthesis and fabrication of nanomaterials were made, its large-scale implementation is still plagued by several issues: (i) coarsening and agglomeration of nanomaterials caused by huge surface energy and (ii) the scale-up synthesis of nanomaterials with homogeneous size distribution, morphology, crystallinity, chemical composition, and microstructure. To tackle the aforementioned problems, many efforts are still required to exploit novel synthesis approaches of nanomaterials, which would expedite the significant development of next-generation nanomaterials.

This chapter also illustrates the nuclei and growth mechanism of nanomaterials. For nanomaterials, they have promising potential for practical application because of their unique morphologies and structures. Therefore, controlling the nuclei and growth processes has been the pinpoint of research. Meantime, the understanding of growth mechanism is also significantly helpful to the large-scale production of nanomaterials. However, we can see that most present researches only involve the synthesis and preparation, and the discussion on the mechanism of nanomaterials is insufficient. In present, some of the explanations for the growth mechanism are reasonable, while some are still doubtful. Thus, a lot of works need to be carried out to exploit the intrinsic mechanisms for various nanomaterials. This chapter mainly lists some typical examples. As the understanding of nanomaterials evolved, the list of examples will continue to be expanded in the future.

Acknowledgments

This work was supported by the Shanghai Municipal Natural Science Foundation (17ZR1414900), the Opening Project of State Key Laboratory of High Performance Ceramics and Superfine Microstructure (SKL201604SIC), the Shanghai Jiao Tong University Medical (Science) Cross Fund (YG2017MS02), the Shanghai Aerospace Science and Technology Innovation Foundation Project, and the National Natural Science Foundation of China (51302169).

References

1. Bhat H.L. *Introduction to Crystal Growth: Principles and Practice.* CRC Press, Boca Raton, FL, 2014.

2. Thanh N.T., Maclean N., Mahiddine S. Mechanisms of nucleation and growth of nanoparticles in solution. *Chemical Review*, 2014, 114: 7610–7630.

3. Nancollas G.H., Purdie N. The kinetics of crystal growth. *Quarterly Reviews*, 1964, 18: 1–20.

4. Morin S.A., Bierman M.J., Tong J., Jin S. Mechanism and kinetics of spontaneous nanotube growth driven by screw dislocations. *Science*, 2010, 328: 476–480.

5. Zhao Q., Xu Z., Yue H., Feng D., Jin Z. Chemical vapor deposition synthesis of near-zigzag single-walled carbon nanotubes with stable tube-catalyst interface. *Science Advances*, 2016, 2: e1501729.

6. Song X. A hierarchical hybrid design for high performance tin based Li-ion battery anodes. *Nanotechnology*, 2013, 24: 205401.

7. Ma T., Yu X., Li H., Zhang W., Cheng X., Zhu W., Qiu X. High volumetric capacity of hollow structured $SnO_2@Si$ nanospheres for lithium-ion batteries. *Nano Letters*, 2017, 17: 3959.

8. Pang J., Mendes R.G., Wrobel P.S., Wlodarski M.D., Ta H.Q., Zhao L., Giebeler L., Trzebicka B., Gemmeng T., Fu L., Liu Z., Eckert J., Bachmatiuk A., Rummeli M.H. Self-terminating confinement approach for large-area uniform monolayer graphene directly over Si/SiO_x by chemical vapor deposition. *ACS Nano*, 2017, 11: 1946–1956.

9. Shi W., Song S., Zhang H. Hydrothermal synthetic strategies of inorganic semiconducting nanostructures. *Chemical Society Reviews*, 2013, 42: 5714–5743.

10. Chen C., Wang H., Han C., Deng J., Wang J., Li M., Tang M., Jin H., Wang Y. Asymmetric flask-like hollow carbonaceous nanoparticles fabricated by the synergistic interaction between soft template and biomass. *Journal of the American Chemical Society*, 2017, 139: 2657–2663.

11. Luo D., Deng Y.P., Wang X., Li G., Wu J., Fu J., Lei W., Liang R., Liu Y., Ding Y. Tuning shell numbers of transition metal oxide hollow microspheres towards durable and superior lithium storage. *ACS Nano*, 2017, 11: 11521–11530.

12. Miao J., Xiao F.X., Yang H.B., SI Y.K., Chen J., Fan Z., Hsu Y.Y., Chen H.M., Zhang H., Liu B. Hierarchical Ni-Mo-S nanosheets on carbon fiber cloth: A flexible electrode for efficient hydrogen generation in neutral electrolyte. *Science Advances*, 2015, 1: e1500259.

13. Wang R., Han M., Zhao Q., Ren Z., Guo X., Xu C., Hu N., Li L. Hydrothermal synthesis of nanostructured graphene/polyaniline composites as high-capacitance electrode materials for supercapacitors. *Scientific Reports*, 2017, 7: 44562.

14. Liu J., Zheng Y., Hong Z., Cai K., Zhao F., Han H. Microbial synthesis of highly dispersed PdAu alloy for enhanced electrocatalysis. *Science Advances*, 2016, 2: e1600858.

15. Liu X., Tan J., Fu J., Yuan R., Wen H., Zhang C. Facile synthesis of nanosized lithium-ion-conducting solid electrolyte $Li_{1.4}Al_{0.4}Ti_{1.6}(PO_4)_3$ and its mechanical nanocomposites with $LiMn_2O_4$ for enhanced cyclic performance in lithium ion batteries. *ACS Applied Materials and Interfaces*, 2017, 9: 11696–11703.

16. Miao R., Dutta B., Sahoo S., He J., Zhong W., Cetegen S.A., Jiang T., Alply S.P., Suib S.L. Mesoporous iron sulfide for highly efficient electrocatalytic hydrogen evolution. *Journal of the American Chemical Society*, 2017, 139: 13604–13607.

17. Wang M., Li Z., Wang C., Zhao R., Li C., Guo D., Zhang L., Yin L. Novel core-shell $FeOF/Ni(OH)_2$ hierarchical nanostructure for all-solid-state flexible supercapacitors with enhanced performance. *Advanced Functional Materials*, 2017, 27, 1701014.

18. Veldhuis S.A., Tay Y.K.E., Bruno A., Dintakurti S.S.H., Bhaumik S., Muduli S.K., Li M., Mathews N., Sum T.C., Mhaisalkar S.G. Benzyl alcohol-treated $CH_3NH_3PbBr_3$ nanocrystals exhibiting high luminescence, stability, and ultralow amplified spontaneous emission thresholds. *Nano Letter*, 2017, 17: 7424–7432.

19. Heintz A.S., Fink M.J., Mitchell B.S. Mechanochemical synthesis of blue luminescent alkyl/alkenyl-passivated silicon nanoparticles. *Advanced Materials*, 2007, 19: 3984–3988.

20. Zhu X., Zhang T., Sun Z., Chen H., Guan J., Chen X., Ji H., Du P., Yang S. Black phosphorus revisited: A missing metal-free elemental photocatalyst for visible light hydrogen evolution. *Advanced Materials*, 2017, 29: 1605776.

21. Wang Y., Fedin I., Zhang H., Talapin D.V. Direct optical lithography of functional inorganic nanomaterials. *Science*, 2017, 357: 385–388.

22. Striccoli M. Photolithography based on nanocrystals. *Science*, 2017, 357(6349): 353.

23. Hondred J.A., Stromberg L.R., Mosher C.L., Claussen J.C. High-resolution graphene films for electrochemical sensing via inkjet maskless lithography. *ACS Nano*, 2017, 11: 9836–9845.

24. Zhou L., Tan Y., Ji D., Zhu B., Zhang P., Xu J., Gan Q., Yu Z., Zhu J. Self-assembly of highly efficient, broadband plasmonic absorbers for solar steam generation. *Science Advances*, 2016, 2: e1501227.

25. Qin J., Qiu G., Jian J., Zhou H., Yang L., Charnas A., Zemlyannov D.Y., Xu C.Y., Xu X., Wu W., Wang H., Ye P.D. Controlled growth of a large-size

2D selenium nanosheet and its electronic and optoelectronic applications. *ACS Nano*, 2017, 11: 10222–10229.

26. Shin D., Zhu T., Huang X., Gunawan O., Blum V., Mitzi D.B. Earth-abundant chalcogenide photovoltaic devices with over 5% efficiency based on a $Cu_2 BaSn(S,Se)_4$ absorber. *Advanced Materials*, 2017, 29: 1601945.

27. Kim J.G., Kang H., Lee Y., Park J., Kim J., Truong T.K., Kim E.S., Yoon D.H., Lee Y.H., Suh D. Carbon-nanotube-templated, sputter-deposited, flexible superconducting NbN nanowire yarns. *Advanced Functional Materials*, 2017, 27: 1701108.

28. Zheng J., Yang R., Xie L., Qu J., Liu Y., Li X. Plasma-assisted approaches in inorganic nanostructure fabrication. *Advanced Materials*, 2010, 22: 1451–1473.

29. Meng J., Nie W., Zhang K., Xu F., Ding X., Wang S., Qiu Y. Enhancing electrochemical performance of graphene fiber-based supercapacitors by plasma treatment. *ACS Applied Materials and Interfaces*, 2018, 10: 13652–13659.

30. Haye E., Yan B., Pires M.D.S., Bocchese F., Job N., Houssiau L., Pireaux J.J. Low-pressure plasma synthesis of Ni/C nanocatalysts from solid precursors: Influence of the plasma chemistry on the morphology and chemical state. *ACS Applied Nano Material*, 2017, 1: 265–273.

31. Powell J.A., Venkatakrishnan K., Tan B. Programmable SERS active substrates for chemical and biosensing applications using amorphous/crystalline hybrid silicon nanomaterial. *Scientific Reports*, 2016, 6: 19663.

32. Ye R., Han X., Kosynkin D.V., Li Y., Zhang C., Jiang B., Marti A.A., Tour J.M. Laser-induced conversion of Teflon into fluorinated nanodiamonds or fluorinated graphene. *ACS Nano*, 2018, 12: 1083–1088.

33. Robinson J.T., Burgess J.S., Junkermeier C.E., Badescu S.C., Reinecke T.L., Perkins F.K., Zalautdniov M.K., Baldwin J.W., Culbertson J.C., Shenhan P.E. Properties of fluorinated graphene films. *Nano Letters*, 2010, 10: 3001–3005.

34. Chou Y., Wu W., Chen L., Tu K. Homogeneous nucleation of epitaxial $CoSi_2$ and NiSi in Si nanowires. *Nano Letters*, 2009, 9: 2337–2342.

35. Suzuki Y., Duran H., Steinhart M., Kappl M., Butt H.J., Floidas G. Homogeneous nucleation of predominantly cubic ice confined in nanoporous alumina. *Nano Letters*, 2015, 15: 1987–1992.

36. Yao Y., Ruckseschel P., Graf R., Butt H.J., Retsch M., Floudas G. Homogeneous nucleation of ice confined in hollow silica spheres. *The Journal of Physical Chemistry B*, 2017, 121: 306–313.

37. Song X., Gao L. Facile synthesis and hierarchical assembly of hollow nickel oxide architectures bearing. *The Journal of Physical Chemistry C*, 2008, 112: 15299–15305.

38. Miyauchi M., Nukui Y., Atarashi D., Sakai E. Selective growth of N-type nanoparticles on P-type semiconductors for Z-scheme photocatalysis. *ACS Applied Materials Interfaces*, 2013, 5: 9770–9776.

39. Wang F., Tang R., William E.B. The trouble with TOPO; identification of adventitious impurities beneficial to the growth of cadmium selenide quantum dots, rods, and wires. *Nano Letters*, 2008, 8: 3521–3524.

40. Kim M., Karol M., Mauro P., Rosaria B., Sotirios C., Mirko P., Sergio M., Liberato M. Influence of chloride ions on the synthesis of colloidal branched CdSe/CdS nanocrystals by seeded growth. *ACS Nano*, 2012, 6: 11088–11096.

41. Waynie M.S., William E.B. Silver chloride as a heterogeneous nucleant for the growth of silver nanowires. *ACS Nano*, 2013, 7: 3844–3853.

42. Zhang L. *Controlled Growth of Nanomaterials*. World Scientific Publishing Co., Inc., Hackensack, NJ, 2007.

43. Lv W., He W., Wang X., Niu Y., Cao H., Dickerson J.H., Wang Z. Understanding the oriented-attachment growth of nanocrystals from an energy point of view: A review. *Nanoscale*, 2014, 6: 2531–2547.

44. Meng F., Stephen A.M., Audrey F., Jin S.J. Screw dislocation driven growth of nanomaterials. *Accounts of Chemical Research*, 2013, 46: 1616–1626.

45. Zhang L., Tang X., Gao W. Growth mechanism of layer-by-layer nanostructured γ-$Na_xCo_2O_4$ hexagonal crystals via a novel protein adsorption method. *Crystal Growth and Design*, 2008, 8: 2489–2492.

46. Liao H.G., Shao Y., Wang C., Lin Y., Jiang Y.X., Sun S.G. TEM study of fivefold twined gold nanocrystal formation mechanism. *Materials Letters*, 2014, 116: 299–303.

47. Langer J.S. Theory of dendritic growth: I. Elements of a stability analysis. In: Pierre P., editor. *Dynamics of Curved Fronts*. San Diego: Academic Press; 1988, pp. 281–287.

48. Yan H., He R., Justin J., Matthew L.J., Richard J.S., Yang P. Dendritic nanowire ultraviolet laser array. *Journal of the American Chemical Society*, 2003, 125: 4728–4729.

49. Mandke M.V., Han S.H., Pathan H.M. Growth of silver dendritic nanostructures via electrochemical route. *CrystEngComm*, 2012, 14: 86–89.

50. Sarma P.V., Patil P.D., Barman P.K., Kini R.N., Shaijumon M.M. Controllable growth of few-layer spiral WS_2. *RSC Advances*, 2016, 6: 376–382.

51. Liu J., Huang Q., Qian Y., Huang Z., Lai F., Lin L., Guo M., Zheng W., Qu Y. Screw dislocation-driven growth of the layered spiral-type SnSe nanoplates. *Crystal Growth and Design*, 2016, 16: 2052–2056.

52. Sang H.O., Matthew F.C., Yaron K., Wayne D.K., Luo W., Manfred R., Christina S. Oscillatory mass transport in vapor-liquid-solid growth of sapphire nanowires. *Science*, 2010, 330: 489–493.

53. Song X., Shen W., Sun Z., Yang C., Zhang P., Gao L. Size-engineerable NiS_2 hollow spheres photo co-catalysts from supermolecular precursor for H_2 production from water splitting. *Chemical Engineering Journal*, 2016, 290: 74–81.

Inkjet Printing of Catalytic Materials

Petros G. Savva and
Costas N. Costa
Cyprus University of Technology

14.1 Introduction

During the last four decades, inkjet printing has become one of the most prevalent printing technologies for home and office desktop printing applications. Advances in inkjet technology allow the printing of full-color, high-resolution photographs of equivalent quality to those from other printing techniques that have prevailed in the recent past. In the commercial printing sector, inkjet technology is being widely used for digital proofing prior to running a print job on a press; for short-run, wide-format digital printing such as posters for outdoor advertising; and for applications that require printing onto non-paper substrates such as rigid display boards. In addition, inkjet printers are being developed for integration with offset presses to print customized information in magazines, such as tailored advertising (Sirringhaus & Shimoda, 2003). Finally, in the general industrial sector, the inkjet technology is considered as the leading technology for printing customized information such as "sell-by" dates and product identification codes, as part of packaging or production processes.

During the last few years, apart from the "typical" printing application of the inkjet technology, a growing number of novel applications have shown up, which require the delivery of small amounts of materials, of variable nature, with specific chemical, biological, electrical, optical, or structural functionalities onto well-defined/confined locations on different types of substrates. Most of these new applications require that such functional materials are processed from a liquid solution, dispersion, or melt. Moreover, many functional materials, such as polymers or large biomolecules, are not consistent with vacuum deposition techniques. In some cases, the requirement for processing from a solution may also be defined by the nature and

properties of the substrate, the need to distribute the materials over a large substrate area, or only to certain locations of the substrate and not to others, or simply to keep the material in a liquid environment at all times, such as for some biological applications (Liu et al., 2015).

This manuscript attempts to summarize the recent advances in the inkjet printing technology and its novel applications, with a particular emphasis on the research conducted, up-to-date, regarding the printing of solid catalytic materials. A detailed report on the specifics of the printing technology itself is presented within the second section (Chapter 2) of the present manuscript, followed by a thorough review of the recent advances in the inkjet printing of catalytic materials (Chapter 3). Finally, the challenges and possible future developments in inkjet printing of catalytic materials are summarized in Chapter 4.

14.2 Basics of Inkjet Printing

Inkjet printing is a type of computerized printing that creates a "hard-copy" image, from a digital version, by driving droplets of ink onto a specific area of a substrate by producing drops from a tank with minimal human involvement (Ko et al., 2007). Practically, there is no limitation as to the type of the substrate. The substrate can be either flexible or rigid. The concept of inkjet printing originated in the 20th century, and the technology was first extensively developed in the early 1950s. This technique has recently attracted increasing interest as a flexible, cost-effective, and reliable method for micro- and nano-fabrication, with plentiful practical applications (Haverinen et al., 2010). Despite its many advantages, the primary cause of inkjet printing problems is considered to be ink drying on the printhead's nozzles, causing the micro- or nanoparticle solutions to dry out and precipitate in the form of a solid block of hardened mass that plugs the microscopic ink passageways. In addition, inkjet printing has also the disadvantage of complex drying behavior to form uniform printed films (Atasheh, 2016). Therefore, the preparation of inkjet inks has attracted much interest and is often the most challenging task for researchers. Table 14.1 shows the main characteristics, requirements, and challenges of inkjet printing.

Although the formation of uniform droplets from a liquid stream through an orifice was introduced in 1833 by Savart (1833) and mathematically explained by Lord Rayleigh (1892) and Strutt & Rayleigh (1878), the inkjet printing technology appeared on the consumer market in the late 1980s, since it had been (and still is) under research for many years (Atasheh, 2016). The core technology for continuous inkjet (CIJ) printing was first patented by Rune Elmqvist who was working for Siemens Elema in Sweden (Elmqvist, 1951). Siemens produced the first commercial inkjet printing devices for medical strip and chart recorders. In the early 1960s, the work of Sweet, from Stanford University, eventually led to the use of continuous drop technology for the production of text (Sweet, 1965). Sweet's experiments demonstrated how a pressure wave, sent to an orifice, can break up a continuous ink stream into droplets of uniform size and shape. In the mid-1970s, Sweet's discovery was utilized by IBM to introduce the first commercial inkjet printer, the IBM 4640 (Buehner, 1977). The drop-on-demand (DoD) inkjet (thermal) technology is claimed to be invented by Hewlett Packard (HP) in 1979. However, it was Canon that actually introduced the first working thermal inkjet printer to the market (1981), a technology then called "bubble jet" (Keeling, 1981).

Ink properties, printhead capability, and substrate suitability were found to be the key factors that define the quality of printed page in inkjet printing. Tremendous effort has been put by researchers, one hand, to control the flow of ink from the printhead onto the substrate and, on the other hand, to prevent nozzles clogging due to dried ink (Atasheh, 2016).

14.2.1 Mechanisms of Drop Formation

Currently, the greatest challenge is to create affordable and reliable inkjet printers that would create high-quality

TABLE 14.1 Main Characteristics, Requirements, and Challenges of the Inkjet Printing Technology

Characteristics (advantages)	Variable film thickness
	Variable printing speed
	High efficiency
	Waste-less technology
	Excellent control of material's micro/nano-structure
	Versatile technology
	Use of different substrates
Requirements	Suitable solvent
	Suitable ink physical properties:
	– Ink viscosity
	– Ink surface tension
	– Ink pH
	– Particles' size
	– Precursors' solubility
Challenges (disadvantages)	Nozzle clogging
	Wetting behavior

printouts (Atasheh, 2016). Inkjet printing technologies are categorized based on the method of droplets formation (injection), into the CIJ and DoD printing systems. The underlying mechanism for CIJ printing is breaking a continuous jet into droplets and was "discovered" first, with the first patent of a working practical Rayleigh break-up inkjet device filed in 1951 (Elmqvist, 1951). The DoD printer, in which actually a jet of liquid is "squirted" and joins in a drop under surface tension-driven forces, was not made until 1977 (Buehner, 1977).

The main characteristics, advantages, and disadvantages of the CIJ and DoD techniques are analyzed in the following sections.

Continuous Inkjet Systems (CIJ)

As implied by the name, the CIJ process starts by the formation of drops from a continuously flowing jet of ink which is forced out of a nozzle under pressure. Figure 14.1 presents schematically a single-jet CIJ printhead system. As discussed above, disturbances at a particular wavelength along the jet can cause the jet to break up into droplets. By imposing a controlled disturbance at a suitable frequency, this break-up can be manipulated and a quite uniform stream of droplets can be produced. Not all the droplets formed during this process are eventually utilized for printing. Certain drops from the jet-stream are selected individually for printing. A usual method of selection is by using electrically conducting ink and charging the droplets inductively as they are forming by having an electrode at a close range and held at an appropriate potential. The droplets then pass through a fixed electric field which deflects the charged droplets by an amount which depends on their charge. Uncharged droplets are recaptured and the ink is being reused, while charged droplets are directed onto the substrate. The level of charge and the characteristics of the electric field control the number of droplets and the position at which they strike the substrate (Atasheh, 2016).

When a stream of droplets is formed in the way described above, it is usual to find that smaller satellite droplets are also formed from the ligament as it parts, along with the principal drop. These satellite droplets will either recombine with the main drop or be deflected to unwanted areas and cause poor printing quality and printer failure. Their final fate depends on the details of the applied disturbance (wave) and the characteristics of the ink (e.g., viscosity, conductivity). Remarkably, there has been some research to create small satellite droplets deliberately and utilize them for higher-resolution printing (Keeling, 1981).

In CIJ printing, usually large-diameter droplets (in the order of some millimeters) are produced, and hence, the print resolution is relatively low. CIJ printing is able to use volatile solvent-based inks, a fact which, on one hand, allows rapid drying and proper adhesion on many substrates but, on the other hand, makes the CIJ a kind of "messy" and environmentally unfriendly technology. Different polymers such as acrylics, cellulose acetate butyrate, nitrocellulose, vinyl chloride co-polymers, and polyvinyl butyrate have

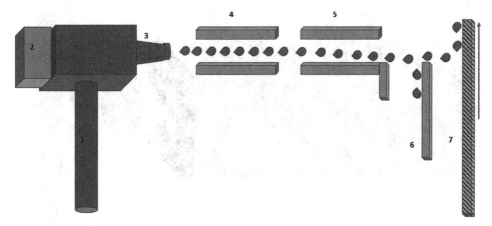

FIGURE 14.1 Schematic representation of a single-jet CIJ printhead system. 1, Ink; 2, piezoelectric crystal; 3, nozzle; 4, electrode; 5, deflector; 6, gutter; and 7, substrate.

been commonly used in solvent-based CIJ inks (Yuan et al., 1997).

Finally, there is another CIJ printing technology known as a "Microdot CIJ printing", in which droplets with different diameters are produced but only the smaller droplets are selectively charged and deflected to the substrate (Yamada et al., 1988).

Drop-on-Demand (DoD) Printing Systems

As mentioned above, the DoD printing technology was invented by Siemens in 1977. In this inkjet technology, droplets are ejected/formed directly from nozzles, by a pressure pulse, only when required. Depending on the method used for the creation of the "pressure pulse", the DoD inkjet printers can be further divided into four main sub-types: thermal (Kobayashi et al., 1981; Buck et al., 1985), piezo-electric (Kyser & Sears, 1976; Stemme, 1973; Zoltan, 1972), electrostatic (Silverbrook, 1998), and acoustic (Hadimioglu et al., 1992; Amemiya et al., 1999).

Thermal Inkjet Printing

Thermal inkjet (TIJ) printers use standard ink cartridge systems and do not require any additional bottles of inks or solvent, making TIJ printers clean and simple to use. This type of printers uses a drop ejection process, storing ink in a cartridge that regulates the pressure of the fluid. Inks are then delivered to the firing chamber and heated with a heating rate of more than $1,000,000°C/s$ by an electric resistor. A 0.1 μm thick film of ink is heated to $300°C$–$400°C$, from which a bubble is formed to expel the ink (Figure 14.2). A droplet breaks away from this bubble causing it to collapse; the firing chamber then refills as the whole process repeats (Le, 1998). The printhead of TIJ printers is usually cheap and has the ability to form small droplets. However, these printers have limitations regarding the use of solvents and other materials in ink formulation. The solvents used must be able to vaporize (e.g., aqueous solution), and in addition, all the materials used in the ink formulations must be able to withstand ultra-high

temperatures. Since the ink components may be functionally damaged due to the high temperatures, TIJ printers are not suitable for polymer printing (Haverinen et al., 2010).

Depending on the location of the heater and the orifice, TIJ printheads can be divided into two types: roof-shooters and side-shooters. In roof-shooter printheads, an orifice is located on top of the heater, while in side-shooter printheads, the heater is located on a side close to the orifice (Le, 1998). Hewlett Packard, Lexmark, and Olivetti use the roof-shooter design in their printheads, whereas the side-shooter design is implemented in Canon and Xerox printheads (Atasheh, 2016).

Piezoelectric Inkjet Printing

Piezoelectric inkjet technology uses piezoelectric material as a key active component within the inkjet printhead (Xaar, 2012). Piezoelectric material exhibits a phenomenon called the piezo effect, i.e., when force is applied to certain (natural) materials, a charge (electricity) is produced. Another effect, the reverse piezo effect, occurs when electricity is applied to the material, causing the material to deform (i.e., to "move"). Piezoelectric printheads incorporate PZT (lead zirconate titanate), which is a manufactured and propelled piezoelectric material (Figure 14.3). All piezo printheads work in practically the same way. The material is deformed in order to "shoot" an ink droplet. The voltage applied for piezo-ceramic deflection, the pulse duration, and the diameter of the nozzle are the critical parameters that define the size of the ink droplets in this type of printing (Atasheh, 2016).

In the piezoelectric inkjet technology, there is no restriction on ink type, while the piezoelectric printheads have a relatively long life. However, the printheads and supplementary components are often quite costly.

Depending on the mechanism of deformation of the piezoelectric element, piezoelectric inkjet printing can be divided into four main modes: (i) bend, (ii) push, (iii) shear, and (iv) squeeze (Xaar, 2012; Le, 1998).

In bend mode printheads, a piece of piezoelectric material is glued to the roof of the ink chamber. When an electric

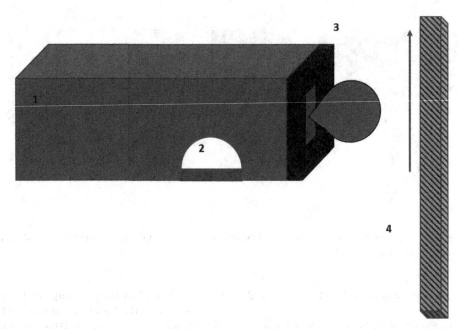

FIGURE 14.2   Schematic representation of a thermal DoD inkjet printhead system. 1, Ink; 2, heating element; 3, nozzle; and 4, substrate.

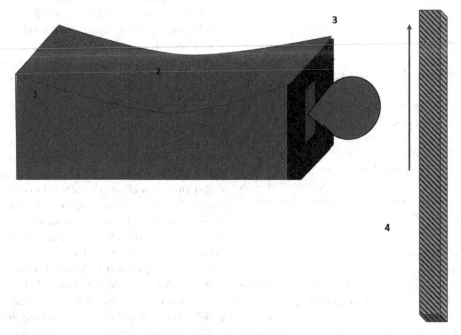

FIGURE 14.3   Schematic representation of a piezoelectric printhead. 1, Ink; 2, piezoelectric ceramic; 3, nozzle; and 4, substrate.

field is applied in the same direction in which the material is poled, the piece of material either lengthens or shrinks. If one side of the material is glued to the top of an ink chamber, the lengthening causes the roof to bend. This creates a pressure change that displaces a volume of ink causing a drop to be ejected from the nozzle (Xaar, 2012).

Shear mode printheads use a single piece of poled PZT for the actuator design. The electric current is applied perpendicularly to the direction in which the material is poled. The deformation creates a shear effect (trapezoid shape) (Xaar, 2012).

With the push mode, also referred to as "bump mode", a piezoelectric element pushes against an ink chamber wall to deform the ink chamber. The electrical field is applied in the poling direction, and the deformation is in the same direction or perpendicular to the poling direction (Wijshoff, 2010).

Finally, a squeeze mode inkjet can be designed with a thin tube of piezo-ceramic surrounding a glass nozzle as in a Gould's impulse inkjet (Johnson & Bower, 1979) or with a piezo-ceramic tube cast in plastic that encloses the ink channel as was implemented in a Seimens PT-80 inkjet printer (Heinzl, 1981).

Commercial printing systems operating in any of the above modes present certain advantages, which are summarized as follows (Le, 1998; Heinzl, 1981):

a. The speed, size, and shape of the droplets can be controlled by adjusting the waveform of voltage applied to the nozzles.
b. Spacing of droplets can be controlled by angling the line of nozzles along the printhead.
c. The droplet formation and impact can be observed by a video camera.
d. The printheads are heat resistance.

Organic and inorganic solutions, suspensions, polymers, and even biomaterials can be printed in all four modes. Since there is no fluid limitation in piezoelectric inkjet printing, some bench-scale piezoelectric inkjet printers are used in research and development, such as the Dimatix (Dimatix, 2017), MicroFab (Hayes et al., 2002), and Microdrop systems (Microdrop, 2018).

The considerable advantages and superior performance of the Dimatix inkjet material printer have been widely and successfully utilized in several studies on printed polymers (Hoth et al., 2007) such as the deposition of poly(3,4-ethylenedioxythiophene)-poly(styrene sulfonate) (PEDOT:PSS) (Han et al., 2011) and a polymeric fullerene blend. In addition, the Dimatix printer has been recently utilized for the synthesis/printing of novel metal-supported catalysts (Chatziiona et al., 2018).

Electrostatic Inkjet Printing

Since its discovery, electrohydrodynamic (EHD) technology has been applied in many areas, such as microcolloid thrusters, inkjet printing, pesticides for crops, and film deposition (Choi et al., 2008). In this technology, the formation of the droplet is induced by electrostatic forces, which are created due to a voltage that is applied between an electrode and the nozzle to attract free charges within the ink to its surface. The charged fluid is then separated from the inkjet head as fine droplets (Figure 14.4). Unlike piezoelectric systems, an EHD inkjet head can produce droplets smaller than the size of the nozzles used. This unique feature distinguishes EHD printing from conventional methods by allowing submicron resolution, since the size of the droplet is not controlled by the nozzle diameter (Park et al., 2007). With the increasing interest in direct writing, inkjet printing has become popular due to its economical and rapid fabrication advantages (Choi et al., 2008). EHD printing has great potential to offer complex and high-resolution printing, since it is able not only to eject fine droplets but also to print more viscous inks. Existing studies of EHD printing have successfully demonstrated the various geometry buildups from dots and continuous lines. However, although electrostatic jetting has been around for a long time, it has been mainly concerned with spraying tiny droplets of a uniform

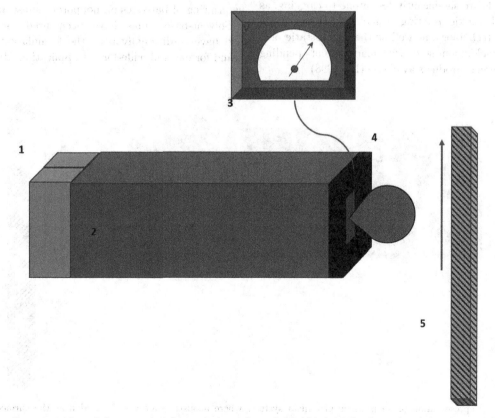

FIGURE 14.4 Representation of electrostatic inkjet printing, where the fluid is initially charged and then separated from the inkjet head as fine droplets. 1, Electrode; 2, ink; 3, voltage supply; 4, nozzle; and 5, substrate.

size for coating applications or spraying a continuous jet for patterning.

To date, there has been little in-depth study of the DoD control of a droplet in micro-dripping mode and patterning (Choi et al., 2008).

Acoustic Inkjet Printing

This technology is based on the development of acoustic droplet ejection (ADE), a method which utilizes a pulse of ultrasound to move low volumes of fluids (typically nanoliters or picoliters) without any physical contact. ADE was first reported, almost a century ago, by Robert Wood and Alfred Loomis (1927). To eject a droplet, a transducer generates and transfers acoustic energy to a source well. When the acoustic energy is focused near the surface of the liquid, a mound of liquid is formed and a droplet is ejected (Figure 14.5) (Amemiya et al., 1999). The diameter of the droplet scales inversely with the frequency of the acoustic energy—higher frequencies produce smaller droplets. Acoustic inkjet printing presents certain advantages, compared to other methods, including (i) the absence of a droplet-defining nozzle, (ii) the absence of a requirement to boil the ink to achieve ejection (which implies a wider latitude of admissible ink properties), (iii) the precision with which uniform small droplets of ink can be ejected, and (iv) the potential for a high degree of integration in the printhead structure, including the use of multiple colors in a single printhead (Hadimioglu et al., 1992).

Due to their unique principle of operation, the acoustic inkjet printheads are significantly less prone to clogging, as compared to other inkjet printing technologies. However, the acoustic inkjet technology, as well as the electrostatic one, is still in the development stage with many patents pending and few commercial products available (Le, 1998).

14.2.2   Ink Composition and Properties

Composition

For succeeding in specific printing applications, the whole printing system should be evaluated, namely, the printhead, the fluid that is jetted from the printhead (inkjet ink), and the substrate onto which the ejected droplets are placed (Hudd, 2009). For specific, novel applications, the requirements need to be established, defining the type of fluid chemistry, which, in turns, directs the selection of the hardware and drives the implementation.

There are currently four main types of inkjet inks: phase-change (Berry & Corpron, 1972), solvent-based (Bhatia & Stallworth, 1986), water-based, (Chandrasekaran et al., 2003), and UV-curable (Noguchi & Shimomura, 2001). Other types of inks have been found less successful, such as oil-based and liquid toner (for electrostatic inkjet technology). Hybrid versions of the four main types also exist (e.g., water-based inks containing some amount of solvent).

Phase-change inks, also known as hot melt, are distributed in solid form and, when introduced into a compatible system, are melted before being inkjet-printed. Advantages of phase-change ink include that they are very fast-drying (solidifying) and environmentally-friendly, and exhibit good opacity. It is also relatively easy to control the quality of the print because they do not tend to spread, due to their fast solidification. Their primary disadvantages are the lack of durability and poor abrasion resistance. Phase-change inks are currently used in applications such as printing of barcodes on nonporous substrates.

Solvent-based inks have been used for many years and have traditionally been the formulation of choice for grand format and wide-format applications due to excellent

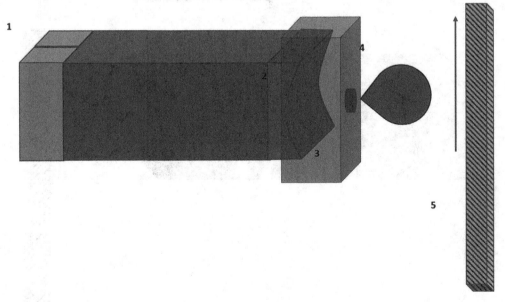

FIGURE 14.5 Representation of an acoustic printhead system, where acoustic energy is focused near the surface of the liquid, causing a mound of liquid to form and then a droplet to be ejected. 1, Piezoelectric; 2, acoustic lens; 3, ink; 4, nozzle; and 5, substrate (Amemiya et al., 1999).

print quality, image robustness, and range of compatible substrates. They are also generally considered to be low-cost inks. Benefits include the ability to adhere to different types of substrates and rapid drying (which is often accelerated by heating). Solvent-based inks can be formulated with either pigments or dyes (or even both). Disadvantages include environmental concerns and a requirement for high maintenance, due to the potential of blocking the printhead nozzles by the dried pigment or dye (Hudd, 2009).

Water-based or aqueous inks are prevalent on the desktop and present the advantage of being relatively inexpensive and environmentally-friendly, but penetration in industrial applications has been relatively slow for a variety of reasons. Water-based inks tend to require porous or specially treated substrates or even lamination to impart durability, since this type of ink tends not to adhere to nonporous substrates.

Additionally, many piezoelectric industrial printheads are incompatible with water-based ink formulations, although this is changing, partly, due to the market demand for systems that can jet water-based biological or food contact fluids.

UV-curable inks are formulations that are designed to remain as a stable liquid until irradiated with a particular wavelength and intensity of light. UV-curable inks and coatings have been used in printing markets for many years, and thanks to recent advancements in the R&D of inkjet printhead and fluid formulation, inkjet is now an established and robust deposition tool for UV-curable fluids. This is not surprising considering the benefits resulting from the partnering of the UV and inkjet technologies.

UV inks are now reliably and successfully employed for a variety of inkjet applications across many different sectors. The benefits achieved by UV, coupled with the flexibility of digital printing, have proved a compelling proposition for many industrial applications and have seen the breadth of UV ink implementations spread from the more traditional wide-format/flatbed sectors into the niche application areas of product coatings, primary package decoration, and labeling. Current limitations are in edible and food contact applications. Disadvantages include cost and facility requirements (space, extraction, power) for the UV curing hardware.

As stated earlier, inkjet printing is a system, which must take into consideration the hardware capabilities, the ink properties, and the interaction with the substrate. Once the requirements are defined and the ink chemistry and inkjet printing technology have been chosen, there are additional considerations including image/information processing, speed, print quality, cost trade-offs, fixed versus scanning heads, and maintenance (Pond et al., 2000). The image is actually a combination of process colors, and therefore, the placement of each ink drop and the order of placement, as well as bleeding issues, play significant roles in the print quality.

For novel materials' deposition applications such as printed functional materials, inkjet system requirements are diverse and can include ultra-high precision substrate handling, drop visualization, and fiducial recognition for printing of multiple layers, not to mention the requirement for inkjet fluids that may incorporate "difficult" ingredients such as nano- or large particles that must remain in suspension, aggressive acids or alkalis, fragile biological materials, magnetic materials, and even radioactive substances (Hudd, 2009).

14.2.3 Ink Properties Requirements

In the printing process, to avoid the agglomeration or precipitation of ink components that would clog the printhead reservoir, the inkjet solutions that can be used for the fabrication of functional metal oxide materials should have specific properties such as particle size, viscosity, surface tension, and density that are equivalent to those of standard inkjet printer inks (Liu et al., 2015). The general property requirements for inks (fluids) to be used in inkjet printers are reported in Table 14.2 (Blazdell et al., 1995; Dong et al., 2006; Oezkol et al., 2010). Among these properties, viscosity and the surface tension must be carefully controlled and adjusted, because they strongly affect the droplets formation and dynamics. Inks with high viscosity reduce the droplet ejection speed, leading to aggregation and agglomeration of the inks in the printhead orifice and/or channels. However, high surface tension makes drop generation difficult and increases the chances for satellite droplets formation rather than a single drop of ideal size. If the surface tension of the ink is too low, it can lead to contamination of the droplets, causing liquid ink to drip towards the substrate, randomly, during the printing process. During the last two decades, several reviews have been published that summarize the principles of ink formulation (Derby, 2010; Sun et al., 1997). The fluid rheological property requirements for printable inks are determined by the physics and the fluid mechanics of the droplet formation process (Maier et al., 2007; Liu et al., 2015). The behavior of fluids can be described by the Reynolds, Weber, and Ohnesorge numbers (Re, We, Oh):

$$\text{Re} = \frac{\nu \rho \alpha}{\eta} \quad (14.1)$$

$$\text{We} = \frac{\nu^2 \rho \alpha}{\gamma} \quad (14.2)$$

$$\text{Oh} = \frac{\sqrt{\text{We}}}{\text{Re}} = \frac{\eta}{(\gamma \rho \alpha)^{\frac{1}{2}}} \quad (14.3)$$

where ρ, γ, and η are the density, surface tension, and dynamic viscosity of the fluid, respectively, α is the characteristic length, and ν is the velocity. Another useful parameter is the Z number (Eq. 14.4), the inverse of the Ohnesorge number (Oh), which is independent of the fluid velocity and can be used to evaluate the printability of an inkjet fluid.

$$Z = \frac{1}{\text{Oh}} \quad (14.4)$$

The latter parameter was defined by Fromm (1984) using a simple model of fluid flow in a drop generator with simplified

TABLE 14.2 Summary of Major Fluid Property Requirements for Inkjet Printing

Printing Method	Sub-Method	Particle Size (μm)	Viscosity (cP)	Surface Tension (dynes/cm)	Density (g/cm³)
Continuous inkjet		<1	1–10	25–70	~1
DoD inkjet	Thermal	<1	5–30	35–70	~1
	Piezoelectric	<1	1–20	35–70	~1

Source: Liu et al. (2015).

geometry and proposed $Z > 2$ as the best condition for the generation of stable droplets. More recent studies further extended the results of Fromm, through numerical simulations, and showed that optimal droplets can be generated for Z values in the range between 1 and 10 (Reis & Derby, 2000). For inkjet fluids with low Z values (highly viscous fluids), drop is prevented from coming out of the nozzle, whereas fluids with high Z values result in multiple satellite droplets that accompany the single defined drop. More recently, researchers have reported a printable range of $4 \leq Z \leq 14$, based on their experimental results by considering characteristics such as single droplet formability, positional accuracy, and maximum allowable jetting frequency (Jang et al., 2009).

Based on the above, it can be concluded that the suitability of a fluid for inkjet printing can be ***roughly*** assessed based on its Z number. However, apart from the Z number, in practice, there are some other restrictions regarding fluid behavior that impose additional limits on practical droplet formation. A minimum Weber number value (We = 4) was proposed, below which it seems that there is not enough fluid flow to overcome the surface tension and generate a small radius droplet (Duineveld et al., 2002). Additionally, Stow and Hadfield observed the onset of splashing that occurred if a droplet hits the substrate at a velocity above a critical threshold value when $We^{1/2}Re^{1/4} > 50$ (Stow & Hadfield, 1981). These limitations define the parameter limits for We and Re, within which DoD inkjet printing is practically possible. The parameters above can therefore be very useful for study and examination of the corresponding properties of novel functional inks.

Ink Compositions for Synthesis of Metal Oxides

Although a successful research has been performed with respect to organic materials, inkjet printing of functional inorganic materials (metal oxides) has had only sporadic success (Liu et al., 2015). Up to date, only a few inorganic materials have been synthesized by inkjet printing, primarily because of the difficulty of preparing functional ink solutions suitable for inkjet printing. The functional inks for inkjet printing of metal oxides are colloidal systems that are typically composed of a metal precursor (1%–30%), a liquid carrier medium (50%–60%), and various additives (1%–10%). The parameters that determine the quality and characteristics of inkjet ink solutions are described below:

i. *Liquid carrier medium*

There are two main types of ink formulation that are based on a liquid carrier medium for an inkjet printing: solvent-based inks and aqueous-based (or water-based) inks. Typically, the liquid carrier medium determines the main properties of the ink (e.g., viscosity, surface tension, density) and is used to disperse the metal precursor. Aqueous-based inks have single or multiple metal colloidal nanoparticles or metal precursor salts or complexes which are dissolved in water. Usually, aqueous-based inks have lower viscosity and higher surface tension values than the corresponding solvent-based systems (Dondi et al., 2012). Their properties can be modified so as to match the requirements of different inkjet printers by the addition of co-solvents, including glycols, glycerin, alcohols, and surfactants. There are three main advantages when using water-based inks: (i) the use of inexpensive source materials, (ii) the low toxicity of produced waste materials, and (iii) the long shelf life of the produced inks due to the prevention of solvent evaporation and dust incorporation (Feys et al., 2012). On the other hand, aqueous-based inks have relatively slow drying rates, particularly when they are deposited on non-paper substrates, e.g., glass, ceramic, or steel. Solvent-based inks are based on organic solvents (most commonly toluene, cyclohexanone, and acetylacetone), as well as a single alcohol or mixtures of different alcohols, e.g., methanol, isopropanol (IPA), ethanol, and ethylene glycol (EG). This type of ink evaporates faster than aqueous-based inks and is thus suitable for the generation of single or multilayer metal oxides on a variety of substrates with various thicknesses for all kinds of applications. Although solvent-based inks are widely used in the field of materials fabrication, they emit volatile organic compounds and are prone to printhead clogging due to the rapid evaporation of the solvent at the nozzles. The major challenge to the application of inkjet printing of metal oxides is therefore the formulation of stable, acceptable inks (Liu et al., 2015).

ii. *Metal species precursors*

Generally, three different types of metal oxide precursors can be identified for jettable ink formulations: metal salt solutions, sol–gel colloidal nanoparticles, and colloidal suspensions of metal oxide nanoparticles. The various precursors in the basic nature of the three types above are likely to lead to differences in the mechanism of fabrication of the material (Liu et al., 2015).

iii. *Colloidal suspensions*

The first method used for the preparation of jettable ink is based on the formulation of stable colloidal suspensions of metal oxide nanoparticles. In a jettable ink formulation, the properties of the suspended metal oxide nanoparticles are only influenced by the particle size and the viscosity, and thus, the inks can be produced in large quantities and dispersed in high concentrations. The critical issues for printability are particle size and the dispersion of the suspensions. In particular, the particle size of the nanoparticles needs to be controlled since small particles tend to agglomerate and eventually lead to clogging of the nozzles and microchannels of the printhead. On the other hand, the particle size in the suspension should be considerably smaller (e.g., 50 times) than the diameter of nozzles and should have a narrow size distribution (Kuscer et al., 2012). In order to prevent aggregation or agglomeration, surfactants or polymeric compounds are added to the suspension, which can successfully be removed by heat treatment (>300°C) or other pretreatment methods (e.g., microwave flash method). Another key issue in the colloidal inks is the homogeneous dispersion of the metal oxide nanoparticles in a liquid with a specific viscosity and surface tension to ensure that the droplets depart successfully from the nozzles and high-quality droplet morphologies are produced. Usually, small amounts of appropriate non-ionic surfactants and glycerol (GC) can be added to colloidal suspensions to modify the properties of the ink solutions, in terms of printability. For successful inkjet printing of a aqueous-based colloidal suspension, two conditions must be fulfilled: (i) an inkjet printer with a relatively large nozzle (50–140 μm) should be used to print the micrometer-sized metal oxide particles, and (ii) the ultrafine powders must be prepared by chemical or mechanical methods, such as microwave synthesis (Arin et al., 2012; Lehnen et al., 2012), hydrothermal synthesis (Andio et al., 2012; Cerna et al., 2013), sol–gel processing (Costa et al., 2012; Huang et al., 2008), ultrasonication mixing (Fasaki et al., 2012), or ball milling/mixing (Dou et al., 2011; Kuscer et al., 2013). In these methods, EG is usually used as a binder to control the mixture's viscosity, an acid (e.g., HCl, poly-acrylic acid, acetic acid, carboxylic acid) is added to regulate pH and thus provide the surface charge of the metal oxide nanoparticles, and a surfactant is added to provide suitable wetting properties and effective stabilization of the nanoparticles. The majority of published studies on the use of printable colloidal suspensions of metal oxide nanoparticles have used an evaporating solvent, mainly

because this system leads to high-density powder compacts (up to 30 wt% solid content) (Jeong et al., 2010) and reduces dimensional changes during subsequent sintering process (Dou et al., 2011). Table 14.2 lists several ink formulations based on colloidal suspensions for inkjet printing and their properties. For a solvent-based suspension ink system, the most commonly used solvent is an alcohol (e.g., ethanol, IPA, and methanol). Addition of small amounts of a dispersive agent and a surfactant to the solution usually enhances the dispersion of the nanoparticles in the mixed solvent and prevents the agglomeration of the suspension. Because colloidal suspensions of metal oxide nanoparticles are usually prone to agglomeration, which affects droplet formation dynamics, the use of precursors of metal salt solutions and sol–gel-based colloidal nanoparticles is recommended.

iv. *Metal salt solutions*

A metal salt solution is the solution that contains dissolved molecules of a metal salt-precursor. This is the simplest, lowest cost process for the preparation of ink formulations. Han et al. (2009) and Lee et al. (2007) used metal halide precursors dissolved in acetonitrile to produce printable inks. After printing, the films were converted into metal oxides by a thermal reaction between the metal halide film and water.

v. *Sol–gel-based colloidal nanoparticles*

The printing "ink" can also use sol–gel-based colloidal nanoparticles as precursors. In this case, the inkjet printing process is consisted of four steps: (i) a homogeneous solution is prepared by the dissolution of suitable metal precursors (usually metal alkoxides) in an organic solvent or a water/organic solvent mixture; (ii) the homogeneous solution is converted into a printable colloidal nanoparticle solution (sol) by treatment with an appropriate chemical reagent (e.g., acid, base, or complexing ligands), which is then the printing "ink"; (iii) a suitable inkjet printer is used to generate printed thin films that are then dried to form a gel layer; (iv) the gel layer is thermally treated to form the oxides. For this type of precursor ink solutions, control of the growth kinetics of the various metal alkoxides is of critical importance to avoid aggregation of the colloidal nanoparticles, which will lead to clogging of the printhead nozzles. Optimal diameters of colloidal nanoparticles can be stabilized, in the 1–100 nm range, by the addition of complexing ligands such as acids. Another issue with sol–gel inks is the hydrolysis and/or condensation of alkoxides or metal chlorides. The dissimilar condensation kinetics and chemistries of the precursors often lead to undesirable macroscopic inorganic/organic phase separations that affect the applications of the functional materials. Strategies used to control the hydrolysis/condensation rates of the precursors for the generation of suitable inkjet inks include the use of stabilizing ligands, nonaqueous media, and controlled hydrolysis by limited addition of water (Fan et al., 2013; Niederberger & Pinna, 2009; Hench & West, 1990).

Representative research regarding the development of multi-component metal oxides based on colloidal nanoparticle precursors and the inkjet printing technique can be found in the study of Liu et al. (2012b). The researchers developed a general, simple, and inexpensive formula for the development of more than 25 different colloidal inks, suitable for inkjet printing synthesis. The ink formula (IF) is consisted of an amphoteric (nonaqueous) solvent, the metal species, and block copolymers. The amphoteric solvents (e.g., alcohols or acetic acid) are used to ensure the appropriate viscosity, density, and surface tension for all metal species. Block copolymers, such as F127 and P123 which can be coassembled with metal ions and nanoparticles, are used as binders and structure-guiding agents to fabricate mesostructured metal oxides (2–50 nm). The main findings of the work of Liu et al. are described in detail in Section 14.3.

14.3 Inkjet Printing of Solid Catalysts

14.3.1 Metal-Supported Catalysts

Inkjet printing has been recently used for the preparation of metal-supported catalyst that can be used in various applications (Singh et al., 2010; Liu et al., 2012a; Zhang et al., 2017; Chatziiona et al., 2018). The use of the method at hand not only reduces the cost for supported catalyst production but also allows the fine-tuning of the parameters that affect behavior of the catalyst in a reaction. At the same time, the time needed for catalyst synthesis is reduced dramatically, compared to the conventional methods used globally.

Liu et al. (2012a) have prepared metal-supported catalyst patterns (Pt/SiO_2) with the use of inject printing for low-temperature catalytic methanol combustion. The catalysts used in their experimental procedures were produced with the use of an inject printer equipped with a piezoelectric printer head, with 70 μm nozzles diameter. The dot distance of Pt, on the SiO_2 substrate, was found to be of 160 μm, while the catalyst was reduced under Ar. Methanol catalytic combustion was performed in a stainless reactor at various temperature. The results obtained showed that methanol conversion increased with temperature reaching the maximum of 85% at 100°C with the use of only 52 μg

of Pt at a total surface area of 5 cm^2 catalyst. Moreover, the researchers also noted the importance of the microdrop drying method on the catalytic morphology and as a result on the overall activity of the catalyst.

Zhang et al. (2017) thoroughly studied the effect of frequency, temperature, and jet height of the DoD inkjet printing method on the particle size of nanomaterials. The microspheres produced were prepared with the use of the spray drying method. The diameter of the nozzles used during the printing process was 160 μm on a piezoelectric printhead which allowed the precise deposition of ammonium nitrate on a DMF (dimethyl formamide)-preimpregnated glass plate.

Different concentrations of the precursor solution were used, and as a result, for each process, the voltage and pressure were adjusted in order to achieve uniform and stable droplets. The researchers showed that the smaller the distance of the nozzle to the substrate, the smaller the deposition position deviation, although the distance cannot be less than 1–2 mm in order to avoid contact between the nozzles and the substrate.

Moreover, Zhang et al. studied the effect of the substrate temperature on the size of the generated particles, at a specific printing frequency (6 Hz), with the use of optical microscopy (Zhang et al., 2017). At room temperature, it has been proved that the deposited particles agglomerate since the evaporation rate of the solvent (DMF) is lower at room conditions. On the other hand, with the gradual increase in the substrate temperature to 80°C, it has been noted that not only the particles size decreases, but also the spreading area increases as well.

The effect of print frequencies was also studied at constant substrate temperatures of 60°C, while the printing frequency used varied in the range of 2–10 Hz. The particle size distribution was determined by an optical microscope.

At 60°C substrate temperature and low print frequency (2 Hz), it was observed that particles agglomerate to form large bract-like structures with a diameter of 10 μm, while with the increase of frequency at 4 and 6 Hz, respectively, particles grow in certain direction forming long stripes. At 10 Hz print frequency, particle size was found to be significantly reduced (1–2 μm) when, at the same time, a uniform distribution was achieved.

Chatziiona et al. (2018) have prepared three 0.1 wt% Pt/Al_2O_3-supported catalysts using three different preparation methods and compared their performance (conversion and selectivity) for the selective catalytic reduction of gaseous NO_x pollutants with the use of $H_2(H_2\text{-SCR})$ under strong oxidizing conditions.

The first catalyst was prepared via the conventional wet impregnation (WI) method, while the second catalyst was prepared using the IF as described by Liu et al. (2012a). The third catalyst was prepared via a novel inkjet printing (NIP) method involving the preparation of a metal and the substrate (Pt and Al_2O_3) precursor ink mixture prior printing. All three catalysts underwent drying, calcination, and pretreatment prior the catalytic reaction at exactly the same conditions. Moreover, temperature-programmed surface reaction (TPSR) experiments as well as scanning electron microscope (SEM) studies were performed to identify how the surface of the three catalysts affects their catalytic behavior towards the H_2-SCR reduction of NO_x.

High-resolution scanning electron microscope (HR-SEM) images taken from all three catalysts revealed that the WI catalyst consisted of large grains of undefined structure, while the surface of the catalyst prepared with the use of the IF method revealed a distinct honey-comb, spongy structure. On the other hand, the catalyst prepared with the method proposed by Chatziiona et al. (NIP) showed a surface structure of distinct spherical particles (nano-spheres 100–300 nm in diameter). H_2 temperature-programmed desorption studies revealed that all catalysts Pt dispersion on the catalytic surface varied between 88% and 91%. As a result, the differences in the catalytic behavior observed can be attributed solely to the structure of the catalytic surface as it derives from the catalyst preparation method used and, in particular, the nature of the contact between the metal and the support (Chatziiona et al., 2018).

Figure 14.6 presents the catalytic activity of the three catalysts towards the H_2-SCR reaction in the 100°C–400°C temperature range. As it can be clearly seen, the NIP catalyst presents the highest NO conversion (83%–91%) at low temperature range (100°C–175°C), while the other two catalysts reduce NO at higher temperatures (150°C–200°C). As a result, the NIP inkjet printing method leads to a catalyst where the Pt particles are in better contact with the surface of the support than the other two catalysts.

Figure 14.7 presents the results obtained by TPSR on the catalysts NIP (a) and WI (b), respectively, following $NO/H_2/O_2$ reaction. The catalysts present significantly different profiles through the whole temperature range examined. Specifically, the WI catalyst does not present any N_2 or NO desorption at temperatures lower than 100°C, while the NIP catalyst desorbs significant amounts of N_2

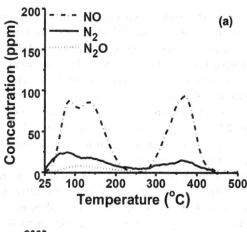

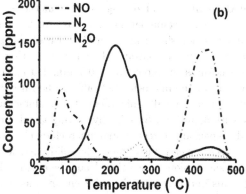

FIGURE 14.7 TPSR curves of NO, N_2O, and N_2 after $NO/H_2/O_2$ reaction on the 0.1 wt% Pt/Al_2O_3 catalyst prepared by inkjet printing (a) and WI (b).

and NO at such low temperatures, revealing the superiority of the preparation method of the NIP method used.

14.3.2 Metal Oxide Catalysts

As mentioned earlier, DoD is a printing technique that has been studied in detail during the last few years and which provides several positive aspects in inkjet printing (IJP), such as uniform catalyst layers (CLs) in microchannels that lead to a steady reactant flow distribution, enhancing the reactor performance. It has been used to deposit either ceramic and metallic particle suspensions (support), or catalytically active materials. This method is widespread in the microchannel reactor technology used in various chemical processes, such as Fischer–Tropsch synthesis and steam reforming. Lee et al. (2013) have demonstrated a DoD IJP method which was used to deposit thin alumina films in microchannels of different geometries and materials. Specifically, four alumina inks were prepared in order to determine the printable ranges, and then, the supports were calcined and Rh was impregnated on the alumina support. Methane steam reforming tests have been carried out to designate the catalytic activity of the prepared materials.

The inkjet printer used (microdrop technology GmbH) was equipped with a MD-K-140 piezoelectric printhead with

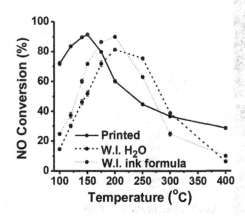

FIGURE 14.6 Figure 14.6 NO conversion versus temperature for three 0.1 wt% Pt/Al_2O_3 catalysts prepared with different synthesis methods (Chatziiona et al., 2018).

100 μm diameter nozzles. Four alumina inks were prepared as follows: Ink1—a commercial stable aqueous nanosuspended aluminum oxide (particle size: 100 nm); Ink2—Ink1 modified with water; Ink3—Ink1 with EG; and Ink4—Ink1 with EG and polyethylene glycol. Figure 14.8 presents the cross-sectional elemental map analysis of a Ink3-coated semi-circular microchannel, after 17 printing repetitions, drying at 343 K for 24 h and calcination at 823 K in air for 6 h. As it can be seen, a uniform thin coating (3 μm) was created over the entire channel.

Figure 14.9 presents the elemental mapping image and the elements' distribution of the cross section of a microchannel foil printed with the use of Ink4. The aluminum map shows the printed layer in the rectangular microchannels. Compared to filling and drying coating methods, it can be seen that the accumulation of the ink in the edges of the channel is not significantly in the case of the printed microchannel. On the contrary, the CL has a rather homogeneous thickness of about 13 ± 0.5 μm. During drying, the geometry of the microchannels affects the rate of vapor removal, and as a result, this affects the agglomeration of the particles deposited. Specifically, during the drying of a colloidal suspension in rectangular microchannels, evaporation increases the concentration of the solid particles in the remaining suspension, which then leads the drying plane to the sidewalls and center of the microchannel, as it can be seen in Figure 14.9.

Moreover, CLs present cracks which are observed to be more intense in microchannels than in plane surfaces, due to capillary forces that are unevenly developed during the drying process. The use of binders in the suspension can reduce the aforementioned phenomena.

Concerning the methane steam reforming reaction and in the case of the catalyst deposited with the use of Ink4, it has been found that the highest methane conversion achieved was 98.9% at 973 K, which was sustained for more than 60 h.

The spatially controlled growth of highly crystalline ZnO nanowires (NWs) was studied by Güell et al. by a catalyst-free vapor-transport method (Güell et al., 2016). A solution of $Zn(NO_3)_2.6H_2O$ and 2-methoxyethanol was stirred for 1 h at 50°C. After filtration (0.45 μm), the solution was used for inkjet printing on SiO_2/Si substrates with the use of a Dimatix DMP-2800 printer. ZnO and graphite powder (1:1 mole ratio) were mixed in a quartz tube which was placed in a chemical vapor deposition (CVD) furnace. Argon was used as a carrier gas, and the temperature of the furnace was raised to 900°C and remained for 30 min, and then the reactor was cooled to room temperature naturally.

Field emission scanning electron microscope (FESEM) images revealed the development of uniform and dense arrays of ZnO NWs with 50 nm diameter and 0.5 μm length, while HRTEM analysis demonstrated the formation of wurtzite structure ($P6_3mc$).

Figure 14.10 represents a schematic illustration of the three steps of the vapor–solid (VS) process followed for the production of ZnO NWs, i.e., (a) inkjet-printing $Zn(NO_3)_2$ (solution in 2-methoxyethanol) arrays on thermally oxidized SiO_2 substrate, (b) heating at 450°C for 1 h in the presence of oxygen and subsequent formation of ZnO

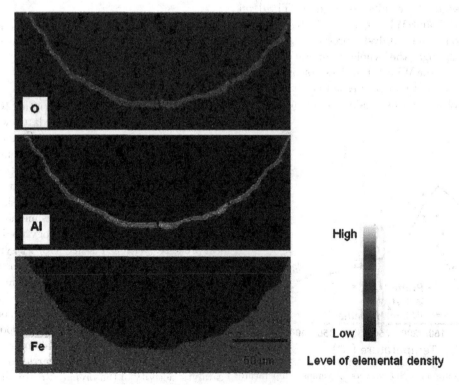

FIGURE 14.8 Elemental mapping images of the cross section of a semi-circular microchannel inkjet-printed alumina (use of Ink3) (Lee et al., 2013).

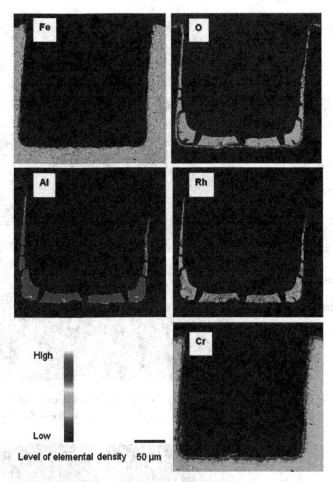

FIGURE 14.9 Cross-sectional elemental mapping images of an inkjet-printed alumina layer in a rectangular microchannel, impregnated with 15 wt% Rh (Lee et al., 2013).

nanocrystallites, and (c) growth of ZnO NWs from a powder mixture of ZnO and graphite at 900°C for 30 min under Ar. The mechanism suggested by the authors regarding the formation of ZnO NWs can be summarized as follows: the thermally reduced ZnO promotes the metallic Zn in the gas phase, and subsequently, due to gradient concentration, Zn diffuses towards the substrate where it reacts with the formed CO at the surface of the ZnO crystallite resulting in the formation of ZnO NWs. The diameter of the printed ZnO drop was found to be around 40 μm while it is further reduced after solvent evaporation during calcination. FESEM images (Figure 14.10f,i) present no substrate growth of ZnO NWs.

HRTEM analysis of the ZnO NWs (Figure 14.11) reveals the important role of the ZnO crystallites that act as promoters for the development of the NWs (the starting point is indicated with an arrow in Figure 14.4a). The lack of catalyst caps on the NWs tips indicates that the growth process was catalyst-free. Figure 14.4b shows the high crystallinity observed in the center of the NWs.

Figure 14.12 compares the Raman spectra obtained on (a) NWs on ZnO (inkjet printing—dark grey), (b) NWs on ZnO thin film (light grey), and (c) ZnO c-plane surface (black). The peaks at 302 and 520 cm^{-1} are attributed to the Si substrate, while the peak at 438 cm^{-1} originates from the optical zone-center phonon mode E_2^{high}. This mode is non-polar, thus excluding carrier concentration as a possible origin of peak position variation. The Raman spectra of NWs exhibit the typical first-order optical Raman modes of ZnO. The peak positions indicate strain-free growth of the ZnO NWs using the suggested inkjet printing, with the E_2^{high} peak position in agreement with the reference substrate.

14.3.3 Inkjet-Assisted Carbon Nanotubes Fabrication

Inkjet printing of catalysts provides the ability to print arbitrary patterns at low cost. On these catalysts, carbon nanotubes (CNTs) can be developed with the use of a carbon containing gas (catalytic chemical vapor deposition—CCVD) which reacts with the catalytic surface leading to separated, high-quality, and perfectly aligned CNTs.

Mansoor et al. prepared catalysts with the use of nickel-containing inks on silicon substrates by inkjet printing, on which CNTs were produced with the use of CCVD (Mansoor et al., 2010). It was found that patterning the catalyst leads to the formation of uniformly well-organized CNTs. In particular, catalysts were printed on Si substrate,

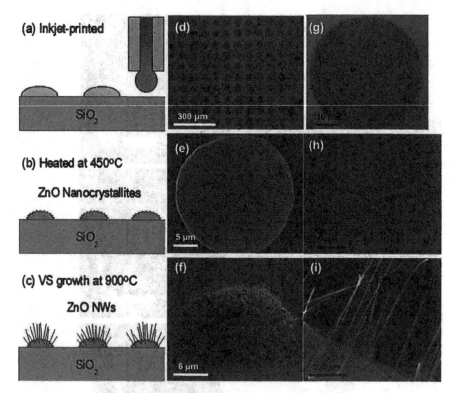

FIGURE 14.10 (a–c) Schematic illustration of the three stages of inkjet-printed growth of ZnO NWs on SiO₂/Si substrate. (d–i) FESEM images of crystallites during the three stages of printing and growth of the NWs (Güell et al., 2016).

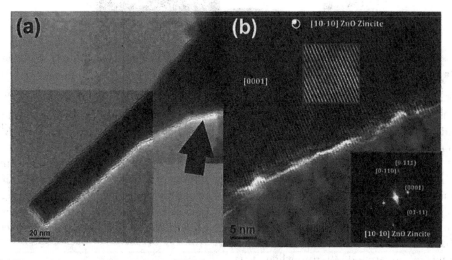

FIGURE 14.11 (a) Transmission electron microscope (TEM) images of a single ZnO NW in [10-10] axis grown with the inkjet-printing technique on a SiO₂/Si substrate and (b) HRTEM characterization of the same ZnO NW (Güell et al., 2016).

with the use of several proportions of ink precursors and solvents (ethanol, water, nickel nitrite, EG, and tetraethyl orthosilicate—TOS). The CNTs produced were characterized by SEM, TEM, and Raman spectroscopy. The inks used had a viscosity range of 10–20 mPas, while the inkjet printer used (DMP 2831) had a 100–5,080 dpi resolution. The matrix which was printed on the Si substrate consisted of a 50 × 50 dots, with a dot diameter of 50 µm ca., while the distance between the dots was 150 µm ca.

Figure 14.13 depicts a simplified diagram of the PFR glass reactor used for the growth of CNTs. A gas mixture

of $N_2/Ar/H_2/CH_4$ and/or acetylene was continuously fed to the reactor.

SEM images obtained on the printed catalysts showed a uniform distribution of the catalyst crystallites (spheres). Cracks and fissures were observed due to the calcination that was conducted at 400°C although the latter can be eliminated by controlling the heat rate during calcination. Moreover, with the use of SEM and TEM microscopy, it was proved that the addition of TOS and EG in the ink formulation reduced the defect ratio of the produced CNTs. Raman spectroscopy showed that the CNTs produced by

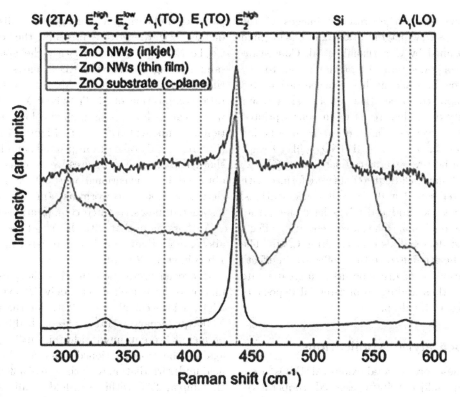

FIGURE 14.12 Comparison of the Raman spectra between (a) NWs on ZnO (inkjet printing), (b) NWs on ZnO thin film, and (c) ZnO c-plane surface (Güell et al., 2016).

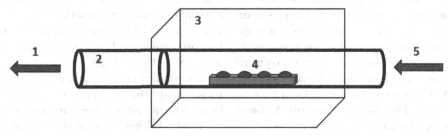

FIGURE 14.13 Schematic representation of the PFR reactor used for CNTs growth. 1, gas mixture outlet; 2, quartz tube reactor; 3, furnace; 4, printed substrate; and 5, gas mixture inlet.

inkjet printing had a high degree of crystallinity resulting in high-quality nanotubes (Mansoor et al., 2010).

The effect of the salt concentration on the viscosity and the printing pattern quality of the prepared CNTs via the CVD method was studied by Chatzikomis et al. with the use of SEM and TEM microscopy (Chatzikomis et al., 2012). The authors used iron precursor salts instead of the most commonly used nickel precursor ones, because they are cheaper and do not possess any environmental potential risk. Specifically, two different iron salts ($Fe(NO_3)_3$ $9H_2O$ and $FeCl_3$) were used to form four different ink solutions (0.1, 0.2, 0.4, and 0.8 M). The two solutions used by Hutchings et al. were a mixture of water and ethanol and the second a model mixture as described by Mansoor et al. (2010) of water (8.3 vol.%), ethanol (50 vol.%), EG (33.3 vol.%), and tetraethyl orthosilicate (TEOS) (8.3 vol.%). After the solvents were mixed with the salts, ultrasonication was performed for 30 min, and viscosity and surface tension tests of the resulted solutions were performed. The inks were deposited on fused quartz substrates (5×5 mm) with the use of a Fujifilm DMP-2831 inkjet printer. The samples underwent calcination at 550°C for 10 min under H_2/Ar gas flow. The reactions for the growth of CNTs were performed at 850°C for 20 min under continuous flow of ethane (0.1 L/min).

It was proved that the increase in salt concentration leads to a higher solution viscosity, although the solvents in the model solutions appeared to have relatively higher viscosity than the water–ethanol mixture solutions, respectively. The observed increase of viscosity was found to be due to the solvation of the highly charged Fe^{3+} ions. On the other hand, surface tension was proved to be unaffected by the salt concentration but was highly affected by the concentration of water in the solution. In particular, the surface tension of the model solution was found to be less than the one observed in the water–ethanol solution.

Table 14.3 shows electron microscopy images of the printed catalyst drops (low magnification) and CNTs (high magnification) obtained by Chatzikomis et al. Concentrations of Fe precursor lower than 0.1 M didn't lead to any sufficient CNT growth, while at high concentration (0.8 M), the ink was sprayed rather than printed as a result of the high viscosity of the ink, resulting in poorly printed dots. Intermediate concentrations lead to the production of the best patterns. In comparison, $Fe(NO_3)_3.9H_2O$ leads to the formation of better patterns compared to $FeCl_3$. As shown in Table 14.3, non-well-defined dots were formed with the use of the latter salt. For 0.1 M salt concentrations, it was observed that the produced CNTs have ring structure, leading to the conclusion that salt concentration affects the printed catalyst dots and also the resulting CNTs. The ring structure can be attributed to the "coffee strain effect" where the evaporation of the drops occurs at a greater rate from its outer rim, thus leading to preferential deposit of particles at the edge of the drop.

14.3.4   Photocatalytic Applications

Multicomponent mesoporous metal oxides (MMMOs) were synthesized by an inkjet printing-assisted cooperative-assembly method (IJP-A) and tested for the photocatalytic reformation of formaldehyde–water solution for hydrogen production (Liu et al., 2012b). In particular, the MMMO synthesis method consisted of three stages: (i) the precursor inks used were prepared via a sol–gel cooperative-assembly system, and their composition included an amphoteric nonaqueous solvent, metal species, and block copolymers coassembled with suitable metal ions and nanoparticles. The last three were used as binders to form a variety of mesoporous metal oxides with high surface area. (ii) A color-management system was developed in order to create a one-to-one output between the percentage of each color in an image as well as the printing volume of each ink. Moreover, a "multichannel combiner" was used to extend the printable space. (iii) The process of evaporation-induced cooperative assembly was used to generate the desired mesoporous structures of the printed libraries. The block-copolymer surfactants were removed with calcination. The oxides (eight different metals) derived have different compositions and mesostructures as they were fabricated with the use of different calcination temperatures and duration (Figure 14.14). Although these particles may have differed in mesoporous structures and composition, when printed, they were found to form a homogeneous distribution of each component, as it can be seen by the TEM and EDX elemental mapping of (a) $Mo_{0.05}V_{0.01}Te_{0.01}Si_{1.0}O_w$ and (b) $Cu_{0.004}TiO_w$ presented in Figure 14.14.

Figure 14.15 summarizes the characterization results of the MMMOs developed with the use of a modified Epson Stylus pro 4,880 inkjet printer and calcined at 350°C. TEM image and elemental mapping (EDX) of IJP-A $Mg_{0.1}Ni_{0.1}Mn_{0.06}Cr_{0.1}Cu_{0.07}Fe_{0.1}Co_{0.1}Ti_{1.0}O_w$ (Figure 14.15a) reveal a uniform X-ray intensity through the particles indicating a homogeneous distribution of the metals. Figure 14.15b presents the optical image of $Co_xFe_yNi_zTi_{1.0}O_w$ library with the space between each sample being 1 mm. SEM-EDX analysis revealed that for the quaternary metal oxides $Mg_xNi_yCu_zTi_{1.0}O_w$, the molar ratio composition of M/Ti (where M is Mg or Ni or Cu) was linear with the programmed color-management deposition of components ($r^2 = 0.99$, Figure 14.15c). TEM images (Figure 14.15d) taken from printed samples of $Cu_xTi_{1.0}O_w$ showed that the structures were consistent with those obtained by conventional methods ranging from ordered (left, $x = 0.005$), to worm-like (middle, $x = 0.05$) and to disordered mesostructured configurations (right, $x = 0.2$).

Authors claim that with the use of the printer mentioned above, more than 1 million materials can be printed in just 1 h (Liu et al., 2012b).

Hydrogen production through the photocatalytic reformation of formaldehyde is widely achieved with the use of TiO_2, a low-cost photocatalyst. For the reaction at hand, 5,000 quaternary metal oxides with different composition of metals were printed ($Mg_xNi_yCu_zTi_{1.0}O_w$) and tested against the aforementioned reaction. Photoreactions were held under irradiation with the use of a 300 W high-pressure Hg lamp at 25°C within a 12-fold parallel system. Hydrogen production was measured with the use of gas chromatography. Further tests were carried out with printed metal oxides containing less metals combined with TiO_2, until the optimal catalyst for the reaction at hand was identified, with regard to the highest hydrogen production rate. The catalyst presenting the best catalytic activity was found to be $Mg_{0.82}Ni_{0.02}Cu_{0.5}Ti_{1.0}O_w$.

As a result, Liu et al. have demonstrated an easy process to accelerate the screening of catalysts for a specific reaction with the use of a new multidimensional group testing approach. Within their work, they rapidly explored the entire candidate catalyst library by using generic analytical tools and reducing the number of experiments performed.

The photocatalytic decomposition of pollutants can be achieved with the use of hierarchical porous photocatalysts that have the ability to recombine photogenerated electrons and holes, thus lowering the photocatalytic degradation efficiency of organic pollutants (Chang et al., 2013). Chang et al. have prepared via inkjet printing hierarchical ZnO nanorod-array films on which Ag was deposited and was used to study its photocatalytic properties towards the photo-induced decomposition of methyl orange. The fabrication process followed by Chang et al. for the development of hierarchical ZnO nanorod-array films is described in Figure 14.16. At first, polymeric barrier ribs were inkjet printed on ITO glass. The ink consisted of 87.5% tri-propylene glycol diacrylate (TPGDA), 10% of pentaerythritol triacrylate (PETA), 2% of photoindicator, and 0.5% of surfactant BYK333. ZnO stripes were placed between the barriers on the indium tin oxide (ITO) glass substrate with electrochemical deposition with the use of a 0.1 M $Zn(NO_3)_2$. ZnO nanorod-array films were developed with the use of hydrothermal and plasma process of a colloid solution

TABLE 14.3 Images of Printed Catalyst Drops and CVD-Grown CNTs

Concentration (M)	FeCl$_3$		Average drop diameter (μm)	Fe(NO$_3$)$_3$		Avg. drop diam.(μm)
	Low magnification	High magnification		Low magnification	High magnification	
0.1			37.91 ± 1.70			31.77 ± 1.58
0.2			60.86 ± 2.80			46.36 ± 1.9
0.4			53.35 ± 2.56			67.92 ± 3.58
0.8			59.10 ± 3.83			50.04 ± 2.43

Source: Chatzikomis et al. (2012).

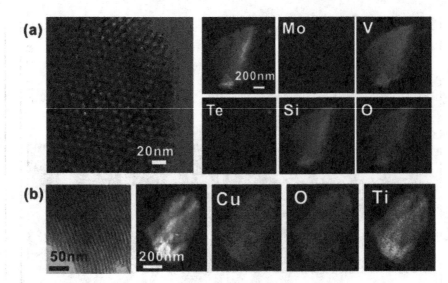

FIGURE 14.14 TEM and EDX elemental mapping of mesostructured (a) $Mo_{0.05}V_{0.01}Te_{0.01}Si_{1.0}O_w$ and (b) $Cu_{0.004}TiO_w$ (Liu et al., 2012).

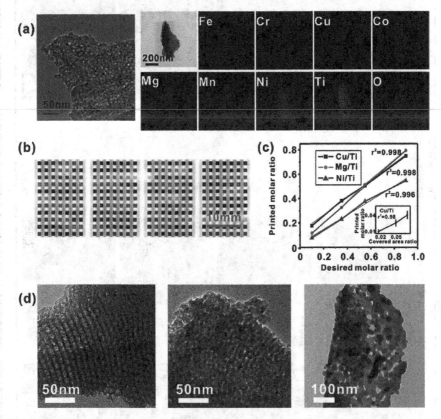

FIGURE 14.15 Surface characterization of MMMOs. (a) TEM image and elemental mapping of inkjet-printed $Mg_{0.1}Ni_{0.1}Mn_{0.06}Cr_{0.1}Cu_{0.07}Fe_{0.1}Co_{0.1}Ti_{1.0}O_w$. (b) Optical image of $Co_xFe_yNi_zTi_{1.0}O_w$ arrays printed on a glass slide. (c) M/Ti molar ratio (M = Cu, Mg, and Ni) of a $Mg_xNi_yCu_zTi_{1.0}O_w$ composite obtained from EDX elemental analyses. Inset: high correlation between actual and desired molar ratio. (d) TEM images of the $Cu_xTi_{1.0}O_w$ samples with a different x value (left, 0.005; middle, 0.05; right, 0.2) (Liu et al., 2012).

$(Zn(CH_3COO)_2$, EtOH, *n*-hexadecyltrimethylammonium hydroxide). Figure 14.17 presents HRTEM images of the polymeric barriers (Figure 14.17a,b), as well as the ZnO nanorods (Figure 14.17c,d). It was found that the width and height of the barriers were 110 and 4.2 μm, respectively.

Images presented in Figure 14.17c,d were taken after the removal of the barriers and present the top view and cross-section of the ZnO porous stripes. The width and height of the ZnO nanorods were found to be 315 and 0.85 μm, respectively, while the surface texture can be tuned by altering

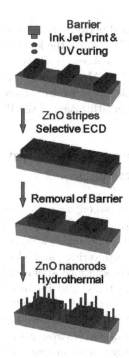

FIGURE 14.16 The fabrication process of making hierarchical ZnO nanorod-array films as proposed by Chang et al. (2013).

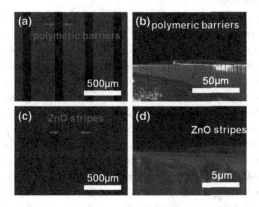

FIGURE 14.17 HRTEM images of the polymeric barriers (a,b) and ZnO stripes (c,d) after the removal of barriers.

the concentration of the precursor and the growth time. The addition of Ag led to an enhanced photocatalytic efficiency towards methyl orange decomposition reaction. This was achieved mainly because the surface area of the printed photocatalyst was enhanced; thus, diffusion and adsorption of organic dye were also enhanced.

14.3.5 Fuel-Cell Applications

Ink jet printing can be used to deposit catalyst materials into gas diffusion layers (GDLs) that are formed in membrane electrode assemblies (MEAs), used in polymer electrolyte fuel cells (PEMFC). Conventional methods cannot achieve an ultra-low loading deposition of metals (<0.5 mg/cm^2); thus, Taylor et al. have developed an IJP method that leads to up to 0.05 mg/cm^2 deposition of Pt on carbon black (CB) anodes (Taylor et al., 2007).

Platinum (II) acetylacetonate was mixed with methanol and CB support in a stainless steel reactor. The mixture was then placed in a sealed sand bath for 30 min at 300°C, conditions that make methanol become a supercritical fluid which caused the reduction of Pt (II) acetylacetonate. The resulting 20 wt% Pt/C catalyst was then dried overnight at room temperature. A Lexmark Z32 thermal inkjet printer (1,200 × 1,200 dpi, drop size 28 pl) was used to print the catalyst on carbon cloth (Toray). Compared to other catalysts studied in the literature, it has been proved that graded catalyst structures performed better than the uniformly distributed catalysts having approximately the same Pt loadings, in terms of electromotive force, power, and current density.

Based on the previous work by Taylor et al. (2007), it is widely accepted that the reduction of Pt loading in the CLs without affecting the overall performance of the cell has drawn the attention of researchers in the fuel-cell industry. Although several Pt deposition methods nowadays are used in the industry, an ultra-low Pt deposition/dispersion cannot be achieved. Even with the use of IJP, a threshold of 0.08 mg/cm^2 Pt loading was found, beyond which the performance of the catalyst was found to be substantially reduced (Shukla et al., 2016).

The CL microstructure is highly dependent on the catalyst ink, since it affects the aggregates size, ionomer interactions, and evaporation rate (Saha, 2013; Holdcroft, 2014; Shukla et al., 2017a; Xu et al., 2010). Shin et al. compared cell performances and concluded that the performance when using a low dielectric constant dispersion medium, when the ionomer is in a colloidal state, is better than that when using constant media with high dielectric constant (Shin et al., 2002). The effect of propylene glycol (PG) and EG on the CLs pore-size distribution, porosity, and performance was studied by Chisaka et al., and it was found that the pore-volume and thickness of the PG sample were higher than that of the EG sample (Chisaka & Daiguji, 2009). However, the electrochemically active area and cell performance were found to be higher with the EG-based CL. Taking into consideration that IJP requires the ink to possess a desired viscosity and surface tension to allow an efficient jetting through the micrometer-sized nozzles, it can be concluded that the faster the evaporation method used, the lower the viscosity of the ink will be and vice versa.

Shukla et al. used a mixture of PG and IPA in order to disperse the catalyst particles and ionomer, using the ink to print directly on a membrane (substrate) (Shukla et al., 2017b). For comparison purposes, EG-based electrodes were also prepared. For the PG-based catalyst ink, a high surface area 50 wt% Pt/C catalyst powder was dispersed in the PG-IPA mixture in a sonication bath. The effect of the PG amount in the mixture was studied by preparing several mixtures of 35, 46, and 58 wt% PG, while IPA volume was adjusted accordingly in order to maintain similar solid fractions in the ink. During sonication, a 5 wt% of solubilized ionomer was added, and in order to determine the optimal ionomer content in the CLs, several inks were produced with an ionomer-to-carbon ratio of 0.5, 0.8, and 1.1. The printer used was a Fujifilm Dimatix-2831 materials printer, and the printed CLs area was found to be of 40 cm^2.

For further electrode microstructure characterization with the use of a laser scanning microscope, samples were printed over polytetrafluoroethylene (PTFE) sheet and NRE-211 membrane. Figure 14.18 presents the surface profiles for the EG- and mixture-based CLs on the two substrates. As it can be seen, the ink recipe does not affect the surface structure as much as the nature of the substrate, as it can be observed by the crack density. PTFE CLs are found to be smooth and without cracks, while the NRE211 CLs appeared to have several cracks. Specifically, the cracking density of PTFE CLs was measured to be only 0.4% for the mixture of PG-IPA (a) and 0.8% for the EG (c), respectively. On the contrary, the cracking density measured on the CLs printed on NRE211 membrane was found to be 6.4% for the PG-IPA mixture (b) and 5.6% for the EG-based CL (d).

It was also found that for a Pt loading of 0.15 mg/cm^2, CLs produced with the PG-IPA mixture had a porosity of 69.5% which is 20% higher than the EG-CLs. Moreover, the mixture-CLs presented a 100 mV improvement in their in situ performance in comparison with the EG-CLs.

The fine-tuning of the material properties of electrodes and membranes is of high importance in order to provide cost-effective materials for PEMFC manufacturing. Saha et al. have studied the structure of cathode catalyst layers (CCLs) in accordance with the preparation method and fuel-cell performance (Saha, 2013). Specifically, they achieved to deposit the catalyst ink directly on a Nafion membrane with high loading accuracy and spatial precision with the use of a printing system capable of tuning the speed and ink ejection characteristics, thus optimizing the printing parameters. Fabrication conditions were studied in relation to the electrochemical performance.

The inks used were prepared by mixing 50 wt% Pt/C with the desired amounts of 5 wt% solubilized Nafion, IPA, and GC. The ionomer-to-carbon ratio (I/C) was kept steady at 0.8. A Fujifilm DPM-2800 Dimatix Materials Printer was used to print the catalyst-coated membrane (CCM), and every layer was dried for 5 min at 50°C prior printing the next layer. After printing, the CCLs were dried, boiled, and dried again in a vacuum oven. Several CCLs were prepared with different Pt loadings (0.075, 0.1, 0.13, and 0.2 mg/cm^2). For comparison purposes, a CCM was also prepared with a conventional method. SEM images of the cross sections of the two differently prepared CCMs are presented in Figure 14.19, i.e., (a) conventional method and (b) IJP method. The electrographs are presented in the back scattering mode, with the contrasting bright layers containing the element with the largest atomic number (Pt). As shown in Figure 14.19, the conventionally prepared CCM presents less homogeneous electrodes (Figure 14.19a) compared to the IJP CCM (Figure 14.18b). Moreover, the delamination observed in Figure 14.19a indicates low adhesion between the membrane and the cathode, unlike the IJP CCM (Figure 14.19b).

The electrochemical performance of both CCMs revealed that for the IJP CCM sample, the lower the Pt loading, the lower performance was observed. However, the 0.2 mg/cm^2 Pt loading of the IJP CCM yields similar performance to the 0.25 mg/cm^2 sample prepared with conventional method, thus indicating better utilization of the Pt catalyst.

14.4 Challenges in Inkjet Printing of Catalysts

Inkjet must be understood as a complete system. Many discipline areas (materials science, chemistry, device physics, production engineering, system integration, software, mechanical engineering, and electronics) must be combined. Customers still face application challenges, and available production tools, still in their infancy, often do not meet all requirements (Hudd, 2009). It is uncommon for systems to work in the laboratory but is not yet ready for a demanding industrial environment.

The main challenges in improving the performance and utilization of inkjet printing are:

i. *Materials*

   Increasing, but still slow, is the development of jettable materials. There is no such thing as a universal ink (Pond et al., 2000). In each case, many issues have to be considered, such as application performance, print quality (e.g., surface wetting), compatibility, drying time/speed, adhesion (sometimes interlayer interactions), "image" robustness (e.g., abrasion resistance), jetting characteristics (viscosity, dynamic surface tension, particle size, compatibility), reliability (volatility, purging, wetting of capillary channels, shelf life), ease of manufacturing (e.g., cost and availability of materials), regulatory, post- or pre-processing requirements (e.g., UV, heating, inert atmospheres).

   An abundant approach to achieving jettable materials is based on adjusting the composition of an existing ink to match the printing requirements of each system. Usually, this approach is not trivial, since non-inkjet formulations usually have very different properties to inkjet fluids (Pond et al., 2000). For example, in order to convert a silk screen printing ink, one would require, among other changes, to significantly decrease the viscosity and particle size in pigment-containing inks.

   Decreasing the viscosity, in the case of large pigment particles, would lead to sedimentation and agglomeration of the nanoparticles. To avoid such an effect would require a submicron pigment size (also important for not clogging the printhead). Such pigments (e.g., metallic, ceramic) are not always available commercially, and in that case, they should be manufactured specifically for the new inkjet ink (Hudd, 2009).

ii. *Feature size*

Another challenge is associated with feature size reduction, especially for sophisticated printing of functional materials, such as printed electronics (Pond et al., 2000). This can be achieved by combined effects of the whole printing system, such as surface treatment of the substrates (see Section 2.2) and achieving unique rheological behavior of the ink.

iii. *Resolution and productivity*

Higher resolution and substrate handling at higher speed is a very demanding task. While approaching the fundamental limits of increased jetting frequency, the productivity needs to be improved in other creative ways. To date, this has been accomplished through increasing the number of nozzles, although this is directly related to increased cost (Hudd, 2009).

iv. *Drop placement accuracy*

Exact drop positioning is uncertain, due to several parameters, such as jet-to-jet variations, sensitivity to nozzle straightness, nozzle and surface wetting, nozzle plate contamination, and drop velocity (Pond et al., 2000). This issue is worse for longer flight paths (throw distance).

In summary, inkjet materials printing, although attractive in concept, is very difficult to implement in practice, especially in very demanding applications such as very high-throughput systems, and printing of novel functional materials. In the case of functional materials (such as metal

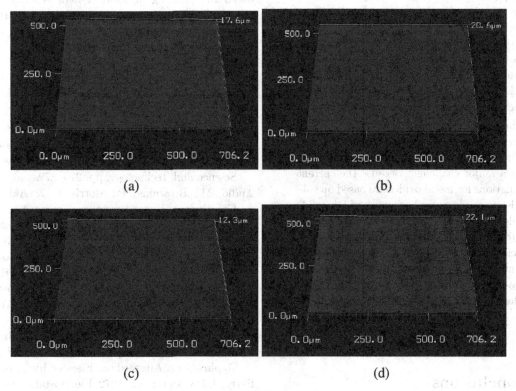

FIGURE 14.18 Scanning laser micrographs of the effect of the substrate on the crack density observed on CLs. (a) PG-IPA CL on PTFE, (b) PG-IPA CL on NRE211, (c) EG CL on PTFE, and (d) EG CL on NRE211 (Shukla et al., 2017).

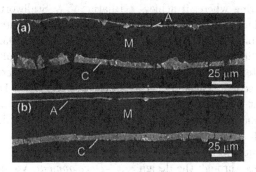

FIGURE 14.19 SEM electrographs of the cross sections of CCMs prepared with (a) conventional method and (b) IJP method used in this research (Saha, 2013).

oxide catalysts), there are two particular obstacles that withhold the widespread application of this technique to the fabrication of functional metal oxides (Liu et al., 2015).

- There are only a few industrial inkjet printers that have been designed to print metal oxides. In addition, challenges such as the fabrication of uniform and homogeneous multi-component structures, coffee staining, and maintaining stability during post-processing must be addressed through improvements in our understanding of nano-ink fluid behavior. These challenges limit the available opportunities to use this technique, especially for newer researchers in this field. Therefore, greater efforts must be focused on the development of a simple, low-cost, energy-efficient, and environmentally-friendly method for ink formulation to produce the variable compositions and structures required for the different applications. The interactions of the droplets and the drop drying processes are clearly important for determination of the quality, accuracy, and properties of the functional metal oxides. However, little systematic work with regard to this aspect exists in the literature to date.

- Current inkjet printing techniques are mostly used to fabricate thin-film metal oxides (Liu et al., 2015). The fabrication of three-dimensional (3D) objects is a major challenge. Because the current ink formulations for metal oxides are based on sol–gel chemistry, the increased amounts of ink solids that would be required for 3D printing would be difficult to achieve with the current formulations. The development of a fast drying process for 3D printing may enable greater complexity in the sol–gel ink system. However, these difficulties will not hinder the exploration of the application of these materials to the fabrication of large 3D device arrays.

14.5 Conclusions

Inkjet printing offers advantages that make it ideal in applications of non-traditional manufacturing where, if achievable, high-resolution and high printing speed would open up the possibilities for competitive production of quality objects by the extension of existing inkjet-based additive manufacturing processes. In this context, the potential for additive manufacture is to be more economical and more environmentally-friendly than current methods. Manufacture by inkjet presents challenges to chemists and materials scientists to find ways to deliver both structural and functional materials in liquid form and subject to the constraints of the process. Solving these challenges through the design of specialized printheads and ink formulation is not made easier by the fact that inkjet technology is still regarded by many experts in the field as "black magic". Although several groups have attempted to model theoretically the hydrodynamical processes occurring in the ink chamber and at the nozzle plate, in many cases, ink formulation and printhead design are still based on empirical experience and trials. A better theoretical understanding of the inkjet process would be of great help to printhead designers and ink formulators. Similarly, a theoretical understanding of the process of spreading and drying of a micro-liquid deposited onto a substrate is challenging because a micro-liquid exhibits specific size effects arising from its high surface-to-volume ratio, such as different drying kinetics, that are not observed for bulk liquids.

The inkjet technology has been classified by many as an emerging manufacturing technique. While the term "desktop manufacturing" is probably inappropriate, nevertheless, the prospect for distributed manufacture by inkjet-based 3D printing at local centers is similar to that of the distributed printing model for graphical products, with potentially far more profound effects both socially and economically.

References

Amemiya, I. et al., 1999. Ink jet printing with focused ultrasonic beams. In: E. Hansen, ed. *Recent Progress in Ink Jet Technologies II*. Springfield, VA: Society for Imaging Science and Technology, pp. 275–279.

Andio, M., Browning, P., Morris, P. & Akbar, S., 2012. Comparison of gas sensor performance of SnO_2 nanostructures on microhotplate platforms. *Sensors and Actuators B: Chemical*, Volume 165, pp. 13–18.

Arin, M. et al., 2012. Deposition of photocatalytically active TiO_2 films by inkjet printing of TiO_2 nanoparticle suspensions obtained from microwave-assisted hydrothermal synthesis. *Nanotechnology*, Volume 23, p. 165603.

Atasheh, S.-G., 2016. Inkjet printing. In: J. Izdebska & S. Thomas, eds. *Printing on Polymers: Fundamentals and Applications*. Amsterdam: Elsevier Inc., pp. 231–246.

Berry, J. & Corpron, G., 1972. Electrostatic printing composition comprising didodecyl sebacate. United States of America, Patent No. 3653932.

Bhatia, Y. & Stallworth, H., 1986. Ink jet printing composition. United States of America, Patent No. 4567213.

Blazdell, P. et al., 1995. The computer-aided manufacture of ceramics using multilayer jet printing. *Journal of Materials Science Letters*, Volume 14, pp. 1562–1565.

Buck, R. et al., 1985. Disposable ink jet head. United States of America, Patent No. 4500895 A.

Buehner, W.L., Hill, J.D., Williams, T.H. & Woods, J.W., 1977. Application of ink jet technology to a word processing output printer. *IBM Journal of Research and Development*, Volume 21, pp. 2–9.

Cerna, M. et al., 2013. Fabrication, characterization and photocatalytic activity of TiO_2 layers prepared by inkjet printing of stabilized nanocrystalline suspensions. *Applied Catalysis B: Environmental*, Volume 138, pp. 84–94.

Chandrasekaran, C., Ahmed, A. & Henzler, T., 2003. Inkjet ink for textiles. European, Patent No. WO 2003076532 A1.

Chang, C. et al., 2013. Hierarchical ZnO nanorod-array films with enhanced photocatalytic performance. *Tin Solid Films*, Volume 528, pp. 167–174.

Chatziiona, V. et al., 2018. Regulating the catalytic properties of Pt/Al_2O_3 through nanoscale inkjet printing. *Catalysis Communications*, Volume 103, pp. 69–73.

Chatzikomis, C., Pattison, S., Koziol, K. & Hutchings, I., 2012. Patterning of carbon nanotube structures by inkjet printing of catalyst. *Journal of Materials Science*, Volume 47, pp. 5760–5765.

Chisaka, M. & Daiguji, H., 2009. Effect of organic solvents on catalyst layer structure in polymer electrolyte membrane fuel cells. *Journal of The Electrochemical Society*, Volume 156(1), pp. B22–B26.

Choi, J. et al., 2008. Drop-on-demand printing of conductive ink by electrostatic field induced inkjet head. *Applied Physics Letters*, Volume 93, p. 193508.

Costa, C., Pinheiro, C., Henriques, I. & Laia, C., 2012. Inkjet printing of sol–gel synthesized hydrated tungsten oxide nanoparticles for flexible electrochromic devices. *ACS Applied Materials and Interfaces*, Volume 4, pp. 1330–1340.

Derby, B., 2010. Inkjet printing of functional and structural materials: Fluid property requirements, feature stability, and resolution. *Annual Review of Materials Research*, Volume 40, pp. 395–414.

Dimatix, 2017. Fujifilm Dimatix Inc. [Online] Available at: www.dimatix.com [Accessed 2017].

Dondi, M., Blosi, M., Gardini, D. & Zanelli, C., 2012. Ceramic pigments for digital decoration inks: An overview. *Ceramic Forum International*, Volume 89, pp. E59–E64.

Dong, H., Carr, W. & Morris, J., 2006. An experimental study of drop-on-demand drop formation. *Physics of Fluids*, Volume 18, p. 072102.

Dou, R., Wang, T., Guo, Y. & Derby, B., 2011. Ink-jet printing of zirconia: Coffee staining and line stability. *Journal of the American Ceramic Society*, Volume 94, pp. 3787–3792.

Duineveld, P. et al., 2002. *Ink-Jet Printing of Polymer Light-Emitting Devices*. San Diego, CA: International Society for Optics and Photonics, pp. 59–67.

Elmqvist, R., 1951. Measuring instrument of the recording type. US Patent, US2566443A.

Fan, J., Liu, X. & Huang, F., 2013. Mesoporous metal oxide catalysts: Rational synthesis and fast exploration. *Chemistry (in Chinese)*, Volume 76(8), pp. 675–683.

Fasaki, I. et al., 2012. Ultrasound assisted preparation of stable water-based nanocrystalline TiO_2 suspensions for photocatalytic applications of inkjet-printed films. *Applied Catalysis A: General*, Volume 411, pp. 60–69.

Feys, J. et al., 2012. Ink-jet printing of $YBa_2Cu_3O_7$ superconducting coatings and patterns from aqueous solutions. *Journal of Materials Chemistry*, Volume 22, pp. 3717–3726.

Fromm, J., 1984. Numerical-calculation of the fluid-dynamics of drop-on-demand jets. *IBM Journal of Research and Development*, Volume 28, pp. 322–333.

Güell, F. et al., 2016. Spatially controlled growth of highly crystalline ZnO nanowires by an inkjet-printing catalyst-free method. *Materials Research Express*, Volume 3, p. 025010.

Hadimioglu, B. et al., 1992. Acoustic ink printing. *IEEE 1992 Ultrasonics Symposium Proceedings*, Tucson, AZ, pp. 929–935.

Han, S.-Y., Lee, D.-H., Herman, G. & Chang, C.-H., 2009. Inkjet printed high mobility transparent oxide semiconductors. *Journal of Display Technology*, Volume 5, pp. 520–524.

Han, W.-S., Hong, J.-M., Kim, H.-S. & Song, Y.-W., 2011. Multi-pulsed white light sintering of printed Cu nanoinks. *Nanotechnology*, Volume 22, p. 395705.

Haverinen, H.M., Myllyla, R.A. & Jabbour, G.E., 2010. Multi-pulsed white light sintering of printed Cu nanoinks. *Nanotechnology*, Volume 22, p. 395705.

Hayes, D. et al., 2002. Inkjet printing in the manufacture of electronics, photonics, and displays. *International Symposium on Optical Science and Technology*, Seattle, Washington DC, International Society for Optics and Photonics, pp. 94–99.

Heinzl, J., 1981. Printing with ink droplets from a multi-nozzle device. In: J. Gaynor, ed. *Advances in Non-Impact Printing Technologies for Computer and Office Applications*, New York: Van Nostrand Reinhold, pp. 1191–1201.

Hench, L. & West, J., 1990. The sol–gel process. *Chemical Reviews*, Volume 90, pp. 33–72.

Holdcroft, S., 2014. Fuel cell catalyst layers: A polymer science perspective. *Chemistry of Materials*, Volume 26(1), pp. 381–393.

Hoth, C., Choulis, S., Schilinsky, P. & Brabec, C., 2007. High photovoltaic performance of inkjet printed polymer: Fullerene blends. *Advanced Materials*, Volume 19, pp. 3973–3978.

Huang, J. et al., 2008. Electrochemical properties of $LiCoO_2$ thin film electrode prepared by ink-jet printing technique. *Thin Solid Films*, Volume 516, pp. 3314–3319.

Hudd, A., 2009. Inkjet printing technologies. In: S. Magdassi, ed. *The Chemistry of Inkjet Inks*. Singapore: World Scientific Publishing Co. Pte. Ltd., pp. 3–11.

Jang, D., Kim, D. & Moon, J., 2009. Influence of fluid physical properties on ink-jet printability. *Langmuir*, Volume 25, pp. 2629–2635.

Jeong, J., Lee, J., Kim, H. & Na, S., 2010. Ink-jet printed transparent electrode using nano-size indium tin oxide particles for organic photovoltaics. *Solar Energy Materials and Solar Cells*, Volume 94, pp. 1840–1844.

Johnson, T. & Bower, K., 1979. Review of the drop on-demand ink-jet with primary emphasis on the gould jet concept. *Journal of Applied Photographic Engineering,* Volume 5(3), p. 174.

Keeling, M., 1981. Ink jet printing. *Physics in Technology,* Volume 12, p. 196.

Kobayashi, H., Koumura, N. & Ohno, S., 1981. Liquid recording medium. United States of America, Patent No. 4243994.

Ko, S. et al., 2007. All ink-jet-printed flexible electronics fabrication on a polymer substrate by low-temperature high-resolution selective laser sintering of metal nanoparticles. *Nanotechnology,* Volume 18, p. 345202.

Kuscer, D., Noshchenko, O., Ursic, H. & Malic, B., 2013. Piezoelectric properties of ink-jet-printed lead zirconate titanate thick films confirmed by piezoresponse force microscopy. *Journal of the American Ceramic Society,* Volume 96, pp. 2714–2717.

Kuscer, D., Stavber, G., Trefalt, G. & Kosec, M., 2012. Formulation of an aqueous titania suspension and its patterning with ink-jet printing technology. *Journal of the American Ceramic Society,* Volume 95, pp. 487–493.

Kyser, E. & Sears, S., 1976. Method and apparatus for recording with writing fluids and drop projection means therefore. United States of America, Patent No. 3946398.

Lee, D.-H., Chang, Y.-J., Herman, G. & Chang, C.-H., 2007. A general route to printable high-mobility transparent amorphous oxide semiconductors. *Advanced Materials,* Volume 19, pp. 843–847.

Lee, S. et al., 2013. Inkjet printing of porous nanoparticle-based catalyst layers in microchannel reactors. *Applied Catalysis A: General,* Volume 467, pp. 69–75.

Le, H., 1998. Progress and trends in ink-jet printing technology. *Journal of Imaging Science and Technology,* Volume 42, pp. 49–62.

Lehnen, T., Zopes, D. & Mathur, S., 2012. Phase-selective microwave synthesis and inkjet printing applications of Zn_2SnO_4 (ZTO) quantum dots. *Journal of Materials Chemistry,* Volume 22, pp. 17732–17736.

Liu, T., Zeng, Z., Wang, X. & Xia, X., 2012a. Application of inkjet printing to fabricate controllable Pt catalyst patterns for low temperature catalytic combustion. *Advanced Materials Research,* Volume 550–553, pp. 257–260.

Liu, X. et al., 2012b. Inkjet printing assisted synthesis of multicomponent mesoporous metal oxides for ultra-fast catalyst exploration. *Nano Letters,* Volume 12, pp. 5733–5739.

Liu, X., Tarn, T.-J., Huanga, F. & Fan, J., 2015. Recent advances in inkjet printing synthesis of functional metal oxides. *Particuology,* Volume 19, pp. 1–13.

Maier, W., Stowe, K. & Sieg, S., 2007. Combinatorial and high-throughput mate-rials science. *Angewandte Chemie International Edition,* Volume 46(32), pp. 6016–6067.

Mansoor, M., Kinloch, I. & Derby, B., 2010. Inkjet printing of catalyst-inks on Si wafers and the subsequent synthesis of carbon nanotubes by chemical vapour deposition. *Key Engineering Materials,* Volume 442, pp. 7–14.

Microdrop, 2018. Microdrop. [Online] Available at: www.microdrop.com.

Niederberger, M. & Pinna, N., 2009. *Metal Oxide Nanoparticles in Organic Solvents.* London: Springer Verlag.

Noguchi, H. & Shimomura, M., 2001. Aqueous UV-curable ink for inkjet printing. *RadTech Report,* Volume 15(2), pp. 22–25.

Oezkol, E., Ebert, J. & Telle, R., 2010. An experimental analysis of the influence of the ink properties on the drop formation for direct thermal inkjet printing of high solid content aqueous 3Y-TZP suspensions. *Journal of the European Ceramic Society,* Volume 30, pp. 1669–1678.

Park, J.-U. et al., 2007. High-resolution electrohydrodynamic jet printing. *Nature Materials,* Volume 6, pp. 782–789.

Pond, S., Wnek, W., Doll, P. & Andreottola, M., 2000. Ink design. In: S. Pond, ed. *Inkjet Technology and Product Development Strategies.* Carlsbad, CA: Torrey Pines Research, pp. 153–210.

Rayleigh, L., 1892. On the instability of a cylinder of viscous liquid under capillary force. *The London, Edinburgh, and Dublin Philosophical Magazine and Journal of Science,* Volume 34, pp. 145–154.

Reis, N. & Derby, B., 2000. Ink jet deposition of ceramic suspensions: Modelling and experiments of droplet formation. *Materials Research Society Symposium,* San Francisco, CA, pp. 117–122.

Saha, M.S. et al., 2013. Characterization of inkjet printed electrodes with improved porosity. *ECS Transactions,* Volume 58(1), p. 797.

Savart, F., 1833. Memoire sur la constitution des veines liquides lancees par des orifices circulaires en mince paroi. *Annales de chimie et de physique,* Volume 53, p. 1833.

Shin, S.-J. et al., 2002. Effect of the catalytic ink preparation method on the performance of polymer electrolyte membrane fuel cells. *Journal of Power Sources,* Volume 106(1), pp. 146–152.

Shukla, S. et al., 2017a. Characterization of inkjet printed electrodes with improved porosity. *ECS Transactions,* Volume 77(11), pp. 1453–1463.

Shukla, S., Bhattacharjee, S., Weber, A.Z. & Secanell, M., 2017b. Experimental and theoretical analysis of ink dispersion stability for polymer electrolyte fuel cell applications. *Journal of The Electrochemical Society,* Volume 164(6), pp. F600–F609.

Shukla, S., Stanier, D., Saha, M. & Stumper, J., 2016. Analysis of inkjet printed PEFC electrodes with varying platinum loading. *Journal of the Electrochemical Society,* Volume 163(7), pp. F677–F687.

Silverbrook, K., 1998. Fax machine with concurrent drop selection and drop separation ink jet printing. United States of America, Patent No. 5781202.

Singh, M., Haverinen, H. & Dhagat, P., 2010. Inkjet printing: Process and its applications. *Advanced Materials,* Volume 22, pp. 673–685.

Sirringhaus, H. & Shimoda, T., 2003. MRS bulletin/November 2003. [Online] Available at: www.cambridge.org/core.

Stemme, N., 1973. Arrangement of writing mechanisms for writing on paper with a colored liquid. United States of America, Patent No. 3747120.

Stow, C. & Hadfield, M., 1981. An experimental investigation of fluid flowresulting from the impact of a water drop with an unyielding dry surface. *Proceedings of the Royal Society of London: Series A*, Volume 373(1755), pp. 419–441.

Strutt, J. W. & Rayleigh, L., 1878. On the instability of jets. *Proceedings of the London Mathematical Society*, Volume 10, pp. 4–13.

Sun, X.-D. et al., 1997. Solution-phase synthesis of luminescent materials libraries. *Advanced Materials*, Volume 9, pp. 1046–1049.

Sweet, R.G., 1965. High frequency recording with electrostatically deflected ink jets. *Review of Scientific Instruments*, Volume 36, pp. 131–136.

Taylor, A. et al., 2007. Inkjet printing of carbon supported platinum 3-D catalyst layers for use in fuel cells. *Journal of Power Sources*, Volume 171, pp. 101–106.

Wijshoff, H., 2010. The dynamics of the piezo inkjet printhead operation. *Physics Reports*, Volume 491, pp. 77–177.

Wood, R. & Loomis, A., 1927. The physical and biological effects of high frequency sound waves of great intensity. *Philosophical Magazine*, Volume 4(22), pp. 417–436.

Xaar, 2012. *A Guide to Industrial Inkjet*, Cambridge: Xaar.

Xu, F. et al., 2010. Investigation of a catalyst ink dispersion using both ultra-small-angle X-ray scattering and cryogenic TEM. *Langmuir*, Volume 26(24), pp. 19199–19208.

Yamada, T., Matsuda, Y., Yoshino, M. & Sakata, M., 1988. Micro-dot ink jet recorder. United States of America, Patent No. 4746928.

Yuan, S. et al., 1999. The development of receiving coatings for inkjet imaging applications. In: E. Hansen, ed. *Recent Progress in Ink Jet Technologies II*, Springfield, VA: Society for Imaging Science and Technology, pp. 378–382.

Zhang, R., Luo, J., Lian, H. & Li, H., 2017. Control of particle size in energetic drop-on-demand inkjet method. *2017 IEEE International Conference on Manipulation, Manufacturing and Measurement on the Nanoscale (3M-NANO)*, Shangai, pp. 335–338.

Zoltan, S., 1972. Pulsed droplet ejecting system. United States of America, Patent No. 3683212.

15

Calix-Assisted Fabrication of Metal Nanoparticles: Applications and Theoretical Insights

Anita R. Kongor,
Manthan K. Panchal, and
Vinod K. Jain
Gujarat University

Mohd Athar
Central University of Gujarat

15.1 Introduction

15.1.1 Design and Fabrication of Nanoparticles: A General Introduction

The term "nano" in the word "nanoparticles" has been originated from a Latin word, meaning "dwarf". Very fine materials with nanometer-scale dimensions; that is, billionth of a meter (1 nm = 10^{-9} m) are defined generally as nanoparticles. These smaller entities do not have any roots of discovery in scientific literature, as because they have been notably used since prehistoric times. For instance, nanoparticles were used by the *Damascans* to create swords with exceptionally sharp edges and the Romans to craft iridescent glassware. Such long history gives us a message of extraordinary inspiration to uplift the research on nanoparticles. There is something special about these nanoparticles which have stimulated an impressive scope of research interests. With novel experimental findings, there has been a great progress in the synthesis and fabrication of nanoparticles.

To achieve specific applications relying on the optical, electronic, or catalytic properties, the control over the size, shape, as well as composition of nanoparticles has to be well accounted. The noble metal nanostructures based on gold and silver have grabbed special attention due to their extraordinary abilities to absorb and scatter light. Preparing metal nanoparticles with well-defined shapes has been a great challenge. Any small changes in the experimental conditions may greatly influence the particle growth. Thus, a space for complexity can arise when working with materials on nanoscale.

The first experimental approach to synthesize metal nanoparticles was in 1857 when Faraday, the father of modern nanoscience, first prepared and characterized colloidal gold nanoparticles in a controlled manner (Faraday 1857). Since then, numerous experimental methods have been developed to synthesize metal nanoparticles, for example, the classical reduction of chloroauric acid in aqueous solution by trisodium citrate. In the 1950s, this synthetic procedure was identified and studied by John Turkevich, a pioneer in the field of catalysis (Turkevich et al. 1951). Different chemical procedures using sodium borohydride, ascorbic acid, trisodium citrate, or alcohols as reducing agents were then reported (Masala and Seshadri 2004). The synthesis of metal nanoparticles was often carried out using a stabilizing agent in order to prevent aggregation and to control the morphology of nanoparticles. A stable bonding is necessary to provide access to durable applications in the field of sensing and catalysis.

The reducing as well as stabilizing properties of one of the most intriguing macrocyclic compound, calixarenes, is undoubtedly an exciting platform to obtain different fabricated metal nanoparticles. It was Stavenset al. who for the first time adopted new strategy to encapsulate nanoclusters of gold using a calix system, resorcinol-derived calix[4]arenes (resorcinarenes) (Stavens et al. 1999). With mesmerizing coordination of two oppositely sized materials (calixarenes being large), there is a possibility to establish some unexplored chemistry with countless and promising applications waiting to be discovered. The chemical functionality of the nanoparticles using calixarene platforms can be tailored accordingly to target a specific application. In this chapter, a correlation between the coordination of experimental and theoretical research for the design of fabricated nanoparticles is also focused. More appropriately, different efficient and accurate methods will be optimized in order to understand more and more about the constantly rising potential of nanoparticles. As per the famous saying: "God has perfect timing: never early, never late. It takes a little patience and a whole lot of faith, but it's worth the wait". Nevertheless, definitely, more newly constructed species of nanoparticles are expected to be generated in near scientific era.

15.2 Synthetic Strategies

15.2.1 Different Synthetic Strategies for the Formation of Nanoparticles

In the past few years, the research and development on the synthesis of nanoparticles have been exploded to extend the prospects of nanoscience with a myriad of applications. No sort of infancy in seeking the knowledge of nanoparticles is witnessed in science laboratories. State-of-the-art methods are now delivered to provide deeper insights to manufacture nanomaterials of varied shapes and size.

With this context, till date, numerous synthetic strategies have been designed to develop materials that have spatial dimensions ranging in nanometric scale. So far, only two broad strategies, *top-down* and *bottom-up* approaches, have been mentioned in the literature for nanofabrication process. It should also be noted that nanoparticles are not bound to be synthesized in laboratories; they are accidentally produced in the environment, for example, combustion of engine in vehicles and cigarettes, and emissions from printing equipment. It is somewhat a challenging task to synthesize pure nanoparticles on a larger scale in terms of economy and sustainability criteria.

The top-down method was first suggested by Feynman in his famous talk, named "There is plenty of room at the bottom" (Hutchings 1960). The "top-down" approach represents step-by-step building of material from bulkier objects to nanometer-sized particles. This method involves cutting of the bulk (precursor) material which is further subdivided into smaller units. On the other hand, the "bottom-up" approach involves building of small and uniform nanoparticles through specific chemical route. Depending on the reproducibility and cost, there are advantages and disadvantages using both the approaches. Comparatively, the "bottom-up" approach holds greater potential to design nanomaterials with practical applications.

Using the techniques developed by solid-state physicists, nanotechnological devices have been prepared by the breakdown of materials. For instance, a bulky block of silicon wafer can be reduced to smaller components by etching, cutting, and slicing down to a desired size or shape. This is known as the "top-down" approach. On the other hand, the synthesis of nanostructures and nanomaterials by making use of supramolecular and biomimetic materials is known as the "bottom-up" approach. The nanomaterials synthesized through the bottom-up method have distinct physico-chemical properties that differ from the bulk material and can stimulate the emergence of novel characteristics (Steed et al. 2007). A schematic representation of these two approaches is shown in Figure 15.1.

The use of both top-down and bottom-up approaches has attracted the researchers in material science, physics, chemistry, and engineering fields because of the promising wide range of applications it heralds. Different application intends the use of particular shape or composition of nanoparticles; thus, the use of a particular approach becomes

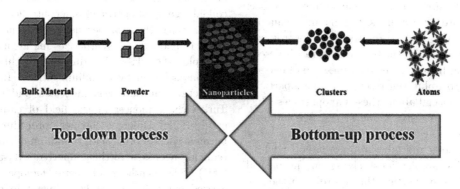

FIGURE 15.1   Representation of Synthetic strategies: top-down and bottom-up approaches.

more specific. More often, the use of a hybrid method involving both the approaches in a simultaneous manner is witnessed.

15.2.2 Types of Top-Down Approach

The "top-down" method of preparing nanoparticles involves high energy processes to physically break the material. Below are some of the top-down approaches used frequently by scientists or industry-related research concerning nanotechnology.

Milling (Mechanical Attrition)

Attrition or milling (Yadav et al. 2012) is a typical top-down method in making nanoparticles and is found to be useful to produce tonnage quantities of materials for various applications, mainly in industries. There are certain advantages of this method such as simplicity and inexpensive cost of equipment. But still, the downside of this approach includes contamination issue due to milling balls and to consolidate the powder product. The milling method includes high energy mills such as vibrating ball mills, low-energy tumbling mills, shaker mills, planetary ball mill, and attrition mill. Using these five types of milling balls, the nanoparticles are produced by high shear and forces during grinding. The implications of such milling process with high shear can impart imperfection on the surface of the nanomaterials. Therefore, the control parameters to be taken into consideration are the size of the milling ball, time duration for milling process, and the material used for milling.

Laser Ablation

Dry and wet laser ablation methods are the two types of laser ablation used so far (Sun et al. 2006, Tsuji et al. 2008, Hassan et al. 2016). This method involves the fabrication of nanoparticles with the use of laser system as an energy source. Depending on the parameters such as source, frequency, and wavelength of the laser, the solid target material is ablated, which means removal of the surface atoms takes place through evaporation. This advanced laser ablation method has an affinity to synthesize morphology-controlled nanoparticles for large-scale production. Moreover, the purity of the nanoparticles is not compromised, due to the absence of any by-products or residue produced in this method. Various kinds of nanoparticles such as semiconductor quantum dots, nanoalloys, nanocarbons, and metal and metal oxide nanoparticles have been prepared using this technique.

Arc Discharge

The technique of arc discharge has received considerable interest with the exploration of novel methods to synthesize nanoparticles (Shi et al. 1999, Zhou et al. 1999). In this method, the word arc refers to luminous electrical discharge in the shape of "arc" between two electrodes. The arc discharge takes place due to electrical breakdown of a gas

when current is applied between two electrodes, resulting in the discharge of plasma. The product depends on the voltage and current applied to the electrode and the material used as cathode/anode. Thus, arc discharge method presents the production of nanoparticles due to arc-assisted breakdown of bulk materials. A high yield of product is difficult to achieve using this method and requires careful control of the experimental conditions.

Carbon nanotubes are the widely reported nanoparticles using the arc discharge method (Shi et al. 1999). The synthetic model is based on the physical properties of the arc discharge plasma where the interplay of the two main components of the bimodal carbon velocity distribution (Maxwellian and directed) controls the process for creating nanotubes in a zone near the cathode surface (Gamaly and Ebbesen 1995).

Nanolithography

Nanolithography is the most widely used standard method to create a large variety of nanostructures with controlled size, shape, and interparticle spacing (Haynes and Van Duyne 2001).This method deals with writing and etching on the surface at microscopic level. Two main microscopic methods involved in lithography techniques are atomic force microscope (AFM) andthe scanning probe microscope (SPM). Lithographic processes can be performed using high-energy electrons (e-beam lithography), ions (i-beam lithography), light (photolithography), or X-ray (X-ray lithography) methods (Madou 2002). New nanolithography technologies such as extreme ultraviolet (EUV) lithography (Jee et al. 2010), "dip-pen" nanolithography (DPN) method (Lu et al. 2012), and Magneto-lithography (Nie et al. 2017) have explored the possibility of better nanoscale features.

15.2.3 Types of Bottom-Up Approach

The bottom-up methods have the ability to miniaturize the building blocks to form the nanostructures in a simpler way. In simple words, atoms and molecules are assembled carefully by controlled chemical reactions to generate small nanoparticles (typical size of 2–10 nm). Thus, this method is efficient when compared to physical methods mentioned into p-down methods.

Chemical Precipitation

Chemical precipitation method involves easy procedure to control the size of the particle and is cost-effective. In this method, solutions of two different ions are mixed under controlled temperature and pressure to generate nano-sized precipitates. Using this method, nanoparticles of metals, alloys, oxides, etc. have been prepared from aqueous or organic solutions. The downside of this method lies with the difficulty to control the distribution of particle size and shape, which limits its applications where indiscriminate particle size distribution is not desired.

Chemical Reduction

As the name suggests, this method involves reduction of ions from its variable oxidation state (Ag^+ or Au^{3+}) to zero oxidation state (Ag^0 or Au^0) with the help of reducing agents. Some commonly used reducing agents are citrate, borohydride, and elemental hydrogen. The major issue is controlling the growth of the nanoparticles and preventing them from agglomerating. This may arise when the chemical reactions are not complete leaving unwanted reactant on the product. Thus, nowadays, interest has been raised for the chemical reduction of inorganic salts in the presence of both reducing as well as stabilizing/coating agent for the preparation of well-dispersed nanoparticles. There exists no chance of aggregation when a stabilizing agent is used; hence, better stability for durable applications can be achieved. By varying the concentrations of reductant and stabilizer, uniformly sized nanoparticles can be synthesized. A better uniformity of the nanoparticles can be used for applications especially in medicine.

Sol–Gel Process

This process involves the development of inorganic networks through the formation of a colloidal suspension (sol) and gelation of the sol to form a network in continuous liquid phase (gel). The four main steps for the sol–gel formation are hydrolysis, condensation, growth of particles, and agglomeration. The main advantages of this technique are low temperature conditions during processing and its versatility. This method has been used for the synthesis of metal oxide nanoparticles and nano-composites. The starting materials used typically to form a solvated precursor (sol) are metal alkoxides. After drying and calcination process at specific temperature, desired nanomaterials can be obtained. The disadvantage of using this method is the segregation of metal ions during thermal decomposition.

Chemical Vapor Deposition

This method deals with the production of thin films on the surfaces. To elaborate, vapor phase precursors are brought into a hot-wall reactor under conditions that favor nucleation of particles in the vapor phase rather than the deposition of a film on the wall. The key feature of chemical vapor synthesis is that it allows formation of doped or multi-component nanoparticles by the use of multiple precursors (Swihart 2003). This technique is widely used industrially.

15.2.4 Merits/Demerits

The most common method employed for the synthesis of nanoparticles is the bottom-up approach because the particles fabricated by this method are extremely scalable.

1. One notable drawback of "top-down" approach is the quality size of the nanoparticles required which has imperfection in the surface with significant crystallographic damage. In addition, these imperfection leads to failure in design and fabrication of nano-devices.

2. The "top-down" techniques tend to be slow and lengthy procedure. Also, it involves the use of expensive equipment to construct two-dimensional nanostructures. In spite of these disadvantages, this approach turns out to achieve bulk production with simpler processes with the desired structures.

3. Top-down approach involves impurities while synthesizing nanoparticles.

4. The size distribution of nanoparticles in top-down approach is broad (10–1,000 nm). In bottom-up approach ultra-fine nanoparticles with narrow size distribution (1–20 nm) can be prepared.

5. Large-scale production is simple in case of top-down approach. Whereas in bottom-up approach, it is difficult.

6. Although nothing is new, "bottom-up" approach plays comparatively important role in the synthesis of nanomaterials. It has been increasingly applied to produce structures of even scale or dimension.

7. Using the bottom-up approaches, there is no fine control of particle size and dispersity and can be limited in the variety of shapes which can be produced. Mostly, spherical-shaped nanoparticles are produced using this method. Nevertheless, there is a need of non-spherical-shaped nanoparticles for biomedical applications such as drug delivery.

15.3 Fabrication of Nanoparticles Using Calixarenes

15.3.1 Introduction

From synthetic chemistry point of view, it is significant to study the different ways for fabricating the surface of a nanoparticle. An ideal fabrication of nanoparticles involves a systematic creation of nanostructures sewed with the flexibility in size, shape, and inexpensive of the substrate being used. Currently, widespread interest has risen on this subject as evident by the burgeoning number of publications and conferences in this topic. Several chemical ligands such as polymers, dendrimers, supramolecular receptors, and ionic liquids are being used for the fabrication of nanoparticles. Among various macrocyclic compounds, calixarenes have devoted much to synthesize and fabricate nanoparticles.

Calixarenes are three-dimensional cup-shaped macrocyclic structure with intrinsic inherent hollow cavity. They can also be described as "anisotropic supramolecular architectures" which are synthesized from relatively simple building blocks. The word "calix" means a cup, and the word "arene" refers to any aromatic building block. The aromatic units may be derived from a phenol, resorcinol, or pyrogallol. The term "calixarenes" was introduced in the year 1978 by D. Gutsche. The nomenclature of calixarene can be done by counting the number (n) of repeating units in the ring and includes it in its name, that is, calix (Turkevich et al.)arene. For example, calix[4]arene comprises four aromatic units in the ring, and calix[6]arene will have six. There are four basic conformations which calixarenes adopt: cone, partial cone, 1,3-alternate, and 1,2-alternate. Calixarenes are structurally characterized by having an upper rim, middle annulus, and lower rim. The head groups are known for molecular encapsulation, whereas the lower rim favors the dispersion stability between the encapsulated complexes (Figure 15.2).

Different substrates can bind with calixarenes within the inherent hydrophobic cavity through non-covalent type of interactions or outside the cavity, that is, by the introduction of suitable functionalized ligand on the upper or lower rim of the calixarene system.

Calixarenes are found to be versatile host molecules playing different roles such as encapsulation, self-assembly, stabilization, and creation of nanoparticles, owing to their supramolecular capacities. Calixarenes have been used in wide areas of applications: in the development of sensor devices, as catalyst in number of chemical reactions, and in subjects of biology, nuclear chemistry, environmental chemistry, and many more.

15.3.2 Types of Calixarenes

Among the class of macrocyclic compounds, calixarenes are quite versatile; herefore, it becomes difficult to categorize them. Based on substitution reactions, regio-selectivity, cyclic moieties, or substituents joined through a bridged atom to form the ring, different number of calixarenes are signified (Agrawal et al. 2009). Figure 15.3a represents the chemical structures of phenol-derived (p-tert-butylcalix[n]arene) and resorcinol-derived (resor-cinarene) calixarenes which have been synthesized. p-Tert-butylcalix[n]arene is synthesized base-induced condensation of p-alkyl phenol with formaldehyde. It is known for its ion binding ability and to form complexes with lanthanides and actinides. The lower rim, phenolic –OH groups of calixarenes show reaction activities of metal–oxygen bonds or H-bond formation. The upper rim of the calixarenes, that is, the opposite side of the phenolic –OH groups,

produces various calixarene derivatives that affect the activities of the phenolic–OH groups (Ling et al. 2003). Calix[4]resorcinarenes are synthesized by the acid-catalyzed condensation of resorcinol with various aliphatic or aromatic aldehydes (Jain and Kanaiya 2011). As shown in Figure 15.3a, resorcinarenes have extra rigidity in their structure, due to methylene bridging between the resorcinol-based oxygen atoms of neighboring units. Also the functional groups at the in-between two –OR positions designate new chemical structures based on modified resorcinarenes.

Other special groups of calixarenes are synthesized in which the linking methylene bridge of calixarenes is replaced by sulfur, oxygen, or nitrogen atom. They are termed as "heteracalixarenes" (Mehta et al. 2015). Figure 15.3b depicts the chemical structure of oxacalix[4]arenes and thiacalix[4]arenes. In the same way, calix[4]pyrroles and calix[4]thiophene are built by replacing the phenolic units with pyrrole and thiophenewhich are shown in Figure 15.3b. They are termed as "heterocalixarenes" (Jain and Mandalia 2007).

Chiral calixarenes have been introduced by Bohmer et al. (1994) based on the absence of a plane of symmetry or an inversion center in the molecule as a whole. "Inherently" chiral calixarenes (Zheng and Luo 2011) are built up of achiral moieties and thus owe their chirality only to the fact that the calixarene molecule is not planar. Several strategies, such as fragment condensation, and regio- and stereoselective functionalization at the lower rim, have been reported so far for the preparation of inherently chiral calixarenes. The other approach is based on the attachment of chiral moieties at the upper or lower rim of calixarene (Sirit and Yilmaz 2009).

15.3.3 Characteristics of Calixarenes

Calixarenes are the key cavity containing building blocks in supramolecular chemistry. They are described to possess three-dimensional, concave surface, deeper/larger cavities, and relatively rigid structure (Rudkevich 2000). Calixarenes undoubtedly possess the potential to be classified as the "third supramolecule", because they can recognize not only metal cations but also organic molecules; large-scale preparation is extremely easy, optically active isomers can be synthesized, and selectivities are generally superior because of the rigid ring structures (Shinkai 1993).

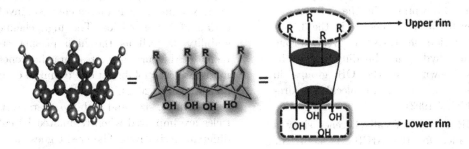

FIGURE 15.2 General representation of calixarenes.

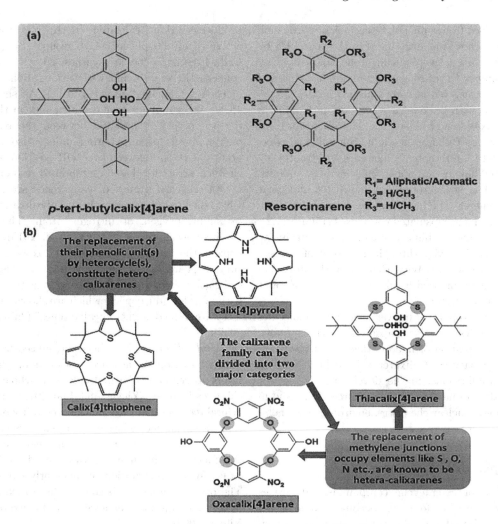

FIGURE 15.3   Types of calixarenes broadly grouped as (a) heteracalixarenes and (b) heterocalixarenes.

Some specific characteristics they possess are

- The calixarenes are characterized by much higher melting points and lower solubility in common organic solvents than their acyclic counterparts. They generally have high chemical stability and thermal stability and low toxicity properties.

- Due to their easy availability and functionalization with selective complexation ability, they are used in supramolecular and host-guest chemistry (Gutsche 1998).

- The calixarenes show concentration-independent OH stretching absorptions in the infrared at unusually low frequency, generally in the 3,150–3,200 cm^{-1} region indicative of very strong intramolecular hydrogen bonding (Gutsche 1983). It is known that the OH groups in calix[4]arenes form strong intramolecular hydrogen bonds (Shinkai 1993).

- The ^{1}H NMR spectra of calix[4]arenes show a *singlet* resonance for the ArCH$_2$Ar methylene protons at high temperature and a pair of doublets at low temperature (Shinkai 1993).

- Their valency can be easily varied at least from one to eight, while the stereochemical orientation of the ligating arms can be properly tuned by shaping, for example, the calix[4]arenemacrocycle in one of its four possible conformations (Baldini et al. 2007).

15.3.4   Properties of Calixarenes-Based Nanoparticles

Apart from being potential host molecules for large variety of guest analytes, the unlimited functionalization properties of calixarenes have attracted chemists involved in the fabrication of nanoparticles. The high chemical and thermal stability as well as the high plasma etching durability increases the current interest in calixarenes as "prospective material for nanofabrication" (Ohnishi et al. 1997, Kaestner and Rangelow 2012).

The shape, size, and stability characteristics of nanoparticles are improved when templated by calixarenes and its different derivatives. The credit goes to both calix system and nanoentities because it is together that they have established new dimensions and scope in chemistry.

Few examples to lay stress upon the properties of calixarenes-fabricated nanoparticles are

1. Size-selective synthesis and stabilization properties: A reduction in the core size of the gold nanoparticles is seen when stabilized with thiolatedcalix[4]arenes (employed as multidentate ligand) (Pescatori et al. 2010). Wei et al. also described that resorcinarenes with extended alkyl chains when fabricated on gold nanoparticles act as surfactants and stabilizers (Balasubramanian et al. 2002). The stability of the nanoparticles dispersions was dependent on the chemisorptive properties of the surfactant head group.

2. Chemical sensing and recognition properties: Filenkoet al. investigated the electrical conductivity modulation of discontinuous gold films (DGF) when coated with calixarenesof different thickness. This property was applied for its application as chemical sensor for different volatile analytes such as water, methanol, ethanol, and isopropanol (Filenko et al. 2005).

3. Bio-analytical properties: The use of calix-capped silver nanoparticles for inhibiting the growth of pathogenic bacteria was revealed (Boudebbouze et al. 2013). The bare nanoparticles decomposed without the capping agent did not at all show antibacterial activity.

15.3.5 Chemistry of Calix-Nanohybrids

With the development of different functionalized calixarenes and calixarene-based nanoparticles, the number of research articles has been increasing enormously. As per the SciFinder source, Figure 15.4 portrays the number of publications per year for the terms, "calix" and "calix-based nano". To the best of our knowledge, calix-based gold, silver, platinum, palladium, ruthenium, cobalt, iron oxide, and titanium oxide nanoparticles have been witnessed in the literature so far. Careful design and modification of nanoparticles using various calixarenes and its compounds

has enabled desired qualitative and quantitative research in almost all fields of contemporary chemistry. In some cases, calixarenes act as template, stabilizer, coating, and/or modifying agent. Figure 15.5 shows an overview of calixarenes using different functional units to fabricate nanoparticles.

Calix-based nanoparticles exhibit physico-chemical properties and reactivities due to the overall composition, structure, and surface characteristics of both calixarenes as well as the metal nanoparticles itself. The presence of calixarene as fabricating material renders enhanced and novel applications in comparison with other un-fabricated nanoparticles. This is due to the improved charge dispersion and stability it provides in terms of pH of the surrounding medium, temperature, and time.

Liu et al. described the "macrocyclic effect" on the formation of nanoparticles depends on the chemical structure of the capping ligand (Liu et al. 2002). Chenet al. confirmed that to promote nucleation, multivalent ligands such as calixarenes are excellent platforms and further provide protection from postnuclear aggregation (Chen et al. 2013). The effect of calixarene cavity–metal interaction was explained by Ha et al. who also further demonstrated the interactions between π-electron-rich calixarene cavities and the surface of gold nanoparticles (Ha et al. 2008).

A mechanistic insight towards the selective recognition ability of calixarene-based gold nanoparticles was demonstrated by Zhonget al. The molecular recognition was based on the strategy of interparticle charge-induced aggregation upon selective capture of metal cations into the nanoparticle-immobilized tert-butylcalixarenes, which produces calorimetric changes for the detection (Yan et al. 2011).

The probability of Mie theory has also been adopted for calixarene-fabricated nanoparticles. According to this, when the distance between two nanoparticles is less than the sum of their radii, the surface plasmon resonance (SPR) band displays a red shift, broadening and decreasing in intensity (Link and El-Sayed 2000). Aggregation of calix-functionalized gold nanoparticles due to detected metal ions was demonstrated by Maity et al. This was further confirmed by experimental results based on transmission electron microscopy (TEM) and dynamic

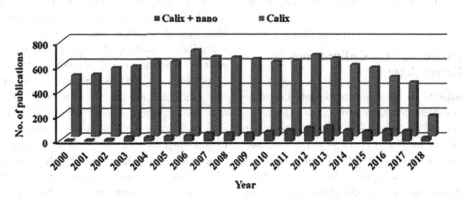

FIGURE 15.4 Evolution of the number of publications according to the keywords "calix" and "calix + nano" between 2000 and 2018.

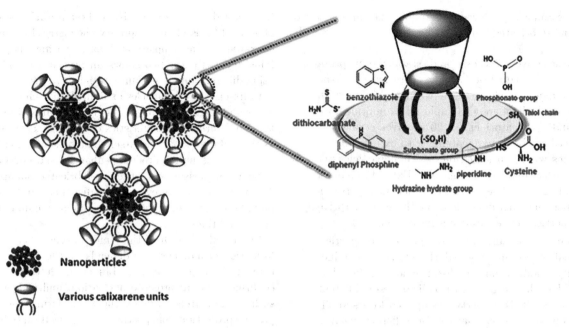

FIGURE 15.5 Representation of calix-fabricated nanoparticles.

light scattering (DLS). A very interesting chemistry behind the selective recognition of mercury ions using calix-functionalized gold nanoparticles was explained by the same group. The detection of mercury has been based on the aggregation of the nanoparticles. Two mechanisms were demonstrated supported by experimental results. In the first mechanism, it was proposed that the calixarene part of the calix-functionalized nanoparticles interacts with the mercury ions, causing aggregation and a red shift in the absorption band of the nanoparticles. Second, the mercury ions displace the calixarene part from the functionalized nanoparticles, leaving the bare nanoparticles to aggregate. Thus, through dipole–dipole interactions, the gold nanoparticles coalesce, and a complex formation takes place between calixarene and mercury ions.

In a nutshell, it is the linkage/functional groups attached on calixarenes which adapt different chemistry with the nanoparticles. Therefore, nanoparticles based on calixarenes can open new vistas for future researchers working in the field of nanoscience.

15.3.6 Fabrication Using Calixarenes as Stabilizing Agents

Fabrication of nanoparticles is important to gain desirable and predictable properties such as electronic, optical, medicinal, and catalytic activities. It is well-known that there are three important factors for the preparation and stabilization of nanoparticles, namely, reaction medium, reducing agent, and capping agent. Among these factors, the stabilizers play a significant role in the dispersion stability as well as to control its formation. In case of calixarenes,

hydrophobic core sandwiched between two functional rims can be easily modified, and such chemical modification is of interest to achieve specificity in applications such as nanodevices, sensors and actuators, smart delivery systems, and many more.

Resorcinol-derived calix[4]arenes were the first reported stabilizing agent used for gold nanoclusters. The large and concave head groups of calixarenes provide suitable encapsulation on the multi-nanometer scale (Stavens et al. 1999). It was concluded that resorcinarene-based surfactants will provide a significant advance in the controlled fabrication of nanometer-scale device. Later, Kim et al. elaborated the chemisorption properties of resorcinarenes for promoting nanoparticle self-organization at aqueous interfaces (Kim et al. 2005). The anchoring of calix[4]arene derivatives on monolayer-protected gold clusters with a core diameter of about 2 nm was reported by Arduini et al. These calixarene hosts were reported as efficient receptors for the recognition of ion pairs (Arduini et al. 2005). The use of phosphonate derivatives of calixarenes as stabilizer for the synthesis of silver nanoparticles was reported by Rastonet al. (Hartlieb et al. 2009). The pyridyl-appended calix[4]arene stabilized silver nanoparticles were reported which can act as a selective colorimetric sensor for Fe^{3+} ions (Zhan et al. 2012). Synthesis of modified calixarenes and its immobilization on magnetic (Fe_3O_4) nanoparticles have also been studied. The grafting of iron oxide nanoparticles with calixarene entities has been important for some application such as quantification of aromatic amines, extraction of ions, and improved catalytic activity. Some examples of calixarene-based nanoparticles with and/or without using external reducing agent have been represented in Table 15.1.

TABLE 15.1 Representation of Calix-Based Nanoparticles

Functionalized Calixarene Derivative	Reducing Agent	Types of Nanoparticles	Media	TEM Size	References
Calix[4]resorcinarene	Calix[4]resorcinarene tetrahydrazide	Gold nanoparticles	Water	11 ± 2 nm	Mishra et al. (2015)
Calix[4]pyrroleocta-hydrazide	Calix[4]pyrroleocta-hydrazide	Gold nanoparticles	Water	8 ± 2 nm	Bhatt et al. (2014)
Octamethoxyresorcin [4]arene tetrahydrazide	Octamethoxyresorcin[4]arene tetrahydrazide	Silver nanoparticles	Water	3–7 nm	Makwana et al. (2016)
Calix[4]arenepolyhydrazide	Calix[4]arenespolyhydrazide	Gold nanoparticles	Water	7 ± 3 nm	Vyas et al. (2012)
Resorcinol-derived calix[4]arenes	–	Gold nanoparticles	–	3–15 nm	Stavens et al. (1999)
25,27-bis(11-thio-1-oxyundecan)-26,28-dihydroxyxycalix[4]arene	–	Gold nanoparticles	THF/water mixture	14 ± 1 nm	Tshikhudo et al. (2005)
Para-sulfonato-calixa[4]arene	Sodium borohydride	Silver nanoparticles	Water	8 ± 1 nm	Xiong et al. (2008)
Para-phosphonated calix[4]arene and para-sulfonatedcalix[4]arene	Light	Silver nanoparticles	Water	18.5 ± 6 nm	Hartlieb et al. (2010)
Calix[4]arene-tetradiazoniumdecorated by four oligo(ethylene glycol) chains	Sodium borohydride	Gold nanoparticles	Water	17 nm	Valkenier et al. (2017)
Thiacalix[4]arenetetrahydrazide	Thiacalix[4]arene tetrahydrazide	Gold nanoparticles	Water	10 ± 2 nm	Darjee et al. (2017)
p-sulfonato-25,27-bis-dimethoxy-26,28 (sulfanylmethoxy)methoxy calix[4]arene	sodium citrate	Gold nanoparticles	Water	28 nm	Sutariya et al. (2016)
p-sulfonato-calix[4]resorcinarene	sodium citrate	Silver nanoparticles	Water	45 ± 10 nm	Menon et al. (2013)
Sulfonato-calixnaphthalene (SCn)	Sodium borohydride	Silver nanoparticles (SCn1:Ag NP), (SCn2:Ag NP), (SCn3:Ag NP)	Water	16.09 ± 0.13 nm 24.21 ± 0.60 nm 40.55 ± 0.41 nm	Valluru et al. (2014)
p-tert-butylcalix[4]arene	Sodium citrate	Silver nanoparticles	Water	~52 nm	Pandya et al. (2013)
Thioester-functionalized calix[8]arene	Hydrogen	Palladium (Pd), Platinum (Pt) and Ruthenium (Ru) nanoparticles	THF	≤1 nm (Pd and Pt nanoparticles) 2–3 nm(Ru nanoparticles)	Huc and Pelzer (2008)
p-phosphonated calix[8]arene	Hydrogen	Pd nanoparticles	Water	1.3–7 nm	Yasin et al. (2013)

15.3.7 Fabrication Using Calixarenes Reducing as well as Stabilizing Agents

Amphiphilic thiacalixarene grafted with poly(ethylene glycol) monomethyl ether (SCa-MPEG550) was used as stabilizer and reductant for the size-controlled preparation of gold nanoparticles in water.By using longer poly(ethylene glycol)to modify the thiacalixarene, the nanoparticles can be stabilized very well due to the improvement of hydrophilicity of the amphiphilic thiacalixarene. The particle sizes of the gold nanoparticles were further confirmed by DLS and TEM techniques. The combination of gold nanoparticles and thiacalixarene can be applied in catalysis and biosensor applications (Tu et al. 2009).

Our research group has established various applications of calixarene-based nanoparticles using resorcinarenes, calix[4]pyrrole, thiacalixarene, and oxacalixarenes. We have used one-pot chemical reduction method using hydrazide derivatives of these calix platforms for the synthesis of gold, silver, palladium, and platinum nanoparticles. Recently, calix-modified gold nanoparticles using octa-O-methoxyresorcin[4]arenetetrahydrazide (OMRTH) as reducing as well as stabilizing agent synthesized via simple one-pot method have been reported (Makwana et al. 2017). The synthesized gold nanoparticles (OMRTH-AuNps) have been used as a selective sensor for copper ions and leucine using fluorescent turn on–off mechanism. Moreover, its application in physics as molecular logic gate was demonstrated. The "Write–Read–Erase–Read" function of the memory element was evaluated for three input combinations of chemical inputs (In Cu and In Leu) integrated through a series of Set/Reset operations involving logic operations to produce an information processing device.

The synthesis and fabrication of platinum nanoparticles (PtNPs) using calix[4]pyrroletetrahydrazide as both reducing and stabilizing agent has also been reported (Kongor et al. 2018). The nanoparticles were explored in the field of catalysis for chemoselective hydrogenation of nitroarenes using molecular hydrogen without the use of external additives or promoters. To understand the binding mode and stabilization role of calixpyrroletetrahydrazide ligand to PtNPs, computational modeling wascarried out for the first time to elaborate the grafting role of calix[4]pyrrole on PtNPs.

To the best of our knowledge, none has explored the use of oxacalixarene hydrazide as reducing as well as stabilizing agent for the fabrication of nanoparticles, except ours. Recently, palladium nanoparticles (PdNps) were synthesized using dihydrazide derivative of oxacalixarene (OXDH)(Panchal et al. 2018). The extreme small size of the nanoparticles (3–4 nm) led to the impressive achievement of using OXDH-PdNps inextensive catalytic applications for carbon–carbon cross-coupling reactions, namely, Suzuki–Miyaura and Mizoroki–Heck reactions. The nano-catalyst could be easily reused for over six consecutive catalytic cycles with excellent quantitative yields.

The use of thiacalix[4]arenetetrahydrazide (TCTH) as both reducing as well as stabilizing agent for the synthesis of gold nanoparticles has also been reported (Darjee et al. 2017). TCTH-AuNps were found tobe selective and sensitive for isoleucine with the detection limit between 1 nm and 1.2 µm based on fluorescence enhancement. TCTH-AuNps also exhibited potential antioxidant activity without using the traditional antioxidant functional group structure-modifying methods.

15.3.8 Applications of Fabricated Calixarene-Based Nanoparticles

Owing to easy accessibility and selective functionalization of calixarenes, these macrocyclic receptors have grabbed great attention for the fabrication of nanoparticles with desired applications. Calixarenes can be converted into different derivatives possessing functional groups such ashydrazide, sulfonates, phosphonates, and dithiocarbamates. These functionalized calixarenes exhibit unique behavior towards nanoparticles (as modifying/templating/reducing/stabilizing agent) which then entails remarkable applications such as catalysis, sensing, pharmaceutical and microbial activities. The applications depend on the size of the calixarene ring and number of donor atoms on the attached functional groups as well as on the conformation of calixarene system. Table 15.2 gives a brief account of different calixarene-fabricated nanoparticles in the field of sensing, catalysis, drug delivery, medicine, etc. with appropriate reference.

15.4 Theoretical Approach for Mechanistic Insight

15.4.1 Introduction

Kelvin's mathematical triumphalism *"When you measure what you are speaking about and express it in numbers, you know something about it, but when you cannot express it in numbers your knowledge about it is of a meagre and unsatisfactory kind"* gave a distinct view to model complex phenomenon via mathematical laws for the predictive extrapolation. Nowadays, it's possible to investigate the structure and properties of materials by the use of computational modeling with quantum mechanical (including *ab initio*) and empirical approaches to understand observed experimental results (Gygi and Galli 2005). Quite often, theoretical approaches address the chemical questions related to molecular geometry and physical properties of the systems. Since past few decades, with the development of computer hardware and methodological efficiencies, there seems a revolutionary increase in the potentials of such approaches to lay down the foundation of chemistry. In the other way, theoretical results can be adjudged before actually performing the experiments to assess their possibilities at the pilot level. Such complimentary trends

TABLE 15.2 Various Applications of Fabricated Calix-Based Nanoparticles

Calix-Based Nanoparticles	Application	References
Calixarene-functionalized gold nanoparticles	Colorimetric detection of copper and lead ions	Gunupuru et al. (2014)
Anionic calix[4]arene-capped silver nanoparticles	Molecular recognition and transport of active pharmaceutical ingredients	Perret et al. (2012)
Calix[4]pyrroleocta-hydrazide based gold nanoparticles	Colorimetric and fluorometric chemosensors for selective signaling of Co (II) ions	Bhatt et al. (2014)
Calixarene-functionalized gold nanoparticles	Sensing of cobalt ion in organic and aqueous medium	Maity et al. (2014)
Calix[4]resorcinarenetetrahydrazide based gold nanoparticles	Turn-off fluorescent sensor for phenylalanine	Mishra et al. (2015)
Thiacalix[4]arenetetrahydrazide based gold nanoparticles	Recognition of isoleucine in aqueous solution and its antioxidant study	Darjee et al. (2017)
Octamethoxyresorcin [4] arenetetrahydrazide based silver nanoparticles	Sensitive and selective turn-off sensor for cadmium ions	Makwana et al. (2016)
p-sulfonato calix[4]arene based gold nanoparticles	Colorimetric and bare-eye detection of iodide in aqueous media	Maity et al. (2015)
Sulfonato-calix [4] resorcinarene-functionalized silver nanoprobe	Ultrasensitive and specific detection of dimethoate	Menon et al. (2013)
Calix [4] arenethiol-functionalized gold nanoparticles	Creatinine sensing via DLS selectivity	Sutariya et al. (2016)
Calix [4] arenethiol-functionalized silver nanoprobe	Selective recognition of ferric ion	Pandya et al. (2013)
calix[4]pyrrole-based platinum nanoparticles	Heterogeneous hydrogenation	Kongor et al. (2018)
Oxacalix[4]arenedihydrazide-based palladium nanoparticles	Heck-type olefination and Suzuki coupling reactions	Panchal et al. (2018)
Calix[4]pyrroletetrahydrazide-based palladium nanoparticles	Catalyst for Mizoroki–Heck reaction	Kongor et al. (2017)
Calixarene-capped silver nanoparticles	Species-dependent binding to serum albumins	Tauran et al. (2013a)
Tetramethoxyresorcinarenetetrahydrazide-based palladium nanoparticles	Antibacterial Pd nanoparticles in CC coupling reactions	Panchal et al. (2016)
Oxacalix[4]arenedihydrazide based palladium nanoparticles	Catalytic application in C–C coupling reactions	Mehta et al. (2016)
Calix[n]arene capped silver nanoparticles	Discriminatory antibacterial effects	Boudebbouze et al. (2013)
Para-sulfonato-calix[4]arene assemblies based silver nanoparticles	Molecular recognition by *para*-sulfonato-calix[4]arene of cytosine	Tauran et al. (2013b)
Calix[n]arenes featuring sulfonated moieties and linked to Ag nanoparticles	Antioxidant activity	Stephens et al. (2015)

are suitable for designing an assignment for the advance material studies. Nowadays, theoretical approach, specifically *computational modeling,* has been used by researchers to substantiate the experimental findings with mechanistic insights. It renders a complete description of chemistry under the shelter of mathematics, physics, and computer science in order to study how exactly a complex system behaves (Davidson 2000). A theoretical prediction can accelerate the research and can help the researcher to find the solution to the problem being studied. For instance, any change in the bond distance and angle in the structure of any molecule (or any change in the actual position of the atom in a molecule) can be interpreted using molecular simulation studies. The key feature of computational studies is that any complex problem can be optimized precisely and solved with different state-of-the-art techniques. Herein, aiming to provide a glimpse in this regard, we have examined the capability of theoretical predictions to address the issues and success reached as of now.

In particular, nanoscale materials are distinctly different from the bulk materials in comprising high surface-to-volume ratio. The resulting complexity of nano-interfaces requires the discrete theoretical treatment by the predictive theories. The fact that only limited experimental technique can probe the nano-interfaces; it generates an immediate attention towards the theoretical approaches to answer the qualitative information. Last few years have witnessed the performance and spatial information given by the simulation and electronic calculations in many cases. For such hybrid systems, the arrangement, composition, conformation, and orientation of nano-interfaces can be accessible by

real simulation. Indeed, this information will help in material and device operations.

It is quite interesting to study the behavior of atoms at nanoscale by considering the surface structure of the nanoparticles (which unit cell it forms) before starting the computational calculations. The physical properties of nanoparticles such as surface charge, band gap, determination of energy of frontier orbitals, HOMO, and LUMO can be well predicted computationally using the density function theory (DFT) studies (Bickelhaupt and Baerends 2007). Moreover, bonding or non-bonding type of interactions between any stabilizing molecule and nanoparticle surface can also be identified using computational methods, that is, molecular dynamics simulation studies. Some examples are illustrated in later section. In the context of the present discussion of the chapter, the theoretical studies have been performed in three broad areas:

- Computational adsorption study of metal surface models
- Dynamics and growth of nanostructures (molecular dynamics calculations)
- Binding energetics and mapping of the electronic populations (surface/atomic charges) to understand the reaction behavior

15.4.2 Computational Adsorption Study of Metal Surface Models

The understanding of the molecule and surface (nano) intermolecular interactions is crucial for device fabrication as well as material science applications. Studies indicate that such interactions can localize the adsorbing molecule and can generate the defects by modifying the charges and morphology of the adsorbed layer. The central aim behind these calculations is to calculate reactivity, stability of the chemisorption as well as the bond strengths, and band structure adhesion of one nanophase to another. The success of the subsequent optimization depends upon the judicious choice of the k-vectors and good initial guesses. Since the molecular adsorption is governed by van der Waals and coulomb-type interactions, it is imperative to treat such forces with great care using the computational methods. For the accurate estimation of the adsorption energetics, the precise treatment of image charges (local atomic charge of a molecule that can polarize conducting metal) and the coulomb contribution (donate/accept charge) at the quantum mechanical level is crucial. For this, charge equilibration model is used that can allow the mimicking of detaching and rebinding of atom to surface.

Gradient-corrected DFT (PW91) methylamine binding study onto the four surfaces of gold nanoparticle and their intersected ridges planes showed that compound can binds to Au surfaces via interaction with a single Au atom. However, this binding was substantially enhanced on the ridges compare to flat surfaces (Pong et al. 2006). Likewise, the adsorption study of C_{60}, ethylene, acetone, formaldehyde, benzene, aniline, and several alkanes has also reported on 111 surface of Ag (Jalkanen and Zerbetto 2006).

Moreover, in another seminal work, combined results of experiments and DFT calculations suggest that adlayers of 1,2,4-triazole adsorb on silver at alkaline and neutral solutions in the form of anion (in parallel oriented to surface). However, in neutral native form in moderately acidic medium, the ring gets tilted to the metal surface. In addition, varying pH values can reverse the switching between anionic and neutral form (Wrzosek et al. 2012).

15.4.3 Binding Energetics and Mapping of the Electronic Populations (Surface/Atomic Charges) to Understand the Reaction Behavior

Stability of the interactions is encoded by the electrostatic interactions that are mediated by the electronic populations at the spatial site of molecule. The 3D charge distribution of molecule gives an idea about how the molecules will interact and communicate with each other? Moreover, such charge-related properties enable to understand the reaction of electron donation/acceptor as observed in various nanoparticle formation reactions using the reductant. Atoms approach each other through the physical vicinity to equilibrate their charges by transferring electron density.

Joshi et al. using DFT calculation showed that indole-2-carboxylic acid capped silver nanoparticle can have large binding energy for Al^{3+} (-135.83 kcal/mol) ion than Ga^{3+} ion (-65.95 kcal/mol). The results were in good

agreement with the observed selective behavior. In a seminal work, DFT calculation explained the formation of curcumin complex with Au^{3+} in various isomeric conformation forms. It was noticed that from the charge and electrostatic potential surface, electron transfer from the curcumin to Au center is responsible for the Au^{3+} reduction to Au. The findings also underline that breakage of intramolecular bonding subjected the more availability of curcumin for gold ions along with water molecules (Singh et al. 2013).

In one of the theoretical calculations (Battistini et al. 2007), the pyrene derivatives were immobilized onto the Au(111) surface through the fully capped alkyl chains (of 4 and 11 carbon). In the predicted structure, the tilt angle for C4 with respect to Au surface was 20; however, this value is half as calculated for the longer alkyl chains (45–48). The authors concluded that the packing of the organic layers between the pyrene moieties is small chain was maximized compare to bigger one (Tam et al. 2007).

15.4.4 Different Software Packages Used for Molecular Modeling Studies

Success of any molecular modeling or the computational prediction is entirely depending upon the quality of the software package employed (Pirhadi et al. 2016). In particular, all such techniques focus on either of these three techniques (Dorsett and White 2000):

i. *Abinitio* or first principle electronic structure methods (relied on quantum mechanics and gives most accurate and consistent predictions)

ii. Semi-empirical Methods: replace some explicit calculation by approximation based on experimental data

iii. Molecular Mechanics: most fast calculations and neglect the electronic picture with limited scope in the present context

It is worth to mention that for very large systems (polymers or inhomogeneous mixtures), only molecular mechanics-based computation is possible due to extensive computation cost required for the electronic structure methods. However, thanks to the recent advances in the computer storage capacity and the computing efficiency that makes it possible to solve the relevant properties of interest in reasonable time. Present following type of predictions is possible for chemical systems such as heats of formation, reaction pathway/mechanism, characteristic of transition states, bond energetics, thermochemical stability, charge distribution (reactive sites), substituent effects, global reactivity descriptors, vibrational frequencies (IR and Raman spectra), electronic transitions (UV/visible spectra), and chemical shift (NMR) (Dorsett and White 2000). Table 15.3 lists the relevant software computational packages with their characteristic feature to study the hybrid nano-molecular systems. Other packages are MOLPRO, NWChem, and GAMESS that also used to perform the calculation, but these are less in use for such applications.

TABLE 15.3 Known Packages Relevant for the Theoretical Study of Nanoparticle-Organic Framework

Package Name	Developer(s)	Application
VASP	University of Vienna	Supports calculation with 3D Periodic with Plane Wave (PW) basis[a]
Gaussian	Carnegie Mellon University Gaussian, Inc.	Most popular electronic structure program, range of calculation (Periodic, Mol. mech, Semi-emp, HF, Post-HF, DFT) with GTO[b] basis
Desmond	D. E. Shaw Research, Schrödinger	Comprehensive GUI Molecular Dynamics code with OPLS[c] force fields
GROMACS	University of Groningen Royal Institute of Technology Uppsala University	Popular Molecular Dynamics GPU accelerated code with variety of force field to handle variety of chemical systems
Spartan	Wavefunction, Inc. & Q-Chem	Supports electronic structure calculation without periodic code for small molecules
TURBOMOLE	TURBOMOLE GmbH	Periodic as well as non-periodic paralleled DFT code (with RI-J method[d]) to handle larger systems
JAGUAR, Material Studio	Schrödinger Inc.	Electronic structure calculations including Hartree–Fock and DFT calculations for structural optimization as well as spectroscopic properties
LAMMPS	Sandia National Laboratories, Temple University	Most efficient free parallel code (coarse grain) for the molecular dynamics for soft and solid-state materials
Accelrys Discovery Studio Visualiser	BIOVIAs, Accelrys	small molecule simulations, QM-MM modeling
HyperChem	Hypercube, Inc.	De facto molecular modeling suite
Quantum ESPRESSO	QUANTUM ESPRESSO Foundation	Open access plane wave package for nanoscale modeling based on DFT, plane wavebasis sets, andpseudopotentials

[a] For periodic systems, plane wave basis is used to handle the 3d periodicity of lattice.

[b] Gaussian-type orbital ($\varphi_a^{GTO} = Nx^a e^{-r\zeta^2}$).

[c] Optimized potential for liquid simulation (OPLS) force field is used to handle potential energy terms for various atom types.

[d] Resolution of identity (RI) is a method to expands the molecular electron density in a set of atom-centered auxiliary functions, leads to tenfold increase in computation

15.4.5 Dynamics and Growth of Calix-Based Nanostructures (Molecular Dynamics Calculations)

Due to large size of supramolecular systems and their frameworks, molecular mechanics methods provide a platform that relies on the analytical expression with potential energy parameters. Combination of such atom specified parameters altogether gives rise to force fields that allow to simulate the systems in order to mimic the growth of nanostructures (Kaminski and Jorgensen 1996, Neria et al. 1996).

Time-dependent properties of the nanomaterial-organic framework (including transport properties) such as self-assembly, nucleation, and growth can be studied with comparable accuracy. To achieve more precision into this tactic, hybrid electronic structure along with molecular dynamics (MD) are used to model combined reaction evolution. Likewise, sometimes it is also worth to use course grain MD to cover the large associative time events and for incorporating the phase equilibria along with solvent effects.

Our group has performed MD simulation of MCPTH-PtNPs to investigate the capping ability of the calixpyrrole against PtNPs. A self-assembly model was developed that suggests the continuous aggregation of MCPTH onto the nanoparticle surface. Results have shown that NH and O engaged in building the intermolecular network of hydrogen bonds along with metal–acceptor interactions while wrapping of MCPTH on PtNPs surface. However, the chemistry of interactions was dominated by amino groups and the pyrrole rings which faces towards the PtNPs in the MD minimized structure (Kongor et al. 2018). For instance, Figure 15.6 shows a schematic representation of calix[4]pyrrole capped around spherical-shaped metal nanoparticles.

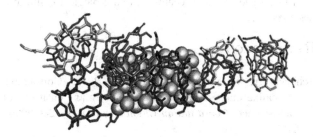

FIGURE 15.6 Example of molecular modeling showing noncovalent interaction of calix[4]pyrrole on surface of metal nanoparticles.

DFT computations have unveiled that modification of the arms of 1,3-conjugates of calix[4]arenes can able to selectively detect the polyaromatic hydrocarbons via naphthylamide moiety as observed in pyrenaldehyde (Bandela et al. 2011). Further, computational studies have deciphered that upon adding HSO_3^- to the AgNPs. Increase in the size of nanoparticle due to aggregation led to the formation of non-spherical one(Acharya et al. 2012).

15.5 Concluding Remarks

Currently, research on design and fabrication of nanoparticles possessing new useful properties have become prominent in the scientific domain. The most exciting challenge seen so far is the development of simple and facile techniques to diminish the use of toxic solvents and sophisticated instrumentation. The most possible way to synthesize stable and durable nanoparticles is to fabricate them. Among various traditional functionalizing compounds used for the fabrication of nanoparticles, calixarenes have achieved a topical position. Different functionalized calixarenes have anchored the surface of nanoparticles as templating, modifying, stabilizing, and reducing agent. The unique and beneficial combination of calix-nanohybrid has been appraised with a number of successful applications. The chemical modification of nanoparticles using calixarenes plays a critical role by modulating their specificity and toxicity properties. It is highly promising that calix-nanohybrid materials may have more and more applications in medicine delivery, sensing, catalysis, and therapy. However, the exact identity and quantity of calixarene molecules attached noncovalently on the surface of different nanomaterials remains a bottleneck challenge. The most reliable concern for this challenging task is the adoption of theoretical calculations based on different computational methods. Recent literature witnesses a continued dedication and efforts to substantiate realistic chemistry to the topic of calixarene-fabricated nanoparticles. Significant assessment on combined experimental and computational approach fulfills the overarching need for the development of advanced calixarene-fabricated nanoparticles coupled with a wide range of applications. It is therefore ensured that the endless efforts on this vital topic shall bring the physical, chemical, and biological communities together.

References

Acharya A, Samanta K, Rao CP, 2012. Conjugates of calixarenes emerging as molecular entities of nanoscience. *Coordination Chemistry Reviews* 256:2096–2125.

Agrawal Y, Pancholi J, Vyas J, 2009. Design and synthesis of calixarene. *Journal of Scientific and Industrial Research* 68(9):745–768.

Arduini A, Demuru D, Pochini A, Secchi A, 2005. Recognition of quaternary ammonium cations by calix[4]arene derivatives supported on gold nanoparticles. *Chemical Communications* (5):645–647.

Balasubramanian R, Kim B, Tripp SL, Wang X, Lieberman M, Wei A, 2002. Dispersion and stability studies of resorcinarene-encapsulated gold nanoparticles. *Langmuir* 18:3676–3681.

Baldini L, Casnati A, Sansone F, Ungaro R, 2007. Calixarene-based multivalent ligands. *Chemical Society Reviews* 36:254–266.

Bandela A, Chinta JP, Hinge VK, Dikundwar AG, Row TNG, Rao CP, 2011. Recognition of polycyclic aromatic hydrocarbons and their derivatives by the 1, 3-dinaphthalimide conjugate of calix[4]arene: Emission, absorption, crystal structures, and computational studies. *The Journal of Organic Chemistry* 76:1742–1750.

Battistini G, Cozzi PG, Jalkanen J-P, Montalti M, Prodi L, Zaccheroni N, Zerbetto F, 2007. The erratic emission of pyrene on gold nanoparticles. *ACS Nano* 2:77–84.

Bhatt KD, Vyas DJ, Makwana BA, Darjee SM, Jain VK, 2014. Highly stable water dispersible calix[4] pyrroleoctahydrazide protected gold nanoparticles as colorimetric and fluorometricchemosensors for selective signaling of Co (II) ions. *SpectrochimicaActa Part A: Molecular and Biomolecular Spectroscopy* 121:94–100.

Bickelhaupt FM, Baerends EJ, 2007. Kohn-Sham density functional theory: Predicting and understanding. *Chemistry Reviews in Computational Chemistry* 15:1–86.

Böhmer V, Kraft D, Tabatabai M, 1994. Inherently chiral calixarenes. *Journal of Inclusion Phenomena and Molecular Recognition in Chemistry* 19:17–39.

Boudebbouze S, Coleman AW, Tauran Y, Mkaouar H, Perret F, Garnier A, Brioude A, Kim B, Maguin E, Rhimi M,2013. Discriminatory antibacterial effects of calix[n]arene capped silver nanoparticles with regard to gram positive and gram negative bacteria. *Chemical Communications(Cambridge)*49:7150–7152. doi:10.1039/c3cc42838a.

Chen Z, Liu J, Evans AJ, Alberch L, Wei A, 2013. Calixarene-mediated synthesis of cobalt nanoparticles: An accretion model for separate control over nucleation and growth. *Chemistry of Materials* 26:941–950.

Darjee SM, Bhatt K, Kongor A, Panchal MK, Jain VK, 2017. Thiacalix[4] arene functionalized gold nanoassembly for recognition of isoleucine in aqueous solution and its antioxidant study. *Chemical Physics Letters* 667:137–145.

Davidson ER, 2000. Computational transition metal chemistry. *Chemical Reviews* 100:351–352, ACS Publications.

Dorsett H, White A, 2000. Overview of molecular modelling and *ab initio* molecular orbital methods suitable for use with energetic materials, Defence Science and Technology Organization Salisbury (Australia).

Faraday M, 1857. X. The Bakerian Lecture: Experimental relations of gold (and other metals) to light. *Philosophical Transactions of the Royal Society of London* 147:145–181.

Filenko D, Gotszalk T, Kazantseva Z, Rabinovych O, Koshets I, Shirshov Y, Kalchenko V, Rangelow IW, 2005. Chemical gas sensors based on calixarene-coated discontinuous gold films. *Sensors and Actuators B: Chemical* 111–112:264–270. doi:10.1016/j.snb.2005.06.053.

Gamaly EG, Ebbesen TW, 1995. Mechanism of carbon nanotube formation in the arc discharge. *Physical Review B* 52:2083.

Gunupuru R, Maity D, Bhadu GR, Chakraborty A, Srivastava DN, Paul P, 2014. Colorimetric detection of Cu^{2+} and Pb^{2+} ions using calix[4] arene functionalized gold nanoparticles. *Journal of Chemical Sciences* 126:627–635.

Gutsche CD, 1983. Calixarenes. *Accounts of Chemical Research* 16:161–170.

Gutsche CD, 1998. *Calixarenes.* Royal Society of Chemistry, Cambridge.

Gygi F, Galli G, 2005. *Ab initio* simulation in extreme conditions. *Materials Today* 8:26–32.

Ha J-M, Katz A, Drapailo AB, Kalchenko VI, 2008. Mercaptocalixarene-capped gold nanoparticles via post-syntheticmodification and direct synthesis: Effect of calixarenecavity-metal interactions. *The Journal of Physical Chemistry C* 113:1137–1142.

Hartlieb KJ, Martin AD, Saunders M,Raston CL, 2010. Photochemical generation of small silver nanoparticles involving multi-functional phosphonatedcalixarenes. *New Journal of Chemistry* 34:1834–1837.

Hartlieb KJ, Saunders M, Raston CL, 2009. Templating silver nanoparticle growth using phosphonatedcalixarenes. *Chemical Communications (Cambridge)*:3074–3076. doi:10.1039/b823067f.

Hassan MS, Taha ZA, Rasheed BG, 2016. Synthesis and modeling of temperature distribution for nanoparticles produced using Nd: YAG lasers. *Journal of Nanotechnology* 2016:8.

Haynes CL, Van Duyne RP, 2001. Nanosphere lithography: A versatile nanofabrication tool for studies of size-dependent nanoparticle optics, *The Journal of Physical Chemistry B* 105:5599–5611, ACS Publications.

Huc V,Pelzer K, 2008. A new specifically designed calix[8] arene for the synthesis of functionalized, nanometric and subnanometricPd, Pt and Ru nanoparticles. *Journal of Colloid and Interface Science* 318:1–4.

Hutchings Jr E, 1960. Engineering and science, volume 23: 5, February 1960. *Engineering and Science* 01 January1960 (online) Available at: http://resolver.caltech.edu/CaltechES:23.5.0. (Accessed 27 September 2013).

Jain V, Kanaiya P, 2011. Chemistry of calix[4] resorcinarenes. *Russian Chemical Reviews* 80:75–102.

Jain VK, Mandalia HC, 2007. The Chemistry of calixpyrroles. *Heterocycles* 71(6): 1261–1314.

Jalkanen J-P, Zerbetto F, 2006. Interaction model for the adsorption of organic molecules on the silver surface. *The Journal of Physical Chemistry B* 110:5595–5601.

Jee H-G, Hwang H-N, Han J-H, Lim J, Shin H-J, Kim YD, Solak HH, Hwang C-C, 2010. Patterning of self-assembled pentacenenanolayers by extreme ultraviolet-induced three-dimensional polymerization. *ACS Nano* 4:4997–5002.

Kaestner M, Rangelow I, 2012. Scanning probe nanolithography on calixarene. *Microelectronic Engineering* 97: 96–99.

Kaminski G, Jorgensen WL, 1996. Performance of the AMBER94, MMFF94, and OPLS-AA force fields for modeling organic liquids. *The Journal of Physical Chemistry* 100:18010–18013.

Kim B, Balasubramanian R, Pérez-Segarra W, Wei A, Decker B, Mattay J, 2005. Self-assembly of resorcinarene-stabilized gold nanoparticles: Influence of the macrocyclicheadgroup. *Supramolecular Chemistry* 17:173–180. doi:10.1080/10610270412331328961.

Kongor A, Panchal M, Mehta V, Bhatt K, Bhagat D, TipreD, Jain VK, 2017. Basketingnanopalladium into calix[4] pyrrole as an efficient catalyst for Mizoroki-Heck reaction. *Arabian Journal of Chemistry* 10:1125–1135.

Kongor A, Panchal M, Athar M, Mehta V, Bhatt K, Jha P, Jain V, 2018. Heterogeneous hydrogenation using stable and reusable calix[4] pyrrole fenced Pt nanoparticles and its mechanistic insight. *Applied Surface Science* 437: 195–201.

Ling J, Shen Z, Zhu W, 2003. Synthesis, characterization, and mechanism studies on novel rare-earth calixarene complexes initiating ring-opening polymerization of 2, 2-dimethyltrimethylene carbonate. *Journal of Polymer Science Part A: Polymer Chemistry* 41:1390–1399.

Link S, El-Sayed MA, 2000. Shape and size dependence of radiative, non-radiative and photothermal properties of gold nanocrystals. *International Reviews in Physical Chemistry* 19:409–453.

Liu J, Ong W, Kaifer AE, Peinador C, 2002. A "macrocyclic effect" on the formation of capped silver nanoparticles in DMF. *Langmuir* 18:5981–5983.

Lu RP, Ramirez AD, Russell SD, 2012. Nanofabrication using dip pen nanolithography and metal oxide chemical vapor deposition, U.S. Patent No. 8,178,429. 15 May 2012.

Madou MJ, 2002. *Fundamentals of Microfabrication: The Science of Miniaturization.* CRC Press, Boca Raton, FL.

Maity D, Bhatt M,Paul P, 2015. Calix[4]arene functionalized gold nanoparticles for colorimetric and bare-eye detection of iodide in aqueous media and periodate aided enhancement in sensitivity. *Microchimica Acta* 182:377–384.

Maity D, Gupta R, Gunupuru R, Srivastava DN, Paul P, 2014. Calix[4] arene functionalized gold nanoparticles: Application in colorimetric and electrochemical sensing of cobalt ion in organic and aqueous medium. *Sensors and Actuators B: Chemical* 191:757–764.

Makwana BA, Vyas DJ, Bhatt KD, Darji S, Jain VK, 2016. Novel fluorescent silver nanoparticles: Sensitive and selective turn off sensor for cadmium ions. *Applied Nanoscience* 6:555–566.

Makwana BA, Vyas DJ, Bhatt KD, Jain VK, 2017. Selective sensing of copper (II) and leucine using fluorescent turn on–off mechanism from calix[4] resorcinarene modified gold nanoparticles. *Sensors and Actuators B: Chemical* 240:278–287.

Masala O, Seshadri R, 2004. Synthesis routes for large volumes of nanoparticles. *Annual Review of Materials Research* 34:41–81.

Mehta V, Panchal M, Kongor A, Panchal U,Jain V, 2016. Synthesis of water-dispersible Pdnanoparticles using a novel oxacalixarenederivative and their catalytic application in C–C coupling reactions. *Catalysis Letters* 146:1581–1590.

Mehta V, Panchal M, Modi K, Kongor A, Panchal U, Jain VK, 2015. The chemistry of nascent oxacalix[n] hetarene (n≥4): A review. *Current Organic Chemistry* 19:1077–1096.

Menon SK, Modi NR, Pandya A, Lodha A, 2013. Ultrasensitive and specific detection of dimethoate using ap-sulphonato-calix[4] resorcinarene functionalized silver nanoprobe in aqueous solution RSC. *Advances* 3:10623–10627.

Mishra DR, Darjee SM, Bhatt KD, Modi KM, Jain VK, 2015. Calix protected gold nanobeacon as turn-off fluorescent sensor for phenylalanine. *Journal of Inclusion Phenomena and Macrocyclic Chemistry* 82: 425–436.

Neria E, Fischer S, Karplus M, 1996. Simulation of activation free energies in molecular systems. *The Journal of Chemical Physics* 105:1902–1921.

Nie Z-Q, Lin H, Liu X-F, Zhai A-P, Tian Y-T, Wang W-J, Li D-Y, Ding W-Q, Zhang X-R, Song Y-L, 2017. Three-dimensional super-resolution longitudinal magnetization spot arrays light. *Science andApplications* 6: e17032.

Ohnishi Y, Fujita J, Ochiai Y, Matsui S, 1997. Calixarenes-prospective materials for nanofabrications. *Microelectronic Engineering* 35:117–120.

Panchal M, Kongor A, Mehta V, Vora M, Bhatt K, Jain V, 2018. Heck-type olefination and Suzuki coupling reactions using highly efficient oxacalix[4] arene wrapped nanopalladium catalyst. *Journal of Saudi Chemical Society* 22:558–568.

Panchal U, Modi K, Panchal M, Mehta V,Jain VK, 2016. Catalytic activity of recyclable resorcinarene-protected antibacterial Pd nanoparticles in CC coupling reactions. *Chinese Journal of Catalysis* 37:250–257.

Pandya A, Sutariya PG, Lodha A, Menon SK, 2013. A novel calix[4] arenethiol functionalized silver nanoprobe for selective recognition of ferric ion with nanomolar sensitivity via DLS selectivity in human biological fluid. *Nanoscale* 5:2364–2371.

Perret F, Tauran Y, Suwinska K, Kim B, Chassain-Nely C, Boulet M, Coleman AW, 2012. Molecular recognition and transport of active pharmaceutical ingredients on anionic calix[4] arene-capped silver nanoparticles. *Journal of Chemistry* 2013: 1–9.

Pescatori L, Boccia A, Ciesa F, Rossi F, Grillo V, Arduini A, Pochini A, Zanoni R, Secchi A, 2010. The effect of ligand denticity in size-selective synthesis of calix[n]arene-stabilized gold nanoparticles: Amultitechnique approach. *Chemistry* 16:11089–11099. doi:10.1002/chem.201001039.

Pirhadi S, Sunseri J, Koes DR, 2016. Open source molecular modelling. *Journal of Molecular Graphics and Modelling* 69:127–143. doi:10.1016/j.jmgm.2016.07.008.

Pong B-K, Lee JY, Trout BL, 2006. A computational study to understand the surface reactivity of gold nanoparticles with amines and DNA. http://hdl.handle.net/1721.1/30380.

Rudkevich DM, 2000. Intramolecular hydrogen bonding in calixarenes. *Chemistry-A European Journal* 6:2679–2686.

Shi Z, Lian Y, Zhou X, Gu Z, Zhang Y, Iijima S, Zhou L, Yue KT, Zhang S, 1999. Mass-production of single-wall carbon nanotubes by arc discharge method1. *Carbon* 37:1449–1453.

Shinkai S, 1993. Calixarenes-the third generation of supramolecules. *Tetrahedron* 49:8933–8968.

Singh DK, Jagannathan R, Khandelwal P, Abraham PM, Poddar P, 2013. In situ synthesis and surface functionalization of gold nanoparticles with curcumin and their antioxidant properties: An experimental and density functional theory investigation. *Nanoscale* 5:1882–1893.

Sirit A, Yilmaz M, 2009. Chiral calixarenes. *Turkish Journal of Chemistry* 33:159–200.

Stavens KB, Pusztay SV, Zou S, Andres RP, Wei A, 1999. Encapsulation of neutral gold nanoclusters by resorcinarenes. *Langmuir* 15:8337–8339.

Steed JW, Turner DR, Wallace K, 2007. *Core Concepts in Supramolecular Chemistry and Nanochemistry*. John Wiley & Sons, Chichester.

Stephens E, Tauran Y, Coleman A,Fitzgerald M, 2015. Structural requirements for anti-oxidant activity of calix[n] arenes and their associated anti-bacterial activity. *Chemical Communications* 51:851–854.

Sun Y-P, Zhou B, Lin Y, Wang W, Fernando KS, Pathak P, Meziani MJ, Harruff BA, Wang X, Wang H, 2006. Quantum-sized carbon dots for bright and colorful photoluminescence. *Journal of the American Chemical Society* 128:7756–7757.

Sutariya PG, Pandya A, Lodha A,Menon SK, 2016. A simple and rapid creatinine sensing via DLS selectivity, using calix[4] arenethiol functionalized gold nanoparticles. *Talanta* 147:590–597.

Swihart MT, 2003. Vapor-phase synthesis of nanoparticles. *Current Opinion in Colloid and Interface Science* 8:127–133.

Tam F, Goodrich GP, Johnson BR, Halas NJ, 2007. Plasmonic enhancement of molecular fluorescence. *Nano Letters* 7:496–501.

Tauran Y, Brioude A, Kim B, Perret F,Coleman AW, 2013a. Anionic calixarene-capped silver nanoparticles show species-dependent binding to serum albumins. *Molecules* 18:5993–6007.

Tauran Y, Rhimi M, Ueno R, Grosso M, Brioude A, Janneau E, Suwinska K, Kassab R, Shahgaldian P, Cumbo A, 2013b. Cytosine: Para-sulphonato-calix[4]arene assemblies: In solution, in the solid-state and on the surface of hybrid silver nanoparticles. *Journal of Inclusion Phenomena and Macrocyclic Chemistry* 77: 213–221.

Tshikhudo TR, Demuru D, Wang Z, Brust M, Secchi A, Arduini A,Pochini A, 2005. Molecular recognition by calix[4] arene-modified gold nanoparticles in aqueous solution. *AngewandteChemie International Edition* 44:2913–2916.

Tsuji T, Thang D-H, Okazaki Y, Nakanishi M, Tsuboi Y, Tsuji M, 2008. Preparation of silver nanoparticles by laser ablation in polyvinylpyrrolidone solutions. *Applied Surface Science* 254:5224–5230.

Tu C, Li G, Shi Y, Yu X, Jiang Y, Zhu Q, Liang J, Gao Y, Yan D, Sun J, Zhu X, 2009. Facile controlled preparation of gold nanoparticles with amphiphilicthiacalix[4]arene as reductant and stabilizer. *Chemical Communications (Cambridge)*: 3211–3213. doi:10.1039/b902033k.

Turkevich J, Stevenson PC, Hillier J, 1951. A study of the nucleation and growth processes in the synthesis of colloidal gold. *Discussions of the Faraday Society* 11: 55–75.

Valkenier H, Malytskyi V, Blond P, Retout M, Mattiuzzi A, Goole J, Raussens V, Jabin I,Bruylants G, 2017. Controlled functionalization of gold nanoparticles with mixtures of calix[4] arenesrevealed by infrared spectroscopy. *Langmuir* 33:8253–8259.

Valluru G, Georghiou PE, Sleem HF, Perret F, Montasser I, Grandvoinnet A, Brolles L,Coleman AW, 2014. Molecular recognition of nucleobases and amino acids by sulphonato-calixnaphthalene-capped silver nanoparticles. *Supramolecular Chemistry* 26:561–568.

Vyas DJ, Makwana BA, Gupte HS, Bhatt KD,Jain VK, 2012. An efficient one pot synthesis of water-dispersible calix[4] arenepolyhydrazide protected gold nanoparticles-A "turn off" fluorescent sensor for Hg [II] ions. *Journal of Nanoscience and Nanotechnology* 12:3781–3787.

Wrzosek B, Cukras J, Bukowska J, 2012. Adsorption of 1, 2, 4-triazole on a silver electrode: Surface-enhanced Raman spectroscopy and density functional theory studies. *Journal of Raman Spectroscopy* 43:1010–1017.

Xiong D, Chen M,Li H, 2008. Synthesis of para-sulfonatocalix [4] arene-modified silver nanoparticles as colorimetric histidine probes. *Chemical Communications* 21:880–882.

Yadav TP, Yadav RM, Singh DP, 2012. Mechanical milling: A top down approach for the synthesis of nanomaterials and nanocomposites. *Nanoscience and Nanotechnology* 2:22–48.

Yan H, Luo J, Xie HM, Xie DX, Su Q, Yin J, Wanjala BN, Diao H, An DL, Zhong CJ, 2011. Cationic recognition by tert-butylcalix[4]arene-functionalized nanoprobes. *Physical Chemistry Chemical Physics: PCCP* 13:5824–5830. doi:10.1039/c0cp02658a.

Yasin FM, Iyer KS,Raston CL, 2013. Palladium nano-carbon-calixarene based devices for hydrogen sensing. *New Journal of Chemistry* 37:3289–3293.

Zhan J, Wen L, Miao F, Tian D, Zhu X, Li H, 2012. Synthesis of a pyridyl-appended calix[4]arene and its application to the modification of silver nanoparticles as an Fe^{3+}colorimetric sensor. *New Journal of Chemistry* 36:656–661. doi:10.1039/c2nj20776a.

Zheng Y-S, Luo J, 2011. Inherently chiral calixarenes: A decade's review. *Journal of Inclusion Phenomena and Macrocyclic Chemistry* 71:35.

Zhou Y, Yu S, Cui X, Wang C, Chen Z, 1999. Formation of silver nanowires by a novel solid: liquid phase Arc discharge method. *Chemistry of Materials* 11: 545–546.

Biofabrication of Graphene Oxide Nanosheets

Badal Kumar Mandal
*Department of Chemistry, Vellore Institute of
Technology, Vellore*

16.1 Introduction

Graphene is an allotrope of carbon and aromatic in nature. Other carbon allotropes contain graphene as the repeating unit, and these are carbon nanotubes (CNTs), charcoal and fullerenes. Single layer of graphite is known as graphene; that is, its structure consists of sp^2-bonded (i.e. covalently bonded) carbon atoms (having a C–C bond length of 0.142 nm) in planar sheets which resembles densely packed honeycomb crystal/hexagonal lattice. Graphene is unstable in nature until about 6,000 atoms but are most stable structures having more than 24,000 carbon atoms held together by weak van der Waals force-type electrostatic attractions. Nowadays, scientists are talking about graphene intensively due to its versatile applications with novel and astonishing properties in different fields. Mostly, it is used as single-molecule gas detection, graphene transistors, integrated circuits, transparent conducting electrodes for the replacement of ITO (indium tin oxide), ultracapacitors, graphene-based bio-devices, reinforcement for polymer nanocomposites, electrical and thermally conductive nanocomposites, antistatic coating and transparent conductive composites, etc. Interestingly, the resistivity of the graphene sheet is lower than silver resistivity which is the least resistivity-containing substance in the world at room temperature ($\sim 10^{-6}$ $\Omega \cdot$cm). It is assumed that phonons are quasi-particles representing the vibrational modes of atoms in the graphene lattice and are responsible for this miraculous observation. Some outstanding properties of graphene include zero band gap (~ 0 to 0.25 eV), high electron mobility capacity at room temperature [~ 200 cm^2/(V·s) compared to Si at RT $\sim 1,400$ cm^2/(V·s), CNTs: ~ 100 cm^2/(V·s), organic semiconductors (polymer, oligomer): <10 cm^2/(V·s)], strong excitation at visible to near-infrared region (NIR), phonon-dominated thermal conductivity, high tensile modulus (~ 130 Gpa compared to structural steel having tensile strength of 0.4 Gpa), high Young's modulus ($\sim 1,100$ Gpa) and high fracture strength (125 Gpa). Also, monolayer graphene is glass-like transparent (97.7% transmittance of white light).

Due to the promising and versatile applications of graphene as well as aqueous solubility, several researchers have modified graphite surface by chemical and heat treatment to its different oxidized as well as reduced forms with required surface functionalization. From the past few years, environmentally friendly methods for reduced graphene oxide (rGO) synthesis have been reported using various biomolecules (Maddinedi and Mandal 2014). For example, the natural biomolecules such as casein (Maddinedi et al., 2014), bovine serum albumin (BSA) (Liu et al., 2010), reducing sugar (Zhu et al., 2010) and L-ascorbic acid (Gao et al., 2010) have been used as reducing and stabilizing agents for graphene synthesis.

16.2 Functionalization of Graphene Oxide

Graphene and reduced GO are mostly applied in nanodevices, actuators and sensors by combing with magnetic nanoparticles (NPs), CNTs, carbon dots, quantum dots, nano-semiconductors, etc. To combine them, surface modification is necessary which could be done by two distinct ways: noncovalent and covalent functionalization of graphene oxide (GO) and rGO. Normally, noncovalent functionalization (NCF) of graphene and rGO is carried out by green routes, and they combine together by a weak interaction force such as Π–Π, van der Waals or electrostatic type (Bagherzadeh and Farahbakhsh, 2015; He et al., 2010). But in the case of covalent functionalization of graphene and rGO interaction of oxygen-containing functional groups such as carboxylic groups on the edges, epoxy and hydroxyl groups on the planes take place with the target matters via covalent bond formation chemically (Singh et al., 2011). During chemical reaction, normally electrophilic addition, nucleophilic substitution, addition and condensation reactions of the target materials take place with the oxygen-containing functional groups of graphene-based materials.

Practically, 2D-graphene appeared in the application field after 1980 when other carbon allotropes such as fullerene, CNTs, and graphene were discovered in addition to graphites and diamonds much before it (Park et al., 2006). Nowadays, graphene-based materials such as GO and reduced GO are playing a key role after modifying its oxygen-containing functional groups, i.e. by modifying its hydrophilic character on the surface for interactions to other entities in creating different molecular architecture, which are highly promising in scientific applications. In majority cases, the edging carboxylic acid groups are activated for anchoring other target molecules, and activation of carboxylic acid groups is carried out by reacting with different chemicals such as 2-(7-aza-1H-benzotriazole-1-yl)-1,1,3,3-tetramethyluronium hexafluorophosphate (HATU), thionyl chloride (SOCl2), N,N dicyclohexylcarbodiimide (DCC) and 1-ethyl-3-(3 dimethylaminopropyl)-carbodiimide (EDC) (Xu et al., 2009; Niyogi et al., 2006; Liu et al., 2008, 2009; Yang et al. 2009; Veca et al., 2009; Mohanty and Berry, 2008) followed by nucleophilic addition of nucleophiles, i.e. amines or alcohols. This results in amides or ester formation by covalent bonding, i.e. covalent GO which could be conjugated with other nanomaterials (NMs) such as metal, metal oxides, quantum dots and magnetic NPs immobilization for advanced applications.

16.2.1 Noncovalent Functionalization (NCF)

GO, which is coined formerly as graphite oxide or graphitic oxide or graphitic acid, is slightly different from graphite oxide. Actually, graphite oxide consists of multilayer of single-carbon-layered material/film, whereas GO consists of few layers of single-carbon thick material. Nowadays,

graphene is prepared easily and cheaply by the reduction of GO using green routes, i.e. by NCF.

In practice, through NCF the interacting species are adhered by the physical adsorption of the desired molecules via hydrophobic, van der Waals, Π–Π, and electrostatic interactions. Out of different target species, organic modifiers are considered universally by the scientific community. Also, researchers have used versatile options such as polymer wrapping, adsorption of organic molecules and composite formation followed by interactions with biomolecules such as deoxyribonucleic acid (DNA) and peptides and secondary plant metabolites.

Similarly, graphene-based materials, i.e. GO and rGO, possess sp^2 networks with Π–Π interactions within conjugated double bond systems and aromatic moieties which can improve the stability as well as dispersibility in aqueous medium without disturbing its electronic conjugation by exfoliation of interlayers, i.e. its scopes of applications in different fields and its conversion to different useful nanostructures. Also, this process improves its biocompatibility and stability in physiological conditions as well as its surface properties such as catalytic activity.

As per literature survey, the first report of NCF over GO surfaces used poly(sodium 4-styrenesulfonate) (PSS) via hydrophobic interaction for surface modification (Stankovich et al., 2006a), while polymer wrapping by using amine-terminated polymers technology was used by others electrostatically to modify GO surfaces (Bai et al., 2009). Another report also used lignin and cellulose derivatives for the hydrophobic modification of GO surfaces (Yang et al., 2010b). Many researchers used Π–Π interaction mode to modify GO surfaces, and for this purpose, different organic compounds such as polyethylene glycol (PEG) (Qi et al., 2010), anionic blue fluorescent coronene derivative (Ghosh et al., 2010), heparin (Lee et al., 2011), thionine (Chen et al., 2011a), porphyrin (Geng and Jung, 2010; Wojcik and Kamat, 2010), zinc phthalocyanines (Malig et al., 2011) and sulfonated copper phthalocyanine (Chunder et al., 2010) were used. A few reports used hydrophilic interaction in modifying GO surface by NCF, and the authors used polyacrylamide (Ren et al., 2010), sodium dodecyl benzene sulfonate (SDBS) (Chang et al., 2010; Zeng et al., 2010), and ionic liquid polymers (Kim et al., 2010) during the modification of GO surfaces by NCF.

In majority of cases, water-dispersible graphene was prepared by sonicating the resultant mixture in an ice-water bath which prevented inter- and intra-π–π stacking of graphene sheets. Basically, NCF was used to prepare surface-functionalized rGO for the electrochemical detection of hydrogen peroxide (Chang et al., 2010; Zeng et al., 2010; Zhang et al., 2010), uric acid (Zhang et al., 2010), variety of optoelectronic devise applications (Yang et al., 2010c), detection of nicotinamide adenine dinucleotide (NADH) (Liu et al., 2009), dispersion in organic solvents like DMF and NMP (Kamada et al., 2011), and sensors (Ren et al., 2010). Also, surface-functionalized rGO by hydrophilic interaction can be highly dispersed in aqueous medium at different pH and has versatile scopes in medicine, biology, nanoelectronics and other aqueous medium-based applications (Ren et al., 2010).

Recently, our group is working on synthesis, stabilization, functionalization and applications of green synthesized rGO and their composites with metals and metal oxides (Kadiyala and Mandal 2016; 2018; Maddinedi et al., 2014, 2015, 2017a, b; Maddinedi and Mandal, 2016) which are categorically classified into NCF. Basically, NCF was used earlier extensively to modify sp² networks of CNTs surfaces (Nakayama-Ratchford et al., 2007; Zhao and Stoddart, 2009).

We used casein protein for the reduction of GO produced from graphite powder by using the modified Hummers' method (Perera et al., 2012). In this process, GO aqueous dispersion (1 mg/mL) was mixed with 1% casein protein, and pH of the mixture was adjusted to 12 using ammonium hydroxide. Then, the mixture was refluxed at 90°C for 7 h on a water bath. The colour change from yellowish brown to black confirmed the reduction completion of GO (Maddinedi et al., 2014). Figure 16.1 shows the casein-mediated synthesized rGO stabilized by casein protein. Also, high-performance liquid chromatography (HPLC) analysis confirmed the stabilization of rGO sheets by casein molecules, but its degraded products, i.e. glutamic acid and aspartic acid, did not provide any role in the stabilization of rGO sheets (Figure 16.2).

Similarly, we prepared rGO from GO using *T. chebula* seed aqueous extract which contains many phytochemicals especially polyphenols that reduced and stabilized rGO sheets (Maddinedi et al., 2015). Figure 16.3 shows the micrographs of rGO obtained by high-resolution-transmission electron microscope (HR-TEM). It clearly indicates that the synthesized rGO had few layers of graphene.

Also, we used *Terminalia bellirica* fruit aqueous extract for the synthesis of rGO sheets (TBG) (Maddinedi and Mandal, 2016) (Figure 16.4). Figure 16.5 shows the proposed stabilization mechanism of TBG using the *T. bellirica* polyphenols. It suggests that π–π interactions between the oxidized polyphenols and rGO sheets which protects the agglomeration of RGO sheets with higher stability than rGO synthesized by chemical method using sodium borohydride, hydrazine and phenylhydrazine due to lack of stabilization. In addition, Ag-rGO nanocomposites synthesized by *T. bellirica* fruit aqueous extract showed very good catalytic activity at a dose of 75 mg/L towards complete reduction of 4-nitrophenol (0.3 mM) to 4-aminophenol in the presence of $NaBH_4$ as reducing agent within 5 min (Kadiyala and Mandal, 2016). Similar type of catalytic reduction of 4-nitrophenol to 4-aminophenol was performed by Ag/rGO nanocomposites which was synthesized by using protein tyrosine (Figure 16.6) (Maddinedi et al., 2017a).

Also, we prepared rGO from GO using tyrosine as reducing and stabilizing agent (Maddinedi et al., 2017a). In this process, GO aqueous dispersion (1 mg/L) was mixed with tyrosine (1%) and was refluxed at 100°C on a water bath for 8 h after adjusting pH 12 with ammonium hydroxide. Then, rGO powder was dispersed by sonication for 1 h. The sonicated rGO dispersion was mixed with 10 mL $AgNO_3$ solution (1 mM) and 2 mL tyrosine solution (1 mM) followed by sonication at 60°C for 10 min to get Ag-decorated rGO (Ag-rGO) nanocomposites. It is clearly observed that Ag NPs are embedded on transparent silk-like rGO surfaces (Figure 16.7).

In another study, we used a thiol group-containing enzyme diastase as reducing and stabilizing agent for the synthesis of Au-decorated rGO nanocomposites for

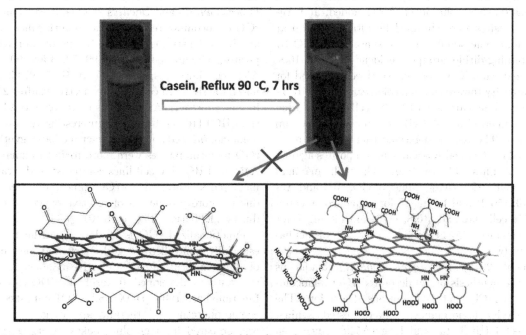

FIGURE 16.1 Casein-mediated rGO. (Reprinted from Maddinedi et al., 2014, with permission from Elsevier.)

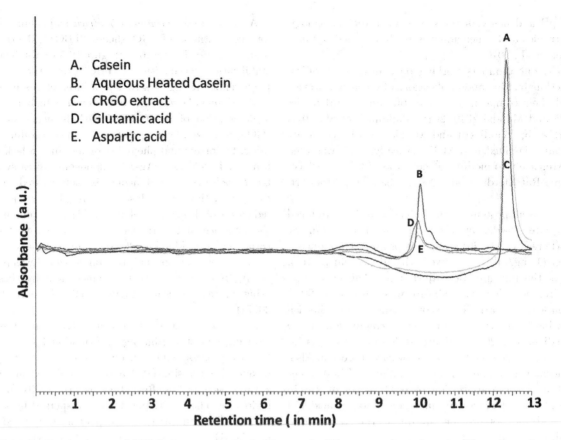

A. Casein
B. Aqueous Heated Casein
C. CRGO extract
D. Glutamic acid
E. Aspartic acid

FIGURE 16.2 Overlay image of HPLC chromatograms of native casein (A), aqueous heated casein (B), casein reduced graphene oxide (CRGO) extract (C), standard glutamic acid (D), and standard aspartic acid (E). The analysis was done by C18 phenomenex column (250×4.6 mm, 5 μm, 100Å). The flow rate was 0.2 mL/min with a mobile phase (70% of methanol and 30% water). The injection volume was 20 μL. (Reprinted from Maddinedi et al., 2014, with permission from Elsevier.)

biomedical applications (Maddinedi et al., 2017b). In brief, 10 mg rGO was added to 10 mL chloroauric acid (HAuCl$_4$, 0.5 mM) solution and dispersed by ultrasonication. Then, 3 mL aqueous diastase solution (1%) was added, and Au-rGO nanocomposites were obtained by ultrasonication at 60°C for 10 min. The stability of the synthesized rGO by diastase containing dithiol groups is evident in Figure 16.8. The synthesized Au-rGO nanocomposites were checked for *in vitro* cytotoxicity towards human colon carcinoma (HCT-116) and human lung carcinoma (A-549) cell lines by using MTT [(3-(4,5-dimethylthiazol-2-yl)-2,5-diphenyltetrazolium bromide)] assay. The results suggested that GO was more toxic compared to rGO and Au-rGo nanocomposites against the tested cell lines (Figure 16.9). The half maximal inhibitory concentration (IC$_{50}$) values of GO, rRG and Au-rRG were 140, 150.21 and 187 mg/L for human colon carcinoma HCT116 cell lines and 116.8, 117.6 and 140 mg/L for human lung carcinoma A549 cell lines after 24 h of incubation, respectively.

Kadiyala et al. (2018) used *Syzygium cumini* seed aqueous extracts for the synthesis of Au-decorated rGO nanocomposites from HAuCl$_4$ and GO aqueous dispersion. The prepared Au/rGO nanocomposites were tested against human colorectal (HCT116) and lung (A549) cancer cell

lines for their *in vitro* cytotoxicity and antibacterial toxicological behaviour towards Gram-negative bacterial strain *Escherichia coli* and Gram-positive bacterial strains *Staphylococcus aureus* and *Bacillus subtilis*. It was found that Au-rGO nanocomposites were toxic to both cancerous cell lines and bacterial strains. The IC$_{50}$ values of Au-rGO nanocomposites, rGO and GO were 51.99, 79.64 and 90.41 mg/L for A549 cell lines, respectively, and 61.57, 97.03 and 107.36 mg/L for HCT116 cell lines, respectively, after 24 h of incubation. Moreover, Au-rGO was more toxic to A459 cell lines than HCT116 cell lines. It is interesting to note that *Syzygium cumini* seed aqueous extract-mediated synthesized Au-rGO nanocomposites were more toxic to human cancerous A459 and HCT116 cell lines compared to diastase enzyme-mediated synthesized Au-rGO nanocomposites. It may be due to more production of reactive oxygen species (ROS) during incubation (Figure 16.10).

Functionalized rGO is used in biomedical applications, especially controlled drug delivery and sustained release of different drugs to the selective regions. In this study, the authors developed doxorubicin (DOX)-loaded BSA-functionalized rGO (DOX-BSA-rGO) nanosheets (NS) for chemo-photothermal therapy applications. In brief, GO was prepared from graphite flakes by using the modified

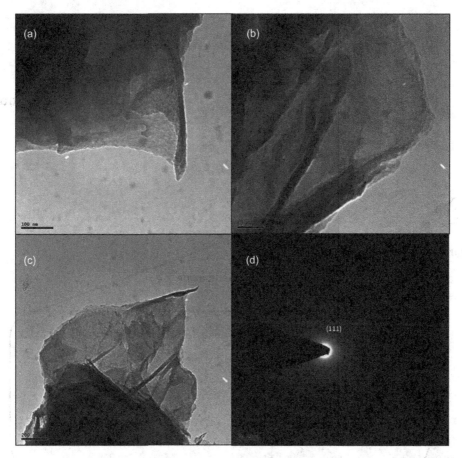

FIGURE 16.3 HR-TEM images of TCG showing the formation of few-layer graphene (a–c) and the selected area electron diffraction (SAED) pattern (d). (Reprinted from Maddinedi et al., 2015, with permission from Elsevier.)

Hummers' method. Then, BSA solution (50 mg/mL) was mixed with GO aqueous dispersion, and pH was adjusted to 12 with NaOH (1 M) at 60°C for 1 h. After stabilization, the solution was centrifuged at 15,000 rpm and washed several times, and precipitates were redispersed in distilled water. Finally, the required concentration of DOX drug solution was mixed with 0.1 mg/mL BSA-rGO, and, after centrifugation, the precipitate of DOX-BSA-rGO was washed with phosphate-buffered saline to remove uncoated drug from the synthesized GO nanocomposites (Cheon et al., 2016).

The exfoliated uniform ultrasmall GO NS (with lateral size of less than 50 nm) were prepared from graphite using the modified Hummers' method. Basically, the synthesized GO was deoxidized twice by following the same procedure with $KMnO_4$ in H_2SO_4 medium. The synthesized ultrasmall GO NS were biocompatible with less/negligible cytotoxicity and higher cellular uptake. Also, the ultrasmall GO NS showed fluorescence which suggested that this NS may be used as the smart nanocarriers for drug delivery and intracellular fluorescent nanoprobe (Zhang et al., 2013).

Targeting drug delivery is a key area in biomedical applications, and hence, highly biocompatible NMs with enhanced drug loading capacity are the focus of the present-day research for the materials scientist throughout the globe. In this study, the authors prepared molybdenum disulfide (MoS_2)/GO nanocomposites for their *in vitro* and *in vivo* efficacy study towards tumour growth of B16 murine melanoma cancer cells in lungs of mice. The authors claimed that MoS_2/GO nanocomposites could be promising in cancer nanotheranostics (Liu et al., 2018).

Separation and catalysis chemistry needs materials with atomic-level pores for attractive effectiveness. In this chapter, the authors prepared functionalized rGO nanoplates or membrane with an interlayer spacing of 3.7 Å. Basically, GO nanoplates were functionalized with N-(trimethoxysilylpropyl) ethylenediamine triacetic acid (EDTA-silane). This GO was functionalized by the silylation to EDTA-GO, while EDTA-rGO was prepared by reducing GO to rGO by hydrazine treatment. The resulting functionalized rGO nanoplates/membranes were not affected severely in wetted state, and the interlayer spacing of the functionalized rGO increased from 3.7 to 4.4 Å while that of GO increased from 9 to 13 Å. Hence, the functionalized rGO nanoplates showed threefold lower permeation to K^+ and Na^+ ions compared to GO nanoplates (Lee et al., 2016). Similar type of graphene-based membrane was reported elsewhere for the desalination of water by modifying pores on the graphene surfaces (Surwade et al., 2015).

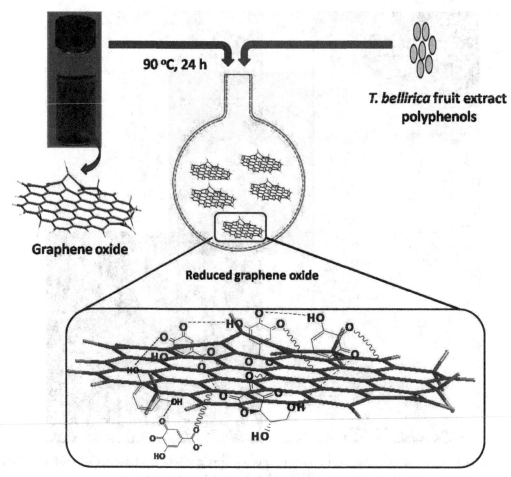

FIGURE 16.4 Schematic illustration of green reduction of GO using *T. bellirica* extract. (Reprinted from Maddinedi and Mandal, 2016, with permission from Bentham Science Publishers.)

16.2.2 Covalent Functionalization

GO consists of sp^2-hybridized carbon atom networks with honeycomb crystal-like structure. Covalent functionalization (CFn) of GO surface results in rehybridization of one or more sp^2-bonded carbon atoms to sp^3-hybridized carbon atoms by chemical transformation which causes substantial loss of electronic configuration on GO surfaces. Chemical transformation of GO surfaces can be done by different techniques such as substitution, addition and condensation reactions. Categorically surface modifications of GO by CFn are done by (i) nucleophilic substitution, (ii) electrophilic substitution, (iii) condensation reactions and (iv) addition reactions. For this purpose, many compounds with different functional groups/moieties, NMs, polymers and biomolecules are interacted with GO surfaces and are bound covalently to prepare CFn GO with modified properties and structures.

Nucleophilic Substitution

GO surfaces contain different oxygen-containing functional groups, i.e. hydroxyl, carboxyl, carbonyl and epoxy groups. Among them, carbon atoms connected to epoxy groups are mostly electron-deficient under high ring strain and behave

as a good electrophilic centre. Hence, nucleophiles with electron-rich atoms, i.e. having lone pair of electrons or negatively charged atomic centres, attack epoxy group of GO surface even at room temperature in aqueous medium and forms nucleophile-modified GO. Normally, different types of nucleophiles such as amino acids, aliphatic and aromatic amines, amine groups containing ionic liquids and polymers, biomolecules with amine functionality, and alcoholic groups containing compounds such as silane, polyphenols, alkaloids and alkoxides can undergo nucleophilic substitution reaction in modifying GO surfaces (Scheme 16.1).

As per literature survey, it is observed that many nucleophiles such as aliphatic amines, i.e., dodecylamine (Stankovich et al., 2010), allylamine (Wang et al., 2009), polyglycerol (Pham et al., 2010) and isocyanate (Stankovich et al., 2006b); aromatic amines, i.e. octadecylamine (Wang et al. 2008), dopamine (Xu et al., 2010), hydrophobins (Laaksonen et al., 2010) and polyallylamine (Park et al., 2009); amino acids such as poly-L-lysine (Shan et al., 2009) and peptides (Cui et al., 2010); and ionic liquid 1-(3-aminopropyl)-3-methylimidazolium bromide (Yang et al., 2009) are successfully modified GO surfaces for effective and selective functionalization towards different applications. Also, it can be used as an intermediate for further

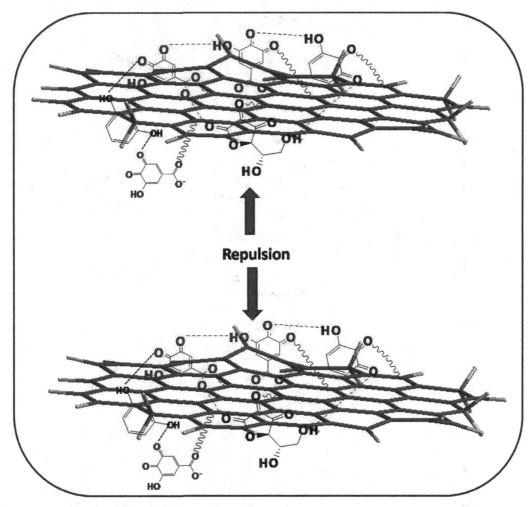

FIGURE 16.5 Proposed stabilization mechanism of TBG using the *T. bellirica* polyphenols. (Reprinted from Maddinedi and Mandal, 2016, with permission from Bentham Science Publishers.)

functionalization, i.e. immobilization of biomolecules (Shan et al., 2009).

Functionalization of GO surfaces by nucleophilic substitution can be done in alkaline as well as aqueous medium. Some researchers carried out functionalization of GO surface by simple solvothermal method (Wang et al., 2009). Among different nucleophiles, isocyanate functionalized GO surfaces by converting carboxylic groups to amide and hydroxyl groups to carbamate esters. As a result, modified GO formed highly stable dispersion in aprotic polar solvent with completely exfoliated GO sheets of approximately 1 nm thickness. Sometimes, these amine and hydroxyl groups on GO surfaces could be used for the attachment of other materials especially polymers to prepare GO-polymer nanocomposites in developing biocompatibility and solubility in suitable solvents.

Electrophilic Substitution Reaction

Electrophiles are positively charged or neutral moiety which are replaced or attracted by a nucleophile, i.e. electron-rich species to form covalent bond. In case of

GO surface, oxygen-containing functional groups bound carbons are slightly positively charged due to higher electronegativity of oxygen atom, and hence, nucleophiles such as species containing amino groups and Π-electron systems (i.e. double/triple bond-containing species) can attack the positively charged carbon centre (i.e., carbenium ion) which may initiate covalent formation followed by functionalization of GO surfaces (Scheme 16.2).

Different nucleophiles such as amines/substituted amines (Bekyarova et al., 2009; Sun et al., 2010), organometallic compounds such as ferrocene (Avinash et al., 2010), and aromatic ring-containing polymers such as polystyrene (Fang et al., 2009) can react with epoxy group of GO surface-forming carbenium ion which further undergo addition reaction via covalent bond formation. Electrophilic substitution of hydrogen atoms by aryl diazonium salts on GO surface is a promising example of GO surface modification with aromatic/substituted aromatic moieties (Lomeda et al., 2008), whereas electrophilic addition of species to double/triple bond system of GO surface occurs followed by atom transfer via radical (Fang et al., 2009) resulting functionalization of GO surfaces.

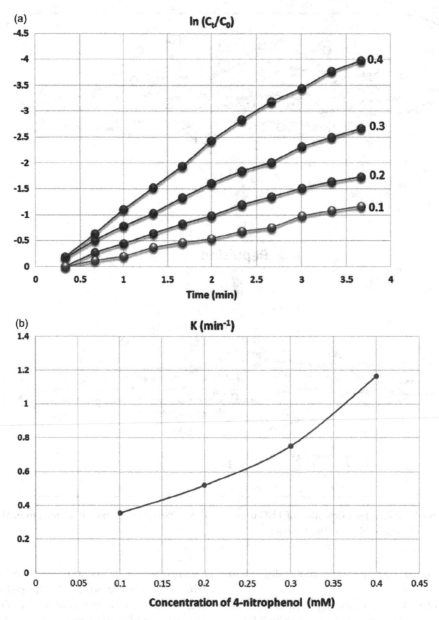

FIGURE 16.6 (a) Plots of ln (C_t/C_0) versus time for the catalytic reduction of 4-nitrophenol with NaBH$_4$ by tyrosine-reduced graphene oxide (TRGO)/silver nanocomposite at different concentration of 4-nitrophenol. (b) Plots showing the relationship between the pseudo-first-order rate constant versus concentration of 4-nitrophenol. An increase in the pseudo-first-order rate constant linearly with the increase in the catalyst concentration and catalyst. (Reprinted from Maddinedi et al., 2017a, with permission from Springer.)

Condensation Reactions

Condensation reaction happens due to loss of small molecules after interaction between functional groups of GO surface with functional groups of the attacking/interacting species. As a result, overall entropy of the resulting product is reduced with covalent bond formation which is the driving force for this surface transformation of GO. This condensation reaction occurs mainly by two ways: esterification between carboxylic acid groups of GO with hydroxyl groups of the interacting species and amide formation among carboxylic acid groups of GO with amino groups of the interacting species. Basically, water-insoluble aromatic

compounds or drugs can be dispersible in aqueous medium after loading them on the amine-functionalized FO surfaces (Liu et al., 2008; Hu et al., 2011).

Among different condensing agents, organic isocyanate and organic diisocyanate are used extensively for functionalization and cross-linking on GO surfaces (Zhang et al., 2009a). In addition, biopolymer chitosan also is used to prepare novel nanocarriers after covalent functionalization of GO surface followed by loading of water-insoluble cancer drug 'Camptothecin' via π–π stacking and hydrophobic interactions. Sometimes, hydroxyl groups of GO surface form ester with carboxylic acid groups of fullereno acetic acid (Zhang et al., 2011) or sulfanilic acid

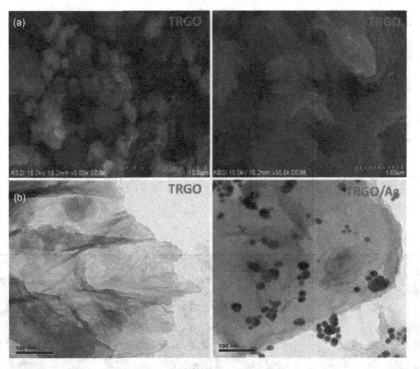

FIGURE 16.7 (a) Field emission scanning electron microscopy (FE-SEM) images of TRGO at different magnifications showed the well-separated layered structures with transparent, thin graphene sheets possessing high surface area. (b) Transmission electron microscopy (TEM) images of TRGO and tyrosine-reduced graphene oxide/silver composite (TRGO/Ag). Images showed the decoration of silver NPs onto the surface of transparent, silk-like graphene sheets. (Reprinted from Maddinedi et al., 2017a, with permission from Springer.)

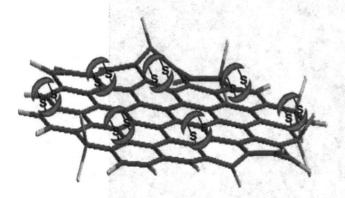

FIGURE 16.8 Possible mechanism of GO reduction and stabilization of the formed DRG with diastase molecules. (Reprinted from Maddinedi et al., 2017b, with permission from Elsevier.)

or cysteamine (Chen et al., 2011b) for sensing/detection of hydrogen peroxides.

Addition Reactions

In organic chemistry by addition reactions, two or more molecules combine to form larger molecule via electrophilic attack or biradical pathway. Interestingly, a nitrene (R–N:) is the nitrogen analogue of a carbene. Having six valence electrons (two for C–N covalent bond, two for two p-orbitals and remaining two as a lone pair of electrons), nitrogen atom is coined as an electrophile. This nitrene is a reactive intermediate and participates in many chemical reactions via (1,3) cycloaddition or biradical pathway to form

functionalized GO surfaces (He and Gao, 2010). In addition, a nitrene intermediate can undergo C–H insertion via an oxime or acetic anhydride formation which results in cyclo compounds (Savarin et al., 2007). This nitrene-based method is used for the functionalization of epitaxial rapheme (Choi et al., 2009). Similarly, rGO surfaces are modified after grafting GO surfaces with maleic anhydride (Hsiao et al, 2010) or polyacetylene (Vadukumpully et al., 2011).

Graphene Nanocomposites

There are different techniques to prepare rapheme/rGO nanocomposites after decorating/drafting inorganic NMs/NPs on 2D rapheme/rGO/GO surfaces. It could be done by noncovalent or covalent functionalization of GO surfaces with novel or improved properties. Actually, decorations of GO surfaces with NMs are done by three different strategies. First, some researchers prepare rapheme/rGO and NMs, and then mix them thoroughly to improve the solubility of rapheme composites in different solvents of interest. This method is known as pre-graphenization technique. Second, other researchers follow different routes; that is, they mix pre-synthesized NPs or its precursor salts with aqueous suspension of GO and then thoroughly mix with a reducing agent to prepare rGO-NP composites. This process is known as post-graphenization technique which prevents aggregation and restacking of prepared rGO during reduction process. Lastly, some researchers mix directly precursor salts and GO with a suitable reducing

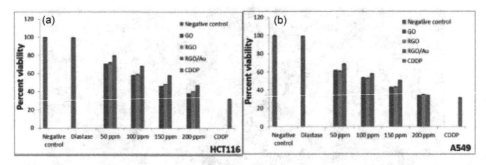

FIGURE 16.9   Cell viability of HCT116 (a), A549 (b) cell lines induced by GO, DRG and DRG/Au. (Reprinted from Maddinedi et al., 2017b, with permission from Elsevier.)

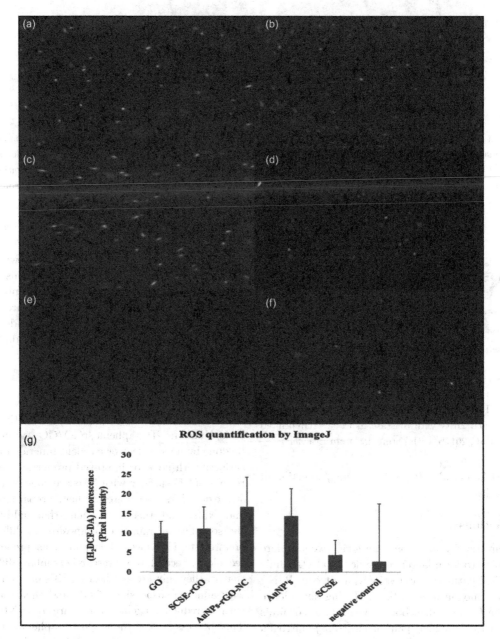

FIGURE 16.10   Fluorescence microscopic images of samples representing the ROS generation. GO (a), *syzygium cumini* seed extract (SCSE)-rGO (b), AuNPs-rGO-NC (c), AuNPs (d), SCSE (e), negative control (f) and ROS quantified data from images shown in (a–f) by using Image J software program (g). (Reprinted from Kadiyala et al., 2018, with permission from Elsevier.)

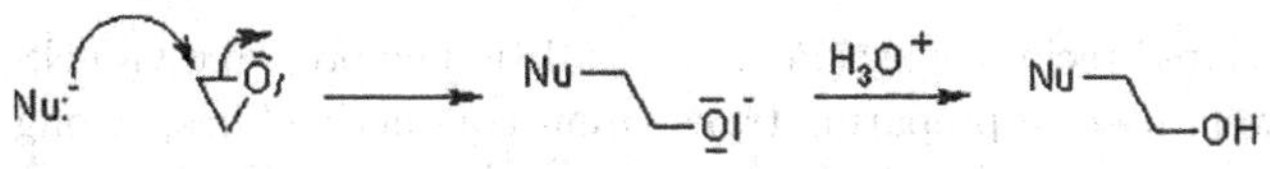

SCHEME 16.1 Mechanism of nucleophilic substitution reaction on GO surfaces.

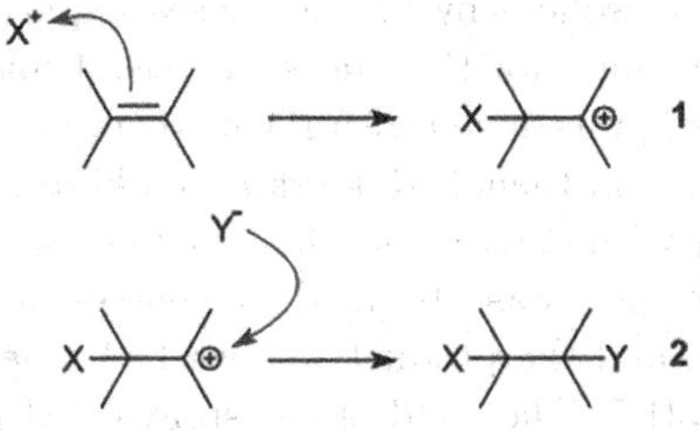

SCHEME 16.2 Mode of electrophilic substitution reaction for functionalization of GO surfaces.

agent in one spot and prepare the rGO-NP composites where reducing acts as both reducing and stabilizing agents in controlling size, coalescence/aggregation and restacking of rGO sheets.

Practically, noncovalent decoration of NPs on 2D-GO surfaces faces severe adherence problem and hence researchers functionalize NPs with suitable functional groups first which can bind covalently to GO surfaces overcoming adherence problem. Also, it helps to decorate and immobilize NPs with proper functional groups for different applications such as photocatalytic, catalytic, energy storage, optoelectronics and sensing applications. Although NCF of GO and its NPs composites formation are relatively easier process, it faces several technical and procedural complications during synthesis and thereafter applications. The fabrication processes face the following uncertain issues: (i) non-uniform decorating of NPs on the GO surface and its interaction with the functional groups existing on the GO surfaces, (ii) whether size and morphology of GO surfaces influence the decoration of NPs, interaction of NPs with functional groups and thereafter properties of the resulting composites and (iii) the mechanism/mode of NPs adherence to GO surfaces. After thorough monitoring and critical analysis, it is found that NCF still possesses the following shortfall: (i) poor dispersity in solvents and (ii) processes that are complex in nature with lower surface coverage density. In addition to these problems and challenges of NCF, many researchers have prepared GO-NPs composites using both noncovalent and covalent functionalization pathways.

16.3 Graphene Metal Nanoparticle Nanocomposites

In this class of nanocomposites, the corresponding metal salts and GO suspension are mixed together with a reducing agent such as hydrazine/sodium borohydride to synthesize the respective GO-metal NPs nanocomposites (Dey and Raj, 2010; Sundaram et al., 2008; Liu et al.,

2011; Yang et al., 2010a; Baby et al., 2010). Among many published articles on the synthesis of graphene-based nanocomposites using noncovalent pathway, some articles on GO-NPs nanocomposites such as GO-Pt (Dey et al., 2010), GO-Pd (Sundaram et al., 2008), GO-Au (Liu et al., 2011), GO-Si (Yang et al., 2010c) and GO-Pt/Au (Baby et al., 2010) are documented in literature.

16.4 Graphene Metal Oxide Nanocomposites

Like metal NP-decorated GO nanocomposites, metal oxide decoration is done by *in situ* synthesis of metal oxide NPs on 2D-GO surfaces. By this process, oxygen-containing groups of GO surfaces control the nucleation/coalescence of metal oxide NPs resulting size and morphology of the synthesized metal oxide NPs. Several metal oxide-grafted GO nanocomposites such as GO-ZnO (Zhang et al., 2009b), GO-SnO$_2$ (Yang et al., 2010c), GO-TiO$_2$ (Wang et al., 2010b), GO-Co$_3$O$_4$ (Wang et al., 2010a), GO-RuO$_2$ (Wu et al., 2010), GO-Mn$_3$O$_4$ (Wang et al., 2010a) and GO-Fe$_3$O$_4$ (Zong et al., 2013) are reported in the literature.

16.5 Graphene Semiconducting Nanoparticles

Solvothermal, sol–gel and UV-assisted reduction and solution mixing processes are used in semiconducting NP-decorated GO sheets. The purpose of this process is to improve the dispersibility/solubility of GO in organic solvents or rGO in aqueous solvent and also to improve conducting properties of the synthesized composite materials. Basically, *ex situ* methods of synthesis is comparatively better due to better control over size and morphology of the embedded/decorated/grafted NPs on GO surfaces with selective desired properties. Available literatures report the following semiconductor-decorated GO sheets such as GO-ZnS (Wang et al., 2010c), GO-CdS (Wang et al., 2010c), GO-CdTe (Dong et al., 2010), GO-CdSe (Geng et al., 2010) and GO-QDs (Pan et al., 2010).

16.6 Miscellaneous Methods for Fabrication of GO Nanosheets

Highly delaminated nickel aluminium-layered double-hydroxide (NiAl-LDHs) NS synthesized by one-step reverse microemulsion were used to synthesize GO/NiAl-LDHs nanocomposites by simple electrostatic self-assembly. The prepared GO nanocomposites with 12 wt% GO showed the best electrochemical properties, i.e. capacitance retention (67%) compared to pure NiAl-LDHs (29%). This improvement may be due to strong interaction between negatively charged GO and positively charged NiAl-LDH NS (Zhang et al., 2017).

Oxygen-containing groups of GO surface can be modified by high-energy electron-beam irradiation on graphite. In this method, the authors controlled the oxygen contents of GO surface by electron beam irradiation, and the modified GO NS were used to remove lead (Pb^{+2}) from aqueous solutions at an initial Pb(II) concentration of 75 mg/L, pH 5.0, 1.5 mg GO for 2 h. The results suggested that the highest adsorption capacity of GO was 636.4 mg/g for Pb(II) at 19.2 kGy of irradiation dosage which indicated that the synthesized GO by EB-irradiated graphite is a promising sorbent for heavy metals in wastewater treatment (Bai et al., 2016).

Microwave irradiation can be used to prepare GO NS (with lateral size of ∼30 nm) without using strong acid in the presence of $KMNO_4$ (1%) (Luo et al., 2016). In a typical method, the GO suspension (0.5 wt%, 100 mL) was stirred for 10 min with 1 g $KMnO_4$ followed by heating with a household microwave oven (2,450 MHz, 700 W) for 5 min. Then, 200 mL deionized water was added to the black-coloured suspension, which was thoroughly washed with dil. HCl to get pure GO NMs (65% yield). After washing step, graphene oxide nanomaterials (GONM) was sonicated for 30 min to prepare GO NS. Finally, 0.2 mg of single-wall carbon nanotubes (SWCNTs) (90%, Chengdu Organic Chemical CO., Ltd) was dispersed in 10 mL of GONS (0.01 wt%) and ultrasonicated at 300 W for 30 min with cooling on an ice-water bath. Afterwards the SWCNTs dispersion was centrifuged at 13,000 rpm to check appearance of any precipitation, but there was no sediment formation indicating GONS's good surfactant behaviour (Luo et al., 2016).

Another study reported the green approach to prepare rGO NS electrochemically (Guo et al., 2009). In this study, three electrode systems, i.e. Pt as counter electrode, calomel electrode as reference electrode, and glassy carbon electrode (GCE) as working electrode, were used for the preparation of rGO from GO electrochemically. The exfoliated GO suspension (5 L) was applied on a GCE or on graphite disk and dried in a vacuum desiccator. The electrochemical reduction of the pretreated exfoliated GO film on GCE was done against a constant potential (−1.5 or −1.3 V *vs* SCE) in 10 mM/L pH 5.0 PBS (K_2HPO_4/KH_2PO_4). The rGO-fabricated GCE was used to study its electrocatalytic activity towards dopamine which showed its superiority over chemically reduced rGO-fabricated GCE. Similarly, bentonite (Bent)-GO thin film coated on GCE was used as a sensor for a neurotransmitter dopamine. Bent-GO-fabricated GCE showed a potential in the peak-peak separation (∼140 mV) which is lower than the peak-peak separation of ∼190 mV at bare GCE. This sensor could be used effectively in analysing pharmaceutical and biological samples containing DA with high recovery (>97.85%) and lower RSD values (<2%) (Tandel et al., 2016). The present synthetic strategy of GO from graphite flakes is based on water electrolytic oxidation of graphite by electrochemical oxidation reaction to overcome the drawbacks of Hummers' method and

related technologies (Pei et al., 2018). The prepared GO NS were used in preparing transparent conductive films, strong papers, and ultra-light elastic aerogels.

Water-soluble single-layer GO was prepared by a green route avoiding environmentally hazardous heavy metals as catalyst and poisonous by-product gases, explosion risk and long reaction times for the said synthesis. Typically, about 125.3 g K_2FeO_4 was mixed with 100 mL concentrated H_2SO_4 (93% purity), and then, 3.13 g graphite (40 μm in size) was added slowly with stirring for 1 h at room temperature. The paste-like product was obtained by centrifugation (10,000 rpm, 3 min), and the product was washed repeatedly with water until pH 7. The yield of the spray-dried powder was about 4.7 g (Peng et al., 2015). The importance of this method relies on liquid crystals of GO which can be processed into films, macroscopic graphene fibres and aerogels.

Another study used very different reducing agent (hydriodic acid with acetic acid, HI-AcOH) for the preparation of rGO from GO. In this method, GO (4.0 g) was dispersed in acetic acid (1.5 L), and then, HI (80.0 mL) was added and stirred constantly at 40°C for 40 h. Finally, the product (yield 2.83 g) was washed with saturated sodium bicarbonate (5 × 100 mL), distilled water (5 × 100 mL) and acetone (2 × 100 mL) followed by vacuum drying at room temperature overnight (Moon et al., 2010).

Tiny rGO sheets were synthesized by a cellbreak sonicator at 14 kHz and $Na_2Ti_3O_7$ nanowires prepared by hydrothermal method (Sun et al, 2017). About 50 mg P25 TiO_2 powder was dissolved in 16 mL NaOH solution (10 mol/L) followed by ultrasonication for 15 min and aged at 200°C for 20 h. The resulting product was rinsed with distilled water until pH 7, and then, 1 mg/mL dispersion was made using ultrasonication for 15 min to fabricate rGO. A required volume (65 mL) of rGO suspension and $Na_2Ti_3O_7$ nanowire suspension was sonicated with 1 μL of hydrazine hydrate for 15 min and stirred further for 3 h to ensure complete fabrication. The prepared composite membrane with 1 wt% rGO showed high absorption capacity of Rhodamine B dyes (1.30×10^{-2} mol/g with respect to GO mass) which was threefold more than earlier reported GO materials-based studies (Jiao et al., 2015; Tiwari et al., 2013; Wu et al., 2013).

GO and nickel ferrite ($NiFe_2O_4$) were prepared to synthesize GO-doped $NiFe_2O_4$ (GO–$NiFe_2O_4$). The as-prepared GO–$NiFe_2O_4$ samples showed good catalytic activity in degrading organic dyes such as methylene blue, rhodamine B and malachite green under visible light irradiation with oxalic acid, and $NiFe_2O_4$ did not show any catalytic activity in degrading the above dyes in the presence of H_2O_2. This further confirmed its photo-Fenton catalytic activity towards the degradation of organic dyes (Liu et al., 2013).

Another interesting application of GO NS is evidenced by Liu et al. (2017). The authors reported that GO NS can be used as good desiccating/adsorptive materials for different preservative applications. The authors found that grape fruits wrapped with GO NS showed delayed mould gathering compared to normal wrapping paper indicating its applications in moisture desiccation and food preservation (Liu et al., 2017).

Recently, single-layer graphene was prepared in bulk quantity in a liquid phase with homogeneous stable dispersion which is a break-through in graphene and graphene-related research (Bepete et al., 2017). In this method, graphenide (negatively charged graphene) was dispersed in tetrahydrofuran with degassed water, and the single-layer graphene was obtained by evaporating the organic solvent. The resulting graphene possessed 400 m/L of developed graphene surface and exhibited a conductivity of up to 32 kS/m.

16.7 Characterization Techniques for GO-Based Materials

A thorough literature survey on characteristics of carbon-based materials shows that graphene-based materials are characterized by different instrumental techniques. Basically, UV-visible, FT-IR, Raman spectroscopy, XPS analysis, XRD, SEM-EDX and TEM-SAED pattern are capable to provide some constructive information to the morphology (i.e. size and shape) and structures of the synthesized NMs.

16.7.1 UV-Vis Studies

The dispersion of the synthesized graphene-based materials is made in double-distilled water or in a suitable solvent, and UV-vis spectroscopy study is done by scanning the dispersion within 200–800 nm of λ_{max}. Normally, GO shows an absorption peak at 230 nm due to π–π^* electron transition of aromatic C–C single bond, while another shoulder peak appears at ~300 nm due to π–π^* electron transition of aromatic C=C double bond and n–π^* electron transition

of aromatic C=O bonds (Figure 16.11). Also, after the reduction of oxygen-containing functional groups of GO surfaces, λ_{max} of the absorption peak is changed which would provide the useful information about the functionalities on GO-based materials surfaces (Maddinedi et al., 2014). In addition, fabrication of metal NPs on rGO surfaces shows their distinct surface resonance band (SPR) in the absorption spectra (Kadiyala et al., 2018; Maddinedi et al., 2017a,b).

16.7.2 Attenuated Total Reflectance-FTIR (ATR-FTIR) Spectroscopy Studies

Basically, Attenuated Total Reflectance-Fourier transform infrared (ATR-FTIR) spectroscopic study is used to find out the functional groups present on the surface of GO-based materials. It can predict the availability of oxygen-containing functional groups present in GO as well as that come from fabricating agent, stabilizing agent, surfactants and other composite-forming agents with specific functional groups. Normally, GO shows specific absorption bands at 3,377 cm^{-1} for carboxylic hydroxyl group stretching vibration and at 1,645 cm^{-1} for carbonyl group stretching vibration. Also, it shows many other characteristics absorption bands of the functional groups attached to the GO surfaces (Figure 16.12) (Kadiyala et al., 2018; Maddinedi et al., 2014).

16.7.3 X-Ray Diffraction (XRD) Studies

X-ray diffraction (XRD) study is carried out to get information on crystalline state, size and phase purity of NMs. It is an ideal technique to investigate the interlayer modifications induced in graphite-related materials and also

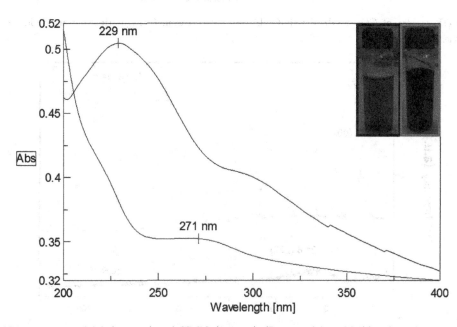

FIGURE 16.11 UV-vis spectra of GO (229 nm) and CRGO (271 nm). (Reprinted from Maddinedi et al., 2014, with permission from Elsevier.)

to identify the crystalline nature of the synthesized material. In addition, one can get d-spacing values from the diffractogram (Figure 16.13). It is evident that GO showed a sharp diffraction peak at 2θ of 11.2° (002 crystalline plane) and rGO synthesized by casein showed a broad diffraction peak at 2θ of 21.8° (222 crystalline plane) (Maddinedi et al., 2014). Moreover, we can calculate mean crystallite diameter using Scherrer's equation as $D = 0.94\lambda/\beta^{1/2}\cos(\theta)$, where D is the crystallite size, θ is the Bragg's angle, λ is the X-ray wavelength (1.5418Å) and $\beta^{1/2}$ is the width of the XRD peak at half height.

16.7.4 Scanning Electron Microscopy (SEM) and Energy Dispersive X-ray (EDX) Studies

Scanning electron microscopic (SEM) study is a great tool to investigate the morphology and microstructure of the as-synthesized nanostructures. Figure 16.7 shows the SEM micrographs of tyrosine-induced rGO (Maddinedi et al., 2017a). This micrograph presents well-separated layered structures with transparent and thin graphene sheets. Also, energy-dispersive X-ray (EDX) analysis is done to

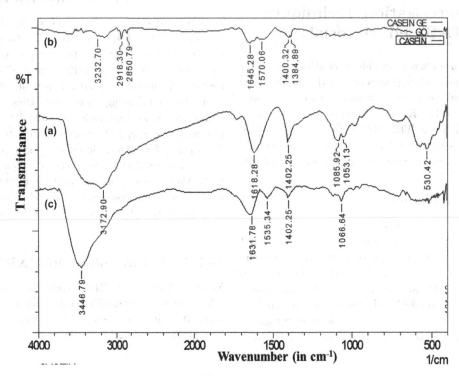

FIGURE 16.12 FTIR spectra of GO (a), dried CRGO (b) and native casein (c). (Reprinted from Maddinedi et al., 2014, with permission from Elsevier.)

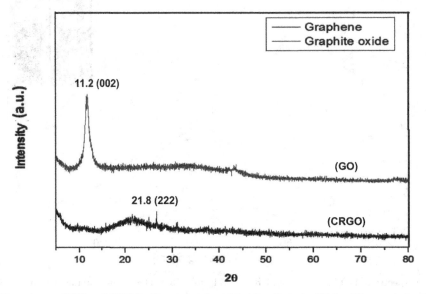

FIGURE 16.13 XRD patterns of GO and CRGO. (Reprinted from Maddinedi et al., 2014, with permission from Elsevier.)

do elemental analysis of the GO (Figure 16.14c) and *T. bellirica*-induced rGO (Figure 16.14d) which can provide the elements present in the NMs as well as the purity of the materials. Figure 16.14 shows the EDX micrograph of GO with 43.2 atom% oxygen and *T. bellirica*-induced rGO with 24.8% oxygen along with remaining carbon atoms.

16.7.5 Transmission Electron Microscopy (TEM) and SAED Studies

TEM analysis is done to find out the morphologies, shapes and sizes of the studied NMs. Figure 16.3 shows the HR-TEM micrograph of *T. chebula*-mediated rGO which suggests ultra-thin silk-like transparent three-layer rGO sheets. The SAED pattern of rGO sheets shows polycrystalline ring pattern with many diffraction spots (Figure 16.3d).

16.7.6 X-Ray Photoelectron Spectroscopy (XPS) Studies

X-ray photoelectron spectroscopy (XPS) is a technique to investigate elemental composition and elemental oxidation states in GO and embedded/fabricated NMs on its surfaces. Figure 16.15 is the C1s XPS spectrum of GO- and diastase-mediated rGO. It shows four distinct peaks related to C=C, C–O, C=O and O–C=O groups at 284.6, 286.7, 287.8 and 289.1 eV. In addition, lowering value of C–O group indicates more reduction of oxygen-containing functional groups in GO to rGO.

16.7.7 Raman Spectroscopy Studies

Raman spectroscopy is mainly used to study the orderness in crystal structures of graphene/rGO as well as degree of reduction. In a typical Raman spectrum of graphene, two

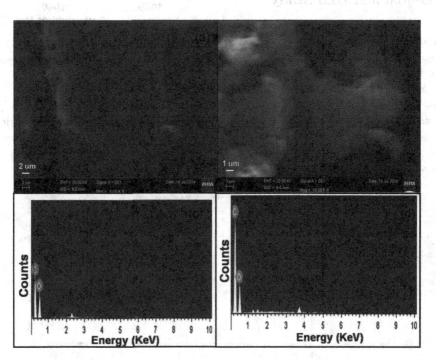

FIGURE 16.14 SEM images at different magnifications (a) 2 μm, (b) 1 μm and EDS spectrum of GO, (c) and TBG (d). (Maddinedi and Mandal, 2016, with permission from Bentham Science Publishers.)

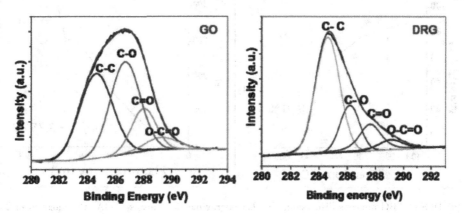

FIGURE 16.15 C1s XPS spectrum of GO and DRG. (Reprinted from Maddinedi et al., 2017b, with permission from Elsevier.)

major peaks appear. One is called G-band (I_G) which arises due to E_{2g} photon emission from Csp^2 atoms at $\sim$1,575 cm^{-1}, and other one is called as D-band (I_D) which arises due to A$_{1g}$ photons activation at $\sim$1,350 cm^{-1} which signifies disorderness on graphene sheets. Basically, D-band/G-band intensity ratio (I_D/I_G) provides useful information on degree of disorder, i.e. defects, ripples and edges in graphene. Figure 16.16 shows Raman spectrum of GO with two strong peaks, i.e. D-band at 1,355 and G-band at 1,598 cm^{-1}, while that of diastase-mediated rGO possesses D-band at 1,325 and G-band at 1,593 cm^{-1}. The increase in I_D/I_G intensity ratio value from 1.47 to 1.85 indicates the reduction GO to rGO with increasing sp^2 domain or orderness which indicates increased functionalization of rGO with diastase molecules (Cui et al., 2011; Guo et al., 2009; Maddinedi et al., 2017b).

16.7.8 Zeta Potential and DLS Study

The zeta potential measurement is done by dynamic light scattering (DLS) study. The stability of GO and rGO suspension/colloidal solution is checked by its zeta potential value which is highly important during its applications in biomedical, catalytic and sensing arena. GO and rGO having oxygen-containing functionalities possess negative charge on the surfaces with a zeta potential of $\sim$46.2 mV (ASTM, 1985). Based on the fabrication of GO/rGO surfaces, zeta

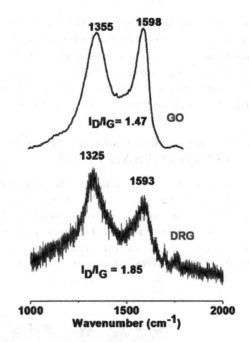

FIGURE 16.16 Raman analysis of GO and DRG. (Reprinted from Maddinedi et al., 2017b, with permission from Elsevier.)

potential value will vary which may help to choose a proper solvent for dispersion as well as fabrication. Figure 16.17 shows a representative DLS spectrum of *T. bellirica*-reduced

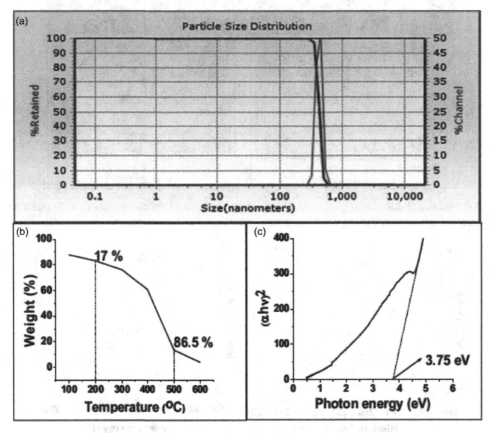

FIGURE 16.17 (a) DLS of TBG aqueous dispersion, (b) thermogravimetric analysis (TGA) thermogram of TBG and (c) optical diffuse reflectance spectra of *T. bellirica*-rGO. (Maddinedi and Mandal, 2016, with permission from Bentham Science Publishers.)

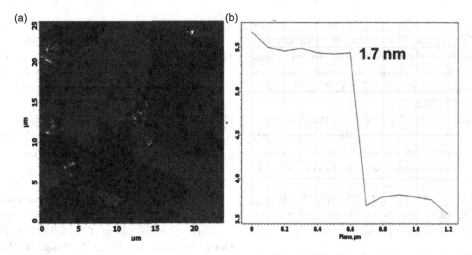

FIGURE 16.18 AFM image (a) and the corresponding height profile (b) of DRG. (Reprinted from Maddinedi et al., 2017b, with permission from Elsevier.)

GO which can provide information on hydrodynamic diameter and particles size distribution.

16.7.9 Atomic Force Microscopic (AFM) Studies

Atomic force microscopic (AFM) studies of graphene-based materials are carried out to find out the thickness of the synthesized graphene-based nanostructures. Figure 16.18 shows a typical AFM image of diastase enzyme-mediated rGO (DRG) (Maddinedi et al., 2017b). It is evident that the thickness of DRG sheet is 1.72 nm; that is, its sheet consists of two layers considering the thickness of single-layered graphene sheet as 0.8 nm (Fan et al., 2008; Novoselov et al., 2004). Also, one can get information on stabilizing agent present on the graphene surface compared to a flat sp^2 carbon atom network (0.335 nm) (Ni et al., 2007).

16.7.10 Thermogravimetric Analysis (TGA)

TGA is normally carried out to check the thermal stability of the synthesized rGO. Sometimes, it can provide good information on desired immobilization/fabrication of biomolecules or surface-modifying agents or functionalizing moieties. Figure 16.17 suggests that the *T. bellirica*-mediated rGO lost 17% weight below 200°C which suggests that all oxygen functional groups were eliminated from its surface at this temperature, while 86.6% weight loss happened below 500°C heating indicating its pyrolysis of all functional groups as well as its carbon ring (Maddinedi and Mandal, 2016).

16.8 Conclusions

Fabrication of GO surfaces is needed for the cutting-age-technological applications. Although the functionalization of GO surface is done by green routes via NCF techniques, it is a challenge to get precise size and selective adhesion on GO surfaces as well as purity of the resultant products. Leaching out of the immobilized/fabricated NMs on GO/rGO surfaces is a major challenge to materials scientists. This can be solved by covalent functionalization of GO surfaces. Also, the required purity is attained by covalent-functionalized rGO nanocomposites for very sophisticated and important applications. Moreover, advanced analytical techniques are in need of the hour for proper characterization of the synthesized graphene-based nanostructures.

Acknowledgment

The author highly acknowledges Vellore Institute of Technology (VIT), Vellore 632014, India, for all sorts of academic help to prepare this review article.

References

ASTM, Standard D4187-92, Zeta potential of colloids in water and waste water, American Society for Testing and Materials, 1985.

Avinash, M.B., Subrahmanyam, K.S., Sundarayya, Y. and Govindaraju, T. 2010. Covalent modification and exfoliation of grapheneoxide using ferrocene. *Nanoscale* 2: 1762–1766.

Baby, T.T., Aravind, S.S.J., Arockiadoss, T., Rakhi, R.B. and Ramaprabhu, S. 2010. Metal decorated graphene nanosheets as immobilization matrix for amperometric glucose biosensor. *Sensor. Actuat. B Chem.* 145: 71–77.

Bagherzadeh, M. and Farahbakhsh, A. 2015. Surface functionalization of graphene. In *Graphene Materials: Fundamentals and Emerging Applications*, eds. A. Tiwari, and M. Syvajarvi, Scrivener Publishing LLC, Beverly, MA, pp. 25–65.

Bai, H., Xu, Y., Zhao, L., Li, C. and Shi, G. 2009. Non-covalent functionalization of grapheme sheets by sulfonated polyaniline. *Chem. Commun.* 0: 1667–1669.

Bai, J., Sun, H., Yin, X., Yin, X., Wang, S., Creamer, A.E., Xu, L., Qin, Z., He, F. and Gao, B. 2016. Oxygen-content-controllable rapheme oxide from electron-beam-irradiated graphite: Synthesis, characterization, and removal of aqueous lead [Pb(II)]. *ACS Appl. Mater. Interfaces* 8: 25289–25296.

Bekyarova, E., Itkis, M.E., Ramesh, P., Haddon, R.C., Berger, C., Sprinkle, M. and de Herr, W.A. 2009. Chemical modification of epitaxial rapheme: Spontaneous grafting of aryl groups. *J. Am. Chem. Soc.* 131: 1336–1337.

Bepete, G., Anglaret, E., Ortolani, L., Morandi, V., Huang, K., Pnicaud, A. and Drummond, C. 2017. Surfactant-free single-layer graphene in water. *Nat. Chem.* 9: 347–352.

Chang, H., Wang, G., Yang, A., Tao, X., Liu, X., Shen, Y. and Zheng, Z. 2010. A transparent, flexible, low-temperature, and solution-processible graphene composite electrode. *Adv. Funct. Mater.* 20: 2893–2902.

Chen, C., Zhai, W., Lu, D., Zhang, H. and Zheng, W. 2011a. A facile method to prepare stable noncovalent functionalized graphene solution with thionine. *Mater. Res. Bull.* 46: 583–587.

Chen, G., Zhai, S., Zhai, Y., Zhang, K., Yue, Q., Wang, L., Zhao, J.S., Wang, H.S., Liu, J.F. and Jia, J.B. 2011b. Preparation of sulfonic-functionalized graphene oxide as ion-exchange material and its application into electro-chemiluminescence analysis. *Biosens. Bioelectron.* 26: 3136–3141.

Cheon, Y.A., Bae, J.H. and Chung, B.G. 2016. Reduced graphene oxide nanosheet for chemo-photothermal therapy. *Langmuir* 32: 2731–2736.

Choi, J., Kim, K.J., Kim, B., Lee, H. and Kim, S. 2009. Covalent functionalization of epitaxial graphene by azidotrimethylsilane. *J. Phys. Chem. C* 113: 9433–9435.

Chunder, A., Pal, T., Khondaker, S.I. and Zhai, L. 2010. Reduced graphene oxide/copper phthalocyanine composite and its optoelectrical properties. *J. Phys. Chem. C* 114: 15129–15135.

Cui, P., Seo, S., Lee, J., Wang, L., Lee, E., Min, M. and Lee, H. 2011. Nonvolatile memory device using gold nanoparticles covalently bound to reduced graphene oxide. *ACS Nano* 5: 6826–6833.

Cui, Y., Kim, S.N., Jones, S.E., Wissler, L.L., Naik, R.R. and McAlpine, M.C. 2010. Chemical functionalization of graphene enabled by phage displayed peptides. *Nano Lett.* 10: 4559–4565.

Dey, R.S. and Raj, C.R. 2010. Development of an amperometric cholesterol biosensor based on graphene–Pt nanoparticle hybrid material. *J. Phys. Chem. C* 114: 21427–21433.

Dong, H., Gao, W., Yan, F., Ji, H. and Ju, H. 2010. Fluorescence resonance energy transfer between quantum dots and graphene oxide for sensing biomolecules. *Anal. Chem.* 82: 5511–5517.

Fan, X., Peng, W., Li, Y., Li, X., Wang, S., Zhang, G. and Zhang, F. 2008. Deoxygenation of exfoliated graphite oxide under alkaline conditions: A green route to graphene preparation. *Adv. Mater.* 20: 4490–4493.

Fang, M., Wang, K., Lu, H., Yang, Y. and Nutt, S. 2009. Covalent polymer functionalization of graphene-nanosheets and mechanical properties of composites. *J. Mater. Chem.* 19: 7098–7105.

Gao, J., Liu, F., Liu, Y., Ma, N., Wang, Z. and Zhang, X. 2010. Environment-friendly method to produce graphene that employs vitamin C and amino acid. *Chem. Mater.* 22: 2213–2218.

Geng, J. and Jung, H.T. 2010. Porphyrin functionalized graphene sheets in aqueous suspensions: From the preparation of graphene sheets to highly conductive graphene films. *J. Phys. Chem. C* 114: 8227–8234.

Geng, X., Niu, L., Xing, Z., Song, R., Liu, G., Sun, M., Cheng, G., Zhong, H., Liu, Z., Zhang, Z., Sun, L., Xu, H., Lu, L. and Liu, L. 2010. Aqueous-processable noncovalent chemically converted graphene-quantum dot composites for flexible and transparent optoelectronic films. *Adv. Mater.* 22: 638–642.

Ghosh, A., Rao, K.V., George, S.J. and Rao, C.N.R. 2010. Noncovalent functionalization, exfoliation, and solubilization of graphene in water by employing a fluorescent coronene carboxylate. *Chem. Eur. J.* 16: 2700–2704.

Guo, H.L., Wang, X.F., Qian, Q.Y., Wang, F.B. and Xia, X.H. 2009. A green approach to the synthesis of graphene nanosheets. *ACS Nano* 3: 2653–2659.

He, F., Fan, J., Ma, D., Zhang, L., Leung, C. and Chan, H.L. 2010. The attachment of Fe_3O_4 nanoparticles to graphene oxide by covalent bonding. *Carbon* 48: 3139–3144.

He, H. and Gao, C. 2010. General approach to individually dispersed, highly soluble, and conductive graphene nanosheets functionalized by nitrene chemistry. *Chem. Mater.* 22: 5054–5064.

Hsiao, M.C., Liao, S.H., Yen, M.Y., Liu, P., Pu, N.W., Wang, C.A. and Ma, C.C. 2010. Preparation of covalently functionalized graphene using residual oxygen-containing functional groups. *ACS Appl. Mater. Interfaces* 2: 3092–3099.

Hu, H., Wang, X., Wang, J., Liu, F., Zhang, M. and Xu, C. 2011. Microwave-assisted covalent modification of graphene nanosheets with chitosan and its electrorheological characteristics. *Appl. Surf. Sci.* 257: 2637–2642.

Jiao, T., Liu, Y., Wu, Y., Zhang, Q., Yan, X., Gao, F., Bauer, A.J.P., Liu, J., Zeng, T.Z. and Li, B. 2015. Facile and scalable preparation of graphene oxide-based magnetic hybrids for fast and highly efficient removal of organic dyes. *Sci. Rep.* 5: 12451.

Kadiyala, N.K. and Mandal, B.K. 2016. *Terminalia bellirica* mediated one-pot green synthesis of reduced graphene oxide-silver nanoparticle (RGO/AgNP) nanocomposite and its catalytic activity towards *p*-nitrophenol reduction in water. *J. Indian Chem. Soc.* 93: 999–1007.

Kadiyala, N.K., Mandal, B.K., Ranjan, S. and Dasgupta, N. 2018. Bioinspired gold nanoparticles decorated reduced graphene oxide nanocomposite using Syzygium cumini seed extract: Evaluation of its biological applications. *Mater. Sci. Eng. C* 93: 191–205.

Kamada, S., Nomoto, H., Fukuda, K., Fukawa, T., Shirai, H. and Kimura, M. 2011. Noncovalent wrapping of chemically modified graphene with π-conjugated disk-like molecules. *Colloid. Polym. Sci.* 289: 925–932.

Kim, T.Y., Lee, H., Kim, J.E. and Suh, K.S. 2010. Synthesis of phase transferable graphene sheets using ionic liquid polymers. *ACS Nano* 4: 1612–1618.

Laaksonen, P., Kainlauri, M., Laaksonen, T., Shchepetov, A., Jiang, H., Ahopelto, J. and Linder, M.B. 2010. Interfacial engineering by proteins: Exfoliation and functionalization of graphene by hydrophobins. *Angew. Chem. Int. Ed.* 49: 4946–4949.

Lee, B., Li, K., Yoon, H.S., Yoon, J., Mok, Y., Lee, Y., Lee, H.H. and Kim, Y.H. 2016. Membrane of functionalized reduced graphene oxide nanoplates with angstrom-level channels. *Sci. Rep.* 6. doi: 10.1038/srep 28052.

Lee, D.Y., Khatun, Z., Lee, J.H., Lee, Y.K. and In, I. 2011. Blood compatible graphene/heparin conjugate through noncovalent chemistry. *Biomacromolecules* 12: 336–341.

Liu, H., Gao, J., Xue, M., Zhu, N., Zhang, M. and Cao, T. 2009. Processing of graphene for electrochemical application: Noncovalently functionalize graphene sheets with water-soluble electroactive methylene green. *Langmuir* 25: 12006–12010.

Liu, J., Fu, S., Yuan, B., Li, Y. and Deng, Z. 2010. Towards a universal "adhesive nanosheet" for the assembly of multiple nanoparticles based on a protein-induced reduction/decoration of graphene oxide. *J. Am. Chem. Soc.* 132: 7279–7281.

Liu, K.P., Zhang, J.J., Wang, C.M. and Zhu, J.J. 2011. Graphene-assisted dual amplification strategy for the fabrication of sensitive amperometric immunosensor. *Biosens. Bioelectron.* 26: 3627–3632.

Liu, R., Gong, T., Zhang, K. and Lee, C. 2017. Graphene oxide papers with high water adsorption capacity for air dehumidification. *Sci. Rep.* 7: 9761–9769.

Liu, S.Q., Xiao, B., Feng, L.R., Zhou, S.S., Chen, Z.G., Liu, C.B., Chen, F., Wu, Z.Y., Xu, N., Oh, W.C. and Meng, Z.D. 2013. Graphene oxide enhances the Fenton-like photocatalytic activity of nickel ferrite for degradation of dyes under visible light irradiation. *Carbon* 64: 197–206.

Liu, Y., Peng, J., Wang, S., Xu, M., Gao, M., Xia, T., Weng, J., Xu, A. and Liu, S. 2018. Molybdenum disulfide/graphene oxide nanocomposites show favorable lung targeting and enhanced drug loading/tumor-killing efficacy with improved biocompatibility. *NPG Asia Mater.* 10: e458–e472.

Liu, Z., Robinson, J.T., Sun, X. and Dai, H. 2008. PEGylated nanographene oxide for delivery of water-insoluble cancer drugs. *J. Am. Chem. Soc.* 130: 10876–10877.

Liu, Z., Robinson, J.T., Sun, X. and Dai, H. 2008. PEGy-lated nanographene oxide for delivery of water-insoluble

cancer drugs. *J. Am. Chem. Soc.* 130: 10876–10877.

Liu, Z.B., Xu, Y.F., Zhang, X.Y., Zhang, X.L., Chen, Y.S. and Tian, J.G. 2009. Porphyrin and fullerene covalently functionalized graphene hybrid materials with large nonlinear optical properties. *J. Phys. Chem. B* 113: 9681–9686.

Lomeda, J.R., Doyle, C.D., Kosynkin, D.V., Hwang, W.F. and Tour, J.M. 2008. Diazonium functionalization of surfactant-wrapped chemically converted graphene sheets. *J. Am. Chem. Soc.* 130: 16201–16206.

Luo, Z.J., Geng, H.Z., Zhang, X., Du, B., Ding, E.X., Wang, J., Lu, Z., Sun, B., Wang, J. and Liu, J. 2016. A time-saving, low-cost, high-yield method for the synthesis of ultrasmall uniform graphene oxide nanosheets and their application in surfactants. *Nanotechnology* 27: 055601 (8pp). doi:10.1088/0957-4484/27/5/055601.

Maddinedi, S.B. and Mandal, B.K. 2014. Low-cost and eco-friendly green methods for graphene synthesis. *Int. J. Nano Sci. Technol.* 3: 46–61.

Maddinedi, S.B. and Mandal, B.K. 2016. Biofabrication of reduced graphene oxide nanosheets using *Terminalia bellirica* fruit extract. *Curr. Nanosci.* 12: 94–102.

Maddinedi, S.B., Mandal, B.K., Nawaz Khan, F.-R. 2017a. High reduction of 4-nitrophenol using reduced graphene oxide/Ag synthesized with tyrosine. *Environ. Chem. Lett.* 15: 467–474.

Maddinedi, S.B., Mandal, B.K., Patil, S.H., Andhalkar, V.V., Ranjan, S. and Dasgupta, N. 2017b. Diastase induced green synthesis of bilayered reduced graphene oxide and its decoration with gold nanoparticles. *J. Photochem. Photobiol. B* 166: 252–258.

Maddinedi, S.B., Mandal, B.K., Vankayala, R., Kalluru, P. and Sreedhara Reddy, P. 2015. Bioinspired reduced graphene oxide nanosheets using *Terminalia chebula* seeds extract. *Spectrochim. Acta, Part A* 145: 117–124.

Maddinedi, S.B., Mandal, B.K., Vankayala, R., Kalluru, P. Tammina, S.K. and Kiran Kumar, H.A. 2014. Casein mediated green synthesis and decoration of reduced graphene oxide. *Spectrochim. Acta, Part A* 126: 227–231.

Malig, J., Jux, N., Kiessling, D., Cid, J.J., Vazquez, P., Torres, T. and Guldi, D.M. 2011. Towards tunable graphene/phthalocyanine-PPV hybrid systems. *Angew. Chem. Int. Ed.* 50: 3561–3565.

Mohanty, N. and Berry, V. 2008. Graphene-based single-bacterium resolution biodevice and DNA transistor: Interfacing graphene derivatives with nanoscale and microscale biocomponents. *Nano Lett.* 8: 4469–4476.

Moon, I.K., Lee, J., Ruoff, R.S. and Lee, H. 2010. Reduced graphene oxide by chemical graphitization. *Nat. Commun.* 1: 73. doi: 10.1038/ncomms1067.

Nakayama-Ratchford, N., Bangsaruntip, S., Sun, X., Welsher, K. and Dai, H. 2007. Noncovalent functionalization of carbon nanotubes by fluorescein-polyethylene glycol: Supramolecular conjugates with pH-dependent absorbance and fluorescence. *J. Am. Chem. Soc.* 129: 2448–2449.

Ni, Z.H., Wang, H.M., Kasim, J., Fan, H.M., Yu, T., Wu, Y.H., Feng, Y.P. and Shen, Z.X. 2007. Graphene thickness determination using reflection and contrast spectroscopy. *Nano Lett.* 7: 2758–2763.

Niyogi, S., Bekyarova, E., Itkis, M.E., McWilliams, J.L., Hamon, M.A. and Haddon, R.C. 2006. Solution properties of graphite and graphene. *J. Am. Chem. Soc.* 128: 7720–7721.

Novoselov, K.S., Geim, A.K., Morozov, S.V., Jiang, D., Zhang, Y., Dubonos, S.V., Grigorieva, I.V. and Firsov, A.A. 2004. Electric field effect in atomically thin carbon films. *Science* 306: 666–669.

Pan, D., Zhang, J., Li, Z. and Wu, M. 2010. Hydrothermal route for cutting graphene sheets into blue-luminescent graphene quantum dots. *Adv. Mater.* 22: 734–738.

Park, M.J., Lee, J.K., Lee, B.S., Lee, Y.W., Choi, I.S. and Lee, S.-G. 2006. Covalent modification of multiwalled carbon nanotubes with imidazolium-based ionic liquids: Effect of anions on solubility. *Chem. Mater.* 18: 1546–1551.

Park, S., Dikin, D.A., Nguyen, S.T. and Ruoff, R.S. 2009. Graphene oxide sheets chemically cross-linked by polyallylamine. *J. Phys. Chem. C* 113: 15801–15804.

Pei, S., Wei, Q., Huang, K., Cheng, H.M. and Ren, W. 2018. Green synthesis of graphene oxide by seconds timescale water electrolytic oxidation. *Nut. Commun.* 9: 145–153.

Peng, L., Xu, Z., Liu, Z., Wei, Y., Sun, H., Li, Z., Zhao, X. and Gao, C. 2015. An iron-based green approach to 1-h production of single-layer graphene oxide. *Nat. Commun.* 6: 5716–5724. doi: 10.1038/ncomms6716.

Perera, S.D., Njiem, N., Chabal, Y., Ferraris, J.P. and Balkus, Jr., K.J. 2012. Deoxygenated graphene oxide for supercapacitor applications: An effective green alternative for chemically reduced graphene. *J. Power Sources* 215: 1–10.

Pham, T.A., Kumar, N.A. and Jeong, Y.T. 2010. Covalent functionalization of graphene oxide with polyglycerol and their use as templates for anchoring magnetic nanoparticles. *Syn. Met.* 160: 2028–2036.

Qi, X., Pu, K.Y., Li, H., Zhou, X., Wu, S., Fan, Q.L., Liu, B., Boey, F., Huang, W. and Zhang, H. 2010. Amphiphilic graphene composites. *Angew. Chem. Int. Ed.* 49: 9426–9429.

Ren, L., Liu, T., Guo, J., Guo, S., Wang, X. and Wang, W. 2010. A smart pH responsive graphene/polyacrylamide complex via noncovalent interaction. *Nanotechnology* 21: 335701.

Savarin, C.G., Gris, C., Murry, J.A., Reamer, R.A. and Hughes, D.L. 2007. Novel intramolecular reactivity of oximes: Synthesis of cyclic and spiro-fused imines. *Org. Lett.* 9: 981–983.

Shan, C., Yang, H., Han, D., Zhang, Q., Ivaska, A. and Niu, L. 2009. Water-soluble graphene covalently functionalized by biocompatible poly-L-lysine. *Langmuir* 25: 12030–12033.

Singh, V., Joung, D., Zhai, L., Das, S., Khondaker, S.I. and Seal, S. 2011. Graphene based materials: Past, present and future. *Prog. Mater. Sci.* 56: 1178–1271.

Stankovich, S., Dikin, D.A., Compton, O.C., Dommett, G.H.B., Ruoff, R.S. and Nguyen, S.T. 2010. Systematic post-assembly modification of graphene oxide paper with primary alkylamines. *Chem. Mater.* 22: 4153–4157.

Stankovich, S., Piner, R.D., Chen, X., Wu, N., Nguyen, S.T. and Ruoff, R.S. 2006a. Stable aqueous dispersions of graphitic nanoplatelets via the reduction of exfoliated graphite oxide in the presence of poly(sodium 4-styrenesulfonate). *J. Mater. Chem.* 16: 155–158.

Stankovich, S., Piner, R.D., Nguyen, S.T. and Ruoff, R.S. 2006b. Synthesis and exfoliation of isocyanate-treated graphene oxide nanoplatelets. *Carbon* 44: 3342–3347.

Sun, Y., Fu, W., Dai, Y., Huang, Y., Zhou, J., Huang, C., Yang, C., Huang, M., Ma, R., Lin, B. 2017. Self-assembly of defect-rich graphene oxide nanosheets with $Na_2Ti_3O_7$ nanowires and their superior absorptive capacity to toxic dyes. *Nanotechnology* 28: 245601 (11pp). doi:10.1088/1361-6528/aa6f7e.

Sun, Z., Kohama, S.I., Zhang, Z., Lomeda, J.R. and Tour, J.M. 2010. Soluble graphene through edge-selective functionalization. *Nano Res.* 3: 117–125.

Sundaram, R.S., Gomez-Navarro, C., Balasubramanian, K., Burghard, M. and Kern, K. 2008. Electrochemical modification of graphene. *Adv. Mater.* 20: 3050–3053.

Surwade, S.P., Smirnov, S.N., Vlassiouk, I.V., Unocic, R.R., Veith, G.M., Dai, S. and Mahurin, S.M. 2015. Water desalination using nanoporous single-layer Graphene. *Nat. Nanotechnol.* 10. doi: 10.1038/NNANO.2015.37.

Tandel, R.D., Pawar, S.K. and Seetharamappa, J. 2016. Synthesis and characterization of bentonite-reduced graphene oxide composite: Application as sensor for a neurotransmitter, dopamine. *J. Electrochem. Soc.* 163: H705–H713.

Tiwari, J.N., Mahesh, K., Le, N.H., Kemp, K.C., Timilsina, R., Tiwari, R.N. and Kim, K.S. 2013. Reduced graphene oxide based hydrogels for the efficient capture of dye pollutants from aqueous solutions. *Carbon* 56: 173–82.

Vadukumpully, S., Gupta, J., Zhang, Y., Xu, G.Q. and Valiyaveettil, S. 2011. Functionalization of surfactant wrapped graphene nanosheets with alkylazides for enhanced dispersibility. *Nanoscale* 3: 303–308.

Veca, L.M., Lu, F., Meziani, M.J., Cao, L., Zhang, P., Qi, G., Qu, L., Shrestha, M. and Sun, Y.P. 2009. Polymer functionalization and solubilization of carbon nanosheets. *Chem. Commun.* 18: 2565–2567.

Wang, G., Wang, B., Park, J., Yang, J., Shen, X. and Yao, J. 2009. Synthesis of enhanced hydrophilic and hydrophobic graphene oxide nanosheets by a solvothermal method. *Carbon* 47: 68–72.

Wang, H., Cui, L.F., Yang, Y., Sanchez Casalongue, H., Robinson, J.T., Liang, Y., Cui, Y. and Dai, H. 2010a. Mn_3O_4-graphene hybrid as a high-capacity anode material for lithium ion batteries. *J. Am. Chem. Soc.* 132: 13978–13980.

Wang, H., Robinson, J.T., Diankov, G. and Dai, H. 2010b. Nanocrystal growth on graphene with various degrees of oxidation. *J. Am. Chem. Soc.* 132: 3270–3271.

Wang, P., Jiang, T., Zhu, C., Zhai, Y., Wang, D. and Dong, S. 2010c. One-step, solvothermal synthesis of graphene-CdS and graphene-ZnS quantum dot nanocomposites and their interesting photovoltaic properties. *Nano Res.* 3: 794–799.

Wang, S., Chia, P.J., Chua, L.L., Zhao, L.H., Png, R.Q., Sivaramakrishnan, S., Zhou, M., Goh, R.G.S., Friend, R.H., Wee, A.T.S. and Ho, P.K.H. 2008. Band-like transport in surface-functionalized highly solution-processable graphene nanosheets. *Adv. Mater.* 20: 3440–3446.

Wojcik, A. and Kamat, P.V. 2010. Reduced graphene oxide and porphyrin. An interactive affair in 2-D. *ACS Nano* 4: 6697–6706.

Wu, T., Chen, M., Zhang, L., Xu, X., Liu, Y., Yan, J., Wang, W. and Gao. J. 2013. Three-dimensional graphene-based aerogels prepared by a self-assembly process and its excellent catalytic and absorbing performance. *J. Mater. Chem. A* 1: 7612–7621.

Wu, Z.S., Wang, D.W., Ren, W., Zhao, J., Zhou, G., Li, F. and Cheng, H.M. 2010. Anchoring hydrous RuO$_2$ on graphene sheets for high-performance electrochemical capacitors. *Adv. Funct. Mater.* 20: 3595–3602.

Xu, L.Q., Yang, W.J., Neoh, K.G., Kang, E.T. and Fu, G.D. 2010. Dopamine-induced reduction and functionalization of graphene oxide nanosheets. *Macromolecules* 43: 8336–8339.

Xu, Y., Liu, Z., Zhang, X., Wang, Y., Tian, J., Huang, Y., Ma, Y., Zhang, X.Y. and Chen, Y. 2009. A graphene hybrid material covalently functionalized with porphyrin: Synthesis and optical limiting property. *Adv. Mater.* 21: 1275–1279.

Yang, H., Shan, C., Li, F., Han, D., Zhang, Q. and Niu, L. 2009. Covalent functionalization of polydisperse chemically-converted grapheme sheets with amine-terminated ionic liquid. *Chem. Commun.* 0: 3880–3882.

Yang, H., Zhang, Q., Shan, C., Li, F., Han, D. and Niu, L. 2010a. Stable, conductive supramolecular composite of graphene sheets with conjugated polyelectrolyte. *Langmuir* 26: 6708–6712.

Yang, Q., Pan, X., Huang, F. and Li, K.2010b. Fabrication of high-concentration and stable aqueous suspensions of graphene nanosheets by noncovalent functionalization with lignin and cellulose derivatives. *J. Phys. Chem. C* 114: 3811–3816.

Yang, S., Feng, X., Ivanovici, S. and Mullen, K. 2010c. Fabrication of graphene-encapsulated oxide nanoparticles: Towards high-performance anode materials for lithium storage. *Angew. Chem. Int. Ed.* 49: 8408–8411.

Zeng, Q., Cheng, J., Tang, L., Liu, X., Liu, Y., Li, J. and Jiang, J. 2010. Self-assembled graphene-enzyme hierarchical nanostructures for electrochemical biosensing. *Adv. Funct. Mater.* 20: 3366–3372.

Zhang, D.D., Zua, S.Z. and Hana, B.H. 2009a. Inorganic-organic hybrid porous materials based on graphite oxide sheets. *Carbon* 47: 2993–3000.

Zhang, H., Peng, C., Yang, J., Lv, M., Liu, R., He, D., Fan, C. and Huang, Q. 2013. Uniform ultrasmall graphene oxide nanosheets with low cytotoxicity and high cellular uptake. *ACS Appl. Mater. Interfaces* 5: 1761–1767.

Zhang, J., Lei, J., Pan, R., Xue, Y. and Ju, H. 2010. Highly sensitive electrocatalytic biosensing of hypoxanthine based on functionalization of graphene sheets with water-soluble conducting graft copolymer. *Biosens. Bioelectron.* 26: 371–376.

Zhang, L., Yao, H., Li, Z., Sun, P., Liu, F., Dong, C., Wang, J., Li, Z., Wu, M., Zhang, C. and Zhao, B. 2017. Synthesis of delaminated layered double hydroxides and their assembly with graphene oxide for supercapacitor application. *J. Alloys Compd.* 711: 31–41.

Zhang, Y., Li, H., Pan, L., Lu, T., Sun, Z. 2009b. Capacitive behavior of graphene-ZnO composite film for supercapacitors. *J. Electroanal. Chem.* 634: 68–71.

Zhang, Y., Ren, L., Wang, S., Marathe, A., Chaudhuri, J. and Li, G. 2011. Functionalization of graphene sheets through fullerene attachment. *J. Mater. Chem.* 21: 5386–5391.

Zhao, Y.L. and Stoddart, J.F. 2009. Noncovalent functionalization of single-walled carbon nanotubes. *Acc. Chem. Res.* 42: 1161–1171.

Zhu, C., Guo, S., Fang, Y. and Dong, S. 2010. Reducing sugar: New functional molecules for the green synthesis of graphene nanosheets. *ACS Nano.* 4: 2429–2437.

Zong, M., Huang, Y., Zhao, Y., Sun, X., Qu, C., Luo, D. and Zheng, J. 2013. Facile preparation, high microwave absorption and microwave absorbing mechanism of RGO–Fe$_3$O$_4$ composites. *RSC Adv.* 3: 23638–23648.

Radio Frequency Magnetron-Sputtered Germanium Nanoislands: Comprehensive Investigations of Growth Parameters

Alireza Samavati and
Ahmad Fauzi Ismail
Universiti Teknologi Malaysia

17.1 Introduction

In 1959, the American Physical Society at California Institute of Technology has witnessed the presentation of principal talk, which was about "things on an ultra-small" by Professor Richard Feynman. The possibility of direct utilization of individual atoms as stronger than artificial chemistry those used at that time was considered. After that talk, world observed the extraordinary developments in nanotechnology. In 1974, the Late Norio who was the researcher in The University of Tokyo used the term "nanotechnology" for the first time to refer to materials at the scale of nanometer.

Recently, design, characterization, production, and application of materials, which involve the manipulation of matter at the smallest scale (from 1 to 100 nm), have been widely used as current meaning of nanotechnology rather than just materials. Three distinct aspects can be considered for the evolution of nanotechnology: indirect, direct, and conceptual. The advanced miniaturization of obtainable technologies, which open new areas of application for those technologies, can be explained by indirect aspect. Direct refers to the application of novel nanoscale artifacts to improve the performance of presented process and materials or for completely novel purposes. Finally, there is a conceptual aspect of nanotechnology, in which all materials and process considered from molecular or even atomic viewpoint especially in living system and biology. Now a few areas of technology are exempt from the advantages of nanotechnology. The information and communication systems such as novel semiconductor and optoelectronic device, environment (filtration), energy (reduction of energy, consumption increasing, the efficiency of energy production nuclear accident cleanup and waste storage), heavy industry (aerospace and catalysis), and consumer goods are some applications of nanotechnology.

Silicon has been used in the solid-state electronics industry as the preferable semiconductor for device applications for more than 60 years. Interestingly, the first transistor was constructed of germanium. In the periodic table, germanium is placed in the fourth column, two levels below carbon, and just below silicon. The lattice constant between Ge and Si is 5%, and germanium has a smaller band gap of 0.7 eV compared to that of 1.17 eV for silicon. Both germanium and silicon are indirect gap materials with otherwise extraordinarily similar band structure. In the solid phase, germanium forms tetrahedrally coordinated covalent bonds and has the diamond crystal structure similar to both carbon and silicon. Germanium has a lower melting point than silicon, 937°C versus 1,410°C. The oxide of germanium is water soluble in contrast to silicon. In the past few years, as heterostructure growing methods improve, interest in germanium among science and commercial has increased consequently, and Ge/Si heterostructure with applications in band gap engineering for optoelectronic devices has attracted a considerable attention.

In 1985, Becker et al. have studied the Ge(111) surface with the Scanning tunneling microscope (STM) for the first time. The microscopic surface topography of germanium layers grown by molecular beam epitaxy (MBE) on Si(111) has been studied by them. A more comprehensive study of the Ge(111) surface structures and their relation to analogous structures on Si(111) was carried out by Becker et al. (1989).

Creating the Ge nanoscale structures consistently and predictably with a high degree of control over size, such as the ability to form monosized particles, shape, and spatial distribution, needs to study many fabrication and characterization techniques. A number of techniques such as MBE (Liu et al., 2002), chemical vapor deposition (CVD) (Thanh et al., 2000; Yam et al., 2000), femtosecond laser ablation (Seo et al., 2006; Riabinina et al., 2006a), and sputtering (Huang et al., 2009; Das et al., 2007) are exercised to produce Ge nanoislands. Desire for controlling the

properties of results originates the motivation for optimizing growth conditions for quantum dot fabrication.

Extensive efforts for fabricating Ge nanoislands having size between 10 and 100 nm using MBE and sputtering technique have been made by Ray et al. (2011). Shape, size, strain and composition evolution, and the optical properties have been investigated by them. Simonsen et al. (1999) produced Ge nanocrystals on Si (001) substrate using electron beam evaporation in which a Gaussian distribution of sizes from 2 to 68 nm was achieved, and the full width at half maximum (FWHM) showed increment with increasing island size. Ge QDs of 8–20 nm were prepared by Fahim et al. (2010) using electron beam evaporation technique, in which the root mean square (RMS) roughness is shown to be highly sensitive to the annealing temperature. Huang et al. (2009) prepared the Ge/Si nanomultilayer by magnetron sputtering and determined the crystalline nature of the film at deposition temperature 300°C. A strong agglomeration and the formation of Ge nanocrystals are observed for the samples annealed beyond 800°C. Radić et al. (2006) characterized Ge thin films grown by magnetron sputtering at different substrate temperatures using atomic force microscopy (AFM) and grazing-incidence small angle X-ray scattering (GISAXS), and found the best island structures at a substrate temperature of 650°C.

Ge nanocrystals are embedded on SiO_2 at temperature below 400°C using RF magnetron sputtering (Zhang et al., 2004). A competitive process between Volmer–Weber (VM) growth and oxidation reaction is used to explain the underlying mechanism. They found that the fabrication of multilayered Ge nanocrystals which exhibited controllable crystallite size with high crystallization quality can be achieved using rf magnetron sputtering with substrate temperature between 350°C and 400°C.

The rapid development of the information society is taken place because of the computation power, and data transmission bandwidth is gradually growing. Current electronic IC on Si platform is facing the limitation of fabrication difficulties, such as thermal dissipation and bandwidth, which are attributed to decreasing the size of transistors. In order to fulfill the growing demand, next-generation integrated circuit on Si platform has been developed. For future optoelectronic hybrid, Si-based light source in the telecommunication wavelengths is one of the most important components. Optoelectronic hybrid integration on Si, which has the advantages of ultra-large bandwidth and low power consumption, is one of the solutions. Complementary metal oxide semiconductor (CMOS)-compatible Si-based light-emitting devices are unavoidable for the optoelectronic integration on Si for the reason that integrating different optical components, such as Si-based light source, optical waveguide circuits, modulator, detector, and electronic integrated circuit for control on a single Si wafer. The indirect band gap origin of Si, makes it unsuitable for light-emitting devices. Among various options such as SiGe quantum wells (Kawaguchi et al., 2002; Nayak et al., 1993), Er doping (Franzo et al., 1994), Si nanocrystals (Pavesi et al., 2000), and Si/SiGe quantum-cascade structures (Lynch et al., 2006; Dehlinger et al., 2000) for light-emitting devices, Si-based Ge quantum dots have great potential, which is fully CMOS-compatible (Dashiel et al., 2002). Strong room-temperature resonant light emission from Ge self-assembled quantum dots shows a promising way to fabricate fully compatible Si-based light-emitting devices (CMOS). It is attracted a huge of attention to the wide applied physics, photonics community, and the IC industry. It will supply the ultimate Si-based light source, and optical interconnections inside Si chips accelerate the increase of the "optical age of silicon".

Controlled growth and fabrication of semiconductor nanostructure by easy and economic method is demanding for optoelectronic application. Despite many techniques, the rf magnetron sputtering technique has added advantage for large-scale and high-density fabrication of germanium nanostructure. We implement this method with different growth condition to examine the role of annealing, variation of substrate temperature, annealing time, and other growth parameters on sample morphology. Several other characterization methods are used for structural and optical behavior. We find the sample quality (shape, size, and microstructure of quantum dots) which is strongly influenced by the growth condition and hence their properties. It is possible to grow high-density large-scale structure by this method, which is detrimental for optoelectronic devices. The mechanism responsible for the growth can be interpreted in terms of nucleation, diffusion, oxygen passivation, quantum confinement, and the formation of core–shell-like structure. The easy and economic nature of our method suitable for the fabrication of varieties of other nanostructures to understand the fundamental physics of nanoscale structure under different growth condition is also important.

Surface morphology of Ge nanoislands deposited on Si(100) substrate, Ge/Si bilayer containing Ge QDs, and Ge/SiO$_2$/Si heterostructure, their characterization using AFM, field emission scanning electron microscopy (FESEM), scanning electron microscopy (SEM), and their optical properties are determined by photoluminescence (PL) and Raman spectroscopy which are important issues. The formation of nanoislands and their characteristics sizes can be estimated by X-ray diffraction (XRD) pattern. The particle size distribution between 8 and 29 nm is desired for practical implementation in devices. Understanding the optical and electron correlation effect in such nanostructure may provide useful information towards the development of nanophotonics. Determining and analyzing the growth processes and the underlying mechanism play a crucial role in the functionality of nanometric structure. The room-temperature visible luminescence and controlling the intensity as well as the size-dependent peak energy contribute significantly in display, laser, and sensing devices. The debatable issues related to the origin of different luminescence peak require careful fabrication and characterization of germanium nanoislands, on which research has been done by a few other scientists. The other important

aspect is making the bilayer of silicon–germanium and Ge/SiO$_2$/Si heteroepitaxial multilayer. Experimental observation requires further verification to modeling and simulation of the opening of band gap due to quantum size effect and surface modification. The results generated from rf sputtering method need to be compared with other techniques.

Nanostructuring of semiconductors is a novel means of developing new electronic and optoelectronic devices. In particular, the discovery of room-temperature visible PL from Si and Ge nanostructures has stimulated much interest in these particular kinds of nanoclusters and in small semiconductor particles. The possibility of tuning the optical response of Si and Ge nanomaterials by modifying their size has become one of the most challenging aspects of recent semiconductor research.

Easy and economic fabrication technique would be developed. The instrumentation for large-scale fabrication has socioeconomic impact. The fundamental physics behind the growth would be understood. The data generated through this research will be published in high impact factor journal, and research data would be presented in conferences, workshops, and seminars. PhD and masters research scholar can be trained using this methodology to pursue their future research. High quality of sample needed for the optoelectronic industries can be supported by using rf magnetron sputtering method. Device would be cheaper and economic. A set of characterization, which we propose, would be able to measure the band gap, right sample structure, and right physics. The extension of this research is that this method is not just limited to germanium, but other semiconductor nanostructures such as silicon and others can also be grown by using this method. This method can be extended and become versatile for nanostructure growth.

17.2 Literature Survey

17.2.1 Introduction

The trend in device-oriented manufacturing towards smaller and smaller dimensions fulfills the requirements of nanofabrication techniques and paves the way to understand the physics of nanometer-scale materials. One of the most complicated and long-term research challenges in the semiconductor industry is growing high-quality films using lattice-mismatched materials (heteroepitaxy). The remarkable changes in physical properties of materials are caused by the conversion of bulk structures into nanoscale. It provides the possibility of observing novel behaviors such as size-dependent structural and optical properties. The influence of the growth procedure and the corresponding mechanism of self-assembly (Stranski–Krastanov) of Ge nanostructure on Si substrate are of particular interest for obtaining highly monodisperse nanostructure for application in semiconductors optoelectronic devices. Magnetron sputtering method due to its high depositional rate and safety currently becomes the most popular commercially viable method for the fabrication of Ge/Si semiconductor heterostructure. The following chapter will discuss the main challenges of the Ge/Si heteroepitaxial system in detail.

17.2.2 Importance of Germanium

The possibility of light emission of silicon-based semiconductor heterostructure makes them quite competitive with such III–V compounds, traditional optoelectronic materials (Lai and Li, 2007; Tong et al., 2004). Studying the Ge/Si heterosystems for designing microwave devices and optoelectronic units has been extended by the most well-known technological centers such as IBM, NEC, Daimler-Benz, Toshiba, and Mitsubishi (Metzger, 1995; Paul, 1998). Ge/Si and InAs/GaAs nanosize islands heterosystems with density of 10^{10}–10^{11} cm^{-2} obtained by discovering the effect of self-organization of ordered arrays resulted in observation of more distinct atomic-like characteristics in the pertinent electron and optical spectra. The pioneering publication in this field is the work of Yakimov et al. (1999), which explained the use of Ge islands as QDs while observing the effects of resonance tunneling and Coulomb blockage. The transition from the 2D film growth to nucleation of 3D islands (the Stranski–Krastanov mechanism) has been studied for a long time. Germanium on silicon heterosystem is an example of S–K growth mode.

Germanium is being considered as a possible semiconductor material because germanium has many advantageous properties compared to silicon. Table 17.1 summarizes many of the basic physical properties of silicon and germanium. Electron mobility is 2.75 times higher than Si, and bulk hole mobility of Ge is four times higher than silicon. As a result, an increase in surface mobility and an ultimate increase in the transistor performance are ideally attributed to increase in bulk mobility. In addition to the increase in mobility, germanium's electron and hole mobilities are more symmetric compared to silicon. This will lead to smaller area devices in a CMOS inverter cell. The larger exciton Bohr diameter for germanium compared with Si enables it to exhibit more strong confinement effect in nanoscale size.

17.2.3 Low-Dimensional Structure Physics

Nanoscale science refers to the study of materials with a dimension of approximately 1–100 nm. This kind of structures belongs to a group called low-dimensional structures (0D–2D). The dimensions are determined by the number of directions that carriers can move freely in the material. For instance, in bulk materials (three-dimensional (3D) structure), electronic carriers move in all three directions. Quantum well is an example of 2D structure in which carriers may be confined in two of the three directions. Further confining the thin film resulted in one-dimensional (1D) system or the quantum wire. The quantum dot (0D) is the final confinement of the material. Figure 17.1 presents the progress for creating the low-dimensional structures. The trend of significant change in the density of states (DOS) as a function of spatial confinement is depicted in Figure 17.2. The confinement process causes interesting changes take place in the allowed carrier energies. In bulk materials, carriers can survive in nearly continuous bands. In contrast, in the case of 0D structure, the carriers are limited to a particular set of completely quantized energy states instead of an existence in the band of permissible energies. The entire quantization of energy states constructs a unique and suitable structure for study. The effect of confinement on the results of energy states in the system is accomplished by solving the eigen-energies of the Schrödinger wave equation for the carriers in a confined space. Hence, the calculation of the relationship between the size of the confined system and changes in the energy levels is made possible.

The basic term for the energy of a confined system is

$$E = \frac{\hbar^2 n^2 \alpha_n^2}{2mL^2} \tag{17.1}$$

where m is equal to the effective mass of the carriers which depends on the degree of confinement. α_n and L are the n zeroes of the spherical Bessel function of order 1 and the confinement dimension, respectively. Consequently, spatially

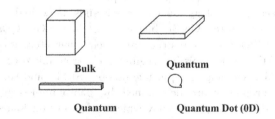

FIGURE 17.1   Low-dimensional structures.

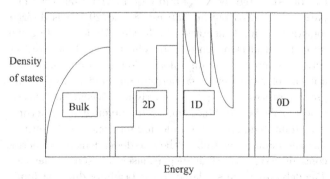

FIGURE 17.2   DOS as a function of energy for a bulk material (3D), quantum well (2D), quantum wire (1D), and quantum dot (0D). (From Bukowski and Simmons, 2002.)

TABLE 17.1   The Physical Properties of Germanium and Silicon

Properties	Bulk Germanium	Bulk Silicon
Atom (cm^3)	4.42×10^{22}	5.0×10^{22}
Crystal structure	Diamond	Diamond
Effective density of states (DOS) in conduction band (cm^3)	1.04×10^{19}	2.08×10^{19}
Effective DOS in valence band (cm^3)	6.0×10^{18}	1.04×10^{19}
Hole mobility (cm^2/V-s)	1,900	471
Electron mobility (cm^2/V-s)	3,900	1,417
Exciton Bohr diameter (nm)	24	9
Band gap energy (eV)	0.67	1.11

confined electronic carriers can shift the band gap of a material towards higher energies.

17.2.4 Exciton

Semiconductor band gap energy is the minimum energy required for optical excitation to form free carriers that means it is impossible for free carriers to be excited below that energy value. However, at low-temperature absorption studies of semiconductors, excitation below the band gap is reported by Pankove (1971). This excitation is taken place when the electron and hole bounded to each other which is called exciton. Because the electron and hole are bound to each other, the energy required for excitation is lower than the band gap energy. The electron orbits a local minimum potential associated with hole, which forms the stable structure. In the case of quantum dot that is spatially confined systems, quantized energy states are attributed to excitons carriers.

The bounding structure of electron and hole is comparable to the hydrogen atom. Coulombic attractive force in the case of hydrogen atom origin binds the electron to a single proton, and it is very similar to behavior of exciton. In fact, a set of hydrogen-like states are produced by orbiting electron around the hole. The Bohr radius is the characteristic dimension that is explained by orbiting electron around nucleus; therefore, in the quantum dot exciton, it is more than the radius of orbiting electron around hole. This distinctive dimension is named as exciton Bohr diameter and fundamentally measures the diameter (or radius) of the exciton. Judgment about the size confinement criteria in materials is based on the exciton Bohr diameter that is the significant parameter of optical behavior of materials. The criterion for confinement must be material-dependent property which is based on the nature of materials.

17.2.5 Confinement Regimes

The exciton Bohr diameter is a measure for evaluating the quantum confinement effects in a specific material. Generally, confinement effects are caused by reducing the material dimension to a size comparable with the exciton Bohr diameter. There are strong and weak confinement regimes that correspond to different energy state levels (Pankove, 1971). The strong and weak states are associated by the level of coupling between the electron and hole in the exciton. The strong confinement case occurs for the smaller dots having larger degree of confinement. In this case, the exciton cannot exist anymore and converts to the free electron and hole states. When the size of crystal is approximately three to ten times more than the exciton Bohr radius of the material, the weak confinement case can happen in a semiconductor nanostructure. This leads to the Coulomb interaction energy having the order of electron and hole sublevel separations. The pair of electron and hole correlating explain the quantization of confinement potential and motion of mass center. In this case, to solve the energy states with the boundary

condition, the parabolic approximation of the Schrödinger equation can be used. The potential is at the radius of the crystal, and the wave function vanishes, $\psi(a_0) = 0$, because a_0 is infinite (Joseph et al., 2000). Separation of variables into center of mass coordinates can make the finding of the solution possible. The energy quantization in a spherical 0D structures is

$$\Delta E_{nl} = \frac{\hbar^2 \alpha_{nl}^2}{2M^* a_0^2} \quad (17.2)$$

where M^* is equal to effective exciton mass which is described as $M^* = m_e^* + m_h^*$. α_{nl} is the n zeroes of the spherical Bessel functions of order l. In this equation, occurrence of energy shift can be observed in a range larger than exciton Bohr radius (weak confinement).

When the size of QDs comes close to the size of exciton (strong confinement), the quantization of the kinetic energy of the carriers is larger than the Coulomb effects. Uncorrelated wave functions of electron and hole are due to the size of quantum dot radius, which is smaller than exciton Bohr radius. The weak confinement technique mentioned earlier is used for solving the energy states in this case. The only difference between strong and weak confinement method is that the electron and hole have independent Bessel wave functions in the former. The shift in the energy is

$$\Delta E_{nl} = \frac{\hbar^2 \alpha_{nl}^2}{2\mu^* a_0^2} \quad (17.3)$$

where μ^* is the effective reduced mass

$$\frac{1}{\mu^*} = \frac{1}{m_e^*} + \frac{1}{m_h^*} \quad (17.4)$$

The presence of the Bessel function term, α_{nl}, and the changes in effective carrier mass are the different variable parameters in the energy expression for the strong confinement, weak confinement cases, and the 3D particle in a box treatment. Electron and hole masses are not similar in different materials. Therefore, the relative shifts in energy are dependent according to the material involved. The blue shift in energy is the consequence of the decrement in the size of crystal (Joseph et al., 2000).

17.2.6 Theory of Lattice Mismatch and Growth Modes

Lattice constant of germanium is 5.6575 Å, while that of silicon is 5.4307 Å as can be inferred from Figure 17.3. Thus, the lattice mismatch between germanium and silicon is (5.6575–5.4307 Å)/5.4307 Å × 100 = 4.1763%. This dissimilarity is very large in the area of semiconductor heteroepitaxy and will have a negative effect on the layer quality. With starting the deposition, the new Ge layer conforms to the lattice spacing of the Si substrate. The lattice constant is reduced, and the Ge layer is now strained. Growing defect-free compressively strained Ge on Si occurs below a certain thickness. People and Bean (1985) have demonstrated that this thickness is around 4–10 nm. Reduction of the strain

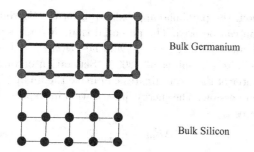

FIGURE 17.3 Lattice mismatch between Si and Ge.

by forming dislocations at the Ge/Si interface takes place by continuing the layer deposition. In addition, formation of 3D islands with rough surface occurs as additional benefit of reducing the elastic strain energy of the film (Fitzgerald, 1991). The next sections will detail the process of islanding and dislocations in germanium on silicon.

A thin liquid layer on a solid substrate in which the gravity can be neglected is considered. Young's equation can be driven from the balance of forces at the liquid–gas (lg), solid–liquid (sl), and solid–gas interfaces (sg):

$$\gamma_{lg} \cos \psi + \gamma_{sl} = \gamma_{sg} \qquad (17.5)$$

For a liquid, the surface tension is the same as the surface energy. In Eq. 17.5, the γ terms and ψ are surface energies and the contact angle, respectively. This equation illustrates that the two growth modes are simply represented in the absence of strain, whether the $\psi = 0°$ is related to wetting overlayer, which corresponds to layer-by-layer growth or $\psi > 0°$ that shows nonwetting overlayer, which corresponds to the formation of 3D islands. Consider two kinds of materials A and B, which present the substrate and the overlayer, respectively. For a microscopic understanding of wetting behavior, the balance of intermolecular forces between A and B is considered. The energies of bond formation for A–A, A–B, and B–B are denoted as Φ_{AA}, Φ_{AB}, and Φ_{BB}, respectively. Wetting occurs, when Φ_{AB} is greater in magnitude than Φ_{AA}. In this state, the system prefers to form A–B bonds over maintaining of A–A bonds. When the opposite condition overcomes, droplet formation occurs

because the system attempts to minimize the contact area between A and B. The energy difference is defined according to the following equation:

$$\Delta = \Phi_{AA} - \Phi_{AB} \qquad (17.6)$$

The wetting condition restates as $\Delta < 0$ and the nonwetting condition as $\Delta > 0$. Equation 17.6 is valid only at equilibrium. We define the vacuum chamber pressure, p and the equilibrium vapor pressure of the material being deposited, p^*, which are different in the growth process. Thus, the Gibbs energy of the system must be accounted for the accompanying pressure dependence.

$$\Delta G = n k_B T \ln \zeta \qquad (17.7)$$
$$\zeta = \frac{P}{P^*}$$

where n and ζ are the number of moles and the degree of supersaturation, respectively. Switching the growth mode of the system potentially becomes possible by controlling over ζ in such a way the Gibbs energy is changed in the system in addition to T, as we shall see below. Three limiting cases of growth mode under equilibrium control are illustrated in Figure 17.4. The conditions of each growth mode are as follows: consider the overlayer A wets the substrate B, thereby the contact angle $\psi = 0°$, so flat layers are expected. Equation 17.7 can be used in the case of the surface Gibbs energy G_s is equal to the surface energy γ, so the Young equation to determine the condition for wetting becomes

$$\gamma_{Bg} \geq \gamma_{AB} + \gamma_{Ag} + C k_B T \ln \zeta. \qquad (17.8)$$

Layer-by-layer growth occurs, if strain can be neglected ($\varepsilon \approx 0$) and Eq. 17.8 represents a simple energetic statement. Layer-by-layer growth is also known as Frank–van der Merve (FM) growth (Frank and van der Merve, 1949). At equilibrium, when $\zeta = 1$, the third term vanishes. For finely balanced systems ($\Delta \approx 0$), under nonequilibrium conditions (system is still under thermodynamic control) for switching the growth mode out of FM growth, the third term plays as an important role. At the other extreme when $\psi > 0°$,

$$\gamma_{Bg} < \gamma_{AB} + \gamma_{Ag} + C k_B T \ln \zeta. \qquad (17.9)$$

FIGURE 17.4 Three modes of growing overlayer B on substrate A which are thermodynamically controlled in the presence of a gas (or more generally a fluid or vacuum). (a) Layer-by-layer growth (FM) of two lattice matched ($\varepsilon_0 = 0$) materials. (b) Layer-plus-island growth (Stranski–Krastanov). The strained wetting layer does not exhibit dislocations at either of its interfaces; however, the islands continuously relax with a lattice deformation in the growth direction. (c) Island growth (VW) of lattice-mismatched ($\varepsilon_0 \neq 0$) materials with dislocations at the interface.

and the overlayer does not wet the substrate. Interactions between the deposited materials are stronger than interactions between substrate material, and the system attempts to minimize the contact area between the deposit and the substrate by forming the 3D islands. Three-dimensional island growth, which is well known as Volmer–Weber growth (Volmer and Weber, 1926), is attributed to a large lattice mismatch.

Stranski–Krastanov is the most commonly observed growth mode that corresponds to neither of these limiting cases and cannot be described in the nonexistence of strain. In Stranski–Krastanov growth (Baskaran, Arvind, and Peter Smereka, 2012), wetting layers of about 1–4 ML first cover the substrate. The overlayer cannot continue to grow in the strained structure because the lattice misfit between deposited material and substrate is more than 2%. At this point, the growth switches from layer-by-layer to the formation of 3D islands. Nature defines what is a small lattice misfit as opposed to a large. Layer-by-layer growth takes place in a small lattice misfit, $\varepsilon_0 < 1.5$. A large lattice misfit, $\varepsilon_0 > 2\%$, does not let the materials grow on the substrate by layer-by-layer growth mode. Ge/Si and InAs/GaAs are the best-known examples of Stranski–Krastanov growth in which the critical coverages for island formation are 3.5 and 1.6 ML, respectively (Priester and Lannoo, 1997). In these cases, the islands are dislocation free and known as coherent islands.

Shchukin et al. (1995) found that under suitable conditions, narrow size distribution of islands can be formed. A change in the growth mode can occur by changing the strain. For instance, Si on Ge(100) grows via a VW mode, whereas Ge on Si(100) grows via a Stranski–Krastanov mode. Ge possesses a lower surface energy than Si as a result, and generally, this inversion of growth modes is taken place; therefore, when the roles of substrate and overlayer are interchanged, Eq. 17.6 changes sign. The growth mode of Ge on Si can be changed from Stranski–Krastanov to layer-by-layer using As and Sb. They act as a surfactant that separates the top surfaces. At the surface, they decrease the energy of a flat film and reduce the formation of islands.

17.2.7 Ostwald Ripening

Liu et al. (2001) in the absence of strain calculate the chemical potential μ_i of island having radius r_i as

$$\mu_i(r_i) = \frac{v\delta}{r_i} \qquad (17.10)$$

where v is the surface area per atom in the island and σ is the island edge (step) energy. Equation (17.10) indicates that the chemical potential of an island is inversely proportional to radius. Therefore, a thin film that does not wet a surface will break up into a distribution of 3D islands. In unstrained system, coarsening of the initial layer/island distribution will tend to form one large island rather than any number of smaller islands.

The communication distance between islands is limited in any real finite system. Once the distance between the contracting islands becomes above the diffusion length, the islands become independent of one another and their accretion is stopped. During accretion, small islands feed material to larger islands. Since the position of islands is random, the number and size of initial neighboring islands are random and the amount of material that can ultimately be accreted into the largest island will vary randomly across the surface. There is no optimum island size, and as a result, size distribution of islands will be broad. This process is known as Ostwald ripening, which was first observed and described for the growth of grains in solution (Ostwald, 1900). Even at equilibrium and with no constraint on diffusion, ripening can be observed under appropriate conditions in the presence of strain.

The chemical potential of a strained island can be written (Liu et al., 2001)

$$\mu_i(r_i) = v \left[\frac{\delta - \alpha}{r_i} - \frac{\alpha}{r_i} \ln \frac{r_i}{a_0} \right] \qquad (17.11)$$

where $\alpha = \frac{4\pi F^2 (1 - v^2)}{\mu}$

F is the misfit strain-induced elastic force monopole along the island edge, ν is the Poisson ratio, μ is the Young modulus, and a_0 is a cutoff length on the order of the surface lattice constant. The strained system exhibits a thermodynamically stable island size that is resistant to further coarsening, and a system of many islands evolves until they reach a radius given by

$$r_0 = a_0 \exp \left(\frac{\delta}{\alpha} \right) \qquad (17.12)$$

At finite temperatures, entropic effects broaden the size distribution into a Gaussian distribution (Priester and Lannoo, 1995) about a mean value that is somewhat different than r_0 (Shchukin et al., 2001). Liu et al. (2001) have shown that further considering the influence of island–island interactions, strain leads not only to the establishment of a preferred size but also to self-organization and performing narrow size distribution.

17.2.8 Ge/Si Island Shapes and Evolution

In the field of self-assembled system, describing the different island shapes and their relative stabilities became one of the most controversial problems in order to derive rational method of growing self-assembled nanostructures. The first Ge/Si island type, which was discovered, is huts (Mo et al., 1990). Medeiros-Ribeiro et al. (1998a) found that they are metastable, and by annealing process, they are transformed into pyramids. On the other hand, pyramids and domes co-exist over a wide range of coverage, growth temperature, and annealing conditions. For over a decade, extensive research has been carried out on the growth and shape transition mechanisms of these two island types (Brunner, 2002; Teichert, 2002; Medeiros-Ribeiro et al., 1998b).

The kinetic roughening process (Ostwald ripening) (Zinke-Allmang et al., 1992) and an equilibrium distribution, which is determined by the appropriate island, energies are the main issues for observing the changes in the island shapes. The equilibrium structures are preferred for practical purposes. They naturally evolve towards controlled uniform sizes and remain stable over the lifetime of devices, which may be orders of magnitude longer than the duration of the deposition process. They can potentially be grown by any deposition method. Furthermore, the growth kinetics are used to determine the island types; thus, for controlling the appropriate growth parameters and stabilizing the structures, then good understanding of the relative rates of the various processes is crucially needed. Obviously, the differences between kinetic and equilibrium processes are readily predicted by simple models. According to classic Ostwald ripening formation of islands with strictly unimodal, albeit not very narrow size distribution, increasing the average particle size with time can be explained (Zinke-Allmang et al., 1992).

In an equilibrium system with several competing interactions, the existence of an optimal particle size (and shape) can be expected. The predictions from simple kinetic or equilibrium models are not in agreement with the actual experimental observations for the Ge/Si system. The next issue was choosing the suitable model extensions, including additional surface processes, which in general tends to obscure the differences between model predictions. Recently, the scientists have extensively investigated the development of the complex equilibrium models for this system (Williams et al., 2000). Some more illustrative examples of the kinetics models are briefly discussed below.

The chemical potential of an island decreases continuously with size because of the decreasing radius of curvature and surface/volume ratio, which are predicted by the classic Ostwald ripening model and its extensions to 3D crystals on surfaces (Giesen, 2001; Zinke-Allmang et al., 1992). Detachment of atoms from smaller islands and attachment to larger ones become more probable; thus, the larger islands grow at the expense of the smaller ones, which means that the average island size enhances and the surface number density of islands decreases with time. Co-existing of bimodal pyramids and domes-like Ge/Si island distribution during growth and annealing (Williams et al., 2000; Medeiros-Ribeiro et al., 1998b) does not correspond to a simple Ostwald ripening. A discontinuous change in the chemical potential of an island during its shape transformation suggests that the Ostwald ripening kinetics should be modified to produce a bimodal distribution (Ross et al., 1998; Daruka et al., 1999). Quantitatively, indications of size-limited behavior for both pyramids and domes are shown in the experimental size distribution (Ross et al., 1998), and thus, the width of these distributions is narrower than expected. Therefore, in a kinetic model, an additional island size-dependent term in the adatom attachment rate is required. The increased strain for large islands at their edges favors detachment and in surrounding substrates leads to flow away from these islands, which have suggested by Williams et al. (2000).

Multiple energy terms such as bulk strain, facet, interface, and edge energies for individual islands (Shchukin et al., 1995; Uemura et al., 2001) should be included in realistic equilibrium models, as well as inter-island elastic interactions and ensemble thermodynamics (Williams et al., 2000). Considering the multiple terms in both kinetic and equilibrium models makes their predictions depend on a number of unknown parameters consequently; comparing with experimental data is difficult. Further increase in complexity is the other practical aspects of this system. For example, above 650°C, all Ge islands are only metastable with respect to SiGe alloying (Kamins et al., 1998), and transition of shapes occurs during growth but disappears after cooling (Ross et al., 1999). The extensive research on the ideal Ge/Si heteroepitaxy shows that although the structures themselves appear to be thermodynamically stable, but their growth is determined as much by kinetic pathways towards the equilibrium, as by the equilibrium configuration itself (Kaganer and Ploog, 2001; Williams et al., 2000). Furthermore, partly annealed configurations may display strong nonequilibrium features, but equilibrium theories emerge with sufficient annealing distributions consistent. Initial island nucleation and 2D–3D transitions are strongly kinetics-dependent. With increasing coverage, strain effects become relevant and island sizes and densities approach equilibrium values, but size-dependent kinetic terms are still significant. As the deposition flux decreases and stops, further annealing brings the system close to the equilibrium configuration, including saturated values of island sizes and density.

17.2.9   Rf Magnetron Sputtering Growth of Ge/Si

In 1990, Mo et al. published a paper in which they have characterized the Stranski–Krastanov formation of metastable 3D Ge on Si (001) that consisting of small clusters by scanning tunneling microscopy. They have a particular facet crystallography and alignment of their principal axes with respect to the substrate.

Four years later, Mosleh et al. (1994) reported the epitaxial growth of Ge on Si(100) by ion beam sputtering in which the effect of deposition temperature on the surface roughness was studied by SEM and in situ Auger electron microscopy. They found that the growth procedure does not agree with layer-by-layer growth mode and the Stranski–Krastanov mode is the right growth procedure. Furthermore, computer calculation using Monte Carlo transport of ions in matter with code number of TRIM was used to support their experimental observation.

Extensive attempt has been made to a systematic study of the morphology of self-organized Ge/Si(001) islands prepared by ultra-high vacuum (UHV) magnetron sputtering epitaxy, based on high-resolution scanning tunneling microscopy measurements. They demonstrated that two main Ge nanoisland families co-exist on the surface: smaller pyramids bound by one type of shallow facets and larger multifaceted domes. Additionally, the proposition of these observations on a possible universal explanation of the Stranski–Krastanov growth mode is discussed with respect to recent theoretical results (Costantini et al., 2005).

In 2006, Radić and co-workers employed magnetron sputtering to form germanium islands on silicon surface held at constant elevated temperature. The GISAXS and AFM data were presented. They concluded that 650°C was the best substrate temperature in order to produce the 4 nm high and 6 nm wide islands. The islands were found in homogenous shape and size. However, in-plane organization was misplaced since the inter-island distance varies more than their size.

Ge thin films epitaxially were grown onto (100) Si substrates via a fast, economic, and industrially scalable deposition techniques such as direct current (DC)-pulsed magnetron sputtering which was proved by the research of Pietralunga et al. (2009). High-quality film-substrate adhesion and roughness at the nanometer scale both at surface and at the Ge/Si interface were observed by detailed investigation on film morphology. They illustrated that the films were single crystalline and slightly misoriented below 0.1°. Defect annihilation was effected by the postdeposition rapid annealing process. Perfect epitaxy at the Si/Ge junction in the nonexistence of interfacial silicon oxide and with sub-nanometer roughness was observed by high-resolution transmission electron microscopy (HRTEM). A weak roughness of ~0.6 nm was obtained both at the film surface and at the Ge-Si interface. The samples confirmed the rectifying electrical performance typical of p-type semiconductors (Pietralunga et al., 2009).

DC-pulsed magnetron sputtering technique was employed to epitaxially produce Ge thin film on Si(100) substrate, which is a promising technique to realize p–n junctions for photodiodes application in the near IR with extended spectral response by Pietralunga et al. (2009). Transmission electron microscopy (TEM) and XRD confirmed that the Ge structure was nanoscale with relaxed epitaxial growth. Additionally, they illustrated that the p-type conductivity, mediated by intragap states, can be explained by the presence of defective region extending in the film from the interface. Extended absorption and photoconductive response in the near IR, which were achieved experimentally, can be explained by narrowing the band gap. They state that this growth technique can also be considered for waveguide photodiode geometries in order to be monolithically integrated into silicon-based photonic circuits.

Recently, Yang et al. (2012) have studied the evolution of Ge/Si islands prepared by ion beam magnetron sputtering. The effect of temperature and Ge coverage on the evolution of nanoislands was studied. Their observations show that the surface diffusion and intermixing between Si and Ge have been effected by the growth temperature. A decrease in island density and an increase in size as consequence of faster surface diffusion by increasing the growth temperature are reported. The interface strain relaxation and redistribution are modified with intermixing. Therefore, the islands having

lower aspect ratios reoccur at higher temperatures. With increasing the amount of Ge, the islands vertically grow faster, which is due to the transferring of additional energy through energy ion bombardment. This leads to the formation of condensed islands that often observed whether low aspect ratio islands or high aspect ratio islands. Furthermore, they reported that due to the intermixing between the Ge islands and the Si substrate, the critical volume of the pyramid-to-dome transition would also increase (Seta et al., 2002).

17.2.10 Epitaxial and Nonepitaxial Ge Nanodots in the Presence of SiO_2 Sub-Layer

As pointed out above, the well-known Stranski–Krastanov mode can explain the growth of Ge overlayers on clean Si surfaces due to slightly larger lattice mismatch between Ge and Si (4.2%), whereas the surface energy of Ge is smaller than that of Si (Mo et al., 1990). On the other hand, the formation of the wetting layer of Ge on the oxidized Si surfaces is slowed down by a smaller surface energy of SiO_2 compared to Ge. As a result, deposition of Ge onto the SiO_2 sub-layer produces the Ge nanodots uniformly cover over the whole surface (Nakayama et al., 2008). Parts of deposited Ge atoms react with SiO_2 as

$$SiO_2(s) + Ge(ad) \rightarrow SiO \uparrow + GeO \uparrow, \qquad (17.13)$$

where silicon monoxide (SiO) and germanium monoxide (GeO) are produced and evaporated at relatively moderate temperature (Shklyaev et al., 2000). This reaction can take away the SiO_2 sub-layer as well as the deposited Ge, thus leading to form the sub-nanometer-sized voids penetrating through the oxide layer. Ge nanodot generation in this case is slower than the Ge/Si because the rate of the void formation will compete in opposition to nucleation of the deposited Ge at the void sites to generate the nanodots. Thus, voids become the nucleation center of the Ge nanodots. In this case, Ge nanodots have a direct contact with Si substrate through the voids, which follows the lattice of the substrate, as schematically shown in Figure 17.5. On the contrary, nonepitaxial growth on the substrate will occur if the sample temperature is not sufficient or/and fast deposition. In those cases, voids cannot be formed and the Ge will deposit on SiO_2 layer without touching the Si substrate. On the other word, Ge nanoislands are separated by SiO_2 overlayer (Figure 17.5) from Si Substrate.

Additionally, transforming from nonepitaxial nanodots into the epitaxial ones by postdeposition annealing is not possible (Shklyaev et al., 2000). TEM is the best equipment for direct observation of the local structure of the epitaxial nanoislands.

17.2.11 Ge/Si Multilayers

The application of Si-based low-dimensional materials like quantum wells in optoelectronic and electronic devices

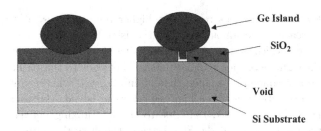

FIGURE 17.5 Schematic diagram of epitaxial and nonepitaxial growth of Ge nanoisland on Si substrate with existence of SiO$_2$ overlayer.

such as hetero-bipolar transistors (Behammer et al., 1996), modulation- doped field-effect transistors (Aniel et al., 2003), strained-CMOS (Mizuno et al., 2005), and resonant tunneling diode devices (Suda et al., 2001) and their compatibility with the current silicon planar technology of integrated circuits have attracted great interest. Among varieties of Si-based integrated-circuit technology, Ge/Si nanomultilayers have several promising applications.

Several attempts have been made to form ordered multilayer structures during the Ge/Si heteroepitaxy arises from inter-layer interactions (Brunner, 2002; Teichert, 2002; Voigtländer, 2001). The importance of the multilayer ordering as a unique self-organization mechanism for self-assembled surface structures is considered here. In a few words, throughout the multilayer Ge/Si heteroepitaxy, nanostructures formed in the preceding layer affect the strain in the subsequent overlayer, which leads to self-alignment of nanostructures between the layers. These layer contacts can lead to enhanced size uniformity and lateral ordering of the nanodots in the topmost layer even when the first layer was randomly nucleated. Almost when the first layer is pre-patterned either artificially or by self-organized template on the substrate, perfect uniformity and lateral ordering can be achieved (Brunner, 2002; Teichert, 2002; Voigtländer, 2001).

Schmidt et al. (1999) employed the TEM microanalysis to reveal that the wetting-layer thickness remained after the onset of the island formation was smaller in upper layers compared to that of the first. They found that the increase in the island size and height is a consequence of decreasing the Ge critical thicknesses in upper layers, which can be explained by elastic strain fields induced by buried layers and mediated by the spacer layers.

Huang et al. (2009) used rf magnetron sputtering to prepare Ge/Si nanomultilayers in different growth conditions. They reported that as-deposited films on substrate at room temperature are amorphous. On the other hand, when the deposition temperature is 300°C, the as-deposited Ge/Si nanomultilayers are crystalline structure. They also observed the Ge/Si nanomultilayers convert into GeSi alloy when the sample is annealed at 500°C. This is attributed to Ge/Si atomic inter-diffusion activated by the thermal annealing. By increasing annealing temperature, Ge atoms are likely to form interstitial atoms or shift to the interface. At the annealing temperature of 700°C, in comparison with

the sample annealed at 600°C, the inter-diffusion augments significantly and the amount of Ge in the germanosilicide phase is reduced. In addition, they reported that the formation of Ge nanocrystal grains is attributed to segregation from the germanosilicide. The strong agglomeration and the formation of Ge nanocrystal beyond 800°C are reported. The effect of agglomeration and nanocrystal formation is well known at higher annealing temperatures because of the strong temperature dependence of the physical transport of atoms.

17.3 Methodology

17.3.1 Growth of Ge Nanoislands in rf Magnetron Sputtering System

The ejection of the surface atoms in the target by series of collisions with high-energy ions is known as sputtering. In radio frequency (RF) magnetron sputtering, 13.6 MHz source is employed to produce the strong electric field target (cathode) and substrate (anode) to accelerate the positive Ar ion to the target and dislodge the target atoms. The powerful magnets are located under cathode in order to confine the plasma to the region closest to the target. This condenses the ion-space ratio, increases the collision rate, and thus improves the deposition rate. The definition of sputtering yield is

$$Y = \frac{\text{Ejected atoms or molecules}}{\text{incident ion}} \quad (17.14)$$

which can be explained as

$$Y = \frac{U_0}{\rho}, \quad (17.15)$$

where U_0 is the energy of sublimation, and ρ is the density. The sputtering yield for Ge varies from 0.4 to 5 corresponding to Ar energy changes from 500 eV to 10 keV (Matsunami et al., 1984).

Argon is usually used due to its general availability in pure form and lack of chemical reaction among the bombarding particles. The chamber pressure used for sputtering is a compromise. Preferably, the chamber pressure should be extremely low for larger mean free path of the material. However, for generating the plasma, collisions between electrons and neutral gas atoms are used. This requires a substantial pressure to take place efficiently. Normally, working pressure is in the range of 0.1–10 Pa (Huang et al., 2009).

Magnetron sputtering has attracted considerable attention for the preparation of samples for fundamental studies and device fabrication in industry due to its lower cost, suitability for large-scale production, high resource usability, and higher deposition rate. Furthermore, most of the elements, alloys, and compounds including insulators can be sputtered and deposited. The sputtering target provides a stable and long-lived vaporization source. A variety of sources are available for sputtering target such as line, rod, or cylinder. There is very little heat radiant in the deposition process; therefore, the target and substrate can be spaced close together which can reduce the dimension of the sputter deposition chamber. However, sputtering technique has some disadvantages such as the deposition non-uniformity of flux distribution, expensive target materials, and damages due to the bombardment of samples by energetic ions. On top of that, oxygen contamination is also one of the critical problems in the sputter deposition. The gas composition must, therefore, be carefully controlled to prevent the poisoning of sputtering target.

Sputtering deposition is widely used to grow thin film metallization on semiconductor materials, coatings on architectural glasses, reflective coating on polymers, magnetic films for storage media, transparent electrically conductive films on glasses and flexible webs, dry-film lubricants, wear-resistant coating on tools, and decorative coatings (Costantini et al., 2005; Das et al., 2007; Jie et al., 2004).

In this work, the chamber consists of substrate and target holders, heater which is located under the substrate, thermocouple for measuring the substrate temperature, and Ar and N_2 flow controller. Ar flow rate is typically in the range of 5–25 sccm. The chamber is evacuated by mean of both rotary and turbo molecular pumps. The rotary vacuum pump is used to obtain the base pressure of 10^{-2} torr which then followed by the turbo molecular pump to reach a pressure of 10^{-6} torr. The substrates are positioned horizontally on top of the target with the distance of 20 cm; therefore, only one side gets deposited. The plasma is generated by radio frequency system.

The following steps are performed to grow Ge nanoislands using rf magnetron sputtering system. The substrate is loaded into chamber after cleaning, and then, the chamber is evacuated to grow Ge nanoislands. Heater is switched on in order to achieve different substrate temperature. Next step is pre-clean sputtering process. The rf power is applied, and the Ar gas starts to flow to the chamber to generate the plasma. Deposition is carried out after opening the shutter. The rf reflection should be adjusted to minimum value during deposition. After deposition, the Ar source is cut off, and heater is switched off. The samples are left to cool to room temperature before taken them out.

17.3.2 Characterization Techniques

Surface morphology, and structural and optical characteristics strongly participate in defining the optical, electrical, and magnetic properties of the samples. Various modern tools have been used to obtain valuable information at the nanoscale level. Thereupon, AFM plays a crucial role to measure the surface morphology of the samples. SEM, FESEM, XRD, and energy dispersive X-ray (EDX) spectroscopy are used for structural analysis. Raman spectroscopy and PL are used for optical characteristics. Multiple characterization instruments are employed to collect important data to improve the reliability of the obtained properties.

Atomic Force Microscopy

The morphology of surface containing nanoislands is important for fundamental physics study and realizing the application of the nanoislands in electronic and photonic devices. AFM is an uncomplicated method without sample preparation and conductivity requirement to study the surface morphology of samples by measuring the size of nanoisland accompanied with RMS surface roughness, number density, and the ratio of grain area. It can be widely used for a variety of samples such as plastics, metals, glasses, semiconductors, and biological samples. Nanoscale 3D image of the surface is provided, by measuring forces between a sharp probe (<10 nm) and the surface at very short distance (0.2–10 nm probe-sample separation). Flexible cantilever supports the probe that is placed at the end of cantilever. The AFM tip touches the surface and records the small force between the probe and the surface, which depends on the stiffness of the cantilever (spring) and the distance between the probe and the sample surface. This force (usually ranging from nN (10^{-9}) to μN (10^{-6}) in the air) can cause the bending of cantilever and then can be calculated by Hooke's law, and the deflection is monitored.

Si_3N_4 and Si are usually used for fabricating the probes having different cantilever lengths, shapes, and materials permit for varied spring constants and resonant frequencies. Probes can be coated with other materials for additional applications such as chemical force microscopy (CFM) and magnetic force microscopy (MFM).

The feedback loop and piezo-electric scanners control the motion of the probe across the surface. The deflection of the probe is typically measured by a beam reflected method. The cantilever is connected to the sensitive photodiode detector by reflecting the semiconductor diode laser. During scanning the sample, the detector measures the bending of cantilever. The surface topography of the sample is generated using measured cantilever deflections.

According to distance between probe and surface, the AFM imaging technique can be divided into three primary imaging modes as shown in Figure 17.6:

- If the probe–surface separation is less than 0.5 nm (contact mode AFM), the dominant interactions are repulsive van der Waals interaction. Therefore, the force between the probe and the sample remains constant. Fast scanning, excellent for rough samples, and used in friction analysis are some advantages of these modes.

- If the probe–surface separation is between 0.5 and 2 nm (intermittent contact AFM), the cantilever is oscillated at its resonant frequency, contacting the surface at the bottom of its swing, and by keeping oscillation amplitude constant, a constant probe–surface interaction is maintained and an image of the surface is obtained. This is excellent for biological samples, and high-resolution images are some benefits.

- Finally, if the probe–surface separation is more than 2 nm (non-contact AFM), attractive van der Waals is the prominent reaction, and probe does not contact the sample surface but oscillates above the adsorbed fluid layer on the surface during scanning (Note: all samples except in a controlled UHV or environmental chamber have some liquid adsorbed on the surface). Some disadvantages of this mode are given as follows: commonly lower resolution, contaminant layer on surface that interferes with oscillation, and UHV is needed to give best imaging increase the tip lifetime.

In comparison with TEM, which is time-consuming to make the sample sufficiently thin for electron transmission, sample preparation is not required for AFM although the AFM image does not demonstrate the actual size of nanoislands.

Using the sharp tip and reconstruction of the actual size can suppress the tip artifact in AFM observations. Carbon nanotube tip is attached to a commercially available probe by group of scientist (Hohmura et al., 2000; Chen et al., 2004) in order to suppress the artifact. Xu and Arnsdorf (1994) reconstructed the actual shape of an object by the observation of a sphere on a flat surface. Broadening of the AFM tip artifact produces two geometric assumptions for spherical and hemispherical objects.

For spherical object,

$$(R + r)^2 = \left(\frac{w}{2}\right)^2 + (R - r)^2 \qquad (17.16)$$

where w is an extra broadening of the width, r is the observed radius of sphere or hemisphere, and R is the radius of tip. It results in

$$w = 4\sqrt{Rr} \qquad (17.17)$$

And for hemispherical object,

$$(R + r)^2 = \left(\frac{w}{2}\right)^2 + R^2 \qquad (17.18)$$

Leads to

$$w = 2\sqrt{2Rr + r^2}$$

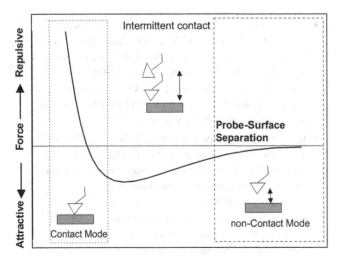

FIGURE 17.6 Force versus probe–surface separation.

The tip artifact for the steep surface objects like spherical and hemispherical shapes originates a larger broadening than that for the object having a moderate surface like pyramid- or dome-like-shaped islands as shown in Figure 17.7 (Shiramine et al., 2007).

X-Ray Diffraction

The cleavage faces of crystals appear to reflect X-ray beams at certain angles of incidence (θ). Two English physicists (W.H. Bragg and his son W.L. Bragg) in 1913 established this fact and explained it as

$$n\lambda = 2d \sin \theta \qquad (17.19)$$

which is called Bragg's law. The diffraction of monochromatic X-rays takes place only at those particular angles of incidence that satisfy the Bragg law. In a crystal, d is the distance between atomic layers. The wavelength of the incident X-ray beam is λ, and n is an integer. XRD is an observation of X-ray wave interference and is direct confirmation for the periodic atomic structure of crystals postulated for several centuries.

XRD can be used for extensive ordered structural solids such as single crystals and polycrystalline in order to measure the average spacing between layers or rows of atoms, determine the orientation of a single crystal or grain, find the crystal structure of an unknown material, and measure the size, shape, and internal stress of small crystalline regions.

Destructive and constructive interferences are just a consequence of the periodic structure of crystal planes. In infinitive period, structure of crystal planes is bulk if the path difference between X-rays scattered from two planes slightly differs from the integral number of wavelength they cancel each other, and only X-rays, which satisfy the Bragg's law, interact and exhibit sharp peak in the XRD spectra.

If the crystal is nanosize (less than 100 nm), then infinitive periodic planes do not exist; thus, complete cancellation does not occur for the scattered X-rays having path difference of non-integer number of wavelength. Therefore, it causes a broadening of the peak in the XRD spectra. Instrumental broadening B_{ins} and microstrain B_{strain} are two factors other than nanosize effect might contribute to the extra line broadening in the XRD peaks. Therefore, the observed peaks $B_{experiment}$ can be presented as

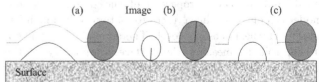

FIGURE 17.7 Schematic views of AFM observations of island (a), sphere (b), and hemisphere (c) on surface. The AFM tip is represented as a sphere having the same radius. (From Shiramine et al., 2007.)

$$B_{\text{experiment}} = B_{\text{ins}} + B_{\text{strain}} + B_{\text{size}} \text{ (for Lorentzian)}$$

(17.20)

and

$$B^2_{\text{experiment}} = B^2_{\text{ins}} + B^2_{\text{strain}} + B^2_{\text{size}} \text{ (for Gaussian)} \quad (17.21)$$

The instrumental broadening B_{ins} can be determined experimentally with a diffraction standard or calculated with the fundamental parameters approach. The separation of the size and the strain effect on the sample broadening is approved as follows:

$B_{\text{strain}}(2\theta) = 4\varepsilon_0 \tan\theta$, where $\varepsilon_0 = \Delta d/d$ (fractional variation in Bragg plane spacing); thus, by plotting the $\sin\theta/\lambda$ as a function of $B\cos\theta/\lambda$ (Williamson–Hall plot), the microstrain can be calculated. The extent of the broadening is given by Scherrer's equation:

$$L = \frac{0.9\lambda}{B_{\text{experiment}} \cos\theta} \quad (17.22)$$

where $B_{\text{experiment}}$ is the observed broadening of diffraction line measured at half its maximum intensity (radians), L is the diameter of crystal particle, λ is the X-ray wavelength, and θ is the diffraction angle of the reflection (Cullity and Stock, 2001).

Raman Spectroscopy

Raman spectroscopy is a technique that the electrons are excited into higher energy states using laser radiation. Some virtual states of energy are existing in which the electrons spend a relatively short time before relaxing. For the excited electrons, relax back to the similar energy state from which they were excited is the most probable relaxation pathway. This signal is indicated by Rayleigh scattering and is blocked by filters. When electrons relax back to a different vibration level, they emit a light in different energy that gives information of the vibrational spectra of the sample that is measured in Raman technique. This gives supplementary information to that of IR measurements. A plan of energy levels involved is given in Figure 17.8.

The obtained spectrum from the Raman measurement can be used for the identification of crystal phases and information about vibrational spectra of the entire molecule or crystal as well as size estimation. Phonons vibration is the quantum of energies in crystals. This is caused by the mixture of vibration of atoms in the lattice that affects the optical and physical properties such as transport of heat, electricity, and sound.

Shift to different energies, changes in the symmetry of the vibration and decay or enhancement of the Raman spectra are attributed to the restriction the size of a nanocrystalline, the phonons and vibrational modes in the small volume of the particle. There are theoretical modelings for analyzing the Raman spectra in which we can get more information especially in the rather tricky experimental achievement.

Atoms moving coherently in the direction of propagation (LA) or perpendicular to the propagation direction (TA) of the lattice out of their equilibrium positions are called acoustic phonons. On the other hand, optical phonons fall out of phase movement of the atoms in the lattice. They can be abbreviated as LO and TO for longitudinal and transverse modes, respectively, which are the same for acoustic phonons. In contrast to acoustic phonon, in the Brillouin zone center, optical phonons do not have zero energy and do not show dispersion near long wavelength (Gouadec and Colomban, 2007). It is because of the positive and negative ions at adjacent lattice that can be excited by infrared radiation, oscillate against one another, and create a time-varying electrical dipole moment. The interaction of optical phonons with light is called infrared active. Raman active optical phonons can also interact indirectly with light through Raman scattering.

Richter et al. proposed a phonon confinement model (PCM) for nanospheres with diameter L (Richter et al., 1981). They multiplied the plane wave of phonon, with wave vector k_0, in an ideal crystal:

$$\phi\left(\vec{k_0}, \vec{r}\right) = u\left(\vec{k_0}, \vec{r}\right) e^{i\vec{k_0}, \vec{r}} \quad (17.23)$$

where u has the same spatial periodicity as the lattice by a Gaussian function:

$$\phi\left(\vec{k_0}, \vec{r}\right) = e^{-\alpha\left(\frac{r}{L}\right)^2} \times u\left(\vec{k_0}, \vec{r}\right) e^{i\vec{k_0}, \vec{r}} \quad (17.24)$$

Assuming $u\left(\vec{k_0}, \vec{r}\right) = u\left(\vec{r}\right)$, Eq. (17.24) is equal to

$$I\left(\vec{v}\right) \propto \int_0^1 dq e^{\frac{k^2_{\text{BZ}}(q-q_0)^2 L^2}{2\alpha}} \frac{1}{\left[\overline{v} - \overline{v}\left(\vec{k}\right)\right]^2 + \left(\frac{\Gamma_0}{2}\right)^2} \quad (17.25)$$

with

$$C\left(\vec{k_0}, \vec{k}\right) = \frac{1}{(2\pi)^3} \int d^3\vec{r} \left(e^{-\alpha\left(\frac{r}{L}\right)^2} \times e^{i\left(\vec{k_0} - \vec{k}\right)\vec{r}}\right) \quad (17.26)$$

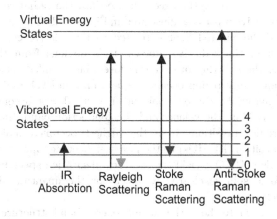

FIGURE 17.8 The energy levels involved in various Raman spectroscopy.

Therefore, in a non-ideal crystal, the wave related to a phonon confined is a superposition of plane waves with a weight of $\left| C\left(\vec{k}_0, \vec{k}\right) \right|^2$. It is known that each wave gives rise to a Lorentzian, and finally, Raman intensity is

$$I\left(\vec{v}\right) \propto \int_{\text{BZ}} d^3\vec{k}\, \frac{\left| C\left(\vec{k}_0, \vec{k}\right) \right|^2}{\left[\overline{v} - \overline{v}\left(\vec{k}\right)\right]^2 + \left(\frac{\Gamma_0}{2}\right)^2} \quad (17.27)$$

In this equation, C and Γ_0 are the Raman coupling coefficient and Lorentzian broadening, respectively. Raman selection rule breaking induced by phonon confinement with a weighed exploration of the dispersion curve is mathematically indicated by Eq. (17.27). k_{BZ} is given as the edge of Brillouin zone (BZ) and q as the reduced wavevector ($q = k/k_{\text{BZ}}$), and with respect to isotropic mode dispersion (Carles et al., 1992), Eq. (17.27) yields

$$I\left(\vec{v}\right) \propto \int_0^1 dq e^{\frac{k_{\text{BZ}}^2 (q-q_0)^2 L^2}{2\alpha}} \frac{1}{\left[\overline{v} - \overline{v}\left(\vec{k}\right)\right]^2 + \left(\frac{\Gamma_0}{2}\right)^2} \quad (17.28)$$

The PCM is not applicable to acoustic modes because their energy is zero at BZ and is rarely used with TO modes because of their low dispersion (Xiong et al., 2003; Tiong et al., 1984). It is completely applied to LO modes, which the dispersion typically is maximum at BZ. On the low-frequency side of the single crystal mode, the integration in Eq. (17.28) for LO illustrates additional contributions; therefore, consequential peaks are asymmetric.

Photoluminescence

PL spectroscopy is a contactless, nondestructive method of probing the electronic structure of materials by emitting the light from the electrons that are absorbed and imparted excess energy, and transferring to an excited energy level, which is called photo-excitation. Photo-excitation causes electrons within a material to move into permitted excited states. These excited electrons return to their equilibrium states by emitting the excess energy into the radiative or nonradiative process as illustrated in Figure 17.9. The quantity of the emitted light is related to the relative contribution of the radiative process. It is obvious from this figure, the splitting of energetic levels into a filled region and an empty region occurs at quantum size level, and in bulk materials, the ensemble of energetic levels becomes dense yielding the valence and the conduction band. Since electronic transitions from the highest occupied molecular orbitals (HOMOs) to the lowest unoccupied molecular orbitals (LUMOs) yield the observed absorption spectrum, the blue shift is clearly explained for the quantum size particles.

Figure 17.10 shows the possible energy band structure of germanium at room temperature. The understanding and accurate description of energy band structures for those

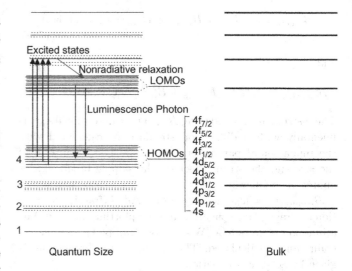

FIGURE 17.9 Molecular orbital model illustrates the quantization of the density and energy levels when going from quantum size to the bulk materials.

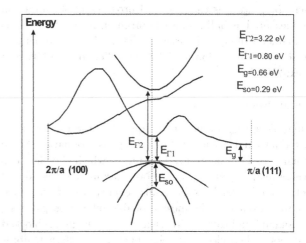

FIGURE 17.10 Energy band structure of germanium. (From Bukowski and Simmons, 2002.)

materials are subjected to stress and shaped as nanostructure is fundamental to explain the experimental data and examined the level of reliability.

The size-dependent band gap energy is theoretically calculated by Brus in which the Schrödinger equation is used for the first excited state considering an effective mass approximation and dielectric solution approach for the kinetic and potential energy, respectively (Brus, 1984). E_{g} in Eq. 17.29 is the first approximation for the energy of the lowest excited state. d is the particle diameter, m_{e}^* and m_{h}^* are the effective mass of the electron and hole, respectively, e is the electron charge, ε is the dielectric constant, α is the polarizability, and s is the spatial position.

$$E_{\text{g}} = \frac{\hbar^2 \pi^2}{2d^2}\left(\frac{1}{m_{\text{e}}^*} + \frac{1}{m_{\text{h}}^*}\right) - \frac{1.8e^2}{d\varepsilon} + \frac{e^2}{d}\sum_{n=1}^{\infty} \alpha_n \left(\frac{s}{d}\right)^{2n} \quad (17.29)$$

$$E_g = 3.337 - \frac{1.35}{d} + \frac{8.47}{d^2} \qquad (17.30)$$

Equation 17.30 gives the numerical values for Brus analysis (Spanhel, 2006).

Field Emission Scanning Electron Microscopy

A FESEM is a microscope that works with electrons instead of light to observe the small structure (as small as 1 nm). High electrical field gradient accelerates the electrons. In order to bombard the object, primary electrons are focused and deflected by electromagnetic lenses in the high vacuum column to produce a narrow scan beam. Consequently, each spot on the object emits the electrons that are called secondary electrons. The surface structure of the object is related to the angle and velocity of these secondary electrons. Secondary electrons are captured by detector and produced an electronic signal. After amplifying the signal and transforming to a monitor, scan-image can be seen.

Energy Dispersive X-Ray Spectroscopy (EDX)

To obtain a localized chemical analysis, focused beam of electrons bombards the sample, and the emitted X-ray spectrums are detected. In X-ray production, typically tens to hundreds of kilovolts of potential is applied to accelerate the electrons in X-ray tube and bombard a target with high-speed electron. The electrons from the inner shells of the atoms can be ejected by bombarding the target with electrons. Electrons dropping down from higher levels and quickly fill those vacancies. The difference between the atomic energy levels of the target atoms causes emitting X-rays with sharply defined frequencies. The X-rays produced by transitions from the $n = 2$ to $n = 1$ levels of targets are called K_α X-rays, and those for the $n = 3$ transition are called K_β X-rays. Transitions to the $n = 2$ or L-shell from $n - 3$ and $n = 4$ are designated as L_α and L_β, respectively. The frequencies of the characteristic X-rays and their corresponding energy can be predicted from the Bohr model, which act as a unique fingerprint of elements.

Simplicity of X-ray spectra establishes the uncomplicated method for qualitative analysis that involves the identification of the lines in the spectrum. Measuring line intensity for each element in the sample determines the concentrations of the present elements (quantitative analysis) and the degree of composition.

Element distribution images can be produced by scanning the beam in a monitor and displaying the intensity of a selected X-ray line. Furthermore, images produced by electrons collected from the sample illustrate surface topography or mean atomic number differences according to the mode selected. The element mapping is done by EDX spectroscopy usually attached with SEM, which is closely related to the electron probe, and SEM is designed primarily for producing electron images. Therefore, there is a considerable overlap in the functions of these instruments.

17.4 Substrate Temperature-Dependent Surface Morphology and Photoluminescence of Germanium Quantum Dots Grown by Radio Frequency Magnetron Sputtering

17.4.1 Introduction

The Ge/Si(100) system among the self-assembled semiconductors nanostructure has generated intense interest since it is the simplest semiconductor heteroepitaxial system that opened new possibilities for optoelectronic and microelectronic applications. The small electron and hole effective masses of bulk germanium (Ge) lead to a significantly larger excitonic Bohr radius ($\sim$24.3 nm) (Gu et al., 2001) implying strong quantum confinement effects in Ge nanocrystals. An enhanced quantum confinement leads to a decrease in indirect band gap transitions and relaxes the selection rules for direct band gap transitions. Thus, Ge nanocrystals are expected to exhibit visible PL with the high quantum efficiency.

The brilliant, tunable fluorescence emission of Ge quantum dots has encouraged their use as nanophotonics applications. In recent years, extensive efforts have been made to fabricate Ge nanocrystals using various chemical and physical methods (Montalenti et al., 2004; Ray et al., 2011; Warner and Tilley, 2006). Alkyl-surface functionalized Ge nanocrystals via metal hydride reduction of nonpolar solutions of GeI_4 at room temperature has been produced by Jonathan et al. (2004). Chiu and Kauzlarich (2006) synthesized Ge nanoparticles using the sodium naphthalide reduction of $GeCl_4$ under varying reaction conditions. Simonsen et al. (1999) produced Ge nanocrystals on Si (001) substrate using electron beam evaporation in which a Gaussian distribution of sizes from 2 to 68 nm was achieved, and the FWHM showed increment with increasing island size. Ge QDs of 8–20 nm were prepared by Fahim et al. (2010) using electron beam evaporation technique, in which the RMS roughness is shown to be highly sensitive to the annealing temperature. Furthermore, the thermal annealing was found to strongly influence the structural, optical ($\sim$0.25 eV shift of the band gap energy), and electrical properties of Ge QDs.

Mestanza et al. (2006) measured the room-temperature PL spectra of Ge nanocrystalline samples on SiO_2 matrix by ion implantation technique and observed a broad blue–violet band at around 3.2 eV (400 nm) originates from germanium-oxygen-deficient centers. The occurrence of a weak peak at around 4 eV is also reported. The temperature-dependent PL peak at 2.05 eV was observed by Sun et al. (2005). A PL red shift from 1.18 to 1.05 eV was found as a function of increasing nanoparticle size from 1.6 to 9.1 nm in the experiment of Riabinina et al. (2006b). Three prominent PL peaks at 2.59, 2.76, and 3.12 nm for Ge nanoparticles synthesized by the inert gas condensation (IGC) method were illustrated by Oku et al. (2000). The PL peaks are related to luminescence that originates from the Ge/GeO$_x$ interface and quantum size effect of Ge clusters. They indicated that the formation of the core–shell structure of Ge and Si with oxide layers is the reason for the blue shift of band gap energy. In a recent communication, we have reported the details of preparation and characterization of Ge nanoislands of ~50 to ~100 nm (Samavati et al., 2012a).

The island size and RMS roughness are found to increase, and the number density is decreased on increasing the annealing temperature. The knowledge of the size and shape distribution is a prerequisite in the understanding of the evolution of islands during growth and predicting the optical properties and surface morphology of such islands. In this chapter, we present the size distribution and surface evolution pattern of Ge nanoislands on Si (100) recorded using AFM and FESEM. The details of the growth behavior and optical properties are investigated by PL spectroscopy with varying substrate temperature.

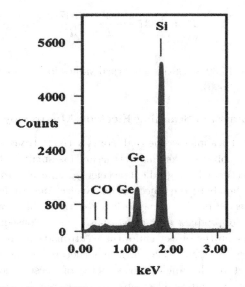

FIGURE 17.11   EDX spectra of sample D.

17.4.2   Experiment

Ge QDs are fabricated in the vacuum chamber pumped with diffusion and rotary pumps at a pressure of ~10 to 3 Pa. The Ge disc (purity 99.99% and 3 in. in diameter) is used as a target. The measured temperature on the rear face of the substrate holder during deposition is room temperature (sample A), 200°C (sample B), 300°C (sample C), and 400°C (sample D). The radio frequency power, Ar flow, and deposition time for all samples are 100 W, 10 sccm, and 180 s, respectively. Before loading the substrates in the sputtering chamber, the oxide layers from the substrate surface are removed by dipping the samples in weak hydrofluoric acid (HF ~ 5%) and then by ultrasonic bath at room temperature for 20 min, and finally dried by blowing nitrogen over them. Atomic force micrographs are recorded, and the room-temperature PL measurement is performed in the visible region using an excitation wavelength of 239 nm. The elemental composition of the Ge QDs is measured by EDX diffraction. The X-ray diffraction (Bruker D8 Advance Diffractometer) using Cu-Kα radiations (1.54 Å) at 40 kV and 100 mA is employed. The 2θ range is 0–60 with a step size of 0.021 and a resolution of 0.011.

17.4.3   Results and Discussion

Figure 17.11 shows the EDX spectra of the sample D to confirm the presence of Ge in addition to the silicon, the oxygen, and the carbon. The occurrences of Ge peaks confirm the existence of nanoislands composed purely of Ge as observed in AFM images. The appearance of the oxygen peak is due to the passivation of the surface dangling bond under exposure of atmospheric oxygen. The FESEM images of the samples A and D (Figure 17.12a,b) clearly show the presence of high-density Ge QDs.

The particle sizes (~8 to ~17 nm) estimated from FESEM are in conformity with the AFM. A schematic illustration

of the pyramidal QDs structure based on structural and optical analyses is presented in Figure 17.13. The surface of the Ge QDs is modeled as covered by thin oxide layers having a thickness of a few nm, and the Ge islands grown at room temperature are considered to have the thickest oxide layer.

The XRD spectra for the 200°C substrate temperature (lower curve) and 400°C substrate temperature (upper curve) samples are shown in Figure 17.14. Two prominent peaks associated with GeO$_2$ core–shell structure and Ge QDs are evidenced. The strongest peak is related to Ge QDs located at 22.2° (Sorianello et al., 2011) whose FWHM is 1.98° (inset). The broadening of the peak describes the quantum size effect of the nanoislands. Upon decreasing the substrate temperature, the peaks become sharper, and the FWHM of each peak decreases indicating the increase of average QDs size. The average size of Ge QDs estimated using Scherrer's formula is ~8 nm.

QDs at different growth temperatures are shown in Figure 17.15. The samples consist of mono-modal distribution of pyramid-shaped islands. The surface clearly exhibits well-resolved regular topography of germanium particle with ultra-small size as evidenced from 3D AFM.

The estimated average size (number density) for the samples A, B, C, and D are ~15 nm (2×10^3 μm^{-2}), ~8 nm (8×10^3 μm^{-2}), ~8.5 nm (14×10^3 μm^{-2}), and ~8 nm (14×10^3 μm^{-2}) respectively. The effect of size reduction of Ge islands with increasing substrate temperature from room temperature to 400°C can be explained in terms of different kinetic mechanism. While uniformity is increased at higher growth temperatures, self-ordering observed in the super-lattice structures may be required to achieve the QDs density and the degree of uniformity needed for their architectures. Undoubtedly, the increase of the growth temperature leads to a high degree of intermixing via thermal diffusion. This, in turn, lowers the strain

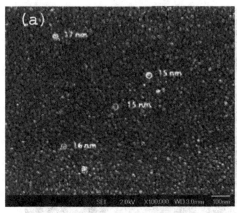

FIGURE 17.12   FESEM image of sample A (a) and D (b).

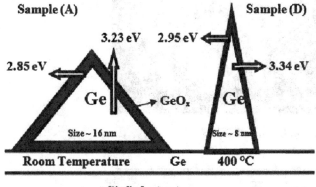

FIGURE 17.13   Schematic diagram of S-K growth mode of Ge QDs on Si substrate at two different substrate temperatures for samples A (room temperature) and D (400°C) responsible for the origin of PL peaks presented in Figure 17.9.

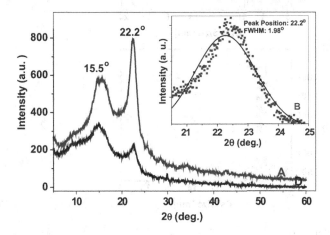

FIGURE 17.14   XRD spectra of samples A and D. The inset shows the de-convolution (Gaussian function) of the strongest peak at $2\theta \sim 22.2°$.

energy that favors nucleation on subsequent islands. At a higher substrate temperature sample (A), the Ge adatoms will diffuse to a longer distance at surface and prefer to produce new nucleation centers with narrow size distribution. A small pyramidal structure that requires higher energy for activation due to entropy maximization would appear.

Results for the substrate temperature-dependent size distribution are illustrated in Figure 17.16. The island size is found to decrease from $\sim$16 to $\sim$8 nm with a decrease of FWHM, thereby producing the narrow size distribution as the substrate temperature increases from room temperature to 400°C. Results for the roughness fluctuation in relation to the variation in the height distribution of the islands corresponding to the samples A, B, C, and D are presented in Figure 17.17a–d, respectively. The roughness variation is quite robust and smooth at 400°C and is expected because the island distribution has a regular pattern. However, as the substrate temperature is decreased, the larger particles are formed and thereby resulted in irregular fluctuation of the height distribution, as indicated clearly in the AFM micrograph.

For determining the sample quality, which eventually decides the optical behavior, the scattering of light and the nature of the surface are important factors. The RMS roughness is a measure of these properties (Mestanza et al., 2006).

The variation of RMS roughness and number density as a function of substrate temperature is shown in Figure 17.18. The corresponding fluctuation in the height distribution (Figure 17.18) expressed in terms of RMS roughness shows a monotonic decrease with substrate temperature. It is clearly seen that beyond 300°C, little change either in the number density (Figure 17.18) or in the roughness is evident. This observation has a direct correlation with the height of the peak in the corresponding normalized distribution obtained from the AFM images. The results obtained might provide a method to control the morphology of QDs, which could be used for tuning the narrow size distribution and ultra-small-sized Ge nanoparticle.

The room-temperature PL spectra for four samples A, B, C, and D are depicted in Figure 17.19. Three peaks appearing at approximately $\sim$2.85, 3.23, and 4.02 eV indicate the interaction between Ge and GeO$_x$, the possibility of the formation of a core–shell structure for the Ge QDs,

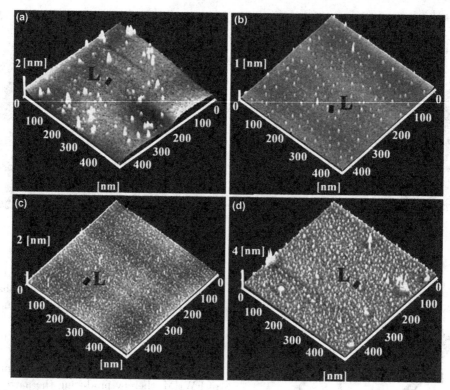

FIGURE 17.15 Three-dimensional AFM images of sample A (a), B (b), C (c), and D (d).

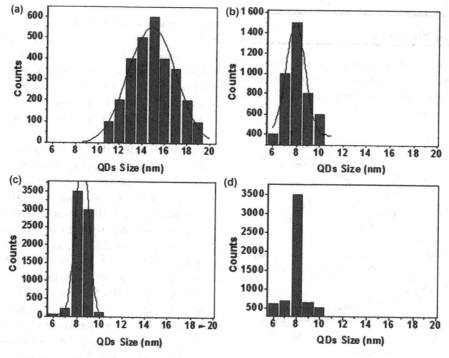

FIGURE 17.16 Size distribution of samples A (a), B (b), C (c), and D (d).

and probably other kind of nanostructures with different symmetries.

The role of surface passivation by atmospheric oxygen is important as it strongly affects the optical behavior. After the deposition, the Ge QDs may be encapsulated by the oxygen layer to form a GeO_x layer, thereby resulting in the formation of the core–shell structure. Configuration of core–shell structures in different thickness and sizes gives rise to a peak and shift in the PL. At the nanoscale, the blue shift in the HOMO-LUMO transition energy gap becomes prominent. The quantum size effect drives the visible PL shift by changing the band gap nature from indirect to direct

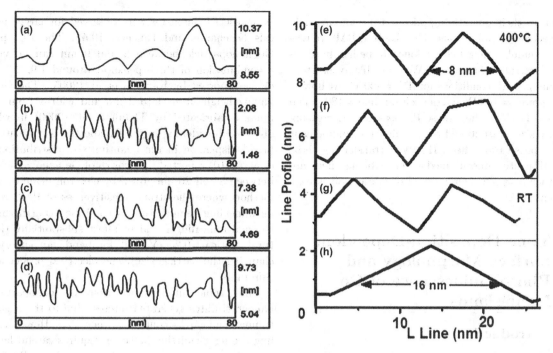

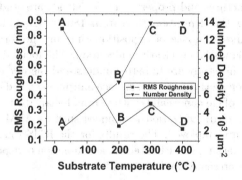

FIGURE 17.17 Height fluctuation of samples A (a), B (b), C (c), and D (d), line scan profile of sample A (e), B (f), C (g), and D (h).

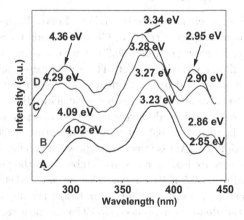

FIGURE 17.18 RMS roughness and number density of samples.

FIGURE 17.19 PL spectra of samples A, B, C, and D.

Four intense peaks around ~3.23, 3.27, 3.28, and 3.34 eV are clearly seen for samples A, B, C, and D, respectively, which are attributed to the presence of Ge QDs. The red shift (~0.11 eV) of the PL peak position for different particle size, which is due to the different substrate temperature, is in close agreement with observations made by others (Riabinina et al., 2006a; Dashiell et al., 2002).

The island-size-dependent shift in the PL peak position at different annealing temperatures is attributed to quantum confinement effects. The maximum PL intensity is obtained for sample D with larger number density, which is explained in terms of the generation of a larger number of photocarriers that contribute to the emission cross section. Our results confirm that the formation of core–shell structures, the presence of mix states, the quantum size, and surface effects are responsible for the visible luminescence in Ge nanostructures. The weak peak at ~2.95 eV observed in sample D presumably originates from the Ge and GeO_x interface. The inner core of these Ge islands consists of finely distributed nanoparticle, and the peak at 3.23 eV may originate from these nanostructures in sample A. The optical properties, the narrow size distribution of nontoxic germanium nanocrystals, and quantum confinement effect reported here can make them excellent candidates for future biological (especially biomedical) applications.

17.4.4 Summary

Ge QDs with precise size and shape distribution are synthesized using rf magnetron sputtering technique. The influence of substrate temperature on the surface morphology and PL response of such QDs is studied. The occurrence of narrow size distribution and light-emitting behavior in the visible

(Takagahara et al., 1992). The increase in substrate temperature causes the formation of smaller islands; the mix state of Ge and the thinner GeO_x interface reaction give rise to a PL peak at around ~2.86, 2.90, and 2.95 eV for samples B, C, and D, respectively.

region confirm their potential application in nanophotonic and optoelectronic applications. The observed RMS roughness and the number densities are found to be highly sensitive to the substrate temperature. We were able to ascertain the optimum growth conditions for TS = 400°C, Ar flow = 10 sccm, rf power = 100 W and deposition time = 180 s. The blue shift ~0.11 eV of the intense PL peak with decreasing nanoparticle size is attributed to the quantum confinement and surface passivation effect. However, transferring these nontoxic QDs into aqueous media and capping them are worth further research. Our method of fabrication is easy and economic.

17.5 Time Deposition-Dependent Surface Morphology and Photoluminescence of Ge Nanoislands

17.5.1 Introduction

The novel means of developing new electronic and optoelectronic devices is by nanostructuring of semiconductors. In particular, the discovery of room-temperature visible PL from Si and Ge nanostructures has stimulated much interest in these particular kinds of nanomaterials and in small semiconductor particles. The possibility of tuning the optical response of Ge nanomaterials by modifying their size has become one of the most challenging aspects of recent semiconductor research. Ge/Si heterostructure is suitable for investigating the growth morphology in two- and 3D growth process of Stranski–Krastanov. Ge nanostructure on Si substrate attracted considerable attention due to the possibility of optoelectronic application. Optical properties of these heterostructured islands can be changed by changing the islands size distribution (Fahim et al., 2010).

In recent years, an intensive research is dedicated towards the growth and optoelectronic characterization of Ge nanoislands on Si substrate employing various deposition conditions for optimization and functionality. The most commonly used fabrication techniques are MBE (Mestanza et al., 2007; Xue et al., 2005), CVD (Seo et al., 2006) and thermal evaporation (Garoufalis, 2009). However, the economic and safer deposition techniques such as RF and DC magnetron sputtering are also exploited for large-scale production of Ge nanoislands (Kartopu et al., 2004). Huang et al. (2009) prepared the Ge/Si nanomultilayer by magnetron sputtering and determined the crystalline nature of the film at deposition temperature 300°C. A strong agglomeration and the formation of Ge nanocrystals are observed for the samples annealed beyond 800°C. Simonsen et al. (1999) produced the Ge nanoparticle on Si(001) using electron beam evaporation in which the height distribution diagram found to be Gaussian and FWHM is increased with increasing island size.

It has been observed in the literature that the Ge and GeO$_2$ nanocrystals luminescence of around 3.1 eV and strongest PL intensity were observed for larger nanocrystals (Zacharias and Fauchet, 1997). The PL spectra of Ge nanoparticle between 2 and 9 nm can be viewed as a combination of three peaks at around 1.9, 2.3, and 3.0 eV as reported by Kartopu et al. (2005). The Ge islands having height from 1 to 2 nm and widths from 40 to 50 nm are fabricated by Thanh et al. (2000) in which the observed broadening in the PL spectra is correlated to the broad dispersion in the quantum size distribution. Fahim et al. (2010) reported the growth of Ge nanoparticle islands between 8 and 20 nm grown by electron beam evaporation method where the strong sensitiveness of RMS roughness on annealing temperature is confirmed. Furthermore, the thermal annealing found to affect substantially the structural, optical (0.25 eV shift for band gap energy due to changing the particle size), and electrical properties of Ge thin films.

The visible luminescence from Ge nanoparticles and nanocrystallites generated interest due to the feasibility of tuning band gap by controlling their sizes. However, controlling the light-emitting behavior requires careful fabrication of such nanostructure. In this work, the radio frequency magnetron sputtering is chosen to deposit Ge/Si islands. The structures and properties of Ge islands are studied by changing the deposition time from 180 to 420 s. Despite few studies, the role of deposition time on the formation of Ge islands is not clearly understood. The focused objectives of this study are to better understand the role of time deposition on the structure, surface morphology, and optical behavior of Ge nanoislands with pyramid-shaped structures. Different characterization techniques such as AFM, EDX, and PL are employed. The results for the RMS roughness, number density, and PL intensity as a function of deposition time are presented and explained.

17.5.2 Experiment

The RF magnetron sputtering (HVC Penta Vacuum) is employed to prepare a series of Ge nanoislands on Si(100) substrate. Single crystal Ge (99.99%) with a diameter of 3 in. is used as sputtering target. The substrate is rinsed to deionized water and dry with nitrogen prior to the placement in the chamber. Weak hydrofluoric acid (HF ~ 5%) is used to remove the oxide layers from the substrate surface and then by ultrasonic bath at room temperature for 20 min. The working gas (Ar) is injected in the vacuum chamber at the base pressure less than 8×10^{-6} mbar, and the pressure is maintained at 6.5×10^{-3} mbar. The samples and chamber are cleaned by ion radiation for 2 min before the deposition. During the deposition process, 350°C substrate temperature, 100 W rf power, and 10 sccm Ar flow are applied. Throughout the growth, the turbo pump remained operational in order to achieve the lower pressure. The islands are grown in different deposition times, 300 s (sample A), 420 s (sample B), 540 s (sample C), and 660 s (sample D). These samples are now subjected to structural characterization. The sizes, density, and RMS roughness of islands are measured with AFM operating in the contact mode. The elemental analysis is carried out using EDX. The PL spectra are recorded under 262 nm excitation with an argon ion laser. All the measurements are performed at room temperature.

17.5.3   Results and Discussion

Figure 17.20 shows the EDX spectra of the sample A (300 s deposition times) and sample D (660 s deposition times) that confirm the presence of Ge in addition to silicon, oxygen, carbon, and a small amount of gold, used as high-resolution coater in gold target for FESEM analysis. The occurrence of Ge peaks supports the existence of nanoislands composed purely of Ge as observed in the AFM. The presence of oxygen is due to the surface oxidation during the sample preparation.

The existence of Ge islands having heights between 2 nm (sample C) to 7 nm (sample B) and width between 12 nm (sample A) to 30 nm (sample D) is clearly depicted in the AFM images. The trend of height fluctuation with increasing deposition time can be explained using the Stranski–Krastanov (S–K) growth mechanism. In the S–K process, 3D square pyramid-shaped Ge islands (Ross et al., 1999) are grown after forming the wetting layer. Following the coarsening process called Ostwald ripening, these pyramidal islands transform to larger multifaceted pyramid-shaped islands (Figure 17.21c,d) by the further increase of deposited material and the collection of more adatoms from the neighboring islands. With continuing the Ge deposition over the Si surface covered with Ge islands as evidenced from Figure 17.21c,d, a large number of relatively smaller islands shrink irrespective of their size and eventually disappear. The theory of self-organized criticality can be used to understand this kind of structural change by achieving the minimally stable free energy configuration of the morphology (Ross et al., 1998). Additionally, the quantum size effect becomes less prominent with decreasing

the number densities and gets reflected in the distribution of islands as indicated in Figure 17.21c. Finally, the distribution of large-sized islands is achieved as depicted in Figure 17.21d by increasing of deposition time to 660 s, which leads the islands structure to evolve in a configuration where a relatively stable and steady-state distribution is achieved. The development of larger Ge islands in the competitive process results in lower number density. At a higher time deposition, the Ge adatoms diffuse to a longer distance and prefer to condense into pre-nucleated islands having lower activation energy rather than producing a new nucleation center that require higher activation energy.

The RMS roughness and the grain area coverage obtained from AFM are presented in Figure 17.22. The observed ratio of grain area shows continuous decrease as a function of increasing deposition time, which is inconsistent with the number density trend. The RMS roughness of sample A increases slightly after continuing the deposition time to 420 s (sample B), which is due to the formation of larger nanoislands and high irregularity associated with the fluctuation of the height and the width distribution. Furthermore, the RMS roughness shows a monotonic decrement as deposition time is further increased to 540 s (sample C). The minimum RMS roughness (sample C) is attributed to the formation of homogenous islands with relatively similar size and smallest height evolving towards more stable state. The variation in the value of RMS roughness is directly correlated to the island height distribution as indicated in Figure 17.21.

The structure of the prepared germanium nanoislands characterized by XRD is depicted in Figure 17.23. Aside from the reflections due to diamond cubic Ge, the XRD pattern exhibits two peaks at $2\theta = 14°$ and $22°$ corresponding to the (220) and (110) planes are in agreement with the report of Sorianello et al. (2011). Scherrer's analysis (inset) reveals the average size is ~9 nm for sample A which is in well consistent with AFM results.

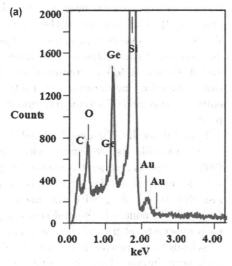 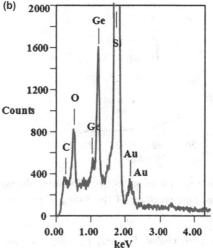

FIGURE 17.20   EDX spectrum of Sample A (a) and D (b).

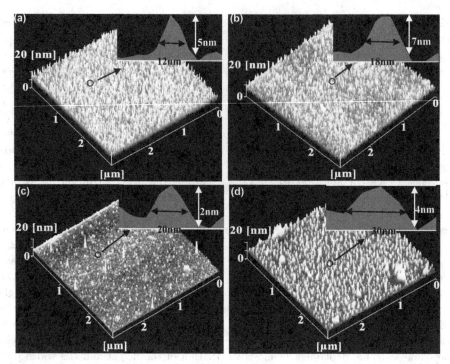

FIGURE 17.21   Three-dimensional AFM images of sample A (a), B (b), C (c), and D (d). Insets show the shape, width, and height of nanoisland.

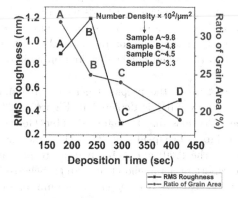

FIGURE 17.22   Plot of RMS roughness (nm) and ratio of grain area.

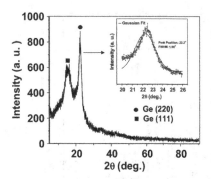

FIGURE 17.23   XRD spectra of sample A. The inset shows the de-convolution (Gaussian function) of the strongest peak at $2\theta \sim 22.2°$.

The room-temperature visible PL and the shift in the band gap energy of Ge nanostructure are studied extensively (Jie et al., 2004; Mestanza et al., 2006). The PL spectra of Ge nanoislands are shown in Figure 17.24. The intense peaks, which are marked as Xc, occurred at $\sim$373.4, $\sim$383.3, $\sim$383.5, and $\sim$396.3 nm, correspond to samples A, B, C, and D, respectively. The existence of Ge island having different sizes leads to shift in the PL peak position. The constant shift in the PL peak position towards the lower band gap energy with increasing islands size (from A to D) confirms the effect of quantum confinement. However, the small size variation ($\sim$2 nm) between samples B and C caused relatively lower band gap shift (0.2 nm). More generally, the appearance of the peaks might have emerged from the photo-generated charge carriers trapped in the shallow states that are tunneled to recombine.

The emission from carrier recombination in shallow traps (sample A) appears at lower wavelength than deep traps (sample D). The increase in the broadening of the emission band gap energy represents the superposition of wide distribution of trap distances and nanoisland sizes (Spanhel and Anderson, 1991) in continuation with the deposition time. The other mechanisms that contribute to the broadening are crystal field interaction, uncertainty, and Doppler process.

From the results (Figures 17.24 and 17.25), we affirm that the Ge nanoislands are excited with high energy photons (262 nm) and the generated carrier interacts with phonon and defect sites in the lattice, and a part of energy is lost due to coupling with the phonon via nonradiative processes. In contrast, the emission energy transforms via its room-temperature PL to violet ones (namely 383 nm). This leads to an increase in the carrier collection probability (enhanced intensity) in the region where the conversion appears.

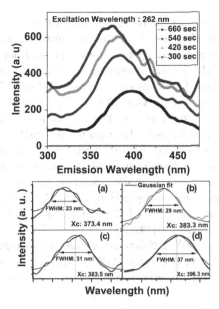

FIGURE 17.24 PL spectra of samples (top) and the Gaussian de-convolution of intense peak of samples A (a), B (b), C (c), and D (d) (bottom).

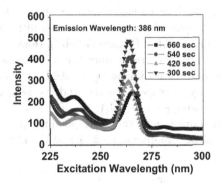

FIGURE 17.25 Excitation spectra of Ge nanoislands in different deposition time.

Figure 17.25 represents the excitation spectra of the Ge nanoislands at λ_{em} = 386 nm in various deposition time. The occurrence of excitation band at the lower wavelength is attributed to the exciton formation.

17.5.4 Summary

Ge nanoislands embedded in Si(100) are grown by easy one step rf magnetron sputtering method. The existence of pyramid-shaped nanoislands on the surface is observed. The surface morphology parameters such as RMS roughness, number density, and the ratio of grain area are studied using AFM. The average size of the nanoislands is estimated from XRD using Scherrer's equation. The optical characteristics of nanoislands are investigated using PL. The appearance of strong and broad visible PL peak in addition to size-dependent band gap shift at room temperature is attributed to the quantum confinement of carriers in Ge nanoislands having a wide distribution of crystallite sizes. Our findings may be useful for nanophotonics.

17.6 Structural and Optical Behavior of Germanium Quantum Dots: Role of Annealing

17.6.1 Introduction

The possibility of tuning the optical response of silicon (Si) and germanium (Ge) nanomaterials by modifying their size has become one of the most challenging aspects of recent semiconductor research. Self-assembled growth of Ge quantum dots using the Stranski–Krastanov (MacLeod et al., 2012) mechanism has attracted extensive attention for use in future nanoelectronic and optoelectronic devices. Ability to control the size and areal density is the prerequisite for using self-assembled QDs in future devices. The possibility for novel optoelectronic devices arises with the miniaturization of device components down to the nanometer length scale, including quantum dots for single-electron transistors, light-emitting diodes, and quantum cellular automata. Transition of the shape and size of nanoislands through on Ostwald ripening process for achieving the energetically stable phase is proved (Liu et al., 2002).

The formation of pyramid (Dilliway et al., 2003)- and multiple steeper facet (dome-like) (Medeiros-Ribeiro et al., 1997)-shaped islands with bi- or tri-modal size distribution has been reported (Capellini et al., 1997; Liu et al., 2002). Several methods have been employed to fabricate the Ge QDs with different sizes such as CVD having a size between 20 and 250 nm (Capellini et al., 1997), rf magnetron sputtering having a size between ~20 and 30nm (Ray and Das, 2005), and MBE with a ~10 nm size (Wang et al., 2012b). Rf magnetron sputtering has received considerable interest because it can offer simplicity, it is economic, and it is suitable for large-scale production and freedom in growing Ge nanostructures (Samavati et al., 2012b). It has been established that quantum confinement can modify the energy gap that results in visible luminescence as experimentally observed. Despite numerous proposed models, experiments and simulations to explain the luminescence, including quantum confinement, surface states, defects in the oxides, core–shell structures, and chemical complexes; however, the mechanism of visible PL and structure transformation is far from being understood.

Mestanza et al. (2006) measured the room-temperature PL spectra of Ge nanocrystalline samples on a SiO_2 matrix by the ion implantation technique and observed a broad blue–violet band at around 3.2 eV (400 nm) which originates from germanium-oxygen deficient centers. The occurrence of a weak peak around 4 eV was also reported. The temperature-dependent PL peak at 2.05 eV was observed by Sun et al. (2005). A PL red shift from 1.18 to 1.05 eV was found as a function of increasing nanodots size from 1.6 to 9.1 nm in the experiment of Riabinina et al. (2006b). Three prominent PL peaks at 2.59, 2.76, and 3.12 nm for Ge nanoparticles synthesized by the IGC method were illustrated by Oku et al. (2000). The PL peaks are related to luminescence that originates from the Ge/GeO$_x$ interface and quantum size effect of the Ge clusters. They indicated that the formation of the core–shell structure of Ge and Si with oxide layers is the reason for the blue shift of band gap energy. All these studies confirm the red shift in the PL peak as observed by us. Three prominent Raman spectroscopy peaks, which are attributed to the Gage optical phonon signal (302 cm^{-1}), Si-Ge interface reaction (403 cm^{-1}), and Si-Si (520 cm^{-1}) atomic vibration, were observed by Liu et al. (2002). A Ge-Ge phonon signal shift from 301.5 to 302.7 cm^{-1} due to the annealing process was obtained in the research of Ray et al. (2011). In this chapter, we report that pyramidal Ge QDs with an average diameter of 8 nm, height 2 nm, and density of 4.4×10^{11} cm^{-2} are fabricated successfully on a Si(100) surface. Rapid thermal annealing (RTA) is carried out. AFM, FESEM, PL, and Raman spectroscopy are used to investigate their detailed morphology and optical properties, respectively.

17.6.2 Experiment

Ge QDs are fabricated in the vacuum chamber pumped with diffusion and rotary pumps at a pressure of $\sim 10^{-3}$ torr. The Ge disc (purity 99.99% and 3 in. in diameter) is used as a target. The measured temperature on the back of the substrate holder during deposition is 400°C, and rf power and Ar flow are 100 W and 10 sccm, respectively (designated as sample A). Before loading the substrates in the sputtering chamber, the oxide layers from the substrate surface are removed by dipping the samples in weak hydrofluoric acid (HF $\sim$ 5%) and then placing the samples in

an ultrasonic bath at room temperature for 20 min; the samples are then dried by blowing nitrogen over them. Prior to deposition, a sputter-etch is used for 10 min to remove the surface contamination on the target. The freshly prepared samples are then annealed by RTA equipment (Anneal sys, As-One 150) at 700°C for 2 min that is designated as sample B. The atomic force microscope (SPI3800) built by Seiko Instrument Inc. (SII) is used and the room-temperature PL measurement (Perkin Elmer Ls 55 Luminescence Spectrometer) is performed in the visible region using an excitation wavelength of 239 nm. The structure and elemental composition of the Ge QDs and their crystal phase are analyzed through field emission scanning electron microscopy (FESEM, JEOL, JSM-6701 F) with an attached EDX diffractometer. Raman spectroscopy is performed using a spectrum GX (NIR, FT-Raman) system with an Nd crystal laser source having a spot size of 1 μm.

17.6.3 Result and Discussion

The AFM images of the structural pattern of the pre-annealed (A) and post-annealed (B) samples are shown in Figure 17.26. The surface clearly exhibits well-resolved regular topography of Ge nanoislands with ultra-small size as evidenced in 3D AFM (Figure 17.26b,f). However, similar interpretation can be drawn from the TEM measurement as reported elsewhere (Singha et al., 2008). The estimated number density, average diameter, and the height for the pre-annealed samples are $\sim 4.41 \times 10^{11}$ cm^{-2}, ~ 8, and ~ 2nm, respectively, and the corresponding quantities for the post-annealed samples are $\sim 2.14 \times 10^{11}$ cm^{-2}, ~ 15 and

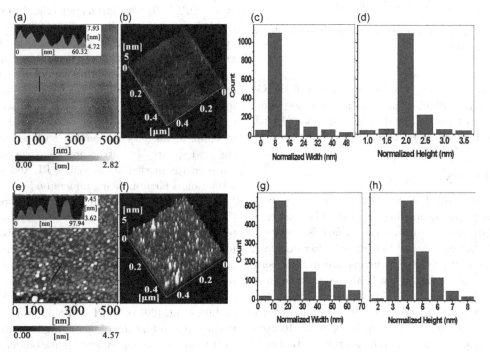

FIGURE 17.26 2D AFM image of pre-annealed Ge QDs (a), 3D AFM image of pre-annealed sample (b), width distribution of pre-annealed sample (c), height distribution of pre-annealed sample (d), 2D AFM image of post-annealed Ge QDs (e), 3D AFM image of post-annealed sample (f), width distribution of post-annealed sample (g), and height distribution of post-annealed sample (h).

~4 nm. The line scan profiles for the pre-annealed and the post-annealed samples are illustrated in the inset of Figure 17.26a,e measured along the indicated black lines. The pre-annealed sample consists of mono-modal distribution of pyramid-shaped islands, and the dome-shaped islands are evidenced for the post-annealed sample. The narrow size distribution with small pyramidal structure of sample A (Figure 17.26c,d) probably occurs due to entropy maximization. The change of density, shape, and size upon heating is due to the diffusion of the Ge adatoms to a longer distance at the surface that preferentially condensate into a developed quantum dots. This, in turn, lowers the barrier energy, minimizes the surface energy, and relaxes the strain that finally transforms the morphology from pyramid- to dome-like-shaped structure (inset Figure 17.26a,e). Additionally, the annealing causes the intermixing of Si into the Ge islands that further decrease the strain energy to form larger islands (Mooney et al., 1993). The origin of multimodal height and width distribution with varying QDs size in the post-annealed sample (Figure 17.26g,h) is due to the rearrangement of the inner structure of Ge/Si. The estimated RMS roughness parameter that measures the sample irregularities is obtained from the AFM topographic scan which is illustrated in Table 17.2. The RMS roughness and the grain size are found to be strongly dependent on the annealing process as shown in Table 17.2. The high irregularity associated with the RMS roughness and the fluctuating width and height distribution as shown in Figure 17.26c,d,g,h are due to the RTA at 700°C for 2 min (Yi et al., 2007).

The insets in Figure 17.27 are the corresponding EDX spectrum of samples A and B that show the presence of the Ge peaks. This also confirms the existence of QDs that purely composed of Ge as observed in the AFM. The Si and the carbon (C) peaks are appeared from the substrate and the supporter carbon tape, respectively. The oxygen (O) peak originates from the oxidization of the sample surface occurs during the sample transfer, atmospheric oxygen intake, and the possibility of the presence of GeO_x. However, the stronger Ge peak in the sample B is related to the existence of larger Ge QDs with higher atomic percentage.

The FESEM images of the pre-annealed and post-annealed samples (Figure 17.27a,b) clearly show the presence of high-density Ge QDs. The larger QDs (~15 nm) in the post-annealed sample (Figure 17.27b) appear from the process of agglomeration. The QDs sizes estimated from FESEM are in conformity with the AFM and XRD results.

The room-temperature PL spectra for two samples A and B are shown in Figure 17.28, and the corresponding values of peak position and FWHM are summarized in Table 17.3. The PL spectra consist of three peaks appeared approximately at ~2.85, ~3.21, and ~ 4.13 eV attributed to the interaction between Ge, GeO_x, the possibility of the formation of core–shell-like structure and the existence of the other kind of Ge nanostructures different from QDs having unknown symmetry. The surface passivation by atmospheric oxygen for both the pre-annealed and rapid thermally annealed sample strongly influences the optical band gap. The position of the PL peak and the shift of the peak are core–shell size dependent. The formation of core–shell structure is primarily due to the postdeposition effect of contamination during the sample transfer from vacuum chamber to PL recorder. These QDs are encapsulated by

TABLE 17.2 The Average Size, RMS Roughness, Number Density, Ge QDs Band Gap Energy, Experimental Ge-Ge Mode Frequencies, and Calculated Ge Composition in Dots

	Pre-annealed (A)	Post-annealed (B)
Average width (nm)	8	18
Average height (nm)	2	4
RMS roughness (nm)	0.83	2.24
Number density ($\times 10^{11}$ cm^{-2})	4.41	2.14
QDs diameter calculated by XRD	6.50	8.50
Band gap energy (eV)	3.26	3.21
W (Ge-Ge) cm^{-1}	305.42	302.15
Ge composition in dots	0.41	0.39

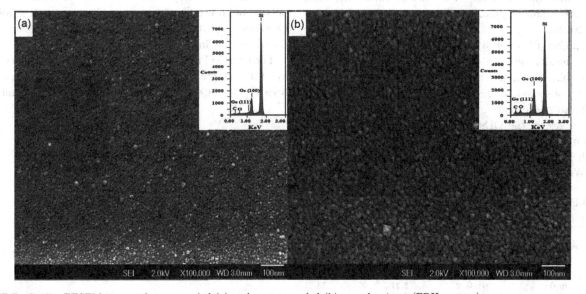

FIGURE 17.27 FESEM image of pre-annealed (a) and post-annealed (b) samples, inset (EDX spectra).

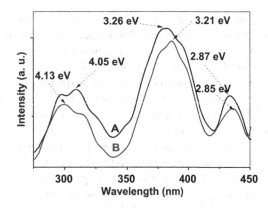

FIGURE 17.28 The PL spectra of pre-annealed (a) and post-annealed (b) samples.

the oxygen layer to form a GeO_x layer that finally gives rise to the formation of the core–shell structure. The thermal annealing process favors the formation of core–shell structures and decides their size. The formation of larger dots, the mix state of Ge, and the thicker GeO_x interface reaction are strong function of annealing process that is responsible for the PL peak at around ∼2.83 eV. Two intense peaks around ∼3.26 and ∼3.21 eV are clearly evidenced for samples A and B respectively, are attributed to the presence of Ge QDs. Moreover, the PL peak that appears at ∼3.1 eV is originated from the Ge/GeO_2 interface-related defects is well known (Pavesi and Ceschini, 1993).

The blue shift in the HOMO-LUMO transition energy gap becomes robust upon nanosizing. The large shift in the PL peak position from 0.67 eV (bulk) To 3.26 eV (QDs) is related to the indirect to direct band gap transition driven by quantum size effect (Takagahara and Takeda, 1992). The red shift of the peaks 2 and 3 (Table 17.3) is about 0.05 and 0.03 eV, respectively, which are in well consistent with the quantum confinement model of carrier recombination in Ge nanostructure. The annealing temperature-dependent red shift of the room-temperature PL peak position observed by us is in conformity with the earlier observations (Mestanza et al., 2006; Sun et al., 2005; Ray et al., 2011; Samavati et al., 2012a).

The higher PL intensity for the sample A having larger number density causes the generation of enhanced photocarriers that contributes to the emission cross section. The PL peak observed at 2.85 eV for sample A is undoubtedly due to the formation of mix states and the core–shell-like structure in the QDs. The annealing process leads to the formation of heavily oxidized QDs as well as bigger core–shell structures responsible for the red shift of the PL peak. We assert that the formation of core–shell structures,

the presence of mix state, the quantum size effect, and surface passivation all play major role for the observation of visible luminescence in Ge QDs.

Figure 17.29 shows the Raman spectra of samples A and B. The vertical dotted line at 300 cm^{-1} represents the optical phonon position for bulk crystalline Ge. The modes related to the Ge-Ge optical phonons in QDs are shifted slightly to higher frequencies with respect to their bulk counterpart. Among several possible physical mechanisms that can cause a Raman shift of optical phonons, the mechanism of the phonon confinement and the presence of strain are the dominant one. The observed Ge-Ge mode shift to the higher frequency is attributed to the induction of compressive strain due to the lattice mismatch of Si and Ge in the lateral directions of the QDs (Riabinina et al., 2006a). For the sample A, three intense peaks at around 305.42 cm^{-1} (Ge-Ge), 409.19 cm^{-1} (Si-Ge), 518.25 cm^{-1} (Si-Si), and a weak peak at around 328.68 cm^{-1} (Ge-O) are clearly evidenced. The difference in the Si-Si peak position ∼1.75 cm^{-1} from the reported value of (Riabinina et al., 2006) may be due to the contamination of the substrate. The appearance of the broad peak at 409.19 cm^{-1} is assigned to the Si-Ge vibration mode arises from Ge-Si bonds on the interface. Furthermore, the sharp and slightly asymmetric Lorentzian peak at 302.15 cm^{-1} corresponds to Ge-Ge vibration in small Ge QDs with phonon confinement. However, the weak peak at 329.75 cm^{-1} suggests that Ge QDs must be present in GeO_x matrix.

Upon annealing, Ge-Ge and Ge-O peak shift to 302.25 and 328.15 cm^{-1}, respectively. A maximum shift ∼5.42 cm^{-1} (from the bulk value) in the Ge-Ge optical phonon frequency of the QDs is observed. Additionally, a small change in frequency ∼3.27 cm^{-1} corresponding to Ge-Ge mode is observed as the samples are annealed. This mode shift is attributed to the different nanodots morphology (pyramid to dome), height-over-base ratio (0.25–0.27), and the degree of atomic inter-diffusion (0.41–0.39). The broadening of the peak is attributed to the phonon confinement and QDs size distribution. Intensity enhancement can be related to a large portion of broken Ge-O bonds enabling a greater number of Ge atoms to participate in the larger QDs formation and at the same time increasing the oxygen vacancies.

The degree of the interface intermixing can be determined by the integrated peak intensity ratio I_{Ge-Ge}/I_{Si-Ge} because the intensity depends on the relative number of

TABLE 17.3 Parameters for the Three Peaks, Which are Observed in the PL Spectra (Figure 17.29)

		Center (eV)	FWHM
Pre-annealed sample	Peak 1	4.05	0.43
(size ∼8 nm)	Peak 2	3.26	0.38
	Peak 3	2.87	0.10
Post-annealed sample	Peak 1	4.13	0.47
(size ∼15 nm)	Peak 2	3.21	0.34
	Peak 3	2.85	0.12

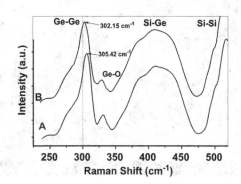

FIGURE 17.29 Raman spectra of samples A and B.

corresponding bonds as are summarized in Table 17.2. Following Mooney et al. (1993), we use the relation for SiGe alloy and Ge quantum dots as

$$\frac{I_{\text{Ge-Ge}}}{I_{\text{Si-Ge}}} \cong B \frac{x}{2(1-x)} \qquad (17.31)$$

where x is the average Ge concentration, and coefficient B is related to the Bose factor and the frequencies of Ge-Ge and Si-Ge optical modes of the alloy. It is found that the coefficient B varies weakly with alloy composition and is determined to be 3.2 (Singha et al., 2008). The values of $I_{\text{Ge-Ge}}/I_{\text{Si-Ge}}$ for samples A and B calculated using Figure 17.29 are ~1.12 and 1.04, respectively, which is in conformity with Liu et al. (2002).

17.6.4 Summary

The influence of annealing temperature on the optical properties and growth morphology of Ge QDs on Si(100) substrate is reported. A simple rf magnetron sputtering method is used to synthesized high-density QDs at large scale. The annealing temperature-dependent shape, size evolution pattern, and QDs size distribution are obtained by AFM and FESEM. The PL and Raman spectroscopy are employed for optical characterization. The average size of the quantum dots is estimated using XRD. The well-known Stranski–Krastanov growth mechanism due to lattice mismatch between Si and Ge is used to explain the formation of pyramid-shape Ge QDs, similar to a more sophisticated MBE growth, albeit at a much higher pressure in our study. FESEM images of post-annealed sample revealed larger QDs size ~15 nm as confirmed by XRD spectra. Room-temperature PL spectra due to direct band transitions for Ge and GeO$_x$ are observed. The emission in the visible wavelength range and PL red shift ~0.05 eV is attributed to quantum confinement of carriers in Ge QDs. Those results suggest that Ge will play a significant role as an enabler for integrating active photonic device on Si. According to the peak position and intensity of Ge-Ge and Ge-Si, the Ge QDs are analyzed. It is found that the Raman shift is remarkably related to the evolution of shape, size, and Ge composition in dots. It is asserted that the rf magnetron sputtering being an accurate technique for the production of large-scale and high-density nanostructure can reveal crucial physical insights of the growth processes. Our results on the growth morphology of Ge QDs are detrimental for optoelectronic devices that can be tuned by controlling the annealing temperature.

17.7 Influence of Ar Flow and rf Power on Growth of Ge/Si(100) Nanoislands

17.7.1 Introduction

The recurring theme of this chapter is to discuss the importance of specific growth parameters such as Ar flow rate and rf power on structural and optical properties of Ge nanoislands. Spontaneous formation of coherent 3D Ge nanoislands on Si substrate due to lattice mismatch of semiconductor heteroepitaxy follows the Stranski–Krastanov growth mode. Typically, the growth begins with a strained 2D wetting layer followed by the formation of 3D islands after deposition of the first few monolayers. In our study, a similar growth mechanism contributes to the formation of Ge nanoislands. The driving force is the dislocations within the islands that reduce the strain. During Ge film growth, a 4.2% lattice mismatch between the Ge layer and the Si substrate ($a_{\text{Ge}} = 5.65$ Å, $a_{\text{Si}} = 5.43$ Å) causes a linear increase in the film elastic energy that drives a layer-to-island transition at a thickness above a critical value depending upon the growth temperature.

The feasibility of light-emitting silicon–germanium heterostructure-based semiconductor materials containing Ge nanostructure received much interest for optoelectronics and nanoelectronics applications. Photo-detectors made of silicon-based photonic materials are competitive with those of conventional optoelectronics materials, such as the III–V compounds. Growth of high-quality small-sized Ge nanoislands in the range of exciton Bohr radius possessing large confinement energies, high density, homogeneity, and low defect density is challenging issue and promising for device fabrication. Controlled growth, optimization of the deposition parameters, and understanding the effect of strain are the solution keys for overcoming these issues.

The structural parameters and surface morphology play significant role in deciding the optical behaviors. Tang et al. (2010) have studied the impact of Ar flow rate on deposition rate, crystallinity, microstructure, and optoelectronic properties of SiGe:H films using H$_2$ and Ar as diluted gases by XRD, Raman, and FTIR. They found that with increasing Ar flow rate, the deposition rate enhances which is attributed to the increased dissociation efficiency of source gases. The film crystallinity is promoted by the addition of Ar effectively which is attributed to the increase in bombardment on growth zone by metastable states Ar, a larger amount of energy from deexcitation of Ar to its ground state and relax the growth zone.

Ge/Si nanomultilayers are prepared by Huang et al. (2009) using rf magnetron sputtering. It is found that increasing the sputtering power causes an increase in surface roughness of the deposited film and the favorable pressure of working gas in their experimental conditions was about 0.6 Pa.

The blue PL band located at ~3.1 eV and a weaker band at ~2.4 eV under the excitation of 325 nm laser are observed by Jie et al. (2004). These bands do not show significant size dependence while their relative intensities are related to the addition of H$_2$ or O$_2$ into the sputtering ambient. Direct exciton radiative recombination in the Ge nanocrystals and defects center in the nanocrystals and surface are reported as the possible origin of blue PL emission.

Shklyaev and Ichikawa (2002) fabricated the Ge nanoisland on Si surfaces deposited with ultrathin SiO$_2$ films in which the observed PL in the high-energy range of 2–3 eV is attributed to the radiative defect states at the Ge-dot/SiO_2 interfaces. As the results suggest, the electron and hole migration from the Si spacer layers to the potential wells of Ge QDs improves the PL intensity.

Argon is often used as sputtering gas in magnetron sputtering. The rate of Ar flow has certain influence on the deposited thin films as reported (Tang et al., 2010; Zhang et al., 2004; Kuo et al., 2011), but the influence on the Ge nanoislands is not studied in detail. To the best of our knowledge, the influence of rf power on growth morphology of Ge nanoislands is not investigated.

17.7.2 Experiment

The RF magnetron sputtering (HVC Penta Vacuum) is employed to prepare a series of Ge nanoislands on Si(100) substrate. Single crystal Ge (99.99%) with a diameter of 3 in. is used as sputtering target. The substrate is rinsed to deionized water and dry with nitrogen prior to the placement in the chamber. Weak hydrofluoric acid (HF ~ 5%) is used to remove the oxide layers from the substrate surface by ultrasonic bath at room temperature for 20 min. The working gas (Ar) is injected in the vacuum chamber at the base pressure less than 8×10^{-6} mbar, and the working pressure is changed from 3×10^{-3} to 7×10^{-3} mbar. The samples and chamber are cleaned by ion radiation for 2 min before the deposition. Throughout the growth, the turbo pump remained operational in order to achieve the lower pressure. The islands are grown in different Ar flow and rf power. The substrate temperature and deposition time are 400°C and 300 s, respectively. The growth parameters are detailed in the Table 17.4. These samples are now subjected to structural and optical characterization.

17.7.3 Results and Discussion

The 3D AFM images of six different samples are illustrated in Figure 17.30. During sputtering process, the sputtering power and the gas pressure significantly effect the deposition rate and the properties of the deposited film. Gas pressure plays an important role in the probability of the

collision between sputtered atom and background gas particles. Increasing the Ar flow rate causes the increment of ambient pressure, decrement of the self-bias potential on the substrate, and decrement of the average energy of the ion bombardment (Huang et al., 2009). In the extremely high pressure with high density of Ar atoms, the number of collisions between the Ar atoms and the sputtered atoms is increased. This causes to lose the energy of sputtered atoms and consequently a decrement in the growth rate. At a lower working pressure, atoms transport at a longer mean free path. Hence, at a higher working-gas pressure, the created film is expected to have a higher deposition rate. However, a lower sticking coefficient can be obtained in a lower working pressure. Thus, an optimal value of the pressure of working gas could be determined, under which the highest deposition rate, number density, and small size distribution occur.

In our experimental conditions, the favorable Ar flow is about 10 sccm, which cause the formation of Ge nanoislands with high density and narrow size distribution as illustrated in Figure 17.30b. Height fluctuation of deposited sample in the lateral diagram (inset Figure 17.30b) shows that the appropriate amount of Ar flow is 10 sccm that allows the formation of dense and homogenous Ge nanoislands. The dependence of deposition rate on deposition pressure in various materials is different, which is attributed to the average number of collisions of the sputtered and background atoms for a certain value of sputtering power. van Hattum et al. (2007) studied the ion and neutral-atom bombardment for the surface growth in magnetron plasma

TABLE 17.4 Details of Growth Parameters at Deposition Time 300 s and Substrate Temperature at 300°C

Sample Name	rf Power	Ar Flow	Pressure $\times 10^{-3}$ torr
		5	3
D	50	7.5	3.5
		10	6.5
		12.5	5.5
A		5	3.5
B	100	7.5	4.5
C		10	5
		12.5	4
		5	4.5
E	150	7.5	5.5
		10	5.5
		12.5	7
		5	2.5
F	250	7.5	3.5
		10	4
		12.5	6

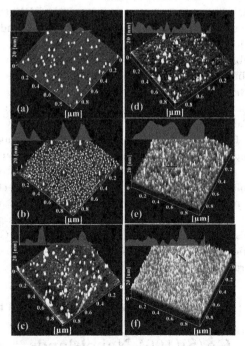

FIGURE 17.30 Three-dimensional AFM images of samples growth at 5 sccm and 100 W (a), 10 sccm and 100 W (b), 12.5 sccm, 100 W (c), 10 sccm and 50 W (d), 10 sccm and 150 W (e), 10 sccm and 250 W, and (f) Ar flow and rf power, respectively (insets: black line scan profile).

sputtering deposition. At lower pressure, the ion mean free path is long, and the intense scattering of the ions does not occur. Therefore, the absence of high-density islands structure on the surface is shown in Figure 17.30a.

The enhancement of pressure diminishes the potential of the plasma, and further increase of the pressure can cause a higher charge-transfer scattering rate. The sputtered atoms with the lower atomic masses may lose energy. The atoms directly become thermalized for a certain sputtering rate and initial kinetic energy of the sputtered particle (Palmero et al., 2007). In comparison with Si and SiO_2, the deposition of Ge is easier. Therefore, the deposition rate of Ge is larger than that of Si. Since this gas scattering is considerably dependent on the atomic mass of the sputtered particles, the gas pressure dependence of the amount of Ge atoms varies from that of the amount of Si and SiO_2 atoms. Figure 17.31 illustrates the histogram for the size distribution of samples A–D. It can be clearly seen that the sample B has narrow size distribution in comparison with others.

The increase of the sputtering power causes an increase of the surface roughness of the deposited islands (Figure 17.30e). At 250 W of the sputtering power, the surface of the deposited film is extremely rough (Figure 17.30f). Thus, for depositing Ge/Si nanoislands, a high sputtering power is not useful. When the sputtering power is 100 W, the surface of the deposited film turns dense and smooth (Figure 17.30c). The formation of the larger atom clusters can occur with an appropriate value of the sputtering power. In fact, after depositing the initial clusters of atoms on the surface, they have sufficient time to diffuse along the growing surface and mutually combine to form larger atomic clusters.

Through increasing the sputtering power, both the plasma excitation and the plasma density increase. Consequently, this increment improves sputtering rate and the nucleation growth of the crystalline thin film. Beyond a certain critical value of sputtering power, the adjacent gas becomes more energetic by transferring the high-energy from particles near the target to the neighbor. Clearly, the surface morphology of Ge nanoislands exhibits a strong dependence on Ar flow rate and rf power. Quantitative analyses of these effects are performed by surface roughness measurement as shown in Figures 17.32 and 17.33. The RMS roughness with increasing rate of Ar flow for varying rf power exhibits a gradual decrement followed by an increment. Furthermore, by increasing the rf power for all Ar flow, the same trend persists. We establish that the surface roughening is dependent on the microstructure and growth process of Ge nanoislands with increasing Ar flow and rf power. Such factors as deposition rate, the type of deposition particle, and the properties of the surface change correspondingly and produced different surface roughness behavior.

Increasing the Ar flow from 10 to 12.5 sccm causes a decrement in the mean free path due to large number of collision with Ar atoms; thus, the sputtered atoms do not have enough kinetic energy to diffuse over the surface and fill the voids. On the other hand, the increasing number of collisions results in an intense bombardment that increases the defects and degrades the sample quality. Figures 17.34 and 17.35 illustrate the number density as a function of Ar flow and rf power, respectively. By increasing the rf power from 100 to 250 W, anti-sputtering of the deposited film takes place.

For relatively higher sputtering power, the kinetic energy of the particles is too high and hence the occurrence of the opposite effect. In addition, large number of defects will appear with intense bombardment and collisions, which

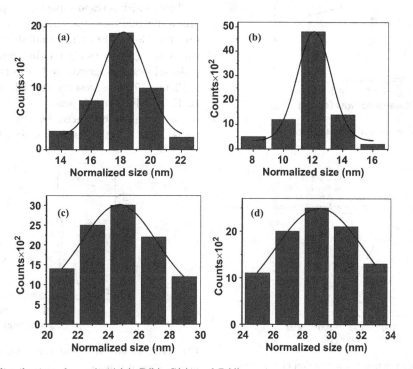

FIGURE 17.31 Size distribution of samples A(a), B(b), C(c), and D(d).

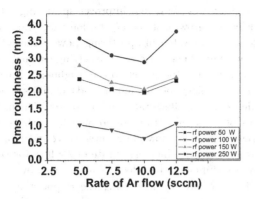

FIGURE 17.32   Ar flow dependent of RMS roughness.

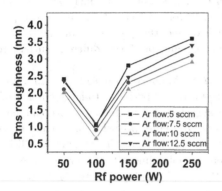

FIGURE 17.33   RMS roughnesses versus rf power.

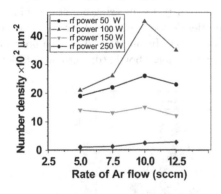

FIGURE 17.34   Number density versus Ar flow.

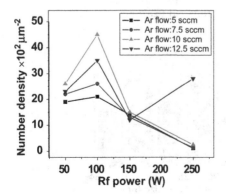

FIGURE 17.35   Number density as a function of rf power.

degrade the film quality by enhancing surface roughness. We optimize the rf power of 100 W to achieve high-quality nanoislands with minimal RMS roughness and precise size distribution.

Figure 17.36 shows the XRD pattern for varying Ar flow and rf power (samples B, C, and D). Reducing the rf power and Ar flow results in higher order reflections of Ge lattice and narrowing of the peaks. The appearance of the intense and narrow peak with decreasing rf power from 250 to 100 W and Ar flow from 15 to 10 sccm, respectively, indicates the growth enhancement of Ge nanocrystal. For both peaks appeared at $2\theta = 24.17$ and $27.2°$, intensities are increased and are associated with a shift from the expected diamond structure of Ge crystal ($23°$ and $27°$) corresponding to the lattice planes (220) and (111), respectively. The shift is more prominent for the (220) plane. This shift is attributed to the changes in the lattice constant of the Ge nanocrystallites from their bulk counterpart (Zhang et al., 1998). The strain field between Ge nanocrystallites and the surrounding matrix causes a change in the lattice constant by shifting the space of the planes. The strain field induced from the high temperature gradient created on samples during the deposition process causes the shift in the XRD peaks.

The broadening of the XRD peaks can originate from three different contributions as mentioned in Section 17.3. Figure 17.37 shows the effect of strain broadening through Williamson–Hall plot for Ge /Si (100) grown at 100 W rf power and 10 sccm Ar flow. The linear plot confirms the existence of strain. It is the effect of strain, which makes slight difference in the estimated crystallite size by Scherrer's formula and AFM.

The strain calculated by Williamson–Hall plot is found to be 3.6%, which is responsible for the peak shift as mentioned above. Furthermore, the measured strain is closed to lattice misfit of Ge and Si (4.2%) that shows the tenability of our controlled growth condition that minimizes the density of dislocation with approximately strained surface morphology.

Figure 17.38 represents the PL spectra of samples A, B, E, and F. The PL properties of Ge nanoisland appear to be strongly influenced by the density and quality of the surface. The strongest visible PL peak is observed at

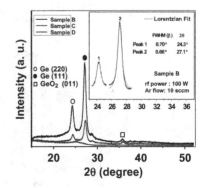

FIGURE 17.36   XRD pattern of Ge/Si nanoislands deposited under different rf power.

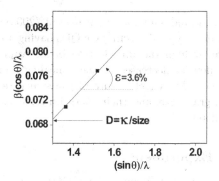

FIGURE 17.37 Williamson–Hall plot for Ge/Si(100) nanoislands grown on 100 W rf power.

blue–violet band at around 3.20 eV for sample B with an average diameter of 12 nm. The origin of this emission can be assigned to the electron–hole recombination within the islands, which can be decomposed into two peaks corresponding to Ge transverse optical (TO) phonon- and no-phonon (NO)-assisted recombination (Yang et al., 2004).

During nanoislands growth, the oxygen deficit defects formed in the silicon and Ge oxide matrix may contribute towards the emission process. The oxidation of Ge nanoislands surface is confirmed by the presence of oxygen peak in the EDX spectra and GeO_2 in XRD pattern that also supports the possibility of the formation of core-shell-like structure. GeO_2 defects have been reported to contain two nonbonding electrons by Ginzburg et al. (1995). The same observation in Ge doped silica glass fibers is made by Gallagher and Osterberg (1993). Such defects form a ground-state singlet level (S0), an excited singlet level (S1), and a triplet level (T1). In this energy scheme,

the blue luminescence of ~3.20 eV may be attributed to T1 → S0 transition. Furthermore, the orange-green PL band centered at ~2.81 eV (440 nm) probably originates from the oxygen deficit-related defects in the silicon oxide matrix. The appearance of relatively weak and broadened PL centered at 3.2 eV for sample F (Figure 17.38f) is attributed to high-density dislocation, defects, and anti-sputtering of the surface. The oxygen passivation and the radiative recombination of excitons in the nanoislands are responsible for the appearance of microstructure in the PL spectra for sample D (Yang et al., 2012).

17.7.4 Summary

Rf magnetron sputtering is employed to grow high-density Ge/Si nanoislands for various values of Ar flow and rf power to optimize the growth behavior. The optimum growth parameters are found to be 10 sccm Ar flow and 100 W rf power to fabricate the high-quality sample with minimum defects and density of dislocation. The XRD spectra show three peaks at $2\theta = 24.17°$, $27.2°$, and $36°$ corresponding to Ge(220), Ge(111), and GeO_2(011), respectively. The strain field between Ge nanocrystallites and the surrounding matrix is responsible for the shift in the peak positions that cause a change in the lattice constant of the nanocrystallites from bulk Ge. The rf power and the rate of Ar flow strongly affect the surface morphology and structural evolution pattern. The optimum size, number density, and RMS roughness are 12 nm, 46×10^2 μm^{-2}, and 0.6 nm, respectively. The room-temperature PL spectra exhibit strong blue-violet band that is attributed to electron–hole recombination from an excited triplet level (T1) to ground-state

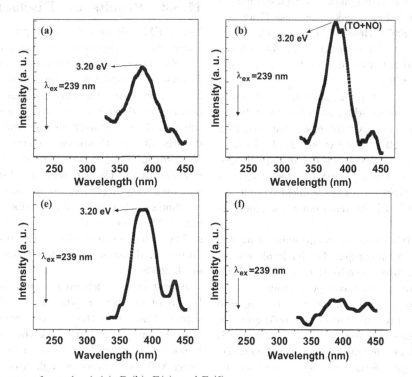

FIGURE 17.38 PL spectra of samples A (a), B (b), E (e), and F (f).

singlet level (S0). It is further established that by carefully controlling the processing parameters, it is possible to achieve high-quality Ge/Si nanoislands useful for optoelectronic applications.

17.8 Radio Frequency Magnetron Sputtering Grown Germanium Nanoislands: Annealing Time-Dependent Surface Morphology and Photoluminescence

17.8.1 Introduction

The Ge/Si(100) self-assembled nanostructures have generated intense interest because of its simplicity and opened new possibilities for optoelectronic and microelectronic applications. Understanding the growth of Ge nanoislands using the Stranski–Krastanov mechanism is topically important and fundamentally challenging for device functionality (Samavati et al., 2012b). Transforming indirect band gap material (bulk Ge) to direct gap material (nano-sized Ge) gives rise to light emission, which is very attractive for full-color display, as well as integrated optoelectronics technology. Among the techniques used to fabricate the low-dimensional semiconductor structures, rf magnetron sputtering has several advantages including low deposition temperature, higher films purity when compared to existing melting techniques, easy fabrication, and low cost (Huang et al., 2009). In recent years, the sputtering method has been widely exploited with the intent of making high-density nanostructures and showed tremendous promise (Ray and Das, 2005; Das et al. 2007). However, a few studies have been focused on the effect of annealing time on surface morphology and optical behavior of nanostructures (Yang et al., 2012; Jie et al., 2004).

It has been established that quantum confinement modifies the energy band gap of low-dimensional semiconductor materials and strongly influences the electronic structure properties (Zhao et al., 2006; Yang et al., 2012). Despite numerous proposed models, to explain the change in PL, due to quantum confinement including surface states, defects in the oxides, core-shell structures, and chemical complexes, the mechanism of visible PL by quantum confinement is still debatable.

Mestanza et al. (2006) recorded room-temperature PL spectra from Ge nanostructure prepared by ion implantation technique and observed broad blue band at around ~3.2 eV (400 nm) in addition to a weak violet peak around ~4 eV (310 nm). The authors suggested that the peaks originate from Ge–oxygen-deficient centers. Sun et al. (2005) grew Ge QDs via MBE that showed a temperature-dependent PL peak at ~2.05 eV. Despite extensive studies on synthesis and characterization, few efforts have been made on the dependence of surface morphology and optical behavior on

annealing time and temperature. Wan et al. (2003) produced ultra-high-density (~10^{11}/cm^2) Ge QDs having a mean size of ~14 nm by high vacuum electron beam evaporation in which the density decreased as the annealing temperature was increased. The exact nature of the structure, density, roughness, grain size, and their control mechanism are not clearly understood.

17.8.2 Experiment

Ge nanoislands were deposited on (100) oriented p-type silicon substrate by rf magnetron sputtering. Prior to deposition, the Si substrate was dipped in 5% hydrofluoric acid to remove the surface native oxide followed by rinsing in deionized water and drying in a flux of nitrogen. The chamber was first evacuated to a base pressure of 2×10^{-5} torr. The deposition was carried out at 300°C substrate temperature, 350 s deposition times, 100 W rf power, and 10 sccm Ar flow. The estimated Ge layer thickness was 50 nm, as measured by spectroscopic reflectometry (Filmetrics). The sample was then subjected to RTA in nitrogen ambient at 600°C for 30 s (sample B), 90 s (sample C) and 120 s (sample D). The as-grown sample was labeled A. A Bruker D8 Advance diffractometer was used to study the structural characteristics of Ge nanoislands using Cu-Kα radiations (1.54 Å) at 40 kV and 100 mA. The 2θ range was set to 0–90° with a step size of 0.021° and a resolution of 0.011°. An AFM built by Seiko Instrument Inc. (SPI3800) was used to study the surface morphology, whereas Perkin Elmer Ls 55 Luminescence Spectrometer probed the PL characteristics of the nanoislands.

17.8.3 Results and Discussion

The AFM micrographs in Figure 17.39 represent the annealing time-dependent evolution of the shape and size distributions of Ge nanoislands. The growth of nanoislands follows the Stranski–Krastanov mode in which the wetting layers first cover the substrate. A lattice misfit ~4.2% between the deposited Ge material and Si substrate hinders the growth of overlayer on the strained structure. Consequently, the growth processes change from layer-by-layer to spontaneous formation of 3D nanoislands (Eaglesham and Cerullo, 1990). The line scan profiles of the as-grown and annealed samples (120 s) are presented in Figure 17.40. The baseline sample consists of a distribution of pyramid-shaped islands that become dome-shaped upon annealing for 120 s. The aspect ratio of islands has been defined by various groups in different manners (Ronda et al., 2000; Mckay et al., 2008).

Aspect ratio is defined as the ratio of the height to the length of the base of the island. The measured average aspect ratios from the line profile scans are 0.35 for the baseline sample and 0.23 for the annealed sample. It is important to note that the size of nanoislands determined using AFM observation is slightly overestimated because of the tip artifact.

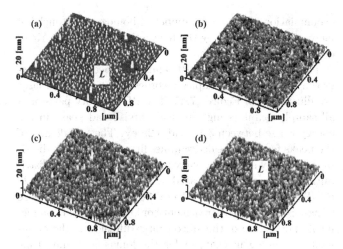

FIGURE 17.39 2D AFM image of as-grown (a) samples annealed for 30 s (b), 90 s (c), and 120 s (d).

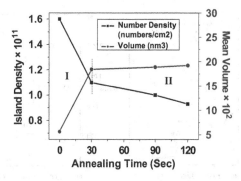

FIGURE 17.41 Island number density (cm^2) and mean island volume (nm^3) with respect to annealing time. The data are calculated from AFM topographic images. The regions marked in the figure are discussed in the text.

The annealed sample at 120 s in Figure 17.40 demonstrates noticeably larger dome-shaped islands. Islands are found to be rough due to heat treatment via structural and compositional change through continuous strain relaxation. Coarsening is a competitive growth process where some islands grow at the expense of others in order to minimize the total surface energy (Kamins et al., 1998). This leads to an increase in the mean island volume with time which results in a simultaneous decrease in the mean areal island density. In a system of pure Ge clusters, coarsening can occur by a combination of different mechanisms, such as Oswald ripening mediated by adatom diffusion between islands and Si inter-diffusion.

The mean island density and volume obtained from AFM micrographs are plotted with respect to annealing time in Figure 17.41. In the sample annealed for 30 s, the Si interdiffusion is a dominant coarsening mechanism allowing the pyramid-shaped islands to re-equilibrate to larger alloyed pyramid-shape islands (Singha et al., 2008). With the incorporation of Si via inter-diffusion into the Ge islands changes the lattice constant of the islands reducing the lattice misfit with respect to the Si substrate, as shown in region I of Figure 17.41. Beyond 30 s of annealing, Oswald ripening is more dominant marked as region II. Larger dome-shaped islands act as sinks and grow at the expense of smaller ones. In addition, the larger islands release their strain very fast via misfit dislocations at the edges of islands.

The RMS roughness and ratio of grain area obtained from AFM are presented in Figure 17.42. The observed ratio of grain area shows continuous decrease as a function of increasing annealing time, which is consistent with the number density trend. Furthermore, the roughness of baseline sample increases slightly after annealing for 30 s because of heat treatment, and the RMS roughness increases due to the formation of larger nanoislands and high irregularity associated with the fluctuating of the height and the width distribution. Thereafter, it decreases rapidly as the annealing time is increased. In addition, there is a gradual change in the morphology of islands to a dome-like form with less than 1.2 nm roughness.

Figure 17.43 shows the EDX spectra of the baseline and 120 s annealed samples to confirm the presence of Ge in addition to silicon, oxygen, and carbon. The Ge peaks present the existence of nanoislands composed purely of Ge as observed in the AFM. By annealing the sample, the Ge content decreased, indicating that some of the Ge vaporized, diffused further inside the Si substrate and intermixed

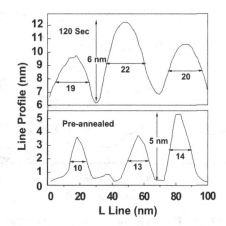

FIGURE 17.40 Line scans profile of as-grown and annealed sample for 120 s.

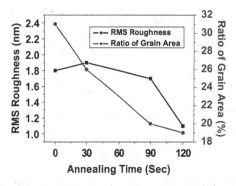

FIGURE 17.42 Plot of RMS roughness (nm) and grain area surface coverage (%) with respect to annealing time.

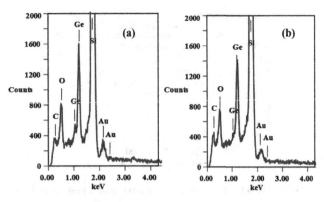

FIGURE 17.43 EDX spectrum of as-grown (a) and 120 s annealed sample (b).

with Si. The presence of oxygen is due to the surface oxidation during the sample preparation and the post-annealed analyses.

The XRD spectra of samples A, C, and D shown in Figure 17.44 confirm the existence of Ge nanoislands on the Si substrate. The noticeable peaks indexed at Ge (111) and (311) clearly indicate the diamond cubic structure of Ge. The most intense peak of Ge appeared at (111) which was used to estimate the size of Ge islands. The peak present at 29° for all the samples can be indexed to the (111) reflection of hexagonal symmetry of GeO_2. The silicon being the same crystal structure as Ge, the peak for Si(100) plane occurs at 69°. It can be inferred from the XRD line shape that the increase in the annealing time causes substantial change in the Ge (111) peak intensity and corresponding FWHM attributed to the growth of the islands size. Interestingly, the small FWHM value signifies a big grain size because there are a small number of grains in one X-ray beam spot. The Scherrer FWHM analysis indicates that the island growth is highly sensitive to the annealing time. The estimated size for the as-grown sample is ~13 nm.

The size of the islands in the sample annealed for 120 s is ~16 nm. The saturation of the Ge (111) peak beyond 90 s implies there is no further increase in the grain size due to annealing. This saturation may be attributed to the

accumulation of the large number of homogeneous nuclei of increasing size formed by the mechanism of annealing time-dependent agglomeration.

The room-temperature PL emission spectra of the as-grown and annealed samples recorded using a 239 nm laser are illustrated in Figure 17.45. The PL intensity pattern of all samples exhibits high intensity and broad peaks in the energy range between 3.19 and 3.29 eV. The broadening of the peaks for the annealed samples fitted to Gaussian distribution clearly shows the increase in nanoislands size distribution. The peaks are located at 3.19, 3.21, 3.23, and 3.29 eV. The PL red shift of ~0.10 eV because of increasing size of nanoisland by heat treatment confirms quantum confinement. Furthermore, the silicon inter-diffusion is the dominant coarsening mechanism for the longer annealing time that causes decrement in the compressive stress; it would then shift the peak to a lower energy value (Samavati et al., 2012b).

The intense blue band centered at 3.29 eV observed by us is different from the PL peak at 3.1 eV under longer wavelength excitation reported by others (Jie et al., 2004; Zacharias and Fauchet., 1997). Since the photon-absorption and photo-excitation take place inside the Ge nanoislands (Jie et al., 2004) and the observed blue emission shows significant size dependence, it is reasonable to exclude the possibility of direct exciton radiative recombination responsible for this emission. However, there may be other mechanisms such as defects-related luminescent centers in the matrix and interface responsible for the occurrence of the weak peak around ~2.84 eV. The PL band at ~3.29 eV shows a significant shift upon annealing, but the ~2.84 eV peak remains unchanged. This observation clearly indicates their different origins. It is important to note that Ge is uniformly

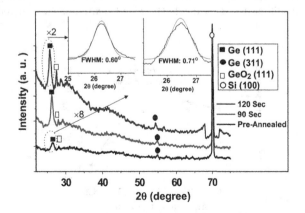

FIGURE 17.44 XRD patterns of as-grown and annealed samples (90 and 120 s). Ge(111) peaks have been magnified and fitted with a Gaussian curve to extract the FWHM value.

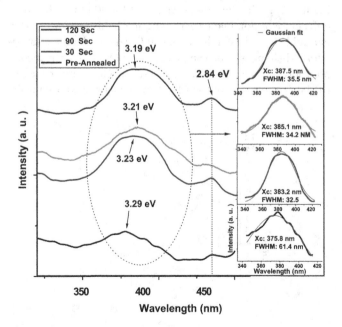

FIGURE 17.45 PL spectra of the samples under 239 nm excitation wavelength (Gaussian de-convolution of the intense peak for each sample is presented in inset).

deposited on the silicon substrate and that the larger Ge nanoislands formed during annealing may have some effect on the observed PL spectra.

During nanoislands growth, the oxygen deficit defects formed in the silicon and Ge oxide matrix may contribute towards the emission process. The oxidation of Ge nanoislands surface is confirmed by the presence of oxygen peak in the EDX spectra and GeO_2 in XRD pattern that supports the possibility of the formation of core–shell-like structure. GeO_2 defects have been reported to contain two nonbonding electrons by Ginzburg et al. (1995). Such defects form a ground-state singlet level (S0), an excited singlet level (S1), and a triplet level (T1). In this energy scheme, the blue luminescence of ~3.29 eV may be attributed to T1 → S0 transition.

Furthermore, the orange-green PL band centered at ~2.84 eV probably originates from the oxygen deficit-related defects in the silicon oxide matrix because the annealing process and the increase in annealing time do not have any influence on this emission. The appearance of microstructure in the PL spectra may be due to oxygen-passivated nanoislands having a broad size distribution. In addition, the near band edge transition of the wide band gap, especially the radiative recombination of the free excitons through an exciton–exciton collision process, may contribute towards this (Yang et al., 2012).

17.8.4 Summary

Ge nanoislands were deposited on Si (100) substrates by rf magnetron sputtering and characterized. The influence of RTA, the surface morphology, and PL properties as studied. The size and shape evolution, RMS roughness, number density, volume of islands, and the ratio of grain area were quantified by AFM. The Scherrer equation was used to estimate the size of the islands using XRD spectra. The increase of the island size was attributed to a coarsening mechanism in Ge/Si (100) system and was found to be highly sensitive to the heat treatment. The room-temperature PL spectra of the baseline sample exhibited strong and broad peaks and were found to shift as the annealing time was increased. This observation was attributed to quantum confinement of carriers in the Ge nanoislands. Our experimental methodology provides a process to control the size and morphology of the Ge nanoislands and hence tune the band gap energy of the material. This is valuable for the development of high-quality nanophotonics based on these structures.

17.9 Optical Properties and Structural Evolution of Self-Assembled Ge/Si Bilayer Containing Ge QDs

17.9.1 Introduction

Growth and optical characterization of high-density low-dimensional semiconductor nanostructures is

challenging issue for device fabrication (Shklyaev and Ichikawa, 2002; Ray and Das, 2005). Multilayers containing Ge quantum dots (QDs) have been fabricated by several groups using rf magnetron sputtering, ion implantation, and DE sputtering in a reactive oxygen environment, CVD with subsequent thermal annealing to induce film crystallization. Among these thin film synthesis methods, rf magnetron sputtering deposition is usually a one-step method (Samavati et al., 2012b), comprising of thin film deposition and post-growth thermal annealing treatment, which is compatible with the conventional integrated-circuit fabrication processes. The strong room-temperature luminescence from Ge QDs is promising for optoelectronics.

17.9.2 Experiment

The radio frequency magnetron sputtering is used to grow Ge/Si single and bilayer structures consist of 10 nm thickness of Ge and 10, 20, 30, and 40 nm thickness of Si space layer as measured by spectroscopic reflectometry (Filmetrics). Thicknesses are controlled by adjusting sputtering power and time, powered by 13.56 MHz RF voltage at 400°C substrate temperature. The growth morphology of Ge/Si heterostructure having Ge QDs is characterized using atomic force microscopy (SPI3800) built by Seiko Instrument Inc. (SII) and XRD (Bruker D8 Advance Diffractometer) using Cu-Kα radiations (1.54 Å) at 40 kV and 100 mA. The 2θ range is 0–60 with a step size of 0.021 and a resolution of 0.011. Raman spectroscopy is performed using a spectrum GX (NIR, FT-Raman) system with an Nd crystal laser source having a spot size of 1 μm. The room-temperature PL measurement (Perkin Elmer Ls 55 Luminescence Spectrometer) is carried out in the visible region using an excitation wavelength of 239 nm.

17.9.3 Results and Discussion

Figure 17.46 shows the 2D AFM images of the top surfaces of the samples in different steps of bilayer formation with

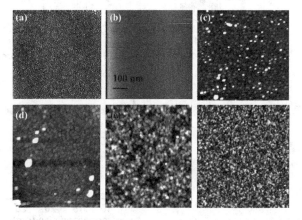

FIGURE 17.46 The AFM micrograph of Ge/Si (a), Si/Ge/Si (b), [Ge/Si]$_{\times 2}$ having 10 nm (c), 20 nm (d), 30 nm (e), and 40 nm (f) Si space layer thickness.

varieties of Si space layers. The size uniformity of the bilayer morphology is significantly improved over the single layer originates from the increasing uniformity of the strain distribution in the upper layer as predicted by the elastic continuum model (Tersoff et al., 1996), evidenced from AFM. The sizes of QDs are reduced from 25 to 16 nm as the Si space layer thickness increased from 10 to 40 nm. However, the size distribution becomes narrow, which is in agreement with the observation of Miura et al. (2000). The increase in Si space layer causes the surface homogeneous due to which the bilayer is more uniform than monolayer.

The Ge QDs size distribution for Ge/Si, Si/Ge/Si, $[Ge/Si(10\ nm)]_{\times 2}$, $[Ge/Si(20\ nm)]_{\times 2}$, $[Ge/Si(30\ nm)]_{\times 2}$, and $[Ge/Si(40\ nm)]_{\times 2}$ is presented in Figure 17.47. The average size of QDs is found to decrease from 25 to 16 nm as the Si space layer thickness is increased from 10 to 40 nm. Furthermore, the FWHM is decreased leading towards the narrow size distribution. The vertical correlation of QDs disappears with increasing Si space layer thickness. This is due to the attenuated strain-field coupling in the thicker Si spacer. In contrast, a decrease in the thickness of space layer may cause a rapid coarsening of the QDs in the subsequent layers and thereby increase the mean size of QDs. These QDs are partially relaxed by the formation of dislocations along the edge of the dot stacks due to the large accumulated strain in the spacer. The strain in the bilayer ($[Ge/Si(10\ nm)]_{\times 2}$) is more relaxed compared to the monolayer as evidenced from Williamson–Hall plot. In addition, the diffusion transport for bilayer is also enhanced with favorable nucleation that leads to the formation of larger QDs. In fact, there are two distinct regimes of sample size distribution associated with monolayer and bilayer as reported earlier by Rodrigues

et al. (2012). The first regime corresponds to monolayer case with smaller dimensions of Ge QDs and the second regime to the bilayer samples having relatively larger dimensions of QDs. The strain associated with the former regime is higher than the latter one.

Eventually, the quality of the sample and the surface morphology determine the structural and optical behaviors. The RMS roughness is a measure of these properties. The variation of RMS roughness, number density, and the average size of Ge QDs obtained from AFM are summarized in Table 17.5. The RMS roughness continuously decreases as the Si space layer thickness increases. The roughness variation is quite robust and smooth for Si/Ge/Si in the absence of any Ge QDs on the surface of this sample.

The roughness fluctuation and the variation in size distribution show the same trend. Moreover, the number density for monolayer is higher than the bilayer, but it increases monotonically with the increase of spacer thickness from 10 to 40 nm.

The XRD spectra of Ge/Si, Si/Ge/Si, and $[Ge/Si(10\ nm)]_{\times 2}$ are presented in Figure 17.48. Single-layered structure displays two prominent peaks attributed to Ge (111) and Ge (220) accompanied by a sharp peak for Si substrate (Sorianello et al., 2011; Jing et al., 2008). Depositing Si on the top of Ge QDs does not affect the nature of Ge peaks, but the Si peak changed to amorphous-like structure. However, bilayering through continuous Ge deposition re-structured the entire XRD pattern by forming poly-oriented structures of the Ge at (004) and (311) planes accompanied by (111) and (220) planes. The appearance of very narrow FWHM and strong intensity associated with the peak Ge (004) may be due to the formation of bulk

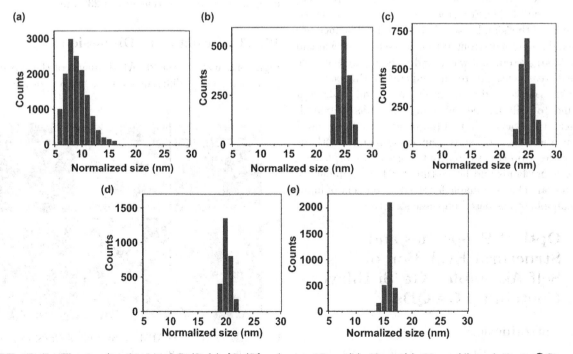

FIGURE 17.47   The size distribution of Ge/Si (a), $[Ge/Si]_{\times 2}$ having 10 nm (b), 20 nm (c), 30 nm (d), and 40 nm € Si space layer thickness.

TABLE 17.5 The RMS Roughness, Number Density, and Average Size of Ge QDs for Different Samples

	RMS Roughness (nm)	Number Density ($\times 10^{11}$ cm^{-2})	Size (nm)
Ge/Si	2.15	3	8
Si/Ge/Si	0.42	–	–
[Ge/Si(10 nm)]$_{\times 2}$	1.95	0.55	25
[Ge/Si(20 nm)]$_{\times 2}$	1.77	0.70	25
[Ge/Si(30 nm)]$_{\times 2}$	1.21	1.35	20
[Ge/Si(40 nm)]$_{\times 2}$	1.05	2.10	16

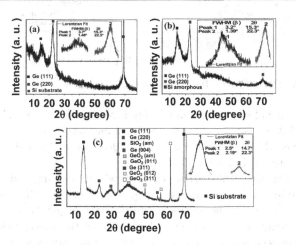

FIGURE 17.48 XRD spectra from Ge/Si (a), Si/Ge/Si (b), and [Ge/Si(10 nm)]$_{\times 2}$ (c).

Ge structure in the sub-layer Ge caused by heat treatment during deposition.

The preferable orientation of Ge to (111) direction gradually enhances the peak intensity of Ge (111) and diminishes the intensity of Ge (220). The presence of GeO$_2$ and SiO$_2$ peaks is attributed to the surface passivation of oxygen arises due to atmospheric exposure during sample transfer or deposition. Finally, the oxygen diffuses in the sub-layer to form GeO$_2$ and SiO$_2$. The transformation of monolayer Ge to bilayer structure reduces the strain, clearly evidenced from the Williamson–Hall plot (Figure 17.49).

The strain field between Ge nanocrystallites and the surrounding matrix leads to the variation in the lattice parameters by changing the inter-spacing of the planes. Consequently, a decreasing strain in the bilayer sample causes the Ge(111) peak shift in the XRD pattern (Zhang et al., 1998). The broadening of the XRD peaks is mainly due to the presence of various crystallite sizes, microstrain, and instrumental effects. The involvement of the strain that causes the broadening can be determined by plotting $\beta\cos\theta$

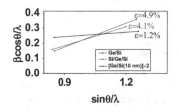

FIGURE 17.49 Williamson–Hall plot of Ge/Si, Si/Ge/Si, and [Ge/Si(10 nm)]$_{\times 2}$.

versus $\sin\theta$, known as a Williamson–Hall plot. Here, β is the line broadening at half the maximum intensity (FWHM). A linear relation between $\beta\cos\theta$ and $\sin\theta$ confirms the contribution from strain in the broadening. On the other hand, if the contribution in broadening appeared from the size variation only, then $\beta\cos\theta$ would have remained constant (Cullity and Stock, 2001).

The Raman spectra of monolayer and bilayer samples containing Ge QDs corresponding to the peak of Ge-Ge optical phonon (marked as circle in the inset) are shown in Figure 17.50. The vertical dotted line is fixed at 300 cm^{-1} indicating the optical phonon position for bulk crystalline Ge. The frequency positions of the Ge-Ge optical phonons are shifted slightly to higher frequencies with respect to their bulk value (300 cm^{-1}) originated from the competitive effect of phonon confinement and strain. The optical phonon branches of the bulk Ge are known to be quadratic and nearly flat at the Brillouin zone center ($k \sim 0$) as described by simple linear chain model. Restricted phonons in a nanocrystal are equal to those vibrations in an infinite crystal whose wave vector is given by $m\pi/d$ with m an integer and d the diameter of the nanocrystal. Therefore, the spatial limitations of superlattices cause a shift of optical phonons towards lower frequency. In the case of dots with very small height $\sim$1.5–2 nm (Liu et al., 2000), the phonon confinement effect was observed that resulted in a shift less than 2 cm^{-1}. It is the effect of strain, which plays the predominant role in shifting the peak towards the higher frequency. Interestingly, the lattice mismatch of Si and Ge leads to a compressive strain on the dots in the lateral directions that induces a Ge-Ge mode shift towards the higher frequency. However, the shift may also change due to intermixing of Si content in the Ge QDs. The Ge-Ge optical phonon frequency induced by a biaxial strain in Ge can be written as

$$\omega = \omega_0 + \frac{1}{2\omega_0}\left[p\varepsilon_{zz} + q\left(\varepsilon_{xx} + \varepsilon_{yy}\right)\right] \qquad (17.32)$$

where $\omega_0 = 0.564 \times 10^{14}$ s^{-1} is the frequency of the Ge zone center LO phonon; $p = -4.7 \times 10^{27}$ s^{-2} and $q = -6.16 \times 10^{27}$ s^{-2} are the Ge deformation potentials; $\varepsilon_{xx} = -0.042$

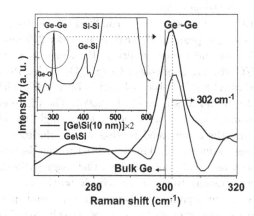

FIGURE 17.50 Raman spectra of [Ge/Si(10 nm)]$_{\times 2}$ and Ge/Si (inset: full range Raman spectra of [Ge/Si(10 nm)]$_{\times 2}$).

and ε_{yy} are the biaxial strain. Here, $\varepsilon_{zz} = -(2C_{12})\varepsilon_{xx}/C_{11}$, $\varepsilon_{xx} = \varepsilon_{yy}$, $C_{11} = 1{,}288$ kbar, $C_{12} = 482.5$ kbar are elastic coefficients (Kwok et al., 1999). The calculated value of ω is 317.4 cm^{-1} for fully strained pure Ge on Si, and the corresponding experimental value is 302 cm^{-1} (Figure 17.49). The difference in theoretical and experimental values suggests that the dots in the samples may not fully be strained as assumed in the theory and can be seen from the Williamson–Hall plot. The microscopic mechanisms such as the presence of threading dislocations and the Ge-Si atomic intermixing play dramatic role causing the strain relaxation of QDs. The signature of the intermixing is pronounced as Ge-Si peak in the full-scale Raman spectra of [Ge/Si (10 nm)]$_{\times 2}$ depicted in the inset of Figure 17.50. The presence of a weak shoulder towards higher energy may be related to the effect of intermixing and strain relaxation processes.

The intensity of peaks depends on the relative number of corresponding bonds; therefore, the degree of the interface intermixing can be determined by the integrated peak intensity ratio $I_{\text{Ge-Ge}}/I_{\text{Ge-Si}}$. The relationship for SiGe alloy and Ge quantum dots is presented by Moony et al. (1993) as

$$\frac{I_{\text{Ge-Ge}}}{I_{\text{Ge-Si}}} \cong 3.2 \frac{x}{2(1-x)} \qquad (17.33)$$

where x is the average Ge concentration. The Ge concentration in quantum dots estimated by this method is 0.46.

The formation of QDs during the deposition of subsequent Ge layer is strongly influenced predominantly by the compressive strain induced by the buried Ge wetting layer and the inhomogeneous strain field due to the partially strain relaxed islands. Furthermore, the intermixing of Ge and Si is greatly enhanced by the heat treatment during deposition. This, in turn, offers alternative pathways to the strain relaxation that results in the reduction of mismatch energy at the phase boundary as shown in Figure 17.49. The above mechanisms cause improved size uniformity and lateral ordering of the nanodots in the topmost layer even when the first layer is randomly nucleated. Almost perfect uniformity and lateral ordering is achieved when the first layer is pre-patterned either artificially or by self-organized template on the substrate.

Figure 17.51 represents the normalized PL spectra consist of a strong blue peak and a weak peak under 239 nm laser excitation in which no significant PL signal is observed from Si/Ge/Si. For single-layer Ge/Si QDs, a blue PL band centered at 3.28 eV is observed and in conformity with Jie et al. (2004) and Mestanza et al. (2006). The PL intensity shows continuous decrement going from single to bilayer Ge/Si, and the same trend persists as the thickness of Si space layer is increased. This is related to the decrease in the concentration of photo-carriers, their aggregations, and the energy transfer to the down layer in particular.

There is no significant change in the band position because the diameter of the dots effectively remains the same although the thickness of Si space layer increased from 10 to 20 nm. The red shift $\sim$0.08 eV in the PL is due to the

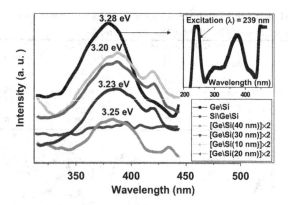

FIGURE 17.51   Room-temperature PL spectra.

change of single layer in to bilayer. Moreover, the blue shift $\sim$0.05 eV is related to the continual increase of the Si space layer thickness that leads to the decrease of QDs size as illustrated in Figure 17.52.

The size of the QDs, the material intermixing, and the strain are major counteracting effects on the PL emission energy. The degree of intermixing depends on the accumulated strain energy. Reduction of the spacer thickness accumulates more strain energy that causes stronger Ge-Si intermixing. The strain relaxation for a thinner spacer is achieved by transferring strain into spacers and material intermixing.

Although reducing the spacer thickness may provide better confinement for the electron in the Si conduction band due to the elastic strain transfer effect, but the predominant role played by material intermixing tends to counteract this effect and degrades the interface sharpness of the dots. The strain transfer mechanism tends to lower the emission energy of QDs. On the other hand, the deformation potential of the Si band which is about three times higher than the Ge band (Schmidt et al., 2000) favors the material intermixing that tends to increase the QD emission energy due to the increased Si content in the QDs. Therefore, the occurrence of similar peak position and FWHM of Ge/Si bilayer with different Si spacer originates from the net competitive effects of sizes, strain, and intermixing.

The occurrence of asymmetric weak PL band (2.9 eV) in bilayer samples is attributed to the existence of carrier in the structural defects state in the GeO$_x$ or SiO$_2$. Therefore,

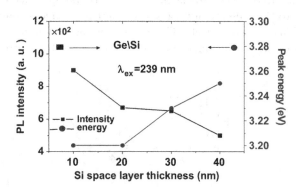

FIGURE 17.52   PL intensity and band gap energy versus Si space layer thickness.

the radiative recombination could involve electrons localized in radiative defect states at the Ge-dot/GeO_x or SiO_2 interfaces (Samavati et al., 2012b; Shklyaev and Ichikawa, 2002).

17.9.4 Summary

The size-dependent shift in the band gap energy associated with the strong room-temperature visible PL is achieved by changing the thickness of Si space layer, attributed to the electron–hole recombination within the dots and the effect of quantum confinement of photo-carriers. The Raman peak for the phonon modes appeared at 302 cm^{-1} is understood in terms of phonon confinement and strain relaxation. Furthermore, our dots are not fully strained on the surface as confirmed by Williamson–Hall plot. We assert controlled bilayering of heterostructure may significantly improve the evolutionary dynamics of growth that may be detrimental for optoelectronic devices.

17.10 Optical Behavior of Self-Assembled High-Density Ge Nanoislands Embedded in SIO_2

17.10.1 Introduction

The phenomenon of self-assembly in nanotechnology has paramount significance in several areas of material processing and fabrication. Understanding the importance of the mechanisms of growth and atomic processes occur in semiconductor nanostructure on the surface in condensed media is challenging. Si and Ge nanostructures among the other kind of semiconductors are receiving much attention due to the reason that in the small-sized particles, the efficiency of optical transitions increased by orders of magnitude resulted in widespread nanophotonic application. In order to overcome the complexity of non-volatile memory devices connected with the continuous miniaturization such as Si-based light-emitting diodes (LEDs) (Horváth, 2006), the dielectric layers with embedded semiconductor nanocrystals are widely studied. Low-dimensional memory devices due to their restricted dimension are not reliable. This is due to the problems connected with defects in tunneling layer and lateral charge loss (Cappelletti, 1998). These kinds of problems can be avoided by employing the semiconductor nanocrystals like Ge and Si embedded in the insulator as charge-storage media (Heng and Finstad, 2005; Horváth, 2006). In order to make use of carrier confinement at room temperature, the size of nanostructures must be comparable to exciton Bohr radius. The typical size of self-organized Ge/Si islands is usually much larger (Schittenhelm, 1995; Chaparro et al., 2000). The growth of Ge islands on a Si surface with a thin SiO_2 coverage results in the formation of ultra-small islands is reported (Shklyaev et al., 2000).

The formation of the self-assembled Ge islands on Si substrates is compressively strained because of the 4% lattice mismatch between Si and Ge. The growth of Ge islands on SiO_2 surfaces is less strained because they are more weakly bound to the surface. The strong strain of Ge islands makes them vulnerable to oxidation, which can act to reduce the strain. Unlike the growth of Ge on the pure silicon surface, the formation of Ge nanoislands on SiO_2 does not follow the S–K growth mode, and it can be explained by VW growth mode (Nikiforov et al., 2005). It proceeds without formation of the wetting layer and the change in the surface lattice of germanium relative to the surface lattice of Si, as in the case of the growth on the pure silicon surface.

Growth process using sputtering consists of two main competing mechanisms. One is the shadowing effect, which leads to preferred growth in the vertical direction. The other is surface diffusion, which tends to smooth the growing surfaces. The morphology evolution of growing surfaces is driven by the competition of these two mechanisms. At the initial stage of deposition, growing a smooth film is possible. The surface stays flat until it reaches a critical thickness and the extra growth leads to a roughening transition (Karunasiri et al., 1989). We report the optical and structural behavior of Ge nanoislands/SiO_2/Si using rf magnetron sputtering technique. Detailed characterization, the mechanism of growth, and the SiO_2 thickness-dependent structural evolution and optical properties are presented and understood.

17.10.2 Experiment

A series of Ge nanoislands are grown on SiO_2/Si(100) employing radio frequency (rf) magnetron sputtering technique (HVC Penta Vacuum). The samples are grown from a polycrystalline SiO_2 and Ge target (99.999% of purity) on Si (100) p-type in argon at 400°C substrate temperature. Initially, a layer of SiO_2 is deposited at three different times of 40, 60, and 90 min in order to achieve 30 (sample A), 40 (sample B), and 50 nm (sample C) thicknesses, respectively. Then, the Ge is sputtered for 420 s under similar condition. After deposition of each layer, the plasma is turned off, and the chamber is evacuated until a pressure around 2×10^{-5} torr is reached. The total gas pressure in the chamber is kept constant at 2×10^{-3} torr during the deposition. The rf power applied on SiO_2 target is 250 W and on Ge target is 100 W. The samples are kept inside the chamber to acquire the room temperature at the end of the growth. The structural details are studied by XRD (Bruker D8 Advance Diffractometer) using Cu-Kα radiations (1.54 Å) at 40 kV and 100 mA. The 2θ range is 0–90 with a step size of 0.021 and a resolution of 0.011. The PL spectra are recorded at room temperature using the (Perkin Elmer Ls 55) luminescence spectrometer under 259 nm excitation sources. Raman spectroscopy is performed using a spectrum GX (NIR, FT-Raman) system with an Nd crystal laser source having spot size of 1 μm with 50 mW laser power.

17.10.3   Results and Discussion

Figure 17.53 represents the AFM images of size 2 μm × 2 μm for the self-assembled high-density nanoislands grown on SiO₂ sub-layer with varying thickness. The corresponding size distribution and the shape of the nanoislands are shown in the inset. The surface of the sample A is found to be inhomogeneous and smooth with lower density of nanoislands. This inhomogeneity and smoothness are due to the lack of nucleation and growth of nanoislands caused by the effect of higher interfacial and strain energy. Furthermore, the formation of Ge nanoislands became more prominent along with an increase in the height and length of pyramid-shaped islands as the thickness of sub-layer increased from 30 to 50 nm. This is related to the enhanced diffusion and the nucleation of Ge atoms at the surface that leads to ordering and lowering the free energy.

The formation of pyramid-shaped islands is attributed to the significantly higher rate of surface reactions between Ge adatoms and the nucleated Ge islands during deposition of Ge in comparison with the reactions between Ge adatoms and SiO₂. This mechanism causes hindrance in the formation of a wetting Ge layer on the SiO₂ surface that causes higher aspect ratio (height divided by base length) ~0.33 of Ge islands. In contrast, the formation of a wetting Ge layer on the Si substrate is more favorable that causes enhanced reaction rate between Ge adatoms and Si substrate, reported to have lower aspect ratio for Ge islands (Liu et al., 2002; Samavati et al., 2012a).

This study reveals that the growth of high-density Ge nanoislands on SiO₂ is more preferable for optoelectronic application than the growth on Si substrate is in good agreement with Kolobov et al. (2001).

The cross section of SEM micrograph, shown in Figure 17.54, confirms the existence of SiO₂ sub-layer having thicknesses 30, 40, and 50 nm on the Si substrate representing samples A, B, and C, respectively.

The SiO₂ sub-layer thickness-dependent RMS roughness of the Ge nanoislands is illustrated in Table 17.6. The RMS roughness of the surface is continuously increased as the thickness of the sub-layer is increased from 30 to 50 nm. However, the surface roughness, the ratio of grain area, and the number density are relatively insensitive to the variation of SiO₂ thickness from 40 to 50 nm. This is a consequence of the reaction of adatoms attachment to the nucleated islands, which dominates over the nucleation through the reaction of individual adatoms.

Figure 17.55 illustrates EDX spectroscopy for samples A and C. The Ge layer started to form island structures as the thickness of SiO₂ bilayer increased from 30 to 50 nm. At 30 nm thickness, the surface is largely smooth with minimal density of Ge islands. However, with the increase of SiO₂ thickness to 50 nm, the island density increases by order of magnitude that results in an effective decrement

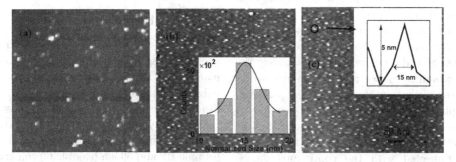

FIGURE 17.53   2D AFM images of Ge nanoislands on SiO₂ sub-layer thickness of 30 nm (a), 40 nm (b), and 50 nm (c). Inset: the size distribution (b) and pyramid-shaped structure (c).

FIGURE 17.54   The cross section of SEM images for the formation of different thicknesses of SiO₂ sub-layer.

TABLE 17.6   The RMS Roughness, the Number Density, and the Ratio of Grain Area as a Function of SiO₂ Thickness

Sample	SiO₂ Thickness (nm)	RMS Roughness (nm)	Number Density (cm⁻²)	Ratio of Grain Area (%)
A	30	~0.12	~2.1 × 10⁹	15
B	40	~0.56	~5.1 × 10¹¹	30
C	50	~0.61	~5.8 × 10¹¹	28

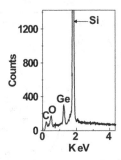

FIGURE 17.55 EDX spectra of samples A and C.

of the Ge content. The observed high Si content is originated from the substrate. It is needless to mention that the occurrence of Au and C peak emerges from the fine conductive coating, the glue used to attach the samples for recording EDX spectra, and FESEM performed together. The recorded FESEM images (not shown here) confirm the presence of Ge nanoislands in the range of 15–20 nm, which is in agreement with XRD and AFM results.

The XRD spectra of samples A and C are presented in Figure 17.56. The sample C with SiO_2 thickness of 50 nm clearly shows the Ge crystallization on silicon oxide sub-layer. In addition, it exhibits poly-oriented structure with several characteristic peaks corresponding to other crystalline planes such as 111, 004, 220, and 311. The peak occurred at 57° can be assigned to the reflection from (012) plane with hexagonal symmetry of GeO_2 nanocrystalline phase (Jing et al., 2008). The peak corresponding to the amorphous structure of SiO_2 sub-layer appeared at lower angles of 20°–30° having FWHM ~4° (Wang et al., 2012b).

For SiO_2 sub-layer thickness ~30 nm and below, the growth of $Ge/SiO_2/Si$ proceeds through layer-by-layer with strain relaxation mediated via nucleation of the misfit dislocations rather than the formation of islands. In contrast, above 30 nm sub-layer thickness, the growth mechanism obeys VW mode. Interestingly, the Ge nanoislands embedded at 50 nm SiO_2 matrix display more complex defect structure. This is due to the presence of misfit dislocations as well as very small pyramidal stacking faults at the Ge/Si interface and threading dislocations propagating all across the Ge thickness. The observed pronounced surface roughening above 30 nm of SiO_2 sub-layer thickness

originates from various kinds of dislocation and stacking faults. The (311) peak is used to estimate the mean nanoislands size by Scherrer's formula (Eq. 17.22). The average size of Ge nanoislands is estimated to be ~13 nm, which tallies with AFM and Raman data.

The room-temperature PL spectra shown in Figure 17.57 exhibit a strong peak centered at 2.9 eV irrespective of the thickness of the SiO_2 sub-layer. The origin of visible luminescence is attributed to the radiative recombination of electrons and holes in the quantum-confined nanostructures that can be explained using Ge nanoislands/SiO_2/Si interface model depicted in Figure 17.57. It is argued that, if the potential well that confines electrons in the conduction band has sufficient depth, the corresponding radiative recombination would involve holes and electrons confined within the Ge islands.

The energy required for the radiative recombination is ~2–3 eV. Meanwhile, the potential barrier at the Ge islands/SiO_2 interface can only confine electrons with energies approximately 1.5–2 eV above the conduction band edge. Therefore, the electron states with such high energies can only exist as structural defects in SiO_2 sub-layer.

This suggests that the radiative recombination does not involve electrons from the conduction band of the Ge or Si, but involves electrons that are localized in the radiative defect states at the Ge nanoislands/SiO_2 interfaces. A similar description for Ge nanocrystals embedded in SiO_2 matrix is suggested by Takeoka et al. (1998).

The room-temperature PL spectra are illustrated in Figure 17.58. The excitation energy-dependent PL spectra suggest the active role of defects because the electron states in defects typically have broader energy distributions (Chambliss and Johnson, 1994; Zinke-Allmang, 1999). Therefore, the lower energy states can be involved through relatively lower excitation energies of the laser beams than the higher one reported earlier (Samavati et al., 2012a). The Ge nanoislands structures in the present case contain a boundary between the Si substrate and the SiO_2 sub-layer that is responsible for the observed weak PL peak ~1.65 eV. Nishikawa et al. (1999) have observed a PL peak around 1.5–1.7 eV originated from SiO_2/Si interface. Furthermore, this weak peak is attributed to the radiative recombination at

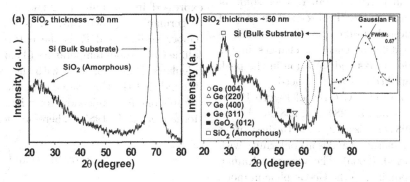

FIGURE 17.56 XRD pattern of Ge nanoislands deposited on 30 and 50 nm SiO_2 thickness, Gaussian fitted of intense Ge peak at (311) orientation plane.

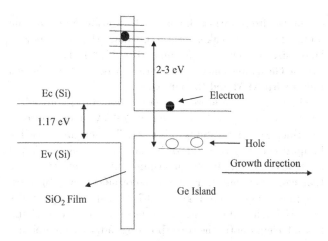

FIGURE 17.57 Schematic band structure of Ge nanoisland/SiO₂/Si. Dotted lines indicate energy levels in the Ge islands and SiO₂ sub-layer, respectively. The energy of 1.17 eV corresponds to the band gap of bulk Si.

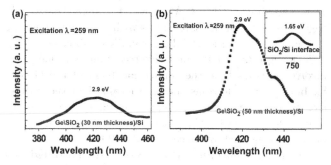

FIGURE 17.58 PL spectra of Ge nanoislands embedded on SiO₂ sub-layer having thicknesses of 30 and 50 nm (inset: the SiO₂/Si interface reaction peak).

the Si/SiO₂ interface, which is much weaker than the Ge nanoislands/SiO₂ interface (inset of Figure 17.58).

The Raman spectra used for the fingerprint of identifying the species and the crystal phases are recorded and presented in Figure 17.59. It contains the information of the vibration (phonon modes) of the entire molecule or the crystal under study. Phonons have important influence on physical, structural, and optical properties. The effect of confinement in low-dimensional system in general and nanocrystalline system in particular can restrict some of the phonons or vibrational modes. This restriction termed as quantization is manifested in different behaviors such as vibrations that shift to different energies, changes in the symmetry of the vibration peaks and variation in the peak intensity that can be vanished or even enhanced.

The Raman spectra of sample C (inset of Figure 17.59) consist of three bands around 278, 289, and 298 cm⁻¹. The band occurred around 298 cm⁻¹ is assigned to the first-order TO phonon mode of Ge nanoisland and is in conformity with the results of Kartopu et al. (2005). The strongest Raman peak (298 cm⁻¹) corresponding to the Ge-Ge phonon mode shows a shift towards lower frequency with respect to the bulk crystalline Ge (∼300 cm⁻¹). The asymmetric bands

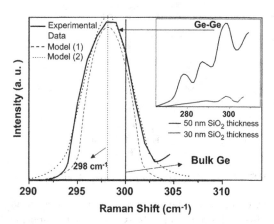

FIGURE 17.59 Raman spectrum around the Ge peak with Gaussian fitted spectrum (inset: the raw data).

appeared between 260 and 290 cm⁻¹ are attributed to the Ge-O phonon mode is inconsistent with the report of Giri and Dhara (2012). Additionally, the peak at 298 cm⁻¹ exhibits two interesting features: first, the spectra possess a well-defined peak with a characteristic shape that is essentially the same as bulk Ge crystals reported by Kartopu et al. (2005). The presence of this band in our spectra confirms the existence of Ge nanoislands in the SiO₂ sub-layer. Second, the spectra are not only wider but also slightly asymmetrical as well as red-shifted compared to the bulk Ge. According to Wellner et al. (2003), the displacement in the position of the Raman peak for Ge nanostructures with respect to the bulk Ge is caused by isotopic effects, phonon confinement, and the effects of stress from the SiO₂ sub-layer on the Ge nanostructures.

Following Fauchet and Campbell (1988), the Raman spectra of the sample C are fitted to the theoretical model to examine the effect of size distribution of the nanocrystals on the spectral line shape. According to this model, the first-order Raman spectrum, $I(w)$ as a function of frequency (w), is given by

$$I(w) = \int \frac{4\pi q^2 \, |C(0,q)|^2}{[w - w(q)]^2 + \left[\frac{\Gamma_c}{2}\right]^2} dq \qquad (17.34)$$

where $w(q)$ is the phonon dispersion curve, vector q is expressed in the units of $2\pi/a$, a is the lattice constant of germanium (0.566 nm), L is the diameter of the Ge nanocrystallite, and Γ_c is the natural line width for crystalline Ge at room temperature (3 cm⁻¹). By comparing the calculated Raman spectra with the experimental results, it is found that the Gaussian confinement function is the most suitable one for semiconductor microcrystallites (Gonzalez-Hernandez et al., 1985). Therefore, $w(r,L)$ is chosen as

$$w(r,L) = \exp\left(\frac{-8\pi^2 r^2}{L^2}\right)$$

and $$|C(0,q)|^2 = \exp\left(\frac{-q^2 L^2}{16\pi}\right)$$

The size distribution profile obtained from the AFM micrographs is used together with Eq. (17.34) to yield

$$I(w) = \int \rho(L)\,dL \times \int \frac{4\pi q^2 \, |C(0,q)|^2}{[w - w(q)]^2 + \left[\frac{\Gamma_c}{2}\right]^2}\,dq \quad (17.35)$$

where $\rho(L)$ is the size distribution of the sample. The model Eq. (17.35) was used by Gonzalez-Hernandez et al. (1985) and Bottani et al. (1996).

The experimental Raman spectra in Figure 17.59 (upper solid curve) are now used to fit with the model Eq. (17.34) (lower dashed curve) and Eq. (17.35) (middle dotted curve) by adjusting the value of L (assuming spherical-shaped nanocrystals), and a best fit is achieved. The dashed line that is obtained for crystal with single size deviates much, while the dotted line generated using Gaussian distribution is closer to the experimental curve. Furthermore, the size variation in the nanocrystals is unable to produce measurable blue shift in the Raman spectra.

The estimated sizes of the Ge nanocrystallites obtained using Eqs. (17.34) and (17.35) are ~10 and 12 nm, respectively. The disagreement between the theoretical and experimental Raman peak frequencies may be due to stress exerted on the Ge nanocrystals.

17.10.4 Summary

High-density Ge nanoislands embedded in various SiO_2 sublayer thicknesses are prepared using rf magnetron sputtering. The formation of Ge nanoislands and their optical behaviors are examined using AFM, XRD, EDX, PL, and Raman spectroscopy. The AFM results for sample C are further used to determine the size distribution. The average size of nanoislands is found to be ~13 nm. The strong PL peak at 2.9 eV is attributed to the radiative recombination of electron and hole from the Ge nanoislands/SiO_2 interfaces. The experimental Raman spectra are fitted to two different models with single and Gaussian size distribution of nanoislands. The model based on size distribution rather fits well with the experimental observation than the model based on phonon confinement. This disagreement is perhaps due to the effect of compressive stress exerted by the SiO_2 network on the Ge nanoislands. We establish that embedding Ge on SiO_2 matrix has strong influence on the growth mechanism through interfacial and strain energy. Our easy sample preparation method and the detailed characterization may contribute towards the fabrication of Ge nanoislands having high density and yet economic.

17.11   Conclusion and Future Works

17.11.1   Conclusion

Nanostructuring germanium, optimizing the growth behavior, and understanding their optical and structural properties in a tunable manner have emerged as one of the challenging issues in optoelectronic applications (importance of the work). Ge nanoislands are grown on Si(100) substrate using radio frequency (rf) magnetron sputtering technique and characterized by various means. The substrate temperature-dependent surface morphology is examined in detail. Quantities such as RMS roughness, number density and the size of islands are found to be influenced strongly by the temperature. The observed very high density ($\sim 10^{11}$ cm^{-2}), narrow size distribution (~ 8 nm), and room-temperature visible PL are promising for their potential application in nanophotonic and optoelectronic devices. The observed intense PL peak and the associated blue shift ~ 0.11 eV with decreasing nanoislands size are attributed to the effect of quantum confinement and the surface passivation.

The influences of annealing temperature (700°C) on growth morphology, and structural and optical properties are also examined. Coarsening mechanism is the prominent reason for increasing the size. The average size of QDs of the post-annealed sample obtained from FESEM and AFM ~ 14 nm is in agreement with XRD data calculated using Scherrer's equation. Ge QDs ~ 7 nm in diameter and ~ 2 nm heights shows strong luminescence at 3.21 eV and a red shift (~ 0.05 eV) which is due to quantum size effect of the Ge QDs. The Raman shift (~ 3.27 cm^{-1}) and the broadening of the peak are attributed to the presence of different island morphology, height-over-base ratio, and phonon confinement of Ge QDs. The mechanism behind the occurrence of PL and Raman peaks and their corresponding shifts are understood.

Now we turn our attention to study the role of deposition time on growth characteristics, namely, size, shape, number density, RMS roughness, and ratio of grain area. Understanding the growth mechanism is the key issue. The mean size of pyramid-shaped nanoislands is found to increase from ~ 12 to ~ 30 nm with rising deposition time. The trend of height and width fluctuation with increasing deposition time is explained using the Stranski–Krastanov (S–K) growth mechanism following Ostwald ripening and self-organized criticality, achieving the minimally stable free energy configuration. PL study revealed that the Ge nanoislands exhibit a broad emission band originates from the electron and hole recombination in the nanoislands accompanied with a weak peak that is related to Ge and GeO_x interface reaction.

It is important to examine the effect of bilayering and Si space layer thickness on overall growth behavior. In this view, $[Ge/Si]_{\times 2}$ heterostructure with Ge quantum dots between ~ 12 and ~ 25 nm is optically characterized. The room-temperature luminescence exhibits size-dependent strong blue-violet broadened peak consists of a red shift as much as 0.05 eV for increasing thickness of Si space layer attributed to the effect of quantum confinement of photo-carriers. The Raman spectra for both the single layer Ge/Si and the bilayer $[Ge/Si]_{\times 2}$ reveal strong Ge-Ge optical phonon mode, shifted to the higher frequency. Furthermore, the first-order feature of Raman spectra from SiGe alloy affirms the presence of intermixing and phase formation at the interface during deposition. Our easy fabrication method may be relevant for the development of nanophotonics.

There is renewed interest to understand the fundamental mechanism of growth of Ge and Si nanostructure on insulator surface useful for the development of low-dimensional memory devices. Defects in tunneling layer and lateral charge loss play paramount role on growth morphology. Employing the semiconductor nanocrystals like Ge and Si embedded in the insulator as charge-storage media is the way of overcoming these problems. Furthermore, the influence of sub-layer thickness on surface evolution has important implication on devices. We achieve high-quality and pyramid-shaped Ge nanoislands embedded in amorphous SiO_2 sub-layer of various thicknesses by controlling the rf power, deposition time, and Ar flow. The estimated size is ~ 13 nm, and the density is $\sim 10^{11}$ cm^{-2}. The AFM micrograph shows the sensitivity of RMS roughness on thickness variation of SiO_2 sub-layer. The formation of nanoislands with high aspect ratio is attributed to the higher rate of surface reaction between Ge adatoms and the nucleated Ge islands than the reaction associated with SiO_2 and Ge. The poly-orientation of Ge nanoislands on SiO_2 (thickness 50 nm) is revealed in the XRD pattern. The observed room-temperature PL peaks at 2.9 and 1.65 eV are attributed to the radiative recombination of electron and hole from the Ge nanoislands/SiO_2 and SiO_2/Si interfaces, respectively. The mean islands sizes are determined by fitting the experimental Raman profile with two different models: the phonon confinement and the size distribution combined with phonon confinement. The latter model yields best fit to the experimental data. We assert that the thickness variation of SiO_2 matrix plays significant role in the growth morphology of Ge nanoislands mediated via the minimization of interfacial and strain energy.

The recurring team of the thesis was to grow Ge nanoislands using easy and economic technique called radio frequency magnetron sputtering. The main aim was to characterize this highly controlled high-density and well-dispread nanoislands for their surface, structural, and optical properties. Different system parameters are used for control growth. We are able to establish that it is possible to grow Ge nanoislands having high-quality, different shape, size, and well-defined morphology. The growth process is optimized, and their physics behind is understood. Our detailed investigation on surface morphology and growth evolution may provide valuable information for further development in nanophotonic.

17.11.2   Recommendation for Future Work

- To future work, one can use growing the Ge nanoislands directly on different substrate materials such as SiO_2 and glasses to compare the morphology and size variation of Ge nanoislands as a function of different substrate materials.

- Higher annealing temperature and time can be applied to observe the surface evolution and structural behavior of Ge nanoislands. It is illustrated that annealing time and temperature cause a change in the size and surface morphology of the samples.

- To grow the Ge/Si multilayer by rf magnetron sputtering and study the effect of number of periods on the surface morphology, quality, and light-emitting behavior of the samples, it is suggested that the number density and optical behavior are effected as going from Ge/Si single layer to bilayer.

Semiconductor QDs are emerging as a novel class of fluorescent probe with high photo-stability for biological and biomedical applications such as *in vivo* bio-molecular and cellular imaging. It is possible that the water-soluble Ge QDs sitting on substrate are removed by mega sonic cleaning process, rolling, sliding, and electrical lifting. Therefore, development of synthetic methods for producing Ge nanocrystals with controlled sizes and modified surfaces, and low toxicity is important for the biological and biomedical applications. Biology and medicine application of water-soluble Ge QDs can be investigated.

- Embedding Ge nanoislands with metal nanoparticles for surface-enhanced Raman and surface plasmon resonance measurement is worth looking.

References

Aniel, F., Enciso-Aguilar, M., Giguerre, L., Crozat, P., Adde, R., Mack, T., Raynor, B. (2003). High performance 100 nm T-gate strained $Si/Si_{0.6}Ge_{0.4}$ n-MODFET. *Solid-State Electronics*, 47(2), 283–289.

Baskaran, Arvind, Peter, S. (2012). Mechanisms of stranski-krastanov growth. *Journal of Applied Physics* 111(4), 044321–044327.

Becker, R. S., Golovchenko, J. A., Swartzentruber, B. S. (1985). Tunneling images of germanium surface reconstructions and phase boundaries. *Physical Review Letters*, 54(25), 2678–2680.

Becker, R. S., Swartzentruber, B. S., Vickers, J. S., Klitsner, T. (1989). Dimer–adatom–stacking-fault (DAS) and non-DAS (111) semiconductor surfaces: A comparison of Ge (111)-c (2×8) to Si (111)-(2×2),-(5×5),-(7×7), and-(9×9) with scanning tunneling microscopy. *Physical Review B*, 39(3), 1633.

Behammer, D., Albers, J. N., König, U., Temmler, D., Knoll, D. (1996). Si/SiGe HBTs for application in low power ICs. *Solid-State Electronics*, 39(4), 471–480.

Bukowski, T. J., Simmons, J. H. (2002). Quantum dot research: Current state and future prospects. *Critical Reviews in Solid State and Material Sciences*, 27(3–4), 119–142.

Bottani, C. E., Mantini, C., Milani, P., Manfredini, M., Stella, A., Tognini, P., Kofman, R. (1996). Raman, optical-absorption, and transmission electron microscopy study of size effects in germanium quantum dots. *Applied Physics Letters*, 69(16), 2409–2411.

Brunner, K. (2002). Si/ge nanostructures. *Reports on Progress in Physics*, 65(1), 27.

Brus, L. E. (1984). Electron–electron and electron-hole interactions in small semiconductor crystallites: The size dependence of the lowest excited electronic state. *The Journal of Chemical Physics*, 80, 4403.

Cappelletti, P. (1998). Flash memory reliability. *Microelectronics Reliability*, 38(2), 185–188.

Capellini, G., Di Gaspare, L., Evangelisti, F., Palange, E. (1997). Atomic force microscopy study of self-organized Ge islands grown on Si (100) by low pressure chemical vapor deposition. *Applied Physics Letters*, 70(4), 493–495.

Carles, R., Mlayah, A., Amjoud, M. B., Reynes, A., Morancho, R. (1992). Structural characterization of Ge microcrystals in Ge_xC_{1-x} films. *Japanese Journal of Applied Physics*, 31, 3511–3518.

Chambliss, D. D., Johnson, K. E. (1994). Nucleation with a critical cluster size of zero: Submonolayer Fe inclusions in Cu (100). *Physical Review B*, 50(7), 5012.

Chaparro, S. A., Zhang, Y., Drucker, J. (2000). Strain relief via trench formation in Ge/Si (100) islands. *Applied Physics Letters*, 76(24), 3534–3536.

Chen, L., Cheung, C. L., Ashby, P. D., Lieber, C. M. (2004). Single-walled carbon nanotube AFM probes: Optimal imaging resolution of nanoclusters and biomolecules in ambient and fluid environments. *Nano Letters*, 4(9), 1725–1731.

Chiu, H. W., Kauzlarich, S. M. (2006). Investigation of reaction conditions for optimal germanium nanoparticle production by a simple reduction route. *Chemistry of Materials*, 18(4), 1023–1028.

Costantini, G., Rastelli, A., Manzano, C., Acosta-Diaz, P., Katsaros, G., Songmuang, R., Kern, K. (2005). Pyramids and domes in the InAs/GaAs (001) and Ge/Si (001) systems. *Journal of Crystal Growth*, 278(1), 38–45.

Cullity, B. D., Stock, S. R. (2001). *Elements of X-Ray Diffraction* (Vol. 3). Upper Saddle River, NJ: Prentice Hall.

Daruka, I., Tersoff, J., Barabási, A. L. (1999). Shape transition in growth of strained islands. *Physical Review Letters*, 82(13), 2753–2756.

Das, K., Goswami, M. L. N., Dhar, A., Mathur, B. K., Ray, S. K. (2007). Growth of Ge islands and nanocrystals using RF magnetron sputtering and their characterization. *Nanotechnology*, 18(17), 175301.

Dashiell, M. W., Denker, U., Muller, C., Costantini, G., Manzano, C., Kern, K., Schmidt, O. G. (2002). Photoluminescence of ultrasmall Ge quantum dots grown by molecular-beam epitaxy at low temperatures. *Applied Physics Letters*, 80(7), 1279–1281.

Dehlinger, G., Diehl, L., Gennser, U., Sigg, H., Faist, J., Ensslin, K., Müller, E. (2000). Intersubband electroluminescence from silicon-based quantum cascade structures. *Science*, 290(5500), 2277–2280.

Dilliway, G. D. M., Bagnall, D. M., Cowern, N. E. B., Jeynes, C. (2003). Self-assembled germanium islands grown on (0 0 1) silicon substrates by low-pressure chemical vapor deposition. *Journal of Materials Science: Materials in Electronics*, 14(5–7), 323–327.

Eaglesham, D. J., Cerullo, M. (1990). Dislocation-free Stranski–Krastanow growth of Ge on Si (100). *Physical Review Letters*, 64(16), 1943–1946.

Fauchet, P. M., Campbell, I. H. (1988). Raman spectroscopy of low-dimensional semiconductors. *Critical Reviews in Solid State and Material Sciences*, 14(S1), S79–S101.

Fitzgerald, E. A. (1991). Dislocations in strained-layer epitaxy: Theory, experiment, and applications. *Materials Science Reports*, 7(3), 87–142.

Frank, F. C., van der Merwe, J. H. (1949). One-dimensional dislocations. I. Static theory. *Proceedings of the Royal Society of London. Series A. Mathematical and Physical Sciences*, 198(1053), 205–216.

Franzo, G., Priolo, F., Coffa, S., Polman, A., Carnera, A. (1994). Room-temperature electroluminescence from Er-doped crystalline Si. *Applied Physics Letters*, 64(17), 2235–2237.

Gallagher, M., Osterberg, U. (1993). Spectroscopy of defects in germanium-doped silica glass. *Journal of Applied Physics*, 74(4), 2771–2778.

Garoufalis, C. S. (2009). Optical gap and excitation energies of small Ge nanocrystals. *Journal of Mathematical Chemistry*, 46(3), 934–941.

Giesen, M. (2001). Step and island dynamics at solid/vacuum and solid/liquid interfaces. *Progress in Surface Science*, 68(1), 1–154.

Ginzburg, L. P., Gordeev, A. A., Gorchakov, A. P., Jilinsky, A. P. (1995). Some features of the blue luminescence in v-Si (1-x) Ge (x)O_2. *Journal of Non-Crystalline Solids*, 183(3), 234–242.

Giri, P. K., Dhara, S. (2012). Freestanding Ge/GeO_2 core-shell nanocrystals with varying sizes and shell thicknesses: Microstructure and photoluminescence studies. *Journal of Nanomaterials*, 2012, 6.

Gouadec, G., Colomban, P. (2007). Raman spectroscopy of nanomaterials: How spectra relate to disorder, particle size and mechanical properties. *Progress in Crystal Growth and Characterization of Materials*, 53(1), 1–56.

Gonzalez-Hernandez, J., Azarbayejani, G. H., Tsu, R., Pollak, F. H. (1985). Raman, transmission electron microscopy, and conductivity measurements in molecular beam deposited microcrystalline Si and Ge: A comparative study. *Applied Physics Letters*, 47(12), 1350–1352.

Gu, G., Burghard, M., Kim, G. T., Dusberg, G. S., Chiu, P. W., Krstic, V., Han, W. Q. (2001). Growth and electrical transport of germanium nanowires. *Journal of Applied Physics*, 90(11), 5747–5751.

Heng, C. L., Finstad, T. G. (2005). Electrical characteristics of a metal–insulator–semiconductor memory structure containing Ge nanocrystals. *Physica E: Low-Dimensional Systems and Nanostructures*, 26(1), 386–390.

Hohmura, K. I., Itokazu, Y., Yoshimura, S. H., Mizuguchi, G., Masamura, Y. S., Takeyasu, K., Nakayama, Y. (2000). Atomic force microscopy with carbon nanotube probe resolves the subunit organization of protein complexes. *Journal of Electron Microscopy*, 49(3), 415–421.

Horváth, Z. J. (2006). Semiconductor nanocrystals in dielectrics: Optoelectronic and memory applications of related silicon-based MIS devices. *Current Applied Physics*, 6(2), 145–148.

Huang, S., Xia, Z., Xiao, H., Zheng, J., Xie, Y., Xie, G. (2009). Structure and property of Ge/Si nanomultilayers prepared by magnetron sputtering. *Surface and Coatings Technology*, 204(5), 558–562.

Jie, Y. X., Wee, A. T. S., Huan, C. H. A., Sun, W. X., Shen, Z. X., Chua, S. J. (2004). Raman and photoluminescence properties of Ge nanocrystals in silicon oxide matrix. *Materials Science and Engineering: B*, 107(1), 8–13.

Jing, C., Hou, J., Xu, X. (2008). Fabrication and optical characteristics of thick GeO_2 sol–gel coatings. *Optical Materials*, 30(6), 857–864.

Jonathan, G. C. (2004). Preparation of alkyl-surface functionalized germanium quantum dots via thermally initiated hydrogermylation. *Chemical Communications*, (4), 386–387.

Joseph, H., Simmons, S., Kelly, S., Potter, S. (2000). Optical properties of semiconductors. In: *Optical Materials*, 191–263.

Kawaguchi, K., Morooka, M., Konishi, K., Koh, S., Shiraki, Y. (2002). Optical properties of strain-balanced SiGe planar microcavities with Ge dots on Si substrates. *Applied Physics Letters*, 81(5), 817–819.

Kamins, T. I., Medeiros-Ribeiro, G., Ohlberg, D. A. A., Williams, R. S. (1998). Dome-to-pyramid transition induced by alloying of Ge islands on Si (001). *Applied Physics A: Materials Science and Processing*, 67(6), 727–730.

Kaganer, V. M., Ploog, K. H. (2001). Energies of strained vicinal surfaces and strained islands. *Physical Review B*, 64(20), 205301.

Kartopu, G., Bayliss, S. C., Hummel, R. E., Ekinci, Y. (2004). Simultaneous micro-Raman and photoluminescence study of spark-processed germanium: Report on the origin of the orange photoluminescence emission band. *Journal of Applied Physics*, 95(7), 3466–3472.

Kartopu, G., Karavanskii, V. A., Serincan, U., Turan, R., Hummel, R. E., Ekinci, Y., Finstad, T. G. (2005). Can chemically etched germanium or germanium nanocrystals emit visible photoluminescence? *Physica Status Solidi (A)*, 202(8), 1472–1476.

Karunasiri, R. P. U., Bruinsma, R., Rudnick, J. (1989). Thin-film growth and the shadow instability. *Physical Review Letters*, 62(7), 788–791.

Kolobov, A. V., Shklyaev, A. A., Oyanagi, H., Fons, P., Yamasaki, S., Ichikawa, M. (2001). Local structure of Ge nanoislands on Si (111) surfaces with a SiO_2 coverage. *Applied Physics Letters*, 78(17), 2563–2565.

Kuo, Y. L., Chen, Y. S., Lee, C. (2011). Growth of 20 mol% Gd-doped ceria thin films by RF reactive sputtering: The O_2/Ar flow ratio effect. *Journal of the European Ceramic Society*, 31(16), 3127–3135.

Kwok, S. H., Yu, P. Y., Tung, C. H., Zhang, Y. H., Li, M. F., Peng, C. S., Zhou, J. M. (1999). Confinement and electron-phonon interactions of the E_1 exciton in self-organized Ge quantum dots. *Physical Review B*, 59(7), 4980.

Lai, W. T., Li, P. W. (2007). Growth kinetics and related physical/electrical properties of Ge quantum dots formed by thermal oxidation of $Si_{1-x}Ge_x$-on-insulator. *Nanotechnology*, 18(14), 145402.

Liu, J. L., Jin, G., Tang, Y. S., Luo, Y. H., Wang, K. L., Yu, D. P. (2000). Optical and acoustic phonon modes in self-organized Ge quantum dot superlattices. *Applied Physics Letters*, 76(5), 586–588.

Liu, J. L., Wan, J., Jiang, Z. M., Khitun, A., Wang, K. L., Yu, D. P. (2002). Optical phonons in self-assembled Ge quantum dot superlattices: Strain relaxation effects. *Journal of Applied Physics*, 92(11), 6804–6808.

Liu, F., Li, A. H., Lagally, M. G. (2001). Self-assembly of two-dimensional islands via strain-mediated coarsening. *Physical Review Letters*, 87(12), 126103.

Lynch, S. A., Paul, D. J., Townsend, P., Matmon, G., Suet, Z., Kelsall, R. W., Ni, W. X. (2006). Toward silicon-based lasers for terahertz sources. *IEEE Journal of Selected Topics in Quantum Electronics*, 12(6), 1570–1578.

MacLeod, J. M., Cojocaru, C. V., Ratto, F., Harnagea, C., Bernardi, A., Alonso, M. I., Rosei, F. (2012). Modified Stranski–Krastanov growth in Ge/Si heterostructures via nanostenciled pulsed laser deposition. *Nanotechnology*, 23(6), 065603.

Matsunami, N., Yamamura, Y., Itikawa, Y., Itoh, N., Kazumata, Y., Miyagawa, S., Tawara, H. (1984). Energy dependence of the ion-induced sputtering yields of monatomic solids. *Atomic Data and Nuclear Data Tables*, 31(1), 1–80.

McKay, M. R., Venables, J. A., Drucker, J. (2008). Kinetically suppressed Ostwald ripening of Ge/Si (100) hut clusters. *Physical Review Letters*, 101(21), 216104.

Medeiros-Ribeiro, G., Bratkovski, A. M., Kamins, T. I., Ohlberg, D. A., Williams, R. S. (1998a). Shape transition of germanium nanocrystals on a silicon (001) surface from pyramids to domes. *Science*, 279(5349), 353–355.

Medeiros-Ribeiro, G., Kamins, T. I., Ohlberg, D. A. A., Williams, R. S. (1998b). Annealing of Ge nanocrystals on Si (001) at 550°C: Metastability of huts and the stability of pyramids and domes. *Physical Review B*, 58(7), 3533.

Metzger, R. A. (1995). Is silicon germanium the new "material of the future". *Compound Semiconductor*, 1(3), 21–28.

Mestanza, S. N. M., Doi, I., Swart, J. W., Frateschi, N. C. (2007). Fabrication and characterization of Ge nanocrystalline growth by ion implantation in SiO_2 matrix. *Journal of Materials Science*, 42(18), 7757–7761.

Mestanza, S. N. M., Rodriguez, E., Frateschi, N. C. (2006). The effect of Ge implantation dose on the optical properties of Ge nanocrystals in SiO_2. *Nanotechnology*, 17(18), 4548.

Mizuno, T., Sugiyama, N., Tezuka, T., Moriyama, Y., Nakaharai, S., Maeda, T., Takagi, S. I. (2005). High-speed source-heterojunction-MOS-transistor (SHOT) utilizing high-velocity electron injection. *IEEE Transactions on Electron Devices*, 52(12), 2690–2696.

Miura, M., Hartmann, J. M., Zhang, J., Joyce, B., Shiraki, Y. (2000). Formation process and ordering of self-assembled Ge islands. *Thin Solid Films*, 369(1), 104–107.

Mo, Y. W., Savage, D. E., Swartzentruber, B. S., Lagally, M. G. (1990). Kinetic pathway in Stranski–Krastanov growth of Ge on Si (001). *Physical Review Letters*, 65(8), 1020–1023.

Montalenti, F., Raiteri, P., Migas, D. B., Von Känel, H., Rastelli, A., Manzano, C., Miglio, L. (2004). Atomic-scale pathway of the pyramid-to-dome transition during Ge growth on Si (001). *Physical Review Letters*, 93(21), 216102.

Mooney, P. M., Dacol, F. H., Tsang, J. C., Chu, J. O. (1993). Raman scattering analysis of relaxed GexSi1-x alloy layers. *Applied Physics Letters*, 62(17), 2069–2071.

Mosleh, N., Meyer, F., Schwebel, C., Pellet, C., Eizenberg, M. (1994). Growth mode of Ge films on Si (100) substrate deposited by ion beam sputtering. *Thin Solid Films*, 246(1), 30–34.

Nakayama, Y., Matsuda, I., Hasegawa, S., Ichikawa, M. (2008). Growth, quantum confinement and transport mechanisms of Ge nanodot arrays formed on a SiO_2 monolayer. *e-Journal of Surface Science and Nanotechnology*, 6, 191–201.

Nayak, D. K., Usami, N., Fukatsu, S., Shiraki, Y. (1993). Band-edge photoluminescence of SiGe/strained-Si/SiGe type-II quantum wells on Si (100). *Applied Physics Letters*, 63(25), 3509–3511.

Nikiforov, A. I., Ulyanov, V. V., Pchelyakov, O. P., Teys, S. A., Gutakovsky, A. K. (2005). Germanium nanoislands formation on silicon oxide surface by molecular beam epitaxy. *Materials Science in Semiconductor Processing*, 8(1), 47–50.

Nishikawa, H., Stathis, J. H., Cartier, E. (1999). Defects in thermal oxide studied by photoluminescence spectroscopy. *Applied Physics Letters*, 75(9), 1219–1221.

Ostwald, W. (1900). Über die vermeintliche Isomerie des roten und gelben Quecksilberoxyds und die Oberflächenspannung fester Körper. *Zeitschrift für Physikalische Chemie*, 34, 495–503.

Palmero, A., Rudolph, H., Habraken, F. H. P. M. (2007). One-dimensional analysis of the rate of plasma-assisted sputter deposition. *Journal of Applied Physics*, 101(8), 083307–83307.

Pankove, J. I. (1971). *Optical Processes in Semiconductors*. New York: Courier Dover Publications.

Paul, D. J. (1998). Silicon germanium heterostructures in electronics: The present and the future. *Thin Solid Films*, 321(1), 172–180.

Pavesi, L., Ceschini, M. (1993). Stretched-exponential decay of the luminescence in porous silicon. *Physical Review B*, 48, 95781–17628.

Pavesi, L., Dal Negro, L., Mazzoleni, C., Franzo, G., Priolo, F. (2000). Optical gain in silicon nanocrystals. *Nature*, 408(6811), 440–444.

People, R., Bean, J. C. (1985). Calculation of critical layer thickness versus lattice mismatch for GeSi/Si strained-layer heterostructures. *Applied Physics Letters*, 47, 322–324.

Pietralunga, S.M., Feré, M., Lanata, M., Piccinin, D., Radnóczi, G., Misják, F., Lamperti, M., Martineli, M., Ossi, P. M. (2009). Heteroepitaxial sputtered Ge on Si (100): Nanostructure and interface morphology. *EPL (Europhysics Letters)*, 88(2), 28005.

Priester, C., Lannoo, M. (1995). Origin of self-assembled quantum dots in highly mismatched heteroepitaxy. *Physical Review Letters*, 75(1), 93–96.

Priester, C., Lannoo, M. (1997). Growth aspects of quantum dots. *Current Opinion in Solid State and Materials Science*, 2(6), 716–721.

Radić, N., Pivac, B., Dubček, P., Kovačević, I., Bernstorff, S. (2006). Growth of Ge islands on Si substrates. *Thin Solid Films*, 515(2), 752–755.

Ray, S. K., Das, K. (2005). Luminescence characteristics of Ge nanocrystals embedded in SiO_2 matrix. *Optical Materials*, 27(5), 948–952.

Ray, S. K., Das, S., Singha, R. K., Manna, S., Dhar, A. (2011). Structural and optical properties of germanium nanostructures on Si (100) and embedded in high-k oxides. *Nanoscale Research Letters*, 6(1), 1–10.

Riabinina, D., Durand, C., Chaker, M., Rowell, N., Rosei, F. (2006a). A novel approach to the synthesis of photoluminescent germanium nanoparticles by reactive laser ablation. *Nanotechnology*, 17(9), 2152.

Riabinina, D., Rosei, F., Chaker, M. (2006b). Structural properties of Ge nanostructured films synthesized by laser ablation. *Journal of Experimental Nanoscience*, 1(1), 83–89.

Richter, H., Wang, Z. P., Ley, L. (1981). The one phonon Raman spectrum in microcrystalline silicon. *Solid State Communications*, 39(5), 625–629.

Ross, F. M., Tersoff, J., Tromp, R. M. (1998). Coarsening of self-assembled Ge quantum dots on Si (001). *Physical Review Letters*, 80(5), 984–987.

Ross, F. M., Tromp, R. M., Reuter, M. C. (1999). Transition states between pyramids and domes during Ge/Si island growth. *Science*, 286(5446), 1931–1934.

Rodrigues, A. D., Chiquito, A. J., Zanelatto, G., Milekhin, A. G., Nikiforov, A. I., Ulyanov, V. V., Galzerani, J. C. (2012). Ge/Si quantum dots superlattices grown at different temperatures and characterized by Raman spectroscopy and capacitance measurements. *Advances in Condensed Matter Physics*, 2012, 1–7.

Ronda, A., Abdallah, M., Gay, J. M., Stettner, J., Berbezier, I. (2000). Kinetic evolution of self-organised SiGe nanostructures. *Applied Surface Science*, 162, 576–583.

Samavati, A, Ghoshal, S. K., Othaman, Z. (2012a). Growth of Ge/Si (100) nanostructures by radio-frequency magnetron sputtering: The role of annealing temperature. *Chinese Physics Letters*, 29(4), 048101.

Samavati, A., Othaman, Z., Ghoshal, S. K., Dousti, M. R., Kadir, M. R. A. (2012b). Substrate temperature dependent surface morphology and photoluminescence of germanium quantum dots grown by radio frequency magnetron sputtering. *International Journal of Molecular Sciences*, 13(10), 12880–12889.

Schmidt, O. G., Eberl, K., Rau, Y. (2000). Strain and band-edge alignment in single and multiple layers of self-assembled Ge/Si and GeSi/Si islands. *Physical Review B*, 62(24), 16715.

Schmidt, O. G., Kienzle, O., Hao, Y., Eberl, K., Ernst, F. (1999). Modified Stranski–Krastanov growth in stacked layers of self-assembled islands. *Applied Physics Letters*, 74(9), 1272–1274.

Schittenhelm, P., Gail, M., Brunner, J., Nutzel, J. F., Abstreiter, G. (1995). Photoluminescence study of the crossover from two-dimensional to three-dimensional growth for Ge on Si (100). *Applied Physics Letters*, 67(9), 1292–1294.

Seo, M. A., Kim, D. S., Kim, H. S., Choi, D. S., Jeoung, S. C. (2006). Formation of photoluminescent germanium nanostructures by femtosecond laser processing on bulk germanium: Role of ambient gases. *Optics Express*, 14(11), 4908–4914.

Shchukin, V. A., Ledentsov, N. N., Hoffmann, A., Bimberg, D., Soshnikov, I. P., Volovik, B. V., Gerthsen, D. (2001). Entropy-driven effects in self-organized formation of quantum dots. *Physica Status Solidi (B)*, 224(2), 503–508.

Shchukin, V. A., Ledentsov, N. N., Kop'ev, P. S., Bimberg, D. (1995). Spontaneous ordering of arrays of coherent strained islands. *Physical Review Letters*, 75(16), 2968–2971.

Shklyaev, A. A., Ichikawa, M. (2002). Effect of interfaces on quantum confinement in Ge dots grown on Si surfaces with a SiO_2 coverage. *Surface Science*, 514(1), 19–26.

Shklyaev, A. A., Shibata, M., Ichikawa, M. (2000). High-density ultrasmall epitaxial Ge islands on Si (111) surfaces with a SiO_2 coverage. *Physical Review B*, 62(3), 1540.

Shiramine, K. I., Muto, S., Shibayama, T., Sakaguchi, N., Ichinose, H., Kozaki, T., Taniwaki, M. (2007). Tip artifact in atomic force microscopy observations of InAs quantum dots grown in Stranski–Krastanow mode. *Journal of Applied Physics*, 101(3), 033527.

Simonsen, A. C., Schleberger, M., Tougaard, S., Hansen, J. L., Larsen, A. N. (1999). Nanostructure of Ge deposited on Si (001): A study by XPS peak shape analysis and AFM. *Thin Solid Films*, 338(1), 165–171.

Singha, R. K., Das, S., Majumdar, S., Das, K., Dhar, A., Ray, S. K. (2008). Evolution of strain and composition of Ge islands on Si (001) grown by molecular beam epitaxy during postgrowth annealing. *Journal of Applied Physics*, 103(11), 114301.

Sorianello, V., Colace, L., Armani, N., Rossi, F., Ferrari, C., Lazzarini, L., Assanto, G. (2011). Low-temperature germanium thin films on silicon. *Optical Materials Express*, 1(5), 856–865.

Spanhel, L. (2006). Colloidal ZnO nanostructures and functional coatings: A survey. *Journal of Sol-Gel Science and Technology*, 39(1), 7–24.

Spanhel, L., Anderson, M. A. (1991). Semiconductor clusters in the sol-gel process: Quantized aggregation, gelation, and crystal growth in concentrated zinc oxide colloids. *Journal of the American Chemical Society*, 113(8), 2826–2833.

Suda, Y., Koyama, H. (2001). Electron resonant tunneling with a high peak-to-valley ratio at room temperature in SiGe/Si triple barrier diodes. *Applied Physics Letters*, 79, 2273–2275.

Sun, K. W., Sue, S. H., Liu, C. W. (2005). Visible photoluminescence from Ge quantum dots. *Physica E: Low-Dimensional Systems and Nanostructures*, 28(4), 525–530.

Takagahara, T., Takeda, K. (1992). Theory of the quantum confinement effect on excitons in quantum dots of indirect-gap materials. *Physical Review B*, 46(23), 15578.

Takeoka, S., Fujii, M., Hayashi, S., Yamamoto, K. (1998). Size-dependent near-infrared photoluminescence from Ge nanocrystals embedded in SiO_2 matrices. *Physical Review B*, 58(12), 7921.

Tang, Z., Wang, W., Wang, D., Liu, D., Liu, Q., Yin, M., He, D. (2010). The effect of Ar flow rate in the growth of SiGe: H thin films by PECVD. *Applied Surface Science*, 256(23), 7032–7036.

Teichert, C. (2002). Self-organization of nanostructures in semiconductor heteroepitaxy. *Physics Reports*, 365(5), 335–432.

Tersoff, J., Teichert, C., Lagally, M. G. (1996). Self-organization in growth of quantum dot superlattices. *Physical Review Letters*, 76(10), 1675–1678.

Thanh, V. L., Yam, V., Zheng, Y., Bouchier, D. (2000). Nucleation and growth of self-assembled Ge/Si (001) quantum dots in single and stacked layers. *Thin Solid Films*, 380(1), 2–9.

Tiong, K. K., Amirtharaj, P. M., Pollak, F. H., Aspnes, D. E. (1984). Effects of As^+ ion implantation on the

Raman spectra of GaAs: Spatial correlation interpretation. *Applied Physics Letters*, 44(1), 122–124.

Tong, S., Liu, F., Khitun, A., Wang, K. L., Liu, J. L. (2004). Tunable normal incidence Ge quantum dot midinfrared detectors. *Journal of Applied Physics*, 96(1), 773–776.

Uemura, H., Saito, Y., Uwaha, M. (2001). Elastic interaction in a two-dimensional discrete lattice model. *Journal of the Physical Society of Japan*, 70(3), 743–752.

van Hattum, E. D., Palmero, A., Arnoldbik, W. M., Rudolph, H., Habraken, F. H. P. M. (2007). On the ion and neutral atom bombardment of the growth surface in magnetron plasma sputter deposition. *Applied Physics Letters*, 91(17), 171501.

Volmer, M., Weber, A. (1926). Keimbildung in übersättigten Gebilden. *Zeitschrift für Physikalische Chemie*, 119, 277–301.

Voigtländer, B. (2001). Fundamental processes in Si/Si and Ge/Si epitaxy studied by scanning tunneling microscopy during growth. *Surface Science Reports*, 43(5), 127–254.

Warner, J. H., Tilley, R. D. (2006). Synthesis of water-soluble photoluminescent germanium nanocrystals. *Nanotechnology*, 17(15), 3745.

Wan, Q., Wang, T. H., Liu, W. L., Lin, C. L. (2003). Ultra-high-density Ge quantum dots on insulator prepared by high-vacuum electron-beam evaporation. *Journal of Crystal Growth*, 249(1), 23–27.

Wang, K. F., Zhang, Y., Zhang, W. (2012a). Morphology and photoluminescence of ultrasmall size of Ge quantum dots directly grown on Si (001) substrate. *Applied Surface Science*, 258(6), 1935–1939.

Wang, Y., Peng, X., Shi, J., Tang, X., Jiang, J., Liu, W. (2012b). Highly selective fluorescent chemosensor for Zn^{2+} derived from inorganic-organic hybrid magnetic core/shell Fe_3O_4/SiO_2 nanoparticles. *Nanoscale Research Letters*, 7(1), 1–13.

Wellner, A., Paillard, V., Bonafos, C., Coffin, H., Claverie, A., Schmidt, B., Heinig, K. H. (2003). Stress measurements of germanium nanocrystals embedded in silicon oxide. *Journal of Applied Physics*, 94(9), 5639–5642.

Williams, R. S., Medeiros-Ribeiro, G., Kamins, T. I., Ohlberg, D. A. (2000). Thermodynamics of the size and shape of nanocrystals: Epitaxial Ge on Si (001). *Annual Review of Physical Chemistry*, 51(1), 527–551.

Xiong, Q., Gupta, R., Adu, K. W., Dickey, E. C., Lian, G. D., Tham, D., Eklunda, P. C. (2003). Raman spectroscopy and structure of crystalline gallium phosphide nanowires. *Journal of Nanoscience and Nanotechnology*, 3(4), 335–339.

Xu, S., Arnsdorf, M. F. (1994). Calibration of the scanning (atomic) force microscope with gold particles. *Journal of Microscopy*, 173(3), 199–210.

Xue, F., Qin, J., Cui, J., Fan, Y. L., Jiang, Z. M., Yang, X. J. (2005). Studying the lateral composition in Ge quantum dots on Si (001) by conductive atomic force microscopy. *Surface Science*, 592(1), 65–71.

Yakimov, A.I, Dvurechenskii, A.V., Proskuryakov Yu Yu., Nikiforov, A.I., Pchelyakov, O.P., Teys, S.A., Gutakovskii, A.K. (1999) *Applied Physics Letters*, 75 1413.

Yam, V., Le Thanh, V., Compagnon, U., Gennser, U., Boucaud, P., Débarre, D., Bouchier, D. (2000). Effect of the bimodal size distribution on the optical properties of self-assembled Ge/Si (001) quantum dots. *Thin Solid Films*, 380(1), 78–81.

Yang, X. Y., Cheng, J. Y., Li, B., Cao, W. Q., Yuan, J., Zhang, D. Q., Cao, M. S. (2012). Micro-nanometer parasitic crystal growth and photoluminescence property of unique screw-cone like Zn_2GeO_4-ZnO by combustion oxidization. *Chinese Physics Letters*, 29(10), 108101.

Yang, Z., Shi, Y., Liu, J., Yan, B., Zhang, R., Zheng, Y., Wang, K. (2004). Optical properties of Ge/Si quantum dot superlattices. *Materials Letters*, 58(29), 3765–3768.

Yi, J. B., Li, X. P., Ding, J., Seet, H. L. (2007). Study of the grain size, particle size and roughness of substrate in relation to the magnetic properties of electroplated permalloy. *Journal of Alloys and Compounds*, 428(1), 230–236.

Zacharias, M., Fauchet, P. M. (1997). Blue luminescence in films containing Ge and GeO_2 nanocrystals: The role of defects. *Applied Physics Letters*, 71(3), 380–382.

Zhang, J. Y., Bao, X. M., Ye, Y. H., Tan, X. L. (1998). Blue and red photoluminescence from Ge implanted SiO films and its multiple mechanism. *Applied Physics Letters*, 73, 1790.

Zhang, W., Li, Y., Zhu, S., Wang, F. (2004). Influence of argon flow rate on TiO_2 photocatalyst film deposited by dc reactive magnetron sputtering. *Surface and Coatings Technology*, 182(2), 192–198.

Zhao, Y. N., Cao, M. S., Jin, H. B., Zhang, L., Qiu, C. J. (2006). Catalyst-free synthesis, growth mechanism and optical properties of multipod ZnO with nanonail-like legs. *Scripta Materialia*, 54(12), 2057–2061.

Zinke-Allmang, M. (1999). Phase separation on solid surfaces: Nucleation, coarsening and coalescence kinetics. *Thin Solid Films*, 346(1), 1–68.

Zinke-Allmang, M., Feldman, L. C., Grabow, M. H. (1992). Clustering on surfaces. *Surface Science Reports*, 16(8), 377–463.

18

Active Scanning Probes in Nanostructure Fabrication

Ivo W. Rangelow and
Matias Holz
Ilmenau University of Technology

Further development of nanotechnology and nanoelectronics is determined not only by the ability to create nanometer-scale features but also by the accuracy of the positioning and alignment to each other. Developments in nanoscience, nanotechnology, and nanoelectronics involve instruments that enable to create and observe objects at the atomic scale. Scanning probe microscopes are capable to image, characterize, measure, and manipulate at atomic scale being an irreplaceable tool in a wide variety of nanofabrication, positioning, and characterization methods. The sharp tip of a micro-mechanical beam, so-called cantilever, is used to scan the surface line by line. The deflection of the cantilever is used to accurately measure the surface topology of the sample. The use of atomically sharp tips enables an ultimately localized interaction at atomic resolution. Often, this technology is called tip-based nanofabrication. With respect to the interaction mechanism, tip-based nanofabrication can be divided into two groups: (i) mechanical or thermomechanical interaction and (ii) voltage-driven additive or ablative physical–chemical interaction. Tip-based nanofabrication can generate features in the so-called "bottom-up" regime. In this case, atoms or molecules are assembled. Tip-based nanofabrication works also in the "top-down" regime, in which features are created into a bulk material.

This chapter provides a description on the development of atomic force microscopy (AFM) nanostructuring and fabrication at the nanometer scale using self-sensing and self-actuation cantilevers (so-called active cantilevers). Attention is paid to the scanning probe techniques and processes that offer high-resolution lithography capabilities for the fabrication of "beyond CMOS" semiconductor devices.

This chapter also reviews the so-called active probes, which have an integrated bending-sensor and actuator. Despite the fact that the active probes are known for more than 30 years, today, the progress in micromachining makes them competitive with regard to the standard probes. The simple operation allows an easy usability in materials science, live science, nanotechnology, and nanolithography.

Scanning probe imaging or lithography is a relatively slow serial process, and despite massive efforts to make it faster, the throughput remains still limited. For this reason, wide industrial applications have hindered its use. An alternative solution is massive parallelization of many active probes working parallel. Such parallel arrays of active cantilevers are now materialized. This increases the productivity and turns out to be a highly attractive technology for many industrial applications. Consequently, active probes become the ultimate solution for next-generation parallel scanning probes.

In addition, scanning-probe-based nanofabrication can be employed in the fabrication of precise templates for the nanoimprint lithography (NIL) and thus open horizons for cost-effective fabrication of many electronical or optical devices.

18.1 Introduction

The finding of the Atomic Force Microscope by Binnig [1] and its realization by Binnig et al. [2] opened up the perspective of surfaces imaging with atomic resolution. Dynamic modes of AFM offer new possibilities for imaging, because of the availability of the various vibrational modes of the probes. In this case, the cantilever is oscillating at fixed amplitude.

Interactions of the tip (attractive or repulsive) that influence as it is moved over a sample cause the cantilever to bend, thereby providing a mechanical means to probe local nanoscale effects. The resonance frequency shift in the oscillation amplitude is used as a feedback control to measure the surface topography. The speediness of the AFM imaging process is the speed at which the active-cantilever tip is scanning over the sample surface. Scanning speed can significantly influence an AFM image quality. High speed can lead to instabilities and artifacts as well as tip/sample damage. Higher AFM image resolution is typically obtained at low scan speeds since the tip has more time to track the surface. Usually, almost all so-called today high-speed AFMs are applied over extremely small scanning areas and demonstrated on very flat sample surfaces at relatively limited resolution. The employment of cantilevers with integrated actuator makes the scanning process possible at higher speeds and the image tasks contents with its adaptability, speediness, and environmental measurement flexibility. More than 30 years, AFM history shows that at the beginning, the AFM was mostly used just for topological controls and metrology. Particularly in the characterization of semiconducting devices, diverse techniques that allow comprehensive microscopic failure investigations become obligatory. In device development, more general methods are usually involved in topographical measurements bat also for the study of mechanical, electrical, thermal, and optical properties. Nowadays, the requirements are endlessly increasing not only for imaging but also for fabrication at the nanoscale.

Cost-effective "on-demand" fabrication of features smaller than 5 nm opens up new possibilities for the realization of many nanoelectronic and optical devices. Driven by the active cantilever technology, a tool for scanning probe lithography (SPL) which is able to image, inspect, align, and pattern features down to the single-digit nano-regime has been developed.

Today, high-resolution lithographic techniques are typically linked with extremely high capital investment in equipment. As a result, fast prototyping of novel devices becomes difficult. The Gaussian electron beam lithography (EBL) is tending towards the application of even higher energetic beams in order to minimize the interaction volume in resist.

Lithography techniques based on high-energy particles are offering high resolution and high throughput. However, using ultraviolet (UV) [3,4] or EBL [5–8] exposure of materials like exfoliated graphene or few-layer MoS_2 and WSe_2 sheets causes significant damages. Since 1982 [9], after the development of the scanning tunnel microscope (STM), SPL methods [10,11] enabled the development of completely new electronic devices. Moreover, in a very short period of time, after demonstration of the atomic force microscope [12], several tip-based lithographic techniques have been demonstrated [13] and an investigation by many authors aimed at a better understanding of the tip-based lithographic interactions at the nanoscale. Exploring different interactions between a sharp tip and a surface SPL can be arranged in couple techniques, all of which tip/sample interactions can result in a highly confined atomic manipulation [14]. Several lithographic tip-based patterning methods have been developed using the following:

1. Force-induced interactions such as plowing [15,16].
2. Nanografting [17,18].
3. Removing atomic hydrogen from the silicon surface using STM in ultra-high vacuum is getting atomic resolution patterning [19–22].
4. Heat-induced thermomechanical interactions [23,26] and even thermomechanical nanolithography capable of fabricating three-dimensional (3D) nanostructures by using self-amplified depolymerization polymers [24,25].
5. Employing thermo-chemical interactions, a high-resolution lithographic technique at high speed, have been shown [13,28–30].
6. The electrochemical modification [31–34] can be used for deposition due to field-induced or current-induced processes.
7. Electric-field-induced field emission (FE) of low-energy electron beam can be used for electron exposure of resist or chemical modification leading to patterning at the nanoscale as a top-down method. Tip-based localized electron beam exposes of electron-sensitive resist, like in standard EBL, have been accomplished in a tabletop system first demonstrated by K. Wilder et al. Nano-patterning occurs when low-energy field-emitted electrons from the tip are locally inducing radiation chemistry into thin film of resist [35–37].

18.2   Electron Field Emission

Emission of electrons induced by an electrostatic field is called FE. FE is dependent on the work function of tip and sample and on the electric field between them and therefore depends only indirectly on the tip-sample distance. The generation of high electric fields is assisted by tips having a small radius of curvature which locally amplifies the external electric field (due to a high density of equipotential surfaces). This field enhancement is described by the ratio between local electric field in the vicinity of the tip and the electric field in a parallel plate configuration [38]. To describe the Fowler–Nordheim [39] emission process as results of the highly localized electric field at the tip, a simulation model based on a two-dimensional (2D) calculation method which combines the computation of the electron trajectories and the electron density of the emitted electrons has been developed [40]. The tip-emitted electrons gain energy from the electric field and are accelerated towards the sample. The following assumptions are used in Lenk's model: Influence of the resist film on the sample surface and additional contaminations, which might occur at ambient conditions and might affect the electric field and current density distribution, is not considered. The electronic energy distribution inside the nano-tip and assumes that the initial electrons have $E_{kin} = 0$ eV and are emitted perpendicular to the tip surface can be neglected. For a total emission current

$I = 30$ pA, approximately 150,000 electrons are emitted in a time of $t = 1$ ms; that is, each electron has around 7 ns before the next electron is emitted. Since the tip-sample distance is in the range of tens of nanometer, the time of flight for an electron with $E_{\text{kin}} = 10$ eV would be below 100 fs. On the other hand, the mean free path in air (ambient conditions) should be well above 100 nm. Therefore, we assume a ballistic transport of electrons between tip and sample. Since the de Broglie wavelength $\lambda = h/p = h/2mE$ [41] of the electrons even for 10 eV is below 0.5 nm, we assume that the electrons behave as classical corpuscles. The 2D calculation model is depicted in Figure 18.3. The tip is characterized by the tip radius 8.5 tip $r =$ nm, the length of the tip (conus) 50 tip $l =$ nm, the opening angle $\gamma = 20°$, and the work function $E = -\nabla\varphi$ for the tip material tungsten. The tip-sample distance d is varied from 10 to 100 nm and the bias voltage $V = 10 \ldots 100$ V, which is applied at the sample. The computation of the electric field $E = -\nabla\varphi$ is based on a 2D solution of Laplace's equation $\nabla\varphi = 0$ for the electrostatic potential φ [42]. This is done by using the partial differential equation (PDE) toolbox of MATLAB®. The boundary conditions are adapted to the model and the applied bias voltage.

The Fowler–Nordheim emission current density is based on the following equation [43]:

$$J = \frac{AE^2}{\varphi} \exp\left(\frac{B}{\varphi^{1/2}} - \frac{C\varphi^{3/2}}{E}\right)$$

Here, A, B, and C are constants taken from [43]. The trajectories of the electrons are computed by the help of a Velocity-Verlet algorithm [44] by using the calculated 2D electric field strength.

In Figure 18.1, the electric field strength distributions (as color-code) for applied bias voltages of 50 V are shown. The tip-sample distance was held constantly at $d = 10$ nm. An identical scale bar for the electric field strength was used for both voltage values. An increase of the electric field by increasing bias voltage is obvious. The enormous field strengths, which appear at the tip apex, are in a range of a few volt per nanometer due to the lightning rod effect.

As expected, a major part of the electrons is emitted closely to the tip apex, where at the smallest distance, the strongest electric field found to be present. Figure 18.1 shows that the electron distribution exceeds the diameter of the nano-tip even though, in principle, no electrons were emitted at the side of the tip. This can be explained by the direction of the emitted electrons. Since the electrons are accelerated in the electric field and the electric field lines are also perpendicular to the tip surface, an initial broadening of the electron distribution is observed. One main difference between the field lines and the trajectories is that the field lines enter the sample perpendicular as well, but the electron trajectories do not. This behavior has its origin in the inertial mass of the electrons. The 2D computational method explains important physical properties related to the FE and the utilization of the emitted electrons for the patterning. The pattern properties are based only on the electron emission and trajectories. It has been showed that the lightning rod effect and the enormous field strengths occur at the nano-tip [40]. The electric field strength exhibits a nonlinear dependency on the tip-sample distance. Here, not only the electron emission is triggered by these high electric fields but also their trajectories are strongly affected. The computed total current clearly shows a stronger nonlinear dependency on the tip-sample distance compared to the electric field strength. This causes difficulties in finding the optimal parameters for the highest resolution patterning, and it clearly proves that the current density depends exponentially on the electric field. The resolution capabilities were studied by assuming a current density threshold value for the lithographic process. In summary, we are able to forecast some patterning properties using our 2D model. The 3D natures of the tip as well as the intrinsic properties of the electrons in the tip are not considered. Additionally, the resist influences the electron distribution on the sample as well. This should also be taken into account for a complete lithography simulation.

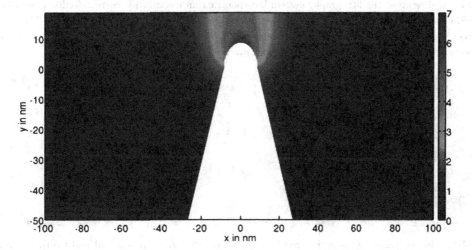

FIGURE 18.1   Calculated electric field strength for a tip-sample distance $d = 10$ nm and an applied bias voltage of 50 V [40].

18.3   Scanning Probe Lithography with Self-Sensing and Self-Actuated Cantilever

In 2010 have been demonstrated a development-less patterning scheme by using electric field, current-controlled SPL on molecular glass resist calixarene [45]. A high non-uniform electric field, which causes a current of low-energy electrons between the tip and sample, is applied for direct patterning of an electron-sensitive resist. These low-energy electrons (<50 eV), field emitted from the scanning probe tip apex, are regulated by the current feedback loop of the SPL system by respective modulations of the tip-sample spacing (Figure 18.2). Depending on the applied exposure conditions such as exposure dose and applied bias voltage, the resist is cross-linked (negative tone patterning) similar to EBL [45–50] or to trigger a direct removal (ablation) reaction (positive-tone patterning) at higher exposure doses. In contrast to EBL, in which high-energy electrons are generating secondary electrons that trigger the lithographic reaction, in FE-SPL is possible to generate directly "lithographically usable" electrons in the secondary electron energy range to carry out the lithographic reaction. Thus, the lithographic processes are more efficient, and practically no proximity effects, which are nominally caused by primary electron scattering and secondary electron generation, are present. By using the direct removal process, no development steps featuring higher resolution capabilities are required. In both lithographic tones, AFM imaging directly after exposure using the same tool and cantilever allows the characterization of the generated features enabling a closed-loop lithography scheme. Moreover, sub-5-nm pattern capability of line and dot features [49], throughput enhancement by combining electric field, current-controlled SPL with electron/EUV lithography, as well as the practical application for pattern transfer by plasma etching have been shown [48].

Based on the thermally actuated, piezoresistive cantilever technology, a compact table-top SPL platform for imaging, alignment, and patterning of features down to a single-digit nano-regime has been developed [48]. We discuss our recent advances with respect to (i) step-and-repeat SPL, (ii) active cantilever for electric-field, current-controlled SPL, and (iii) lithographic demonstrations showing sequential read–write cycle patterning combining positive and negative tone writing, patterning over large areas and over existing topography as well as writing of diverse typical features such as line and dots. The general setup of the applied electric field, current-controlled SPL is shown in Figure 18.3. The application of such a kind of active cantilever allows easier

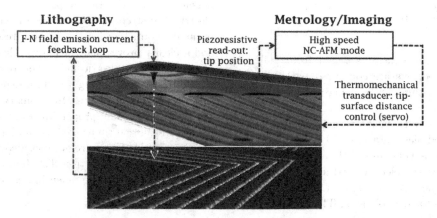

FIGURE 18.2   Principal setup of the lithography system incorporating an electron FE current feedback loop for SPL and a force feedback loop for AFM imaging. Fast switching between either mode (imaging and lithography) is possible. Thus, the same nanoprobe is used for both direct writing of nano-features using a low-energy FN-field-electron emission from the tip of an active cantilever and AFM imaging for pre- and post-inspection, as well as for pattern overlay alignment [49].

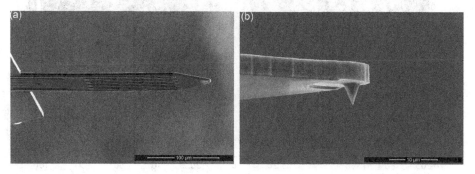

FIGURE 18.3   SEM side-view image of the active cantilever probe. (a) Piezoresistive read-out and (b) thermomechanical actuation are integrated. The silicon tip at the end of the cantilever is 5.6 μm high and has 5 nm tip radius [50].

system integration, higher processing speeds, and higher flexibility compared with conventional passive cantilever approaches.

Self-sensing, self-actuating probes have the potential to overcome the cantilever down seizing limitations optical read-out. Typically, a piezoresistivity-based deflection sensor is integrated onto the cantilever beam itself. The so-called active scanning probe technology integrates both bending read-out as well as actuation capability. In terms of the technological development of AFM systems, the cantilever evolved from a simple passive deflection element to a complex microelectromechanical system through integration of functional groups, such as piezoresistive detection sensors and thermomechanical actuators. Thereby, the deflection sensor is carefully thermally isolated from the actuator and designed for minimum electronic crosstalk. Active cantilevers allow the measurement of surface topology and related surface characteristics, whereby the integrated actuation enables the control of the tip position. Here, the system is even capable of measuring the thermomechanical noise of the cantilever. The progress of active cantilever development has resulted in a significant downscaling of the cantilever system, which is linked to a reduction in cantilever mass and increase of resonance frequency (f_c)-to-spring constant (k_c) ratio. One of our main research targets has been to further extend self-sensing and self-actuating probe capabilities to achieve the requirements of high-speed imaging. For high-speed AFM imaging, we follow the two approaches described earlier: optimizing the active cantilevers and the developing high-speed, large-range nanopositioning stages with appropriate control. In this context, active cantilever with higher resonance frequency, higher bandwidth, and low spring constant will be in the future further developed. To improve the force sensitivity and measurement throughput (speed), we increased the ratio of the resonance frequency to spring constant. In other words, we aimed to fabricate softer and smaller probes, which feature faster imaging dynamics and reduced probe–sample interaction forces while maintaining a high sensitivity. Furthermore, it has been shown that, even with static thermal time constant of 30 µs, high resonance frequencies of up to 4 MHz can be addressed through thermomechanical actuation. Thermal actuation, due to simple implementation and extraordinary dynamic behavior, is an excellent candidate for self-actuated high-frequency cantilever sensors. Due to tailoring of the material properties of the probes, a static displacement in range of 10 µm can be realized. An important advantage in the performance of the FE-SPL technology arises from the fact that the same active cantilever is used for both direct low-energy Fowler–Nordheim electron exposure of the resist using FE from the tip and fast noncontact AFM. To enable both modes, FE-SPL and AFM to be used, two control feedback loops are implemented in the FE-SPL tool (Figure 18.2). The closed-loop FE-SPL tool uses the standard dynamic amplitude modulation feedback loop for topographic imaging (AFM).

Thereby, the height of the cantilever is regulated by a scanner on the basis of the difference of the oscillation amplitude to a predefined amplitude set-point. The imaging is performed routinely with atomic resolution and employed for pre- and post-inspection. The contact of the tip with the sample blunts atomically sharp tip, and to overcome this problem, the noncontact operation mode is employed where the active cantilever is oscillated close to the sample surface without "touching" it. This enables precise pattern overlay alignment and feature stitching. For FE-SPL mode, a second independently working feedback loop is used, which regulates the height of the cantilever to maintain the current set-point. To control the electron emission current required for the lithography feedback loop, a two-stage high-precision current-to-voltage (IV) converter with subsequent amplification stage was developed. The low-noise preamplifier is characterized by a transfer function of 5 V/nA and 0.07 pA noise level at 1.5 kHz bandwidth.

18.4 Scanning Probe Lithography Instrument

The basic system consists of a top scanner mounted on a cross-beam suspension providing high mechanical stability. The sample is placed onto a coarse positioning bottom stage to enable an increased patterning area of up to 150×150 mm^2 size (6 in. wafer). Two different scanners could be applied: one providing a 200×200 µm^2 scanning area. The additionally integrated bottom stage enables a step-and-repeat functionality within an active area of 150×150 mm^2. The resolution of the readout signal of the bottom-stage movement is 10 nm, and the average movement error measured for the set of 100 µm steps is 80 nm. This accuracy of the bottom positioning stage is sufficient, because the integrated AFM imaging functionality is applied for fine feature alignment and stitching. An improved mechanical design of the mainframe shaped into a cross-beam construction had been realized. The system base and suspension beam are made of massive blocks of aluminum alloy to minimize thermal drift effects. Herein, the mechanical stability was simulated by finite element methods and optimized to make the SPL's Z-loop short and most robust, especially due to the integration of a large bottom coarse positioning stage. In contrast to the previous setup, we decided to go away from a one-arm setup towards a cross-beam (bridge)-like construction (Figure 18.4), in which the optical navigation unit is separated from the SPL mainframe. Thus, the stiffness of the mainframe is not weakened by the heavy optical microscope structure, and mechanical noises and thermal drift effects are significantly reduced. In addition, due to the bridge setup, an increased movement area of the coarse positioning unit is created, enabling a patterning range of up to 100×100 mm^2. In the center of the cross-beam, the main SPL unit is mounted, which consists of a small coarse approach motor moving a linear translation stage

(a)   (b)

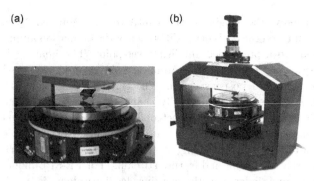

FIGURE 18.4   Field-emission scanning probe lithography (FE-SPL) tool. (a),(b) FE-SPL tool developed in the Rangelow group, including camera for optical navigation (150×150 mm^2), bottom stage for step-and-repeat operation and a granite cross-beam suspension holding the ($x = 200$ μm; $y = 200$ μm; $z = 20$ μm) top-scanner with the active cantilever and control electronics.

on which the XYZ top piezoscanner unit is mounted. Onto that top scanner, the basic scan head including the mechanical cantilever clamping mechanism as well as the first-stage amplifying electronics is fitted. To improve the dynamics of the top-scanner system, the total mass, which has to be moved during scanning, was reduced. This includes the trimming of the electronic adaption, the minimization of the electronic board thickness, the optimization of the cabling length, as well as the reduction of the mass of the mechanical cantilever clamping part (care has to be taken not to weaken the cantilever holder structure). The field-emission current can be controlled by changing the distance between tip and sample or the voltage between cantilever tip and substrate. Since the tip-sample distance is below 100 nm and the mean free path of the electrons in air is >100 nm, the system can be operated at ambient conditions without the requirement for a vacuum environment.

The emitted electrons are used to expose electron-sensitive resist layers with typical thicknesses in the range of 5–50 nm. The energy of the electrons is in the range of several 10 eV, which is comparable to the relevant chemical reactions of the resist. Proximity effects, as observed for EBL due to the generation of secondary electrons, are practically negligible. FE-SPL provides a self-development positive tone mode, for which resist material is removed without the requirement of a development step (Figure 18.5) [50]. A negative tone (cross-linking) can be obtained after development of the exposed structures. The exposure dose, i.e., the lithography current divided by the writing speed, acts as a switch between the different lithographic tones. The most frequently selected resist for FE-SPL is calixarene, offering high-resolution patterning due to its small molecule size and low surface roughness. Furthermore, for non-conducting samples, the conducting resist P3HT (an organic poly(3-hexythiophene-2,5-diyl resist) can be used, which, in this case, enables a conducting pathway for the emission current. Thus, there are no restrictions for the applicability of FE-SPL regarding the conductivity of the samples. Resistless lithography is enabled if FE-SPL is used to oxidize the

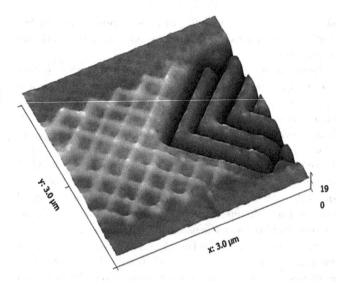

FIGURE 18.5   Demonstration of switching of lithographic tones by FE-SPL. AFM image is showing a single corner feature array patterned in negative and positive tones (without development). The exposure dose for negative tone was 44 nC/cm. The same lines were overexposed (secondary exposure of the same lines) with 120 nC/cm. This is resulting in a positive lithographic tone [51].

sample (e.g., Cr, W). For high-resolution lithography and imaging, tip shape and tip radius become crucial. In both cases, resolution is determined primarily by the sharpness of the tip.

The tip radius of curvature is approximately 5 nm enhancing the resolution capability for imaging and lithography significantly. Furthermore, resolution depends on the exposure dose; that is, increasing the exposure dose results in larger features. Cantilever deflection and oscillation amplitude can be precisely controlled by dissipated electrical power within resistors on top of the cantilever. In this way, resonance and static cantilever deflection can be excited. The deflection amplitude of the cantilever vibration as well as the static bending of the cantilever is determined by acquisition of the output signal of integrated piezoresistive deflection sensors. These form an integrated Wheatstone bridge at the point of highest internal mechanical stress, enabling deflection detection at low thermal noise.

Hence, there are compelling reasons to establish new Si-based and CMOS-compatible devices that can work at these dimensions. Fortunately, at dimensions <5 nm, quantum effect devices such as SE devices in silicon look increasingly attractive. To observe these single-electron effects, small silicon islands with diameters less than 10 nm have to be generated to produce quantum dots (QDs). These can be obtained by direct lithographic definition of the silicon island or by the point contact approach, whereby the point contacts are defined and are then reduced in size by geometric controlled oxidation. Here, the second approach is followed. The SE devices were made on SOI wafer with a highly (~10^{20} cm^{-3}) n-doped, ultrathin (12 ± 1 nm) silicon layer on top of a 25 nm thick buried oxide layer. Contact pads and an area for device definition were defined by optical lithography and etching. Point contacts were defined as trenches into a 9–15 nm thick resist layer (AZ Barli or calixarene resists) by FE-SPL. The resist pattern was transferred into silicon using a SF_6/O_2 plasma with gas flows of 10 and 2.5 sccm, respectively, using cryo-etching with the sample cooled to $-120°C$ and a process pressure of 5 mtorr. AFM measurements were made prior to and post etching to control and optimize the process. After resist removal, a geometrically controlled oxidation step took place for QD definition [50]. Today, NIL could enable a fast and reproducible process chain for the routine fabrication of nanoelectronic, *"beyond CMOS"* devices for the next-generation electronics. Thereby, due to the merits of ambient condition operation, imaging and metrological capabilities, and throughput, FE-SPL could be the shortest way to get access to diverse quantum and single-electron (SE) devices without employing capital equipment. Thus, the miniaturization of electronic devices to the single-digit nanometer range and the possible use of SETs as a sensor of Q-bits in quantum computing become more realistic (Figure 18.6). Due to the high resolution, writing speed is limited but can be enhanced

by using the mix-and-match lithography [27] Due to the high resolution, writing speed is limited but can be enhanced by using the mix-and-match lithography. Furthermore, FE-SPL written and reactive ion etching at cryogenic temperatures can be used for cost-effective fabrication of templates for the NIL, thus resulting in a high-throughput process chain for nanodevices, plasmonic structures (Figure 18.7), and nanoelectronic devices fabrication.

18.5 Cantilevers Parallelization for High-throughput Operation

By the integration of the cantilever deflection detection and actuation within the beam itself, an array operation without the requirement of the up-scaling of external components (e.g., laser-based read-out and Z-piezo scanning unit) is enabled. In this manner, the architecture of the measurement and control electronics can be simplified so that its integration with various metrological or proximal probe lithography systems is enabled. Thermomechanical

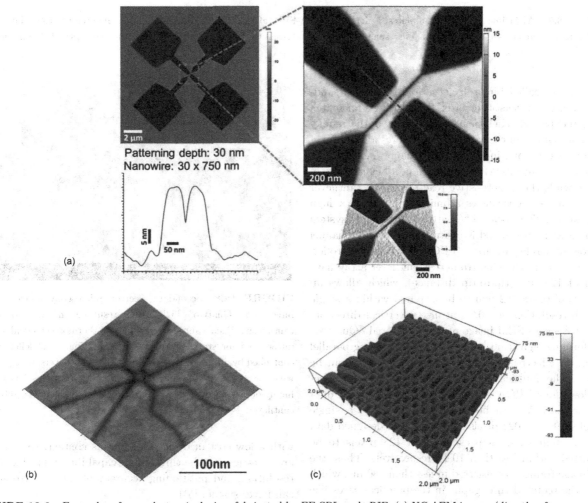

FIGURE 18.6 Examples of nanoelectronic devices fabricated by FE-SPL and cRIE. (a) NC-AFM images (directly after exposure) of a nanowire device patterned by multi-step FE-SPL in positive tone in 30 nm thick calixarene resist. (b) Fast-AFM image of SE transistors with nano-sized silicon island as QD obtained after etching. (c) NC-AFM image of plasmonic resonators [51].

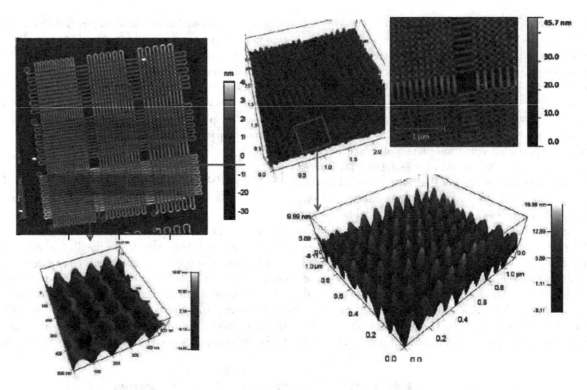

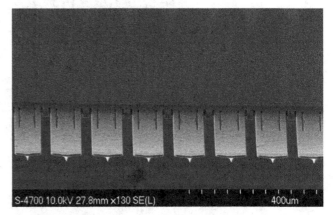

FIGURE 18.7 AFM images of NIL-templates, generated by FE-SPL, after pattern transfer by cryogenic etching. Here, lines with different pitches were crossed to form plasmonic resonators. The initial resist thickness was about 10 nm, and the depth of etching reaches 100 nm.

actuation, also called bimorph actuation, requires a sandwich structure of at least two materials featuring a mismatch of thermal expansion coefficients. The employ of thermomechanical actuation on every cantilever ensures autonomy of operation of each one in static and dynamic modes. This enables a simultaneous AFM operation of all cantilevers in an array, while the piezoresistive read-out of the cantilever bending routinely guarantees atomic resolution at a high imaging speed. The so-called "Quattro" is cantilever system containing four probes and is controlled by a multichannel field-programmable gate array (FPGA) 125 MHz controller. The cantilevers in the Quattro active cantilever array have a pitch of 125 μm (tip-to-tip distance), which allows an image size of 0.5 × 0.2 mm to be acquired within a single scan with resolution of <0.2 nm in the vertical direction. Figure 18.8 shows SEM image of two integrated "Quattro" arrays forming eight cantilevers in one array. Using parallel imaging, an effective scanning speed of 5.6 mm/s can be achieved. The multichannel, scalable controller architecture allows four FPGA channels to scan and collect data simultaneously. A data buffer of 128 Mbits for a single frame of 4,096 × 1,024 pixels is applied. The designed data transfer system allows a packet size of 128 pixels to be transmitted within less than 10 ls, respectively. Thus, the entire image frame is transferred in less than 280 ms, which exceeds the required throughput in the practical cases like critical dimension-metrology and inspection. In this article, the authors are presenting the concept of the system, which combines imaging, metrology, and lithography capabilities

FIGURE 18.8 Secondary electron microscopy image of two integrated "Quattro" [52] arrays resulting in array of eight cantilevers. Each cantilever includes a thermomechanical transducer used for static and dynamic deflection. The deflection is controlled by a piezoresistive readout. The cantilevers are equipped with 6.8 μm high tips having sub-10 nm tip radius formed by highly doped silicon. The distance (pitch) between the neighboring cantilevers is of 125 μm.

with a low cost of ownership. In this context, the authors are investigating the throughput capability, reproducibility, resolution, and positioning accuracy of the Quattro active cantilever system [51].

Due to the small mass of the deflected cantilever, the control of the tip displacement can be carried out with direct current (DC) and alternating current (AC). Each cantilever of the array is excited individually by applying a voltage to its actuator. The excitation signal consists of a static DC component and a dynamic AC sine-wave component:

$$v(t) = V_{\text{DC}} + V_{\text{AC}} \sin(\omega t),$$

where ω is the AC voltage frequency. The dissipated heat in the thermomechanical actuator causes a cantilever deflection, which linearly depends on the heating power, given by the following equation:

$$p(t) = \frac{v(t)^2}{R_{\text{actuator}}} = \frac{1}{R_{\text{actuator}}} \left(2V_{\text{DC}}V_{\text{AC}} \sin(\omega_0 t) \right.$$
$$\left. - \frac{V_{\text{AC}}^2}{2} \cos(2\omega_0 t) + \left(V_{\text{DC}}^2 + \frac{V_{\text{AC}}^2}{2}\right)\right).$$

Parabolic dependence of the cantilever bending on the actuator current is achieved, since the dissipated power is proportional to the square of the applied (sinus) voltage. A nonlinear parabolic dependence is solved by a linear approximation scheme implemented in the AFM control unit. In that case, the DC voltage applied to the heater is used for a static bending control of each cantilever, whereas the AC amplitude is independent from the DC bending. Typically, in conventional AFM systems, the z-piezo stack actuator tracks the topography in the z-direction and needs to be much faster than the highest spatial frequency of the sample. In case of "Quattro" cantilever array operation, the thermal actuator is used for both an AC excitation of the cantilever in resonance and applying a feedback signal for topography.

Thereby, the low-frequency actuation acting as feedback signal for topography tracking, combined with the high-frequency actuation for a resonant excitation, is serving as an unique capability of the Quattro array for parallel operation. The amplitude in the range of 3 μm is used both for actuation at the resonance frequency and as feedback signal for topography tracking. The resonant frequency actuation of the cantilever operation in amplitude modulation mode is several nanometers peak-to-peak and can be adjusted individually for each cantilever in the array in order to achieve a consistent setup of all cantilever in the array.

The oscillation amplitude of each cantilever is measured as AC voltage at the output of piezoresistive bridge. This voltage serves the Z-feedback signal, and the driving DC voltage applied to the thermomechanical actuator keeps the distance between the tip and surface constant. The AC voltage at the output of piezoresistive bridge is also used as the topographic information signal. A direct digital synthesis (DDS) unit generates the AC part of the actuator driving signal running on $\omega_0/2$ (ω_0 is fundamental resonance frequency of the cantilever). This part is summed up with the DC voltage, which is setting the cantilever bending. Thereby, the DC components are not affecting the amplitude of the mechanical oscillation or vice versa. In accordance, the DC-bending controller sets a DC voltage applied to the actuator. Each individual cantilever in the array gives a Z-actuation range of about 6 μm (pick-to-pick). The complete Z-feedback function is aimed not only

to track the topography details but also to compensate the overall system drift caused by sample slope, temperature drift, etc. The latter one can be up to five times larger than the Z-range of the cantilever. For that reason, a second feedback loop is implemented, while the first feedback tracks the topography and determines the DC voltage set-point for the second feedback, and keeps it constant. The output of the second feedback controller is used as a compensation signal for the Z-piezo scanner. Both feedback-regulation loops are implemented in the FPGA controller. Since four channels are required corresponding to each cantilever of the array, in total, four feedbacks controllers are implemented and synchronized.

Due to applied control scheme, the deflection of every cantilever is regulated independently resulting in large imaging fields (Figure 18.9). The application of a resonance noncontact imaging technology reduces the force interaction acting at the cantilever's nano-tip. As a result, the probe lifetime is increased significantly, which enables high-throughput measurements. The piezoresistive deflection detection scheme, combined with the cantilever design, allows a precise closed-loop and tip deflection measurement in a quantitative manner. We also presented high-performance FPGA control electronics, providing all needed functionality in order to drive and collect data from four active cantilevers simultaneously and also a data transfer system, whose capacity and speed exceed 100 times the requirements for the four-channel parallel operation. With respect to other commercially available AFM probes,

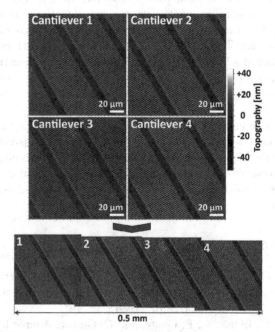

FIGURE 18.9 Merging of four AFM images into a single image obtained with four cantilevers as "Quattro" array. The merged scan covers a total image width of 0.5 mm. The Si test structure consists of 45 μm long and 14 nm high lines. The scanning speed was 10 lines/s at 1,028 pixels per line. The data size of the image was 256 Mbits [52].

the application of Quattro cantilever arrays could be a cost-effective solution in the case of industrial applications for imaging of the large surface structures. For further enhancement of the imaging throughput capability, we are working on (i) the improvement of the cantilever scan speed, (ii) the downscaling of the active cantilever size, and (iii) the up-scaling of the cantilever number based on the thermomechanically driven self-sensing probe concept.

18.6 Summary

In this chapter, we have introduced the basics of AFM with active cantilevers and described how it has been used in single nano-digit capable of manufacturing. Scanning proximal lithography (SPL) provides benefits over EBL in damage-free exposure, better resolution, low complexity, and the ability to much better alignment accuracy and stitching.

The results demonstrate that by adapting fabricating techniques <10 nm structures needed for SETs and quantum computing devices in silicon can be defined using FE-SPL with an enhanced yield. The strengths, limitations, and restrictions of this manufacturing technique are presented, along with the current and potential applications. The self-actuated piezoresistive cantilever allows easy system integration and significant reduction the weight of the scanning head. Hence, the microscope provides better controllability and significant higher scan speeds, with the potential for full lithographic and metrological automation. For FE-SPL, the implemented AFM mode is enabling an in situ inspection capability, a quantitative mapping at atomic resolution. As with most scientific developments, the technological impact of the SPL is difficult to predict. New patterning technologies have to be competitive with today's industry's enormous investments. The closed-loop lithography scheme was applied to sequentially write positive and negative tone features.

Due to unique combination of exposure and imaging with the same cantilever, each single feature is aligned separately with the highest precision and inspected after patterning.

A cost-effective solution to increase the throughput of tip-based nanofabrication is the use of parallel scanning probe arrays.

References

1. G. Binnig, Atomic force microscope and method for imaging surfaces with atomic resolution, U.S. Patent No. 4724318 (1986).

2. G. Binnig, C. F. Quate, and C. Gerber, Atomic force microscope, *Phys. Rev. Lett.* 56, 930 (1986).

3. R. Zhang, T. Chen, A. Bunting, and R. Cheung, Optical lithography technique for the fabrication of devices from mechanically exfoliated two-dimensional materials, *Microelectron. Eng.* 154, 62–68 (2016). doi: 10.1016/j.mee.2016.01.038.

4. G. Imamura and K. Saiki, Modification of graphene/SiO_2 interface by UV-irradiation: Effect on electrical characteristics, *ACS Appl. Mater. Interfaces* 7, 2439–2443 (2015).

5. T. Lehnert, O. Lehtinen, G. Algara–Siller, and U. Kaiser, Electron radiation damage mechanisms in 2D $MoSe_2$, *Appl. Phys. Lett.* 110 (2017). doi: 10.1063/1.4973809.

6. K. Elibol, et al., Atomic structure of intrinsic and electron-irradiation-induced defects in $MoTe_2$, *Chem. Mater.* 30, 1230–1238 (2018). doi: 10.1021/acs.chemmater. 7b03760.

7. C. Jannik et al., Accurate measurement of electron beam induced displacement cross sections for single-layer graphene, *Phys. Rev. Lett.* 108, 196102 (2012); *Erratum Phys. Rev. Lett.* 110, 239902 (2013). doi: 10.1021/nn4044035.

8. R. Zan et al., Control of radiation damage in MoS_2 by graphene encapsulation, *ACS Nano* 7, 10167–10174 (2013).

9. G. Binnig and H. Rohrer, Scanning tunneling microscopy, *Helv. Phys. Acta* 55, 726–735 (1982).

10. M. Ringger, H. R. Hidber, R. Schlogel, P. Oelhafen, and H. J. Guntherodt, Nanometer lithography with the scanning tunneling microscope, *Appl. Phys. Lett.* 46, 832 (1985). doi: 10.1063/1.95900.

11. D. Abraham, H. Mamin, E. Ganz, and J. Clarke, Surface modification with tine scanning tunnelling microscope, *IBM J. Res. Dev.* 30, 492 (1986). doi: 10.1147/rd.305.0492.

12. A. Dagata, J. Schneir, H. H. Harary, C. J. Evans, M. T. Postek, and J. Bennett, Modification of hydrogen-passivated silicon by a scanning tunneling microscope operating in air, *Appl. Phys. Lett.* 56, 2001 (1990); doi: 10.1063/1.102999.

13. A. A. Tseng (ed.), *Tip-Based Nanofabrication.* Springer, New York (2011). ISBN: 978-1-4419-9899-6.

14. D. M. Eigler and E. K. Schweizer, Positioning single atoms with a scanning tunnelling microscope, *Nature* 344, 524 (1990). [A. J. Heinrich, C. P. Lutz, J. A. Gupta, and D. M. Eigler, Molecule cascades, *Science* 298, 1381–1387 (2002)].

15. F. Pérez-Murano, G. Abadal, N. Barniol, and X. Aymerich, Nanometer-scale oxidation of Si(100) surfaces by tapping mode atomic force microscopy, *J. Appl. Phys.* 78, 6797 (1995). doi: 10.1063/1.360505.

16. M. Muller et al., Controlled structuring of mica surfaces with the tip of an atomic force microscope by mechanically induced local etching, *Surf. Interface Anal.* 36, 189 (2004).

17. S. Xu and G. Y. Liu, Nanometer-scale fabrication by simultaneous nanoshaving and molecular self-assembly, *Langmuir* 13, 127 (1997). doi: 10.1021/la962029f.

18. O. Custance, R. Perez, and S. Morita, Atomic force microscopy as a tool for atom manipulation, *Nat. Nanotechnol.* 4, 803 (2009).

19. J. W. Lyding, T. C. Shen, J. S. Hubacek, J. R. Tucker, and G. C. Abeln, Nanoscale patterning and oxidation of H-passivated Si(100) 2 × 1 surfaces with an ultrahigh vacuum scanning tunneling microscope, *Appl. Phys. Lett.* 64, 2010 (1994). doi: 10.1063/1.111722.

20. J. N. Randall et al., Ding, atomic precision lithography on Si, *J. Vac. Sci. Technol. B* 27, 2764 (2009).

21. M. Fuechsle et al., A single-atom transistor, *Nature Nano* 7, 242–246 (2012).

22. T. M. Arruda et al., Toward quantitative electrochemical measurements on the nanoscale by scanning probe microscopy: Environmental and current spreading effects, *ACS Nano* 7, 8175–8182 (2013). doi: 10.1021/nn4034772.

23. H. J. Mamin and D. Ruga, Thermomechanical writing with an atomic force microscope tip, *Appl. Phys. Lett.* 61(8), 1003 (1992). doi: 10.1063/1.108460.

24. D. Pires et al., Nanoscale three-dimensional patterning of molecular resists by scanning probes, *Science* 328, 732 (2010).

25. S. Alexander et al., An atomic-resolution atomic-force microscope implemented using an optical lever, *J. Appl. Phys.* 65, 164 (1989).

26. K. M. Carroll et al., Speed dependence of thermochemical nanolithography for gray-scale patterning, *ChemPhysChem* 15, 2530–2535 (2014). doi: 10.1002/cphc.201402168.

27. D. Wang et al., Thermochemical nanolithography of multifunctional nanotemplates for assembling nano-objects, *Adv. Funct. Mater.* 19, 3696–3702 (2009). doi: 10.1002/adfm.200901057.

28. W.-K. Lee and P. E. Sheehan, Scanning probe lithography of polymers: Tailoring morphology and functionality at the nanometer scale, *Scanning* 30, 172–183 (2008). doi: 10.1002/sca.20084.

29. Y. S. Chi and I. S. Choi, Dip-pen nanolithography using the amide-coupling reaction with interchain carboxylic anhydride- terminated self-assembled monolayers, *Adv. Funct. Mater.* 16, 1031–1036 (2006). doi: 10.1002/adfm.200500796.

30. Robert Szoszkiewicz, Takashi Okada, Simon C. Jones, Tai-De Li, William P. King, Seth R. Marder and Elisa Riedo, High-Speed, Sub-15 nm Feature Size Thermochemical Nanolithography, *Nano Lett.* 7(4),1064–1069, 2007. doi: 10.1021/nl070300f.

31. J. A. Dagata, T. Inoue, J. Itoh, and H. Yokoyama, Understanding scanned probe oxidation of silicon, *Appl. Phys. Lett.* 73, 271 (1998).

32. R. V. Martinez, N. S. Losilla, J. Martinez, Y. Huttel, and R. Garcia, Patterning polymeric structures with 2 nm resolution at 3 nm half pitch in ambient conditions, *Nano Lett.* 7, 1846 (2007).

33. M. Rolandi, I. Suez, H. J. Dai, and M. J. Frechet, Dendrimer monolayers as negative and positive tone resists for scanning probe lithography, *Nano Lett* 4(5), 889–893 (2004).

34. J. D. Torrey et al., Scanning probe direct-write of germanium nanostructures, *Adv. Mater.* 22, 4639–4642 (2010).

35. K. Wilder, H. T. Soh, A. Atalar, and C. F. Quate, Hybrid atomic force/scanning tunneling lithography, *J. Vac. Sci. Technol., B* 15, 1811 (1997).

36. K. Wilder, C. F. Quate, B. Singh, and D. F. Kyser, Electron beam and scanning probe lithography: A comparison, *J. Vac. Sci. Technol., B* 16, 3864 (1998). doi: 10.1116/1.590425.

37. K. Wilder, C. F. Quate, D. Adderton, R. Bernstein, and V. Elings, Fabrication and lateral electronic transport measurements of gold nanowires, *Appl. Phys. Lett.* 73, 2527 (1998).

38. A. J. l. Fèbre, L. Abelmann, and J. C. Lodder. Field emission at nanometer distances for high-resolution positioning, *J. Vac. Sci. Technol., B* 26(2), 724–729 (2008).

39. R. H. Fowler and L. Nordheim, Electron emission in intense electric fields, *Proc. R. Soc. Lond., A* 119, 173 (1928).

40. S. Lenk, M. Kästner, C. Lenk, T. Angelov, Y. Krivoshapkina and I. W. Rangelow, 2D Simulation of Fowler- Nordheim Electron Emission in Scanning Probe Lithography, *Journal of Nanomaterial & Molecular Nanotechnology.* 5(7) (2016). doi: 10.4172/2324-8777.1000210.

41. L. De Broglie, The wave nature of the electron. *Nobel Lect.* (1929).

42. J. D. Jackson, *Klassische Elektrodynamik.* Walter de Gruyter, Berlin (1983). [W. Nolting, *Grundkurs Theoretische Physik 3, Elektrodynamik.* Springer, Berlin (2007)].

43. L. Nordheim, The effect of the image force on the emission and reflexion of electrons by metals. *Proc. R. Soc. Lond., A* 121, 626 (1928).

44. L. Verlet, Computer "experiments" on classical fluids. I. thermodynamical properties of Lennard-Jones molecules. *Phys. Rev.* 159, 98 (1967).

45. I. W. Rangelow, Tz. Ivanov, Y. Sarov, A. Schuh, A. Frank, H. Hartmann, J.-P. Zöllner, D. Olynick, V. Kalchenko, Nanoprobe maskless lithography, *Proc. SPIE*, 7637, 76370V-1-10, (2010). doi: 10.1117/12.852265.

46. J. Fujita et al., Nanometer-scale resolution of calixarene negative resist in electron beam lithography, *J. Vac. Sci. Technol. B* 14, 4272–4276 (1996).

47. H. H. Solak et al., Photon-beam lithography reaches 12.5 nm half-pitch resolution, *J. Vac. Sci. Technol. B* 25, 91–95 (2007).

48. M. Kaestner et al., Electric field scanning probe lithography on molecular glass resists using self-actuating, self-sensing cantilever, *Proc. SPIE* 9049, 90490C (2014).

49. I. W. Rangelow et al., Pattern-generation and pattern-transfer for single-digit nano devices, *JVST B* 34(6), 06K202 (2016).

50. I. W. Rangelow et al., Active scanning probes: A versatile toolkit for fast imaging and emerging nanofabrication, *JVST B* 35, 06G101 (2017). doi: 10.1116/1.4992073.

51. I. W. Rangelow et al., Field-emission scanning probe lithography with self-actuating and self-sensing cantilevers for single digit nanodevices, *Proc. SPIE 10584, Emerging Patterning Technologies* 2018, 1058406.

52. A. Ahmad et al., Large area fast-AFM scanning with active "Quattro" cantilever arrays, *J. Vac. Sci. Technol. B* 34, 06KM03 (2016). doi: 10.1116/1.4967159.

Nanoscale Electrocrystallization: Eco-Friendly and Site-Selective Nanofabrication of Organic Nanocrystals Based on Electrochemistry

Hiroyuki Hasegawa
Shimane University

19.1 Introduction

In modern nanofabrication, reduction of environmental burden is required. It is important to save resources and energy not only in products but also during their manufacture. Electron-beam, ion-beam, sputtering, and high-energy equipment, even those that involve the use of vacuum, consumes large amounts of energy in the fabrication of electronic devices [1–3]. Therefore, the needs for low-cost and low-environmental-burden processes are gradually increasing.

Organic materials have been extensively applied for the fabrication of electronic devices from this viewpoint [4–8]. As advantages of organic electronic devices, low weight, flexibility, and low-cost processability are often cited; however, many techniques such as vapor deposition require ultra-high-vacuum equipment [8]. Further, poor molecular orientation of the obtained film and grain boundary are responsible for performance degradation. Meanwhile, as bottom-up technology, studies are being conducted to control alignment at the level of one molecule to fabricate nanodevices [9,10]. In such nanofabrication, it is still challenging to easily arrange molecules with good controllability. Therefore, a technique that can efficiently arrange high-performance nanodevices is required.

One such promising technique is the electrochemical technique. The features of the electrochemical technique are

1) It is one of the liquid-phase processes and can be implemented using a small amount of electric power with a simple device.

2) As the oxidation–reduction potential is peculiar to molecules, it is possible to select reactive molecules.

3) As the product amount is proportional to the quantity of electricity flowing, it is possible to control the amount to be produced.

4) As the product generally deposits on the electrode surface after reacting on the electrode, it is possible to control the location at which the material is produced.

Consequently, the electrochemical technique has potential for nanofabrication. There has been research on nanofabrication employing electrochemical techniques. For example, electrochemical atomic layer epitaxy, electrochemical nanolithography using scanning probe microscopy (SPM), and electrochemical micromachining were identified as fine nanofabrication methods [11–15]. In both methods, the processability, which has not been realized in methods other than the electrochemical technique, was highly realized.

We developed *nanoscale electrocrystallization*, which is a technique to site-selectively form nanoscale single crystals between two electrodes [16–18], for nanofabrication. Assuming application to electronic devices, where materials capable of highly efficient electron transfer with high orientation should be fabricated, the *single crystals*, in which the molecules are originally efficiently arranged, were reduced to the nanoscale and then made to crosslink between the electrodes. Thus, it was possible to efficiently construct high-performance devices. The features of nanoscale electrocrystallization are as follows:

a) Simplicity: The method is based on bulk electrocrystallization, which is a liquid process. The device configuration is simple and does not require a vacuum device.

b) Ecological process: Materials can be precisely fabricated in the gap between two electrodes. Masks or patterning are not required after fabrication.

c) On-demand fabrication: Once the gap is prepared at the desired position, the nanocrystal is assembled at that location. At this time, both electrodes are crosslinked with a nanocrystal.

d) High efficiency: The obtained nanomaterial comprises single crystals. The molecular orientation enables highly efficient electron transfer, via self-assembly of the molecules.

In subsequent sections, we describe nanoscale electrocrystallization, which is nanofabrication based on the electrochemical method.

19.2   Nanoscale Electrocrystallization

The *nanoscale electrocrystallization* is based on electrocrystallization (hereinafter referred to as *bulk electrocrystallization*). Electrocrystallization is known as one of the methods for producing electrically conductive organic single crystals of visible size [19,20]. A direct current (DC) of several microamperes is mainly applied between two separated, platinum electrodes, and then, a single crystal mainly grows on either electrode where the electrochemical reaction occurs, in about several days to several weeks. There are several examples in which a nanocrystal is produced by an electrocrystallization. First, the group of Penner et al. has succeeded in growing nanocrystals of tetraselenafulvalene (TTF) bromide by DC electrolysis from platinum nanoparticles prepared on a substrate in 2001 [21]. After that, our first report (2003) showed that nanocrystals of phthalocyanine have grown on the electrode surface at DC. Here, when alternating current (AC) was used, it was reported for the first time that it selectively grows between the electrodes and showed the possibility of nanofabrication [16–18]. In 2006, the group of Yamamoto et al. reported crystal growth by DC electrolysis of 2,5-dimethyl-N,N'-dicyanoquinonediimine complexes and detailed electronic properties at the nanoscale [22]. In 2007, Caro et al. reported nanowires of molecule-based charge transfer salts of $Per_2[Au(mnt)_2]$ and $(EDT-TTFVO)_4(FeCl_4)_2$ using DC electrolysis. (Per: perylene; mnt^{2-}: maleonitriledithiolate; EDT-TTFVO: ethylenedithiotetrathiafulvalenoquinone-1,3-dithiolemethide) [23]. In 2008, the group of Kobayashi et al. reported nanocrystal

growth of tetramethyltetraselenafulvalene complex by DC electrolysis using a glass nanovolume cell [24]. In 2010, the group of Liu et al. reported the growth and properties of various nanocrystals in DC electrolysis using carbon nanotube electrodes [25]. Thus, most studies on nanocrystal growth have used electrocrystallization, mainly DC electrolysis, like in bulk electrocrystallization. When aiming for site-selective crystal growth via DC electrolysis, however, extra processing is necessary to prevent undesired crystals from growing at the electrode and/or to remove them after their occurrence. In addition, it should be noted that in the case where the nanocrystal must be crosslinked in order to monitor the electronic characteristics by a two-terminal method, it is necessary to design the electrode pattern so that the nanocrystals bridge between the two electrodes.

Although nanoscale electrocrystallization is based on electrocrystallization, some improvements have been made to the equipment (Figure 19.1). First, from the viewpoint of handling and application, electrolysis was performed on the substrate, and the electrode for electrolysis was prepared on the substrate. A one-compartment cell was also newly developed according to the shape of the substrate [26]. These modifications make the equipment different from that used in bulk electrolysis. By providing the electrode on the substrate, nanocrystals can be directly used as a device after electrolysis.

The electrochemical conditions for preparing nanocrystals were as follows. First, a solution of the starting material was prepared, which was filtered through a syringe filter to fill the cell in order to prevent the entry of fine particles. A substrate on which the two electrodes were fabricated was then set into the cell. Generally, in bulk electrocrystallization, it is necessary to collect the obtained

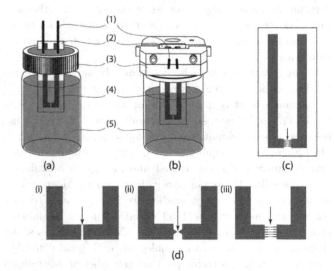

FIGURE 19.1 Electrochemical cells for nanoscale electrocrystallization. (a) Is a prototype and (b) is a commercially available model [26]: (1) electric terminal, (2) electrode substrate, (3) cap (substrate holder), (4) electrode, and (5) sample solution. (c) Is an example of the electrode substrate. (d) Three electrode shapes, (i) parallel, (ii) tapered, and (iii) multi-tip, were prepared to investigate the effect of electrode shape.

crystals; hence, the crystals are grown on an electrode on the side where an electrochemical reaction occurs using DC. On the other hand, in nanoscale electrocrystallization, since the nanocrystal formation is performed on the substrate, it is advantageous that whole substrate with as-grown nanocrystals can be utilized for devices. Furthermore, the formation position of the nanocrystals was controlled by using not only DC but also AC. The electrolysis time ranged from several tens of seconds to several minutes. Although the applied voltage differed depending on the material, nanocrystals were obtained at 0.5 to several volts in the case of DC, and 0.1 to several tens volts (peak-to-peak) in the case of AC, with 0.5–15.0 kHz of frequency. As the obtained nanocrystals were generally insoluble in organic solvents, after electrolysis, the substrate was dipped into a pure solvent to wash it. The size of these nanocrystals was controlled by adjusting the electrolysis conditions, to widths of 100 nm to several hundred nanometers and lengths of several micrometers, with a minimum width of 14 nm. Structure analysis indicated that this minimum width corresponds to the width of the 14 constituent molecules [27].

In nanoscale electrocrystallization, a redox reaction occurs, in which electrons are exchanged between the electrode and molecule during electrolysis. Consequently, three types of nanocrystals are formed according to the amount of exchanged charge (Figure 19.2). For example, considering donor molecules, three types of nanocrystals are possible, depending on the degree of oxidation of the molecules. One is a partially oxidized nanocrystal, where one electron is removed per two or more molecules; the second is a completely oxidized nanocrystal, where one electron is oxidized per molecule; and the third is the band insulator-type nanocrystal, which is described later. In a partially oxidized nanocrystal, where some molecules constituting the conductive path are oxidized, electrons are lost during the oxidation reaction; that is, holes are doped. Thus, a highly conductive nanocrystal can be obtained. In a completely oxidized nanocrystal, each of the molecules constituting the conductive path is oxidized by one electron and many carriers are doped; however, in the completely oxidized state, energy is necessary for electron transfer. This state is known as the Mott insulator [19,20,28,29]. Thus,

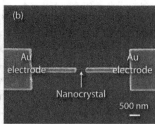

FIGURE 19.3 SEM images of nanocrystal fabrication. With DC, the nanocrystals were formed all over the anode surface when donor molecules were utilized (a). However, using AC, nanocrystals were selectively formed only in the gap between the two electrodes (b).

a low-conductivity nanocrystal can be obtained. Furthermore, by designing the molecule, a band insulator-type nanocrystal that suppresses the oxidation of the π-electron system responsible for electric conduction can be achieved when we utilize a certain kind of metal complex.

In the case of DC, the randomly oriented nanocrystals are formed on the entire electrode surface, at the side where the electrochemical reaction occurs (e.g., the anode in the case of the donor molecule, Figure 19.3a), whereas when using AC, the nanocrystal is selectively grown only on the portion where the two electrodes are closest to each other; further, the two electrodes are connected by nanocrystals. The nanocrystals thus obtained can be observed by scanning electron microscopy (SEM) as well as transmission electron microscopy (TEM), and the crystal structure can be determined by analyzing the selected area electron diffraction (SAED) pattern. Analysis of the diffraction spot revealed that a π-stack was formed along the long-axis direction of the nanocrystal in most cases; moreover, the nanocrystal had high conductivity in the long-axis direction.

These nanocrystals were formed from the site where the molecules that were oxidized at the electrode as a result of the electrochemical reaction reached saturation (in the case of donor molecules). Therefore, when DC is used, nanocrystals grow on the electrode surface at the side where the electrochemical reaction occurs. Although there are still uncertainties regarding the cause for site-selective growth when using AC, electrode reactions alternately occur due to AC, so that the point where the two electrodes are closest to each other will reach saturation at the earliest. This is considered to be one of the factors for selective growth. In addition, as the potential gradient is the steepest at the place where the two electrodes are closest to each other, the mobility of the ions becomes maximum, which is another reason for selective growth.

19.3 Nanodevice Fabrication Using Nanoscale Electrocrystallization

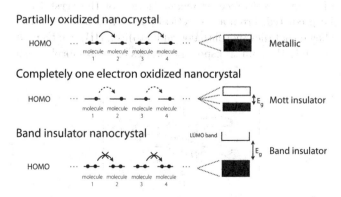

FIGURE 19.2 Three electronic states which can be obtained by nanoscale electrocrystallization (in the donor case).

As nanocrystals could be obtained as described in the preceding paragraph, device fabrication could be performed via the site-selective growth of nanocrystals in the gap between the two electrodes by AC electrolysis and crosslinking with the nanocrystal (Figure 19.4). Electrodes used in the nanoscale electrocrystallization were mainly formed on a silicon substrate with an oxide layer or a glass substrate, by photolithography and electron-beam lithography. Electrochemically stable platinum and gold were used as the electrode material, and a small amount of titanium or chromium was appropriately utilized as an adhesion layer.

During nanoscale electrocrystallization, efficient nanofabrication was achieved by using nanocrystals crosslinked between two electrodes as a device, together with an electrode substrate. For a wide variety of organic materials, the electrolysis conditions (current, voltage, and frequency), which can be modified as per the material, were studied. It was found that the crystal width was small and the selectivity was high for a high frequency of several kilohertz, low voltage, and short electrolysis time. Thus, after optimizing the electrolysis conditions to improve the results, it was possible to selectively fabricate nanocrystals in the gap for any material (Figure 19.5). Among the materials, tetramethyltetraselenafulvalene(TMTSF),

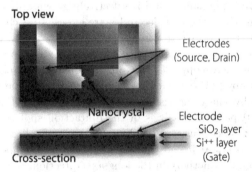

FIGURE 19.4 Device fabrication using nanoscale electrocrystallization.

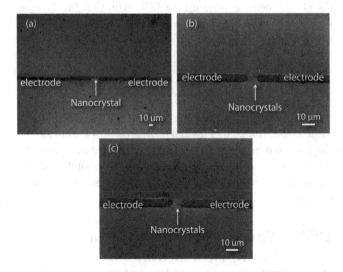

FIGURE 19.5 Examples of device fabrication (SEM images). Nanocrystals were bridged between two electrodes in each sample. (a) TMTSF-based nanocrystal, (b) TTF-based nanocrystal, and (c) phthalocyanine-based nanocrystal.

which is a component in organic superconductors; metal phthalocyanines (MPc); TTF; and tetracyanoquinodimethane (TCNQ), exhibited a tendency to form needle-shaped (wire-like) nanocrystals. In addition to these compounds, the nanoscale electrocrystallization of bis(ethylenedithio)tetrathiafulvalene (BEDT-TTF) yielded nanocrystals. However, in this case, two-dimensional platelet crystals grew in the gap.

19.4 Magnetic Field Effect Devices

The basis of modern information and communication technology is electronic devices. In order to process a large amount of information at ultra-high speeds, it is essential that a very large-scale integrated energy-efficient electronic device is used. In addition to utilizing the charge of electrons, such an electronic device is expected to be based on spintronics, which is an emerging technology using the spin of electrons and related magnetic moment.

With a few exceptions, organic molecules generally have no localized spin due to their closed-shell electronic structure; nevertheless, many organometallic complexes have spins derived from metal elements. When an electronic device with a magnetic field effect is considered, the interaction between the π-electron system and the localized spin is important in determining the operation of the device, because a π-electron system responsible for electrical conduction exists. Dicyano(phthalocyaninato)iron (III) anion, $[Fe(Pc)(CN)_2]^-$ (Figure 19.6), is known to have a localized spin at the central iron (III) atom. While the π electrons of the phthalocyanine ligand contribute to electrical conduction through the π-stack of the crystal, the spin is localized at the center of the conduction path. In this situation, the conduction electrons interact with the localized spins, so that the material also shows magnetoresistance. Matsuda and Inabe et al. successfully synthesized a highly conductive bulk single crystal of dicyano(phthalocyaninato)iron (III), which exhibited negative giant magnetoresistance and high magnetic anisotropy [30,31]. The unique properties of bulk crystals were attributed to the design of their molecular units. Therefore, if the molecular orientation is determined, the anisotropy of magnetoresistance or magnetization of the crystal can be predicted. From a theoretical point of view, it has been shown that dicyano(phthalocyaninato)iron (III) can be used to study the spin transport properties of its isomorphs such

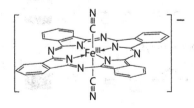

FIGURE 19.6 Dicyano(phthalocyaninato)iron (III) anion, $[Fe(Pc)(CN)_2]^-$.

as dicyano(phthalocyaninato)cobalt (III) [32]. However, as this complex cannot be processed in the bulk crystal form, it is not applied to actual devices. In order to solve this problem, nanoscale electrocrystallization was utilized to fabricate conductive nanocrystals of tetraphenylphosphonium (TPP) salt dicyano(phthalocyaninato)iron (III), TPP·[Fe(Pc)(CN)$_2$].

The formation of nanocrystals depends on the appropriate combination of frequency, amplitude of voltage, and duration of electrolysis. Electrolysis at 1.5–3.0 V AC over a frequency range of 2.0–10.0 kHz using a square waveform for 5–30 min resulted in the controlled formation of nanocrystals around the gap. The size of the nanocrystals was dependent on the electrolysis conditions; typically, the nanocrystals were sized 50 nm to several hundred nanometers in width and just about several micrometers in length. The crystal morphology was the same as that of the non-magnetic nanocrystal TPP·[Co(Pc)(CN)$_2$]$_2$; in fact, these two types of nanocrystals were isomorphic. Analysis of the SAED pattern of the nanocrystal revealed that it had the same structure as the tetragonal lattice of the partially oxidized bulk crystal TPP·[Fe(Pc)(CN)$_2$]$_2$. The SAED pattern indicated that the π-π stacking direction was

parallel to the long axis of the nanocrystal. As the nanocrystals grew in the source–drain direction, smooth electrical conduction can be expected in that direction.

If the electrochemical electrodes can be considered as the source and drain electrodes, a two-terminal device can be easily obtained by the one-pot process described in this study. Therefore, we simply measured the electrical properties of the as-grown nanocrystals formed between the two electrodes. For this analysis, we used multi-tip electrodes separated by a 5 μm gap, with nanocrystals present in each gap. The current–voltage (i–V) curves of the nanocrystals at room temperature (around 295 K) and at 20 K are shown in Figure 19.7a,b, respectively. The activation energy determined using the resistance at these temperatures (1.49 × 10^5 Ω and 1.48 × 10^{11} Ω) was estimated as 0.02 eV. This is comparable to that of the bulk crystal of TPP·[Fe(Pc)(CN)$_2$]$_2$ [30].

The Fe nanocrystal also exhibited a negative giant magnetoresistance, while the bulk crystal of its Co analog did not exhibit any negative magnetoresistance above 6 K [34]. An enhancement of the current depending on the magnetic field was observed (Figure 19.7c). On comparison with the zero-field resistivity, the negative giant magnetoresistance

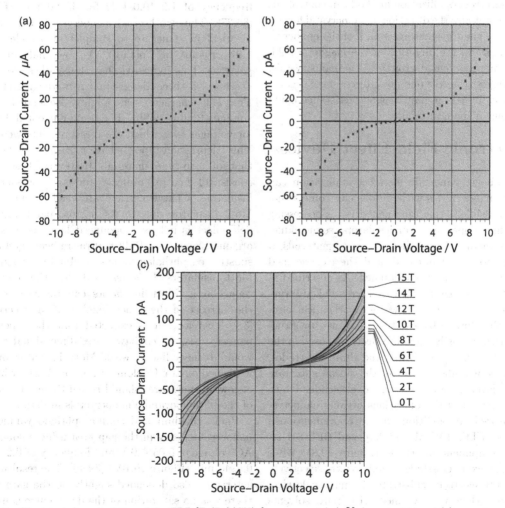

FIGURE 19.7 Electrical characteristics of the TPP·[Fe(Pc)(CN)$_2$]$_2$ nanocrystal. i–V characteristics at (a) room temperature and (b) 20 K. (c) Magnetic field dependence of the i–V characteristics. (Ref. [33], Reproduced by permission of The Royal Society of Chemistry.)

$([R(B)-R(0)]/R(0))$ under a magnetic flux density of 15 T at 20 K was -56%, which is smaller than that of the bulk crystal, -90% [35]. This negative giant magnetoresistance of the nanocrystal is an essential property, although it has been observed in several batches. The difference between the bulk and the nanocrystals can also be due to the difference in the measuring method. The bulk crystal was examined by a four-probe method, while the nanocrystals prepared in this study were measured by a two-probe method. In addition, the positive magnetoresistance of the thin-film electrodes in the nanocrystal-based device should also be considered. However, the intrinsic negative giant magnetoresistance of the nanocrystal and bulk forms of the material should be comparable.

An organic nanocrystal exhibiting strong correlation between its localized spin and conduction π electrons was successfully fabricated using nanoscale electrocrystallization. For the first time, negative giant magnetoresistance was observed in the organic nanocrystals. This giant magnetoresistance, originating due to the interaction between the localized spins and the conduction electrons at the center of the conducting path of the nanocrystals, was reduced to -56% of its original value at a magnetic flux density of 15 T. Our nanoscale electrocrystallization method can control not only the position but also the direction of nanocrystal formation. The negative giant magnetoresistance of the nanocrystals was dependent on the direction of the magnetic field, because of the highly oriented growth of the nanocrystals. These results suggest that the one-pot fabrication of organic spintronic devices based on giant magnetoresistance could become prevalent in the near future.

19.5 Electronic Field Effect Devices

As mentioned above, nanoscale electrocrystallization can produce organic nanocrystals with various electronic structures, and it is possible to construct nanocrystals possessing various electronic properties from insulators, semiconductors, as well as metals. Using these types of materials, a transistor structure was fabricated, and the electric field effect of these nanocrystals was examined. As described in Section 19.3, it is possible to fabricate a crosslinked structure with nanocrystals between two electrochemical electrodes by AC; therefore, a back-gate-type transistor structure can be fabricated easily when the electrodes used in the electrolysis are reused as the source and drain electrodes, and the highly doped silicon layer of the silicon substrate with an oxide layer is used as a gate electrode. Thus, the field-effect characteristics of various organic nanocrystals were measured. In addition to the abovementioned phthalocyanines, TTF, TMTSF, TCNQ, and BEDT-TTF were used as component materials (Figure 19.8). After the nanocrystals were crosslinked and grown selectively in the gap on the electrode substrates by nanoscale electrocrystallization, they were washed with pure solvents and dried, and then evaluated in vacuum and in the atmosphere.

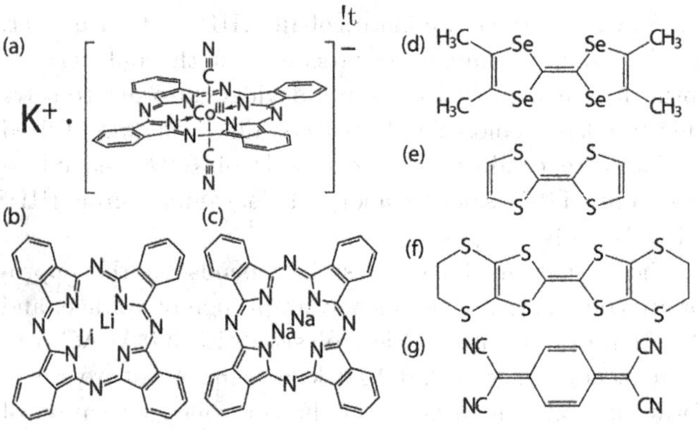

FIGURE 19.8 Component materials of nanocrystals for nanoscale electrocrystallization. (a) K·[Co(Pc)(CN)$_2$], (b) lithium phthalocyanine, (c) sodium phthalocyanine, (d) TMTSF, (e) TTF, (f) BEDT-TTF, and (g) TCNQ.

In the case of phthalocyanines, for the nanoscale electrocrystallization of a potassium salt of dicyano(phthalocyaninato)cobalt(III), K·[Co(Pc)(CN)$_2$] in methanol, nanocrystals were formed only in the gap using a square-waveform AC voltage of 1.5–2.5 V and frequency of 1.0–10.0 kHz for 45–60 min (Figure 19.9a) [36,37]. The synthesized nanocrystals were found to be identical in structure to completely one-electron-oxidized bulk crystals of Co(Pc)(CN)$_2$, by analyzing the SAED patterns. This indicated that the synthesized nanocrystals possessed a three-dimensional π-π stacking structure [37]. The i–V characteristics slightly depended on the gate voltage (Figure 19.9b). The drain current did not saturate for voltages over which the measurements were performed. This depression-type behavior seems to be similar to that of a p-type material. However, these nanocrystals consist of neutral radicals and are considered as Mott insulators. Charge carriers are easily generated within a Mott insulator due to its narrow band gap. Thus, the field effect is not as pronounced here, as it is in ordinary organic transistors. Bulk neutral radical crystals of axially substituted phthalocyanines exhibit high conductivities as Mott insulators. It is assumed that the on-site Coulomb repulsion in Mott insulators can be lowered because of the increase in the dimensionality of the electronic system [38]. Therefore, it is expected that these neutral radical nanocrystals, which have three-dimensional π-π stacking, would behave like a weak Mott insulator owing to the weakened on-site Coulomb repulsion. Further investigations at low temperatures should reveal the electronic properties of the phthalocyanine nanocrystals in detail.

Planar lithium and sodium phthalocyanines also gave nanocrystals only in the gap area using a square-waveform AC voltage of 0.05–5.0 V and frequency of 0.2–10.0 kHz for 0.5–45 min (Figure 19.10a) [39,40]. The resulting i–V characteristics also depended slightly on the gate voltage, and there was no saturation of the drain current in the voltage regions over which these measurements were performed (Figure 19.10b). In the case of sodium phthalocyanine, the

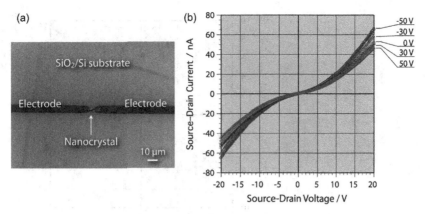

FIGURE 19.9   (a) SEM image of the neutral radical phthalocyanine nanocrystal formed in the gap by nanoscale electrocrystallization and (b) i–V characteristics of the nanocrystals for various gate voltages. (Ref. [37], Reproduced by permission of The Royal Society of Chemistry.)

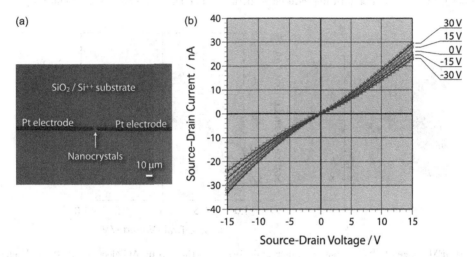

FIGURE 19.10   (a) SEM image of sodium phthalocyanine nanocrystal formed in the gap and (b) i–V characteristics of the nanocrystals for various gate voltages. (Ref. [40], Reproduced by permission of The Royal Society of Chemistry.)

characteristics appeared similar to those of an n-type material, in that the current increased with increasing gate voltage, though lithium phthalocyanine exhibited p-type characteristics. As phthalocyanines are generally p-type materials, these results seem to be unreasonable. It is considered that these nanocrystals also consist of neutral radicals, thereby resulting in a Mott insulator. These electronic states may then affect the electrical characteristics.

In the case of TTF and TMTSF derivatives, we site-selectively synthesized a TTF-based nanocrystal only in the gap between the two electrodes [41,42]. To synthesize the partially oxidized TTF nanocrystal, tetrabutylammonium nitrate was used as the electrolyte to introduce the counter anion. The TMTSF-based nanocrystals were also first site-selectively synthesized using nanoscale electrocrystallization. Tetrabutylammonium perchlorate and tetrabutylammonium nitrate were utilized as the electrolyte to introduce the counter anion. The site-selective formation of nanocrystals was observed using a square-waveform AC voltage of 1.0–1.5 V (peak-to-peak) and a frequency of 5.0–10.0 kHz for 5–10 min in TTF salts, and using a square-waveform AC

voltage of 0.5–2.0 V (peak-to-peak) and a frequency of 2.0–5.0 kHz for 0.5–15 min in TMTSF salts (Figures 19.11a and 19.12a, respectively).

The structures of the synthesized nanocrystals were found to be identical to those of the partially oxidized bulk crystals of TTF·(NO$_3$)$_{0.55}$ and (TMTSF)$_2$·ClO$_4$, respectively [43,44], from the analysis of the SAED patterns. This result also indicates that the most electrically conductive π-π stacking direction was parallel to the long axis of the nanocrystals. Because the nanocrystals were grown in the source–drain direction, smooth electrical conduction is expected when we fabricate the device using this method.

The i–V characteristics of the obtained nanocrystals indicate reasonable conductivity. Here, we could also readily obtain a two-terminal device via a one-pot process when the two electrochemical electrodes were reused as source and drain electrodes. The electrical properties were measured using the as-grown nanocrystals that bridged the two electrodes. The i–V characteristics of this device exhibited linear behavior, suggesting ohmic contact between the

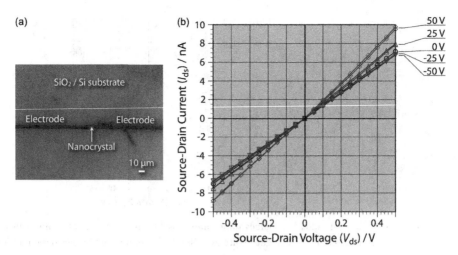

FIGURE 19.11 (a) SEM image of TTF-based nanocrystals formed in the gap by AC electrolysis. (b) i–V characteristics of the nanocrystals for various gate voltages. (Ref. [41], Reproduced by permission of The Royal Society of Chemistry.)

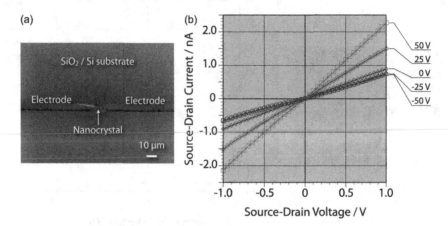

FIGURE 19.12 (a) SEM image of TMTSF-based nanocrystals formed in the gap by AC electrolysis. (b) i–V characteristics of the nanocrystals for various gate voltages. (Ref. [42], Reproduced by permission of Elsevier.)

electrodes and nanocrystals (Figures 19.11b and 19.12b). The field effects on the i–V characteristics were weak in both samples. Therefore, the i–V plots slightly depended on the gate voltage (V_g). In the range of the drain voltage measured, the drain current did not saturate and increased with increasing V_g. However, the weak gate voltage dependence appears similar to n-type behavior. As the band structures of both crystals are considered to be metallic, this result might appear inconsistent. Such an anomalous behavior is generally attributed to the contact resistance and/or charge injection barriers. In addition, we should consider anion sites in these crystals, which do not exist in ordinary single-component organic transistor materials. In the previous section, we determined the structures of both nanocrystals by using SAED. The anions in these nanocrystals were disordered; they were defects in TTF·(NO$_3$)$_{0.55}$ and randomly oriented in (TMTSF)$_2$·ClO$_4$. Consequently, these anions will be able to change their position or rotate in the crystal upon application of external fields, such as an electric field. This bias of the distribution or the orientation of anions in the crystal could also cause the anomalous i–V characteristics. It was reported that the electrochemical

doping of ions can modify the electronic properties [45]. Therefore, understanding the displacement of ions in the crystal could enable the development of new devices with such unusual properties. However, sufficient data on the field effect of cation radical salts, including on vacancies, are unavailable even though there are many reports of organic field-effect transistors consisting of single-component crystals. Further investigations of these materials are needed to comprehensively understand the electrical properties of nanocrystals based on organic ion radical salts.

Electrolysis was carried out using TCNQ. The difference in the crystal growth mode between DC and AC was found to be the same as before; growth occurred only on the electrode (cathode) side where the electrochemical reaction occurred in the DC condition. Electrolysis was performed for several minutes at an amplitude of 0.5–1.5 V (peak-to-peak) and a frequency of 0.5–10 kHz; hence, growth with better controllability was possible. However, the resulting nanomaterial was bundle-shaped with fine fibers, unlike before. SAED was performed for these nanofibers; however, we could not analyze the structure. The electronic characteristics of this material were unique, as seen in a thin film of Cu-TCNQ [46,47]. A switching effect was also observed in the nanofibers. This will be reported elsewhere in the near future.

During evaluation of the electronic characteristics, many materials were observed in which no field effect appeared; however, the field effect was observed in the TTF, TMTSF, and phthalocyanine systems, despite the on/off ratio being insufficient. Although many of these materials possess a metallic band structure, some exhibit the electric field effect. The mechanism of the effect is currently not clear; however, there is a possibility that not only the effect of the interface but also a narrow bandwidth peculiar to the organic material and a low-dimensional electron system may be involved. In addition, defects and/or disorder of the counter component of the nanocrystals may affect the electric characteristics. Interesting results were obtained in the subsequent elucidation of the physical properties of the materials for organic electronics.

19.6 Complete Device Fabrication under Ambient Conditions

As nanoscale electrocrystallization is performed under ambient temperature and pressure, it has very low environmental burden. Instead of the lithography technique, which typically requires a high-vacuum device and sputtering device, a printing technique such as an inkjet method is used to prepare the electrode pattern. Then, a device structure based on the nanocrystal is constructed by nanoscale electrocrystallization. Consequently, all processes can be carried out in the atmosphere. Therefore, we studied the fabrication of an electrode substrate using a material printer and its application to nanoscale electrocrystallization [48]. Using a nanometal ink (ULVAC) of gold and indium tin oxide (ITO) as an electrode material, an electrode was printed with the material printer DMP-2831 (FujiFilm Dimatix) following which nanoscale electrocrystallization was performed. The finest electrodes printed had a gap of approximately 20 μm, and this gap was independent of the material of the nanometal ink used. The finest electrodes printed had a gap of approximately 20 μm, which was independent of the material of the nanometal ink used. For the process, the tetraphenylphosphonium (TPP$^+$) salt of the dicyanocobaltphthalocyanine anion ([Co(Pc)(CN)$_2$]$^-$), namely, TPP·[Co(Pc)(CN)$_2$], which is the first material employed for nanoscale electrocrystallization [16,17], was used. The process yielded partially oxidized nanocrystals of TPP·[Co(Pc)(CN)$_2$]$_2$; in these crystals, one could observe the one-dimensional π-π overlapping of partially oxidized [Co(Pc)(CN)$_2$]$^{0.5-}$ along the long axis [18,27]. The nanocrystals were shaped like square pillars and had a wire-like morphology and were thus suitable for forming a bridging structure.

As a result, all electrode materials successfully formed a crosslinked structure, demonstrating the possibility of fabricating nanodevices via an *all-atmospheric process*. In the case of AC electrocrystallization, nanocrystals were formed only within the gap between the two electrodes. In particular, for parallel electrodes, nanocrystals were formed throughout the gap between the electrode tips (Figure 19.13a), while they were formed only in the narrowest part of the gap between the electrode tips for tip electrodes (Figure 19.13b). The obtained nanocrystals ranged from a hundred to several hundred nanometers in width, and five to several tens of micrometers in length. We also investigated the effects of the type of substrate and electrode material used on nanocrystal formation by performing nanoscale electrocrystallization using ITO electrodes fabricated on glass substrates. It was observed that nanocrystals could be grown in the same manner (Figure 19.14).

The electronic transport properties of the as-grown bridging nanocrystals were measured by reusing the two electrochemical electrodes. As stated above, under AC conditions, nanocrystals only grew within the gap between

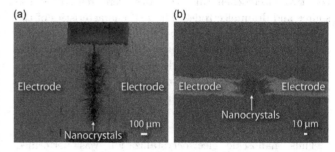

FIGURE 19.13 SEM images of nanocrystals formed in the gap using printed electrodes. (a) Parallel and (b) tip electrodes. (Ref. [48], Reproduced by permission of Elsevier.)

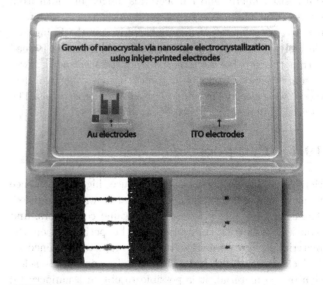

FIGURE 19.14 Nanocrystals fabricated on glass substrate by nanoscale electrocrystallization using an inkjet-printed electrode. The ITO electrode is shown on the right and the gold electrode on the left. In both cases, it was possible to fabricate nanocrystals within the gap between the two electrodes. (Ref. [48], Reproduced by permission of Elsevier.)

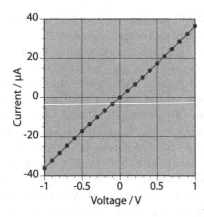

FIGURE 19.15 *i–V* characteristics of the nanocrystals formed on the printed electrode substrates. (Ref. [48], Reproduced by permission of Elsevier.)

the two electrodes, thereby forming a bridging structure. Therefore, the samples acted as two-terminal devices when the two electrochemical electrodes were reused as the source and drain electrodes. The *i–V* characteristics of the nanocrystals are shown in Figure 19.15. The *i–V* behavior is linear and ohmic in nature. Hence, we believe that the nanocrystals can be used as switching devices under the effect of an external field, such as a magnetic field, as mentioned in previous sections; however, they would be unsuitable for use in field-effect transistors owing to their metallic nature.

Thus, nanoscale electrocrystallization was performed successfully using inkjet-printed electrodes. The method used in this study allows the fabrication of nanocrystal-based devices completely under ambient conditions. The synthesized micro- and nanocrystals ranged in width from hundreds of nanometers to several micrometers and were one to several tens of micrometers in length. Using AC electrolysis, micro- and nanocrystals could also be formed within the gap between the two printed electrodes. Two-terminal devices could be obtained because the two electrodes were electrically connected by the microcrystals. The devices exhibited ohmic behavior.

19.7   Perspective

Herein, a simple method to produce high-performance nanocrystals using a simple apparatus was introduced. These nanocrystals are based on organic conductors, and nanoscale electrocrystallization can be performed on the material as long as it is an organic conductor obtained by *bulk* electrocrystallization. Therefore, by appropriate selection of the material, it is possible to obtain a nanocrystal with the desired electronic characteristics. In addition, it is important in applications that the nanocrystal be grown so as to serve as a bridge between the two electrodes, and can be utilized as a nanodevice after washing and drying. For example, if a wiring pattern with a gap on the substrate is prepared beforehand, it is possible to fabricate nanocrystals

in the gap only by applying an electric current through the wiring in the solution (Figure 19.16). When a separate gate electrode is prepared, a circuit incorporating nanocrystal transistors is completed.

Nanodevice fabrication technology based on nanoscale electrocrystallization has the following characteristics:

1) Manufacturable in an *all-atmospheric process*, including the electrode.

2) Energy is conserved because high-vacuum/high-energy equipment is not required.

3) As the device can be manufactured with pinpoint precision, resources are saved.

Although it was still insufficient as a transistor in terms of the electronic characteristics, nanoscale electrocrystallization was able to produce a material in which the field effect could be observed despite being a partially carrier-doped material, and interesting results were obtained as fundamental physical properties. Currently, we are working on improving the characteristics by designing materials and the device structure.

Recently, in the material design of organic devices, it has been realized that not only the grain size but also the molecular orientation, crystal structure, and band structure of the microcrystal are indispensable for performance improvement and characterization analysis. Hence, we consider that designing molecules and devices efficiently by utilizing the inherent properties of organic conductors that have been studied for more than half a century is one effective means. Comprehensive material design, including control of the aggregate state, is vital, together with the shape and orbital energy of one molecule. Hence, applying this method to organic conductor nanocrystals would help improve the efficiency of organic devices.

Finally, nanoscale electrocrystallization is an energy-saving process, as shown by its *all-atmospheric* fabrication; moreover, the solvents and starting materials at the time of preparation can be reused by distillation and recrystallization. Furthermore, as the nanocrystal is formed only where required, extra material is not wasted, like that in spin coating or vapor deposition. Considering these facts, nanoscale electrocrystallization is

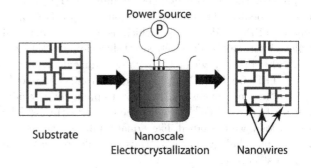

FIGURE 19.16   Application example of device fabricated using nanoscale electrocrystallization. (Ref. [27], Reproduced by permission of Elsevier.)

a low-environmental-burden production method which does not waste resources, as compared to conventional manufacturing methods. The author hopes that this method will become a next-generation manufacturing process with its low environmental burden and energy as well as resource efficiency.

Acknowledgments

This work was previously a research project under *Search for Nanomanufacturing Technology and Its Development* in the Precursory Research for Embryonic Science and Technology (PRESTO) program, Japan Science and Technology Agency (JST). The project was accomplished in Kobe Research Institute of the National Institute of Information and Communications Technology (NICT). Work on the magnetic field-effect devices was performed in collaboration with Professor Hiroyuki Tajima in the University of Hyogo and Professor Masaki Matsuda in Kumamoto University, when they were in the Institute of Solid-State Physics (ISSP), the University of Tokyo. The author gratefully acknowledges all who contributed to this work.

References

1. Sze, S. M. 2012. *Semiconductor Devices: Physics and Technology*, 3rd ed. New York: Wiley.
2. Zant, P. V. 2014. *Microchip Fabrication: A Practical Guide to Semiconductor Processing*, 6th ed. New York: McGraw-Hill.
3. Campbell, S. A. 2013. *Fabrication Engineering at the Micro- and Nanoscale*, 4th ed. New York: Oxford University Press.
4. Cantatore, E. 2013. *Applications of Organic and Printed Electronics: A Technology-Enabled Revolution*. New York: Springer Science + Business Media.
5. Wong, W. S. and Salleo, A. 2009. *Flexible Electronics: Materials and Applications*. New York: Springer Science + Business Media.
6. Köhler, M. and Fritzsche, W. 2007. Nanotechnology: An introduction to nanostructuring techniques, 2nd Completely Revised ed. Chapter 5: 149–210, *Nanotechnical Structures*. Weinheim: Wiley.
7. Petty, M. C. 2007. *Molecular Electronics: From Principles to Practice*. Chichester: Wiley.
8. Klauk, H. 2006. *Organic Electronics: Materials, Manufacturing, and Applications*. Weinheim: Wiley.
9. Iniewski, K. 2010. *Electronic Properties of Organic Conductors*. New York: McGraw-Hill.
10. Baldea, I. 2016. *Molecular Electronics: An Experimental and Theoretical Approach*. Boca Raton, FL: Taylor & Francis.
11. Gregory, B. W. and Stickney, J. L. 1991. Electrochemical atomic layer epitaxy (ECALE). *J. Electroanal. Chem. Interfacial Eletrochem.* 300: 543–61.
12. Herrero, E., Buller, L. J. and Abruña, H. D. 2001. Underpotential deposition at single crystal surfaces of Au, Pt, Ag and other materials. *Chem. Rev.* 101: 1897–930.
13. Gewirth, A. A. and Niece, B. K. 1997. Electrochemical applications of in situ scanning probe microscopy. *Chem. Rev.* 97: 1129–62.
14. Tseng, A. A., Notargiacomo, A. and Chen, T. P. 2005. Nanofabrication by scanning probe microscope lithography: A review. *J. Vac. Sci. Tech., B* 23: 877–94.
15. Bhattacharyya, B. 2015. *Electrochemical Micromachining for Nanofabrication, MEMS and Nanotechnology*. Oxford: Elsevier.
16. Hasegawa, H., Kubota, T. and Mashiko, S. 2003. Fabrication of molecular nanowire using an electrochemical method. *Thin Solid Films* 438–39: 352–55.
17. Hasegawa, H., Kubota, T. and Mashiko, S. 2003. An electrochemical fabrication of nanowire comprised of phthalocyanine. *Synth. Met.* 135–36: 763–64.
18. Hasegawa, H., Kubota, T. and Mashiko, S. 2005. Site-selective fabrication of conducting molecular nanowires based on electrocrystallization. *Electrochim. Acta* 50: 3029–32.
19. Ishiguro, T., Yamaji, K. and Saito, G. 1998. *Organic Superconductors*, 2nd ed. Berlin: Springer-Verlag.
20. Mori, T. 2016. *Electronic Properties of Organic Conductors*. Tokyo: Springer Japan.
21. Favier, F., Liu, H. and Penner, R. M. 2001. Size-selective growth of nanoscale tetrathiafulvalene bromide crystallites on platinum particles. *Adv. Mater.* 13: 1567–70.
22. Yamamoto, H. M., Ito, H., Shigeto, K., Tsukagoshi, K. and Kato, R. 2005. Direct formation of micro-/nanocrystalline 2,5-dimethyl-N,N'-dicyanoquinonediimine complexes on SiO_2/Si substrates and multiprobe measurement of conduction properties, *J. Am. Chem. Soc.* 128: 700–1.
23. Savy, J.-P., de Caro, D., Faulmann, C., Valade, L., Almeida, M., Koike, T., Fujiwara, H., Sugimoto, T., Fraxedas, J., Ondarçuhu T. and Pasquier, C. 2007. Nanowires of molecule-based charge-transfer salts. *New J. Chem.* 31: 519–27.
24. Kobayashi, K., Koyama, H., Ishikura, K. and Mitsui, T. 2008. Growth and observation of micro-organic crystals in two-dimensional glass nanovolume cell. *Appl. Phys. Lett.* 93: 143114.
25. Ren, L., Xian, X., Yan, K., Fu, L., Liu, Y., Chen, S. and Liu, Z. 2010. A general electrochemical strategy for synthesizing charge-transfer complex micro/nanowires. *Adv. Func. Mater.* 20: 1209–23.
26. Commercially available from Iwata Glass Industrial Co., Ltd. as *Nanowire Fabrication Kit*. www.iwataglass.com.
27. Hasegawa, H., Ueda, R., Kubota, T. and Mashiko, S. 2006. Multiple simultaneous fabrication of

molecular nanowires using nanoscale electrocrystallization. *Thin Solid Films* 499: 289–92.

28. Pierrot, M. 1990. *Structure and Properties of Molecular Crystals*. Amsterdam, Tokyo: Elsevier.

29. Mott, N. F. 1990. *Metal–Insulator Transitions*, 2nd ed. London: Taylor & Francis.

30. Matsuda, M., Naito, T., Inabe, T., Hanasaki, N., Tajima, H., Otsuka, T., Awaga, K., Narymbetovd, B. and Kobayashi, H. 2000. A one-dimensional macrocyclic π-ligand conductor carrying a magnetic center. Structure and electrical, optical and magnetic properties of TPP[Fe(Pc)(CN)$_2$]$_2$ {TPP = tetraphenylphosphonium and [Fe(Pc)(CN)$_2$] = dicyano(phthalocyaninato)iron(III)}. *J. Mater. Chem.* 10: 631–6.

31. Hanasaki, N., Matsuda, M., Tajima, H., Ohmichi, E., Osada, T., Naito, T. and Inabe, T. 2006. Giant negative magnetoresistance reflecting molecular symmetry in dicyano(phthalocyaninato)iron compounds. *J. Phys. Soc. Jpn.* 75: 033703.

32. Hasegawa, H., Naito, T., Inabe, T., Akutagawa, T. and Nakamura, T. 1998. A highly conducting partially oxidized salt of axially substituted phthalocyanine. Structure and physical properties of TPP[Co(Pc)(CN)$_2$]$_2$ {TPP=tetraphenylphosphonium, [Co(Pc)(CN)$_2$]= dicyano(phthalocyaninato)cobalt(III)}. *J. Mater. Chem.* 8: 1567–70.

33. Hasegawa, H., Matsuda, M. and Tajima, H. 2013. Giant negative magnetoresistance in an organic nanocrystal: Site-selective device fabrication by nanoscale electrocrystallization. *J. Mater. Chem. C* 1: 6416–21.

34. Hanasaki, N., Tajima, H., Matsuda, M., Naito, T. and Inabe, T. 2001. Giant negative magnetoresistance of one-dimensional conductor TPP[Fe(Pc)(CN)$_2$]$_2$. *Synth. Met.* 120: 797–8.

35. Hanasaki, N., Tajima, H., Matsuda, M., Naito T. and Inabe, T. 2000. Giant negative magnetoresistance in quasi-one-dimensional conductor TPP[Fe(Pc)(CN)$_2$]$_2$: Interplay between local moments and one-dimensional conduction electrons. *Phys. Rev. B* 62: 5839–42.

36. Hasegawa, H., Noguchi, Y., Ueda, R., Kubota, T. and Mashiko, S. 2008. Organic Mott insulator-based nanowire formed by using the nanoscale-electrocrystallization. *Thin Solid Films.* 516: 2491–4.

37. Hasegawa, H. 2013. Site-selectively fabricated phthalocyanine neutral radical nanocrystals: Structure and electrical properties. *J. Mater. Chem. C* 1: 7890–5.

38. Morimoto, K. and Inabe, T. 1995. Conducting neutral radical crystals of axially substituted phthalocyanine: Crystal structures, dimensionality and electrical conductivity. *J. Mater. Chem.* 5: 1749–52.

39. Hasegawa, H. 2014. Site-selective electrochemical fabrication of lithium phthalocyanine nanocrystals. *Sci. Adv. Mater.* 6: 1548–52.

40. Hasegawa, H. 2013. Fabrication of sodium phthalocyanine nanocrystals using nanoscale electrocrystallization. *New J. Chem.* 37: 2271–4.

41. Hasegawa, H. 2015. Tetrathiafulvalene-based nanocrystals: Site-selective formation, device fabrication, and electrical properties. *J. Mater. Chem. C* 3: 8986–91.

42. Hasegawa, H. 2016. Current-dependent formation of Bechgaard salt nanocrystals using nanoscale electrocrystallization for the fabrication of organic nanodevices. *Mater. Today Commun.* 7: 11–5.

43. Kathirgamanathan, P., Mazid, M. A. and Rosseinsky, D. R. 1982. The highly conductive nonstoichiometric tetrathiafulvalene nitrate: Composition, conductivity, and structure. *J. Chem. Soc. Perkin Trans.* 2: 593–6.

44. Bechgaard, K., Carneiro, K., Rasmussen, F. B. and Olsen, M. 1981. Superconductivity in an organic solid. Synthesis structure, and conductivity of bis(tetramethyltetraselenafulvalenium)perchlorate, (TMTSF)$_2$ClO$_4$. *J. Am. Chem. Soc.* 103: 2440–2.

45. Inagawa, M., Yoshikawa, H., Yokoyama, T. and Awaga, K. 2009. Electrochemical structural transformation and reversible doping/dedoping of lithium phthalocyanine thin films. *Chem. Commun.* 3389–91.

46. Potember, R. S. and Poehler, T. O. 1979. Electrical switching and memory phenomena in Cu-TCNQ thin films. *Appl. Phys. Lett.* 34: 405–7.

47. Iwasa, Y., Koda, T., Tokura, Y., Koshihara, S., Iwasaka, N. and Saito, G. 1989. Switching effect in organic charge transfer complex crystals. *Appl. Phys. Lett.* 55: 2111–3.

48. Hasegawa, H. 2017. Inkjet printing and nanoscale electrocrystallization: Complete fabrication of organic microcrystals-based devices under ambient conditions. *Appl. Mater. Today* 9: 487–92.

Bio-Inspired Graphene-Derived Membranes

Enlai Gao, Xiangzheng Jia,
Han Shui, and Ruishan Li
Wuhan University

20.1 Introduction

Over millions of years of evolution, biological tissues have realized perfect unification of the structures and functions. Many of them, such as bamboo, tooth, bone and nacre, share similar layered structures made of inorganic and organic constituents, which endow biomaterials with remarkable high mechanical performance and multifunction. One of the most used examples is nacre, which consists of 95 vol.% inorganic aragonite ($CaCO_3$) platelets ($\sim$200–500 nm thick) glued by 5 vol.% adhesive ductile biopolymers (proteins and polysaccharides, $\sim$10–50 nm thick). These natural composites exhibit synergistic strengthening/toughening through staggered arrangement of inorganic platelets and organic layers (Wegst et al., 2015). Inspired by the hierarchical structures and fascinating properties of natural nacre (Figure 20.1a,b) (Cheng et al., 2015; Gao et al., 2017b; Wan et al., 2016), artificial materials are likely to achieve unprecedented properties by mimicking the hierarchical structures of nacre to assemble building blocks. Choosing building blocks with outstanding mechanical properties is the first key to fabricate high-performance artificial nacres, which reminds us of graphene, a two-dimensional planar with monolayer of carbon atoms arranged in a hexagonal lattice that offers unsurpassable stiffness ($\sim$1 TPa), strength ($\sim$130 GPa) and resilience ($\sim$20% strain to failure) (Lee et al., 2008). Hence, bio-inspired graphene-derived membranes with layered structures are assembled by the building blocks of graphene and its derivatives (e.g., graphene oxides, GO) (Figure 20.1c,d) (An et al., 2011; Compton et al., 2012; Kotov et al., 1996; Qiu et al.,

2012; Stankovich et al., 2006; Zhao et al., 2013). Although load transfer through native interfaces with van der Waals interaction between graphene nanoplatelets is weak, which usually leads to the pull-out failure (Gong et al., 2010; Kis et al., 2004; Salvetat et al., 1999), significant advances in interfacial engineering by various cross-links and functional groups at the molecular scale have enabled the optimum design and fabrication of the graphene-derived membranes as reinforcing composites (An et al., 2011; Bekyarova et al., 2013; Compton et al., 2012; Gao et al., 2011; Hartmann and Fratzl, 2009; Liu et al., 2011, 2012, 2013; Mark and Erman, 2007; Park et al., 2008; Wojtecki et al., 2011).

In addition to reinforcing composites, another promising application of graphene-derived membranes is the selective mass transport. More specifically, the two-dimensional structures and engineerable physicochemical properties of graphene and its derivatives such as GO offer exciting platforms in the design of molecular-sieving membranes by stacking the building blocks layer by layer (LBL) (Abraham et al., 2017; Chen et al., 2017; Joshi et al., 2014; Qin and Buehler, 2015; Sealy, 2017; Xie et al., 2016). Typically, 40%–60% surface area of GO nanoplatelets is not oxygen-rich functionalized, and fast molecular permeation through the graphene-derived membranes is believed to occur in the nearly-frictionless pristine graphene nanochannels among oxide areas (Loh et al., 2010; Wilson et al., 2009). The size of nanochannels of graphene-derived membranes is characterized by the interlayer spacing h between basal planes of GO nanoplatelets, or the void spacing $\delta = h - t$, where t is the interlayer spacing of graphite (3.4 Å), which quantifies the size exclusion effect in the selective mass transport. For

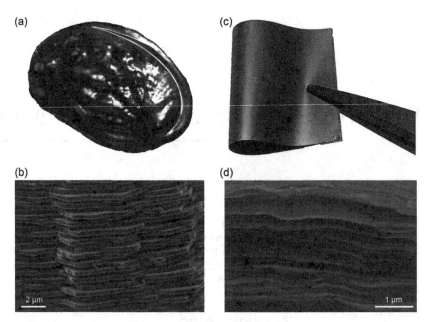

FIGURE 20.1 (a) Optical image of the nacre. (b) Optical image of the graphene membrane. (c) The electron micrograph of the nacre, showing the staggered arrangement of CaCO₃ platelets. (d) The electron micrograph showing the cross section of a graphene-derived membrane with stacked GO nanoplatelets. (Reprinted with permission from Gao and Xu, 2018. Copyright 2018 The Royal Society of Chemistry.)

example, the void spacing to reject sodium ions from water should be lower than the hydrated diameter of Na^+, 7.2 Å (Abraham et al., 2017; Mi, 2014). Therefore, a few methods have been developed to control the interlayer spacing of graphene-derived membranes, including partial reduction (Liu et al., 2015b), capillary compression (Cheng et al., 2016; Yang et al., 2013), ultraviolet reduction of GO-titania hybrid (Sun et al., 2015), hydrogen bond-/coordinative bond (HB/CB)-assisted cross-linking (Hung et al., 2014; Lin et al., 2016) and strain engineering (Gao and Xu, 2018). However, graphene-derived membranes disintegrate upon hydration due to the electrostatic repulsion between GO nanoplatelets with negative charge and swell as immersed in aqueous solutions (Figure 20.2) (Yeh et al., 2014; Zheng et al., 2017). It is thus a challenge to control the interlayer spacing at a specific value in nanoscale for the selective mass transport while maintaining the structural stability of membranes (Chen et al., 2017; Mi, 2014).

In this chapter, we present the vast studies and the latest groundbreaking advances of graphene-derived membrane in both theories and experiments over the past decades. First, the synthesis of GO and the preparation of graphene-derived membrane are briefly reviewed. Then, the multilevel structures, mechanical properties, and separation features are summarized. Finally, the challenges and opportunities of applications in reality are particularly discussed.

20.2 Assembly Approaches

Because of the sp^2-hybridization structures, graphene could not be melting-processed. In addition, it's difficult to directly disperse graphene into solvents, which limits the

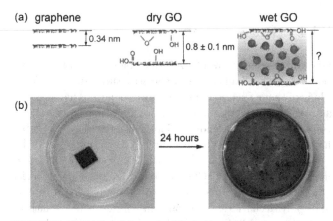

FIGURE 20.2 (a) Schematic illustration of interlayer spacings of graphene, dry GO and wet GO. (Reprinted with permission from Zheng et al., 2017. Copyright 2017 American Chemical Society.) (b) GO (Teflon) readily disintegrates in water. (Reprinted with permission from Yeh et al., 2014. Copyright 2015 Springer Nature.)

manners to assemble graphene into macroscopic materials massively. GO, with oxidation functional groups such as epoxide and hydroxyl that endow it with unique surface properties and solubility, comes into researchers' sights as the precursor of graphene and building block of macroscopic membranes. Briefly, Brodie (1859) first synthesized GO, and the method was improved by Staudenmaier (1898), Hofmann and König (1937), and Hummers and Offeman (1958). Afterwards, a lot of modified Hummers' methods and other novel methods were developed. Benefit from the excellent solubility, graphene-derived membranes can be prepared via various fluid assembly approaches. Here, three approaches, including filtration-assisted assembly,

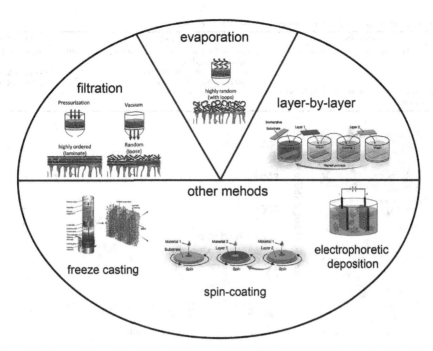

FIGURE 20.3 Representative assembly methods: filtration, evaporation (Reprinted with permission from Tsou et al., 2015. Copyright 2014 Elsevier), LBL, spin-coating (Richardson et al., 2015), freeze casting (Reprinted with permission from Wegst et al., 2015. Copyright 2014 Springer Nature); and electrophoretic deposition (Reprinted with permission from Chavez-Valdez et al., 2013. Copyright 2013 American Chemical Society).

evaporation-assisted assembly and LBL methods, as shown in Figure 20.3, are introduced.

20.2.1 Filtration-Assisted Assembly

Filtration, including vacuum filtration and pressure filtration, is straightforward and easy-operable, as well as effective for assembly at the meantime. Under vacuum filtration, GO nanoplatelets stack regularly on the surface of filter paper by the fluid field, which form membranes after drying, and it can be peeled from the substrate and become self-supported membrane when its thickness reaches micron level. Dikin et al. (2007) took the lead to use vacuum filtration method to prepare GO membranes in 2007.

20.2.2 Evaporation-Assisted Assembly

In the evaporation-assisted assembly, the GO dispersion is placed in an open container, and the membranes are obtained by evaporating solvent. First, GO nanoplatelets quickly transform into membranes as the solvent evaporates at the gas–liquid interface and gas–liquid interface transforms into gas–solid–liquid interface as solvent further evaporates. Then, GO nanoplatelets in the solvents are gradually captured by the solid phase and stabilize over time as a result of evaporation of solvent molecules and the thermal movement of GO nanoplatelets. Finally, self-supported GO membranes gradually emerge as uniform, dense, layered membranes (Shao et al., 2014). The thickness of GO membranes can be controlled via changing the concentration and volume of the solution. However, the evaporation process may take a long time, which makes this method less effective.

20.2.3 Layer-by-Layer Assembly

Nowadays, LBL assembly is gradually becoming a prevalent approach to prepare membranes. This method is able to change membrane thickness to exert nanometer control and widely used for coating planar. The LBL assembly methods can be divided into five categories: (i) immersive, (ii) spin, (iii) spray, (iv) electromagnetic and (v) fluidic assembly (Richardson et al., 2015). About 20 years ago, Kotov et al. (1996) had successfully prepared a self-assembled graphite oxide via LBL method and then reduced it, showing an excellent chemical stability to resist acid and alkali. Hu and Mi (2014) prepared GO membrane for water purification via this method and found that the membrane can remain its structure in low ionic concentration solution, showing great separation features. Additionally, the GO membrane thickness can be controlled by changing the cycles of LBL deposition.

Besides aforementioned approaches, many other methods have been developed to fabricate GO membranes, including drop-casting (Shen et al., 2016a), dip-coating (Wang et al., 2016a), spin-coating (Robinson et al., 2008), spray-coating (Nair et al., 2012), freeze casting (Wegst et al., 2015), hydrogel casting (Xiong et al., 2015) and electrophoretic deposition (Chavez-Valdez et al., 2013). To investigate the influence of various methods, Tsou et al. (2015) prepared

TABLE 20.1   Assembly Approaches for Graphene-Derived Membranes

Methods	Classification	Advantages	Disadvantages
Filtration	Vacuum filtration, pressure filtration	Straightforward, easy operation, well-layered structure	Difficult to scale up
Evaporation	–	Easy operation, scale-up	Time-consuming, disordered
LBL	–	Controllable thickness	Time-consuming, difficult to scale up
Others	Casting/coating-assisted, electrophoretic deposition	–	–

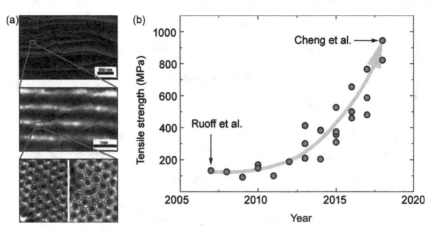

FIGURE 20.4   (a) The multilevel structures of graphene-derived membranes: cross section with ruga structures, layered structures, high-resolution transmission electron microscopy (TEM) images of graphene with SW defect and vacancy. (The high-resolution TEM images are reprinted with permission from Meyer et al., 2008. Copyright 2008 American Chemical Society.) (b) The advancement of tensile strength of graphene-based membranes from 2007 to 2018. (The data are summarized in Table 20.2.)

GO membranes through pressure filtration, vacuum filtration and evaporation, and found that the orientation of the nanoplatelets changed dramatically, showing highly ordered, random and highly random structures, respectively. Table 20.1 summarizes the features of these methods.

20.3   Multilevel Structures

Unlike traditional materials such as metal, ceramic and polymer, graphene-derived membranes are bottom-up assembled using nanostructures with large slenderness ratio by non-covalent interfaces, showing clear multilevel structures (Figure 20.4a). More specifically, for graphene, during the growth and post-processing, the van der Waals forces among the interfaces assemble the carbon nanostructures into microsheets of layered structures with regular arrangement. Furthermore, the microsheets intertwine together to form a higher level of ruga structures, including wrinkles, creases, ridges and folds of nanoplatelets and microsheets. Therefore, the graphene-derived membranes feature multilevel structures. Inspired by the classification of protein into multilevel structures, we break down the microstructural complexity of graphene-derived membrane into three levels: the primary building block structures (individual graphene nanoplatelets), secondary layered structures (microsheets with regular layered nanoplatelets) and tertiary ruga structures (the graphene-derived membrane as an assembly of the ruga microsheets). This microstructural hierarchy of graphene-derived membranes is reminiscent of the three-level model of carbon nanotube fibers (individual nanotubes

as building blocks, the close-packed bundles, and their inter-connected networks) (Gao et al., 2018a), which can improve our understanding of the multiscale nature of carbon nanostructured assemblies, their evolution under

TABLE 20.2   The Key Mechanical Properties of Graphene-Derived Membranes

Years	Young's Modulus (GPa)	Tensile Strength (MPa)	Toughness (MJ m^{-3})	References
2007	32	133	–	Dikin et al. (2007)
2008	28.1	125.8	–	Park et al. (2008)
2009	33.3	91.9	–	Park et al. (2009)
2010	–	170	–	Pei et al. (2010)
2010	–	148.3	–	Putz et al. (2010)
2011	30.4	101	–	Gao et al. (2011)
2012	–	188.9	–	Li et al. (2012)
2013	–	156.8	3.91	Cheng et al. (2013)
2013	84.8	178.9	–	Tian et al. (2013)
2013	103.4	209.9	–	Tian et al. (2013)
2013	–	300	2.8	Hu et al. (2013)
2013	20.1	412	–	Jalili et al. (2013)
2014	14.03	120.5	–	Liu et al. (2014b)
2014	–	204.9	4.0	Cui et al. (2014)
2014	26.6	384.3	–	Kim et al. (2014)
2014	–	382	7.5	Zhang et al. (2014)
2015	26.2	100.5	–	Yeh et al. (2014)
2015	35.4	142.2	–	Lam do et al. (2015)
2015	–	235.3	6.9	Wan et al. (2015a)
2015	–	356	7.5	Ming et al. (2015)
2015	–	374.1	9.2	Gong et al. (2015)
2015	–	526.7	17.7	Wan et al. (2015b)
2015	–	309	8.26	Tan et al. (2015)
2016	–	314.6	9.8	Duan et al. (2016)
2016	169	655	–	Xiong et al. (2016a)
2016	105	460	2.1	Wang et al. (2016b)
2016	120	500	–	Xiong et al. (2016b)
2017	–	765	15.64	Wen et al. (2017)
2017	–	480.5	11.8	Shahzadi et al. (2017a)
2017	–	586.6	12.1	Shahzadi et al. (2017b)
2018	–	419.4	11	Gao et al. (2018b)
2018	–	821	20	Wan et al. (2018a)
2018	–	944.5	20.6	Wan et al. (2018b)

mechanical loads, as well as their impacts on the macroscopic performance. For example, it was recently found that modulating the wavy ruga structures in graphene-derived membranes yields finely-tuned negative Poisson's ratios (Wen et al., 2019).

20.3.1 Building Block Structures

The first-level structure, the building block of graphene, is two-dimensional sp^2-hybridization carbon atoms arranged in covalent honeycomb lattice. The p orbit of the carbon atoms overlaps each other and forms π bond, and the weak localized π electronics can move freely between the adjacent carbon atoms, which endows it with unique electron transport properties and excellent electrical conductivity. The free-standing graphene has intrinsic ripples because of the thermal fluctuations, with a wavelength of 10–15 nm and amplitude of about 1 nm. The three-dimensional ripple structure improves the flexural rigidity of graphene and endows it with great stability. Indeed, the arrangement of atoms or molecules in most crystalline materials, however, is usually not perfect. The regular patterns are interrupted by various crystallographic defects—for graphene specifically, these defects appear to be the deviation from the perfect hexagonal arrangement of atoms: Stone–Wales (SW) defects, vacancies, functional groups as well as the combinations of them. In fact, some defects often occur during the growth and assembly process, which drastically reduce the strength and stiffness of graphene. Because of the nature of its low-dimensional structure, the atoms of graphene are all exposed outside, which means its specific surface area is far larger than bulk materials. Thus, the defects can be characterized by experiments conveniently, including vacancy defect, SW defect, dislocation and grain boundary (GB). It is essential to understand how the defects influence the mechanical properties of graphene from a fundamental standpoint for the rational design of high modulus and strength of graphene-derived membranes.

Vacancy defect, that is, missing of atoms, can reduce the strength and stiffness of graphene (Zandiatashbar et al., 2014). Dewapriya and Rajapakse (2014) investigated the tensile strength of defective graphene through molecular dynamics (MD) simulations and an atomistic model with various vacancy concentrations, strain rates and temperatures. The failure strain remains constant, while the strength dramatically reduces with the increase of vacancy concentration, which indicates that vacancies result in the reduction of stiffness of graphene. Besides, with the increasing of defects, graphene shows brittle-to-ductile transition (Carpenter et al., 2013; Lanqing et al., 2013), which provides the possibilities for the tailoring of its mechanical properties. On the other hand, SW defects are reconstructed from hexagon to pentagon-heptagon arrangement of carbon atoms, in which the number of atoms remains constant. Wang et al. (2012) simulated two types of SW defects created by 90° rotation of C–C bonds in different directions and found that the fracture strength decreases

with the temperature increasing. Furthermore, SW defects significantly reduced the fracture strength of graphene up to 45.3%. He et al. (2014) investigated the effects of orientation and the tilting angles of SW defects on the mechanical properties of graphene using MD simulations and found that when there is more than one SW defect, the mechanical properties of graphene depend on the tilting angle of SW defects—the tensile strength decreases with the increase of tilting angle. Besides, when the tilting angle is smaller than 25°, the strength of graphene with multiple SW defects is even higher than that of graphene with single SW defect. Though SW defect is simply believed to be an in-plane transformation of lattice, theoretical results have showed something different. Ma et al. (2009) predicted an out-of-plane wave-like structure of SW defects that extend several nanometers via density functional theory and quantum Monte Carlo. Compared with in-plane situation, the out-of-plane displacement further minimizes the potential energy of the system. In addition, dislocation formed by pentagon-heptagon defects is observed in experiments. Hashimoto et al. (2004) *in situ* observed single dislocation of monolayered graphene by high-resolution TEM. Yazyev and Louie (2010) constructed dislocation in graphene characterized by arbitrary Burgers vectors and investigated thermodynamic and electronic properties via first-principles calculations. The result indicated that the dislocations as intrinsic defects in graphene could be used for engineering graphene and functional devices. A series of dislocations placed head-to-tail, that is, GB, is observed to significantly decrease the strength of graphene. Song et al. (2013) studied GBs with finite size by MD and found that triple GB junctions play a significant role in the nucleation of cracks. Besides, the larger the GB is, the more the strength reduces, which suggests a way of improving the strength of defective graphene by controlling the size of GB. However, graphene with larger tilting angle of GBs and denser defects shows a great increase in overall strength (Yi et al., 2013), which can be explained as that compressive and tensile stresses overlap and partially cancel each other.

20.3.2 Layered Structures

The second-level structure, also known as the layered sheet, consisting of stacked two-dimensional building blocks, is similar to biomaterials such as teeth, bones and nacres. The load transfer among graphene nanoplatelets is usually insufficient because of the weak interaction of van der Waals forces, which leads to the failure of interlayer slippage (Gong et al., 2010; Kis et al., 2004; Salvetat et al., 1999). With the technological advances in modifying graphene with functional groups and the adoption of various cross-linking mechanisms (An et al., 2011; Bekyarova et al., 2013; Compton et al., 2012; Gao et al., 2011; Liu et al., 2011, 2012, 2013; Park et al., 2008), such as multimodality, sacrificial and self-healing bonds (Hartmann and Fratzl, 2009; Liu et al., 2011; Mark and Erman, 2007), it is possible to design the assembly method of graphene-derived membrane that enhances the mechanical properties by interfacial engineering at the molecular level. Recently, through oxidation (Dimiev and Tour, 2014), reduction (Faucett and Mativetsky, 2015) and irradiation (Trevethan et al., 2013) treatments, graphene nanoplatelets can be cross-linked by covalent bond, CB and HB, and the effect of interlayer cross-links was studied by the experiment (Gao et al., 2011; Park et al., 2008) and the simulation (Medhekar et al., 2010; Wang et al., 2013). These treatments can introduce some defects such as vacancies and adatoms in graphene (Jin et al., 2013; Kis et al., 2004; Salvetat et al., 1999), which substantially improve the interfacial load transfer capacity and thus elevate the overall mechanical properties.

20.3.3 Ruga Structures

The LBL structures of graphene-derived membranes are usually not well controlled, since the synthesis and assembly methods of graphene-derived membranes are mainly self-assembly processes. In these processes, ruga structures, including wrinkles, creases, ridges and folds of nanoplatelets and microsheets, are formed during the building blocks synthesizing, gathering, concentrating and forming membranes. Therefore, ruga structures are typical microstructural features of graphene-derived materials (Figure 20.4a) and partially responsible for the weakening of the nanoplatelets and the local interfacial load transfer. More specifically, the formation of ruga structures breaks down the load transfer path within the graphene nanoplatelets, which can be considered as the reduction of nanoplatelet size l, and could reduce the overall performance of graphene-derived membranes. Qin et al. (2014) studied how wrinkles affected the specific area of graphene by MD and found that the high specific surface area of graphene can only be affected up to 2% regardless of loading conditions, geometry and defects. Kunz et al. (2013) presented a wrinkling method to achieve mechanically homogenous single GO nanoplatelets, where the wrinkles become more homogeneous with the increase of number of layers. Zhu et al. (2012) conducted atomic force microscope (AFM) study on the evolution of ruga structures in graphene and found that the wrinkles reached a maximum height before folding, and tall ripples collapsed into narrow standing wrinkles under van der Waals forces, similar to large-diameter nanotubes. In addition, the ruga structures can be used to modulate the mechanical properties of graphene-derived membranes. Xiao et al. (2017) found that the GO nanoplatelets well dispersed in good solvents can collapse to hierarchically wrinkled conformations as triggered by poor solvents, and the hierarchical ruga structures endowed graphene-derived membranes with a remarkably large failure strain up to 23%.

20.4 Mechanical Performance

The sp^2-hybridization network endows graphene with extraordinary mechanical properties. Through nanoindentation with AFM, Lee et al. (2008) measured the Young's modulus and intrinsic strength of monolayer graphene as high as 1 TPa and 130 GPa, respectively, which indicated that graphene is the strongest material ever measured. Since then, many experiments and simulations have been reported on the monolayered graphene, in which the Young's modulus as well as the strength is ultrahigh (Faccio et al., 2009; Frank et al., 2007; Gomez-Navarro et al., 2008; Gupta et al., 2005; Huang et al., 2011; Kalosakas et al., 2013; Koenig et al., 2011; Kudin et al., 2001; Liu et al., 2007; Meo et al., 2006; Sakhaee-Pour et al., 2008; Van Lier et al., 2000; Zakharchenko et al., 2009; Zhao et al., 2009). Counterintuitively, compared with monolayered graphene, graphene-derived membranes show a great reduction of both Young's modulus and strength (1–2 orders of magnitude lower than that of monolayered graphene), which greatly limits the applications of graphene-derived membranes.

20.4.1 Experimental Advances

Extensive studies have been conducted to enhance the mechanical properties of graphene-derived membranes based on strategies of interfacial engineering and building block improving, and remarkable progress has been achieved. For example, the tensile strength of graphene-derived membranes has been improved from on the order of 0.1 GPa to on the order of 1 GPa during the last decade (Figure 20.4b). More specifically, the abundant functional groups on the surface of GO provide the platform for the interfacial engineering, including hydrogen bonding, ionic bonding, π–π bonding and covalent bonding. Besides, synergistic effect by various interfacial engineering plays a key role in strengthening and toughening graphene-derived membranes (Figure 20.5).

To be specific, hydrogen bonding is easily introduced in the process of preparing graphene-derived membranes. For example, Dikin et al. (2007) reported a Young's modulus of 32 GPa and tensile strength of 133 MPa of GO membranes with hydrogen bonding between nanoplatelets with water molecules via vacuum-assisted filtration, and the modulus and water content of GO membranes showed a negative correlation. Pei et al. (2010) reduced GO membranes into the highly conductive reduced GO (rGO) membranes by hydroiodic acid and exhibited an enhanced tensile strength of 170 MPa and a toughness of 2.98 MJ m^{-3}. Another effective way is to cross-link GO nanoplatelets by polymers with hydroxyl groups or oxygen groups. Recently, graphene-derived membranes were fabricated with various polymers cross-linked by hydrogen bonding including GO-poly(vinyl alcohol) (PVA) (Li et al., 2012; Putz et al., 2010), GO-poly(methyl methacrylate) (PMMA) (Putz et al., 2010) and GO-silk fibroin (SF) (Hu et al., 2013). The rGO-PVA prepared through solution-casting method by Li et al. (2012) demonstrated a high tensile strength of 188.9 MPa and toughness of 2.52 MJ m^{-3}. Hu et al. (2013) also prepared rGO-SF through electrochemical reduction, showing a pretty high tensile strength of 300 MPa and toughness of 2.8 MJ m^{-3} at low content of silk.

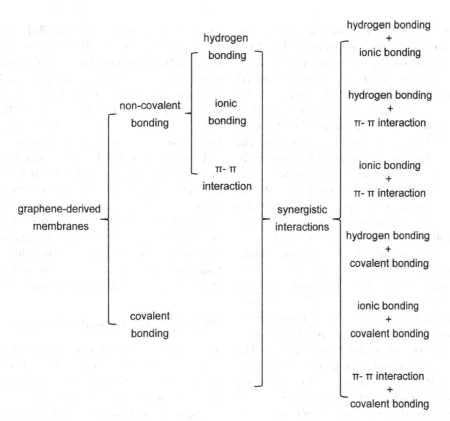

FIGURE 20.5 The interfacial engineering of graphene-derived membranes.

For biological materials, a small amount of metal ions in the protein matrix plays a significant role in improving the mechanical properties. Similarly, ionic bonding is also an efficient way to enhance the mechanical properties of graphene-derived membranes. Park et al. (2008) first modified GO nanoplatelets with Mg^{2+} and Ca^{2+}, and studied the ionic bonding reaction mechanism via Fourier transform infrared spectroscopy, X-ray photoelectron spectroscopy and Raman spectra. They found that significant enhancement in mechanical stiffness (10%–200%) and fracture strength (~50%) of graphene-derived membranes can be achieved upon modification with a small amount (<1 wt%) of Mg^{2+} and Ca^{2+}, and the main contribution to mechanical enhancement of the membranes is attributed to forming bonds between ions and the carboxylate groups on the edges. Lam et al. (2015) impregnated a small amount of Zn via atomic layer deposition. The GO-Zn membranes demonstrated Young's modulus of 35.4 GPa and strength of 142.2 MPa, which were 20% and 27% higher than the pristine GO membranes, respectively. It should be noted that metal ions not only can be added in GO solution but also may be introduced during the assembly process. Yeh et al. (2014) introduced Al^{3+} to GO nanoplatelets during the vacuum filtration assembly using anodized aluminum oxide (AAO) as the filter discs. The GO (AAO) membranes demonstrated a Young's modulus of 26.2 ± 4.6 GPa and tensile strength of 100.5 ± 19.2 MPa, which were higher than 7.6 ± 1.1 GPa and 86.9 ± 8.9 MPa of the pristine GO membranes (Teflon as the filter discs), respectively.

Conjugated molecules such as pyrene derivatives can build π–π bonding between adjacent graphene nanoplatelets due to its sp^2-hybridization (Liu et al., 2014a; Xu et al., 2008). A large number of studies so far demonstrated enhanced electrical conductivity and mechanical properties of graphene-derived membranes through introducing π–π bonding. Xu et al. (2008) cross-linked graphene nanoplatelets with 1-pyrenebutyrate (PB), and the rGO-PB showed enhanced mechanical properties, with Young's modulus of 4.2 GPa and tensile strength of 8.4 MPa. Zhang et al. (2015) functionalized polyethylene glycols (PEG) with phenyl, pyrene and di-pyrene on the both sides of PEG chain. The rGO with adjacent nanoplatelets cross-linked through strong π–π interaction with modified PEG demonstrated a dramatic enhancement of tensile strength from 15 to 45 MPa. In addition, it was found that both purine and pyrimidine bases from DNA can be applied to cross-link graphene nanoplatelets with π–π bonding. Patil et al. (2009) used DNA in the fabrication of graphene sheets and found that graphene-derived membranes can be produced through co-assembly of negatively charged DNA-rGO and positively charged cytochrome c (Cytc). Recently, Wan et al. (2018b) sequentially bridged rGO with π–π bonding, which demonstrated an extraordinary tensile strength of 668.5 MPa.

Covalent bonding is relatively stronger than abovementioned cross-links. Gao et al. (2011) introduced glutaraldehyde (GA) molecules to construct covalent bonding between the GO nanoplatelets. The GA-treated GO membranes demonstrated an average Young's modulus of 30.4 GPa and strength of 101 MPa, which were 190% and 60% higher than as-prepared GO membranes, respectively. The large enhancement may depend on the intermolecular acetalization of aldehyde groups of GA molecules and hydroxyl groups on the nanoplatelets. Besides, 10,12-pentacosadiyn-1-ol (PCDO), a kind of long-chain molecule, is applied to cross-link GO nanoplatelets via covalent bonding between alcohol groups of PCDO and carboxylic acids on the nanoplatelet surface. The long chain of PCDO results in huge slippage upon mechanical loading and thus endows the membranes with an extraordinary toughness of 2.52 MJ m^{-3} and tensile strength of 106.6 MPa. The rGO-PCDO achieved a higher toughness of 3.91 MJ m^{-3} and tensile strength of 156.8 MPa (Cheng et al., 2013). In addition, it is efficient to cross-link GO nanoplatelets covalently with multifunctional polymers. Park et al. (2009) reported GO membranes cross-linked by poly(acrylic acid) (PAA) with Young's modulus of 33.3 ± 2.7 GPa and tensile strength of 91.9 ± 22.4 MPa. Tian et al. (2013) fabricated GO membranes cross-linked with poly(etherimide) (PEI), a kind of polymer with rich amino groups. The maximum Young's modulus and tensile strength reached 84.8 GPa and 178.9 MPa, which were 163.4% and 50.3% higher than those of uncross-linked GO membranes, respectively. Furthermore, GO-diamine-fiber membranes were fabricated and displayed Young's modulus from 17.3 to 26.6 GPa and tensile strength from 275.1 to 384.3 MPa depending on the length of diamine fiber (Kim et al., 2014).

Natural nacre exhibits synergistic strengthening and toughening from various interfacial interactions (Wegst et al., 2015), which provides inspiration of interfacial engineering in enhancing the mechanical properties of graphene-derived membranes by combining different cross-links. Cui et al. (2014) prepared GO membranes cross-linked by dopamine (DA) that self-polymerized into long-chain poly(dopamine) (PDA) and formed both hydrogen and covalent bonding with GO nanoplatelets through evaporation assembly process. The GO-PDA with 5% DA demonstrated the tensile strength of 204.9 MPa and toughness of 4.0 MJ m^{-3}. Wan et al. (2015b) fabricated GO-chitosan (CS) membranes with an impressive tensile strength of 526 MPa and toughness of 17.7 MJ m^{-3}, which was attributed to the synergistic effect of hydrogen and covalent bonding. In addition, Wan et al. (2018a) combined ionic bonding (Cr^{3+}) and π–π bonding (pyrene end groups) to enhance the mechanical properties, showing a high tensile strength of 821 MPa and toughness of 20 MJ m^{-3}. Recently, Wan et al. (2018b) sequentially bridged graphene (SBG) with various ratios of π–π bonding and covalent bonding. The maximum tensile strength reached ultrahigh 944.5 MPa, which was one of the highest reported values of tensile strength for graphene-derived membranes.

Synergistic effect has been found not only in cross-links but also in building blocks. Combination of one-dimensional fibers such as single-walled nanotubes (SWNTs)

(Shin et al., 2012), double-walled nanotubes (DWNTs) (Gong et al., 2015) or cellulose nanocrystals (CNC) (Xiong et al., 2016a), and two-dimensional platelets such as molybdenum disulfide (MoS$_2$) (Wan et al., 2015a) has been found to make a contribution to enhance the mechanical properties of GO membranes. Wan et al. (2015a) demonstrated a synergistic effect of GO and MoS$_2$ through vacuum filtration self-assembly process with thermoplastic polyurethane (TPU), and the rGO-MoS$_2$-TPU showed the tensile strength of 235.3 MPa and toughness of 6.9 MJ m^{-3}. Furthermore, synergistic toughening combining both building blocks and interfacial interactions has been achieved. Gao et al. (2018b) chose CNC and Cd^{2+} to toughen GO membranes. The synergistic effect from one-dimensional CNC and two-dimensional GO, the hydrogen bonding from CNC and the ionic bonding from Cd^{2+} contributed to the high tensile strength of 419.4 MPa and toughness 11 MJ m^{-3}.

20.4.2 Theoretical Advances

The relationship between mechanical properties and layered structures can be captured by shear-lag model. A few theoretical studies have been conducted on the effect of building blocks and interlayer cross-links on the stiffness, strength and toughness of the layered materials. Gao et al. (2003) established a tensile-shear chain (TSC) model. Zhang et al. (2010) expanded the model to a non-uniform or even random platelet alignment. Although widely used for layered nanocomposites, TSC model failed to predict the mechanical properties of graphene-derived membranes for ignoring the tensile deformation of the nanoplatelets, which, however, is significant for atom-thick building blocks in graphene-derived membranes. Therefore, Liu et al. (2012) proposed a deformable tension-shear chain (DTSC) model that transfers the tensile loads acting on the membranes into in-plane tensile deformation of the platelets and shear deformation of adjacent interface. By virtue of the DTSC model (Figure 20.6a), the overall mechanical and structural responses of the graphene-derived membranes can be analyzed. The in-plane displacements of the two adjacent platelets are denoted as $u_1(x)$ and $u_2(x)$. For the two graphene nanoplatelets in the representative volume element (RVE), the equilibrium equations are

$$D\frac{\partial^2 u_1(x)}{\partial x^2} = 2G\frac{u_1(x) - u_2(x)}{h_0} \qquad (20.1)$$

$$D\frac{\partial^2 u_2(x)}{\partial x^2} = 2G\frac{u_2(x) - u_1(x)}{h_0} \qquad (20.2)$$

And the boundary conditions are

$$D\frac{\partial u_1(0)}{\partial x} = 0, D\frac{\partial u_1(l)}{\partial x} = F_0, D\frac{\partial u_2(0)}{\partial x} = F_0,$$
$$\text{and } D\frac{\partial u_2(l)}{\partial x} = 0.$$

Here F_0 is the applied tensile load, l is the lateral size of the RVE, and D, G and h_0 are the tensile stiffness of the platelets, the shear modulus of interlayer cross-links

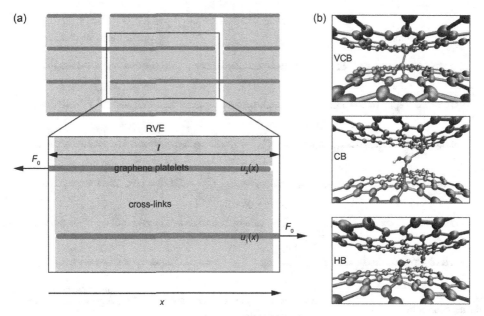

FIGURE 20.6 (a) DTSC model of graphene-derived membranes. The RVE containing two sheets (1 and 2) is denoted by the dashed box. (b) Detailed atomic structures of the cross-links, which include the VCBs, the divalent atom (magnesium)-assisted CBs and the HBs formed between two hydroxyl groups. (Reprinted with permission from Gao et al., 2017a. Copyright 2017 American Chemical Society.)

and the initial interlayer spacing. Based on Eqs. 20.1 and 20.2, the modulus, strength and toughness can be derived for theory analysis and optimum design (Gao et al., 2017a; Liu and Xu, 2014; Liu et al., 2012). Gao et al. (2017a) proposed stiffening, strengthening and toughening strategies for graphene-derived membranes through optimizing the concentration of interlayer cross-links on the basis of first-principles calculations and DTSC model. They considered three types of interlayer cross-linking mechanisms reported in experimental work, including vacancy-induced covalent bond (VCB), the divalent atom (magnesium)-assisted CB and the HB formed between two hydroxyl groups (Figure 20.6b). For VCB, two carbon atoms in the adjacent graphene nanoplatelets are covalently bonded, as well as leaving a vacancy defect in each platelet. For CB, a divalent metal atom (Mg) in the interlayer spacing bridges the hydroxyl groups bonded to the platelets. For HB, two hydroxyl groups on the top and bottom platelets interact through HBs. The first-principles calculation demonstrated that introducing cross-links can enhance the interlayer shear resistance while reducing the in-plane tensile resistance. Therefore, an optimum cross-link density to maximize the overall tensile stiffness and tensile strength appears, which are captured by the DTSC model. The optimum Young's modulus, tensile strength and strain to failure of graphene-derived membranes are predicted, which indicates that there is plenty of room for improvements of mechanical performance compared with present experiment advances. In addition, Liu and Xu (2014) explored the strategies of strengthening and toughening of graphene-derived membranes by introducing multimodal cross-links (including self-healable cross-links)-based DTSC model and MD simulation. The theoretical and simulated results

demonstrated that multimodal cross-links can effectively transfer tensile load to enhance the strength while self-healable cross-links can improve the toughness. However, it should be noted that DTSC model is developed for layered materials, and it neglects the rich ruga structures of graphene-derived membranes, as introduced in Section 20.3.3. More theoretical efforts should be paid on the mechanics of ruga structures, which may provide a thorough understanding of the structure–property relation of graphene-derived membranes, and bridge the huge gap of mechanical properties of graphene-derived membranes between present theory (DTSC) prediction and experiment reports.

20.5 Separation Features

GO is found to be an excellent nanoscale material to construct separation membranes for its two-dimensional structure, thickness (only one atom thick), and great stability. We could simply believe that the interlayer nanochannels between GO membranes allow ionic hydrates and molecules that have a smaller size than the interlayer spacing to pass through while the larger ones staying at the entrance (Figure 20.7a). For example, Joshi et al. (2014) reported that ions smaller in size than the void spacing of GO can permeate in the graphene-derived membranes at a speed that is several orders of magnitude faster than that through simple diffusion. Size exclusion appears to be the dominant separation mechanism. Compared with typically commercial polymeric membranes, the channel-size distribution of graphene-derived membranes is narrower and thus more advantageous for precise sieving. Therefore,

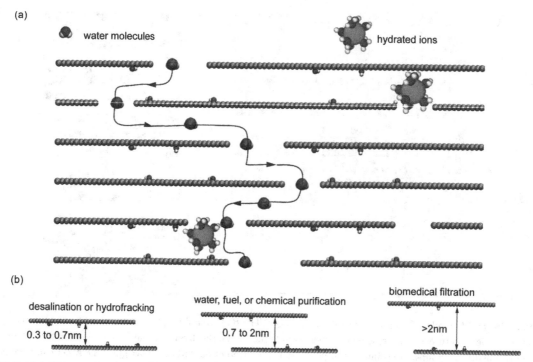

FIGURE 20.7   (a) Water molecules quickly permeate through the graphene-derived membrane while hydrated ions are blocked. (b) The separation capability of the graphene-derived membrane can be tunable by adjusting the interlayer spacing.

a broad spectrum of GO membranes to precisely separate target ions and molecules within a specific size range could be made by adjusting the interlayer spacing between GO nanoplatelets (Figure 20.7b). More specifically, by partially reducing GO or covalently bonding the stacked GO nanoplatelets, the GO spacing could be less than 0.7 nm for desalination (sieving the hydrated ions from water). By inserting large, rigid chemical groups or soft polymer chains (e.g., polyelectrolytes) between GO nanoplatelets, an enlarged GO spacing (0.7–2 nm) can be achieved for applications in water purification and reclamation, and pharmaceutical and fuel separation. If even larger sized nanoparticles or nanofibers are used as spacers, GO membranes with larger spacing (>2 nm) may be assembled for precise separation of large biomolecules and small waste molecules in biomedical applications (e.g., artificial kidneys and dialysis) (Mi, 2014). In addition, the performance of GO membranes, however, depends on more other factors, for example, interfacial chemistry of GO. Therefore, not only the interlayer spacing should be precisely controlled, interfacial engineering of GO membranes by functional groups is also important. In addition, GO membranes would react with water, that is to say, hydration in aqueous, which makes the control of interlayer spacing become a challenge. The separation in aqueous is different from that in dry gas, since hydrogen bonding, hydrophobic interaction, electrostatic interaction and numerous hydrated ions in aqueous should be considered (Ma et al., 2017). Herein, the separation can be briefly divided into two types: liquid separation and gas separation.

20.5.1   Liquid Separation

Sieving ions and molecules from solution is of great importance in a wide range of fields such as water desalination and energy production. The free oxidized surfaces of graphene in graphene-derived membranes exhibit frictionless walls and allow water molecules flow extremely fast, leading to the efficient sieving. Research on graphene-derived membranes for water purification has been excited since the unimpeded water permeation through GO membranes was found (Figure 20.8) (Nair et al., 2012). More specifically, by assuming that water flows in the nanochannels of graphene-derived membranes as a classical liquid, Nair et al. (2012) described the flux J of water in graphene-derived membranes with the Poiseuille's law:

$$J = \frac{\delta^3 \Delta P}{12 L^2 \eta l} \tag{20.3}$$

where δ is the void spacing of adjacent nanoplatelets, ΔP is the differential pressure, L is the average lateral size of the graphene nanoplatelets, η is the water viscosity and l is the length of the nanochannel. The calculation indicated an enhanced water flow by a factor of a few hundred compared with the classical flow. The huge gap between experimental measurements and the theoretical equation indicated that the flow velocity of the water in the nanochannel surface was not zero, and a slip length of 10–100 nm was predicted to explain the enhanced water flow. However, it should be remarked here that the slip flow is a controversial theory without direct proof. As counterexamples, Huang et al. (2013b) also measured 4–6 orders of magnitude higher water flux in the filtration

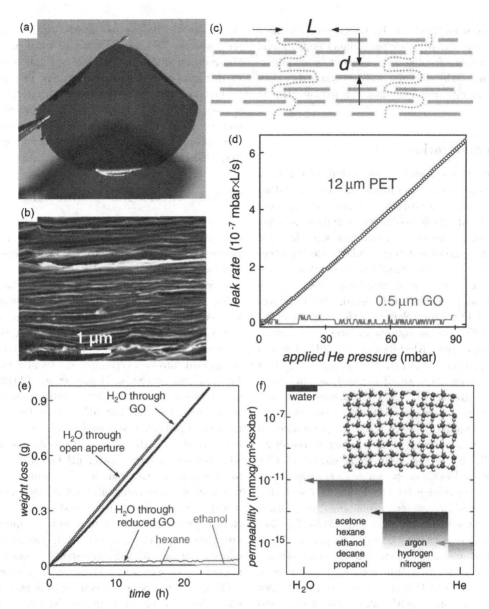

FIGURE 20.8 (a) Optical image of a 1-mm-thick GO membrane peeled off of a Cu foil. (b) Electron micrograph of the film's cross section. (c) Schematic view for possible permeation through the laminates. (d) Examples of He-leak measurements for a freestanding sub-micrometer-thick GO membrane and a reference PET membrane (normalized per square centimeter). (e) Weight loss for a container sealed with a GO film. No loss was detected for ethanol, hexane, etc., but water evaporated from the container as freely as through an open aperture. The measurements were carried out at room temperature in zero humidity. (f) Permeability of GO membranes with respect to water and various small molecules (arrows indicate the upper limits set by our experiments) (Nair et al., 2012).

experiments compared with the theoretical prediction by Poiseuille's law. However, they demonstrated that water flow through the nanochannels between GO nanoplatelets exhibits a classic viscous feature with negligible boundary slip. The rich porous structures of GO, including enlarged interlayer spacing, wide channels formed from ruga structures, holes and inter-edge spaces, were regarded as the main cause of fast flow of water across graphene-derived membranes. In addition, Wei et al. (2014a,b) revealed the breakdown of fast water transport in GO nanochannels by MD due to the side-pinning effect by water confined among oxidized regions. Therefore, a more reasonable explanation

appears to be that water molecules prefer to go through the defects of graphene-derived membranes, leading to smaller L, explaining the higher flux calculated from Eq. 20.3. Besides, the water purification performance of the graphene-derived membranes was evaluated in two main aspects, including retention of organic dyes and salts rejection. In general, the graphene-derived membranes were proven to be efficient for rejecting organic dyes with the rejection of more than 85%. While the rejections of monovalent or divalent salts were relatively lower than 50%. This is because the interlayer spacings of graphene-derived membranes are within the molecular size of organic dyes while too large for salts rejection. Therefore, creating more thin porous structure and reducing the channel length would be effective approaches for flow enhancement in graphene-derived membranes, achieving much higher water flux than the commercial membranes with larger rejection (Huang et al., 2013b).

20.5.2   Gas Separation

Graphene-derived membranes are considered as the new-generation gas separation membranes (Liu et al., 2015a) and thus have been extensively studied. In general, the nanochannels of graphene-derived membranes allow the permeation of gas molecules smaller in size than the interlayer spacing of the nanoplatelets while hindering larger molecules. Indeed, even the small gas molecules cannot permeate into the interlayer of graphitic layers with a spacing of 0.34 nm. On the other hand, the interlayer spacing between GO nanoplatelets is in the range from 0.7 to 1.2 nm, depending on the content of intercalated water and the oxidation degree (Klechikov et al., 2015; Rezania et al., 2014; Talyzin et al., 2014). Nair et al. (2012) first showed that no gases (even small helium) can permeate through fully dry graphene-derived membranes. However, helium appeared to permeate and water unprecedentedly fast transported through the hydrated graphene-derived membrane as it was exposed to humidified atmosphere, which was attributed to the hydrophilic nature of GO. More specifically, the interlayer spacing of GO nanoplatelets increases with the increasing relative humidity due to the increasing of intercalated water, explaining the unusual permeation behaviors of molecules (Figure 20.8). In addition, various factors, such as the size, shape, oxidation degree, stacking manner and defects, influence the intrinsic gas transport properties of graphene-derived membranes. For example, Li et al. (2013) prepared ultrathin and thick graphene-derived membranes with thicknesses of ~1.8 and ~18 nm, respectively, and they found that the ultra-thin graphene-derived membranes exhibited much higher gas permeability than the thick ones, indicating that the gas permeability of graphene-derived membranes strongly depends on their thickness. Besides, the gas permeation path in graphene-derived membranes is different compared with traditional homogenous membranes, since the nanochannels are formed in layered structures. Kim et al. (2013) studied the gas permeabilities of ultrathin GO supported on microporous polymer membranes, as well as those of freestanding thick and dry membranes by applying a high pressure. Small gas molecules permeated through the freestanding thick graphene-derived membranes with various low permeation rates due to the narrow interlayer spacing at dry state. The order of gas permeabilities ($He > H_2 > CO_2 > O_2 > N_2 > CH_4$) is consistent with the kinetic diameters of these gases, indicating that the molecular sieving appeared to be the dominated mechanism. They also prepared GO membranes from GO with various sizes by adjusting the sonication time of the

suspensions. The gas permeabilities increased remarkably with average GO size decreasing, which was attributed to the reduced diffusional pathways. Therefore, the presence of intrinsic or post-treated selective defects can be used to enhance the molecular-sieving properties of graphene-derived membranes. In addition, the permeability and selectivity of membranes can be influenced and controlled by the separation conditions such as pressure and temperature (Xu and Zhang, 2016).

20.5.3   Interfacial Engineering

As preparing separation membranes, various kinds of factors may affect the separation efficiency. For example, the permeability would be affected if the ions or molecules block the pores on the surface of GO or nanochannels between platelets (Xu and Zhang, 2016). The main challenge using graphene-derived membranes for separation is how to preciously control the interlayer spacing and interfacial structures to break the permeability-selectivity trade-off. In general, the strategies to control interlayer spacing can be divided into two types: physical and chemical methods. For physical methods, Huang et al. (2013a) found that permeability and selectivity of separation membranes are sensitive to the pH. For pH $\leq$ 2, water can hardly pass through the graphene-derived membranes. For 2 < pH $\leq$ 6, the pore size on the surface of membranes would increase with the increasing of water pH for electrostatic repulsion force among GO nanoplatelets decreasing, which caused the permeability increasing and selectivity decreasing. For 6 < pH $\leq$ 8, the permeability and selectivity would hardly change for the negative charges of GO nanoplatelets nearly not changing. And for pH $\geq$ 9, the interlayer spacing would decrease with the increase in ion concentration, which caused the permeability decreasing. In addition, strain engineering and temperature can be also used to control the interlayer spacing of graphene-derived membranes. Gao and Xu (2018) cross-linked GO nanoplatelets with covalent interlayer cross-links and applied tensile strain to the membranes, which resulted in the rotation of cross-links and further led to the change of the interlayer spacing, as shown in Figure 20.9. Liu et al. (2017) reported membranes with a negative temperature-response coefficient that were capable of gating tunable water and precisely separating small molecule. On the other hand, chemical methods are also used to engineer the structure of graphene-derived membranes. Yuan et al. (2017) proved that the GO-COOH membranes showed a higher performance in salt rejection and water permeability. It is because the addition of carboxyl groups increases the negativity of membranes, resulting in the improvement of the salt rejection, and the carboxyl groups on the membranes increase the number of wrinkles, causing enlarged nanochannels and high water flux. In addition, Shen et al. (2016b) showed that separation membranes with subnanometer interlayer spacing could be prepared by removing oxygenated functional groups on GO nanoplatelets. Meanwhile, this kind of membranes has

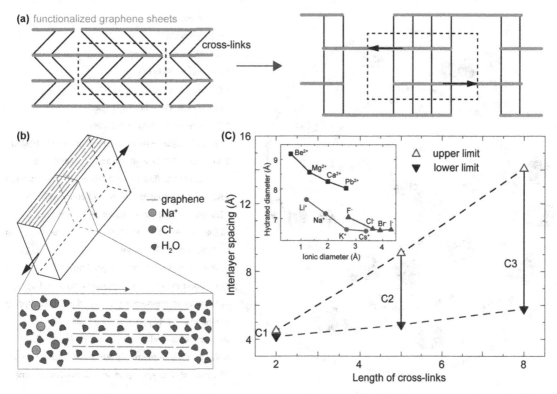

FIGURE 20.9 (a) DTSC model and RVE of graphene-derived membranes, before and after tensile loading is applied. The interlayer spacing changes with tensile strain applied. (b) Size-sieving mechanism in water desalination. (c) A summary of the hydrated ionic diameters and the range of strain-controlled interlayer spacing in the graphene-derived membranes, where the upper and lower limits correspond to the extreme values of interlayer spacing engineered through strain. The critical size range in molecular sieving is covered by the strain-engineering approach. (Reprinted with permission from Gao and Xu, 2018. Copyright 2018 The Royal Society of Chemistry.)

a performance of highly selective gas transport. Besides, inserting some specific chemical groups such as copper hydroxide nanostrands (Huang et al., 2013b), dicarboxylic acids (Jia and Wang, 2015) or poly-(vinylpyrrolidone) (PVP) (Li et al., 2014) into GO membranes by covalently cross-linking can also modulate the separation performance of graphene-derived membranes (Ma et al., 2017).

20.5.4 Remarks on the Separation Features

Graphene-derived membrane, with excellent performance of mechanical strength (Liu et al., 2012), chemical stability (Gao, 2015), antibacterial activity (Hu et al., 2010) and antifouling properties (Dikin et al., 2007), is believed to be one of the next generation of ultrathin, high-flux and energy-efficient membranes for precise ionic and molecular sieving in aqueous solution, with applications in numerous important fields. But limited by the assembly and post-treated technologies, the permeability and selectivity of graphene-derived membranes cannot be totally precisely controlled in reality. More efforts are needed for thorough understanding of the separation mechanisms of ions and molecules besides size exclusion. Further research is needed to address the specific issues considering various exciting yet challenging applications in desalination, water treatment and energy production, as well as in biomedical and pharmaceutical

fields. For gas separation, even in ultrathin graphene-derived membranes, gas permeabilities are still low because of high tortuosity and limited interlayer spacing, and more efforts should be paid on tailoring the interlayer spacing, flake size and the interfacial chemistry of GO to break the permeability–selectivity trade-off. In addition, the chemical or thermal stability of layered structures as well as their ruga structures should be further considered for practical applications.

20.6 Conclusions and Perspectives

This chapter provides a comprehensive review of the state of the art of research in graphene-derived membranes, encompassing their assembly approaches, multilevel structures, mechanical properties, structure–property models and separation features. Briefly, numerous assembly approaches of graphene-derived membranes have been introduced, and the Young's modulus, tensile strength and toughness have been summarized from experiments, theoretical analysis and numerical simulations. Besides, the separation of ions and molecules of graphene-derived membranes has been discussed, and various methods including interfacial engineering and interlayer spacing control have been applied to break the permeability-selectivity trade-off of graphene-derived membranes.

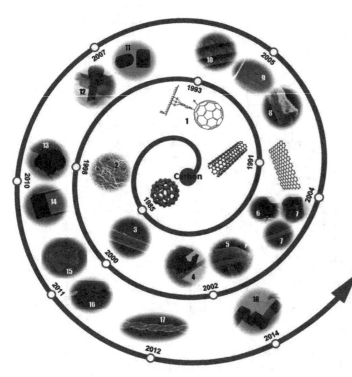

1985 - Kroto et al. C_{60}

1991 - Iijima CNT

2004 - Geim et al. Graphene

1 Mirkin et al. SAM of C_{60}

2 Smalley et al. Bucky paper

3 Poulin et al. Wet-spun CNT fiber

4 Fan et al. Array-drawn CNT fiber

5 Wu and Ajayan et al. FCCVD synthesized CNT fiber

6 Xie et al. FCCVD synthesized CNT membrane

7 Windle et al. CVD aerogel derived CNT fiber/membrane

8 Kozlov and Baughman et al. Wet-spun CNT fiber

9 Baughman et al. Array-drawn CNT membrane

10 Cao and Ajayan et al. CVD-grown aligned CNT foam

11 Yodh et al. CNT aerogel

12 Ruoff et al. Free-stranding graphene-derived membrane

13 Cao and Wu et al. CVD-grown random CNT sponge

14 Shi et al. Hydrothermally prepared graphene hydrogel

15 Gao et al. Wet-spun graphene fiber

16 Cheng et al. CVD-grown graphene foam

17 Qu et al. Hydrothermallly fabricated graphene fiber

18 Gao et al. Wet-spun graphene-derived membrane

FIGURE 20.10   Growth ring of assembly of sp^2 nanocarbons. (Reprinted with permission from Li et al., 2015. Copyright 2015 American Chemical Society.)

As a kind of graphene assembly, preliminary studies have shown that bio-inspired graphene-derived membranes can potentially transfer the superb properties of individual graphene to the macro-structural level using simpler bottom-up assembly approaches than those for fabricating traditional high-performance and multifunctional membranes. During the past decade, research on graphene-derived membranes has been a focus of the research of sp^2 nanocarbons (Figure 20.10). Considerable progress has been achieved over the past few years in the developing of graphene-derived membranes to improve both performance and stability. However, there are still some challenges for practical applications due to the limited performance improvement of graphene-derived membrane compared with its building blocks. For example, the graphene-derived membrane shows great reduction of mechanical performance compared with that of graphene, which makes it less competitive with commercial high-performance materials. Considering the non-uniformity of the building blocks in terms of size and shape, and the functional groups distributed on the interfaces, future research efforts on the precisely controlled assembly technologies are fully required before the potential applications of graphene-derived membranes can be realized.

Acknowledgments

E.G. acknowledges the support from the starting-up funding of Wuhan University.

References

Abraham, J., Vasu, K. S., Williams, C. D., et al. 2017. Tunable sieving of ions using graphene oxide membranes. *Nat. Nanotech.*, 12, 546.

An, Z., Compton, O. C., Putz, K. W., Brinson, L. C. and Nguyen, S. T. 2011. Bio-inspired borate cross-linking in ultra-stiff graphene oxide thin films. *Adv. Mater.*, 23, 3842.

Bekyarova, E., Sarkar, S., Wang, F., et al. 2013. Effect of covalent chemistry on the electronic structure and properties of carbon nanotubes and graphene. *Acc. Chem. Res.*, 46, 65.

Brodie, B. C. 1859. On the atomic weight of graphite. *Philos. Trans. R. Soc. Lond.*, 149, 249.

Carpenter, C., Maroudas, D. and Ramasubramaniam, A. 2013. Mechanical properties of irradiated single-layer graphene. *Appl. Phys. Lett.*, 103, 013102.

Chavez-Valdez, A., Shaffer, M. S. and Boccaccini, A. R. 2013. Applications of graphene electrophoretic deposition. A review. *J. Phys. Chem. B*, 117, 1502.

Chen, L., Shi, G., Shen, J., et al. 2017. Ion sieving in graphene oxide membranes via cationic control of interlayer spacing. *Nature*, 550, 380.

Cheng, C., Jiang, G., Garvey, C. J., et al. 2016. Ion transport in complex layered graphene-based membranes with tuneable interlayer spacing. *Sci. Adv.*, 2, e1501272.

Cheng, Q., Duan, J., Zhang, Q. and Jiang, L. 2015. Learning from nature: Constructing integrated graphene-based artificial nacre. *ACS Nano*, 9, 2231.

Cheng, Q., Wu, M., Li, M., Jiang, L. and Tang, Z. 2013. Ultratough artificial nacre based on conjugated cross-linked graphene oxide. *Angew. Chem.,* 52, 3750.

Compton, O. C., Cranford, S. W., Putz, K. W., et al. 2012. Tuning the mechanical properties of graphene oxide paper and its associated polymer nanocomposites by controlling cooperative intersheet hydrogen bonding. *ACS Nano,* 6, 2008.

Cui, W., Li, M., Liu, J., et al. 2014. A strong integrated strength and toughness artificial nacre based on dopamine cross-linked graphene oxide. *ACS Nano,* 8, 9511.

Dewapriya, M. A. N. and Rajapakse, R. K. N. D. 2014. Molecular dynamics simulations and continuum modeling of temperature and strain rate dependent fracture strength of graphene with vacancy defects. *J. Appl. Mech.,* 81, 081010.

Dikin, D. A., Stankovich, S., Zimney, E. J., et al. 2007. Preparation and characterization of graphene oxide paper. *Nature,* 448, 457.

Dimiev, A. M. and Tour, J. M. 2014. Mechanism of graphene oxide formation. *ACS Nano,* 8, 3060.

Duan, J., Gong, S., Gao, Y., et al. 2016. Bioinspired ternary artificial nacre nanocomposites based on reduced graphene oxide and nanofibrillar cellulose. *ACS Appl. Mater. Interfaces,* 8, 10545.

Faccio, R., Denis, P. A., Pardo, H., Goyenola, C. and Mombru, A. W. 2009. Mechanical properties of graphene nanoribbons. *J. Phys. Condens. Matter,* 21, 285304.

Faucett, A. C. and Mativetsky, J. M. 2015. Nanoscale reduction of graphene oxide under ambient conditions. *Carbon,* 95, 1069.

Frank, I. W., Tanenbaum, D. M., Van der Zande, A. M. and McEuen, P. L. 2007. Mechanical properties of suspended graphene sheets. *J. Vac. Sci. Technol. B,* 25, 2558.

Gao, E., Cao, Y., Liu, Y. and Xu, Z. 2017a. Optimizing interfacial cross-linking in graphene-derived materials, which balances intralayer and interlayer load transfer. *ACS Appl. Mater. Interfaces,* 9, 24830.

Gao, E. and Xu, Z. 2018. Bio-inspired graphene-derived membranes with strain-controlled interlayer spacing. *Nanoscale,* 10, 8585.

Gao, E., Lu, W. and Xu Z. 2018a. Strength loss of carbon nanotube fibers explained in a three-level hierarchical model. *Carbon,* 138, 134.

Gao, H., Ji, B., Jager, I. L., Arzt, E. and Fratzl, P. 2003. Materials become insensitive to flaws at nanoscale: Lessons from nature. *Proc. Natl. Acad. Sci.,* 100, 5597.

Gao, H. L., Chen, S. M., Mao, L. B., et al. 2017b. Mass production of bulk artificial nacre with excellent mechanical properties. *Nat. Commun.,* 8, 287.

Gao, W. 2015. The chemistry of graphene oxide. In: *Graphene Oxide* (ed. W. Gao). Springer: Berlin, pp. 61–95.

Gao, Y., Liu, L. Q., Zu, S. Z., et al. 2011. The effect of interlayer adhesion on the mechanical behaviors of macroscopic graphene oxide papers. *ACS Nano,* 5, 2134.

Gao, Y., Xu, H. J. and Cheng, Q. F. 2018b. Multiple synergistic toughening graphene nanocomposites through cadmium ions and cellulose nanocrystals. *Adv. Mater. Interfaces,* 5, 1800145.

Gomez-Navarro, C., Burghard, M. and Kern, K. 2008. Elastic properties of chemically derived single graphene sheets. *Nano Lett.,* 8, 2045.

Gong, L., Kinloch, I. A., Young, R. J., et al. 2010. Interfacial stress transfer in a graphene monolayer nanocomposite. *Adv. Mater.,* 22, 2694.

Gong, S., Cui, W., Zhang, Q., et al. 2015. Integrated ternary bioinspired nanocomposites via synergistic toughening of reduced graphene oxide and double-walled carbon nanotubes. *ACS Nano,* 9, 11568.

Gupta, S., Dharamvir, K. and Jindal, V. K. 2005. Elastic moduli of single-walled carbon nanotubes and their ropes. *Phys. Rev. B,* 72, 165428.

Hartmann, M. A. and Fratzl, P. 2009. Sacrificial ionic bonds need to be randomly distributed to provide shear deformability. *Nano Lett.,* 9, 3603.

Hashimoto, A., Suenaga, K., Gloter, A., Urita, K. and Iijima, S. 2004. Direct evidence for atomic defects in graphene layers. *Nature,* 430, 870.

He, L., Guo, S., Lei, J., Sha, Z. and Liu Z. 2014. The effect of stone-thrower–wales defects on mechanical properties of graphene sheets: A molecular dynamics study. *Carbon,* 75, 124.

Hofmann, U. and König, E. 1937. Untersuchungen über graphitoxyd. *Z. Anorg. Allg. Chem.,* 234, 311.

Hu, K., Tolentino, L. S., Kulkarni, D. D., et al. 2013. Written-in conductive patterns on robust graphene oxide biopaper by electrochemical microstamping. *Angew. Chem.,* 52, 13784.

Hu, M. and Mi, B. X. 2014. Layer-by-layer assembly of graphene oxide membranes via electrostatic interaction. *J. Mem. Sci.,* 469, 80.

Hu, W., Peng, C., Luo, W., et al. 2010. Graphene-based antibacterial paper. *ACS Nano,* 4, 4317.

Huang, H., Mao, Y., Ying, Y., et al. 2013a. Salt concentration, pH and pressure controlled separation of small molecules through lamellar graphene oxide membranes. *Chem. Commun.,* 49, 5963.

Huang, H., Song, Z., Wei, N., et al. 2013b. Ultrafast viscous water flow through nanostrand-channelled graphene oxide membranes. *Nat. Commun.,* 4, 2979.

Huang, M., Pascal, T. A., Kim, H., Goddard, W. A. and Greer, J. R. 2011. Electronic-mechanical coupling in graphene from in situ nanoindentation experiments and multiscale atomistic simulations. *Nano Lett.,* 11, 1241.

Hummers, W. S. and Offeman, R. E. 1958. Preparation of graphitic oxide. *J. Am. Chem. Soc.,* 80, 1339.

Hung, W.-S., Tsou, C.-H., De Guzman, M., et al. 2014. Cross-linking with diamine monomers to prepare composite graphene oxide-framework membranes with varying d-spacing. *Chem. Mater.,* 26, 2983.

Jalili, R., Aboutalebi, S. H., Esrafilzadeh, D., et al. 2013. Scalable one-step wet-spinning of graphene fibers and yarns from liquid crystalline dispersions of graphene oxide: Towards multifunctional textiles. *Adv. Funct. Mater.*, 23, 5345.

Jia, Z. Q. and Wang, Y. 2015. Covalently crosslinked graphene oxide membranes by esterification reactions for ions separation. *J. Mater. Chem. A*, 3, 4405.

Jin, E. Z., He, J. Y., Sheng, K. X., et al. 2013. Electron-irradiation-induced reinforcement of reduced graphene oxide papers. *Acta. Mater.*, 61, 6466.

Joshi, R. K., Carbone, P., Wang, F. C., et al. 2014. Precise and ultrafast molecular sieving through graphene oxide membranes. *Science*, 343, 752.

Kalosakas, G., Lathiotakis, N. N., Galiotis, C. and Papagelis, K. 2013. In-plane force fields and elastic properties of graphene. *J. Appl. Phys.*, 113, 134307.

Kim, H. W., Yoon, H. W., Yoon, S. M., et al. 2013. Selective gas transport through few-layered graphene and graphene oxide membranes. *Science*, 342, 91.

Kim, Y. S., Kang, J. H., Kim, T., et al. 2014. Easy preparation of readily self-assembled high-performance graphene oxide fibers. *Chem. Mater.*, 26, 5549.

Kis, A., Csanyi, G., Salvetat, J. P., et al. 2004. Reinforcement of single-walled carbon nanotube bundles by inter-tube bridging. *Nat. Mater.*, 3, 153.

Klechikov, A., Yu, J., Thomas, D., Sharifi, T. and Talyzin, A. V. 2015. Structure of graphene oxide membranes in solvents and solutions. *Nanoscale*, 7, 15374.

Koenig, S. P., Boddeti, N. G., Dunn, M. L. and Bunch, J. S. 2011. Ultrastrong adhesion of graphene membranes. *Nat. Nanotech.*, 6, 543.

Kotov, N. A., Dekany, I. and Fendler, J. H. 1996. Ultrathin graphite oxide-polyelectrolyte composites prepared by self-assembly: Transition between conductive and non-conductive states. *Adv. Mater.*, 8, 637.

Kudin, K. N., Scuseria, G. E. and Yakobson, B. I. 2001. C_2F, BN, and C nanoshell elasticity from ab initio computations. *Phys. Rev. B*, 64, 235406.

Kunz, D. A., Feicht, P., Godrich, S., et al. 2013. Space-resolved in-plane moduli of graphene oxide and chemically derived graphene applying a simple wrinkling procedure. *Adv. Mater.*, 25, 1337.

Lam do, V., Gong, T., Won, S., et al. 2015. A robust and conductive metal-impregnated graphene oxide membrane selectively separating organic vapors. *Chem. Commun.*, 51, 2671.

Lanqing, X., Ning, W. and Yongping, Z. 2013. Mechanical properties of highly defective graphene: From brittle rupture to ductile fracture. *Nanotechnology*, 24, 505703.

Lee, C., Wei, X., Kysar, J. W. and Hone, J. 2008. Measurement of the elastic properties and intrinsic strength of monolayer graphene. *Science*, 321, 385.

Li, H., Song, Z., Zhang, X., et al. 2013. Ultrathin, molecular-sieving graphene oxide membranes for selective hydrogen separation. *Science*, 342, 95.

Li, W., Zhang, Y., Xu, Z., et al. 2014. Self-assembled graphene oxide microcapsules with adjustable permeability and yolk-shell superstructures derived from atomized droplets. *Chem. Commun.*, 50, 15867.

Li, Y. Q., Yu, T., Yang, T. Y., Zheng, L. X. and Liao, K. 2012. Bio-inspired nacre-like composite films based on graphene with superior mechanical, electrical, and biocompatible properties. *Adv. Mater.*, 24, 3426.

Li, Z., Liu, Z., Sun, H. and Gao, C. 2015. Superstructured assembly of nanocarbons: Fullerenes, nanotubes, and graphene. *Chem. Rev.*, 115, 7046.

Lin, X., Shen, X., Sun, X., et al. 2016. Graphene oxide papers simultaneously doped with Mg^{2+} and Cl^- for exceptional mechanical, electrical, and dielectric properties. *ACS Appl. Mater. Interfaces*, 8, 2360.

Liu, F., Ming, P. M. and Li, J. 2007. Ab initio calculation of ideal strength and phonon instability of graphene under tension. *Phys. Rev. B*, 76, 064120.

Liu, G., Jin, W. and Xu, N. 2015a. Graphene-based membranes. *Chem. Soc. Rev.*, 44, 5016.

Liu, H., Wang, H. and Zhang, X. 2015b. Facile fabrication of freestanding ultrathin reduced graphene oxide membranes for water purification. *Adv. Mater.*, 27, 249.

Liu, J., Wang, N., Yu, L. J., et al. 2017. Bioinspired graphene membrane with temperature tunable channels for water gating and molecular separation. *Nat. Commun.*, 8, 2011.

Liu, L., Gao, Y., Liu, Q., et al. 2013. High mechanical performance of layered graphene oxide/poly(vinyl alcohol) nanocomposite films. *Small*, 9, 2466.

Liu, Y., Yuan, L., Yang, M., et al. 2014a. Giant enhancement in vertical conductivity of stacked CVD graphene sheets by self-assembled molecular layers. *Nat. Commun.*, 5, 5461.

Liu, Y. L., Xie, B. and Xu, Z. P. 2011. Mechanics of coordinative crosslinks in graphene nanocomposites: A first-principles study. *J. Mater. Chem.*, 21, 6707.

Liu, Y. L., Xie, B., Zhang, Z., Zheng, Q. S. and Xu, Z. P. 2012. Mechanical properties of graphene papers. *J. Mech. Phys. Solids*, 60, 591.

Liu, Y. L. and Xu, Z. P. 2014. Multimodal and self-healable interfaces enable strong and tough graphene-derived materials. *J. Mech. Phys. Solids*, 70, 30.

Liu, Z., Li, Z., Xu, Z., et al. 2014b. Wet-spun continuous graphene films. *Chem. Mater.*, 26, 6786.

Loh, K. P., Bao, Q., Eda, G. and Chhowalla, M. 2010. Graphene oxide as a chemically tunable platform for optical applications. *Nat. Chem.*, 2, 1015.

Ma, J., Alfè, D., Michaelides, A. and Wang, E. 2009. Stone-wales defects in graphene and other planar sp^2-bonded materials. *Phys. Rev. B*, 80, 033407.

Ma, J., Ping, D. and Dong, X. 2017. Recent developments of graphene oxide-based membranes: A review. *Membranes*, 7, 7030052.

Mark, J. E. and Erman, B. 2007. *Rubberlike Elasticity: A Molecular Primer*. Cambridge University Press: Cambridge.

Medhekar, N. V., Ramasubramaniam, A., Ruoff, R. S. and Shenoy, V. B. 2010. Hydrogen bond networks in graphene oxide composite paper: Structure and mechanical properties. *ACS Nano,* 4, 2300.

Meo, M. and Rossi, M. 2006. Prediction of Young's modulus of single wall carbon nanotubes by molecular-mechanics based finite element modelling. *Compos. Sci. Technol.,* 66, 1597.

Meyer, J. C., Kisielowski, C., Erni, R., et al. 2008. Direct imaging of lattice atoms and topological defects in graphene membranes. *Nano Lett.,* 8, 3582.

Mi, B. 2014. Materials science. Graphene oxide membranes for ionic and molecular sieving. *Science,* 343, 740.

Ming, P., Song, Z. F., Gong, S. S., et al. 2015. Nacre-inspired integrated nanocomposites with fire retardant properties by graphene oxide and montmorillonite. *J. Mater. Chem. A,* 3, 21194.

Nair, R. R., Wu, H. A., Jayaram, P. N., Grigorieva, I. V. and Geim, A. K. 2012. Unimpeded permeation of water through helium-leak-tight graphene-based membranes. *Science,* 335, 442.

Park, S., Dikin, D. A., Nguyen, S. T. and Ruoff, R. S. 2009. Graphene oxide sheets chemically cross-linked by polyallylamine. *J. Phys. Chem. C,* 113, 15801.

Park, S., Lee, K. S., Bozoklu, G., et al. 2008. Graphene oxide papers modified by divalent ions-enhancing mechanical properties via chemical cross-linking. *ACS Nano,* 2, 572.

Patil, A. J., Vickery, J. L., Scott, T. B. and Mann, S. 2009. Aqueous stabilization and self-assembly of graphene sheets into layered bio-nanocomposites using DNA. *Adv. Mater.,* 21, 3159.

Pei, S. F., Zhao, J. P., Du, J. H., Ren, W. C. and Cheng, H. M. 2010. Direct reduction of graphene oxide films into highly conductive and flexible graphene films by hydrohalic acids. *Carbon,* 48, 4466.

Putz, K. W., Compton, O. C., Palmeri, M. J., Nguyen, S. T. and Brinson, L. C. 2010. High-nanofiller-content graphene oxide–polymer nanocomposites via vacuum-assisted self-assembly. *Adv. Funct. Mater.,* 20, 3322.

Qin, Z. and Buehler, M. J. 2015. Nonlinear viscous water at nanoporous two-dimensional interfaces resists high-speed flow through cooperativity. *Nano Lett.,* 15, 3939.

Qin, Z., Taylor, M., Hwang, M., Bertoldi, K. and Buehler, M. J. 2014. Effect of wrinkles on the surface area of graphene: Toward the design of nanoelectronics. *Nano Lett.,* 14, 6520.

Qiu, L., Liu, J. Z., Chang, S. L., Wu, Y. and Li, D. 2012. Biomimetic superelastic graphene-based cellular monoliths. *Nat. Commun.,* 3, 1241.

Rezania, B., Severin, N., Talyzin, A. V. and Rabe, J. P. 2014. Hydration of bilayered graphene oxide. *Nano Lett.,* 14, 3993.

Richardson, J. J., Bjornmalm, M. and Caruso, F. 2015. Technology-driven layer-by-layer assembly of nanofilms. *Science,* 348, aaa2491.

Robinson, J. T., Zalalutdinov, M., Baldwin, J. W., et al. 2008. Wafer-scale reduced graphene oxide films for nanomechanical devices. *Nano Lett.,* 8, 3441.

Sakhaee-Pour, A., Ahmadian, M. T. and Naghdabadi, R. 2008. Vibrational analysis of single-layered graphene sheets. *Nanotechnology,* 19, 085702.

Salvetat, J. P., Briggs, G. A. D., Bonard, J. M., et al. 1999. Elastic and shear moduli of single-walled carbon nanotube ropes. *Phys. Rev. Lett.,* 82, 944.

Sealy, C. 2017. Graphene sieve ions out water filtration. *Nano Today,* 14, 1.

Shahzadi, K., Mohsin, I., Wu, L., et al. 2017a. Bio-based artificial nacre with excellent mechanical and barrier properties realized by a facile in situ reduction and cross-linking reaction. *ACS Nano,* 11, 325.

Shahzadi, K., Zhang, X., Mohsin, I., et al. 2017b. Reduced graphene oxide/alumina, a good accelerant for cellulose-based artificial nacre with excellent mechanical, barrier, and conductive properties. *ACS Nano,* 11, 5717.

Shao, J. J., Lv, W. and Yang, Q. H. 2014. Self-assembly of graphene oxide at interfaces. *Adv. Mater.,* 26, 5586.

Shen, J., Zhang, M., Liu, G., Guan, K. and Jin, W. 2016a. Size effects of graphene oxide on mixed matrix membranes for CO_2 separation. *AIChE J.,* 62, 2843.

Shen, J., Zhang, M. C., Liu, G. P. and Jin, W. Q. 2016b. Facile tailoring of the two-dimensional graphene oxide channels for gas separation. *RSC Adv.,* 6, 54281.

Shin, M. K., Lee, B., Kim, S. H., et al. 2012. Synergistic toughening of composite fibres by self-alignment of reduced graphene oxide and carbon nanotubes. *Nat. Commun.,* 3, 650.

Song, Z., Artyukhov, V. I., Yakobson, B. I. and Xu, Z. 2013. Pseudo hall-petch strength reduction in polycrystalline graphene. *Nano Lett.,* 13, 1829.

Stankovich, S., Dikin, D. A., Dommett, G. H., et al. 2006. Graphene-based composite materials. *Nature,* 442, 282.

Staudenmaier, L. 1898. Verfahren zur darstellung der graphitsure. *Ber. Dtsch. Chem. Ges.,* 31, 1481.

Sun, P. Z., Chen, Q., Li, X. D., et al. 2015. Highly efficient quasi-static water desalination using monolayer graphene oxide/titania hybrid laminates. *NPG Asia Mater.,* 7, e162.

Talyzin, A. V., Hausmaninger, T., You, S. and Szabo, T. 2014. The structure of graphene oxide membranes in liquid water, ethanol and water-ethanol mixtures. *Nanoscale,* 6, 272.

Tan, Z., Zhang, M., Li, C., Yu, S. and Shi, G. 2015. A general route to robust nacre-like graphene oxide films. *ACS Appl. Mater. Interfaces,* 7, 15010.

Tian, Y., Cao, Y., Wang, Y., Yang, W. and Feng, J. 2013. Realizing ultrahigh modulus and high strength of macroscopic graphene oxide papers through crosslinking of mussel-inspired polymers. *Adv. Mater.,* 25, 2980.

Trevethan, T., Dyulgerova, P., Latham, C. D., et al. 2013. Extended interplanar linking in graphite formed from vacancy aggregates. *Phys. Rev. Lett.,* 111, 095501.

Tsou, C. H., An, Q. F., Lo, S. C., et al. 2015. Effect of microstructure of graphene oxide fabricated through different self-assembly techniques on 1-butanol dehydration. *J. Mem. Sci.,* 477, 93.

Van Lier, G., Van Alsenoy, C., Van Doren, V. and Geerlings, P. 2000. Ab initio study of the elastic properties of single-walled carbon nanotubes and graphene. *Chem. Phys. Lett.*, 326, 181.

Wan, S., Fang, S., Jiang, L., Cheng, Q. and Baughman, R. H. 2018a. Strong, conductive, foldable graphene sheets by sequential ionic and π bridging. *Adv. Mater.*, 30, e1802733.

Wan, S., Hu, H., Peng, J., et al. 2016. Nacre-inspired integrated strong and tough reduced graphene oxide-poly(acrylic acid) nanocomposites. *Nanoscale*, 8, 5649.

Wan, S., Li, Y., Mu, J., et al. 2018b. Sequentially bridged graphene sheets with high strength, toughness, and electrical conductivity. *Proc. Natl. Acad. Sci.*, 115, 5359.

Wan, S., Li, Y., Peng, J., et al. 2015a. Synergistic toughening of graphene oxide-molybdenum disulfide-thermoplastic polyurethane ternary artificial nacre. *ACS Nano*, 9, 708.

Wan, S., Peng, J., Li, Y., et al. 2015b. Use of synergistic interactions to fabricate strong, tough, and conductive artificial nacre based on graphene oxide and chitosan. *ACS Nano*, 9, 9830.

Wang, C., Wang, L. F. and Xu, Z. P. 2013. Enhanced mechanical properties of carbon nanotube networks by mobile and discrete binders. *Carbon*, 64, 237.

Wang, J., Zhang, P., Liang, B., et al. 2016a. Graphene oxide as an effective barrier on a porous nanofibrous membrane for water treatment. *ACS Appl. Mater. Interfaces*, 8, 6211.

Wang, M. C., Yan, C., Ma, L., Hu, N. and Chen, M. W. 2012. Effect of defects on fracture strength of graphene sheets. *Comp. Mater. Sci.*, 54, 236.

Wang, Y., Ma, R., Hu, K., et al. 2016b. Dramatic enhancement of graphene oxide/silk nanocomposite membranes: Increasing toughness, strength, and Young's modulus via annealing of interfacial structures. *ACS Appl. Mater. Interfaces*, 8, 24962.

Wegst, U. G., Bai, H., Saiz, E., Tomsia, A. P. and Ritchie, R. O. 2015. Bioinspired structural materials. *Nat. Mater.*, 14, 23.

Wei, N., Peng, X. and Xu, Z. 2014a. Breakdown of fast water transport in graphene oxides. *Phys. Rev. E*, 89, 012113.

Wei, N., Peng, X. and Xu, Z. 2014b. Understanding water permeation in graphene oxide membranes. *ACS Appl. Mater. Interfaces*, 6, 5877.

Wen, Y., Wu, M., Zhang, M., Li, C. and Shi, G. 2017. Topological design of ultrastrong and highly conductive graphene films. *Adv. Mater.*, 29, 1702831.

Wen Y., Gao E., Hu Z., Xu T., Lu H., Xu Z. and Li C. 2019. Chemically modified graphene films with tunable negative Poisson's ratios. *Nat. Commun.*, 10, 2446.

Wilson, N. R., Pandey, P. A., Beanland, R., et al. 2009. Graphene oxide: Structural analysis and application as a highly transparent support for electron microscopy. *ACS Nano*, 3, 2547.

Wojtecki, R. J., Meador, M. A. and Rowan, S. J. 2011. Using the dynamic bond to access macroscopically responsive structurally dynamic polymers. *Nat. Mater.*, 10, 14.

Xiao, Y., Xu, Z., Liu, Y., et al. 2017. Sheet collapsing approach for rubber-like graphene papers. *ACS Nano*, 11, 8092.

Xie, Q., Xin, F., Park, H. G. and Duan, C. 2016. Ion transport in graphene nanofluidic channels. *Nanoscale*, 8, 19527.

Xiong, R., Hu, K., Grant, A. M., et al. 2016a. Ultrarobust transparent cellulose nanocrysal-graphene membranes with high electrical conductivity. *Adv. Mater.*, 28, 1501.

Xiong, R., Hu, K., Zhang, S., Lu, C. and Tsukruk, V. V. 2016b. Ultrastrong freestanding graphene oxide nanomembranes with surface-enhanced Raman scattering functionality by solvent-assisted single-component layer-by-layer assembly. *ACS Nano*, 10, 6702.

Xiong, Z., Liao, C., Han, W. and Wang, X. 2015. Mechanically tough large-area hierarchical porous graphene films for high-performance flexible supercapacitor applications. *Adv. Mater.*, 27, 4469.

Xu, Q. and Zhang, W. 2016. Next-generation graphene-based membranes for gas separation and water purifications. Chapter 2. In: *Advances in Carbon Nanostructures* (ed. A. Silva and S. A. C. Carabineiro). IntechOpen: London, pp. 39–61.

Xu, Y., Bai, H., Lu, G., Li, C. and Shi, G. 2008. Flexible graphene films via the filtration of water-soluble noncovalent functionalized graphene sheets. *J. Am. Chem. Soc.*, 130, 5856.

Yang, X., Cheng, C., Wang, Y., Qiu, L. and Li, D. 2013. Liquid-mediated dense integration of graphene materials for compact capacitive energy storage. *Science*, 341, 534.

Yazyev, O. V. and Louie, S. G. 2010. Topological defects in graphene: Dislocations and grain boundaries. *Phys. Rev. B*, 81, 195420.

Yeh, C. N., Raidongia, K., Shao, J., Yang, Q. H. and Huang, J. 2014. On the origin of the stability of graphene oxide membranes in water. *Nat. Chem.*, 7, 166.

Yi, L. J., Yin, Z. N., Zhang, Y. Y. and Chang, T. C. 2013. A theoretical evaluation of the temperature and strain-rate dependent fracture strength of tilt grain boundaries in graphene. *Carbon*, 51, 373.

Yuan, Y. Q., Gao, X. L., Wei, Y., et al. 2017. Enhanced desalination performance of carboxyl functionalized graphene oxide nanofiltration membranes. *Desalination*, 405, 29.

Zakharchenko, K. V., Katsnelson, M. I. and Fasolino, A. 2009. Finite temperature lattice properties of graphene beyond the quasiharmonic approximation. *Phys. Rev. Lett.*, 102, 046808.

Zandiatashbar, A., Lee, G. H., An, S. J., et al. 2014. Effect of defects on the intrinsic strength and stiffness of graphene. *Nat. Commun.*, 5, 3186.

Zhang, J., Xu, Y., Cui, L., et al. 2015. Mechanical properties of graphene films enhanced by homo-telechelic functionalized polymer fillers via π-π stacking interactions. *Compos. Part A*, 71, 1.

Zhang, M., Huang, L., Chen, J., Li, C. and Shi, G. 2014. Ultratough, ultrastrong, and highly conductive graphene films with arbitrary sizes. *Adv. Mater.*, 26, 7588.

Zhang, Z. Q., Liu, B., Huang, Y., Hwang, K. C. and Gao, H. 2010. Mechanical properties of unidirectional nanocomposites with non-uniformly or randomly staggered platelet distribution. *J. Mech. Phys. Solids*, 58, 1646.

Zhao, H., Min, K. and Aluru, N. R. 2009. Size and chirality dependent elastic properties of graphene nanoribbons under uniaxial tension. *Nano Lett.*, 9, 3012.

Zhao, X., Xu, Z., Zheng, B. and Gao, C. 2013. Macroscopic assembled, ultrastrong and H_2SO_4-resistant fibres of polymer-grafted graphene oxide. *Sci. Rep.*, 3, 3164.

Zheng, S., Tu, Q., Urban, J. J., Li, S. and Mi, B. 2017. Swelling of graphene oxide membranes in aqueous solution: Characterization of interlayer spacing and insight into water transport mechanisms. *ACS Nano*, 11, 6440.

Zhu, W., Low, T., Perebeinos, V., et al. 2012. Structure and electronic transport in graphene wrinkles. *Nano Lett.*, 12, 3431.

Nanoscale Shape Control

Zhaohui Wu
Changsha University

Wei Wu
Wuhan University

21.1 The Development of Shape Controlling of Materials in Nanoscale

Nowadays, nanomaterials including inorganic and organic nanomaterials with controlling and fascinating shapes have attracted numerous research interesting in extensive application fields, such as environmental pollution degradation, energy storage and production, electronic display, and sensors.[1-4] Because the unique shape of diverse materials in nanoscale not only presents large surface area, intrinsic quantum-confined electrons, and facet-dependent anisotropic properties but also exhibits unique photoelectrochemical properties, such as light adsorption, electronic anisotropy, energy band structure, and high surface activity.[5-8] Generally, the controlled shape of nanomaterials includes zero-dimensional (0D), one-dimensional (1D), two-dimensional (2D), and three-dimensional (3D), which is available to achieve through numerous strategies. Recently, "bottom-up" approach, based on the assembled ionic, atomic, or molecular units through various reaction processes, is the preferential strategy to synthesize lots of nanomaterials with defined shape. During the synthesis process, multiple shapes in nanoscale are established, providing nanoparticles with diverse structure, composition, and surface properties. Typical bottom-up approach involving many wet-chemical methods, such as precipitated methods, preformed-seed-mediated growth method, polyol approach, template approach, electrochemical synthesis and photochemical synthesis routes, have been developed.[9-12] Through all the methods, the thermodynamics and kinetics are referred in the synthesized process, which are the essential parameter to control the final shape of nanomaterials.[13] The intrinsic parameters for shape controlling of nanomaterials through synthesis process are described. Additionally,

a great deal of creative and frequently used mechanisms and vital factors affected the eventual shape of nanoparticles are also presented, expecting to provide valuable information on shape controlling of nanoparticles.

21.2 Classical and Non-classical Nucleation and Growth Theory for Shape Controlling of Materials

21.2.1 Classical Nucleation and Growth Theory

The classical nucleation and growth theory has been described by Mullin since 1961 and supplemented by other researchers continuously, in which nucleation is a process whereby the second phase generated from first phase.[14-22] Typically, solid particles as the second phase generated from initial liquid or gaseous phase. Normally, the nucleation process can be divided into primary nucleation and secondary/heterogeneous nucleation, which are referred to the generation of solid nuclei from homogenous supersaturation without or with other solid particles, respectively. The occurrence of heterogeneous nucleation and secondary nucleation generated on the presented surfaces is much easier than primary nucleation owing to its lower energy barrier.

In terms of homogenous nuclei, the formation process is considered as a thermodynamical process, driving by the supersaturation of bulk solution and deciding by total free energy (ΔG) of nanoparticles. It is defined as the sum of bulk free energy $\Delta G_v \left(\Delta G_v = \frac{-2\gamma}{r} = \frac{-2k_B T \ln(S)}{v} \right)$ and surface free energy, as shown in Eq. 21.1.[23]

$$\Delta G = 4\pi r \gamma - \frac{4}{3}\pi r^3 \Delta G_v \qquad (21.1)$$

where ΔG_ν is the free energy change of transformation to unit volume of particles. r, γ, T, k_B, ν, and S are the radius of formed particle, surface energy, temperature, Boltzmann's constant, molar volume, and supersaturation ratio of bulk solution, respectively. S defined as the ratio of monomer concentration C to equilibrium monomer concentration C^* in the crystals ($S = C/C^*$) is the driven force for initial nucleation and growth. The stable nuclei are produced when S is larger than 1. At this point, the critical value of ΔG and critical radius of existed nuclei is set to be zero ($d(\Delta G_{\text{crit}})/dr = 0$), deriving the corresponding minimum radius of nuclei without dissolution and critical free energy ΔG_{crit} in Eqs. 21.2 and 21.3.

$$r_{\text{crit}} = \frac{-2\gamma}{\Delta G_v} = \frac{2\gamma v}{k_B T \ln S} \tag{21.2}$$

$$\Delta G_{\text{crit}} = \frac{4}{3}\pi\gamma r_{\text{crit}}^2 = \Delta G_{\text{crit}}^{\text{homo}} \tag{21.3}$$

During nucleation process, nucleation rate of formed nuclei N is written in a form of Arrhenius reaction velocity equation as follows:

$$\frac{dN}{dt} = A\exp(-\Delta G_{\text{crti}}/k_B T) = A\exp\left(\frac{-16\pi\gamma^3 v^2}{3k_B^3 T(\ln S)^2}\right) \tag{21.4}$$

where A is a pre-exponential factor. Apparently, the nucleation rate is determined by the supersaturation, surface free energy, and temperature.[24] That is, higher nucleation rate is facilitated under higher temperature, higher concentration of precursor monomer, and lower critical energy barrier, breeding a large number of nuclei with small size (Figure 21.1).[25–27]

After nucleation, subsequent growth of existed stable nuclei is the crucial step to decide the shape of

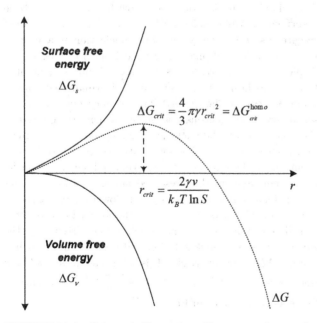

FIGURE 21.1 Schematic illustration of free energy diagram for nucleation.

nanoparticles.[28–32] Growth process involves the deposition of elementary (including atoms, molecules, assemblies, or particles) on the presented nanoparticles in a medium state (plasma, melt, solution, gel, etc.) through the provided sites on particles surface. The density of existed sites on preformed nuclei surface together with the incorporated kinetics plays the crucial role to determine the growth rate of nanoparticles.

In classical growth theory, surface reaction and monomer's diffusion to particle surface are the two involved mechanisms. According to the Fick's first law of diffusion, the growth of nanoparticles is controlled by the two limiting factors neither by diffusion nor by surface reaction, and then increasing the particles size with time. Diffusion-limited or reaction-limited process with different concentration of precursor monomer determines the shape of nanoparticles through the growth rate. Within high concentration of precursor monomer solution, the growth rate is controlled by the diffusion-limited process.[33–35] Then, precursor monomers are precipitated immediately onto the surface of nanoparticles through the bulk reaction medium and solvent.[20] Nevertheless, in the case of reaction-limited growth process, when concentration of precursor monomer is low and growth of particles is greatly limited by surface reaction of monomers.

Although both diffusion-limited and reaction-limited growth is driven by the concentration of precursor monomer, their role for deciding shape and size of final particles is different. That is, the diffusion-limited growth is desirable process for production of nanoparticles with monodispersity, while reaction-limited growth determines the final shape of nanoparticles. During the process of diffusion-limited growth, adsorption of organic ligands or surfactants on the surface of performed nanoparticles introducing the diffusion barrier is a flexible and effective approach to control the shape with monodisperse size. Typically, various pristine, heterostructure, or Z-scheme complex nanomaterials are able to form through the classical nucleation and growth theory, and the resulted shapes of diverse materials are involved all dimensional.

21.2.2 Non-classical Nucleation and Growth Theory

Recently, non-classical nucleation and growth theory and process for shape-controlled nanomaterials have drawn great deal of attentions. Different from classical crystal growth via ion-by-ion addition, non-classical aggregation-based crystal growth emphasizes a particle-by-particle pathway for single-crystal formation. And, the reported non-classical nucleation and growth theory refers to as orientation attachment, reverse growth, Kirkendall effect, aggregation/agglomeration, self-assembling, and so on.[36–42] However, the mechanisms of these non-classical growth theories are different from each other significantly. Therefore, the resulted shapes of nanoparticles are also possessed dramatically difference.

Usually, aggregation frequently happens among particles, which is a common and complex phenomenon for collision of particles. A common and acceptable explanation for aggregation is that aggregation is caused by high surface free energy of their tiny nanoscale size (Figure 21.2). To form the aggregated particles, a Brownian or flow motion and a net interparticle attractive force are the vital roles to form aggregated particles with irregular shape. However, aggregation of particles is intended to avoid during the colloid science process due to the uncontrollable morphology, undesirable properties, and unsatisfied applications in the past decades. To avoid the aggregation, coating with foreign capping agents and/or changing the surface charges of performed nanoparticles is the effective and tailored route.

Normally, in classical growth theory, the polyhedral morphology is highly related to the surface free energy and independent of the interior structure of particles. However, this theory is subverted by the reverse growth, in which there is competition between crystal growth and aggregation of particles at early stage of crystal growth. Reverse growth of particles takes place when the aggregation or disordered crystals is presented, resulting in the growth of crystals from the surface to interiors core (Figure 21.3).[42] During this reversed growth process, some surfactants or polymers such

as ethylamine and the biopolymer chitosan play the dominant role, resulting in the formation of hollow cubic crystals. Additionally, dissolution of some amorphous crystals occurred on the surface of resulted particles also can derive the reverse growth of crystals, resulting in the formation of spherical particles with reduced size.[43]

"Oriented attachment" involves spontaneous self-organization of adjacent particles and continuous growth of the self-organized particles at initial step. Then, the self-organized particles share a common crystallographic orientation and join at a planar interface. During this oriented attachment process, the rotation for perfect lattice matching and fusion among the collisional or agglomerated particles are the vital steps. Recently, the orientation attachment process has been observed directly through the observation of iron oxyhydroxide nanoparticles on high-resolution transmission electron microscopy (HRTEM) using a fluid cell. According to their observation, perfect lattice match occurs among some particles when particles undergo continuous rotation and interaction to each other. When their distance is less than 1 nm, a sudden jump and contact of one smaller particle to another one, the lateral atom-by-atom addition initiates at the contact point of attached particles.[44] This interface elimination proceeds at a rate consistent with the curvature dependence of Gibbs free energy. Based on measured translational and rotational accelerations, strong and highly direction-specific interaction is the direct driving force for particles growth via orientation attachment. Additionally, orientation attachment also has been found to be a significant mechanism in controlling growth to obtain various anisotropic nanostructures, including 3D complex-shaped nanostructures by "building blocks" (architecture structure), core–shell.[45,46]

Self-assembling is another common strategy to regulate the shape of nanoparticles, and static self-assembling is the frequently used route, which involves atomic, ionic, and

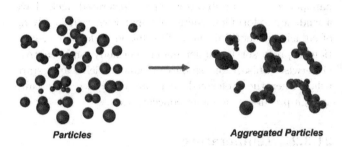

Particles   **Aggregated Particles**

FIGURE 21.2 Schematic illustration of aggregation among nanoparticles.

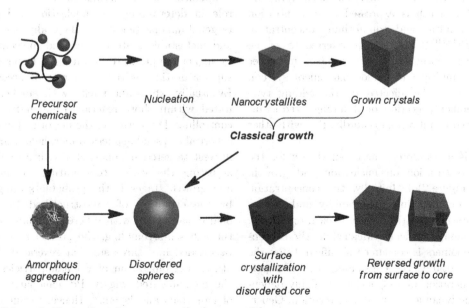

FIGURE 21.3 Schematic illustration of reverse growth of nanoparticles.

molecular or colloidal crystals system with a global or local equilibrium but without energy dissipation.[47] Furthermore, non-covalent or weak covalent interactions (electrostatic, van der Waals, hydrophobic interactions, hydrogen and coordination bonds) are the important parameters for the self-assembly. Therefore, diverse ligands and surfactants are employed to facilitate the recognition among special component for reconfiguring different ordered structure during self-assembly process. For instance, flower-like Bi_2WO_6 with decorated Cu_2O nanodots was synthesized via an interfacial self-assembly method. During this assembling process, $Cu(OH)_2$ with negative charge is able to assemble onto the presented Bi_2WO_6 nanosheets with the assistance of positive charge of cetyltrimethyl ammonium bromide (CTAB). After adding ascorbic acid, the $Cu(OH)_2$ is reduced into Cu_2O nanodots onto the surface of Bi_2WO_6 flower.[48] However, self-assembling of nanoparticles on substrates is largely determined by multiple factors including monodispersity, shape and surface-adsorption feature of performed nanoparticles, type of substrate, and reaction conditions (concentration of presented sites, ambient temperature and humidity, etc.). Although it is worth noting that there is difference between orientation attachment and self-assembling for shape controlling of nanoparticles that self-assembly is controllable process by managing the ligands or bonds among particles, rather than the uncontrollable process of orientation attachment which is highly depending on the reaction process and conditions.

21.3 Typical Parameters for Shape Controlling of Materials in Nanoscale

21.3.1 Supersaturation

According to the classical nucleation theory, supersaturation is regarded as the driven force for nucleation and growth of particles, which is expressed as concentration difference between initial and equilibrium concentration ($\Delta C = C_b - C_r$).[24,49,50] It also can be referred to as the precursor monomer concentration or precursor monomer ratio in single or multiple precursor monomers system, respectively, which can be affected by the adding type of precursor monomer (injection or dumping), reductant concentration, or reductant ratio, in coordination with other ions or agents.

LaMer theory is a widespread accepted theory for the influence of supersaturation on nucleation and growth, as presented in Figure 21.4. Initially, the concentration of bulky small monomers increases rapidly and crosses metastable zone until reach the "burst nucleation" point, breeding a large number of nuclei. Meanwhile, the concentration of free monomers is consumed significantly by the generated nuclei. Notably, the consuming rate of supersaturation and generation rate of nuclei are highly dependent on the thermodynamic and kinetic of reaction. Generally, higher supersaturation, higher temperature, or faster

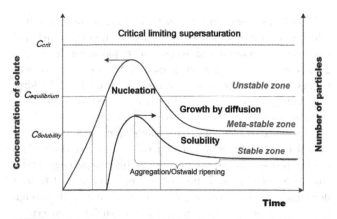

FIGURE 21.4 Scheme of LaMer theory for nucleation and growth process.

increased temperature rate contribute to fast nucleation rate and a larger number of nuclei with smaller size. Then, the generated nuclei also consume the remained supersaturation to grow up under diffusion of monomers during growth period. During this process, aggregation/agglomeration, or Ostwald ripening, self-assembling, and other mechanisms are able to take place. Thus, the eventual shape of generated nanoparticles is highly dependent on the reaction conditions. Generally, 0D spherical, pseudo-spherical, or other isotropic nanoparticles are produced when the generated nuclei have a tendency to form low-energy nanoparticles with supplying of sufficient energy at the bulk solution or low concentration of precursor monomer under thermodynamic control. Otherwise, anisotropic shapes of nanoparticles are established when the preformed nuclei are controlled kinetically at high precursor monomer concentration.

21.3.2 Temperature

Based on the classical nucleation and growth theory, temperature as a thermodynamic factor plays a direct role in determining the nucleation and growth rate of targeted nanoparticles.[51–54] Typically, energetic movement and metastable state of ions and molecules are supplied by higher temperature, resulting in faster consume rate of supersaturation or reduction rate of precursor monomer. Eventually, abundant nuclei with smaller size are facilitated within short reaction time under thermodynamic controlling. Furthermore, the performed nanoparticles with spherical or pseudo-spherical morphology are fabricated. To harvest targeted nanoparticles with an anisotropic shape, adjusting the varied temperature for initial nucleation and growth stages is the probable strategy. As observed by the evolution of nanostructured Pd on the hexagonal c-plane GaN with different amount of Pd deposition through self-assembling, the growth regimes were able to be divided into three steps at several distinctive temperatures: agglomeration of Pd nanoparticles, coalescence of the agglomerated wiggly Pd nanoparticles, and growth of nanovoids dend layers.[55] Hence, through non-isothermal temperature controlling to adjust the thermodynamics state of nanoparticles at different growth stage is probably the appealing strategy to obtain the diverse anisotropic shape of resulted nanoparticles.[56] For instance, varying different temperature at early growth stage of Au, the shapes of resulted Au plates were able to adjust from well-defined, star-like, shield-like, and other novel polygonal because these unique Au plates grew from the (111) plane along <211>, <110>, and other high-index directions by altering the growth temperature.[57]

21.3.3 Seeds, Templates, and Defects

The existed seeds or templates are the feasible sites for further nucleation because the provided sites are able to reduce the energy barriers for adding precursor monomers onto the performed surface of the seeds or templates. The components of seeds and templates can be as same or different as the eventual particles, which are assigned as homogenous or heterogeneous structured nanoparticles. However, there is significant difference over shapecontrolling of nanoparticles when using seeds- and templates-mediated route.[58] That is, the shape of prepared nanoparticles is determined by the employed templates, whereas the shape of resulted nanoparticles is independent on the shape of used seeds.[59-61] Additionally, the size of employed seeds should be much smaller than that of templates.[59]

Generally, defects caused by some extreme conditions during synthesis process (such as heating, calcination in different gaseous environments, and ionic liquid conditions) are avoided during the regulation of the shape of nanoparticles.[62-64] However, recently, many unique shapes of nanoparticles are produced through defects on the surface of particles, such as the screw morphology of nanoparticles. Additionally, the defects generated on the special facet of nanoparticles provide polarized surface and sites, contributing to the reduction of some metal ions and formation of doping morphology onto nanoparticles. For example, the (001) facets of BiOCl with high density of oxygen were able to provide fast generation and multiple amount of oxygen vacancies under microwave irradiation.

The produced oxygen vacancies on the (001) facet of BiOCl nanosheet are able to reduce Ag selectively, and then, the Ag nanoparticles are generated at the surface of BiOCl, which presented higher stability and reactivity over sodium pentachlorophenate oxidation and Cr (VI) reduction under visible light irradiation.[65]

21.3.4 Functional Groups

Generally, a bunch of small nanoparticles with undesirable shapes are intended to aggregate due to their higher surface energy.[51,52,66,67] Employing functional groups stemmed from some surfactants, additives, or solvent is the desirable strategy to alter the shape of nanoparticles in bulk solution. During the synthesis process, the functional groups are regarded to recognize and absorb onto the special surface of growing nanoparticles through the static-electric adsorption or covalent bonding, contributing to the growth of some certain facet of nanoparticles. Furthermore, the adsorbed layer onto the surface of nanoparticles also protects the formed nanoparticles against aggregation in the bulk solution state. Generally, the most frequently used functional groups are halide ions; primary or secondary amine groups; and hydroxyl, carbonyl, and thiol groups, which are usually stemmed from small organic/inorganic molecules, polymers, and solvents, resulting in diverse unique morphology of nanoparticles. Poly(acrylic acid) (PAA), poly(vinylpyrrolidone) (PVP), poly(ethylene glycol) (PEG), poly(allylamine hydrochloride) (PAH), polyetherimide (PEI), and poly(vinyl alcohol) (PVA) with linear and branched structures are the frequently used polymers.[68-73] Small molecules including cetyltrimethylammonium chloride/bromide (CTAC/B), sodium dodecyl sulfate (SDS), octadecylamine (ODA), mannitol, citric acid, oleic acid and/or oleylamine, and trioctylphosphine oxide (TOPO) are used during the synthesis process.[74-77] In terms of solvents, ethylene glycol (EG), glycerol (G), N,N-dimethylformamide (DMF), ionic liquid, and so on are the common solvents.[78,79] The scheme of functional groups controlling over the shape of different nanoparticles is illustrated in Figure 21.5.[13]

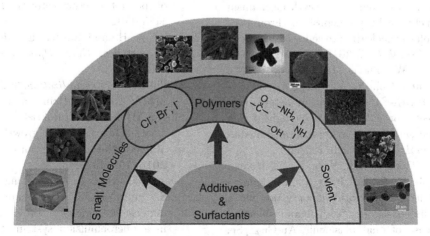

FIGURE 21.5 Schematic of additive/surfactants for the shape control of nanoparticles.

21.4   Prospective

Recently, there are many novel strategies and outstanding techniques, such as chemical vapor deposition (CVD), salt melting, exfoliation assisting with ultrasonication, in situ growth, electrolysis with ultrasonication, and fiber spinning, for shape-controlled nanoparticles.[80−83] However, during these processes, the nucleation and growth controlling over the tolerable shape of nanoparticles probably is different from the known mechanisms. Hence, the further exploration of clear nucleation and growth mechanism during different synthesis process is still the emergency issue.

References

1. Yadav, G. G., Wei, X., Huang, J., Gallaway, J. W., Turney, D. E., Nyce, M., Secor, J., and Banerjee, S., A conversion-based highly energy dense Cu^{2+} intercalated Bi-birnessite/Zn alkaline battery, *J. Mater. Chem. A*, **2017**, 5, 15845–15854.

2. Kim, S., Dong, W. J., Gim, S., Sohn, W., Park, J. Y., Yoo, C. J., Jang, H. W., and Lee, J. L., Shape-controlled bismuth nanoflakes as highly selective catalysts for electrochemical carbon dioxide reduction to formate, *Nano Energy*, **2017**, 39, 44–52.

3. Bhattacharjee, A. K., Stochastic kinetics reveal imperative role of anisotropic interfacial tension to determine morphology and evolution of nucleated droplets in nematogenic films, *Sci. Rep. UK*, **2017**, 7, 40059.

4. Garcia-Tunon, E., Feilden, E., Zheng, H., D'Elia, E., Leong, A., and Saiz, E., Graphene oxide: An all-in-one processing additive for 3D printing, *ACS Appl. Mater. Inter.*, **2017**, 9, 32977–32989.

5. Zhu, C. B., Chen, Z. W., Zhong, C. F., and Lu, Z. Y., Facile synthesis of $BiFeO_3$ nanosheets with enhanced visible-light photocatalytic activity, *J. Mater. Sci. Mater. El.*, **2018**, 29, 4817–4829.

6. Zhou, D., Shu, H. B., Hu, C. L., Jiang, L., Liang, P., and Chen, X. S., Unveiling the growth mechanism of MoS_2 with chemical vapor deposition: From two-dimensional planar nucleation to self-seeding nucleation, *Cryst. Growth Des.*, **2018**, 18, 1012–1019.

7. Wang, Q., Wang, W., Zhong, L. L., Liu, D. M., Cao, X. Z., and Cui, F. Y., Oxygen vacancy-rich 2D/2D $BiOCl$-g-C_3N_4 ultrathin heterostructure nanosheets for enhanced visible-light-driven photocatalytic activity in environmental remediation, *Appl. Catal. B*, **2018**, 220, 290–302.

8. Zou, Y., Sun, C., Gong, W. B., Yang, X. F., Huang, X., Yang, T., Ju, W. B., and Jiang, J., Morphology-controlled synthesis of hybrid nanocrystals via a selenium-mediated strategy with ligand shielding effect: The case of dual plasmonic $Au-Cu_{2-x}Se$, *ACS Nano*, **2017**, 11, 3776–3785.

9. Sun, L. C., Zhang, Q. F., Li, G. G., Villarreal, E., Fu, X. Q., and Wang, H., Multifaceted gold-palladium bimetallic nanorods and their geometric, compositional, and catalytic tunabilities, *ACS Nano*, **2017**, 11, 3213–3228.

10. Panfilova, E. V., Khlebtsov, B. N., Burov, A. M., and Khlebtsov, N. G., Study of polyol synthesis reaction parameters controlling high yield of silver nanocubes, *Colloid J.*, **2012**, 74, 99–109.

11. Zhao, S., Zhang, Y. W., Zhou, Y. M., Qiu, K. B., Zhang, C., Fang, J. S., and Sheng, X. L., Reactable polyelectrolyte-assisted preparation of flower-like $Ag/AgCl/BiOCl$ composite with enhanced photocatalytic activity, *J. Photochem. Photobio. A*, **2018**, 350, 94–102.

12. Thota, S., Zhou, Y. D., Chen, S. T., Zou, S. L., and Zhao, J., Formation of bimetallic dumbbell shaped particles with a hollow junction during galvanic replacement reaction, *Nanoscale*, **2017**, 9, 6128–6135.

13. Wu, Z. H., Yang, S. L., and Wu, W., Shape control of inorganic nanoparticles from solution, *Nanoscale*, **2016**, 8, 1237–1259.

14. Griffiths, H., Mechanical crystallization, *Trans. J. Soc. Chem. Ind.*, **1925**, 44, 7T–18T.

15. Saema, W. C., Crystal-size distribution in mixed suspensions, *AIChE J.*, **1956**, 2, 107–112.

16. Mullin, J. W., *Crystallization*. London: Butterworth-Heinemann, **1961**, ix, 268 p.

17. Van Hook, A., *Crystallization: Theory and Practice*. ACS Monograph Series. New York: Reinhold Publishing Corporation, **1961**, 325 p.

18. Bamforth, A. W., *Industrial Crystallization*. Chemical and Process Engineering Series. London: L. Hill, **1965**, xxiv, 361 p.

19. Nývlt, J., *Industrial Crystallisation from Solutions*. London: Butterworth-Heinemann, **1971**, 9, 189 p.

20. Mullin, J. W., *Crystallisation*, 2d ed. International Scientific Series. Cleveland: CRC Press, **1972**, 480 p.

21. Hong, S-D and Cebe, P., Crystallization behaviour of poly(ether-ether-ketone), *Polymer*, **1986**, 27, 1183–1192.

22. Adair, J. H. and Suvaci, E., Morphological control of particles, *Curr. Opin. Colloid In.*, **2000**, 5, 160–167.

23. Mullin, J. W., *Crystallization*, 3rd ed. Oxford; Boston, MA: Butterworth-Heinemann, **1993**, xiv, 527 p.

24. Singh, M. R. and Ramkrishna, D., Dispersions in crystal nucleation and growth rates: Implications of fluctuation in supersaturation, *Chem. Eng. Sci.*, **2014**, 107, 102–113.

25. Zhang, J., Sun, L. D., Jiang, X. C., Liao, C. S., and Yan, C. H., Shape evolution of one-dimensional single-crystalline ZnO nanostructures in a microemulsion system, *Cryst. Growth Des.*, **2004**, 4, 309–313.

26. Mikami, T., Takayasu, Y., and Hirasawa, I., PEI-assisted preparation of Au nanoparticles via reductive crystallization process, *Chem. Eng. Res. Des.*, **2010**, 88, 1248–1251.

27. Bian, B. R., Xia, W. X., Du, J., Zhang, J., Liu, J. P., Guo, Z. H., and Yan, A., Growth mechanisms and size control of FePt nanoparticles synthesized using Fe(CO)$_x$ (x < 5)-oleylamine and platinum(II) acetylacetonate, *Nanoscale*, **2013**, 5, 2454–2459.

28. Johnson, C. J., Dujardin, E., Davis, S. A., Murphy, C. J., and Mann, S., Growth and form of gold nanorods prepared by seed-mediated, surfactant-directed synthesis, *J. Mater. Chem.*, **2002**, 12, 1765–1770.

29. Cozzoli, P. D., Kornowski, A., and Weller, H., Low-temperature synthesis of soluble and processable organic-capped anatase TiO$_2$ nanorods, *J. Am. Chem. Soc.*, **2003**, 125, 14539–14548.

30. Jana, N. R., Gram-scale synthesis of soluble, near-monodisperse gold nanorods and other anisotropic nanoparticles, *Small*, **2005**, 1, 875–882.

31. Yong, K. T., Sahoo, Y., Swihart, M. T., and Prasad, P. N., Shape control of CdS nanocrystals in one-pot synthesis, *J. Phys. Chem. C*, **2007**, 111, 2447–2458.

32. Grzelczak, M., Perez-Juste, J., Mulvaney, P., and Liz-Marzan, L. M., Shape control in gold nanoparticle synthesis, *Chem. Soc. Rev.*, **2008**, 37, 1783–1791.

33. Dawn, A., Mukherjee, P., and Nandi, A. K., Preparation of size-controlled, highly populated, stable, and nearly monodispersed Ag nanoparticles in an organic medium from a simple interfacial redox process using a conducting polymer, *Langmuir*, **2007**, 23, 5231–5237.

34. Xie, S. F., Zhang, H., Lu, N., Jin, M. S., Wang, J. G., Kim, M. J., Xie, Z. X., and Xia, Y. N., Synthesis of rhodium concave tetrahedrons by collectively manipulating the reduction kinetics, facet-selective capping, and surface diffusion, *Nano Lett.*, **2013**, 13, 6262–6268.

35. Shayeganfar, F., Eskandari, Z., Tabar, M. R. R., and Sahimi, M., Molecular dynamics simulation of formation and growth of CdS nanoparticles, *Mol. Simulat.*, **2014**, 40, 361–369.

36. Wolff, A., Hetaba, W., Wissbrock, M., Loffler, S., Mill, N., Eckstadt, K., Dreyer, A., Ennen, I., Sewald, N., Schattschneider, P., and Hutten, A., Oriented attachment explains cobalt ferrite nanoparticle growth in bioinspired syntheses, *Beilstein J. Nanotechnol.*, **2014**, 5, 210–218.

37. Yu, Y., Zhang, Q. B., Yao, Q. F., Xie, J. P., and Lee, J. Y., Guiding principles in the galvanic replacement reaction of an underpotentially deposited metal layer for site-selective deposition and shape and size control of satellite nanocrystals, *Chem. Mater.*, **2013**, 25, 4746–4756.

38. Baldan, A., Review progress in Ostwald ripening theories and their applications to the gamma - precipitates in nickel-base superalloys: Part II: Nickel-base superalloys, *J. Mater. Sci.*, **2002**, 37, 2379–2405.

39. Galbraith, S. C., Schneider, P. A., and Flood, A. E., Model-driven experimental evaluation of struvite nucleation, growth and aggregation kinetics, *Water Res.*, **2014**, 56, 122–132.

40. Grogan, J. M., Rotkina, L., and Bau, H. H., In situ liquid-cell electron microscopy of colloid aggregation and growth dynamics, *Phys. Rev. E*, **2011**, 83, 061405.

41. Gibert, M., Abellan, P., Martinez, L., Roman, E., Crespi, A., Sandiumenge, F., Puig, T., and Obradors, X., Orientation and shape selection of self-assembled epitaxial Ce$_{1-x}$Gd$_x$O$_{2-y}$ nanostructures grown by chemical solution deposition, *CrystEngComm*, **2011**, 13, 6719–6727.

42. Zhou, W. Z., Reversed crystal growth: Implications for crystal engineering, *Adv. Mater.*, **2010**, 22, 3086–3092.

43. Chandrasekar, G., Mougin, K., Haidara, H., Vidal, L., and Gnecco, E., Shape and size transformation of gold nanorods (GNRs) via oxidation process: A reverse growth mechanism, *Appl. Surf. Sci.*, **2011**, 257, 4175–4179.

44. Li, D., Nielsen, M. H., Lee, J. R., Frandsen, C., Banfield, J. F., and De Yoreo, J. J., Direction-specific interactions control crystal growth by oriented attachment, *Science*, **2012**, 336, 1014–1018.

45. 45. Yan, Y. H., Zhou, Z. X., Cheng, Y., Qiu, L. L., Gao, C. P., and Zhou, J. G., Template-free fabrication of α- and β-Bi$_2$O$_3$ hollow spheres and their visible light photocatalytic activity for water purification, *J. Alloy. Compound.*, **2014**, 605, 102–108.

46. Yu, H. G., Zhu, Z. F., Zhou, J. H., Wang, J., Li, J. Q., and Zhang, Y. L., Self-assembly and enhanced visible-light-driven photocatalytic activities of Bi$_2$MoO$_6$ by tungsten substitution, *Appl. Surf. Sci.*, **2013**, 265, 424–430.

47. Garcia-Perez, U. M., Sepulveda-Guzman, S., Martinez-de la Cruz, A., and Peral, J., Selective synthesis of monoclinic bismuth vanadate powders by surfactant-assisted co-precipitation method: Study of their electrochemical and photocatalytic properties, *Inter. J. Electrochem. Sc.*, **2012**, 7, 9622–9632.

48. Liu, L., Ding, L., Liu, Y. G., An, W. J., Lin, S. L., Liang, Y. H., and Cui, W. Q., Enhanced visible light photocatalytic activity by Cu$_2$O-coupled flower-like Bi$_2$WO$_6$ structures, *Appl. Surf. Sci.*, **2016**, 364, 505–515.

49. Sandhya, S., Sureshbabu, S., Varma, H. K., and Komath, M., Nucleation kinetics of the formation of low dimensional calcium sulfate dihydrate crystals

in isopropyl alcohol medium, *Cryst. Res. Technol.*, **2012**, 47, 780–792.

50. Jeong, S., Jho, Y., and Zhou, X., Micro-structural change during nucleation: From nucleus to bicontinuous morphology, *Sci. Rep. UK*, **2015**, 5, 15955.

51. El-Kass, M., Ladj, R., Mugnier, Y., Le Dantec, R., Hadji, R., Marty, J. C., Rouxel, D., Durand, C., Fontvieille, D., Rogalska, E., and Galez, C., Temperature-dependent adsorption of surfactant molecules and associated crystallization kinetics of noncentrosymmetric $Fe(IO_3)_3$ nanorods in microemulsions, *Mater. Res. Bull.*, **2013**, 48, 4431–4437.

52. Chu, H. B., Li, X. M., Chen, G. D., Zhou, W. W., Zhang, Y., Jin, Z., Xu, J. J., and Li, Y., Shape-controlled synthesis of CdS nanocrystals in mixed solvents, *Cryst. Growth Des.*, **2005**, 5, 1801–1806.

53. Baidakov, V. G. and Tipeev, A. O., Nucleation of liquid droplets and voids in a stretched Lennard-Jones fcc crystal, *J. Chem. Phys.*, **2015**, 143, 31.

54. Sui, M., Li, M. Y., Pandey, P., Zhang, Q. Z., Kunwar, S., and Lee, J., Morphological, structural and optical evolution of Ag nanostructures on c-plane GaN through the variation of deposition amount and temperature, *Metals and Materials International*, **2018**, 24, 337–350.

55. 55. Sui, M., Kunwar, S., Pandey, P., Zhang, Q. Z., Li, M. Y., and Lee, J., Fabrication and determination of growth regimes of various Pd NPs based on the control of deposition amount and temperature on c-plane GaN, *J. Mater. Res.*, **2017**, 32, 3593–3604.

56. Sharma, K. P., Choudhury, C. K., Srivastava, S., Davis, H., Rajamohanan, P. R., Roy, S., and Kumaraswamy, G., Assembly of polyethyleneimine in the hexagonal mesophase of nonionic surfactant: Effect of pH and temperature, *Journal of Physical Chemistry B*, **2011**, 115, 9059–9069.

57. Zhu, J. J., Kan, C. X., Li, H. C., Cao, Y. L., Ding, X. L., and Wan, J. G., Synthesis and growth mechanism of gold nanoplates with novel shapes, *J. Cryst. Growth*, **2011**, 321, 124–130.

58. Sun, Y. G. and Xia, Y. N., Shape-controlled synthesis of gold and silver nanoparticles, *Science*, **2002**, 298, 2176–2179.

59. Perera, S. D., Zhang, H. T., Ding, X. Y., Nelson, A., and Robinson, R. D., Nanocluster seed-mediated synthesis of $CuInS_2$ quantum dots, nanodisks, nanorods, and doped Zn-CuInGaS2 quantum dots, *J. Mater. Chem. C*, **2015**, 3, 1044–1055.

60. Lu, H. F., Ren, X. G., Sha, W. E. I., Chen, J. J., Kang, Z. W., Zhang, H. X., Ho, H. P., and Choy, W. C. H., Experimental and theoretical investigation of macro-periodic and micro-random nanostructures with simultaneously spatial translational symmetry and long-range order breaking, *Sci. Rep. UK*, **2015**, 5, 7876.

61. Ye, X. C., Hickey, D. R., Fei, J. Y., Diroll, B. T., Paik, T., Chen, J., and Murray, C. B., Seeded growth of metal-doped plasmonic oxide heterodimer nanocrystals and their chemical transformation, *J. Am. Chem. Soc.*, **2014**, 136, 5106–5115.

62. Meng, F., Morin, S. A., Forticaux, A., and Jin, S., Screw dislocation driven growth of nanomaterials, *Accounts Chem. Res.*, **2013**, 46, 1616–1626.

63. Cui, Z. K., Li, S. L., Zhou, J. Q., Zhang, J. L., Ge, S. X., and Zheng, Z., Preparation and optical properties of spherical Bi_2S_3 nanoparticles by in situ thermal sulfuration method, *Nano*, **2015**, 10, 1550021.

64. Wang, Z. R., Wang, F., Peng, Y., Zheng, Z. Y., and Han, Y. L., Imaging the homogeneous nucleation during the melting of superheated colloidal crystals, *Science*, **2012**, 338, 87–90.

65. Li, H. and Zhang, L. Z., Oxygen vacancy induced selective silver deposition on the –001″ facets of BiOCl single-crystalline nanosheets for enhanced Cr(VI) and sodium pentachlorophenate removal under visible light, *Nanoscale*, **2014**, 6, 7805–7810.

66. Green, A. E., Chiang, C. Y., Greer, H. F., Waller, A., Ruszin, A., Webster, J., Niu, Z. Y., Self, K., and Zhou, W. Z., Growth mechanism of dendritic hematite via hydrolysis of ferricyanide, *Cryst. Growth Des.*, **2017**, 17, 800–808.

67. Statt, A., Virnau, P., and Binder, K., Finite-size effects on liquid-solid phase coexistence and the estimation of crystal nucleation barriers, *Phys. Rev. Lett.*, **2015**, 114, 026101.

68. Yang, T. I., Huang, Y. C., Yang, S. C., Yeh, J. M., and Peng, Y. Y., Effect of hydroxyapatite particles on the rheological behavior of poly(ethylene glycol)-poly(lactic-co-glycolic acid) thermosensitive hydrogels, *Mater. Chem. Phys.*, **2015**, 152, 158–166.

69. Xing, R. M., Xu, F. L., Liu, S. H., and Niu, J. Y., Surfactant-free fabrication of Fe_3O_4 nanospheres with selective shape, *Mater. Lett.*, **2014**, 134, 71–74.

70. Liu, M., Jin, H. Y., Uchaker, E., Xie, Z. Q., Wang, Y., Cao, G. Z., Hou, S. E., and Li, J. Y., One-pot synthesis of in-situ carbon-coated Fe_3O_4 as a long-life lithium-ion battery anode, *Nanotechnology*, **2017**, 28, 155603.

71. Cazia, V., Piras, R., Ardu, A., Musinu, A., Saba, M., Bongiovanni, G., and Mattoni, A., Atomistic modeling of morphology and electronic properties of colloidal ultrathin Bi_2S_3 nanowires, *J. Phys. Chem. C*, **2015**, 119, 16913–16919.

72. Liu, Y. Y., Wang, G. Z., Dong, J. C., An, Y., Huang, B. B., Qin, X. Y., Zhang, X. Y., and Dai, Y., A bismuth based layer structured organic-inorganic hybrid material with enhanced photocatalytic activity, *J. Colloid Interf. Sci.*, **2016**, 469, 231–236.

73. Choudhury, C. K. and Roy, S., Structural and dynamical properties of polyethylenimine in explicit water at different protonation states: A molecular dynamics study, *Soft Matter*, **2013**, 9, 2269–2281.

74. Tsao, Y. C., Rej, S., Chiu, C. Y., and Huang, M. H., Aqueous phase synthesis of Au-Ag core-shell nanocrystals with tunable shapes and their optical and catalytic properties, *J. Am. Chem. Soc.*, **2014**, 136, 396–404.

75. Feng, J., Wang, Z. F., Shen, B., Zhang, L. M., Yang, X., and He, N. Y., Effects of template removal on both morphology of mesoporous silica-coated gold nanorod and its biomedical application, *RSC Adv.*, **2014**, 4, 28683–28690.

76. Nopwinyuwong, A., Kitaoka, T., Boonsupthip, W., Pechyen, C., and Suppakul, P., Effect of cationic surfactants on characteristics and colorimetric behavior of polydiacetylene/silica nanocomposite as time-temperature indicator, *Appl. Surf. Sci.*, **2014**, 314, 426–432.

77. Okunaka, S., Tokudome, H., Hitomi, Y., and Abe, R., Preparation of fine particles of scheelite-monoclinic phase $BiVO_4$ via an aqueous chelating method for efficient photocatalytic oxygen evolution under visible-light irradiation, *J. Mater. Chem. A*, **2016**, 4, 3926–3932.

78. Ye, L. Q., Jin, X. L., Liu, C., Ding, C. H., Xie, H. Q., Chu, K. H., and Wong, P. K., Thickness-ultrathin and bismuth-rich strategies for BiOBr to enhance photoreduction of CO_2 into solar fuels, *Appl. Catal. B*, **2016**, 187, 281–290.

79. Becerro, A. I., Criado, J., Gontard, L. C., Obregon, S., Fernandez, A., Colon, G., and Ocana, M., Bifunctional, monodisperse $BiPO_4$-based nanostars: Photocatalytic activity and luminescent applications, *Cryst. Growth Des.*, **2014**, 14, 3319–3326.

80. Ertorer, E., Avery, J. C., Pavelka, L. C., and Mittler, S., Surface-immobilized gold nanoparticles by organometallic CVD on amine-terminated glass surfaces, *Chem. Vap. Depos.*, **2013**, 19, 338–346.

81. Xu, H., Wu, J. X., Feng, Q. L., Mao, N. N., Wang, C. M., and Zhang, J., High responsivity and gate tunable graphene-MoS_2 hybrid phototransistor, *Small*, **2014**, 10, 2300–2306.

82. Ludwig, T., Guo, L. L., McCrary, P., Zhang, Z. T., Gordon, H., Quan, H. Y., Stanton, M., Frazier, R. M., Rogers, R. D., Wang, H. T., and Turner, C. H., Mechanism of bismuth telluride exfoliation in an ionic liquid solvent, *Langmuir*, **2015**, 31, 3644–3652.

83. Ambrosi, A., Sofer, Z., Luxa, J., and Pumera, M., Exfoliation of layered topological insulators Bi_2Se_3 and Bi_2Te_3 via electrochemistry, *ACS Nano*, **2016**, 10, 11442–11448.

Index

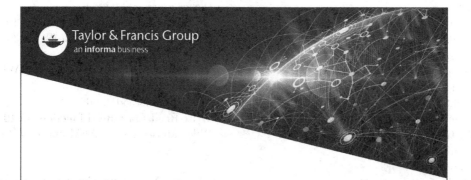